D0646938

HEATING, COOLING, LIGHTING

SECOND EDITION

HEATING, COOLING, LIGHTING

Design Methods for Architects

Norbert Lechner

JOHN WILEY & SONS, INC.

New York • Chichester • Weinheim • Brisbane • Singapore • Toronto

Front cover photo shows the north terminal at Ronald Reagan Washington National Airport in Washington, D.C. (Architect: Cesar Pelli. Photo: Jeff Goldberg/Esto.)

All other photographs that are not otherwise credited were provided by the author.

Library of Congress Cataloging-in-Publication Data:

Lechner, Norbert
    Heating, cooling, lighting : design methods for architects / by Norbert Lechner.—2nd ed.
        p. cm.
    Includes bibliography and index.
    ISBN 0-471-24143-1 (cloth : alk. paper)
    1. Heating. 2. Air conditioning. 3. Lighting. I. Title

    TH7222 .L33 2000
    697—dc21
                                                            00-033008

Printed in the United States of America.
10 9 8 7 6

*This book is dedicated
to a more sustainable
and just world*

CONTENTS

FOREWORD xiii

PREFACE xv

ACKNOWLEDGMENTS xvii

1

HEATING, COOLING, AND LIGHTING AS FORM-GIVERS IN ARCHITECTURE 1

1.1 Introduction 2
1.2 Vernacular and Regional Architecture 2
1.3 Formal Architecture 4
1.4 The Architectural Approach 7
1.5 Dynamic Versus Static Buildings 8
1.6 Energy and Architecture 9
1.7 Architecture and Mechanical Equipment 9
1.8 Conclusion 9

2

SUSTAINABLE DESIGN 11

2.1 Easter Island: Learning from the Past 12
2.2 Sustainable Design 12
2.3 Reuse, Recycle, and Regenerate by Design 14
2.4 The Green Movement 14
2.5 Population and Sustainability 15
2.6 Growth 16
2.7 Exponential Growth 16
2.8 The Amoeba Analogy 16
2.9 Production Versus Efficiency (Conservation) 18
2.10 Sustainable-Design Issues 18
2.11 Climate Change 19
2.12 The Global Greenhouse 21
2.13 The Ozone Hole 21
2.14 Energy Sources 22
2.15 Ancient Greece: A Historical Example 22
2.16 Nonrenewable Energy Sources 22
2.17 Renewable Energy Sources 26
2.18 Hydrogen 33
2.19 Conclusion 34

3

BASIC PRINCIPLES 37

3.1 Introduction 38
3.2 Heat 38
3.3 Sensible Heat 38
3.4 Latent Heat 39
3.5 Evaporative Cooling 39
3.6 Convection 40
3.7 Transport 40
3.8 Energy-Transfer Mediums 41
3.9 Radiation 41
3.10 Greenhouse Effect 42
3.11 Equilibrium Temperature of a Surface 43
3.12 Mean Radiant Temperature 44
3.13 Heat Flow 44
3.14 Heat Sink 45
3.15 Heat Capacity 45
3.16 Thermal Resistance 45
3.17 Heat-Flow Coefficient 46
3.18 Time Lag 46
3.19 Insulating Effect of Mass 46
3.20 Energy Conversion 47
3.21 Combined Heat and Power 48
3.22 Fuel Cells 48
3.23 Embodied Energy 49
3.24 Summary 49

4

THERMAL COMFORT 51

4.1 Biological Machine 52
4.2 Thermal Barriers 53
4.3 Metabolic Rate 55
4.4 Thermal Conditions of the Environment 56
4.5 The Psychrometric Chart 56
4.6 Dew-Point and Wet-Bulb Temperatures 59
4.7 Heat Content of Air 60
4.8 Thermal Comfort 61
4.9 Shifting to the Comfort Zone 62

4.10 Clothing and Comfort 64
4.11 Strategies 65

5

CLIMATE 67

5.1 Introduction 68
5.2 Climate 68
5.3 Microclimate 71
5.4 Climatic Anomalies 73
5.5 Climate Regions of the United States 74
5.6 Explanations of the Climatic Data Tables 75
5.7 Design Strategies 116

6

SOLAR GEOMETRY 125

6.1 Introduction 126
6.2 The Sun 126
6.3 Elliptical Orbit 126
6.4 Tilt of the Earth's Axis 127
6.5 Consequences of the Altitude Angle 128
6.6 Winter 128
6.7 The Sun Revolves Around the Earth! 128
6.8 Sky Dome 129
6.9 Determining Altitude and Azimuth Angles 131
6.10 Solar Time 131
6.11 Horizontal Sun-Path Diagrams 131
6.12 Vertical Sun-Path Diagrams 133
6.13 Sun-Path Models 134
6.14 Solar Site-Evaluation Tools 135
6.15 Sun Machines 136
6.16 Sundials for Model Testing 137
6.17 Integrating Sun Machine & Sun Emulator 138
6.18 Summary 139

7

PASSIVE SOLAR 141

7.1 History 142
7.2 Solar in America 142
7.3 Solar Hemicycle 144
7.4 Latest Rediscovery of Solar 145
7.5 Passive Solar 146
7.6 Direct-Gain Systems 147
7.7 Design Guidelines for Direct-Gain Systems 150
7.8 Example 152
7.9 Trombe Wall Systems 152
7.10 Design Guidelines for Trombe Wall Systems 157

7.11 Example 157
7.12 Sun Spaces 158
7.13 Balcomb House 159
7.14 Sun-Space Design Guidelines 160
7.15 Comparison of the Three Main Passive Heating Systems 162
7.16 General Consideration for Passive Solar Systems 162
7.17 Heat-Storage Materials 165
7.18 Other Passive Heating Systems 166
7.19 Summary 169

8

PHOTOVOLTAICS AND ACTIVE SOLAR 171

8.1 Introduction 172
8.2 The Almost-Ideal Energy Source 172
8.3 History of Photovoltaics (PV) 172
8.4 The Photovoltaic Cell 174
8.5 Types of Photovoltaic Systems 175
8.6 Balance of System Equipment 176
8.7 Building-Integrated Photovoltaics (BIPV) 176
8.8 Orientation and Tilt 179
8.9 Roofs Clad with Photovoltaics 179
8.10 Facades Clad with Photovoltaics 180
8.11 Glazing and Photovoltaics 180
8.12 Photovoltaic Shading Devices 182
8.13 Photovoltaics: Part of the Second Tier 182
8.14 Sizing a Photovoltaic System 183
8.15 Finding Photovoltaic Array Size for a Stand-Alone Building by the Short Calculation Method 184
8.16 Design Guidelines 185
8.17 The Promise of Photovoltaics 186
8.18 The Cost-Effectiveness of Active Solar Applications 186
8.19 Active Solar Swimming Pool Heating 188
8.20 Solar Hot-Water Systems 189
8.21 Solar Hot-Air Collectors 191
8.22 Designing an Active Solar System 193
8.23 Active/Passive Solar Systems 195
8.24 Preheating of Ventilation Air 196
8.25 The Future of Active Solar 197
8.26 Conclusion 197

9

SHADING 201

9.1 History of Shading 202
9.2 Shading 207
9.3 Orientation of Shading Devices 210
9.4 Movable Shading Devices 212
9.5 Shading Periods of the Year 216

9.6 Horizontal Overhangs 219
9.7 Shading Design for South Windows 220
9.8 Design Guidelines for Fixed South Overhangs 220
9.9 Design Guidelines for Movable South Overhangs 222
9.10 Shading for East and West Windows 223
9.11 Design of East and West Horizontal Overhangs 225
9.12 Design of Slanted Vertical Fins 226
9.13 Design of Fins on North Windows 226
9.14 Design Guidelines for Eggcrate Shading Devices 227
9.15 Special Shading Strategies 228
9.16 Shading Outdoor Spaces 231
9.17 Using Physical Models for Shading Design 234
9.18 Glazing as the Shading Element 237
9.19 Interior Shading Devices 239
9.20 Shading Coefficient and Solar Heat Gain Coefficient (SHGC) 240
9.21 Conclusion 243

10
PASSIVE COOLING 245

10.1 Introduction to Cooling 246
10.2 Historical and Indigenous Use of Passive Cooling 246
10.3 Passive Cooling Systems 254
10.4 Comfort Ventilation Versus Night Flushing 255
10.5 Basic Principles of Air Flow 255
10.6 Air Flow Through Buildings 259
10.7 Example of Ventilation Design 265
10.8 Comfort Ventilation 266
10.9 Night-Flush Cooling 269
10.10 Radiant Cooling 270
10.11 Evaporative Cooling 273
10.12 Earth Cooling 275
10.13 Dehumidification with a Desiccant 277
10.14 Conclusion 278

11
SITE DESIGN AND COMMUNITY PLANNING 279

11.1 Introduction 280
11.2 Site Selection 283
11.3 Solar Access 285
11.4 Shadow Patterns 289
11.5 Site Planning 292
11.6 Solar Zoning 296
11.7 Physical Models 299
11.8 Wind and Site Design 301

11.9 Plants and Vegetation 308
11.10 Landscaping 315
11.11 Community Design 322
11.12 Cooling Our Communities 322
11.13 Conclusion 322

12
LIGHTING 325

12.1 Introduction 326
12.2 Light 328
12.3 Reflectance/Transmittance 330
12.4 Color 331
12.5 Vision 335
12.6 Perception 336
12.7 Performance of a Visual Task 339
12.8 Charactericstics of the Visual Task 340
12.9 Illumination Level 341
12.10 Brightness Ratios 342
12.11 Glare 344
12.12 Equivalent Spherical Illumination 347
12.13 Activity Needs 348
12.14 Biological Needs 351
12.15 Light and Health 352
12.16 The Poetry of Light 352
12.17 Rules for Lighting Design 355
12.18 Career Possibilities 356
12.19 Conclusion 356

13
DAYLIGHTING 359

13.1 History of Daylighting 360
13.2 Why Daylighting? 364
13.3 The Nature of Daylight 365
13.4 Conceptual Model 367
13.5 Illumination and the Daylight Factor 368
13.6 Light Without Heat? 369
13.7 Cool Daylight 370
13.8 Goals of Daylighting 371
13.9 Basic Daylighting Strategies 372
13.10 Basic Window Strategies 375
13.11 Advanced Window Strategies 379
13.12 Window Glazing Materials 383
13.13 Top Lighting 384
13.14 Skylight Strategies 384
13.15 Clerestories, Monitors, and Light Scoops 389
13.16 Special Daylighting Techniques 394
13.17 Translucent Walls and Roofs 397
13.18 Electric Lighting as a Supplement to Daylighting 399
13.19 Physical Modeling 400
13.20 Conclusion 404

14
ELECTRIC LIGHTING 407

14.1 History of Light Sources 408
14.2 Light Sources 409
14.3 Incandescent Lamps 410
14.4 Discharge Lamps 412
14.5 Fluorescent Lamps 412
14.6 High-Intensity Discharge Lamps (Mercury, Metal-Halide, and High-Pressure Sodium) 416
14.7 Comparison of the Major Lighting Sources 417
14.8 New Light Sources 418
14.9 Lighting Fixtures (Luminaires) 419
14.10 Lenses, Diffusers, and Baffles 419
14.11 Lighting Systems 421
14.12 Remote-Source Lighting Systems 424
14.13 Visualizing Light Distribution 425
14.14 Architectural Lighting 427
14.15 Maintenance 430
14.16 Switching and Dimming 430
14.17 Rules for Energy-Efficient Electric Lighting Design 430
14.18 Conclusion 431

15
THE THERMAL ENVELOPE: KEEPING WARM AND STAYING COOL 433

15.1 Background 434
15.2 Heat Loss 435
15.3 Heat Gain 437
15.4 Solar Reflectivity (Albedo) 438
15.5 Compactness, Exposed Area, and Thermal Planning 439
15.6 Insulation Materials 443
15.7 Insulating Walls, Roofs, and Floors 447
15.8 Windows 452
15.9 Movable Insulation 455
15.10 Insulating Effect from Thermal Mass 456
15.11 Earth Sheltering 457
15.12 Moisture Control 463
15.13 Infiltration and Ventilation 466
15.14 Appliances 468
15.15 Conclusion 468

16
MECHANICAL EQUIPMENT FOR HEATING AND COOLING 471

16.1 Introduction 472
16.2 Heating 472
16.3 Thermal Zones 475
16.4 Heating Systems 475
16.5 Electric Heating 476
16.6 Hot-Water (Hydronic) Heating 478
16.7 Hot-Air Systems 481
16.8 Cooling 484
16.9 Refrigeration Cycles 484
16.10 Heat Pumps 486
16.11 Geo-Exchange 487
16.12 Cooling Systems 489
16.13 Air Conditioning for Small Buildings 493
16.14 Air Conditioning for Large Multistory Buildings 496
16.15 Design Guidelines for Mechanical Systems 504
16.16 Air Supply (Ducts and Diffusers) 505
16.17 Ventilation 509
16.18 Energy-Efficient Ventilation Systems 510
16.19 Air Filtration and Odor Removal 512
16.20 Special Systems 512
16.21 Integrated and Exposed Mechanical Equipment 514
16.22 Conclusion 518

17
CASE STUDIES 521

17.1 Introduction 522
17.2 Real Goods Solar Living Center 522
17.3 The Urban Villa 530
17.4 The Emerald People's Utility District Headquarters 533
17.5 Hood College Resource Management Center 535
17.6 Colorado Mountain College 538
17.7 Gregory Bateson Building 543
17.8 Hongkong Bank 549
17.9 Commerzbank 554
17.10 Phoenix Central Library 558

APPENDIX A
Horizontal Sun-Path Diagrams 563

APPENDIX B
Vertical Sun-Path Diagrams 565

APPENDIX C

Sun Machine 569

C.1 Construction of Sun Machine 569
C.2 Directions for Initial Set-Up 571
C.3 Directions for Use 572
C.4 Alternate Mode of Use of the Sun
       Machine 572

APPENDIX D

**Methods for Estimating the Height of Trees,
Buildings, Etc.** 573

D.1 Proportional-Shadow Method 573
D.2 Similar-Triangle Method 573
D.3 45-Degree Right-Triangle Method 574
D.4 Trigonometric Method 574
D.5 Tools for Measuring Vertical Angles 575

APPENDIX E

Sundials 577

APPENDIX F

Sun-Path Models 579

F.1 Directions for Constructing a Sun-Path
       Model 579
F.2 Directions for Creating Other Orthographic
       Projections 584

APPENDIX G

**Computer Software Useful for the Schematic
Design Stage** 585

G.1 UCLA Energy Design Tools 585
G.2 Energy Scheming 588

APPENDIX H

Site Evaluation Tools 589

H.1 The Sun Locator 589
H.2 Do-It-Yourself Solar Site Evaluator 589
H.3 Parts List 590
H.4 Construction Process 590
H.5 Using the Solar Site Evaluator 593

APPENDIX I

**Educational Opportunities in Energy-
Conscious Design** 595

APPENDIX J

**Resources (Books, Journals, Videos, and
Organizations)** 601

J.1 Books 601
J.2 Journals 601
J.3 Videos 601
J.4 CD-ROMs 602
J.5 Organizations 602

BIBLIOGRAPHY 605

INDEX 611

FOREWORD

Professor Lechner's book differs from most of its predecessors in several important respects: (1) he deals with the heating, cooling, and lighting of buildings, not as discrete and isolated problems, but in the holistic sense of being integral parts of the larger task of environmental manipulation; (2) he deals with these subjects not merely from the engineer's limited commitment to mechanical and economic efficiency but from the much broader viewpoint of human comfort, physical and psychic well being; (3) he deals with these problems in relation to the central paradox of architecture—how to provide a stable predetermined internal environment in an external environment that is in constant flux across time and space; and finally, (4) he approaches all aspects of this complex subject from a truly cultural—as opposed to a narrowly technological—perspective.

This attitude toward contemporary technology is by no means hostile. On the contrary, Professor Lechner handles them competently and comprehensively. But he never loses sight of the fact that the task of providing a truly satisfactory enclosure for human activity is that of the *building as a whole*. He points out, quite correctly, that until the last century or so, the manipulation of environmental factors was, of necessity, an architectural problem. It was the building itself—and only incidentally any meager mechanical equipment which the period happened to afford—that provided habitable space. To illustrate this point, he makes continuous and illuminating analysis to both high-style and vernacular traditions, to show how sagaciously the problems of climate control were tackled by earlier, prescientific, premechanized societies.

This is no easy-to-read copy book for those designers seeking short cuts to glitzy post-modern architecture. On the contrary, it is a closely reasoned, carefully constructed guide for architects (young *and* old) who are seeking an escape route from the energy-wasteful, socially destructive cul-de-sac into which the practices of the past several decades has led us. Nor is it a Luddite critique of modern technology; to the contrary, it is a wise and civilized explication of how we must employ it if we in the architectural field are to do our bit towards avoiding environmental disaster.

JAMES MARSTON FITCH
Hon. AIA, Hon. FRIBA

In memory of James Marston Fitch, architect, historian, professor,
preservationist, and architectural theorist,
1909–2000

In this completely reworked second edition, the focus of the original 1991 edition remains: to provide the appropriate knowledge at the level of complexity needed at the schematic design stage. In the years since the first edition was published we have moved from a shortage of information to a flood because of the Internet. This book will aid the designer because it presents the information in a concise, logical, and useful arrangement.

Energy conscious design and sustainability were an implicit part of the first edition, while now sustainable design is the central issue. In the new Chapter 2, sustainability is discussed as it relates to heating, cooling, and lighting. Chapter 2 covers the impact of growth, climate change, and the use of renewable energy resources. Also new is Chapter 8 which covers Photovoltaics and Active Solar. All chapters have been revised, while a few have also been totally reorganized. There are also many other new features, including hundreds of new diagrams and photos.

The following new features have been added to increase the reference value of the book. Every chapter now ends with a list of "Key Ideas" for review and quick reference. Many sideboxes consisting of mathematical analyses and examples have been added. Six color plates were included to better explain the phenomenon of color in lighting design. S.I. units and conversion factors are given where appropriate. Each chapter ends with a list of resources consisting of some or all of the following: books, journals, organizations, videos, and web pages.

Although minor changes were made on just about every page, the following are important additions noted by chapter. Chapter 3 (Basic Principles) now also covers fuel cells and hydrogen as a fuel. Chapter 4 (Thermal Comfort) now includes a much expanded discussion of the psychrometric chart. To Chapter 5 (Climate) were added a discussion on urban heat islands and a map of daily temperature ranges. Chapter 8 (Photovoltaics and Active Solar) discusses in detail building integrated, Photovoltaic systems. Chapter 9 (Shading) now also discusses advanced glazing systems such as electrochromic and photochromic glazing, as well as the new shading factor called the "Solar Heat Gain Coefficient." A completely new table of "Common Shade Trees" was added to Chapter 11 (Site Design).

Major revisions and additions were made to the three lighting chapters. In Chapter 12 (Lighting) the discussion on light and health has been expanded, and a section on career possibilities in the lighting field has been added.. Chapter 13 (Daylighting) was completely revised and reorganized, and a discussion of such factors as "Visible Transmittance" and "Light-to-Solar-Gain Ratio" was added. The Electric Lighting Chapter (14) discusses new light sources such as induction lamps, sulfur lamps, and light-emitting diodes. Chapter 15 (Thermal Envelope) now discusses albedo, structural insulated panels (SIPs), insulated concrete forms (ICFs), and an expanded section on moisture control. New topics in Chapter 16 (Mechanical Equipment) include displacement ventilation, ductless split systems, and geo-exchange systems, while the section on thermal storage is expanded. The last Chapter (17) consists of nine case studies, four of which are new: The Real Goods Solar Living Center in California, The Urban Villa in the Netherlands, The Commerzbank in Germany, and The Phoenix Central Library in Arizona.

Most of the appendixes have also changed. Vertical sun path diagrams are now included as Appendix B. "Methods for Estimating the Height of Objects," covered in Appendix D, has been completely revised and expanded. A new "Do-it-yourself Site Evaluation Tools" section was added as Appendix H. For students who want to continue their studies in sustainable heating, cooling, and lighting, Appendix I (Educational Opportunities in Energy Conscious Design) will be a valuable resource. Lastly, Appendix J lists current resources, including: books, journals, videos, CD-ROMs, and organizations.

This book focuses on the schematic design stage, where the key decisions are made. The graph below points out how the earliest decisions have the greatest impact on a project. A building's cost and environmental impact are mainly established at the schematic design stage. The most basic decisions of size, orientation, and form often have the greatest impact on the resources required both during construction and operation.

The information in this book is presented to support the three-tier approach to the heating, cooling, and lighting of buildings. The first tier is load avoidance. Here the need for heating, cooling, and lighting is minimized by the design of the building itself. The second tier consists of using natural energies as much as is practical. This tier is also accomplished mainly by the design of the building itself. The third and last tier uses mechanical equipment to satisfy the needs not provided for by the first two tiers.

With the knowledge and information presented in this book, the first two tiers can provide most of the thermal and lighting needs of a building and, as a consequence, the mechanical equipment of the third tier will be substantially smaller and will use much less energy than is typical now, thereby resulting in more sustainable buildings. Since tiers 1 and 2 are the domain of the architect, the role of the engineer at the third tier is to provide only that heating, cooling, and lighting that the architect could not.

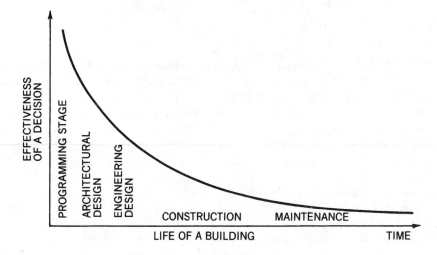

Acknowledgments

Many people helped create this book—too many to mention all. Above all, I would like to thank my wife Judy, my sons Walden and Ethan, and my mother Leni for their help and understanding.

Special thanks also to Paul C. Brandt and Auburn University for the support required for this long project. Special appreciation is also due to the following: readers and reviewers—William Bobenhausen, Murray Milne, Michael Swimmer, Eugene Pauncz, and Lorna Wiggins; research and editorial assistance— Judith V. Lechner; artists—Daniel C. Ly, Andy Ballard, Keith Myhand, Charles Carr, Troy Batson, Blayne Rose, and Keith Pugh; typists— Rosetta Massingale, Darlene Kenny, Valerie Samuel, and Margaret Wright.

For the Second Edition, the author would like to thank Prof.Russell Leslie of Rensellear Polytechnic Institute and the Lighting Research Center for his thorough review of the three lighting chapters; Prof. Robert Aderholdt of Auburn University for reviewing the Mechanical Equipment Chapter; Prof. Richard Kenworthy of Auburn University for reviewing the Site and Community Design Chapter; and Roger Hill, John Stevens, and Earl Rush of the Sandia National Laboratory for reviewing the PV and Active Solar Chapter. Prof. Kenworthy also created for this book Table 11.9: Common Shade Trees. Executive Editor Alex Wilson of the Environmental Building News allowed me to use an abbreviated version of his editorial on Easter Island; the abbreviated version was written by Dr. Eugene Goldwater. Most of the new artwork and line drawings were created by Cormac Phalen. The many generous contributors of images are too numerous to mention here, but they are given credit in the appropriate captions.

I would also like to thank the following people for their help. Prof. Emerita Joan Nist, Auburn University, and Librarian Emerita Lorna Wiggins, Auburn University, helped with the proofreading. The typing and secretarial work was done by my very capable sons, Walden and Ethan Lechner. Ethan also helped with proofreading and editing. And especially I want to thank my wife, Dr. Judy Lechner, Auburn University, not only for her technical help with the index and bibliography, but for her encouragement and patience without which this book would not exist.

NORBERT LECHNER
Auburn University
October 2000

HEATING, COOLING, LIGHTING

Heating, Cooling, and Lighting as Form-Givers in Architecture

"Two essential qualities of architecture [commodity and delight], handed down from Vitruvius, can be attained more fully when they are seen as continuous, rather than separated, virtues.

… In general, however, this creative melding of qualities [commodity and delight] is most likely to occur when the architect is not preoccupied either with form-making or with problem-solving, but can view the experience of the building as an integrated whole —…."

John Morris Dixon

Please note that figure numbers are keyed to sections. Gaps in figure numbering result from sections without figures.

1.1 INTRODUCTION

Until about 100 years ago, the heating, cooling, and lighting of buildings was the domain of architects. Thermal comfort and lighting were achieved with the design of the building and a few appliances. Heating was achieved by a compact design and a fireplace or stove, cooling by opening windows to the wind and shading them from the sun, and lighting by windows, oil lamps, and candles.

By the 1960s, the situation had changed dramatically. It had become widely accepted that the heating, cooling, and lighting of buildings were accomplished mainly by mechanical equipment as designed by engineers.

Our consciousness has been raised as a result of the energy crisis of 1973. It is now recognized that the heating, cooling, and lighting of buildings are best accomplished by *both* the mechanical equipment and the design of the building itself. Some examples of vernacular and regional architecture will show how architectural design can contribute to the heating, cooling, and lighting of buildings.

1.2 VERNACULAR AND REGIONAL ARCHITECTURE

One of the main reasons for regional differences in architecture is the response to climate. If we look at buildings in hot and humid climates, in hot and dry climates, and in cold climates, we find they are quite different from one another.

In hot and dry climates, one usually finds massive walls used for their time-lag effect. Since the sun is very intense, small windows will adequately light the interiors. The windows are also small because during the daytime the hot outdoor air makes ventilation largely undesirable. The exterior surface colors are usually very light to minimize the absorption of solar radiation. Interior surfaces are also light to help diffuse the sunlight entering through the small windows (Fig. 1.2a).

Since there is usually little rain, roofs can be flat and, consequently, are available as additional living and sleeping areas during summer nights. Outdoor areas cool quickly after the sun sets because of the rapid radiation to the clear night sky. Thus, roofs are more comfortable than the interiors, which are still quite warm from the daytime heat stored in the massive construction.

Even community planning responds to climate. In hot and dry climates, buildings are often closely clustered for the shade they offer one another and the public spaces between them.

In hot and humid climates, we find a very different kind of building. Although temperatures are lower, the high humidity creates great discomfort. The main relief comes from moving air across the skin to increase the rate of evaporative cooling. Although the water vapor in the air weakens the sun, direct solar radiation is still very undesirable. The typical antebellum house (see Fig. 1.2b) responds to the humid climate by its use of many large windows, large overhangs, shutters, light-colored walls, and high ceilings. The large windows maximize ventilation, while the overhangs and shutters protect from both solar radiation and rain. The light-colored walls minimize heat gain.

Since in humid climates nighttime temperatures are not much lower than daytime temperatures, massive construction is not an advantage. Buildings are, therefore, usually made of lightweight wood construction. High ceilings permit larger windows and permit the air to stratify. As a result, people inhabit the lower and cooler air layers. Vertical ventilation through roof monitors or high windows not only increases ventilation but also exhausts the hottest air layers first. For this reason, high gabled roofs without ceilings are popular in many parts of the world that have very humid climates (Fig. 1.2c).

Buildings are sited as far apart as possible for maximum access to the cooling breezes. In some of the humid regions of the Middle East, wind scoops are used to further increase the natural ventilation through the building (Fig. 1.2d).

In mild but very overcast climates, like the Pacific Northwest, buildings open up to capture all the daylight possible. In this kind of climate, the use of "bay" windows is quite common (Fig. 1.2e).

Figure 1.2a Massive construction, small windows, and light colors are typical in hot and dry climates, as in this Saudi village. It is also common, in such climates, to find flat roofs and buildings huddled together for mutual shading. (Drawing by Richard Millman.)

FIGURE 1.2b In hot and humid climates, natural ventilation from shaded windows is the key to thermal comfort. This Charleston, SC, house uses covered porches and balconies to shade the windows, as well as to create cool outdoor living spaces. The white color and roof monitor are also important in minimizing summer overheating.

FIGURE 1.2c In hot and humid climates, such as Sumatra, Indonesia, native buildings are often raised on stilts and have high roofs with open gables to maximize natural ventilation.

FIGURE 1.2d When additional ventilation is desired, wind scoops can be used, as on this reconstructed historical dwelling in Dubai. Also note the open weave of the walls to further increase natural ventilation. (Photograph by Richard Millman.)

FIGURE 1.2e Bay windows are used to capture as much light as possible in such a mild but very overcast climate as is found in Eureka, CA.

FIGURE 1.2f In cold climates, compactness, thick wooden walls, and a severe limit on window area were the traditional ways to stay warm. In very cold climates, the fireplace would be either on the inside of the exterior wall or in the center of the building.

And finally, in a predominantly cold climate we see a very different kind of architecture again. In such a climate, the emphasis is on heat retention. Buildings, like the local animals, tend to be very compact, to minimize the surface-area-to-volume ratio. Windows are few because they are weak points in the thermal envelope. Since the thermal resistance of the walls is very important, wood rather than stone is usually used (Fig. 1.2f). Because hot air rises, ceilings are kept very low (often below 7 feet). Trees and landforms are used to protect against the cold winter winds. In spite of the desire for views and daylight, windows are often sacrificed for the overpowering need to conserve heat.

1.3 FORMAL ARCHITECTURE

Not only vernacular structures but also buildings designed by the most sophisticated architects have responded to the needs for environmental control. After all, the Greek portico is simply a feature to protect against the rain and sun (Fig. 1.3a). The repeating popularity of classical architecture is based not only on aesthetic but also on practical grounds. There is hardly a better way to shade windows, walls, and porches than with large overhangs supported by colonnades or arcades (Fig. 1.3b).

The Roman basilicas consisted of large high-ceilinged spaces that were very comfortable in hot climates during the summer. Clerestory windows were used to bring daylight into these central spaces. Both the trussed roof and groin-vaulted basilicas became prototypes for Christian churches (Fig. 1.3c).

One of the Gothic builders' main goals was to maximize the window area for a large fire-resistant hall. By means of an inspired structural system, they sent an abundance of day-light through stained glass (Fig. 1.3d).

The need for heating, cooling, and lighting has also affected the work of the twentieth-century masters, such as Frank Lloyd Wright. The Marin County Court House emphasizes the importance of shading and daylighting. To give most offices access to daylight, the building consists of lin-ear elements separated by a glass-covered atrium (Figs. 1.3e and 1.3f). The outside windows are shaded from the direct sun by an arcade-like overhang (Fig. 1.3g). Since the arches are not structural, Frank Lloyd Wright shows them hanging from the building.

FIGURE 1.3a The classical portico has its functional roots in the sun- and rain-protected entrance of the early Greek megaron. (Maison Carée, Nimes, France.)

FIGURE 1.3b The classical revival style was especially popular in the South because it was very suitable for hot climates.

FIGURE 1.3c Roman basilicas and the Christian churches based on them used clerestory windows to light the large interior spaces. The Thermae of Diocletian, Rome (302 A.D.), was converted by Michelangelo into the church of Saint Maria Degli Angeli. (Photograph by Clark Lundell.)

FIGURE 1.3d Daylight was given a mystical quality as it passed through the large stained-glass windows of the Gothic cathedral. (Photograph by Clark Lundell.)

FIGURE 1.3e White surfaces reflect light down to the lower levels. The offices facing the atrium have all-glass walls.

FIGURE 1.3f The Marin County Court House, California, designed by Frank Lloyd Wright, has a central gallery to bring daylight to interior offices.

FIGURE 1.3g The exterior windows of the Marin County Court House are protected from the direct sun by an arcade-like exterior corridor.

Le Corbusier also felt strongly that the building should be effective in heating, cooling, and lighting itself. His development of the "brise soleil" will be discussed in some detail later. A feature found in a number of his buildings is the **parasol roof,** an umbrella-like structure covering the whole building. A good example of this concept is the "Maison d' Homme," which Le Corbusier designed in glass and painted steel (Fig. 1.3h).

Today, with no predominant style guiding architects, revivalism is common. The buildings in Fig. 1.3i use the classical portico for shading. Such historical adaptations can be more climate responsive than the "international style," which often ignores the local climate. Buildings in cold climates can continue to benefit from compactness, and buildings in hot climates still benefit from massive walls and light exterior surfaces. Looking to the past in one's locality will lead to the development of a new and suitable regional style.

1.4 THE ARCHITECTURAL APPROACH

The design of the heating, cooling, and lighting of buildings is accomplished in three tiers (Fig. 1.4). The first tier is the architectural design of the building itself to minimize heat loss in the winter, to minimize heat

FIGURE 1.3h The "Maison d'Homme" in Zurich, Switzerland, demonstrates well the concept of the parasol roof. The building is now called "Center le Corbusier." (Photograph by William Gwinn.)

FIGURE 1.3i These Postmodern buildings promote the concept of "regionalism" in that they reflect a previous and appropriate style of the Southeast.

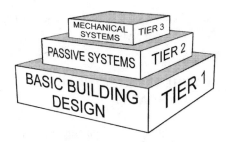

Figure 1.4 The three-tier approach to the design of heating, cooling, and lighting systems produces comfortable, energy-efficient, economical, and sustainable buildings.

gain in the summer, and to use light efficiently. Poor decisions at this point can easily double or triple the size of the mechanical equipment and energy eventually needed. The second tier involves the use of natural energies through such methods as passive heating, cooling, and daylighting systems. The proper decisions at this point can greatly reduce the unresolved problems from the first tier. Tiers one and two are both accomplished by the architectural design of the building. Tier three consists of designing the mechanical equipment using mostly nonrenewable energy sources to handle the loads that remain after tiers one and two have

reduced the loads as much as possible. Table 1.4 shows the design considerations that are typical at each of these three tiers.

The heating, cooling, and lighting design of buildings always involves all three tiers whether consciously considered or not. Unfortunately, in the recent past, minimal demands were placed on the building itself to affect the indoor environment. It was assumed that it was primarily the engineers at the third tier who were responsible for the environmental control of the building. Thus, architects, who were often indifferent to the heating, cooling, and lighting needs of buildings, sometimes designed buildings that were at odds with their environment. For example, buildings with large glazed areas were designed for very hot or very cold climates. The engineers were then forced to design giant, energy-guzzling heating and cooling plants to maintain thermal comfort. Ironically, these mostly glass buildings had their electric lights on during the day when daylight was abundant because they were not designed for quality daylighting. The size of the mechanical equipment can be seen as an indicator

of the architect's success, or lack thereof, in using the building itself to control the indoor environment.

When it is consciously recognized that each of these tiers is an integral part of the heating, cooling, and lighting design process, the buildings are better in several ways. The buildings are often less expensive because of reduced mechanical-equipment and energy needs. Frequently, the buildings are also more comfortable because the mechanical equipment does not have to fight such giant thermal loads. Furthermore, the buildings are often more interesting because some of the money that is normally spent on the mechanical equipment is spent instead on the architectural elements. Unlike hidden mechanical equipment, features, such as shading devices, are a very visible part of the exterior aesthetic.

Proper attention to tiers one and two can easily cut the size of the mechanical equipment by 50 percent, and with extra attention as much as 90 percent. In certain climates, some buildings can even be designed to use no mechanical equipment at all. The Lovins' home/office, which maintains *full* comfort high in the Rocky Mountains, has no mechanical equipment at all.

1.5 DYNAMIC VERSUS STATIC BUILDINGS

Contemporary buildings are essentially static with a few dynamic parts, such as the mechanical equipment, doors, and sometimes operable windows. On the other hand, intelligent buildings adapt to their changing environments. This change can occur continuously over a day as, for example, a movable shading device that extends when it is sunny and retracts when it is cloudy. Alternately, the change could be on an annual basis where a shading device is extended during the summer and retracted in the winter, much like a deciduous

TABLE 1.4 THE THREE-TIER DESIGN APPROACH

	Heating	Cooling	Lighting
Tier 1	*Conservation*	*Heat avoidance*	*Daylight*
Basic Building Design	1. Surface-to-volume ratio 2. Insulation 3. Infiltration	1. Shading 2. Exterior colors 3. Insulation	1. Windows 2. Glazing type 3. Interior finishes
Tier 2	*Passive solar*	*Passive cooling*	*Daylighting*
Natural Energies and Passive Techniques	1. Direct gain 2. Trombe wall 3. Sunspace	1. Evaporative cooling 2. Convective cooling 3. Radiant cooling	1. Skylights 2. Clerestories 3. Light shelves
Tier 3	*Heating equipment*	*Cooling equipment*	*Electric light*
Mechanical and Electrical Equipment	1. Furnace 2. Ducts 3. Fuels	1. Refrigeration machine 2. Ducts 3. Diffusers	1. Lamps 2. Fixtures 3. Location of fixtures

tree. The dynamic aspect can be modest, as in movable shading devices, or it can be dramatic, as when the whole building rotates to track the sun (Figs. 9.15c to 9.15e). Not only will dynamic buildings perform much better than static buildings, but they also will provide an exciting aesthetic, the **aesthetic of change.** Numerous examples of dynamic buildings are included throughout the book, but most will be found in the chapters on shading, passive cooling, and daylighting.

1.6 ENERGY AND ARCHITECTURE

The heating, cooling, and lighting of buildings is accomplished by either adding or removing energy. Consequently, this book is about the manipulation and use of energy. In the 1960s, the consumption of energy was considered a trivial concern. For example, buildings were sometimes designed without light switches because it was believed that it was more economical to leave the lights on—continuously. Also, the most popular air-conditioning equipment for larger buildings was the "terminal reheat system," in which the air was first cooled to the lowest level needed by any space, then reheated as necessary to satisfy the other spaces. The double use of energy was not considered an important issue.

Buildings now use about 35 percent of all the energy consumed in

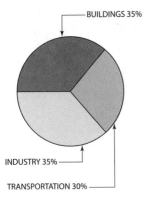

BUILDINGS 35%

INDUSTRY 35%

TRANSPORTATION 30%

FIGURE 1.6 The major energy-consuming sectors of the United States.

the United States (Fig. 1.6). Clearly then, the building industry has a major responsibility in the energy picture of this nation. Architects have both the responsibility and the opportunity to design in an energy-conserving manner.

The responsibility is all the greater because of the effective life of the product. Automobiles last only about ten years, and so any mistakes will not burden society too long. Most buildings, however, have a useful life of at least fifty years. The consequences of design decisions now will be with us for a long time.

Unfortunately, the phrase **energy conservation** has negative connotations. It makes one think of shortages and discomfort. Yet architecture that conserves energy can be comfortable, sustainable, humane, and aesthetically pleasing. It can also be less expensive than conventional architecture. Operating costs are reduced because of lower energy bills, and first costs are often reduced because of the smaller heating and cooling equipment that is required. To avoid the negative connotations, the more positive and flexible phrases of **energy-efficient design** or **energy-conscious design** have been adopted to describe a concern for energy conservation in architecture. Energy-conscious design yields buildings that minimize the needs for expensive, polluting, and nonrenewable energy. Because of the benefit to planet Earth, such design is now frequently called **sustainable** or **green.** The importance of energy consciousness is discussed in more detail in the next chapter.

1.7 ARCHITECTURE AND MECHANICAL EQUIPMENT

The following design considerations have impact on both the appearance and the heating, cooling, and lighting of a building: compactness (surface-area-to-volume ratio), size and location of windows, and the nature of the

building materials. Thus, when architects start to design the appearance of a building, they simultaneously start the design of the heating, cooling, and lighting. Because of this inseparable relationship between architectural features and the heating, cooling, and lighting of buildings, we can say that the environmental controls are **form-givers** in architecture.

It is not just tiers one and two that have aesthetic impact. The mechanical equipment required for heating and cooling is often quite bulky, and because it requires access to outside air, it is frequently visible on the exterior. The lighting equipment, although less bulky, is even more visible. Thus, even tier three is interconnected with the architecture, and, as such, must be considered at the earliest stages of the design process.

The plumbing and electrical wiring systems do not have this same form-giving and integral relationship with architecture. Since these systems are fairly small, compact, and flexible, they are easily buried in the walls and ceilings. Thus, they require little or no attention at the schematic design stage and are not discussed in this book.

1.8 CONCLUSION

The heating, cooling, and lighting of buildings is accomplished not just by mechanical equipment, but mostly by the design of the building itself. The design decisions that affect these environmental controls have, for the most part, a strong effect on the form and aesthetics of buildings. Thus, through design, architects have the opportunity to simultaneously satisfy their need for aesthetic expression and to efficiently heat, cool, and light buildings. Only through architectural design can buildings be heated, cooled, and lit in an environmentally responsible way. The importance of that is explained in the next chapter on sustainability.

KEY IDEAS OF CHAPTER 1

1. Both vernacular and formal architecture were traditionally designed to respond to the heating, cooling, and lighting needs of buildings.
2. Borrowing appropriate regional design solutions from the past (e.g., the classical portico for shade) can yield environmentally responsive buildings.
3. It is a twentieth-century development that only the engineers with their mechanical and electrical equipment respond to the environmental needs of buildings. Architects resolved these needs in the past, and they can again be important players in the future.
4. The heating, cooling and lighting needs of buildings should be designed by the three-tier approach:

   TIER ONE: the basic design of the building form and fabric (by the architect)
   TIER TWO: the design of passive systems (mostly by the architect)
   TIER THREE: the design of the mechanical and electrical equipment (by the engineer).

5. Buildings use about 35 percent of all the energy consumed in the United States.
6. Currently, the dynamic mechanical equipment responds to the continually changing heating, cooling, and lighting needs of a building. There are both functional and aesthetic benefits when the building itself is more responsive to the environment (e.g., movable shading devices). Buildings should be dynamic rather than static.
7. There is great aesthetic potential in energy-conscious architecture.

Resources

FURTHER READING

(See Bibliography in back of book for full citations. The list includes valuable out-of print books.)

Duly, C. *The Houses of Mankind.*
Banham, R. *The Architecture of the Well-Tempered Environment.*
Brown, G. Z., and M. DeKay. *Sun, Wind, and Light: Architectural Design Strategies.*
Fathy, H. *Natural Energy and Vernacular Architecture.*
Fitch, J. M., and W. Bobenhausen. *American Building—The Environmental Forces That Shape It.*
Fitch, J. M. *The Architecture of the American People.*
Fitch, J. M. *Shelter: Models of Native Ingenuity.*

Heschong, L. *Thermal Delight in Architecture.*
Konya, A. *Design Primer for Hot Climates.*
Nabokov, P., and R. Easton. *Native American Architecture.*
Olgyay, V. *Design with Climate: Bioclimatic Approach to Architectural Regionalism.*
Rapoport, A. *House Form and Culture.*
Rudofsky, B. *Architecture Without Architects: A Short Introduction to Non-Pedigreed Architecture.*
Rudofsky, B. *The Prodigious Builders.*
Stein, R. G. *Architecture and Energy.*
Taylor, J. S. *Commonsense Architecture: A Cross-Cultural Survey of Practical Design Principles.*

PAPERS

Knowles, R. "On Being the Right Size," http://www-rcf.usc.edu/~rknowles
Knowles, R. "Rhythm and Ritual," http://www-rcf.usc.edu/~rknowles
Knowles, R. "The Rituals of Place," http://www-rcf.usc.edu/~rknowles

SUSTAINABLE DESIGN

"Sustainable development is development that meets the needs of the present without compromising the ability of future generations to meet their own needs."

The United Nations World Commission on Environment and Development, the Brundtland Report, 1987

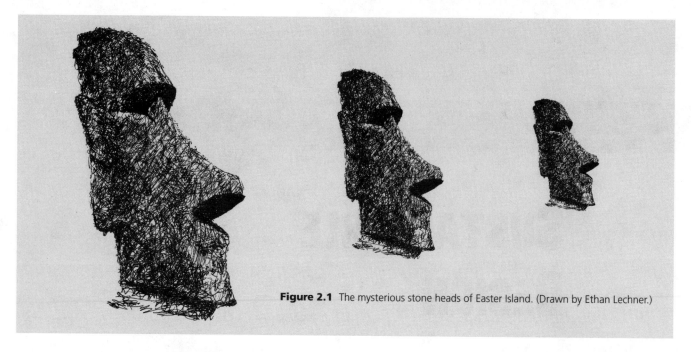

Figure 2.1 The mysterious stone heads of Easter Island. (Drawn by Ethan Lechner.)

2.1 EASTER ISLAND: LEARNING FROM THE PAST

Easter Island has long mystified archaeologists. When the tiny remote island, 2,000 miles (3,200 km) from the nearest continent, was "discovered" on Easter day in 1722, about 200 mammoth stone statues, some more than 30 feet tall and weighing more than 80 tons, stood on the island.

The island was a biological wasteland. Except for introduced rats and chickens, there were no animal species higher than insects and only a few dozen plant species—mostly grasses and ferns—and nothing more than 10 feet in height. There was no obvious way that the island's 2,000 or so inhabitants could have transported and hoisted the huge statues.

Based on an analysis of ancient pollen, researchers have now established that Easter Island was a very different place when the Polynesians first arrived around 400 A.D. In fact, it was a subtropical paradise, rich in biodiversity. The Easter Island palm grew more than 80 feet tall and would have been ideal for carving into canoes for fishing, as well as equipment for erecting statues. In addition to the rich plant life, there were at least twenty-five species of nesting birds.

We now believe that Easter Islanders exploited their resources to the point that they exterminated all species of higher animals and many species of plants. The island's ecosystem might have been destroyed in a cascading fashion; as certain birds were eliminated, for example, trees dependent on those birds for pollination could no longer reproduce. Denuded of forests, the land eroded, carrying nutrients out to sea.

Researchers believe that the island population had grown to a peak of 20,000 that lived in a highly organized structure. But as food (or the ability to get it) became scarce, this structure broke down into warring tribal factions. By 1722, the island's population had dropped to 2,000.

Why didn't the Easter Islanders see what was happening? Jared Diamond, in the August 1995, *Discover* magazine, suggests that the collapse happened "not with a bang but a whimper." Their means of making boats, rope, and log rollers disappeared over decades or even generations and either they didn't see what was happening or couldn't do anything about it.

Will humanity as a whole do better with planet Earth than the Polynesian settlers did with their Easter Island paradise? Many politicians and talk-show hosts claim that there are no limits to growth—that environmental doomsayers are wrong. But Easter Island shows us that limits are real. Let's not wait until it is too late to come to grips with these limits.

Shortened by permission from Alex Wilson, Editor and Publisher, *Environmental Building News (EBN).* * The full article appeared in *EBN*, Volume 4, Number 5, September/October 1995.

Environmental Building News is a monthly newsletter for architects and builders committed to improving the sustainability of buildings and the built environment (see Appendix J for address).

2.2 SUSTAINABLE DESIGN

In the long run, sustainable design is not an option but a necessity. Earth, with 6 billion people, is rapidly approaching the same stress that 20,000 people caused to Easter Island. We are literally covering the planet Earth with people (see Fig 2.2a). We are depleting our land and water resources; we are destroying biodiversity; we are polluting the land, water, and air; and we are changing the climate with potentially catastrophic results.

In the short termm, it *seems* that we do not have to practice sustainable design, but that is only true if we ignore the future. We are using up resources and polluting the planet without regard to the needs of our children and our children's children (Fig. 2.2b).

The World Congress of Architects in Chicago, June, 1993, said:

Sustainability means meeting the needs of the current generation without compromising the ability of future generations to meet their own needs. A sustainable society restores, preserves, and enhances nature and culture for the benefit of all life present and future; a diverse and healthy environment is intrinsically valuable and essential to a healthy society; today's society is seriously degrading the environment and is not sustainable.

Figure 2.2a Nighttime lights of the continental United States as viewed from satellites clearly show how people are filling up our land. (The image is derived from data from the Defense Meteorological Satellite Program (DMSP) and can be viewed at www.ngdc.noaa.gov/dmsp.)

Figure 2.2b Where a mountain once stood, now a colossal hole exists. Human beings are literally moving mountains to feed their appetite for resources. For a sense of scale, note the trains on the terraces on the far side. The tunnel at the bottom of this open-pit copper mine in Utah is for the trains to take the ore to smelters beyond the mountains.

Many ways exist to describe sustainable design. One approach urges using the four Rs:

REDUCE
REUSE
RECYCLE
REGENERATE.

This book will focus on the first "R," **reduce.** Although the word "reduce" might evoke images of deprivation, it applies primarily to the reduction of waste and extravagance. For example, American houses have more than doubled in size since 1950, and since families are now smaller, the increase in size per person is 2.8. Is that really necessary? Consider how inefficient a conventionally built home is, when a demonstration home in Lakeland, Florida, wastes 80 percent less energy (FSEC, 1998). Proven techniques in the areas of heating, cooling, and lighting can easily reduce energy use in commercial buildings by 50 percent, and with a little effort 80-percent reductions are possible. We have the knowledge, tools, and materials to tremendously cut energy use by designing more efficient buildings. The most advanced example is the Lovins' home/office in Snowmass, Colorado, which is more than 99 percent passively heated. Although the primary focus of this book is "Reduce" by design, the building industry can also make use of the other three sustainability techniques, which will be briefly discussed in the next section.

2.3 REUSE, RECYCLE, AND REGENERATE BY DESIGN

Fig. 2.3a shows a sight much too common, a building being demolished. Instead, it should be renovated and **reused.** An excellent example is the Audubon Society headquarters, which is a classic case study in the "green" renovation of an old New York City office building. Both a book and a video of this project are available (see Resources at end of chapter).

Even if the building in Fig. 2.3a could not be saved, it could still be **recycled.** By a process of deconstruction, it can be taken apart, and its component parts could be either recycled (concrete, steel, lumber, etc.) or reused (windows, doors, bricks, etc.). Instead, most buildings end up as landfill with their resources and embodied energy (see Section 3.23) completely lost.

The fourth "R," **regenerate,** deals with the fact that much of the Earth has already been degraded and needs to be restored. Since little is known on how to restore the Earth, the Center for Regenerative Studies was established at Cal Poly Pomona through the pioneering work of John T. Lyle (see Fig. 2.3b). Participating students from Cal Poly Pomona reside on-site to learn and investigate how to live a sustainable and regenerative lifestyle. Both the landscape and the architecture of the Center—for instance, the use of fruit-bearing plants as shading elements—were carefully designed to demonstrate and explore green techniques (Fig. 2.3c).

2.4 THE GREEN MOVEMENT

The issues related to sustainability are so all-encompassing that many feel that a different word should be used. The word **green** is often used because its connotations are flexible and it symbolizes nature, which truly is sustainable. For the same reason, many use the word **ecological.** Still others prefer the phrase **environmentally responsible.** The words might be different, but the goals are the same.

The world community is increasingly aware of the seriousness of our situation, and many important steps have been taken. The most successful so far has been the Montreal Protocol of 1987, through which the world agreed to rapidly phase out chlorofluorocarbons, which are depleting the ozone layer, thereby exposing the planet to more harmful ultraviolet

Figure 2.3a Buildings marked for demolition should be either reused through renovation or recycled through the process of deconstruction.

Figure 2.3b The Center for Regenerative Studies at Cal Poly Pomona was established to teach and explore how to restore the planet. The buildings are all oriented to the south with few if any windows facing east or west. The roofs also face south to support active solar hot water collectors and future photovoltaic panels.

Figure 2.3c The Center for Regenerative Studies. (Photo by Walter Grondzik.)

radiation. Because the danger was very clear and imminent, the world's resolve was swift and decisive.

Other important gatherings have addressed the need for environmental reform. In 1992, the largest gathering of world leaders in history met at the Earth Summit in Rio de Janeiro to endorse the principle of sustainable development. The American Institute of Architects has set up the Committee on the Environment to help architects understand the problems and responses needed for creating a sustainable world. In 1997, many countries met in Kyoto, Japan, to agree on concrete measures to address global warming.

William McDonough and Michael Braungart, two creative individuals who are addressing the challenge, have proposed the "Next Industrial Revolution" (McDonough and Braungart, 1998), when eco-effectiveness will govern. They say that the next Industrial Revolution will be based on three principles: **waste equals food** (i.e., all waste produced must be used as the raw materials of some process), **respect for ecological and human diversity**, and **the use of solar energy.**

2.5 POPULATION AND SUSTAINABILITY

In 1999, the population of the earth was 6 billion. There are various estimates on the rate of population growth, as seen in Fig. 2.5. It is appropriate to ask how many people the Earth can hold. The answer to that question depends on the response to further questions. Is the capacity of

the Earth to be sustainable, and what is to be the standard of living?

The sustainable population or **carrying capacity** of the Earth might already have been exceeded. Assuming that we can make the changes necessary to sustain the present population requires a high level of optimism. Technology is not an automatic solution since we have seen over and over how technological fixes have created unexpected new problems (e.g., the ozone hole and global warming). Another solution often mentioned is that people should eat less meat since that is an inefficient way to feed people. This brings us to the second point: what is the desired affluence level for the people of the Earth?

Paul Ehrlich and John Holden proposed the following relationship:

$$I = P \times A \times T$$

where I = Environmental impact
P = Population
A = Affluence per person
T = Technology

This relationship clearly shows that the greater the population, the greater the impact on the environment. The relationship also shows that the greater the affluence of a soci-

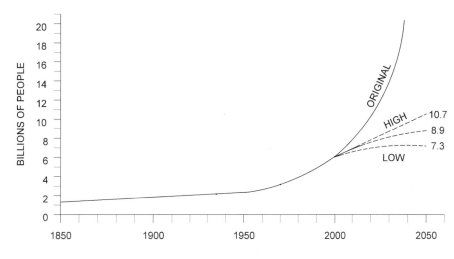

Figure 2.5 Various population projections are given since actual growth is impossible to predict with any certainty. (After "World Population Prospects: The 1998 Revision," New York, December, 1998.)

ety the greater the impact on the environment. For example, a person who drives a car affects the environment much more than a person who walks. It should also be noted that for a given impact on the environment, the greater the population, the lower must be the affluence. Thus, the higher a standard of living we want, the greater is the need to stop population growth.

Technology also has a great impact on the environment. A person today will have a much greater impact on the environment than a person a couple of centuries ago did. There were no automobiles, air travel, air conditioning, endless electrical appliances, abundant electrical lighting, etc. Although some technology is less wasteful than other technology, all technology affects the environment.

Trying to create a high standard of living for the inhabitants of the world without population control "is as though one attempted to build a 100 story skyscraper from good materials, but one forgot to put in a foundation." (Bartlett, 1997).

2.6 GROWTH

As we have seen, both the growth of population and affluence place stress on the planet. There are environmental impacts from the growing use of petroleum, wood, concrete, water, and just about everything else. How is it, then, that we generally think positively about growth? Most politicians get elected by promising growth. Most communities think a 5-percent growth is a great idea, but do

they realize that with a steady 5-percent growth per year, the community will double in size every fourteen years? The **doubling time** for any fixed growth rate is easy to determine. See Sidebox 2.6.

Growth is popular for several reasons: many people make a good living based on growth, we generally think bigger is better, and we don't fully understand the long-term consequences of growth.

Let us look to nature for guidance on what kind of growth we want. Most living things grow until they mature. In nature, unlimited growth is seen as pathological. As the environmental writer Edward Abbey noted: "Growth for the sake of growth is the ideology of the cancer cell." Nature suggests that growth should continue until a state of maturity is reached, whereupon the focus should be on improving the quality and not the quantity.

A steady growth *rate* does not mean steady growth. This misconception is a major reason for our inability to properly plan for the future. For example, if the world population continued growing at its 1.9-percent rate (small rate?) from 1975, it will grow to a size where there will be one person for every square meter (approximately a square yard) of dry land on the earth in only 550 years (Bartlett, 1978). This is an example of the power of exponential growth.

2.7 EXPONENTIAL GROWTH

Since this book is about heating, cooling, and lighting, let us look at the growth of energy consumption over

the last 10,000 years (Fig. 2.7). As in all exponential curves, growth is very slow for a very long time. Then, all of a sudden, the growth becomes very rapid and then almost instantly out of control. Because the implications of exponential growth are almost sinister, it is important to take a closer look at this concept.

We have a very good intuitive feel for linear, straight-line relationships. We know that if it takes one minute to fill one bucket of water, it will take five minutes to fill five such buckets. We do not, however, have that kind of intuitive understanding of nonlinear (exponential) relationships. Yet some of the most important developments facing humankind involve exponential relationships. Population, resource depletion, and energy consumption are all growing at an exponential rate, and their graphs look very much like Figure 2.7.

2.8 THE AMOEBA ANALOGY*

Suppose a single-celled amoeba split in two once every minute. The growth rate of this amoeba would be exponential, as Fig. 2.8a illustrates. If we graph this growth, it yields the exponential curve seen in Fig. 2.8b. Now let us also suppose we have a certain size bottle (a resource) that it would take the reproducing amoeba ten hours to fill. In other words, if we put one amoeba into the bottle and it splits every minute, then in ten hours the bottle will be full of amoeba, and all the space will be used up.

Question: How long will it take for the amoeba to use up only 3 percent of the bottle?

A. 18 minutes (3 percent of 10 hours)
B. about one hour
C. about 5 hours
D. about 8 hours
E. 9 hours and 55 minutes

---

*Based on the work of Albert A. Bartlett (Bartlett, 1978).

SIDEBOX 2.6

To determine the doubling time for any fixed rate of growth, use the following equation:

$$T_2 = 70/G$$

where:

T_2 = Doubling time
G = Growth rate in percent

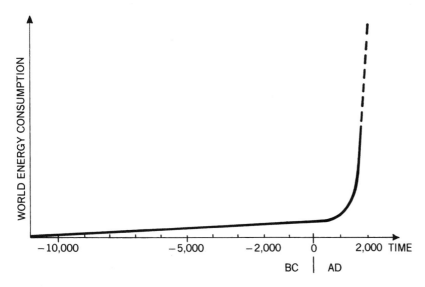

Figure 2.7 The exponential growth in world energy consumption and population growth are very similar.

Time	Percent of the bottle filled
10:00	100 percent
10:01	200 percent
10:02	400 percent

The amoeba gained only two more minutes by finding three more bottles. Obviously, it is hopeless to try to supply the resources necessary to maintain exponential growth at its later stages. What, then, is the solution?

In nature, there is no such thing as limitless exponential growth. For example, the growth of the amoeba actually follows an **"S" curve.** Although the growth starts at an exponential rate, it quickly levels off, as seen in Fig. 2.8c. The amoeba not only run out of food, but also poison themselves with their excretions. Since humans are not above nature, they cannot support exponential growth very long either. If people do not control growth willingly, nature will take over and reduce growth by such timeless measures as pollution, shortages, famine, disease, and war.

Until 1973, the growth of energy consumption followed the exponential curve A in Fig. 2.8d. Then, with the beginning of the energy crisis of 1973, energy consumption moved into an "S"-shaped curve. Initially, the shortages and later the implementation of efficiency strategies dramatically reduced growth. Our attitude to the growth of energy consumption

Since the amoeba double every minute, let us work backward from the end.

Time	Percentage of bottle used up
10:00	100 percent
9:59	50 percent
9:58	25 percent
9:57	12 percent
9:56	6 percent
9:55	3 percent Answer

For the amoeba, the space in the bottle is a valuable resource. Do you think the average amoeba would have listened to a doomsayer who at 9 hours and 55 minutes predicted that the end of the "bottle space" is almost upon them? Certainly not—they would have laughed. Since only 3 percent of the precious resource is used up, there is plenty of time left before the end.

Of course, some enterprising amoeba went out and searched for more bottles. If they found three more bottles, then they increased their resource to 400 percent of the original. Obviously, that was a way to solve their shortage. Or was it?

Question: How much additional time was bought by the 400-percent increase?

Answer: Since the amoeba double every minute, the following table tells the sad tale.

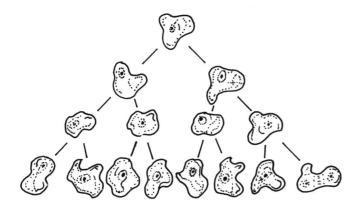

Figure 2.8a The exponential growth of amoeba.

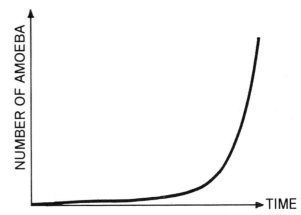

Figure 2.8b The theoretical exponential growth of amoeba.

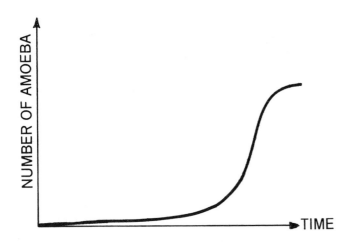

Figure 2.8c The actual growth of an amoeba colony.

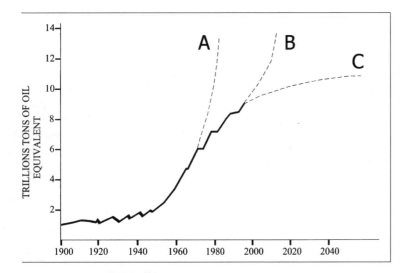

Figure 2.8d Alternate paths for future world energy consumption: (A) historical trend not taken because of the 1973 energy crisis; (B) trend if old wasteful habits return; (C) trend if conservation and efficiency continue to guide our policies. (After: "State of the World, 1999," Fig. 2.5, A Worldwatch Institute Report.)

will determine whether we will follow another dangerous exponential curve B or a more sensible growth pattern, such as that indicated by curve C.

2.9 PRODUCTION VERSUS EFFI-CIENCY (CONSERVATION)

The laws of exponential growth make it quite clear that we can match energy production with demand only if we limit the growth of the demand. In addition, it turns out that efficiency (conservation) is more attractive than increasing the supply from both an economic and an environmental point of view. The Harvard School of Business published a major report called *Energy Future* (Stobaugh and Yerkin, 1979), which clearly presented the economic advantages of conservation (efficiency). The report concluded that conservation combined with the use of solar energy is the best solution to our energy problem. In the following discussion of the energy sources, it will become clear

that almost every other source of energy involves some environmental costs.

The economic advantage of efficiency is demonstrated by the following example. The Tennessee Valley Authority (TVA) was faced with an impending shortage of electrical energy required for the economic growth of the valley. The first inclination was to build new electric generating plants. Instead, a creative approach showed that efficiency would be significantly less expensive. The TVA loaned its customers the money required to insulate their homes. Although the customers had to repay the loans, their monthly bills were lower than before because the reduced energy bills more than compensated for the increase due to loan payments. Because of reduced consumption due to efficiency, the TVA had surplus low-cost electricity to sell, the customers paid less to keep their homes warm, and everyone had a better environment because no new power plants had to be built. Efficiency is a strategy where everyone wins.

2.10 SUSTAINABLE-DESIGN ISSUES

Creating a sustainable green building involves all aspects of design, which is more than one book can discuss. There is, however, an important subset of issues that is discussed here, namely energy (Fig 2.10).

Heating, cooling, and lighting are all accomplished by moving energy into or out of a building. As mentioned in the previous chapter, buildings use about 35 percent of all the energy consumed in the United States. Because of global warming, air pollution, and energy-resource depletion, the energy subset of all the sustainability issues is probably the most urgent to address.

The highly regarded *Environmental Building News* has printed what it

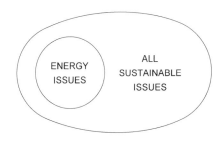

Figure 2.10 Energy issues are probably the most important subset of all sustainable (green) issues because of global warming.

believes are the eleven most important sustainable design issues. They are reproduced below. Note that the first issue is **"Save Energy — Design and build energy-efficient buildings."** Although this book covers only the first issue, the whole list is reproduced because it is succinct yet comprehensive.

Priority List for Sustainable Building*

1. **Save Energy:** Design and build energy-efficient buildings.
2. **Recycle Buildings:** Utilize existing buildings and infrastructure instead of developing open space.
3. **Create Community:** Design communities to reduce dependence on automobiles and to foster a sense of community.
4. **Reduce Material Use:** Optimize design to make use of smaller spaces and utilize materials efficiently.
5. **Protect and Enhance the Site:** Preserve or restore local ecosystems and biodiversity.
6. **Select Low-Impact Materials:** Specify low-environmental-impact, resource-efficient materials.
7. **Maximize Longevity:** Design for durability and adaptability.

*Reprinted by permission from the *Environmental Building News*. See the September/October, 1995, issue for a more thorough discussion of these issues, and see Appendix J for more information about *EBN*.

8. **Save Water:** Design buildings and landscapes that are water-efficient.
9. **Make the Buildings Healthy:** Provide a safe and comfortable indoor environment.
10. **Minimize Construction and Demolition Waste:** Return, reuse, and recycle job-site waste, and practice environmentalism in your business.
11. **"Green Up" Your Business:** Minimize the environmental impact of your own business practices, and spread the word.

2.11 CLIMATE CHANGE

It has been previously mentioned that energy issues are related to global warming. The 1995 report of the Intergovernmental Panel on Climate Change (IPCC), based on the work of 2,500 scientists from more than 60 countries, concluded that "the balance of evidence suggests that there is discernible human influence on global climate." The report also noted that the Earth has already been warmed 2°F (see Fig. 2.11a upper graph), and its best guess is that the air tempera-

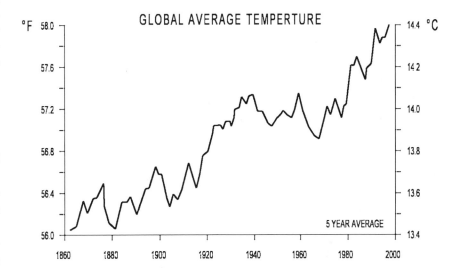

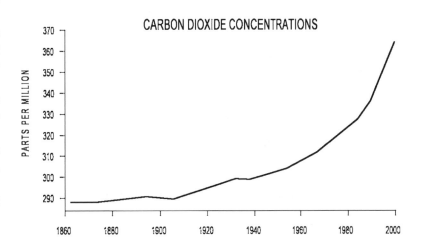

Figure 2.11a The upper graph represents the increase of the global average temperature, and the lower graph represents the increase of carbon dioxide (CO_2) during the same time period. (Sources: temperature data from Goddard Institute for Space Studies; carbon dioxide data from Scripps Institution of Oceanography.)

ture will rise another 2°F to 6.5°F in the next 100 years.

The cause of the global warming is no mystery when we look at the increase of the **greenhouse** gas, carbon dioxide (CO_2), over the same time period (see Fig. 2.11a lower graph). Humanity is also heating the planet by producing methane, nitrous oxide, and some other minor greenhouse gases. Most of the heating, however, is due to the carbon dioxide produced from burning the fossil fuels of coal, oil, and natural gas. The greenhouse effect will be explained in the next section.

Small increases in global temperatures can have serious effects besides deadly hotter summers. Precipitation patterns will change with a corresponding disruption in agriculture; some of the world's poorest and most populated regions will be losers. There will be more droughts in some areas and floods in others. Diseases that thrive in warmer climates, such as malaria, will spread over more of the globe, and species extinction will have a further negative impact on the present ecology. And perhaps most important, there will be a rise in the sea level. A rise of 1 to 3 feet is almost certain from the expansion of water as it warms. A 3-foot rise could flood an area of Bangladesh where 70 million people live, and Miami Beach will be below sea level. Much greater rises are possible from the melting of the icecaps. A 10- to 30-foot rise is possible if there is significant melting of the Greenland and Antarctic ice sheets. Such a rise would submerge all of southern Florida and 25 percent of Louisiana.

One reason for the uncertainty in the rate and extent of global warming is that change could be like a toggle switch. As you move the switch, nothing happens until a certain point is reached, then suddenly the switch makes contact. For example, as the permafrost of the arctic tundra melts, the thick layer of organic material decomposes, giving off carbon dioxide and methane (natural gas), both of which are powerful greenhouse gases. Thus, the Earth heats up faster and more permafrost is melted, and so on and so on.

How likely are the patterns described above? The hottest year on record had been 1997, and then it was 1998. The insurance industry paid out four times as much for weather-related disasters in the decade of the 1990s as it did in the decade of the 1980s. One large insurance company has estimated that the number of natural catastrophes has tripled since the 1960s. This pattern fits with the scientific models of global warming, and we can expect it to get worse as the Earth continues to warm. Just as it takes a long time to heat up, the Earth will take a long time to cool down after we start reducing the carbon dioxide in the air.

We know that global warming is a reality and that serious problems wil occur. Just because we don't know how severe the problems will be doesn't mean we shouldn't take action *now*. As the eminent physicist Albert A. Bartlett said, "We must recognize that it is not acceptable to base our national future on the motto 'When in doubt, gamble'."

Because of our tremendous appetite for energy, each American produces more CO_2 than anyone else (Fig. 2.11b), and because of our large population, the United States is also the country that produces the most carbon dioxide. Reducing our dependence on the carbon-dioxide-producing fossil fuels has more benefits than just reducing global warming. In the short run, it will reduce our dependency on foreign oil, and in the long run, it will prepare us for the day when fossil fuels are finally used up.

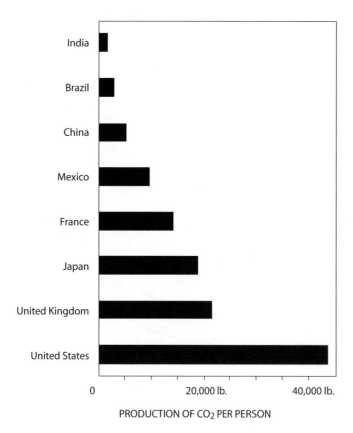

PRODUCTION OF CO_2 PER PERSON

Figure 2.11b Americans must take a lead in reducing carbon-dioxide emissions because we produce far more carbon dioxide per person than the citizens of any other country. (After *Consumer Reports,* September, 1996.)

2.12 THE GLOBAL GREENHOUSE

The greenhouse gases of the atmosphere act as a one-way radiation trap. They allow most of the solar radiation to pass through to reach to the Earth's surface, which when heated radiates increased amounts of heat back into space. The greenhouse gases, however, block the escape of much of this heat radiation (see Fig. 2.12). The present average temperature is the result of the existing level of water vapor and other greenhouse gases mentioned above. As a result of these gases, the Earth is about 60°F (35°C) warmer than it would be without these gases. When more greenhouse gases are added to the atmosphere, the equilibrium temperature increases and the Earth gets warmer.

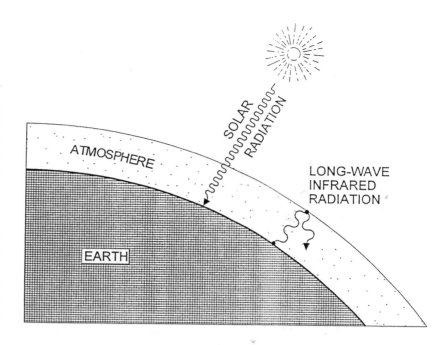

Figure 2.12 The atmosphere acts as a horticultural greenhouse by allowing most of the solar radiation to enter but blocks part of the long-wave infrared radiation from leaving the planet.

2.13 THE OZONE HOLE

The ozone hole is another example of a critical undesired change to the atmosphere. The cooling of buildings has led indirectly to a hole in the ozone layer that protects the Earth from most of the sun's harmful ultraviolet radiation (Fig. 2.13). The chlorofluorocarbon (CFC) molecules that were invented to provide a safe, inert refrigerant for air conditioners have turned out to have a tragic flaw, inertness, which ironically was considered its major virtue. When these molecules escaped from air conditioners or were released as propellants in spray cans, they survived to slowly migrate to the upper atmosphere, which contains ozone. There, the CFCs deplete the protective ozone layer for an estimated fifty years before they themselves are destroyed. Consequently, the problems will be with us long after we eliminate all CFCs on the surface.

The 1987 Montreal Protocol, which the United States wholeheartedly supports, requires countries to phase out the production of CFCs.

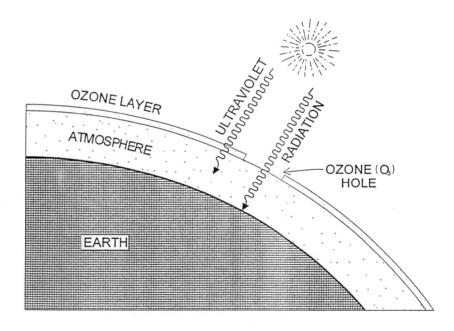

Figure 2.13 The depletion of the ozone layer allows greater amounts of the sun's harmful ultraviolet radiation to reach the Earth's surface.

Although this is a classic example of how technological solutions can be the source of new problems, it is also a good example of how world cooperation based on sound science can quickly respond to a serious problem.

Regretfully, international cooperation is not succeeding in controlling greenhouse emissions. Progress is slow for various reasons, one of which is the shortsighted policies of some fossil-fuel and transportation industries.

2.14 ENERGY SOURCES

Which energy sources are available to power our buildings, and which are sustainable? We can divide all of the sources into the two main categories of renewable and nonrenewable:

I. Renewable
   A. Solar
   B. Wind
   C. Biomass
   D. Hydroelectric
   E. Geothermal

II. Nonrenewable
   A. Fossil fuels
      1. Oil
      2. Natural gas
      3. Coal
   B. Nuclear
      1. Fission
      2. Fusion?

Fig. 2.14 shows that we are using mostly nonrenewable energy sources. This is an unfortunate condition because not only are we using up these sources, but these are the very ones causing pollution and global warming. We must switch as quickly as possible from the nonrenewable to the renewable sources. Before we look at each source in regard to its ability to power buildings sustainably, let's look at a brief example in the history of energy use in buildings.

2.15 ANCIENT GREECE: A HISTORICAL EXAMPLE

The role of energy in buildings had largely been ignored in recent history until the "Energy Crisis of 1973," when some of the leading members of the Organization of Petroleum Exporting Countries (OPEC) suddenly raised prices and set up an embargo on oil exports to the United States. The resulting energy shortages made us realize how dependent we were (and still are) on unreliable energy sources. We began thinking about how we use energy in buildings.

Before the energy crisis, a discussion of Ancient Greek architecture would not have even mentioned the word "energy." The Ancient Greeks, however, became aware of energy issues as the beautiful, rugged land on which they built their monuments became scarred and eroded by the clearing of trees to heat their buildings. The philosopher Plato said of his country: "All the richer and softer parts have fallen away and the mere skeleton of the land remains."

The Ancient Greeks responded to the problem partly by using solar energy. Socrates, another philosopher, thought that this was important enough to compel him to explain this method of designing buildings. According to Xenophon, Socrates said: "In houses that look toward the south, the sun penetrates the portico in winter, while in summer the path of the sun is right over our heads and above the roof so that there is shade" (see Fig. 2.15). Socrates continued talking about a house that has a two-story section: "The section of the house facing south must be built lower than the northern section in order not to cut off the winter sun" (Butti and Perdin, 1980).

2.16 NONRENEWABLE ENERGY SOURCES

When we use nonrenewable energy sources, we are much like the heir living it up on his inheritance with no thought of tomorrow until one day he finds that the bank account is empty. Two major categories of nonrenewable energy sources exist: fossil fuels and nuclear.

Fossil Fuels

For hundreds of millions of years, green plants trapped solar energy by the process of photosynthesis. The accumulation and transformation of these plants into solid, liquid, and gaseous states are what we call the **fossil fuels:** coal, oil, and gas. When we burn these, we are actually using the solar energy that was stored hundreds of millions of years ago. Because of the extremely long time required to convert living plants into fossil fuels, in effect they are depletable or nonrenewable energy sources. The fossil-fuel age started around 1850 and will last at most a few centuries. The finite nature of the fossil age is clearly illustrated by the graph of Fig. 2.16a.

Most air pollution and smog are a result of the burning of fossil fuels (see Fig. 2.16b). The use of fossil fuels also causes acid rain and, most important of all, global warming.

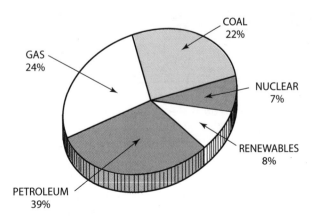

Figure 2.14 Energy consumption by source in the United States. Note that 92 percent is from nonrenewable sources.

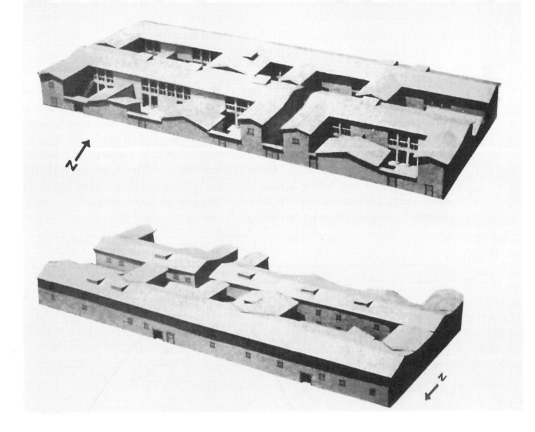

Figure 2.15 Solar buildings were considered "modern" in Ancient Greece. Olynthian apartments faced south to capture the winter sun. Note the few and small northern windows. (From *Excavations at Olynthus, Part 8: The Hellenic House.* © Johns Hopkins University Press, 1938).

Figure 2.16a The age of fossil fuels in the longer span of human history. (After Hubbert.)

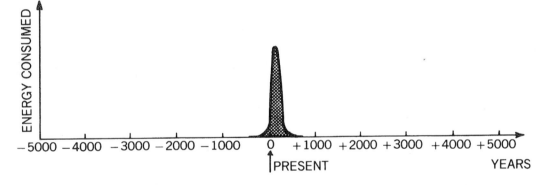

Figure 2.16b Air pollution covering New York City is not like this every day. Often nature blows the pollution away to places like Connecticut or New Jersey.

23

Natural Gas

Natural gas, which is composed primarily of methane, is a very clean and convenient source of energy. With the extensive pipeline system that exists, natural gas can be delivered to most of the populated areas of the United States. Once burnt at the oil well as a waste byproduct, it is in great demand today. Unfortunately, most of the easily obtained gas is already out of the ground. Most of the new sources come from wells as deep as 15,000 feet, and even these supplies are limited. Gas will, therefore, be a much more expensive fuel in the future. Since gas is also a valuable raw material for fertilizer and other chemicals, it will soon be too valuable to burn. We are importing natural gas in the form of **liquefied natural gas** (LNG). To maintain the gas in liquid form, it is shipped in tankers at -260°F. There is some concern about safety because if such a tanker exploded and ignited in a busy harbor, the devastation would be similar to that caused by a small nuclear bomb.

Oil

The most useful and important energy source today is certainly oil. But the world supply is limited and will be mostly depleted by the end of this century. Our domestic supply will run out even sooner. The United States is already importing more than 50 percent of its oil needs. Since oil is also important in making lubricants, plastics, and other chemicals, it, like natural gas, will soon become too valuable to burn.

Long before we run out of oil, its price will rise. Research has shown that the demand becomes greater than the supply when half of the resource has been extracted. This is also the point where peak production occurs. Best estimates for the peaking of world oil production is sometime during the second decade of this century (Fig. 2.16c). Demand will exceed production for several reasons. Since

much of the easily obtainable oil has already been pumped out of the ground, we are now forced to either drill deeper, under water (Fig. 2.16d), or in almost inaccessible places, such as the north slope of Alaska. We are also forced to use lower grades of petroleum, which either increase refinery costs or increase air pollu-

tion. Unconventional sources of oil, such as tar sands, oil shale, and coal liquefaction, are not economically viable at present and will be very expensive if and when used. Besides all of the other problems with this energy source, there is the likelihood of more ecologically devastating oil spills.

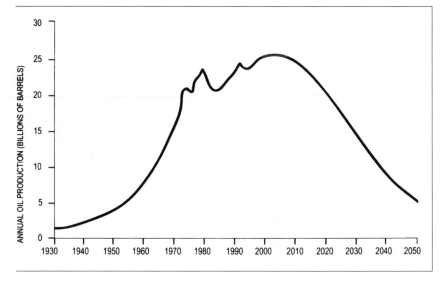

Figure 2.16c The price of oil is expected to increase significantly after production peaks, which is expected somewhere between 2010 and 2020. (After Campbell and Laherrère, 1998).

Figure 2.16d Oil platforms for drilling under water are very expensive, and, therefore, the oil extracted there is also more expensive.

Coal

By far, the most abundant fossil fuel we have is coal. Although there is enough coal in the United States to last us well over 100 years, significant problems are associated with its use. The difficulties start with the mining. Deep mining is dangerous to miners in two ways. First, there is the ever-present danger of explosions and mine cave-ins. Second, in the long run there is the danger of severe respiratory ailments due to the coal dust. If the coal is close enough to the surface, strip mining might be preferred. Less dangerous to people, strip mining is quite harmful to the land. Reclamation is possible but expensive. Much of the strip mining will occur in the western United States, where the water necessary for reclamation is a very scarce resource.

Additional difficulties result because coal is not convenient to transport, handle, or use. Since coal is a rather dirty fuel to burn and a major cause of acid rain, its use will be restricted to large burners, where expensive equipment can be installed to reduce the air pollution. Even if coal were burned "cleanly," it would still continue to produce carbon dioxide and global warming.

All of these difficulties add up to making coal an inconvenient, expensive, and harmful source of energy.

Although plentiful, it will not be the answer to our energy problems.

Nuclear Fission

In fission, certain heavy atoms, such as uranium-235, are split into two middle-size atoms and in the process also give off neutrons and an incredible amount of energy (Fig. 2.16e). During the 1950s it was widely believed that "electricity produced from nuclear energy will be too cheap to meter."

Even with huge governmental subsidies because of nuclear energy's defense potential, this dream has not become reality. In fact, just the opposite has happened. Nuclear energy has become one of the most expensive and least desirable ways to produce electricity. One important factor in the decline of nuclear power is that the public is now hesitant to accept the risks. Another nuclear-power accident, such as the one that took place at Chernobyl in the Soviet Union in 1986, would spread deadly radiation over a large area. The Three Mile Island nuclear accident in the United States (Pennsylvania) in 1979 might have been just as serious if the very expensive containment vessel had not been built. More than twenty years later, the reactor is still entombed with a billion-dollar cleanup bill. The safety features needed to prevent accidents have made the plants uneconomical.

Another problem facing the nuclear-power industry is the shortage of uranium-235, which is very rare and will be depleted soon. The main hope was to use the much more plentiful U-238 isotope by turning it into a fuel. This requires the construction of breeder reactors that turn U-238 into the fissionable element **plutonium.** Plutonium, appropriately named for the god of the underworld, is a very horrendous material. It is extremely toxic and ideal for making atomic bombs. Four pounds of plutonium, evenly distributed, is enough to kill all the people on earth.

The overall efficiency of the power plants has not been as high as had been hoped. The initial cost of a nuclear power plant is high, the operating efficiency is low, and the problem of disposing of radioactive nuclear waste has still not been solved. The net effect of all these difficulties is that no new nuclear power plants have been ordered by the electric-power industry in the United States since 1978, and none might be ordered in the future.

Nuclear Fusion

When two light atoms fuse to create a heavier atom, energy is released by the process called **fusion** (Fig. 2.16f). This is the same process that occurs in the sun and stars. It is quite unlike **fis-**

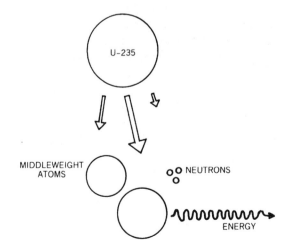

Figure 2.16e Nuclear fission is the splitting apart of a heavy atom.

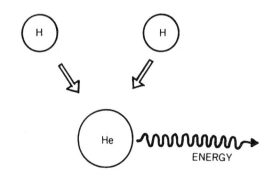

Figure 2.16f Nuclear fusion consists of the union of very light atoms. For example, the fusion of two hydrogen atoms yields an atom of helium as well as a great deal of extra energy.

sion, a process through which atoms decay by coming apart.

Fusion has many potential advantages over fission. Fusion uses hydrogen, the most plentiful material in the universe, as its fuel. It produces much less radioactive waste than fission. It is also an inherently much safer process because fusion is self-extinguishing while fission is self-exciting, meaning it has a tendency to start a chain reaction.

All the advantages, however, do not change the fact that a fusion power plant does not yet exist, and we have no guarantee that we can ever make fusion work economically. Even the optimists do not expect fusion to supply significant amounts of power anytime soon.

Considering the shortcomings, perhaps the best nuclear power plant is the one 93 million miles away — the sun.

2.17 RENEWABLE ENERGY SOURCES

The following sources all share the very important assets of being renewable and of not contributing to global warming. Solar, wind, hydro-electric, and biomass are renewable because they are all variations of solar energy. Of the renewable energy sources, only geothermal energy does not depend on the sun.

Solar Energy

The term **solar energy** refers to the use of solar radiation in a number of different ways. The primary solar utilization methods are all discussed at some depth in this book:

- Passive solar (Chapter 7)
- Photovoltaics and active solar (Chapter 8)
- Daylighting (Chapter 13)

The word "solar" is also used for some specialty applications not dis-

cussed here, such as solar thermal, which is used for the generation of electricity.

In one year, the amount of solar energy that reaches the surface of the Earth is 10,000 times greater than all the energy of all kinds humanity uses in that period. Why, then, aren't we using solar energy? This question can be explained partly by the technical problems involved. These technical problems stem from the diffuseness, intermittent availability, and uneven distribution of solar energy.

Many solutions to the technical problems exist, but they all add to the costs of what would otherwise be a free energy source. However, the clever techniques developed for using solar energy have reduced the collection and storage costs to the point where some form of solar energy is economical in most situations.

Besides being renewable, solar energy has other important advantages. It is exceedingly kind to the environment. No air, water, land, or thermal pollution results. Solar energy is also very safe to use. It is a decentralized source of energy available to everyone everywhere. With its use, individuals are less dependent on brittle or monopolistic centralized energy sources, and countries are secure from energy embargoes. Japan and Switzerland have embarked on ambitious solar programs in order to become more energy-independent, while the United States is depending more and more on foreign oil, already importing more than 50 percent.

The nontechnical problems facing the acceptance of solar energy are primarily a result of people's beliefs that it is unconventional, looks bad, does not work, is futuristic, etc. On the contrary, in most applications, such as daylighting or the use of sun spaces, solar energy can add special delight to architecture. Interesting aesthetic forms are a natural product

of solar design (see Fig. 11.6e). Solar energy promises not only to benefit the nation's energy supply and reduce global warming, but also to enrich its architecture.

Photovoltaic Energy

If one were to imagine the ideal energy source, it might well be photovoltaic (PV) cells. They are made of the most common material on Earth — silicon. They produce the most flexible and valuable form of energy — electricity. They are very reliable — no moving parts. They do not pollute in any way — no noise, no smoke, no radiation. And they draw on an inexhaustible source of energy — the sun.

Over the last twenty years, the cost of PV electricity has been declining steadily, and it is in the process of becoming competitive with conventional electricity. PV electricity is already competitive for installations that are far from the existing power grid.

The greatest potential lies with **building-integrated photovoltaics** (BIPV) both for our energy future and for architecture (see Fig. 8.3b).

Wind Energy

The Ancient Persians used wind power to pump water. Windmills first came to Europe in the twelfth century. They were successfully used over the centuries for such tasks as grinding wheat and pumping water. More than 6 million windmills and wind turbines have been used in the United States over the last 150 years (Fig. 2.17a). Windmills were used primarily to pump water on farms and ranches (Fig. 2.17b). Wind turbines also produced electricity for remote areas before rural electrification in the 1930s.

Today, wind turbines are having a revival because they can produce clean, renewable energy at the same cost as conventional energy. Where

Figure 2.17a This windmill in Colonial Williamsburg, VA, was used to grind wheat.

Figure 2.17b Windmills still pump water on some ranches and farms, while large commercial size wind generators use the same power of the wind to produce clean, pollution free electricity. One commercial size wind turbine like the one shown here can provide clean power for approximately 600 people. (Courtesy Enron Wind Corp.)

wind is plentiful and electricity is expensive, wind power is often the best source of energy. All over the world, giant wind turbines and windfarms are generating electricity for the power grid (Fig. 2.17c).

Small wind turbines can be an excellent source of electricity for individual buildings where the wind resource is sufficient, which is a function of both velocity and duration. Figure 2.17d shows where favorable wind conditions can be found in the United States. Of course, local conditions are critical, and a local survey should be made for obstacles or especially favorable conditions. Mountaintops, mountain passes, and shorelines are often good locations in all parts of the country. For economic reasons, minimum annual average wind speeds of 9 miles per hour (mph), or 4 meters per second (m/s), are needed. See the end of the chapter for information on wind-resource availability.

Since the power output of a wind turbine is proportional to the *cube* of the wind speed (see Sidebox 2.17), a windy site is critical, and there is a great incentive to raise the turbine as

Figure 2.17c Utility size wind farm in the Delaware Mountains, near El Paso, Texas. (Courtesy Enron Wind Corp.)

27

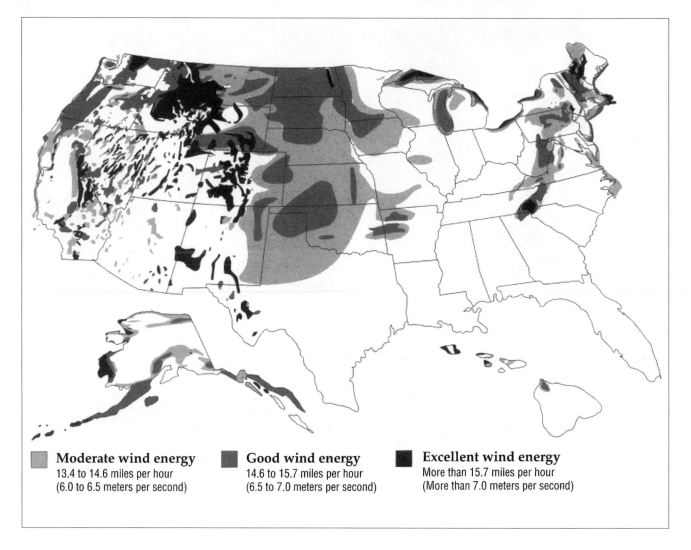

Moderate wind energy
13.4 to 14.6 miles per hour
(6.0 to 6.5 meters per second)

Good wind energy
14.6 to 15.7 miles per hour
(6.5 to 7.0 meters per second)

Excellent wind energy
More than 15.7 miles per hour
(More than 7.0 meters per second)

Figure 2.17d This map of the United States, including Alaska and Hawaii, gives general information on the average wind resources that are available. (From DOE/GO-10097-374, FS 135, January, 1997.)

SIDEBOX 2.17

The power produced by a wind machine is proportional to the cube of the wind speed and the square of the rotor radius.

$$P \approx V^3 \times D^2$$

where

P = power output
V = air speed
D = rotor diameter

For example, the power output will be 8 (2^3) times as large if the wind speed doubles, or stated another way, a 12.6-mph wind will yield twice the power of a 10-mph wind.

Also, if the rotor diameter is doubled (2^2), the power output will be 4 times as large.

high into the air as possible to reach higher wind speeds. Most often, wind machines are supported on towers (Fig. 2.17e), but they can also be placed on buildings. Towers from 80 to 120 feet high are very common.

The power output of a wind turbine is also proportional to the square of the length of the rotor blades (Fig. 2.17f). A 6.6-foot-diameter rotor is enough to power a television, while a 66-foot-diameter rotor can generate enough electricity for 500 Californians or 1,000 Europeans.

The greatest disadvantage of wind power is that it is not constant. In power-grid-connected installations, the intermittent nature of the wind is not a problem because the grid acts as a giant storage battery to supply power when the wind is not blowing.

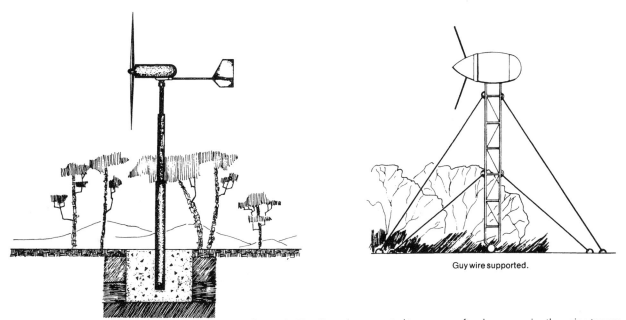

Freestanding pipe tower.

Guy wire supported.

Figure 2.17e Guy-wire-supported towers are often less expensive than pipe towers but require more land area. (From *Wind Power for Farms, Homes, and Small Industry*, U.S. Department of Energy, 1978.)

In stand-alone systems, a large battery would be needed. It is found, however, that **hybrid systems,** in which wind power is combined with PV cells, are very efficient because they complement each other. In winter, there is less sun but more wind, while in summer, the PV cells generate more electricity than the wind turbine. Similarly, on stormy days, there is less sun but more wind. Thus, wind turbines and PV cells are frequently used together, as shown in Fig. 8.5c. Also, see Sections 8.5 and 8.6 for the typical electrical systems and balance of system equipment that is needed.

Small wind turbines do make some noise, but most people do not find the noise to be very objectionable, and wind machines are being designed to be less noisy. Wind turbines do not interfere with television reception. In certain locations, wind machines have killed some birds. Although this is a concern, it is a minor problem when compared with about 57 million birds that are killed each year in collisions with cars and another 97 million birds that die in collisions with plate-glass windows.

Although some wind farms are spread over large areas of land, the land can still be used for crops and grazing. This is not the case for hydro-power, where the land behind the dam is flooded. Also, wind farms require only about one-fifth the amount of land that hydro-power requires.

There is also some concern about the aesthetic impact because by their very nature, wind machines must be high in the air. There have been few complaints about actual installations, and the author believes there is inherent beauty in a device that produces renewable, nonpolluting energy.

Biomass Energy

Photosynthesis stores solar energy for later use. This is how plants solve the problems of diffuseness and intermittent availability, which are associated with solar energy. This stored energy can be turned into heat or electricity, or converted into such fuels as methane gas, alcohol, and hydrogen. Because **biomass** is renewable and because in modern large-scale burners its use is relatively pollution-free,

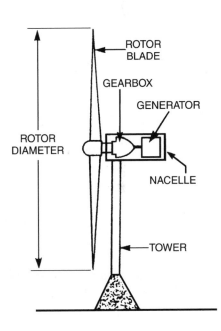

Figure 2.17f The essential components of a wind turbine. (Drawing from DOE/CE-0359P.)

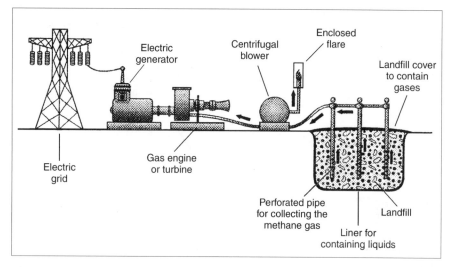

Figure 2.17g Landfill gas can be collected to generate electricity. (From the Texas State Energy Conservation Office—Fact Sheet No. 16.)

Figure 2.17h Saving biomass to replenish the soil makes sense at both the macro and micro level. This sign was found in a residential development outside of Houston, TX.

biomass is an attractive energy source. Two major sources of biomass exist: (1) plants grown specifically for their energy content; and (2) organic waste from agriculture, industry, or consumers (garbage). For the United States to get all of its energy from growing biomass, we would have to use about 1.1 million square miles of land. Because this is equivalent to the present cultivated land area of the United States, we would be forced to choose between food and fuel. The world's population is already too great to switch food farms into fuel farms. But land now not used and organic waste products could still supply significant amounts of energy.

As William McDonough, architect, author, and former Dean of the School of Architecture at the University of Virginia, said, "Waste is food."

Burning biomass instead of fossil fuels can reduce the problem of global warming because biomass is "**carbon-neutral.**" When growing, plants remove the same amount of carbon dioxide from the atmosphere that is returned when the biomass is burned. Thus, over time, there is no net change in the carbon-dioxide content of the atmosphere.

Agricultural waste, such as straw; industrial waste, such as wood chips; or consumer waste, such as garbage, can be burned for process heat or the production of electricity. Some agricultural waste products can be fermented into ethanol alcohol and used as a liquid fuel. All kinds of organic waste can be digested and turned into methane gas. Landfills release methane into the atmosphere where it causes global warming. Instead of harming the planet, the landfill could be easily adapted to produce methane for supplying natural-gas pipelines, or it can generate electricity locally (Fig. 2.17g). Sewage can be used in a similar way.

Wood used to heat houses is an example of biomass energy. Large-scale burning of wood in fireplaces or stoves, however, is not desirable because of the low efficiency and large amount of air pollution produced. Fireplaces are *very* inefficient (see Section 16.2), while wood stoves are more polluting.

Biomass is a desirable source of energy, but it is limited in magnitude because plants are needed to produce food and other products, such as lumber. Furthermore, using all biomass for energy will deprive the soil of decaying material that is needed for the next generation of plants (see Fig. 2.17h).

Hydro-Electric Energy

The use of water power, also called **hydro-power** or **hydro-electricity,**

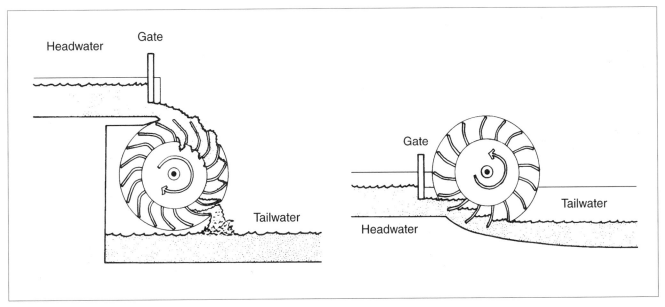

Figure 2.17i *Left:* An overshot water wheel. *Right:* An undershot wheel. (From *Building Control Systems* by Vaughn Bradshaw. 2nd edition. © John Wiley & Sons, Inc., 1993.)

has an ancient history: watermills were already popular in the Roman Empire. The **overshot wheel** (Fig. 2.17i left) was found to be the most efficient, but it required at least a 10-foot fall (head) of the water. When there was little vertical fall in the water but sufficient flow, an **undershot wheel** (Fig. 2.17i right) was found to be best. Today, compact turbines are driven by water delivered in pipes.

The power available from a stream is a function of both head and flow. **Head** is the pressure developed by the vertical fall of the water, often expressed in pounds per square inch. **Flow** is the amount of water that passes a given point in a given time as, for example, cubic feet per minute. The flow is the result of both the cross section and velocity of a stream or river.

Since the power output is directly proportional to both head and flow, different combinations of head and flow will work equally well. For example, a very small hydro-power plant can be designed to work equally with 8 feet of head and a flow of 100 cubic feet per minute, or 16 feet of head and a flow of 50 cubic feet per minute.

Figure 2.17j Hydro-electric dams produce pressure (head), and some also store water and, therefore, energy for later use.

Today, water power is used almost exclusively to generate electricity. The main expense is often the dam that is required to generate the head and store water to maintain an even flow (Fig. 2.17j). One advantage of hydro-power over some other renewable sources is the relative ease of storing energy. The main disadvantage of hydro-electricity is that large areas of land must be flooded to create the storage lakes. This land is most frequently prime agricultural land and is often highly populated. Another disadvantage is the disturbance of the local ecology, for example, where fish cannot reach their spawning grounds. For this reason, many existing dams in the United States are being demolished.

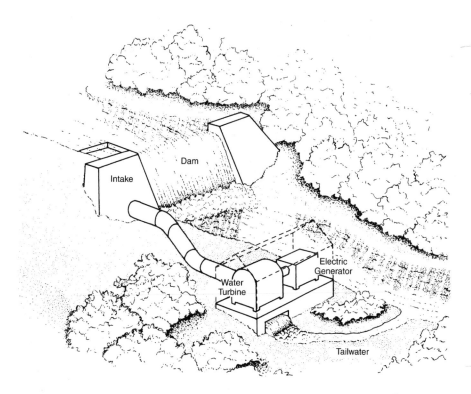

Figure 2.17k A simple, small-scale hydro-electric system. (From *Building Control Systems* by Vaughn Bradshaw. 2nd edition. © John Wiley & Sons, Inc., 1993.)

Fig. 2.17k illustrates a simple, small-scale hydro-electric system. The dam generates the required head, stores water, and diverts water into the pipe leading to the turbine located at a lower elevation. Modern turbines have high rotational speeds (rpm) so that they can efficiently drive electric generators.

All but the smallest systems require dams, which are both expensive and environmentally questionable. Very small systems are known as micro-hydro-power and can use the "run of the river" without a dam. The site must still have an elevation change of at least 3 feet in order to generate the minimum head required. Of course, the more head (elevation change), the better.

About 5 percent of the energy in the United States is supplied from falling water. At present, we are using about one-third of the total hydro-electric resource available. Full use of this resource is not possible because some of the best sites remaining are too valuable to lose. For example, it would be hard to find anyone who would want to flood the Grand Canyon or Yosemite Valley behind hydro-electric dams. Most Americans now see our scenic rivers and valleys as great resources to be protected. Hydro-electric energy will continue to be a reliable but limited source for our national energy needs.

Geothermal Energy

The term **geothermal** has been used to describe two quite different energy systems: (1) the extraction of heat originating from deep in the earth; and (2) the use of the ground just below the surface as a source of heat in the winter and a **heat sink** in the summer. To eliminate confusion, the second system is now called by the much more descriptive name **geo-exchange.**

Geothermal energy is available where sufficient heat is brought near the surface by conduction, bulges of magma, or circulation of ground water to great depths (Fig 2.17l). In a few places, like Yellowstone, hot water and steam bring the heat right to the surface. Other such sites, like the geysers in northern California and the Hatchobaru power station in Japan (Fig. 2.17m), use this heat to generate electricity. In some places like Iceland, geothermal energy is also used to heat buildings. Although surface sites are few in number, there is a tremendous resource of hot rock energy at depths of 5 to 10 miles. Unfortunately, current technology and economics do not enable us to tap this vast heat reservoir.

Geo-Exchange

The low-grade thermal energy at the normal temperature of shallow ground can be extracted by a heat pump to heat buildings or domestic hot water (heat pumps are explained in Section 16.10). This same heat pump can use the ground as a heat sink in the summer. Since the ground is warmer than the air in winter and cooler than the air in summer, a **ground-source heat pump** is much more efficient than normal **air-source heat pumps.** Also, since electricity is used to pump heat and not create it, a geo-exchange heat pump is three to four times more efficient than resistance electric heating.

The use of geo-exchange heat pumps can significantly reduce our consumption of energy and the corresponding emission of pollution and greenhouse gases. Reductions of 40 percent over air-source heat pumps and reductions of 70 percent compared to electric-resistance heating and standard air-conditioning equipment are feasible. See Chapter 16 for a more detailed discussion of this excellent system.

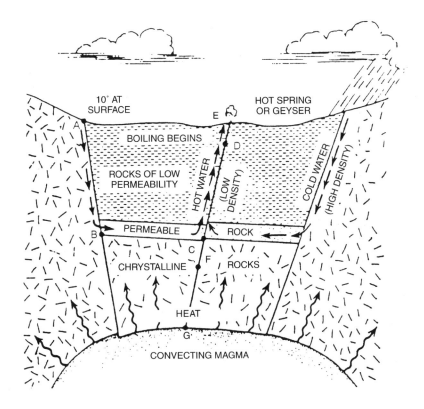

Figure 2.17l At present, the primary source of geothermal energy comes from water circulating deep into the earth. (From "Sourcebook on the Production of Electricity from Geothermal Energy," DOE.)

2.18 HYDROGEN

Although hydrogen is *not* a source of energy, it might play an important role in a sustainable economy. Hydrogen is the ideal nonpolluting fuel because when it is burned, only water is produced. It does not contribute to global warming.

Hydrogen is abundant, but all of it is locked up in compounds, such as water (H_2O). To produce free hydrogen, energy is needed to break the chemical bonds. Although several methods exist for producing hydrogen, if the required energy is from a nonrenewable source like natural gas, hydrogen production is not sustainable. On the other hand, if the process uses renewable energy sources, such as biomass, solar, or wind energy, we have a truly clean sustainable fuel.

Hydrogen is a good match for the intermittent sources of solar and wind, whose main weakness is energy storage. When excess electricity is produced, it can be used to produce

Figure 2.17m The Hatchobaru Geothermal Power Station in Japan. (Courtesy of Kyushu Electric Power Co., Inc.)

hydrogen from water by electrolysis. Later, hydrogen can be used to generate pollution-free electricity in fuel cells, which are explained in Section 3.22. It can also be used to fire building furnaces or automobile engines.

The efficient and economical storage of hydrogen remains a technical problem. High-pressure tanks are heavy and expensive. To store hydrogen as a liquid, one must cool it to −423°F (−253°C). A more efficient solution might be to store the hydrogen in chemical compounds called **hydrides.**

Hydrogen has the potential to become a clean renewable fuel to power our cars and buildings, but since it is not a source of energy, we must still develop the renewable energy sources described above.

2.19 CONCLUSION

If we are looking for insurance against want and oppression, we will find it only in our neighbors' prosperity and goodwill and, beyond, that, in the good health of our worldly places, our homelands. If we were sincerely looking for a place of safety, for real security and success, then we would begin to turn to our communities—and not the communities simply of our human neighbors, but also of the water, earth, and air, the plants and animals, all the creatures with whom our local life is shared.

—*Wendell Berry, Author*

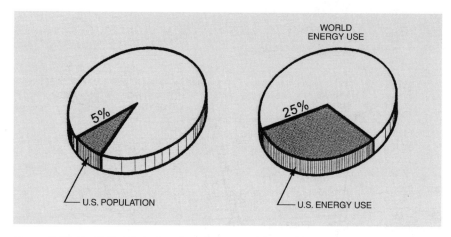

Figure 2.19 The United States has about 5 percent of the world's population, but it consumes about 25 percent of the world's energy.

We are not achieving safety by the way we use energy. We are damaging the environment, changing the climate, and using up our nonrenewable energy sources at a phenomenal rate (Fig. 2.19). Our present course is not sustainable.

Since buildings use more than one-third of all energy consumed, the building-design community has both the responsibility and the opportunity to make major changes in energy use. The amount of energy a building consumes is mainly a function of its design.

Since the energy crisis of 1973, many fine buildings have shown us that buildings can be both energy-efficient and aesthetically successful. As Bob Berkebile, one of our most environmentally responsive architects, has said, "If a building makes animals or people or the planet sick, it's not beautiful and it's not good design" (Wylie, 1994).

The following chapters present the information and design tools needed to create aesthetic, energy-conscious buildings. The goal is to *reduce* the amount of energy that buildings need, using the three-tier approach: design of the building itself, use of passive systems, and finally, efficient mechanical systems.

Since heating, cooling, and lighting are a consequence of energy manipulation, it is important to understand certain principles of energy. The next chapter reviews some of the basic concepts and introduces other important relationships between energy and objects.

KEY IDEAS OF CHAPTER 2

1. We are squandering the Earth's riches without regard to the needs of future generations.
2. Sustainability can be achieved by implementing the 4 "Rs": reduce, reuse, recycle, and regenerate.
3. Sustainable design is also known as green, ecological, and environmentally responsible design.
4. The greater the population, the more difficult it is to achieve sustainability.
5. The greater the affluence, the more difficult it is to achieve sustainability.
6. Limitless growth is the enemy of sustainability.
7. Because many important phenomena, such as energy consumption, are exhibiting exponential growth, and because people do not have a good understanding of the implications of exponential growth, improper decisions are being made about the future.
8. Sustainability can be achieved only if we design and build energy-efficient buildings.
9. The massive use of fossil fuels is causing global warming and climate change.
10. At present, most of our energy comes from nonrenewable and polluting energy sources, such as coal, oil, gas, and nuclear energy.
11. We must switch to the renewable non-polluting energy sources, such as solar, wind, biomass, hydro-power, and geothermal energy.
12. Geo-exchange heat pumps have great potential for energy conservation.
13. Although not a source of energy, hydrogen has the potential to be the clean fuel of the future.
14. As architect Bob Berkabile said, "If a building makes animals or people or the planet sick, it's not beautiful and it's not good design."

References

Bartlett, A. "Forgotten Fundamentals of the Energy Crisis." *American Journal of Physics,* September, 1978, 46 (9).

Bartlett, A. "Reflections on Sustainability, Population Growth, and the Environment — Revisited. *Renewable Resources Journal.* Winter, 1997. 6–23.

Butti, K., and J. Perdin. *A Golden Thread.* New York: Van Nostrand Reinhold, 1980.

Campbell, C., and J. Laherrère. "The End of Cheap Oil." *Scientific American,* March, 1998, 278 (3).

FSEC. "Air-conditioning use cut by 80 percent in Lakeland research home, Cocoa, FL," *Solar Collector,* June/July 1998, 23 (5).

McDonough, W., and M. Braungart. "The Next Industrial Revolution." *The Atlantic Monthly,* October, 1998, 282 (4): 82–92.

Stobaugh, R., and D. Yerkin. *Energy Future — Report of the Energy Project at the Harvard Business School.* New York: Random House, 1979.

Wylie, A. "Form with a Function." *America West Airlines Magazine,* April, 1994, 57–61.

Resources

FURTHER READING
(See Bibliography in back of book for full citations. The list includes valuable out-of-print books.)

Barnett, D. L., and W. D. Browning. *A Primer on Sustainable Building.*

Bartlett, A. A. "Reflections on Sustainability, Population Growth, and the Environment — Revisited." *Renewable Resources Journal.* 6–23.

Campbell, C. J., and J. H. Leherrère. "The End of Cheap Oil."

Climate Change: State of Knowledge.

Commoner, B. *The Politics of Energy.*

Crosbie, M. J. *Green Architecture: A Guide to Sustainable Design.*

Crowther, R. *Ecologic Architecture.*

Environmental Building News. See Appendix J.

Daniels, K. *The Technology of Ecological Building.*

Ehrlich, P. R. *The Population Explosion.*

Gelbspan, R. *The Heat is On: The High Stakes Battle Over Earth's Threatened Climate.*

Gipe, P. *Wind Energy Comes of Age.*

Gipe, P. *Wind Power for Home and Business: Renewable Energy for the 1990's and Beyond.*

Gore, A. *Earth in the Balance.*

Guiding Principles of Sustainable Design.

Hawken, P. *The Ecology of Commerce: A Declaration of Sustainability.*

Hawken, P., A. Lovins, and L H. Lovins. *Natural Capitalism: Creating the Next Industrial Revolution.*

Jones, D. L. *Architecture and the Environment: Bioclimatic Building Design.*

Lovins, A. B. "Soft Energy Paths."

Lovins, A. B., and L. H. Lovins. "Brittle Power: Energy Strategy for National Security."

Lyle, J. T. *Regenerative Design for Sustainable Development.*

McIntyre, M. "Solar Energy: Today's Technologies for a Sustainable Future."

Miller, B. *Buildings for a Sustainable America Case Studies.*

National Audubon Society and Croxton Collaborative, Architects. *Audubon House: Building the Environmentally Responsible, Energy-Efficient Office.*

Pearson, D. *The New Natural House Book: Creating a Healthy, Harmonious, and Ecologically Sound Home.*

Potts, M. *The New Independent Home: People and Houses that Harvest the Sun.*

Recommendations to the United Nations Commission on Sustainable Development: Final Report on the NGO Renewable Energy Initiative.

Scheer, H. *A Solar Manifesto: The need for a total solar energy supply—and how to achieve it.*

Singh, M. *The Timeless Energy of the Sun for Life and Peace with Nature.*

Van der Ryn, S., annotator. *Designing Sustainable Systems: A Reader,* 1993, Vols. 1 and 2. Contains many hard-to-find key source articles. [Contact Ecological Design Institute—see Appendix J.]

Van der Ryn, S. *The Toilet Papers: Recycling Waste and Conserving Water.* Ecological Design Press, 1995.

Van der Ryn, S., and P. Calthorpe. *Sustainable Communities: A New Design Synthesis for Cities, Suburbs, and Towns.*

Van der Ryn, S., and S. Cowan. *Ecological Design.*

Wilson, E. O. "Is Humanity Suicidal: We're Flirting with the Extinction of Our Species." *New York Times Magazine.*

World Commission on Environment and Development. *Our Common Future.*

Zeiher, L. C. *The Ecology of Architecture: A Complete Guide to Creating the Environmentally Conscious Building.*

CD-ROM

(See Appendix J in back of book for full citations and ordering information.)

Green Building Advisor—A CD-ROM that guides and suggests approaches to sustainable design. It offers a great variety of design strategies to the user.

VIDEOS

(See Appendix J in back of book for full citations and ordering information.)

Affluenza. KCTS Television

Arithmetic, Population, and Energy. Dr. Albert A. Bartlett. 65 minutes.

Building Green: Audubon House. 28 minutes.

Case Studies in Sustainable Design. 1 hour, 17 minutes.

Ecological Design—Inventing the Future.

Environmental Architecture. 30 minutes.

Keeping the Earth: Religious and Scientific Perspectives on the Environment. 27 minutes.

World Population. ZPG.

ORGANIZATIONS

(See Appendix J in back of book for full citations and ordering information.)

American Hydrogen Association
Internet: (E-mail) aha@getnet.com; www.clean-air.org

American Solar Energy Society
www.csn.net.solar/factbase.htm
National group for renewable-energy education.

Caddet Center for Renewable Energy
www.caddet.co.uk
Provides summaries of renewable-energy projects around the world.

CREST—Center for Renewable Energy and Sustainable Technology
http://www.crest.org
Comprehensive educational source for renewables.

Energy Efficiency & Renewable Energy Clearinghouse (EREC)
http://eren.doe.gov/erec/factsheets
Provides free general and technical information on topics and technologies pertaining to energy efficiency and renewable energy, including photovoltaic systems, solar energy, and solar radiation.

National Renewable Energy Laboratory
www.nrel.gov
Good source for detailed information on renewables.

National Pollution Prevention Center for Higher Education
734/764-1412, fax: 647-5841, nppc@umich.edu

Union of Concerned Scientists
www.ucsusa.org

Wind Information

Wind energy data can be obtained from:

American Wind Energy Association (AWEA)
122 C Street, NW, 4th Floor
Washington, DC 20001
(202) 383-2500
Fax (202) 383-2505
www.econet.org/awea

National Climatic Data Center
151 Patton Avenue, Room 120
Ashville, NC 28801
(704) 271-4800
Fax (704) 271-4876
E-mail < orders@ncdc.noaa.gov >
< http://www.ncdc.noaa.gov >

Other Wind Information

A Siting Handbook for Small Wind Energy Conversion Systems. Battelle Pacific Northwest Laboratory, National Technical Information Service, U.S. Department of Commerce, 5285 Port Royal Road, Springfield, VA 22161, 1980.

Park, J. *The Wind Power Book.* Palo Alto, CA: Chesire Books, 1981. (out of print)

Gipe, P. *Wind Power for Home & Business: Renewable Energy for the 1990s and Beyond.* Chelsea Green Publishing Company, P.O. Box 130, Route 113, Post Mills, VT 05058-0130, 1993.

Wind Energy Resource Atlas of the U.S. Battelle Pacific Northwest Laboratories. Available from the American Wind Energy Association (see above) or the National Technical Information Service (see Appendix J).

BASIC PRINCIPLES

"If we are anything, we must be a democracy of the intellect. We must not perish by the distance between people and government, between people and power.... And that distance can only be conflated, can only be closed, if knowledge sits in the homes and heads of people with no ambition to control others, and not up in the isolated seats of power."

J. Bronowski
The Ascent of Man, 1973 (p. 435)

Please note that figure numbers are keyed to sections. Gaps in figure numbering result from sections without figures.

3.1 INTRODUCTION

The heating, cooling, and lighting of buildings are accomplished by adding or removing energy. A good basic understanding of the physics of energy and its related principles is a prerequisite for much of the material in the following chapters. Consequently, this chapter is devoted to both a review of some rather well-known concepts as well as an introduction to some less familiar ideas such as mean radiant temperature, time lag, the insulating effect of mass, and embodied energy.

3.2 HEAT

Energy comes in many forms and most of these are used in buildings. Much of this book, however, is concerned with energy in the form of heat, which exists in three different forms:

1. Sensible heat — can be measured with a thermometer
2. Latent heat — the change of state or phase change of a material
3. Radiant heat — a form of electromagnetic radiation

3.3 SENSIBLE HEAT

The random motion of molecules is a form of energy called **sensible heat.** An object whose molecules have a larger random motion is said to be hotter and to contain more heat (see Fig. 3.3a). This type of heat can be measured by a thermometer and is, therefore, called sensible heat. If the two objects in Fig. 3.3a are brought into contact, then some of the more intense random motion of the object on the left will be transferred to the object on the right by the heat-flow

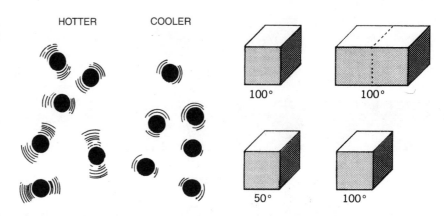

Figure 3.3a Sensible heat is the random motion of molecules and temperature is a measure of the intensity of that motion.

Figure 3.3b The amount of sensible heat is a function of both temperature and mass. In each case, the blocks on the right have more sensible heat content than the blocks on the left.

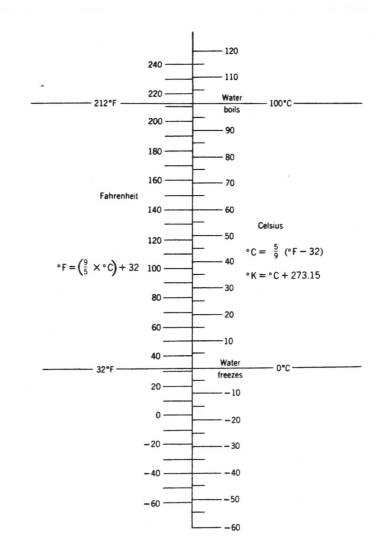

$$°F = \left(\frac{9}{5} \times °C\right) + 32$$

$$°C = \frac{5}{9}\,(°F - 32)$$

$$°K = °C + 273.15$$

Figure 3.3c Conversion: Fahrenheit–Celsius degrees. (From *Mechanical and Electrical Equipment for Buildings,* 9th edition, Stein and Reynolds, ©2000 John Wiley & Sons, Inc.)

mechanism called **conduction.** The molecules must be close to each other in order to collide. Since in air the molecules are far apart, air is not a good conductor of heat. A vacuum allows no conduction at all.

Temperature is a measure of the intensity of the random motion of molecules. We cannot determine the heat content of an object just by knowing its temperature. For example, in Fig. 3.3b (top), we see two blocks of a certain material that are both at the same temperature. Yet the block on the right will contain twice the heat because it has twice the mass.

The mass alone cannot determine the heat content either. In Fig. 3.3b (bottom), we see two blocks of the same size, yet one block has more heat content because it has a higher temperature. Thus, sensible heat content is a function of both mass and

temperature. It is also a function of heat capacity, which is discussed in Section 3.15.

In the United States, we still use the British Thermal Unit (BTU) as our unit of heat. The rest of the world, including Great Britain, uses the System International (S.I.) unit called the joule or calorie. The amount of heat required to raise 1 pound of water 1°F is called a BTU.

	U.S. System	SI. System
Heat	BTU	joule or calorie
Temperature*	degree Fahrenheit (°F)	degree Celsius (°C)

*For conversions see Fig. 3.3c.

3.4 LATENT HEAT

By adding 1 British Thermal Unit (BTU) of heat to a pound of water its temperature is raised 1°F. It takes,

however, 144 BTU to change a pound of ice into a pound of water and about 1000 BTU to change a pound of water to a pound of steam (Fig. 3.4). It takes very large amounts of energy to break the bonds between the molecules when a change of state occurs. "Heat of fusion" is required to melt a solid and "heat of vaporization" is required to change a liquid into a gas. Notice also that the water was no hotter than the ice and the steam was not hotter than the water, even though a large amount of heat was added. This heat energy, which is very real but cannot be measured by a thermometer, is called latent heat. In melting ice or boiling water, sensible heat is changed into latent heat.

Latent heat is a compact and convenient form for storing and transferring heat. However, since the melting and boiling points of water are not always suitable, we use other materials such as "Freon," which has the melting and boiling temperatures necessary for refrigeration machines.

3.5 EVAPORATIVE COOLING

When sweat evaporates from the skin, a large amount of heat is required. This heat of vaporization is drawn from the skin, which is cooled in the process. This sensible heat in the skin is turned into the latent heat of the water vapor.

As water evaporates, the air next to the skin becomes humid and eventually even saturated. The moisture in the air will then inhibit further evaporation. Thus, either air motion to remove this moist air or very dry air is required to make evaporative cooling efficient (Fig. 3.5).

Buildings can also be cooled by evaporation. Water sprayed on the roof can dramatically reduce its temperature. In dry climates, air entering buildings can be cooled with water sprays. Such techniques will be described in Chapter 10.

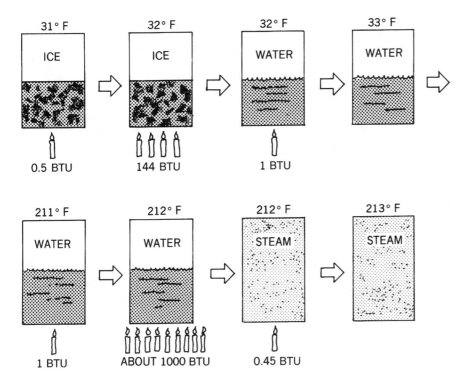

Figure 3.4 Latent heat is the large amount of energy required to change the state of a material (phase change), and it cannot be measured by a thermometer. The values given here are for 1 pound of water, ice, or steam.

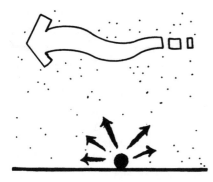

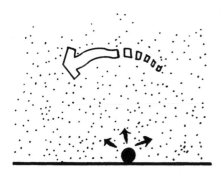

Figure 3.5 The rate of evaporative cooling is a function of both humidity and air movement. Evaporation is rapid when the humidity is low and air movement is high. Evaporation is slow when the humidity is high and air movement is low.

3.6 CONVECTION

As a gas or liquid acquires heat by conduction, the fluid expands and becomes less dense. It will then rise by floating on top of denser and cooler fluid as seen in Fig. 3.6a. The resulting currents transfer heat by the mechanism called **natural convection.** This heat-transfer mechanism is very much dependent on gravity and, therefore, heat never convects down. Since we live in a sea of air, natural convection is a very important heat-transfer mechanism.

Natural convection currents tend to create layers that are at different temperatures. In rooms hot air collects near the ceiling and cold air near the floor (Fig. 3.6b). This stratification can be an asset in the summer and a liability in the winter. Strategies to deal with this phenomenon will be discussed throughout this book.

A similar situation occurs in still lakes where the surface water is much warmer than deep water (Fig. 3.6b).

A different type of convection occurs when the air is moved by a fan or by the wind, or when water is moved by a pump (Fig. 3.6c). When a heated fluid (gas or liquid) is circulated between hotter and cooler areas, heat will be transferred by the mechanism known as **forced convection.**

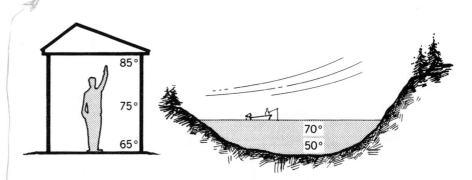

Figure 3.6a Natural-convection currents result from differences in temperature.

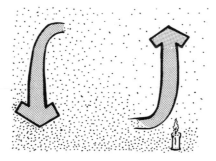

Figure 3.6b Stratification results from natural convection unless other forces are present to mix the air or water.

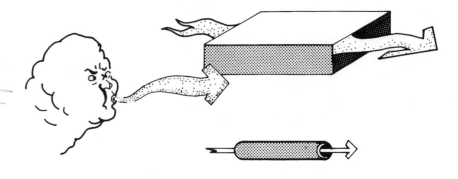

Figure 3.6c Forced convection is caused by wind, fans, or pumps.

3.7 TRANSPORT

In the eighteenth and nineteenth centuries, it was common to use warming pans to preheat beds. The typical warming pan, as shown in Fig. 3.7, was about 12 inches in diameter and about 4 inches deep, and it had a long wooden handle. It was filled with hot embers from the fireplace, carried to the bedrooms, and passed between the sheets to remove the chill. In the early twentieth century, it was common to use hot-water bottles for the same purpose. This transfer of heat by moving material is called **transport.** Because of convenience, forced convection is much more popular today for moving heat around a building than is "transport."

Figure 3.7 Warming pans and hot-water bottles were popular in the past to transport heat from the fireplace or stove to the cold beds.

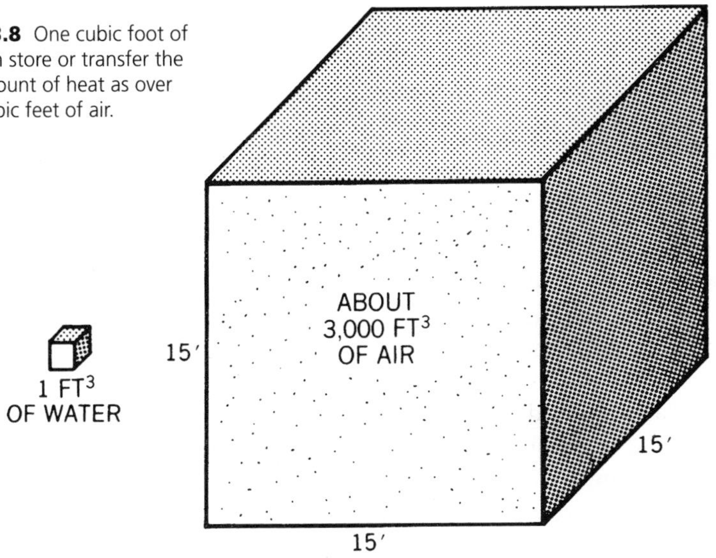

Figure 3.8 One cubic foot of water can store or transfer the same amount of heat as over 3,000 cubic feet of air.

ABOUT 3,000 FT³ OF AIR

15′

15′

15′

1 FT³ OF WATER

3.8 ENERGY-TRANSFER MEDIUMS

In both the heating and cooling of buildings, a major design decision is the choice of the energy-transfer medium. The most common alternatives are air and water. It is, therefore, very valuable to understand the relative heat-holding capacity of these two materials. Because air has both a much lower density and specific heat than water, much more of it is required to store or transport heat. To store or transport equal amounts of heat, we need a volume of air about 3000 times greater than water (Fig. 3.8).

3.9 RADIATION

The third form of heat is radiant heat. It is the part of the electromagnetic spectrum called **infrared.** All bodies facing an air space or a vacuum emit and absorb radiant energy continuously. Hot bodies lose heat by radiation because they emit more energy than they absorb (Fig. 3.9a). Objects at room temperature radiate in the infrared region of the electromagnetic spectrum, while objects hot enough to glow radiate in the visible part of the spectrum. Thus, the wavelength or frequency of the radiation emitted is a function of the temperature of the object.

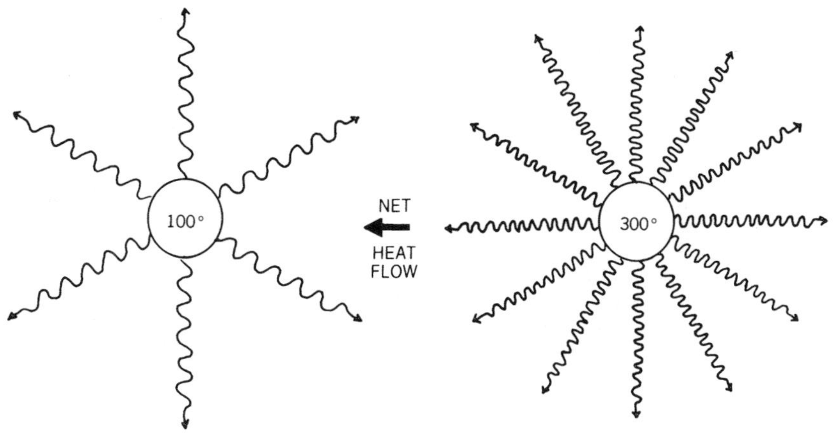

100°

NET HEAT FLOW

300°

Figure 3.9a Although all objects absorb and emit radiant energy, there will be a net radiant flow from warmer to cooler objects.

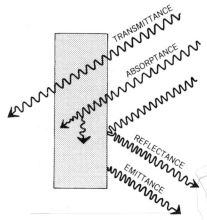

Figure 3.9b Four different types of interactions are possible between energy and matter.

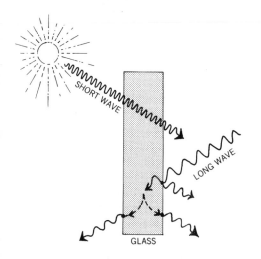

Figure 3.9c The type of interaction depends not only on the nature of the material but also on the wavelength of the radiation.

Radiation is not affected by gravity, so a body will radiate down as much as up. Radiation is, however, affected by the nature of the material with which it interacts and especially the surface of the material. The four possible interactions, as illustrated in Fig. 3.9b, are as follows:

1. Transmittance—the situation in which the radiation passes through the material
2. Absorptance—the situation in which the radiation is converted into sensible heat within the material

3. Reflectance—the situation in which the radiation is reflected off the surface
4. Emittance—the situation in which the radiation is given off by the surface, thereby reducing the sensible heat content of the object. Polished metal surfaces have low emittance, while most other materials have high emittance.

The type of interaction that will occur is not only a function of the material but also the wavelength of the radiation. For example, glass interacts very differently with solar radiation (short wavelength) than with thermal radiation (long-wave infrared), as is shown in Fig. 3.9c. Glass is mostly transparent to short-wave radiation and opaque to long-wave radiation. The long-wave radiation is mostly absorbed, thereby heating up the glass. Much of the absorbed radiation is then reradiated from the glass inward and outward. The net effect is that some of the long-wave radiation is blocked by the glass. The greenhouse effect, explained below, is partly due to this property of glass and most plastics used for glazing. Polyethelene is the major exception, since it is transparent to infrared radiation.

3.10 GREENHOUSE EFFECT

The concept of the **greenhouse effect** is vital for understanding both solar energy and climate. The greenhouse effect is partly due to the fact that the type of interaction that occurs between a material and radiant energy depends on the wavelength of that radiation.

Figure 3.l0a illustrates the basic concept of the greenhouse effect. The short wave solar radiation is able to easily pass right through the glass whereupon it is absorbed by indoor objects. As these objects warm up, they increase their emission of radiation in the long-wave portion of the

Figure 3.10a The greenhouse effect is partly a consequence of the fact that glazing transmits short-wave but blocks long-wave radiation.

electromagnetic spectrum. Since glass is opaque to this radiation, some of the energy is trapped. The glass has created, in effect, a heat trap and the indoor temperature begins to rise.

To get a better understanding of this very important concept, let us look at the graphs in Fig. 3.l0b. First look at the top graph, which describes the behavior of glass with respect to radiation. The percentage transmission is given as a function of wavelength. Notice that glass has a very high transmission for radiation between 0.3 and 3 μm, and zero transmission for radiation above and below that "window." This means that glass is transparent to short-wave radiation and opaque to long-wave radiation. It is very "selective" in what may pass through. The graphs together show that the part of the electromagnetic spectrum for which glass is transparent corresponds to solar radiation, and the part for which glass is opaque corresponds to the long-wave heat radiation given off by objects at room temperature (see lower graph of Fig. 3.l0b). The solar radiation enters through the glass and is absorbed by objects in the room. These objects heat up and then increase their reradiation in the long-wave infrared part of the spectrum. Since glass is opaque to this radiation, a part of the energy is trapped, and the room heats up. This is one of the mechanisms of the

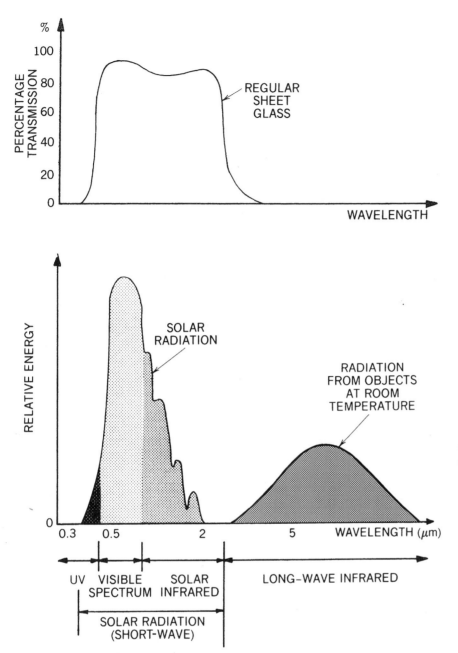

Figure 3.10b These two graphs are aligned vertically by wavelength. Notice that glass has about 85 percent transmission for short-wave (solar) radiation and about 0 percent transmission for long-wave radiation.

greenhouse effect. The other major mechanism of the greenhouse effect is the obvious fact that the glazing stops the convective loss of hot air. This mechanism together with the "selective" quality of glass form a very effective heat trap.

A few additional observations can be made with regard to these graphs.

The first is that solar energy consists of about 5 percent ultraviolet radiation, 45 percent visible light, and 50 percent infrared radiation. To differentiate this infrared from that given off by objects at room temperature, the phrases "short-wave" and "long-wave' are added, respectively. Similarly, ultraviolet radiation has

shorter and longer wavelengths. The portion of the ultraviolet spectrum that causes sunburns is blocked by glass but the part that causes colors to fade is not.

3.11 EQUILIBRIUM TEMPERATURE OF A SURFACE

Understanding the heating, cooling, and lighting of buildings requires a fair amount of knowledge of the behavior of radiant energy. For example, what is the best color for a solar collector, and what is the best color for a roof to reject solar heat in the summer? Figure 3.11 illustrates how surfaces of different color and finish interact with radiant energy. To understand why a black metal plate will get much warmer in the sun than a white metal plate, we must remember that materials vary in the way they emit and absorb radiant energy. The balance between the absorptance and emittance will determine how hot the plate will get, the **equilibrium temperature.** Black has a much higher equilibrium temperature than white because it has a much higher absorptance factor. However, black is not the ideal collector of radiant energy because of its high emissivity. Its equilibrium temperature is suppressed because it reradiates much of the energy it has absorbed.

To increase efficiency in solar collectors, a type of **selective surface** was developed. One particular type has the same high absorptance as black but is more stingy in emitting radiation. Its equilibrium temperature is, therefore, very high.

White is the best color to minimize heat gain in the summer because it is not only a poor absorber but also a good emitter of any energy that is absorbed. Thus, white does not like to collect heat, and a very low equilibrium temperature results. This low surface temperature minimizes the heat gain to the material below the surface.

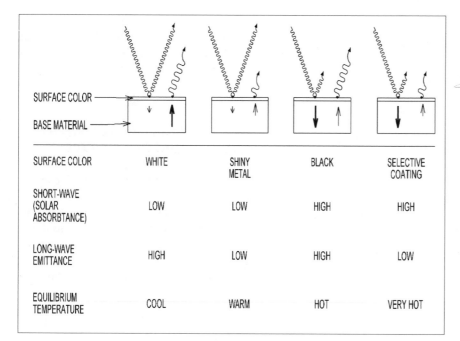

Figure 3.11 The equilibrium temperature is a consequence of both the absorptance and the emittance characteristics of a material.

Polished-metal surfaces, such as shiny aluminum can be used as radiant insulators because they neither absorb nor emit radiation readily. For this reason, aluminum foil is sometimes used inside of walls as a radiant barrier. However, the equilibrium temperature of a polished-metal surface is higher than white because the metal does not like to emit whatever it has absorbed. Although both white and polished metals reflect about the

same percentage of sunlight, white is a much better emitter of heat radiation and so will be cooler in the sun than a polished-metal surface.

3.12 MEAN RADIANT TEMPERATURE

To determine if a certain body will be a net gainer or loser of radiant energy, we must consider both the tempera-

ture and the exposure angle of all objects that are in view of the body in question. The **mean radiant temperature (MRT)** describes the radiant environment for a point in space. For example, the radiant effect on one's face by a fireplace (Fig. 3.12) is quite high because the fire's temperature at about 1000°F more than compensates for the small angle of exposure. A radiant ceiling can have just as much of a warming effect but with a much lower temperature (90°F) because its large area creates a large exposure angle. The radiant effect can also be negative as in the case of a person standing in front of a cold window.

Walking toward the fire (Fig. 3.12) would increase the MRT, while walking toward the cold window would result in a lower MRT because the relative size of the exposure angles would change. Many a "cold draft" near large windows in winter is actually a misinterpretation of a low MRT. The significant effect MRT has on thermal comfort is further explained in the next chapter.

3.13 HEAT FLOW

Heat flows naturally from a higher temperature to a lower temperature, but not necessarily from more heat to less heat. To better understand this,

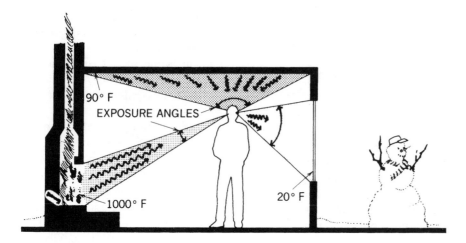

Figure 3.12 The mean radiant temperature (MRT) at any point is a result of the combined effect of a surface's temperature and angle of exposure.

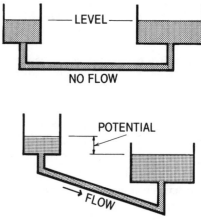

Figure 3.13 A water analogy shows how temperature and not heat content determines heat flow.

we can consider a water analogy. In this analogy, the height between different levels of water represents the temperature difference between two heat sources and the volume of water represents the amount of heat.

Since both reservoirs are at the same level as shown in Fig. 3.13 (top), there is no flow. The fact that there is more water (heat) on one side than the other is of no consequence.

If, however, the levels of the reservoirs are not the same, then flow occurs as indicated in Fig. 3.13 bottom. Notice that this occurs even when the amount of water (heat) is less on the higher side. Just as water will flow only down, so heat will flow only from a higher temperature to a lower temperature.

To get the water to a higher level some kind of pump is required. Heat, likewise, can be raised to a higher temperature only by some kind of "heat pump," which works against the natural flow. Refrigeration machines, the essential devices in air conditioners or freezers, pump heat from a lower to a higher temperature. They will be explained in some detail in Chapter 16.

3.14 HEAT SINK

It is easy to see how transporting hot water to a room also supplies heat to the room. It is not so obvious, however, to see how supplying chilled water cools the room. Are we supplying "coolth"? This imaginary concept only confuses and should not be used. A correct and very useful concept is that of a **heat sink**. In Fig. 3.14 (top) the room is cooled by the chilled water that is acting as a heat sink. The chilled water gets warmer while the room gets cooler.

Often the massive structure of a building acts as a heat sink. Many massive buildings feel comfortably cool on hot summer days, as in Fig. 3.14 (bottom). During the night, these buildings give up their heat by con-

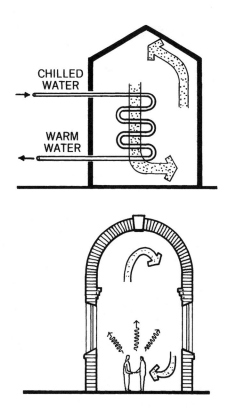

ROMAN VAULTS

Figure 3.14 The cooling effect of a heat sink can result from a cold fluid or from the mass of the building itself.

vection to the cool night air and by radiation to the cold sky—thus recharging their heat sink capability for the next day. However, in very humid regions the high nighttime temperatures prevent effective recharging of the heat sink, and, consequently, massive buildings are not helpful.

3.15 HEAT CAPACITY

The amount of heat required to raise the temperature of a material 1°F is called the **heat capacity** of that material. The heat capacity of different materials varies widely, but in general heavier materials have a higher heat capacity. Water is an exceptional material in that it has the highest heat capacity even though it is a middleweight material. In architecture we

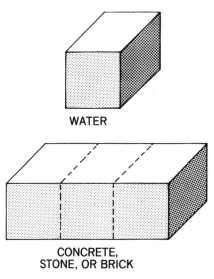

WATER

CONCRETE, STONE, OR BRICK

Figure 3.15 Since 1 cubic foot of water can hold the same amount of heat as about 3 cubic feet of concrete, the volumetric heat capacity of concrete is one-third that of water.

are usually more interested in the heat capacity per volume than in the heat capacity per pound, which is more commonly known as **specific heat.** Each volume in Fig. 3.15 has the same heat capacity. Also note again the dramatic difference in heat capacity between air and water as shown in Fig. 3.8. This clearly indicates why water is used so often to store or transport heat. See Table 7.17A for the heat capacity of various common materials.

3.16 THERMAL RESISTANCE

The opposition of materials and air spaces to the flow of heat by conduction, convection, and radiation is called **thermal resistance.** By knowing the resistance of a material, we can predict how much heat will flow through it and can compare materials with each other. The thermal resistance of building materials is largely a function of the number and size of air spaces that they contain. For example, 1 inch of wood has the same thermal resistance as 12 inches of concrete mainly because of the air spaces created by the cells in the

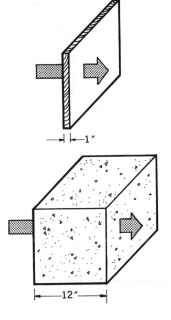

Figure 3.16 The heat flow is equal through the two materials because the thermal resistance of wood is twelve times as great as that of concrete.

SIDEBOX 3.16

*units of thermal resistance or R-value = $\dfrac{ft^2 \times °F}{BTU/h}$

where BTU/h = heat flow per hour

or in SI, R-value = $\dfrac{m^2 \times °C}{W}$

where

m = meter
°C = degrees Celsius
W = watts
1W = 3.412 BTU/h

wood (Fig. 3.16). However, this is true only under steady-state conditions, i.e., where the temperature across a material remains constant for a long period of time. Under certain dynamic temperature conditions, 12 inches of concrete can appear to have more resistance to heat flow than 1 inch of wood. To understand this, we must consider the concept of time lag, explained in Section 3.18. Because the units of thermal resistance are complex and hard to remember, technical literature frequently gives the thermal resistance in terms of "R-Value*" (see Sidebox 3.16).

3.17 HEAT-FLOW COEFFICIENT

Much of the technical literature describes the thermal characteristics of wall or roof systems in terms of the heat-flow coefficient "U" rather than the total thermal resistance. Because the heat-flow coefficient is a measure of heat flow (conductance), it is the reciprocal of thermal resistance (see Sidebox 3.17).

3.18 TIME LAG

Consider what happens when two walls with **equal heat resistance,** 12 inches of concrete or 1 inch of wood, are first exposed to a temperature difference. Let's say that the temperature is 100°F on one side and 50°F on the other side of both the concrete and the wooden walls. Heat will flow through the concrete, but the initial heat to enter will be used to raise the temperature of the massive material. Only after the wall has substantially warmed up can heat exit the other side. On the other hand, this delay in heat conduction is very short for 1 inch of wood because of its low heat capacity. This delayed heat-flow phenomenon is known as **time lag.**

This concept can be understood more easily by means of a water analogy in which the flow of water through an in-line storage tank represents the thermal capacity of a material (Fig. 3.18). The small tank represents 1 inch of wood (small heat capacity) and the large tank represents 12 inches of concrete (large heat capacity). After four hours, water (heat) is flowing through the system with the small capacity but not through the system with the high capacity. Thus, high-capacity materials have a greater time lag than low-capacity materials. Also note the time-lag phenomenon ends when the storage tank is full. Under steady-state conditions there is no time lag.

3.19 INSULATING EFFECT OF MASS

If the temperature difference across a massive material fluctuates in certain specific ways, then the massive material will act as if it had high thermal resistance. Let us consider a massive concrete house in the desert on a summer day. A wall of this building is

SIDEBOX 3.17

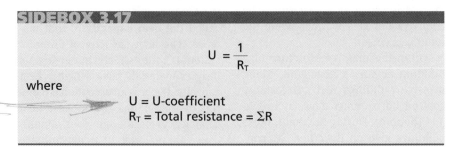

$$U = \frac{1}{R_T}$$

where

U = U-coefficient
R_T = Total resistance = ΣR

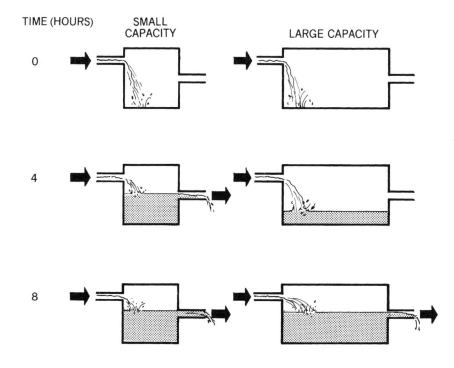

TIME (HOURS) SMALL CAPACITY LARGE CAPACITY

0

4

8

Figure 3.18 This water analogy of time-lag illustrates how high capacity delays the passage of water. Similarly, high heat capacity delays the transmission of heat. This is similar to heat flow through either 1 inch of wood (small capacity) or 12 inches of concrete (high capacity).

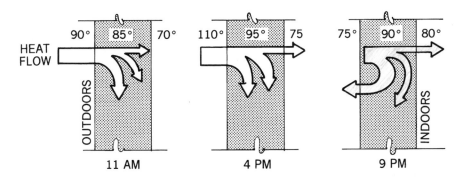

HEAT FLOW OUTDOORS INDOORS

90° 85° 70°   110° 95° 75   75° 90° 80°

11 AM   4 PM   9 PM

Figure 3.19 The insulating effect of mass is most pronounced in hot and dry climates in the summer.

shown at three different times of day (Fig. 3.19). At 11 A.M. the indoor temperature is lower than the outdoor temperature and heat will flow inward. However, most of this heat is diverted to raising the temperature of the wall.

At 4 P.M. the outdoor temperature is very high. Although some heat is now reaching the indoors, much heat is still being diverted to further raising the temperature of the wall.

However, at 9 P.M. the outside temperature has declined enough to be below the indoor and especially the wall temperatures. Now most of the heat that was stored in the wall is flowing outward without ever reaching the interior of the house. In this situation, the time lag of the massive material "insulated" the building from the high outdoor temperatures. It is important to note that the benefits of time lag occur only if the outdoor temperature fluctuates. Also the larger the daily temperature swing, the greater will be the insulating effect of the mass. Thus, this insulating effect of mass is most beneficial in hot and dry climates during the summer. This effect is not helpful in cold climates where the temperature remains consistently below the indoor temperature, and only slightly helpful in humid climates where the daily temperature range is small. In *very* humid climates, the thermal mass can be a liability and should be avoided.

3.20 ENERGY CONVERSION

The First Law of Thermodynamics states that energy can be neither created nor destroyed, only changed in form. But while energy is never lost, the Second Law of Thermodynamics states that its ability to do work can decline. For example, high-temperature steam can generate electricity with a steam turbine, while the same amount of heat in the form of warm water cannot perform this task. Electricity is a very high-grade, valuable energy form, and to use it to purposely generate low grade heat is a terrible waste. Sunlight is another high-grade energy source. It should be used to light a building before it turns into heat instead of turning light into heat directly.

Figure 3.20 shows the conversion of a fossil fuel into electricity. The low efficiency (approximately 30 percent) is a consequence of the large number of conversions required. Thus, electrical energy should not be used when a better alternative is available. For example, heating directly with fossil fuels can be more than 80 percent efficient.

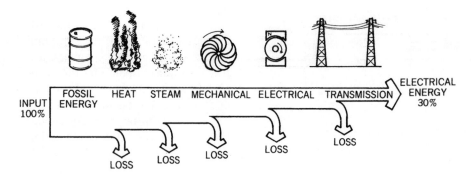

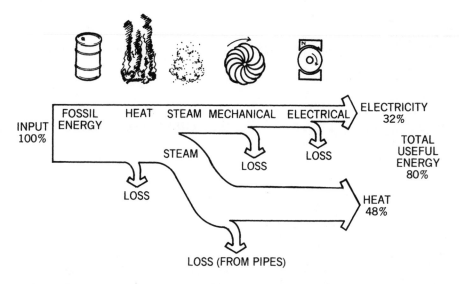

Figure 3.20 In the conversion of fossil fuel into electricity, about 70 percent of the original energy is lost.

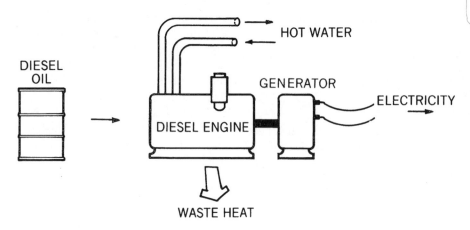

Figure 3.21a Because combined-heat-and-power (CHP) systems generate electricity at the building site, they are able to utilize much of the heat normally wasted.

Figure 3.21b Packaged CHP units are self-contained and easily integrated into a building. Other fuels, such as natural gas, can also be used.

3.21 Combined Heat and Power

Combined heat and power (CHP), also known as **cogeneration,** can greatly reduce the energy losses in producing electricity. Through the generation of electricity at the building site, efficiencies up to 80 percent are possible. Heat, normally wasted at the central power plant, can be used for domestic hot-water or space heating (Fig. 3.21a). Also overland electrical-transmission losses are almost completely eliminated. Compact and fairly maintenance-free packaged combined heat and power units are commercially available for all sizes of buildings (Fig. 3.21b).

3.22 FUEL CELLS

Combined heat and power (CHP) can be even more efficient if a **fuel cell** is used to generate the electricity. Because fuel cells are safe, clean, noiseless, low-maintenance, and compact, they can be placed in any building. Thus, there are no transmission losses, and the waste heat can be used (Fig. 3.22).

Fuel cells are powered with hydrogen that combines with oxygen in the air to form water, electricity, and heat. No emissions pollute the air or cause global warming. No flue is needed.

A "green" high-rise building, 4 Times Square in New York City, uses two fuel cells, located on the fourth floor, to generate a significant portion of the electrical load. Since natural gas is used to create the hydrogen, some carbon dioxide is produced. However, much less carbon dioxide is produced than in conventional buildings because of the high efficiency of the fuel cells. Fuel cells will have their greatest potential to create sustainable buildings when fueled with hydrogen made from renewable sources of energy, such as wind or solar energy.

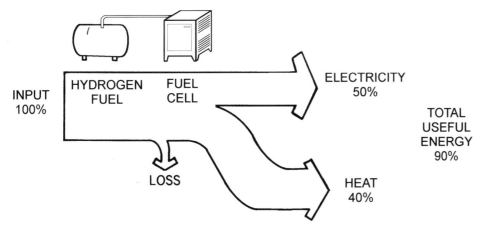

Figure 3.22 Because fuel cells use hydrogen to directly generate electricity and useful heat right inside buildings, about 90 percent of the original energy can be utilized. Fuel cells run off of non-polluting hydrogen.

3.23 EMBODIED ENERGY

Most discussions of energy and buildings are concerned with the use and operation of a building. It is now recognized that it can take large quantities of energy to construct a building. This **embodied energy** is a result of both the construction machinery and the energy required to make and transport the materials. For example, aluminum embodies four times as much energy as steel and about twelve times as much as wood. The embodied energy in a modern office building is about the same as the amount of energy the building will consume in fifteen years.

Much of the embodied energy can be saved when we recycle old buildings. Thus, conservation of energy is another good argument for adaptive reuse and historic preservation (Fig. 3.23).

3.24 SUMMARY

The basic principles described in this chapter will be applied throughout this book. Many of these ideas will make more sense when their applications are mentioned in the forthcoming chapters. It will often prove useful to refer back to these explanations, although more detailed explanations will be given when appropriate. Special concepts, such as those related to lighting, will be explained when needed.

Figure 3.23 A large amount of embodied energy can be saved when existing buildings are reused. (From a poster, copyright 1980 by National Trust for Historic Preservation.)

KEY IDEAS OF CHAPTER 3

1. Sensible heat is the type of heat that can be measured with a thermometer. Dry air has only sensible heat.

2. Heat energy absorbed or given off as a substance changes phase is called latent heat. It is also called heat of vaporization and heat of fusion, and it cannot be measured with a thermometer. Air has latent heat when it has water vapor.

3. Heat is transferred by conduction, convection, radiation, and transport.

4. Stratification of temperatures results from natural convection.

5. Water can hold about 3000 times as much heat as an equal volume of air. Therefore, we say that water has a much greater heat capacity than air.

6. Matter and energy interact in four ways:
   a. Transmittance
   b. Absorptance
   c. Reflectance
   d. Emittance

7. The greenhouse effect traps heat by allowing most short-wave radiation to be admitted, while blocking most long-wave radiation from leaving.

8. The equilibrium temperature of an object sitting in the sun is a result of the relative absorptance and emittance characteristics of the exposed surface.

9. The mean radiant temperature (MRT) describes the radiant environment. An object will simultaneously gain radiation from hotter objects and lose radiation to cooler objects.

10. A cooler object is a potential heat sink. Chilled water or a massive building cooled overnight can act as a heat sink to cool the interior of a building.

11. Thermal resistance is a measure of a material's resistance to heat flow by the mechanisms of conduction, convection, and radiation.

12. Time lag is the phenomenon describing the delay of heat flow through a material. Massive materials have more time lag than light materials.

13. Under certain dynamic temperature conditions, the time lag of massive materials can resist heat flow.

14. The Second Law of Thermodynamics tells us that there will be a loss of usable energy every time energy is converted from one form to another. As a consequence, heating a home directly with gas can be 90 percent efficient, while heating with resistance electricity, which was generated by gas, is only 30 percent efficient.

15. Fuel cells have the potential to efficiently generate electricity and supply useful heat as a byproduct right inside buildings with little or no pollution.

THERMAL COMFORT

"Thermal Comfort — that condition of mind which expresses satisfaction with the thermal environment."

ASHRAE Standard 55-66

Please note that figure numbers are keyed to sections. Gaps in figure numbering result from sections without figures.

4.1 BIOLOGICAL MACHINE

The human being is a biological machine that burns food as a fuel and generates heat as a byproduct. This metabolic process is very similar to what happens in an automobile where gasoline is the fuel and heat is also a significant byproduct (Fig. 4.la). Both types of machines must be able to dissipate the waste heat in order to prevent overheating (Fig. 4.lb). All of the heat-flow mechanisms mentioned in Chapter 3 are employed to maintain the optimum temperature.

All warm-blooded animals, and humans in particular, require a very constant temperature. Our bodies try to maintain an interior temperature of about 98.6°F, and any small deviation creates severe stress. Only 10 to 15 degrees higher or 20 degrees lower can cause death. Our bodies have several mechanisms to regulate the heat flow to guarantee that the heat loss equals the heat generated and that thermal equilibrium will be about 98.6°F.

Some heat is lost by exhaling warm moist air from the lungs, but most of the body's heat flow is through the skin. The skin regulates the heat flow by controlling the amount of blood flowing through it. In summer the skin is flushed with blood to increase the heat loss, while in winter little blood is allowed to circulate near the surface and the skin becomes an insulator. The skin temperature will, therefore, be much lower in winter than in the summer. The skin also contains sweat glands that control body heat loss by evaporation.

Hair is another important device to control the rate of heat loss. Although we no longer have much fur, we still have the muscles that could make our fur stand upright for extra thermal insulation. When we get gooseflesh, we see a vestige of the old

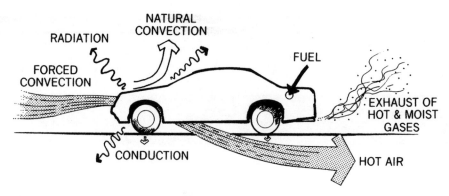

Figure 4.1a Methods of dissipating waste heat from an automobile.

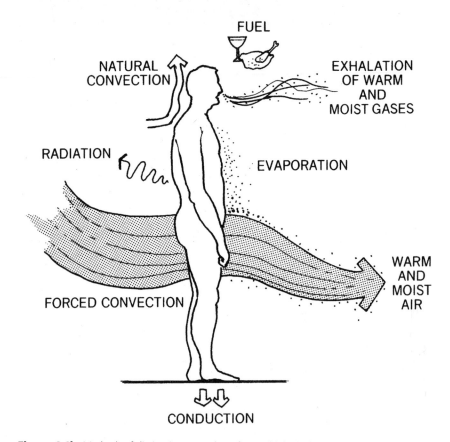

Figure 4.1b Methods of dissipating waste heat from a biological machine.

mechanism. After some days of exposure the body can acclimatize to very high or low temperatures. Changing the total amounts of blood is one important mechanism, with more blood produced under warmer conditions. Excessive heat loss is called **hypothermia,** while insufficient heat loss is **hyperthermia.**

The graph in Fig. 4.1c shows how the effectiveness of our heat loss

mechanisms vary with the ambient temperature. Curve 1 represents the heat generated by a person at rest as the ambient temperature changes. Curve 2 represents the heat lost by conduction, convection, and radiation. Since the heat loss by these mechanisms depends on the temperature difference, it is not surprising that the heat loss decreases as the ambient temperature increases. When the

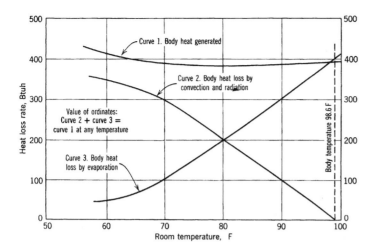

Figure 4.1c The way heat is lost from a body depends on the ambient temperature. This chart assumes the person is at rest and that the relative humidity is 45 percent. (From *Mechanical and Electrical Equipment for Buildings,* 9th edition, Stein and Reynolds, ©2000 John Wiley & Sons, Inc.)

ambient temperature reaches the body temperature (98.6°F), no heat loss by conduction, convection, and radiation will occur. Fortunately, another heat-loss mechanism does not depend purely on ambient temperature. Heat loss by evaporation actually works better at higher temperatures. Curve 3 (Fig. 4.1c) represents the heat lost by evaporation as the ambient temperature changes with the relative humidity fixed at 45 percent.

4.2 THERMAL BARRIERS

If we could all live in the Garden of Eden, it would be easy for our body mechanisms to control heat flow. The real world, however, places our bodies under almost constant thermal stress. Any barrier as thin as the skin will have great difficulty in maintaining a constant temperature in a widely changing environment. Consequently, additional barriers are needed to achieve thermal comfort. Clothing, though it acts as an extra skin, is not always sufficient for thermal comfort. Buildings provide a milder environment for the clothed human being. In the drafty buildings of previous ages still more barriers were needed. The canopy bed was one solution (Fig. 4.2a). In modern buildings we come close to recreating the thermal aspects of the Garden of Eden.

This concept of progressive barriers promises to be continued. There was a serious suggestion, for example,

Figure 4.2a The concept of multiple barriers is very appropriate for thermal comfort. (From *Mansions of England in Olden Time* by Joseph Nash.)

Figure 4.2b The geodesic dome of the U.S. Pavilion, Expo 67, Montreal, protects the interior structures from sun, wind, and rain.

Figure 4.2c The Galleria Vittorio Emanuel, Milan, Italy, completed 1877, protects both the street and buildings. (Photograph by Clark Lundell.)

to enclose the new capital of Alaska in a pneumatic membrane structure and thereby greatly reduce the thermal stress on the buildings inside. Pneumatic structures are ideal for this purpose because they can enclose very large areas at reasonable cost. The U.S. Pavilion for Expo 67 in Montreal, Canada, used a different structural system for the same purpose. Figure 4.2b shows the geodesic dome that created a microclimate within which thermally fragile structures were built. Vents and shades were used to control this microclimate (see Fig. 9.15a).

More modest but quite common are the sheltered streets of our modern enclosed shopping malls, which had their beginnings in such projects as the Galleria Vittorio Emanuele in Milan, Italy, completed in 1877 (Fig. 4.2c). The Crystal Palace, built for the Great Exhibition of 1851 in London (Fig. 4.2d), was the ancestor of both the Galleria and the modern expo pavillion mentioned above. With an area of 770,000 feet^2 it created a new microclimate in a large section of Hyde Park.

Figure 4.2d The Crystal Palace for the Great Exhibition of 1851 created a benign microclimate in a London park. (Victoria and Albert Museum, London.)

4.3 METABOLIC RATE

To maintain thermal equilibrium, our bodies must lose heat at the same rate at which the metabolic rate produces it. This heat production is partly a function of outside temperature but mostly a function of activity. A very active person generates heat at a rate more than six times that of a reclining person. Table 4.3 shows the heat production related to various activities. For a better intuitive understanding, the equivalent heat production in terms of 100-watt lamps is also shown.

TABLE 4.3 BODY HEAT PRODUCTION AS A FUNCTION OF ANATOMY

	Activity	Heat Produced (btu/hour)		Watts
	Sleeping	300	💡	100
	Light work	600	💡 💡	200
	Walking	900	💡 💡 💡	300
	Jogging	2400	💡 💡 💡 💡 💡 💡 💡 💡	800

4.4 THERMAL CONDITIONS OF THE ENVIRONMENT

To create thermal comfort, we must understand not only the heat dissipation mechanisms of the human body but also the four environmental conditions that allow the heat to be lost. These four conditions are:

1. Air temperature (°F)
2. Humidity
3. Air velocity (feet/minute)
4. Mean radiant temperature (MRT)

All of these conditions affect the body simultaneously. Let us first examine how each of these conditions affects the rate of heat loss in human beings by themselves.

1. *Air temperature* The air temperature will determine the rate at which heat is lost to the air, mostly by convection. Above 98.6°F, the heat flow reverses and the body will gain heat from the air. The comfort range for most people (80 percent) extends from 68°F in winter to 78°F in summer. The range is this large mostly because warmer clothing is worn in the winter.
2. *Relative humidity* Evaporation of skin moisture is largely a function of air humidity. Dry air can readily absorb the moisture from the skin, and the resulting rapid evaporation will effectively cool the body. On the other hand, when the relative humidity (RH) reaches 100 percent, the air is holding all the water vapor it can and cooling by evaporation stops. For comfort the RH should be above 20 percent all year, below 60 percent in the summer and below 80 percent in the winter. These boundaries are not very precise, but at very low humidity levels there will be complaints of dry noses, mouths, eyes, and skin, and increases of respiratory illnesses. Static electricity and shrinkage of wood are also problems caused by low humidities.

   High humidity not only reduces the evaporative cooling rate, but also encourages the formation of skin moisture (sweat), which the body senses as uncomfortable. Furthermore, mildew growth is frequently a serious problem when the humidity is high.
3. *Air velocity* Air movement affects the heat-loss rate by both convection and evaporation. Consequently, air velocity has a very pronounced effect on heat loss. In the summer, it is a great asset and in the winter a liability. The comfortable range is from about 20 to about 60 feet/minute (fpm). From about 60 to about 200 fpm, air motion is noticeable but acceptable depending on the activity being performed. Above 200 fpm (2 mph), the air motion can be slightly unpleasant and disruptive (e.g., papers are blown around.) A **draft** is an undesirable local cooling of the human body by air movement, and it is a serious thermal comfort problem. See Table 10.8 for a more detailed description of how air velocity affects comfort.

   In cold climates, **windchill factors** are often given on weather reports because they better describe the severity of the cold than is possible with temperatures alone. The windchill factor is equal to the still-air temperature that would have the same cooling effect on a human being as does the combined effect of the actual temperature and wind speed.

   Although air movement from a breeze is usually desirable in the summer, it is not in very hot and dry climates. If the air is above 98.6°F, it will heat the skin by convection while it cools by evaporation. The higher the temperature, the less the total cooling effect.
4. *Mean radiant temperature* When the MRT differs greatly from the air temperature, its effect must be considered. For example, when you sit in front of a south-facing window in the winter you might actually feel too warm, even though the air temperature is a comfortable 75°F. This is because the sun's rays raised the MRT to a level too high for comfort. As soon as the sun sets, however, you will probably feel cold even though the air temperature in the room is still 75°F. This time the cold window glass lowered the MRT too far, and you experience a net radiant loss. It is important to realize that the average skin and clothing temperature is around 85°F, and this temperature determines the radiant exchange with the environment. In general, the goal is to maintain the MRT close to the ambient air temperature.

The psychrometric chart described in the next section is a powerful tool for understanding how the combination of temperature and humidity affect comfort.

4.5 THE PSYCHROMETRIC CHART

A useful and convenient way to understand some of the interrelationships of the thermal conditions of the environment is by means of the psychrometric chart (Fig. 4.5a). The horizontal axis describes the temperature of the air, the vertical axis describes the actual amount of water vapor in the air called **humidity ratio** or **absolute humidity,** and the curved lines describe the **relative humidity (RH)**. The diagram has two boundaries that are absolute limits. The bottom edge describes air that is completely dry (0 percent RH), and the upper curved boundary describes air that is completely saturated with water vapor (100 percent RH). The upper boundary is curved because as air gets warmer, it can hold more water vapor. Even if we know how much water vapor is already in the air, we cannot predict how much more it can hold unless we also know the temperature of the air. The RH is

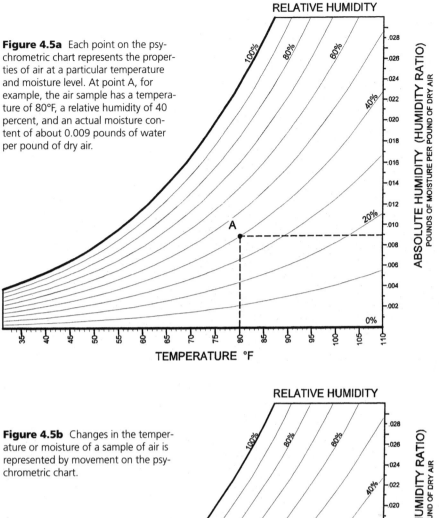

Figure 4.5a Each point on the psychrometric chart represents the properties of air at a particular temperature and moisture level. At point A, for example, the air sample has a temperature of 80°F, a relative humidity of 40 percent, and an actual moisture content of about 0.009 pounds of water per pound of dry air.

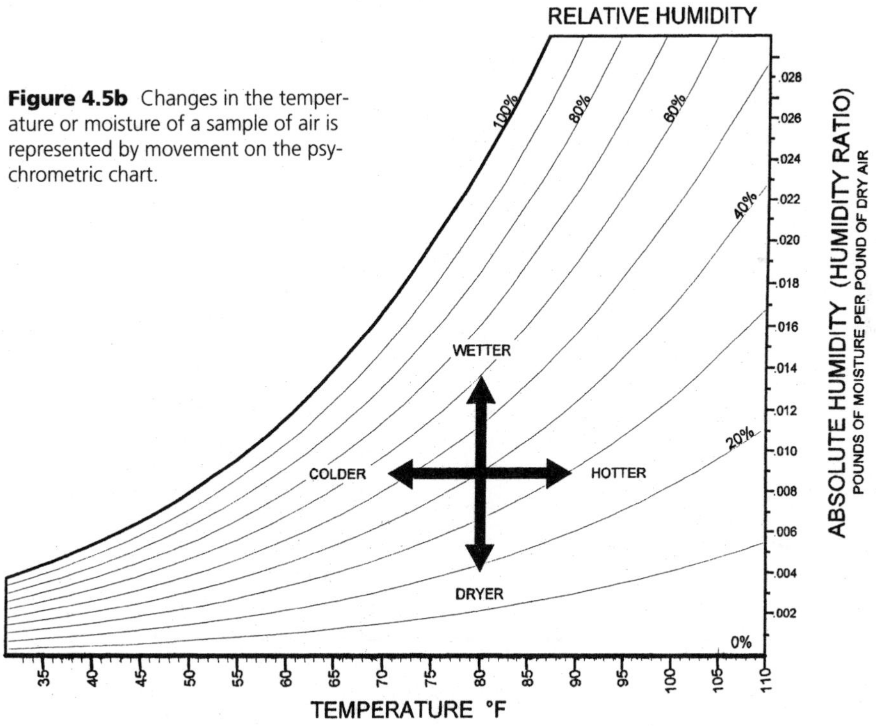

Figure 4.5b Changes in the temperature or moisture of a sample of air is represented by movement on the psychrometric chart.

affected by changes in either the moisture in the air or the temperature of the air.

Every point on the psychrometric chart represents air at a particular temperature and moisture level (Fig. 4.5a). Moving vertically up indicates that moisture is being added to that air sample (see Fig. 4.5b), while a downward motion on the graph represents water vapor removal (dehumidification). Movement to the right indicates that the air sample is being heated, and similarly, movement to the left indicates cooling of the air. Thus, if a sample of air at 80°F and 40 percent RH (Point A) is cooled to 60°F, it will move horizontally to the

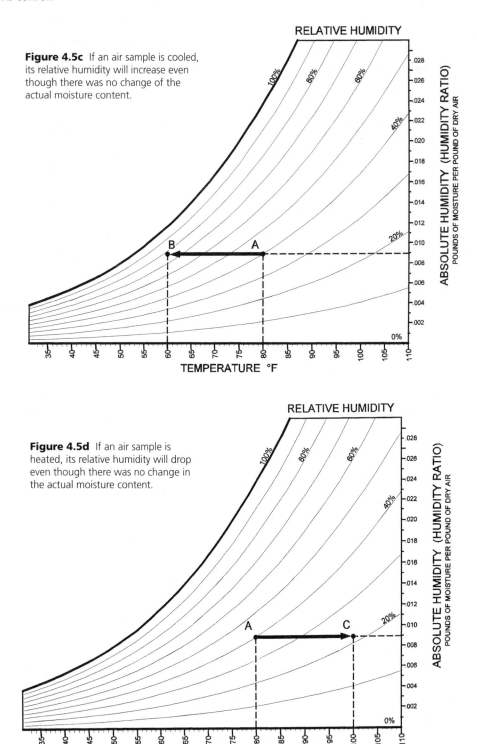

Figure 4.5c If an air sample is cooled, its relative humidity will increase even though there was no change of the actual moisture content.

Figure 4.5d If an air sample is heated, its relative humidity will drop even though there was no change in the actual moisture content.

left on the psychrometric chart to point B (Fig. 4.5c). Its relative humidity, however, has increased to about 78 percent even though there was no change in the actual moisture content of the air (i.e., no vertical movement on the chart). The RH increased because cool air can hold less moisture than warm air, and the same moisture level is now a larger percentage of what air can hold at that cooler temperature. On the other hand, if the air at point A is heated to 100°F (point C in Fig. 4.5d), then its relative humidity will be about 22 percent. The relative humidity changed because warm air can hold more moisture than cool air, and the same moisture level is now a smaller percentage of what the air can hold at that higher temperature.

4.6 DEW POINT AND WET-BULB TEMPERATURES

What would happen to air that is at 80°F and 40 percent RH, if it were cooled to 53°F? Look at point A in Fig. 4.6a. As the air is cooled the RH keeps increasing until it is 100 percent at about 53°F (point D). This is a special condition called the dew point temperature. At this point the air is fully saturated (100 percent RH) and cannot hold any more moisture. Any cooling beyond this point results in condensation where some of the water comes out of solution in the air. This phenomenon is also seen in rain, snow, fog, hoarfrost, and the "sweating" of a cold glass of water.

If the above sample of air is cooled beyond 53°F to 40°F, it will reach point E on the psychrometric chart (Fig. 4.6a). Although its RH is still 100 percent, its absolute humidity (humidity ratio) has decreased. Note the downward movement on the psychrometric chart from a humidity ratio of about 0.009 to about 0.0055 pounds of water per pound of dry air. Consequently, about 0.0035 pounds of water per pound of dry air was removed from the air when it was cooled from 80°F to 40°F. We can say that the air was **dehumidified.**

The dew-point temperature (DPT) is also an indication of how much moisture is in the air at any point. The higher the DPT, the more moisture. Thus, the DPT can be used to describe the actual amount—not the relative amount—of moisture in the air. Weather reports often give the DPT to describe the actual moisture content of the air.

Another way to describe the amount of moisture in an air sample is by giving its **wet-bulb temperature.** The wet-bulb temperature is determined by slinging two thermometers side by side through the air. One thermometer has its bulb covered with a wet sock. If this **sling psychrometer** is slung around in dry air, the temperature of the wet-bulb thermometer will drop significantly below the tempera-

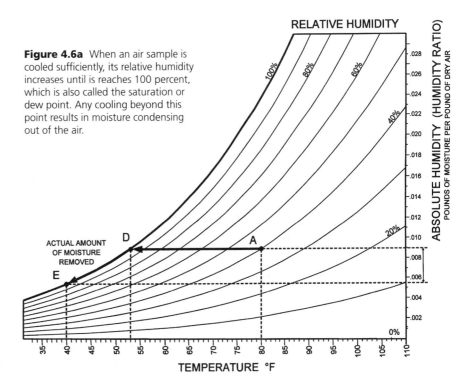

Figure 4.6a When an air sample is cooled sufficiently, its relative humidity increases until is reaches 100 percent, which is also called the saturation or dew point. Any cooling beyond this point results in moisture condensing out of the air.

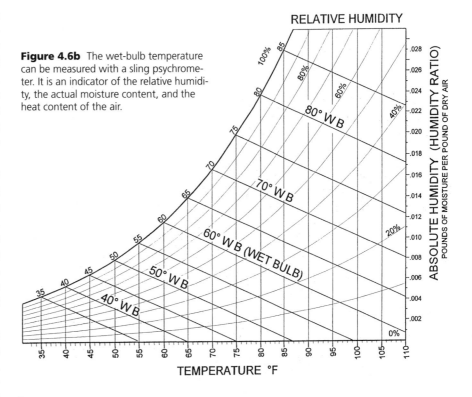

Figure 4.6b The wet-bulb temperature can be measured with a sling psychrometer. It is an indicator of the relative humidity, the actual moisture content, and the heat content of the air.

ture of the dry-bulb thermometer because of the large evaporation of water. Similarly, if the air is humid, the wet-bulb temperature will drop only a little. And, of course, if the air is at 100 percent RH, no evaporation will take place, and the wet-bulb and dry-bulb temperatures will be the same. Fig. 4.6b shows how at 100 percent RH, the wet-bulb temperatures (slanted lines) and the dry-bulb temperatures (vertical lines) are the same.

4.7 HEAT CONTENT OF AIR

The psychrometric chart can also be used to describe the sensible-, latent-, and total-heat content of an air sample. The total-heat **(enthalpy)** scale is a standard part of the psychrometric chart and is shown in Fig. 4.7a. Note that an upward movement on the chart not only increases the moisture content but also the latent-heat content. This is not a surprise if you remember that water vapor is latent heat. Also, note that a movement to the right increases not only temperature but also the sensible-heat content of an air sample. This also is not a surprise because temperature is an indicator of sensible-heat content.

Fig. 4.7b shows air that is being both heated and humidified. Thus,

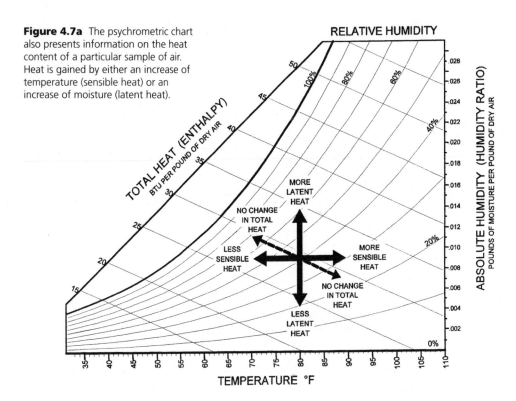

Figure 4.7a The psychrometric chart also presents information on the heat content of a particular sample of air. Heat is gained by either an increase of temperature (sensible heat) or an increase of moisture (latent heat).

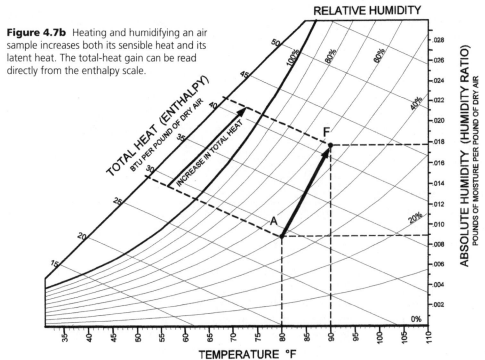

Figure 4.7b Heating and humidifying an air sample increases both its sensible heat and its latent heat. The total-heat gain can be read directly from the enthalpy scale.

when the air reaches point F, it has both more sensible and latent heat than it had at point A. The total increase in BTUs per pound of dry air can be read directly from the enthalpy scale.

In Fig. 4.7c, we see a sample of air that is being cooled by the evapora-

tion of water. If the air is humidified to 80 percent RH, the moisture content will increase and the temperature will decrease. Since the loss of sensible heat equals the gain in latent heat, the total-heat content is the same for point G as it was for point A. Note

that there is no change on the total-heat scale. A change on the psychrometric chart that does not result in a change of total-heat content is called an **adiabatic change.** This is an important and common phenomenon since this is what happens in evaporative cooling, in which the evaporation of water converts sensible heat to latent heat and the total-heat content remains the same. Thus, although the air becomes cooler, it also becomes more humid.

4.8 THERMAL COMFORT

Thermal comfort occurs when body temperatures are held within narrow ranges, skin moisture is low, and the body's effort of regulation is minimized (after ASHRAE 1997). Certain combinations of air temperature, relative humidity (RH), air motion, and mean radiant temperature (MRT) will result in what most people consider thermal comfort. When these combinations of air temperature and RH are plotted on a psychrometric chart, they define an area known as the **comfort zone** (Fig. 4.8a). Since the psychrometric chart relates only temperature and humidity, the other two factors (air motion and MRT) are held fixed. The MRT is assumed to be near the air temperature, and the air motion is assumed to be modest.

It is important to note that the given boundaries of the comfort zone are not absolute, because thermal comfort also varies with culture, time of year, health, the amount of fat an individual carries, the amount of clothing worn, and, most important, physical activity. The American Society for Heating, Refrigerating and Air Conditioning Engineers (ASHRAE) defined thermal comfort as "that condition of mind which expresses satisfaction with the thermal environment." While conditions required for thermal confort vary from person to person, the comfort zone

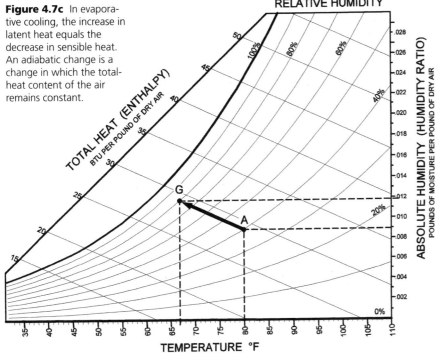

Figure 4.7c In evaporative cooling, the increase in latent heat equals the decrease in sensible heat. An adiabatic change is a change in which the total-heat content of the air remains constant.

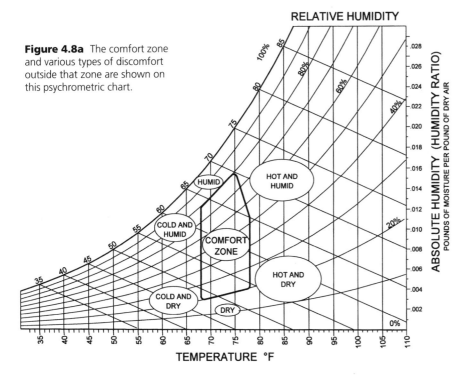

Figure 4.8a The comfort zone and various types of discomfort outside that zone are shown on this psychrometric chart.

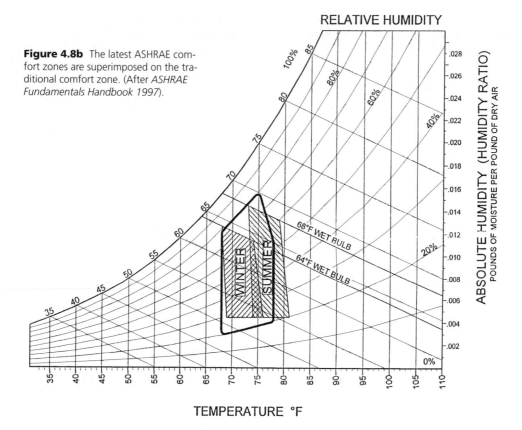

Figure 4.8b The latest ASHRAE comfort zones are superimposed on the traditional comfort zone. (After *ASHRAE Fundamentals Handbook 1997*).

should be the goal of the thermal design of a building because it defines those conditions that *most* people in our society find comfortable. Fig. 4.8b shows the latest ASHRAE comfort zones superimposed on the traditional comfort zone of Fig. 4.8a. Since the differences are small, this book continues to use the traditional zone with the caveat that the left side of the zone is more appropriate for winter and the right side, for summer.

Whenever possible, additional controls should be made available for the occupants of a building so that they can create the thermal conditions that are just right for them. Portable fans and heaters, numerous thermostats, and operable windows are devices people can use to fine-tune their environment. Mechanical equipment systems are now commercially available that allow individual thermal control at each work station.

The chart in Fig. 4.8a also indicates the type of discomfort one experiences outside of the comfort zone. These discomfort zones correspond to different climates. For example, the Southwest has a summer climate that is hot and dry, found on the lower right of the psychrometric chart (Fig. 4.8a). Unfortunately, very few climates have a sizable portion of the year in the comfort zone.

The following discussion shows how the comfort zone shifts when certain variables that had been held constant are allowed to change.

4.9 SHIFTING OF THE COMFORT ZONE

The comfort zone will shift on the psychrometric chart, if we change some of the assumptions made above. In Fig. 4.9a the shift of the comfort

zone is due to an increase in the mean radiant temperature (MRT). Cooler air temperatures are required to compensate for the increased heating from radiation. Likewise, a low MRT would have to be offset by an increase in the air temperature. For example, a room with a large expanse of glass must be kept warmer in the winter and cooler in the summer than a room with more modest window area. The large window area creates a high MRT in the summer and a low MRT in the winter. For every 3-degree increase or decrease in MRT, the air temperature must be adjusted 2 degrees in the opposite direction. Window shading (Chapter 9) and insulation (Chapter 15) can have tremendous effects on the MRT.

In Fig. 4.9b the shift of the comfort zone is due to increased air velocity. The cooling effect of the air motion is offset by an increase in the air tem-

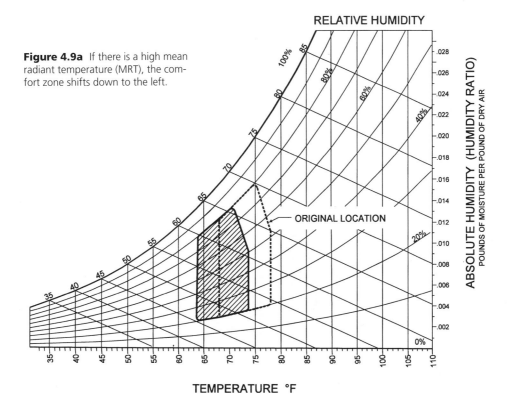

Figure 4.9a If there is a high mean radiant temperature (MRT), the comfort zone shifts down to the left.

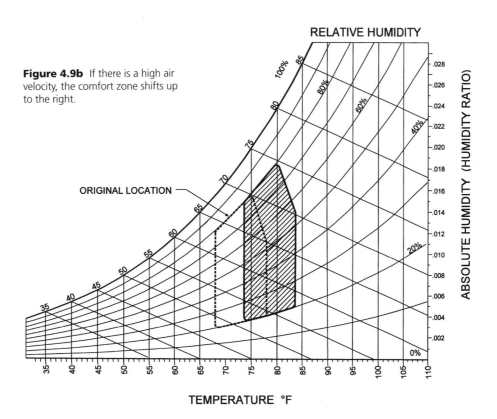

Figure 4.9b If there is a high air velocity, the comfort zone shifts up to the right.

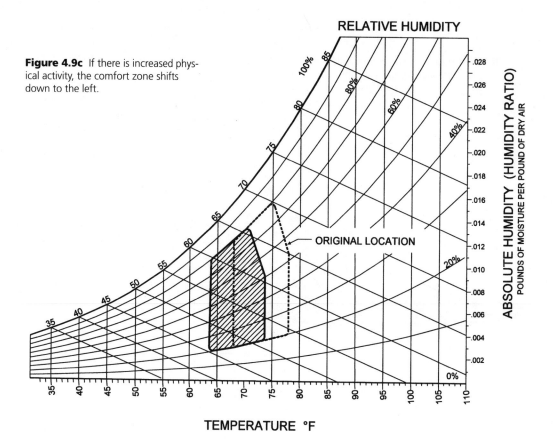

Figure 4.9c If there is increased physical activity, the comfort zone shifts down to the left.

perature. We usually make use of this relationship in the reverse situation. When the air temperature is too high for comfort, we often use air motion (i.e., open a window or turn on a fan) to raise the comfort zone so that it includes the higher air temperature. Chapter 10 will explain how air movement can be used for passive cooling.

Lastly, there is a shift of the comfort zone due to physical activity. Cooler temperatures are required to help the body dissipate the increased production of heat. Gymnasiums, for example, should always be kept significantly cooler than classrooms. Thus, the comfort zone shifts down to the left when physical activity is increased (Fig. 4.9c).

4.10 CLOTHING AND COMFORT

Unfortunately, an architect cannot specify the clothing to be worn by the occupants of his building. Too often, fashion, status, and tradition in clothing work against thermal comfort. In some extremely hot climates women were—in a few places still are—required to wear black veils and robes that completely cover their bodies. Unfortunately, some of our own customs are almost as inappropriate. A three-piece suit with a necktie can get quite hot in the summer. A miniskirt in the winter is just as unsuitable. Clothing styles should be seasonal indoors as well as out-

doors because indoor temperatures vary to some extent with outdoor temperatures. We could save countless millions of barrels of oil if men wore three-piece suits only in the winter and women wore miniskirts only in summer.

The following story is a case in point. Because of the energy crisis that started in 1973, President Carter mandated energy-saving temperatures in government buildings. One such building was the U.S. Capitol, which got quite warm in the summer with the high thermostat setting. Although the members of Congress were all extremely hot, they nevertheless voted to maintain the traditional dress code that was more appropriate for cooler temperatures.

The insulating properties of clothing have been quantified in the unit of thermal resistance called the **clo.** In winter a high clo value is achieved by clothing that creates many air spaces either by multiple layers or by a porous weave. If wind is present, then an outer layer that is fairly airtight but permeable to water vapor is required.

In summer a very low clo value is, of course, required. Since it is even more important in the summer than the winter that moisture can pass through the clothing, a very permeable fabric should be used. Cotton is especially good because it acts as a wick to transfer moisture from the skin to the air. Although wool is not as good as cotton in absorbing moisture, it is still much better than most manmade materials. Also loose, billowing clothing will promote the dissipation of both sensible and latent (water-vapor) heat.

4.11 STRATEGIES

Much of the rest of this book discusses the various strategies that have been developed to create thermal comfort in our buildings. The version of the psychrometric chart shown in Fig. 4.11 summarizes some of these strategies. If you compare this chart with Fig. 4.8a, you will see the relationship between strategies and discomfort (climate) conditions more clearly. For example, the strategy of evaporative cooling (the lower-right area in Fig. 4.11) corresponds with the hot and dry discomfort zone (the lower-right area in Fig. 4.8a). The diagram shows that internal heat gains from sources such as machines, people, and lights are sufficient to heat the building in slightly cool conditions. Also, when the climate conditions are to the right of the **shade line,** the sun should be prevented from entering the windows. This line as well as all the boundaries of the various zones shown in the diagram are not precisely fixed but should be considered as fuzzy limits.

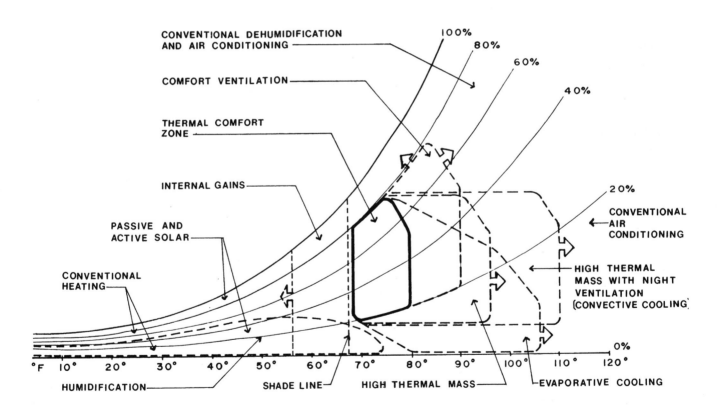

Figure 4.11 Summary of design strategies as a function of ambient conditions (climate). (From *Psychrometric-Bioclimatic Chart,* copyright by Baruch Givoni and Murray Milne.)

KEY IDEAS OF CHAPTER 4

1. For thermal comfort, the body must eliminate waste heat by means of conduction, convection, radiation, and evaporation.

2. The amount of waste heat produced is mostly a function of the physical activity being performed.

4. Four factors of the environment together determine how easily the body can eject the waste heat. Their comfort ranges are:
   a) Air temperature (68°F to 78°F)
   b) Relative humidity (20 percent to 80 percent)
   c) Air velocity (20 fpm to 60 fpm)
   d) Mean radiant temperature (MRT; near air temperature)

4. A combination of certain values of these four factors result in what is called thermal comfort, which can be represented by the comfort zone on various charts (e.g., psychrometric chart).

5. When one or more of the four factors of the environment is somewhat outside the in-comfort range, the remaining factors can be adjusted up or down to compensate, thereby restoring thermal comfort.

6. The psychrometric chart describes the combined effect of temperature and its coincident humidity.

7. A certain set of temperatures and coincident humidities is called the comfort zone on the psychrometric chart.

8. The psychrometric-bioclimatic chart shows which architectural design strategies are appropriate for different climates as determined by temperatures and their coincident humidities.

Resources

FURTHER READING
(See Bibliography in back of book for full citations. This list includes valuable out-of-print books.)

ASHRAE Handbook of Fundamentals, 1997.
Flynn, J. E., et al. *Architectural Interior Systems: Lighting, Acoustics, Air Conditioning.*
Givoni, B. *Man, Climate and Architecture.*
Givoni, B. *Climate Considerations in Building and Urban Design.*
Stein, B., and J. Reynolds. *Mechanical and Electrical Equipment for Buildings.* A general resource.

ORGANIZATIONS
(See Appendix J for full citation.)

American Society of Heating, Refrigerating, and Air-Conditioning Engineers (ASHRAE).

5

CLIMATE

"We must begin by taking note of the countries and climates in which homes are to be built if our design for them are to be correct. One type of house seems appropriate for Egypt, another for Spain…one still different for Rome.…It is obvious that design for homes ought to conform to diversities of climate."

Vitruvius
Architect, first century BC

Please note that figure numbers are keyed to sections. Gaps in figure numbering result from sections without figures.

5.1 INTRODUCTION

As the quote by Vitruvius indicates, designing buildings in harmony with their climates is an age-old idea. To design in conformity with climate, the designer needs to understand the microclimate of the site, since all climatic experience of both people and buildings is at this level. Besides adjusting the building design to the climate, it is also possible, to a limited extent, to adjust the climate to the needs of the building. We cannot agree with Mark Twain when he said, "Everyone talks about the weather but no one does anything about it." It is easy to see how man changes the microclimate by such acts as replacing farmland and forest with the hard and massive materials of cities, irrigating a desert and making it a humid area, and constructing high-rise buildings to form windy canyons. Unfortunately, these changes in the microclimate are rarely beneficial since they are usually done without concern for the consequences.

Most serious, however, are the changes we are making to the macroclimate. As was discussed more fully in Chapter 2, large-scale burning of fossil fuels is increasing the amount of carbon dioxide in the air. Carbon dioxide, like water vapor, is transparent to solar energy but not to the long-wave radiation emitted by the earth's surface. Thus, the ground and atmosphere are heated by the phenomenon known as the **greenhouse effect.** The heating of the earth might create very undesirable changes in the world's climates. Also, various chemicals are depleting the ozone layer, and large-scale cutting of tropical forests might also be creating worldwide changes in climate.

To properly relate buildings to their microclimate and to make beneficial changes in that microclimate, we must first understand the basics of climate.

5.2 CLIMATE

The climate or average weather is primarily a function of the sun. The word "climate" comes from the Greek "klima," which means the slope of the earth in respect to the sun. The Greeks realized that climate is largely a function of sun angles (latitude) and, therefore, they divided the world into the tropic, temperate, and arctic zones.

The atmosphere is a giant heat machine fueled by the sun. Since the atmosphere is largely transparent to solar energy, the main heating of the air occurs at the earth's surface (Fig. 5.2a). As the air is heated, it rises and creates a low-pressure area at ground level. Since the surface of the earth is not heated equally, there will be both relatively low- and high-pressure areas with wind as a consequence.

A global north-south flow of air is generated because the equator is heated more than the poles (Fig. 5.2b). This global flow is modified by both the changes in season and the rota-

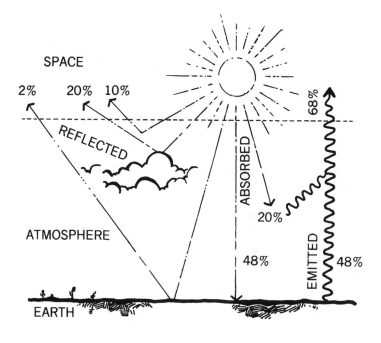

Figure 5.2a The atmosphere is mainly heated by contact with the solar heated ground. On an annual basis, the energy absorbed by the earth equals the energy radiated back into space. In the summer, there is a gain while in the winter an equal loss.

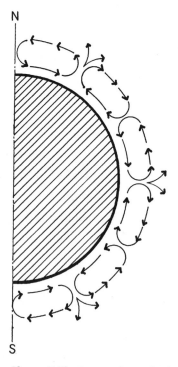

Figure 5.2b Because the earth is heated more at the equator than at the poles, giant global convection currents are generated.

Figure 5.2c The rotation of the earth deflects the north-south air currents by an effect known as the Coriolis force. (From *Wind Power for Farms, Homes and Small Industry,* by the U.S. Department of Energy, 1978.)

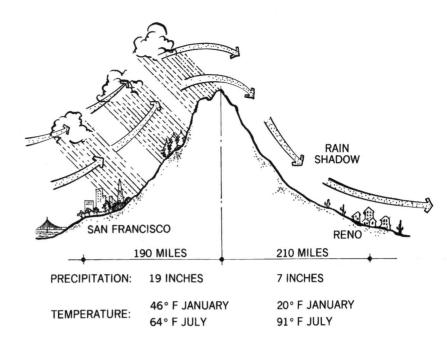

	SAN FRANCISCO	RENO
	190 MILES	210 MILES
PRECIPITATION:	19 INCHES	7 INCHES
TEMPERATURE:	46° F JANUARY	20° F JANUARY
	64° F JULY	91° F JULY

Figure 5.2d In certain cases, mountain ranges cause rapid changes from relatively wet and cool to hot and dry climates. (From *American Buildings: 2: The Environmental Forces That Shape It* by James Marston Fitch, copyright James Marston Fitch, 1972.)

tion of the earth (Fig. 5.2c). Another major factor affecting winds and, therefore, climate is the uneven distribution of land masses on the globe. Because of its higher heat capacity, water does not heat up or cool down as fast as the land. Thus, temperature changes over water tend to be more moderate than over land, and the farther one gets from large bodies of water the more extreme are the temperatures. For example, the annual temperature range on the island of Key West, Florida, is about 24°F, while the annual range inland at San Antonio, Texas, only slightly further north, is about 56°F. These water-land temperature differences also create pressure differences that drive the winds.

Mountain ranges not only block or divert winds but also have a major effect on the moisture content of the air. A good example of this important climatic phenomenon is the American West. Over the Pacific Ocean solar radiation evaporates water, and the air becomes quite humid. The westerlies blow this moist air over land where it is forced up over the north-south mountain ranges (Fig. 5.2d). As the air rises, it cools at a rate of about 3.6°F for every 1,000 feet. When the temperature drops the relative humidity increases until it reaches 100 percent, the saturation point. Any additional cooling will cause moisture to condense in the form of clouds,

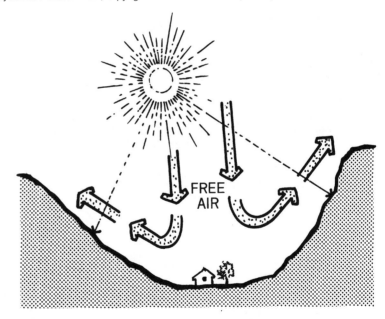

Figure 5.2e During the day, the air moves up the mountain sides.

rain, or snow. On the far side of the mountains, the now drier air falls and, consequently, heats up again. As the temperature increases, the relative humidity decreases and a rainshadow is created. Thus, a mountain ridge can be a sharp border between a hot, dry and a cooler, wetter climate.

Mountains also create local winds that vary from day to night. During the day, the air next to the mountain surface heats up faster than free air at the same height. Thus, warm air moves up along the slopes during the day (Fig. 5.2e). At night, the process is reversed: the air moves down the

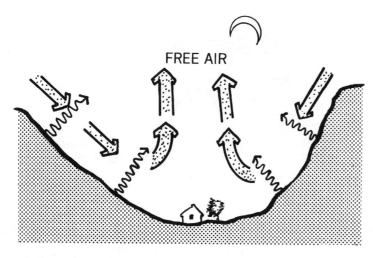

FREE AIR

Figure 5.2f At night, the land cools rapidly by radiation and the air currents move down the mountain sides.

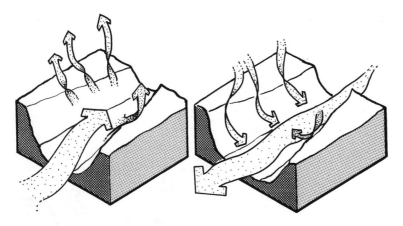

Figure 5.2g The effects described in Fig. 5.2e and 5.2f are greatly magnified in narrow sloping valleys. During the day, strong winds blow up the valley; at night, the winds reverse.

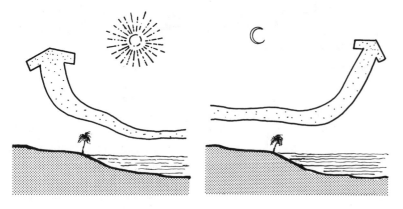

Figure 5.2h The temperature differences between land and water create sea breezes during the day and land breezes at night.

slopes because the mountain surface cools by radiation more quickly than the free air (Fig. 5.2f). In narrow valleys, this phenomenon can create very strong winds up along the valley floor during the day and down the valley at night (Fig. 5.2g).

A similar day-night reversal of winds occurs near large bodies of water. The large heat capacity of water prevents it from heating or cooling as fast as land. Thus, during the day the air is hotter over land than over water. The resultant pressure differences generate sea breezes (Fig. 5.2h). At night, the temperatures and air flows reverse. In the late afternoon and early morning, when the land and sea are at the same temperature, there is no breeze. Furthermore, at night the breezes are weaker than during the day because the temperature differences between land and water are smaller.

The amount of moisture in the air has a pronounced effect on the ambient temperature. In dry climates, there is little moisture to block the intense solar radiation from reaching the ground, and, thus, summer daytime temperatures are very high—over 100°F. Also at night there is little moisture to block the outgoing long-wave radiation; consequently, nights are cool and the diurnal temperature range is high—more than 30°F (Fig. 5.2i). On the other hand, in humid and especially cloudy regions, the moisture blocks some solar radiation to make summer daytime temperatures much more moderate—below 90°F. At night, the outgoing long-wave radiation is also blocked by the moisture, and consequently, temperatures do not drop much (Fig. 5.2j). The diurnal temperature range is, therefore, small—below 20°F. It should be noted that water has a much stronger blocking effect on radiation when it is in the form of droplets (clouds) than in the form of a gas (humidity).

The various forces in the atmosphere interact to form a large set of diverse climates. Later in the chapter there will be a description of seventeen different climate regions in the United States.

5.3 MICROCLIMATE

For a number of reasons, the local climate can be quite different from the climate region in which it is found. If buildings are to relate properly to their environment, they must be designed for the microclimate in which they exist. The following factors are mainly responsible for making the microclimate deviate from the macroclimate:

1. *Elevation above sea level:* The steeper the slope of the land, the faster the temperature will drop with an increase in elevation. The limit, of course, is a vertical ascent, which will produce a cooling rate of about 3.6°F per 1,000 feet.
2. *Form of land:* South-facing slopes are warmer than north-facing slopes because they receive much more solar radiation (Fig. 5.3a). For this reason ski slopes are usually found on the north slopes of mountains, while vineyards are located on the south slopes (Fig. 5.3b). South slopes are also protected

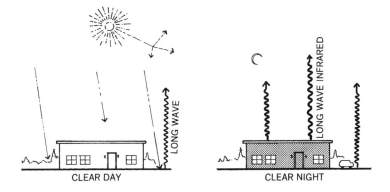

Figure 5.2i Since dry climates have little moisture to block radiation, daytime temperatures are high and night temperatures are low. The diurnal temperature range is, therefore, large.

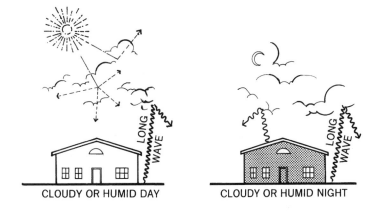

Figure 5.2j Water in the form of humidity and especially in the form of clouds blocks both solar and long wave radiation. Thus, in humid or cloudy climates, the daytime temperatures are not as high and night temperatures are not as low. The diurnal range is, therefore, small.

Figure 5.3a The north side (south slope) of this east–west road in Maryland is several weeks further into spring than the south side (north slope), where the snow melts much more slowly.

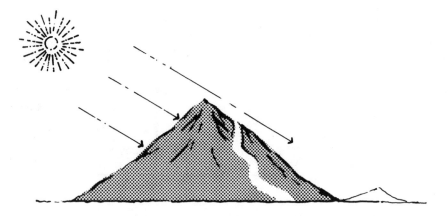

Figure 5.3b South-facing slopes can receive more than 100 times as much solar radiation as north-facing slopes.

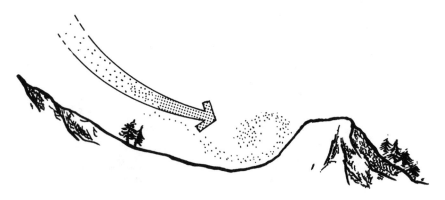

Figure 5.3c Since cool air is heavier than warm air, it drains into low-lying areas forming pools of cold air.

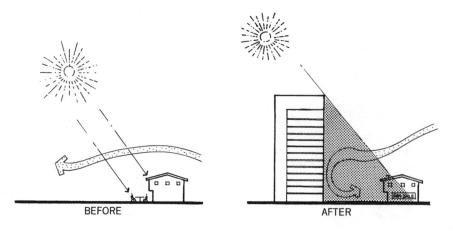

Figure 5.3d A delightfully sunny and wind-protected southern exposure can be turned into a cold windy microclimate by the construction of a large building to the south.

from the cold winter winds that usually come from the north. West slopes are warmer than east slopes because the period of high solar radiation corresponds with the high ambient air temperatures of the afternoon. Low areas tend to collect pools of cold heavy air (Fig. 5.3c). If the air is also moist, fog will frequently form. The fog, in turn, reflects the solar radiation so that these areas remain cool longer in the morning.

3. *Size, shape, and proximity of bodies of water:* As mentioned before, large bodies of water have a significant moderating effect on temperature, they generate the daily alternating land and sea breezes, and they increase the humidity.

4. *Soil types:* The heat capacity, color, and water content of soil can have a significant effect on the microclimate. Light-colored sand can reflect large amounts of sunlight, thereby reducing the heating of the soil, and, thus, the air, but at the same time greatly increases the radiation load on people or buildings. Because of their high heat capacity, rocks can absorb heat during the day and then release it again at night. The cliff dwellings of the Southwest benefited greatly from this effect (see Fig. 10.2m).

5. *Vegetation:* By means of shading and transpiration, plants can significantly reduce air and ground temperatures. They also increase the humidity whether or not it is already too high. In a hot and humid climate, the ideal situation is to have a high canopy of trees for shade, but no low plants that could block the breeze. The stagnant air from low trees and shrubs enables the humidity to build up to undesirably high levels. In cold climates, plants can reduce the cooling effect of the wind. Vegetation can also reduce noise and clean the air of dust and certain other pollutants.

6. *Manmade structures:* Buildings, streets, and parking lots, because

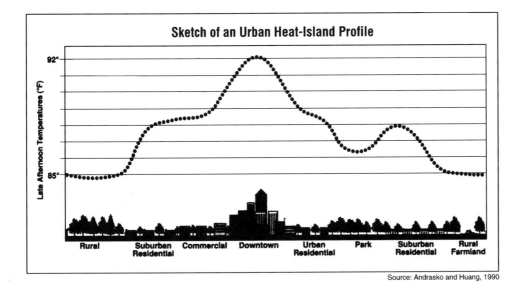

Figure 5.3e A sketch of a typical urban heat-island profile. This profile of a hypothetical metropolitan area shows temperature (°F) changes correlated to the density of development and trees. (Reproduced from *Cooling Our Communities: A Guidebook on Tree Planting and Light-Colored Surfacing*, LBL-31587, published by Lawrence Berkeley National Laboratory, 1992.)

Source: Andrasko and Huang, 1990

of their number and size, have a very significant effect on the microclimate. The shade of buildings can create a cold north-like orientation on what was previously a warm southern exposure (Fig. 5.3d). On the other hand, buildings can create shade from the hot summer sun and block the cold winter winds. Large areas of pavement, especially dark-colored asphalt, can generate temperatures as high as 140°F. The heated air then migrates to overheat adjacent areas as well.

In large cities, the combined effect of all the man-made structures results in a climate significantly different from the surrounding countryside (see Color Plate 6 on color insert). The **annual mean** temperature will usually be about 1.5°F warmer while the winter minimum temperature may be about 3°F higher. In summer, cities can be 7°F warmer than rural areas and are, therefore, sometimes known as **heat islands** (Fig. 5.3e). Solar radiation, however, will be about 20 percent lower due to the air pollution, and the relative humidity will be about 6 percent less because of the reduced amount of vegetation. Although the overall wind speed is about 25 percent lower, very high local wind speeds often occur in the urban canyons.

5.4 CLIMATIC ANOMALIES

Radical variations in the climate of a region are possible under certain conditions. One of the most famous climate anomalies is found in Lugano,

Figure 5.4 The combination of a low elevation, south-facing slopes, high mountains to the north, and a large lake to the south creates a subtropical climate in Lugano, Switzerland, even though it is as far north as Quebec.

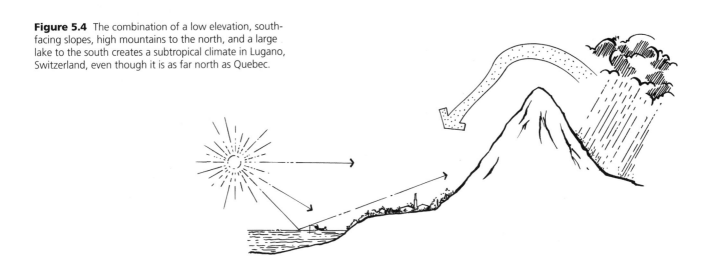

Switzerland. Although Lugano has the same latitude as Quebec (47°), the climates are as different as if Lugano were 1,500 miles farther south.

This unusually warm climate in a northern region is largely a result of the unique geography of the area. Lugano is located where the southern slope of the Alps meets a large lake (Fig. 5.4). It is thus fully exposed to both the direct winter sun and that reflected off the lake. The water also has a moderating effect on sudden temperature changes. The Alps protect the area from cold and wet winter winds. Those winds that do get across the Alps are dried and heated just as are the winds crossing the Sierra Nevadas in California (Fig. 5.2d). And, lastly, the climate in Lugano is so unusually warm because of the low elevation of the land. Meanwhile, cold alpine climates are only a few miles away.

Less dramatic variations in the microclimates of a region are quite common. It is not unusual to find in rather flat country that two areas only a few miles apart have temperature differences as great as 30°F. Suburban areas are often more than 7 degrees cooler during the day and more than 10 degrees cooler at night than urban areas. Even a distance of only 100 feet can make a significant difference. The author has noted very dramatic temperature differences on his half-acre lot. Consequently, he relaxes in one part of the garden in the summer and a different part in the cooler seasons.

Very localized variations in the microclimate are especially obvious in the spring. When the snow melts, it does so in irregular patches. The warm areas are also the first to see the green growth of spring. Areas only a few feet apart might be two or more weeks apart in temperature. These variations are not hard to understand, when one considers the fact that in New York on December 21 a south wall receives 108 times as much solar radiation as a north wall. A designer must not only know the climate of the region but also the specific microclimate of the building site.

5.5 CLIMATE REGIONS OF THE UNITED STATES

No book could ever describe all of the microclimates found in the United States. Designers must, therefore, use the best available published data and modify it to fit their specific site. The National Oceanic and Atmospheric Administration (NOAA) collects and publishes extensive weather and climatic data. See the end of this chapter for a listing of NOAA and other

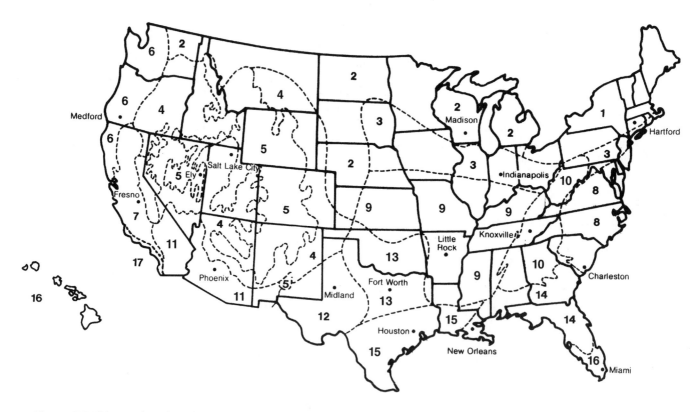

Figure 5.5 This map shows how the United States is divided into the seventeen climate regions used in this book. A description of each climate region can be found in the Climatic Data Tables below.

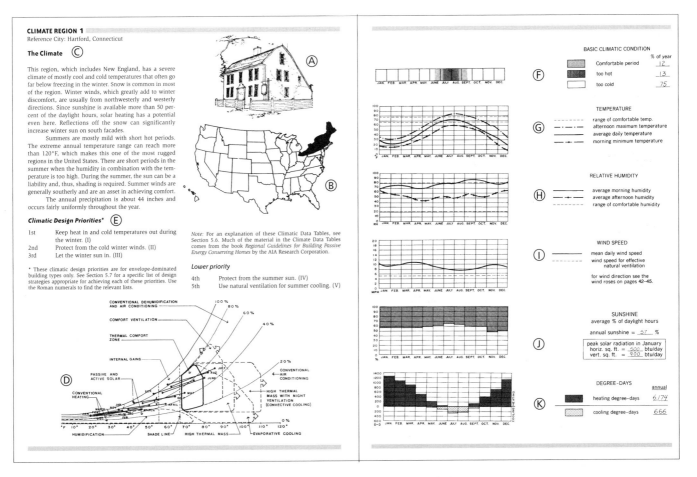

Figure 5.6a Key to the Climatic Data Tables. Each climate is described by two facing pages of information.

publications on climate. Since the information is usually not arranged in a convenient form for architects to use, some appropriate climate data in a graphic format is included in this book.

When the United States is divided into only a few climate regions, the information is too general to be very useful. On the other hand, when too much information is presented it often becomes inaccessible. As a compromise, this book divides the United States into seventeen climate regions (Fig. 5.5). This subdivision system and much of the climatic information are based on material from the book *Regional Guidelines for*

Building Passive Energy Conserving Homes produced by the AIA/Research Corporation.

The remainder of this chapter describes these seventeen climate regions. Included with the climate data for each region is a set of specific climate related design priorities appropriate for envelope-dominated buildings, such as homes and small institutional and commercial buildings. Specific design strategies that respond to those climatic design priorities are given at the end of the chapter (see Section 5.7 on page 116).

Some words of caution are very important here. The following climatic data should be used only as a start-

ing point. As much as possible, corrections should be made to account for local microclimates. For building sites near the border between regions, the climatic data for the two regions should be interpolated. The borders should be considered as fuzzy lines rather than the sharp lines shown in Fig. 5.5.

5.6 EXPLANATIONS OF THE CLIMATIC DATA TABLES

Each of the seventeen climatic regions is described by a **climatic data table** consisting of two facing pages (Fig. 5.6a), and each part marked with a circled uppercase letter is described below.

(A) *Sketch:* The drawing is a representative example of a residential building appropriate for the particular climate region.

(B) *Climate Region:* The climate of the region is represented by the climatic data for the reference city. The darkened portion of the map represents the particular region for which the data are given.

(C) *The Climate:* This section of the Climatic Data Table provides a verbal description of the climate.

(D) *The Psychrometric–Bioclimatic Chart:* This chart defines the climate in relationship to thermal comfort and the design strategies required to create thermal comfort.

See Section 4.11 for an explanation of the psychrometric–bioclimatic chart.

The climate of the region is presented on this chart by means of straight lines, each of which represents the temperature and humidity conditions for one month of the year. Each line is generated by plotting the monthly normal daily maximum and minimum temperatures with their corresponding relative humidities. The line connecting these two points is assumed to represent the typical temperature and humidity conditions of that month (Fig. 5.6b). The twelve monthly lines represent the annual climate of that region.

This method of presenting the climate has several advantages. It graphically defines in one diagram both

temperature and humidity for each month of the year. This is important because thermal comfort is a function of their combined effect. It shows how severe or mild the climate is by the relationship of the twelve lines to the comfort zone. It also shows which design strategies are appropriate for the particular climate. For example, the chart for Climate 7 (Fresno, California) indicates a hot and dry summer climate for which evaporative cooling is an appropriate strategy.

(E) *Climatic Design Priorities:* For each climate, a set of design priorities is given for "envelope-dominated buildings," such as residences and small office buildings. "Internally dominated buildings," such as large office buildings, are less affected by

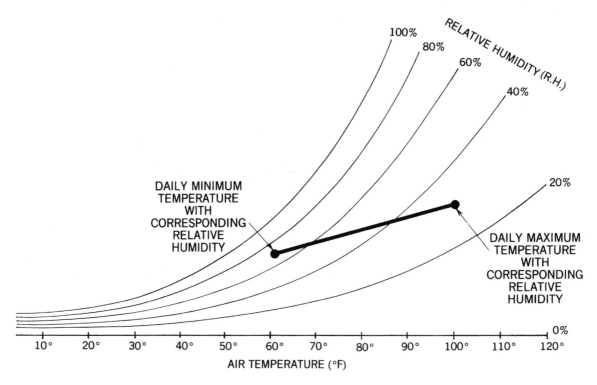

Figure 5.6b On the psychrometric–bioclimatic chart the climate of a region for any one month is represented by a line.

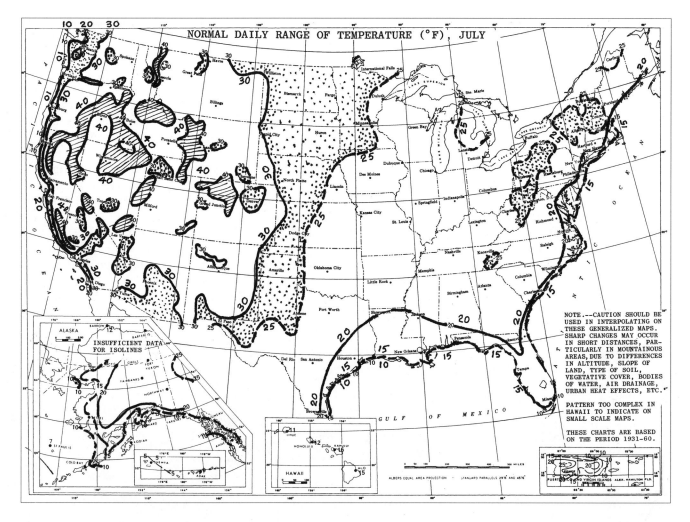

Figure 5.6c July normal, daily temperature ranges. (From *Climate Atlas of the U.S.,* National Oceanic and Atmospheric Administration (NOAA) 1983.)

climate and have much smaller heating and much greater cooling needs. They usually also have a greater need for daylighting. Consequently, these priorities are not directly applicable for "internally dominated building" types.

The priorities are listed in descending order of importance. The designer should start with the first priority and include as many as possible. Note that the words summer and winter are used to refer to the overheated and underheated periods of the year and not necessarily to the calendar months.

(F) *Basic Climate Condition:* This chart shows the periods of the year when the combined effect of temperature and humidity makes the climate either too hot, too cold, or just comfortable. The chart offers a quick answer to the question of what the main thrust of the building design should be: whether the design should respond to a climate that is mainly too hot, too cold, both too hot and too cold, or mostly comfortable.

(G) *Temperature:* The temperatures given are averages over many years. Although occasional temperatures are much higher and lower than the averages shown, most designs are based on normal rather than extreme conditions. The vertical distance between the afternoon maximum and morning minimum temperatures represents the diurnal (daily) temperature range. The horizontal dashed lines define the comfort zone.

Daily temperature ranges can also be obtained from the map of Fig. 5.6c.

These values are critical in choosing the appropriate passive cooling techniques explained in Chapter 10.

(H) *Humidity:* Even when the absolute moisture in the air remains fairly constant throughout the day, the relative humidity will vary inversely with the temperature. Since hot air can hold more moisture than cold air, the relative humidity will generally be lowest during the afternoons when the temperatures are the highest. Early in the morning when temperatures are at their lowest, the relative humidity will be at its highest.

The horizontal dashed lines define the comfort range for relative humidity. However, even humidity levels within the comfort zone can be excessive if the coincident temperature is high enough. Thus, the psychrometric–bioclimatic chart is a better indicator of thermal comfort than this chart.

(I) *Wind:* This chart shows the mean daily wind speeds in an open field at the reference city. The dashed line indicates the minimum wind speed required for effective natural ventilation in humid climates.

For wind direction, see the wind roses shown on maps of the United States in Figs. 5.6d, 5.6e, 5.6f, and 5.6g. A wind rose shows the percentage of time the wind blows from the sixteen compass points or it was calm. Each notch represents 5 percent of the time, and the number in the center circle represents the percentage of time there was no wind (calm). Maps of wind roses are included here for four critical months: the coldest (January), the hottest (July), and two transition months (April and October). Maps are available for every month in the *Climate Atlas of the United States.*

It is extremely important to note that local wind direction and speed

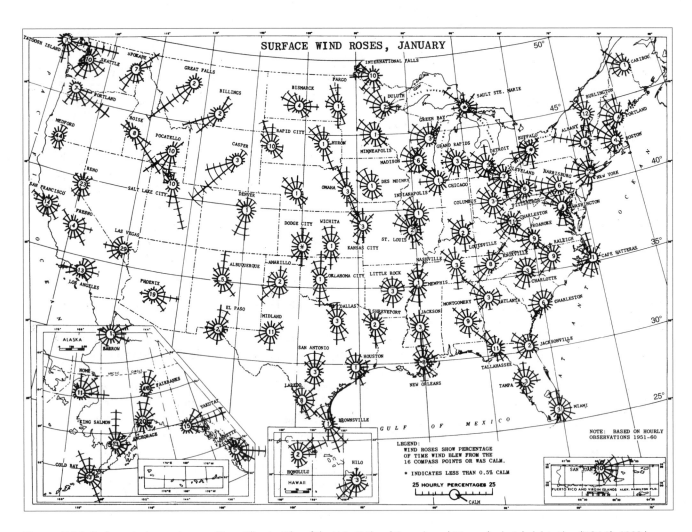

Figure 5.6d Surface wind roses, January. (From *Climate Atlas of the U.S.,* National Oceanic and Atmospheric Administration (NOAA), 1983.)

can be very different from those at the weather station. All wind charts should, therefore, be used with great care.

(J) *Sunshine:* This chart shows the percentage of the daylight hours of each month that the sun is shining. These data are useful for solar heating, shading, and daylighting design. The charts show that direct sunlight is plentiful in all climates. Since there are about 4,460 hours of daylight in a year, these percentages indicate that there are more than 2,000 hours of sunshine even in the cloudiest climate. Thus, direct sunshine is a major design consideration in all climates.

TABLE 5.6 HOURS OF DAYLIGHT PER DAY*

	Latitude		
Month	30°N	40°N	50°N
January	10:25	9:39	8:33
February	11:09	10:43	10:07
March	11:58	11:55	11:51
April	12:53	13:15	13:45
May	13:39	14:23	15:24
June	14:04	15:00	16:21
July	13:54	14:45	15:57
August	13:14	13:46	14:30
September	12:22	12:28	12:39
October	11:28	12:28	12:39
November	10:39	9:59	9:04
December	10:14	9:21	8:06

*Values given are for the fifteenth day of each month.

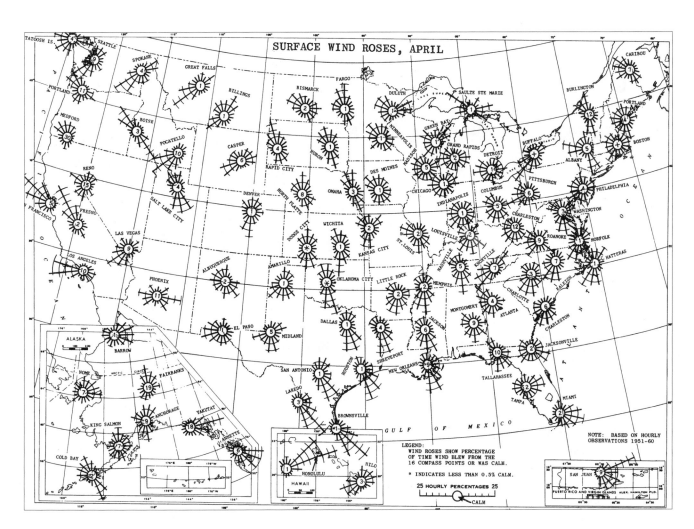

Figure 5.6e Surface wind roses, April. (From *Climate Atlas of the U.S.,* National Oceanic and Atmospheric Administration (NOAA), 1983.)

To determine the average number of sunshine hours for a representative day for any month, multiply the percentage sunshine from the chart by the number of daylight hours in Table 5.6. Since the number of daylight hours varies with latitude, the table contains values for 30 degrees, 40 degrees, and 50 degrees north latitude.

The annual percentage of sunshine is also given at the right of the chart. Along with the data on sunshine, some solar-radiation data are also given (enclosed in rectangle). These data can give a quick estimate of the peak amount of solar heating that can be expected during one day in January on either a horizontal or vertical square foot.

(K) *Degree-Days:* "Degree-days" are a good indicator of the severity of winter and summer. Although the concept was developed to predict the amount of heating fuel required, it is now also used to predict the amount of cooling energy required. The **degree-days** shown here are for the typical base of 65°F. The difference between the average temperature of a particular day and 65°F is the number of degree-days for that day. The chart shows the total number of degree-days for each month with heating degree-days above the zero line and cooling degree-days below. Thus, it is easy to visually determine both the length and depth of the heating and cooling periods and the relative size of each period. The yearly totals are given in numerical form.

Degree-Day Rules of Thumb

Heating Degree-Days (HDDs)

1. Areas with more than 5,500 HDDs per year are characterized by long cold winters.

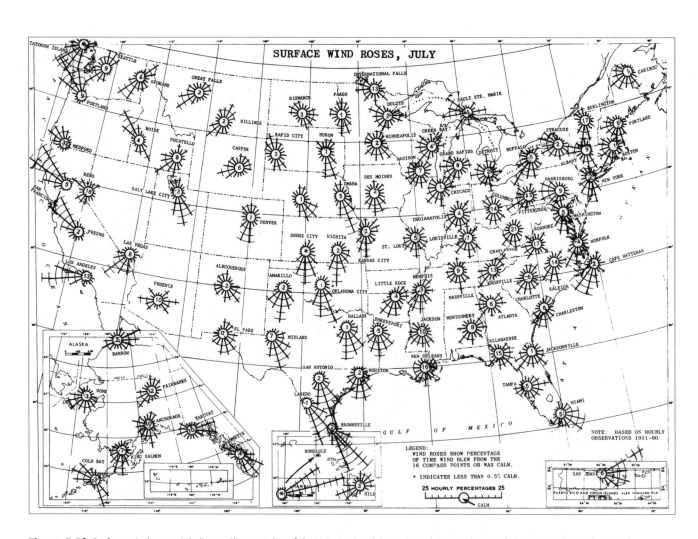

Figure 5.6f Surface wind roses, July (From *Climate Atlas of the U.S.,* National Oceanic and Atmospheric Administration (NOAA), 1983.)

2. Areas with fewer than 2,000 HDDs per year are characterized by very mild winters.

Cooling Degree-Days (CDDs)

1. Areas with more than 1,500 CDDs per year are characterized by long hot summers and substantial cooling requirements.
2. Areas with fewer than 500 CDDs per year are characterized by mild summers and little need for mechanical cooling.

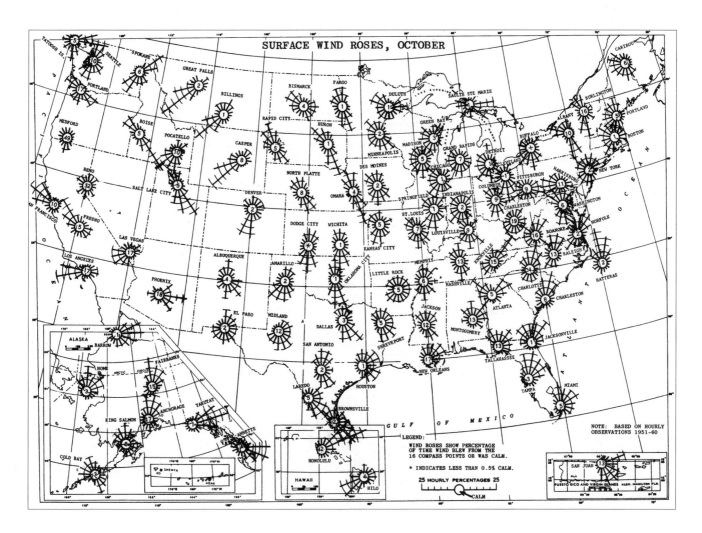

Figure 5.6g Surface wind roses, October. (From *Climate Atlas of the U.S.,* National Oceanic and Atmospheric Administration (NOAA), 1983.)

CLIMATE REGION 1

Reference City: Hartford, Connecticut

The Climate

This region, which includes New England, has a severe climate of mostly cool and cold temperatures that often go far below freezing in the winter. Snow is common in most of the region. Winter winds, which greatly add to winter discomfort, are usually from northwesterly and westerly directions. Since sunshine is available more than 50 percent of the daylight hours, solar heating has a potential even here. Reflections off the snow can significantly increase winter sun on south facades.

Summers are mostly mild with short hot periods. The extreme annual temperature range can reach more than 120°F, which makes this one of the most rugged regions in the United States. There are short periods in the summer when the humidity in combination with the temperature is too high. During the summer, the sun can be a liability and, thus, shading is required. Summer winds are generally southerly and are an asset in achieving comfort.

The annual precipitation is about 44 inches and occurs fairly uniformly throughout the year.

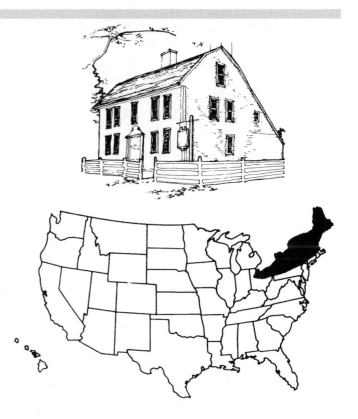

Climatic Design Priorities*

1st Keep heat in and cold temperatures out during the winter. (I)
2nd Protect from the cold winter winds. (II)
3rd Let the winter sun in. (III)

* These climatic design priorities are for envelope-dominated building types only. See Section 5.7 for a specific list of design strategies appropriate for achieving each of these priorities. Use the Roman numerals to find the relevant lists.

Note: For an explanation of these Climatic Data Tables, see Section 5.6. Much of the material in the Climate Data Tables comes from the book *Regional Guidelines for Building Passive Energy Conserving Homes* by the AIA Research Corporation.

Lower priority

4th Protect from the summer sun. (IV)
5th Use natural ventilation for summer cooling. (V)

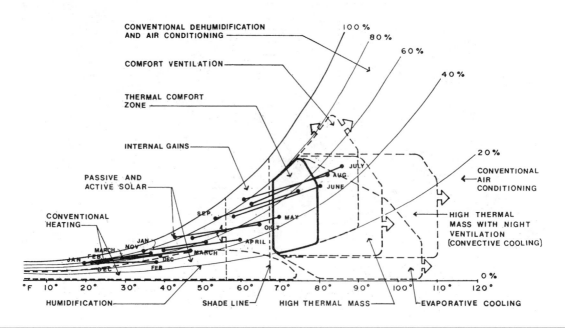

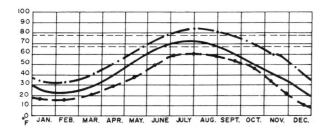

JAN. FEB. MAR. APR. MAY. JUNE JULY AUG. SEPT. OCT. NOV. DEC.

BASIC CLIMATIC CONDITION

		% of year
▨	Comfortable period	12
▨	too hot	13
☐	too cold	75

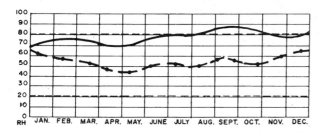

TEMPERATURE

– – – – – – – range of comfortable temp.
– · – · – · – afternoon maximum temperature
———————— average daily temperature
– ● – ● – ● – morning minimum temperature

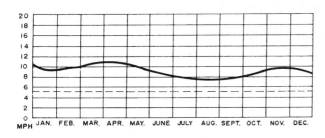

RELATIVE HUMIDITY

———————— average morning humidity
– ● – ● – ● – average afternoon humidity
– – – – – – – range of comfortable humidity

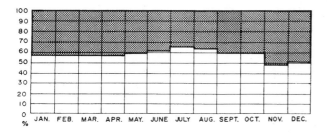

WIND SPEED

———————— mean daily wind speed
– – – – – – – wind speed for effective
natural ventilation

for wind direction see the
wind roses on pages 42–45.

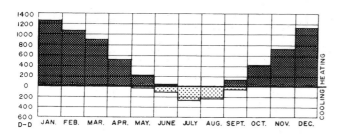

SUNSHINE
average % of daylight hours

annual sunshine = 57 %

> peak solar radiation in January
> horiz. sq. ft. = 500 btu/day
> vert. sq. ft. = 900 btu/day

DEGREE-DAYS

		annual
▨	heating degree-days	6,174
▨	cooling degree-days	666

CLIMATE REGION 2

Reference City: Madison, Wisconsin

The Climate

The climate of the northern plains is similar to that of region 1 but is even more severe because this inland region is far from the moderating effect of the oceans. The main concern is with the winter low temperatures, which are often combined with fairly high wind speeds.

Although summers are very hot, they are less of a concern because they are short. The sun is an asset in the winter and a liability in the summer.

The annual precipitation of about 31 inches occurs throughout the year, but summer months receive over twice as much as winter months.

Climatic Design Priorities*

1st Keep heat in and cold temperatures out in the winter. (I)
2nd Protect from the cold winter winds. (II)
3rd Let the winter sun in. (III)

* These climatic design priorities are for envelope-dominated building types only. See Section 5.7 for a specific list of design strategies appropriate for achieving each of these priorities. Use the Roman numerals to find the relevant lists.

Note: For an explanation of these Climatic Data Tables, see Section 5.6. Much of the material in the Climate Data Tables comes from the book *Regional Guidelines for Building Passive Energy Conserving Homes* by the AIA Research Corporation.

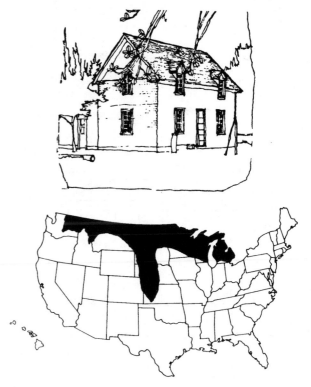

Lower priority

4th Use thermal mass to flatten day-to-night temperature swings in the summer. (VII)
5th Protect from the summer sun. (IV)
6th Use natural ventilation for summer cooling. (V)

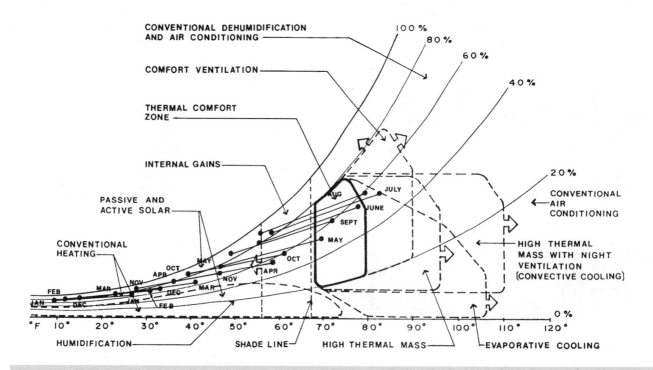

JAN. FEB. MAR. APR. MAY. JUNE JULY AUG. SEPT. OCT. NOV. DEC.

BASIC CLIMATIC CONDITION

		% of year
░	Comfortable period	12
▓	too hot	12
☐	too cold	76

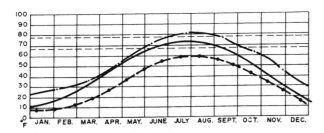

TEMPERATURE

— — — — —	range of comfortable temp.
— · — · —	afternoon maximum temperature
——————	average daily temperature
—•—•—•	morning minimum temperature

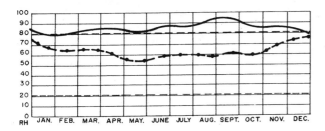

RELATIVE HUMIDITY

——————	average morning humidity
—•—•—•	average afternoon humidity
— — — —	range of comfortable humidity

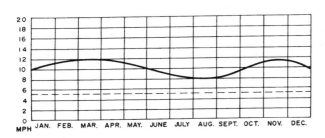

WIND SPEED

| —————— | mean daily wind speed |
| — — — — | wind speed for effective natural ventilation |

for wind direction see the
wind roses on pages **42–45**.

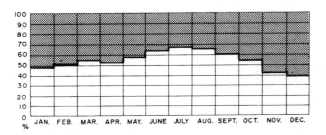

SUNSHINE
average % of daylight hours

annual sunshine = 54 %

peak solar radiation in January
horiz. sq. ft. = 600 btu/day
vert. sq. ft. = 1100 btu/day

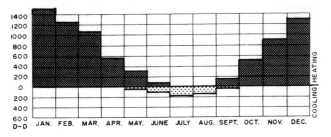

DEGREE-DAYS

		annual
▓	heating degree-days	7,642
░	cooling degree-days	467

CLIMATE REGION 3

Reference City: Indianapolis, Indiana

The Climate

This climate of the Midwest is similar to that of regions 1 and 2, but it is somewhat milder in winter. Cold winds, however, are still an important concern. The mean annual snowfall ranges from 12 to 60 inches. There is some potential for solar energy in the winter since the sun shines more than 40 percent of the daylight hours.

Significant cooling loads are common since high summer temperatures often coincide with high humidity. Winds are an asset during the summer.

The annual precipitation is about 39 inches and occurs fairly uniformly throughout the year.

Climatic Design Priorities*

1st Keep heat in and cold temperatures out in the winter. (I)
2nd Protect from the cold winter winds. (II)
3rd Let the winter sun in. (III)

* These climatic design priorities are for envelope-dominated building types only. See Section 5.7 for a specific list of design strategies appropriate for achieving each of these priorities. Use the Roman numerals to find the relevant lists.

Note: For an explanation of these Climatic Data Tables, see Section 5.6. Much of the material in the Climate Data Tables comes from the book *Regional Guidelines for Building Passive Energy Conserving Homes* by the AIA Research Corporation.

Lower priority

4th Keep hot temperatures out during the summer. (VIII)
5th Protect from the summer sun. (IV)
6th Use natural ventilation for summer cooling. (V)

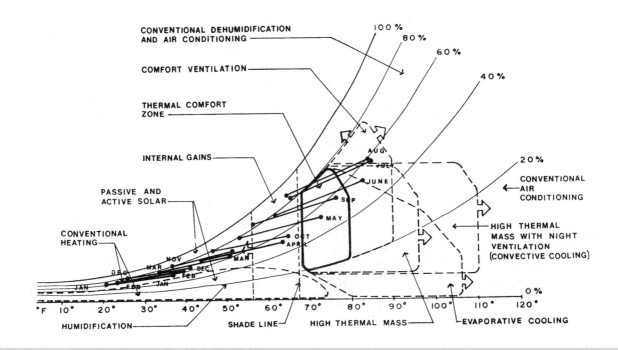

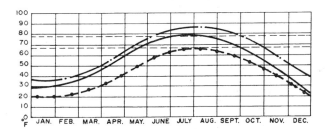

BASIC CLIMATIC CONDITION

		% of year
	Comfortable period	_14_
	too hot	_20_
	too cold	_66_

TEMPERATURE

– – – – – range of comfortable temp.
– · – · – afternoon maximum temperature
———— average daily temperature
– • – • – morning minimum temperature

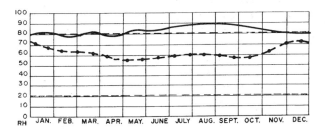

RELATIVE HUMIDITY

———— average morning humidity
– • – • – average afternoon humidity
– – – – – range of comfortable humidity

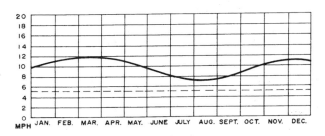

WIND SPEED

———— mean daily wind speed
– – – – – wind speed for effective
natural ventilation

for wind direction see the
wind roses on pages 42–45.

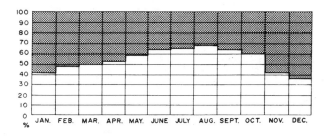

SUNSHINE
average % of daylight hours

annual sunshine = _55_ %

peak solar radiation in January
horiz. sq. ft. = _500_ btu/day
vert. sq. ft. = _800_ btu/day

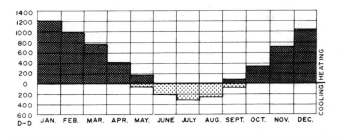

DEGREE-DAYS

		annual
	heating degree-days	_5,650_
	cooling degree-days	_988_

CLIMATE REGION 4

Reference City: Salt Lake City, Utah

The Climate

This is the climate of the Great Plains, intermountain basin, and plateaus. It is a semiarid climate with cold windy winters and warm dry summers. Winters are very cold with frequent but short storms alternating with sunny periods.

Summer temperatures are high, but the humidity is low. Thus, the diurnal temperature range is high and summer nights are generally cool. There is much potential for both passive heating and cooling.

The annual precipitation is about 15 inches and occurs fairly uniformly throughout the year, but spring is the wettest season.

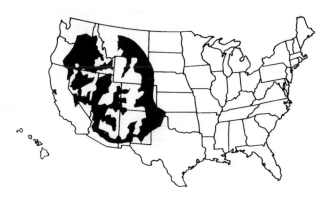

*Climatic Design Priorities**

1st   Keep the heat in and the cold temperatures out during the winter. (I)

2nd Let the winter sun in. (III)

3rd   Protect from the cold winter winds. (II)

* These climatic design priorities are for envelope-dominated building types only. See Section 5.7 for a specific list of design strategies appropriate for achieving each of these priorities. Use the Roman numerals to find the relevant lists.

Note: For an explanation of these Climatic Data Tables, see Section 5.6. Much of the material in the Climate Data Tables comes from the book *Regional Guidelines for Building Passive Energy Conserving Homes* by the AIA Research Corporation.

Lower priority

4th   Use thermal mass to flatten day-to-night temperature swings in the summer. (VII)

5th   Protect from the summer sun. (IV)

6th   Use evaporative cooling in the summer. (IX)

7th   Use natural ventilation for summer cooling. (V)

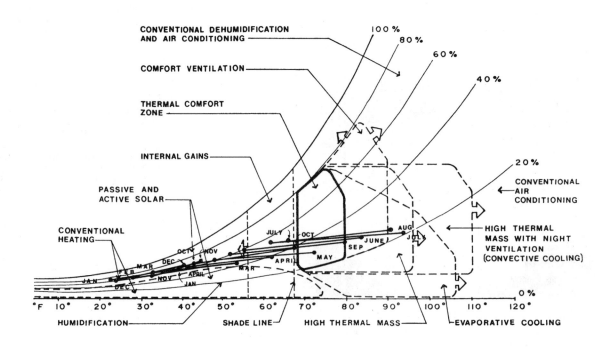

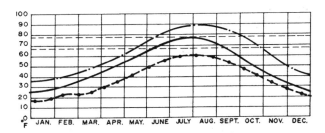

JAN. FEB. MAR. APR. MAY. JUNE JULY AUG. SEPT. OCT. NOV. DEC.

BASIC CLIMATIC CONDITION

		% of year
▦	Comfortable period	12
▓	too hot	11
☐	too cold	77

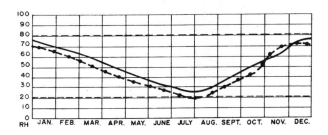

TEMPERATURE

– – – – – – range of comfortable temp.
–·–·–·– afternoon maximum temperature
———— average daily temperature
—•—•— morning minimum temperature

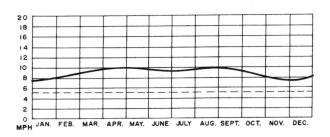

RELATIVE HUMIDITY

———— average morning humidity
—•—•— average afternoon humidity
– – – – – – range of comfortable humidity

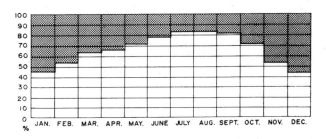

WIND SPEED

———— mean daily wind speed
– – – – – – wind speed for effective
        natural ventilation

for wind direction see the
wind roses on pages **42–45**.

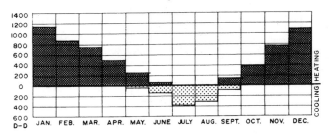

SUNSHINE
average % of daylight hours

annual sunshine = __66__ %

peak solar radiation in January
horiz. sq. ft. = __700__ btu/day
vert. sq. ft. = __1100__ btu/day

DEGREE-DAYS

		annual
▓	heating degree-days	5,802
▦	cooling degree-days	981

CLIMATE REGION 5

Reference City: Ely, Nevada

The Climate

This is a high, mountainous, and semiarid region above 7,000 feet in southern latitudes and above 6,000 feet in northern latitudes. It is a mostly cool and cold climate. Snow is plentiful and might remain on the ground for more than half the year. The temperature and, thus, the snow cover vary tremendously with the slope orientation and elevation. Heating is required most of the year. Fortunately, sunshine is available more than 60 percent of the winter daylight hours.

Summer temperatures are modest, and comfort is easily achieved by natural ventilation. Summer nights are quite cool because of the high diurnal temperature range.

The annual precipitation is about 9 inches and occurs fairly uniformly throughout the year.

*Climatic Design Priorities**

1st   Keep the heat in and the cold temperatures out during the winter. (I)
2nd Let the winter sun in. (III)
3rd   Protect from the cold winter winds. (II)
4th   Use thermal mass to flatten day-to-night temperature swings in the summer. (VII)

* These climatic design priorities are for envelope-dominated building types only. See Section 5.7 for a specific list of design strategies appropriate for achieving each of these priorities. Use the Roman numerals to find the relevant lists.

Note: For an explanation of these Climatic Data Tables, see Section 5.6. Much of the material in the Climate Data Tables comes from the book *Regional Guidelines for Building Passive Energy Conserving Homes* by the AIA Research Corporation.

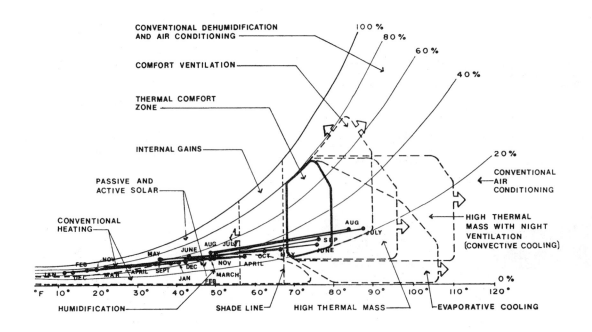

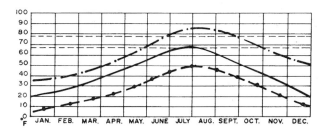

JAN. FEB. MAR. APR. MAY. JUNE JULY AUG. SEPT. OCT. NOV. DEC.

BASIC CLIMATIC CONDITION

		% of year
	Comfortable period	8
	too hot	0
	too cold	92

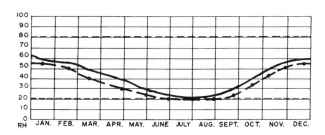

TEMPERATURE

– – – – – – range of comfortable temp.
– · – · – · – afternoon maximum temperature
————— average daily temperature
—•—•—•— morning minimum temperature

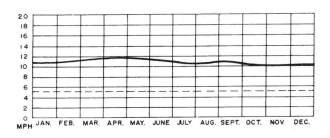

RELATIVE HUMIDITY

————— average morning humidity
—•—•—•— average afternoon humidity
– – – – – – range of comfortable humidity

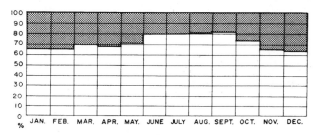

WIND SPEED

————— mean daily wind speed
– – – – – – wind speed for effective
          natural ventilation

for wind direction see the
wind roses on pages 42–45.

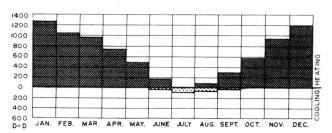

SUNSHINE
average % of daylight hours

annual sunshine = 73 %

peak solar radiation in January
horiz. sq. ft. = 900 btu/day
vert. sq. ft. = 1600 btu/day

DEGREE-DAYS

		annual
	heating degree-days	7,700
	cooling degree-days	192

CLIMATE REGION 6 ▬▬▬▬▬▬▬▬▬▬▬▬▬▬▬▬▬

Reference City: Medford, Oregon

The Climate

The northern California, Oregon, and Washington coastal region has a very mild climate. In winter the temperatures are cool and rain is common. Although the skies are frequently overcast, solar heating is still possible because of the small heating load created by the mild climate. The high relative humidity is not a significant problem because it does not coincide with high summer temperatures.

The region has a large variation in microclimates because of changes in both elevation and distance from the coast. In some areas the winter winds are a significant problem. A designer should, therefore, obtain additional local weather data.

The annual precipitation is about 20 in. but most of it occurs in the winter months. The summers are quite dry and sunny.

*Climatic Design Priorities**

1st Keep the heat in and the cold temperatures out during the winter. (I)
2nd Let the winter sun in (mostly diffused sun because of the clouds). (III)
3rd Protect from the cold winter winds. (II)

* These climatic design priorities are for envelope-dominated building types only. See Section 5.7 for a specific list of design strategies appropriate for achieving each of these priorities. Use the Roman numerals to find the relevant lists.

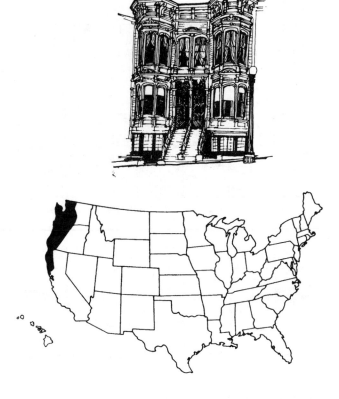

Note: For an explanation of these Climatic Data Tables, see Section 5.6. Much of the material in the Climate Data Tables comes from the book *Regional Guidelines for Building Passive Energy Conserving Homes* by the AIA Research Corporation.

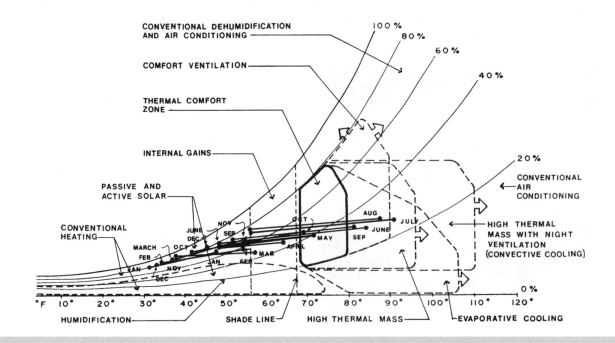

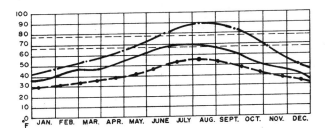

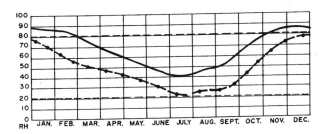

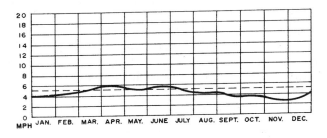

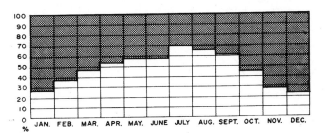

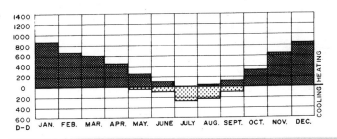

BASIC CLIMATIC CONDITION

		% of year
▨	Comfortable period	_13_
▉	too hot	_8_
☐	too cold	_79_

TEMPERATURE

– – – – – range of comfortable temp.
– · – · – afternoon maximum temperature
——— average daily temperature
—•—•— morning minimum temperature

RELATIVE HUMIDITY

——— average morning humidity
—•—•— average afternoon humidity
– – – – – range of comfortable humidity

WIND SPEED

——— mean daily wind speed
– – – – – wind speed for effective natural ventilation

for wind direction see the wind roses on pages 42–45.

SUNSHINE
average % of daylight hours

annual sunshine = _47_ %

peak solar radiation in January
horiz. sq. ft. = _300_ btu/day
vert. sq. ft. = _550_ btu/day

DEGREE-DAYS

		annual
▨	heating degree-days	_4,798_
▨	cooling degree-days	_645_

CLIMATE REGION 7

Reference City: Fresno, California

The Climate

This region includes California's Central Valley and parts of the central coast. Winters are moderately cold with most of the annual rain of about 11 inches falling during that period. Winter sunshine, nevertheless, is plentiful.

Summers are hot and dry. The low humidity causes a large diurnal temperature range, and, consequently, summer nights are cool. Rain is rare during the summer months.

Since spring and fall are very comfortable, and much of the rest of the year is not very uncomfortable, outdoor living is very popular in this region.

Because of the varying distances to the ocean, significant changes in microclimate exist. Near the coast the temperatures are more moderate in both winter and summer. Neither winter nor summer dominates the climate of this region.

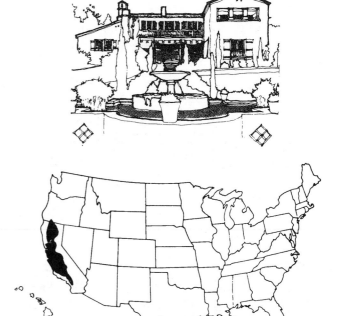

*Climatic Design Priorities**

1st   Keep the heat in and the cold temperatures out during the winter. (I)
2nd   Keep hot temperatures out during the summer. (VIII)
3rd   Let the winter sun in. (III)
4th   Protect from the summer sun. (IV)
5th   Use thermal mass to flatten day-to-night temperature swings during the summer. (VII)
6th   Use natural ventilation for cooling in the spring and fall. (V)
7th   Use evaporative cooling in the summer. (IX)
8th   Protect from the cold winter winds. (II)

*These climatic design priorities are for envelope-dominated building types only. See Section 5.7 for a specific list of design strategies appropriate for achieving each of these priorities. Use the Roman numerals to find the relevant lists.

Note: For an explanation of these Climatic Data Tables, see Section 5.6. Much of the material in the Climate Data Tables comes from the book *Regional Guidelines for Building Passive Energy Conserving Homes* by the AIA Research Corporation.

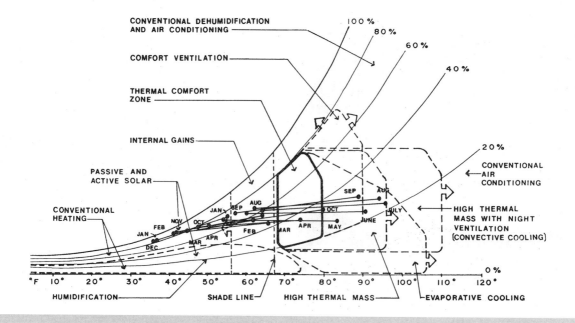

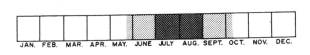

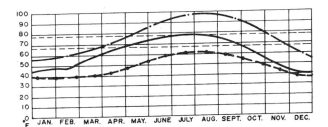

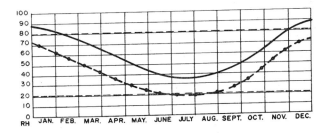

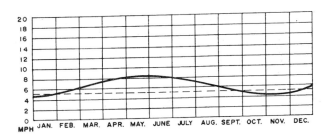

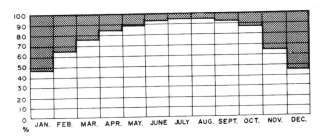

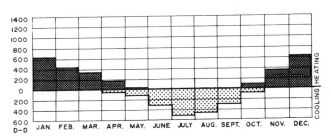

BASIC CLIMATIC CONDITION

		% of year
	Comfortable period	_21_
	too hot	_17_
	too cold	_62_

TEMPERATURE

– – – – – range of comfortable temp.
– · – · – afternoon maximum temperature
——— average daily temperature
—•—•— morning minimum temperature

RELATIVE HUMIDITY

——— average morning humidity
—•—•— average afternoon humidity
– – – – – range of comfortable humidity

WIND SPEED

——— mean daily wind speed
– – – – – wind speed for effective
            natural ventilation

for wind direction see the
wind roses on pages **42–45**.

SUNSHINE
average % of daylight hours

annual sunshine = _78_ %

peak solar radiation in January
horiz. sq. ft. = _600_ btu/day
vert. sq. ft. = _1050_ btu/day

DEGREE-DAYS

		annual
	heating degree-days	_2,647_
	cooling degree-days	_1,769_

CLIMATE REGION 8

Reference City: Charleston, South Carolina

The Climate

This Mid-Atlantic-coast climate is relatively temperate with four distinct seasons. Although summers are very hot and humid and winters are somewhat cold, spring and fall are generally quite pleasant. Summer winds are an important asset in this hot and humid climate.

The annual precipitation is about 47 inches and occurs fairly uniformly throughout the year. Summer, however, is the wettest season with thunderstorms common during that period. Tropical storms are an occasional possibility.

Climatic Design Priorities*

1st   Keep the heat in and the cold temperatures out during the winter. (I)
2nd Use natural ventilation for summer cooling. (V)
3rd   Let the winter sun in. (III)
4th   Protect from the summer sun. (IV)

Lower priority

5th   Protect from the cold winter winds. (II)
6th   Avoid creating additional humidity during the summer. (X)

*These climatic design priorities are for envelope-dominated building types only. See Section 5.7 for a specific list of design strategies appropriate for achieving each of these priorities. Use the Roman numerals to find the relevant lists.

Note: For an explanation of these Climatic Data Tables, see Section 5.6. Much of the material in the Climate Data Tables comes from the book *Regional Guidelines for Building Passive Energy Conserving Homes* by the AIA Research Corporation.

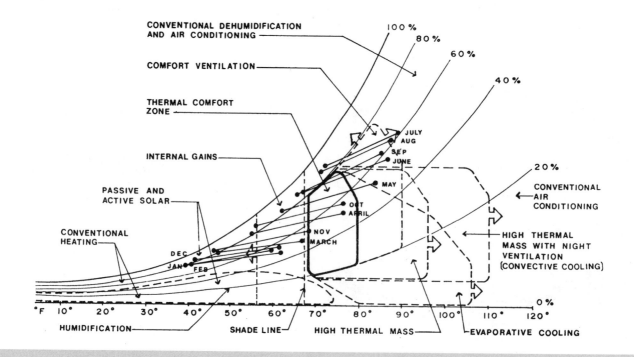

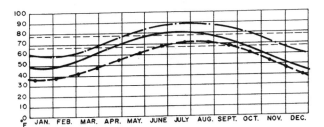

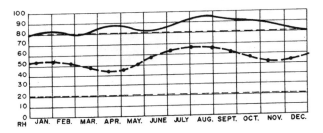

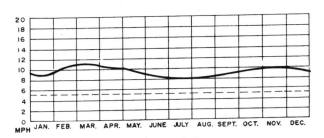

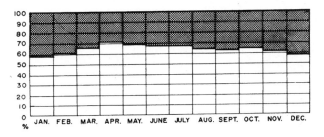

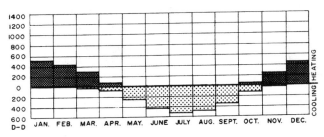

BASIC CLIMATIC CONDITION

		% of year
▒	Comfortable period	12
▓	too hot	42
☐	too cold	46

TEMPERATURE

— — — — — range of comfortable temp.
— · — · — afternoon maximum temperature
———— average daily temperature
—•—•— morning minimum temperature

RELATIVE HUMIDITY

———— average morning humidity
—•—•— average afternoon humidity
— — — — range of comfortable humidity

WIND SPEED

———— mean daily wind speed
— — — — wind speed for effective
         natural ventilation

for wind direction see the
wind roses on pages 42–45.

SUNSHINE
average % of daylight hours

annual sunshine = 65 %

peak solar radiation in January
horiz. sq. ft. = 900 btu/day
vert. sq. ft. = 1300 btu/day

DEGREE-DAYS

		annual
▓	heating degree-days	1,868
▒	cooling degree-days	2,304

CLIMATE REGION 9 ▬▬▬▬▬▬▬▬▬

Reference City: Little Rock, Arkansas

The Climate

This climate of the Mississippi Valley is similar to that of region 8 except that it is slightly more severe in both summer and winter due to the distance from the oceans. Winters are quite cold with chilling winds from the northwest. Summers are hot and humid with winds often from the southwest.

The annual precipitation is about 49 inches and occurs fairly uniformly throughout the year.

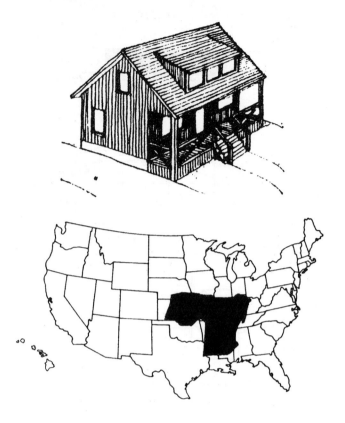

*Climatic Design Priorities**

1st Keep the heat in and cold temperatures out during the winter. (I)
2nd Let the winter sun in. (III)
3rd Use natural ventilation for summer cooling. (V)
4th Protect from the cold winter winds. (II)
5th Protect from the summer sun. (IV)
6th Avoid creating additional humidity during the summer. (X)

*These climatic design priorities are for envelope-dominated building types only. See Section 5.7 for a specific list of design strategies appropriate for achieving each of these priorities. Use the Roman numerals to find the relevant lists.

Note: For an explanation of these Climatic Data Tables, see Section 5.6. Much of the material in the Climate Data Tables comes from the book *Regional Guidelines for Building Passive Energy Conserving Homes* by the AIA Research Corporation.

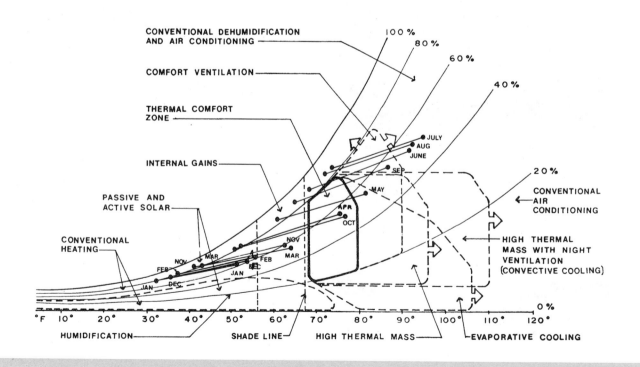

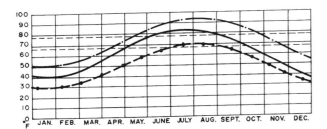

BASIC CLIMATIC CONDITION

		% of year
░	Comfortable period	13
▓	too hot	35
☐	too cold	52

TEMPERATURE

- – – – – – range of comfortable temp.
- – · – · – afternoon maximum temperature
- ──────── average daily temperature
- ──•──•── morning minimum temperature

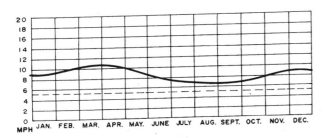

RELATIVE HUMIDITY

- ──────── average morning humidity
- ──•──•── average afternoon humidity
- – – – – – range of comfortable humidity

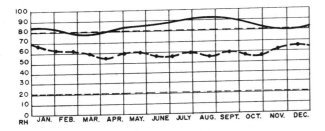

WIND SPEED

- ──────── mean daily wind speed
- – – – – – wind speed for effective natural ventilation

for wind direction see the wind roses on pages 42–45.

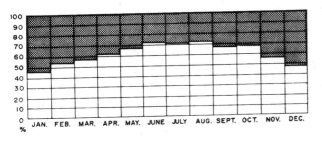

SUNSHINE
average % of daylight hours

annual sunshine = 62 %

peak solar radiation in January
horiz. sq. ft. = 600 btu/day
vert. sq. ft. = 900 btu/day

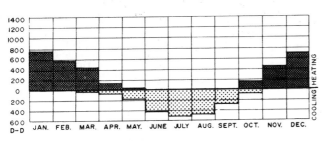

DEGREE–DAYS

		annual
▓	heating degree-days	3.152
░	cooling degree-days	2.045

CLIMATE REGION 10 ▬▬▬▬▬▬▬▬▬▬

Reference City: Knoxville, TN

The Climate

The climate of Appalachia is relatively temperate with a long and pleasant spring and fall. Winters are quite cool with a significant chilling effect from the wind. Temperatures are somewhat cooler at the northern end of this region. Snow is also more common at the northern end although it does occur fairly frequently at higher elevations at the southern end of the region.

Summers are hot and somewhat humid. However, the humidity is low enough to allow a fair amount of night cooling, and, thus, the diurnal temperature range is fairly high. There is also a fair amount of wind available for cooling in the summer.

The annual precipitation is about 47 inches and occurs rather uniformly throughout the year.

Climatic Design Priorities*

1st   Keep the heat in and the cold temperatures out in the winter. (I)
2nd   Use natural ventilation for summer cooling. (V)
3rd   Let the winter sun in. (III)
4th   Protect from the summer sun. (IV)
5th   Protect from the cold winter winds. (II)
6th   Avoid creating additional humidity during the summer. (X)

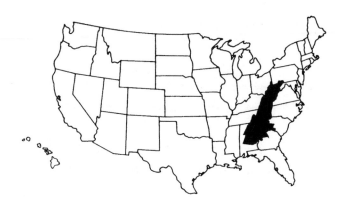

*These climatic design priorities are for envelope-dominated building types only. See Section 5.7 for a specific list of design strategies appropriate for achieving each of these priorities. Use the Roman numerals to find the relevant lists.

Note: For an explanation of these Climatic Data Tables, see Section 5.6. Much of the material in the Climate Data Tables comes from the book *Regional Guidelines for Building Passive Energy Conserving Homes* by the AIA Research Corporation.

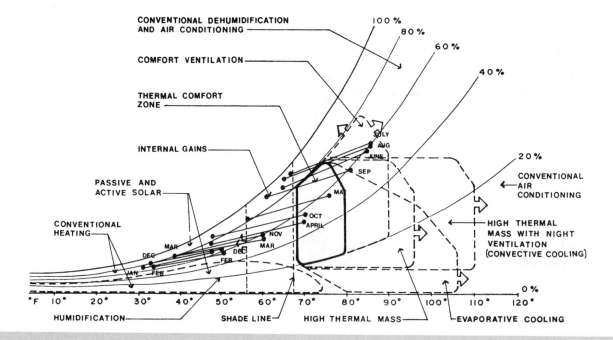

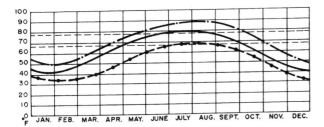

JAN. FEB. MAR. APR. MAY. JUNE JULY AUG. SEPT. OCT. NOV. DEC.

BASIC CLIMATIC CONDITION

		% of year
▨	Comfortable period	16
▓	too hot	28
☐	too cold	56

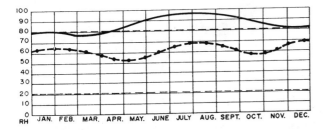

TEMPERATURE

- - - - - - - range of comfortable temp.
- · - · - · - afternoon maximum temperature
———— average daily temperature
—•—•— morning minimum temperature

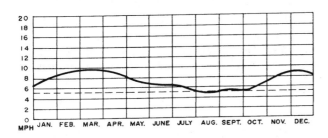

RELATIVE HUMIDITY

———— average morning humidity
—•—•— average afternoon humidity
- - - - - - - range of comfortable humidity

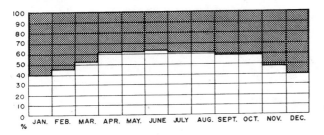

WIND SPEED

———— mean daily wind speed
- - - - - - - wind speed for effective natural ventilation

for wind direction see the wind roses on pages 42–45.

SUNSHINE
average % of daylight hours

annual sunshine = 55 %

peak solar radiation in January
horiz. sq. ft. = 600 btu/day
vert. sq. ft. = 800 btu/day

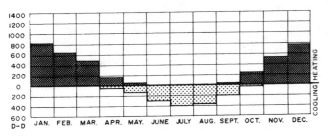

DEGREE-DAYS

		annual
▓	heating degree-days	3,658
▨	cooling degree-days	1,449

CLIMATE REGION 11

Reference City: Phoenix, Arizona

The Climate

The climate of the Southwest desert regions is characterized by extremely hot and dry summers and moderately cold winters. The skies are clear most of the year with an annual sunshine of about 85 percent.

Since summers are extremely hot and dry, the diurnal temperature range is very large, and, consequently, nights are quite cool. The humidity is below the comfort range much of the year. Summer overheating is the main concern for the designer.

The annual precipitation of about 7 inches is quite low and occurs throughout the year. April, May, and June, however, are the driest months, while August is the wettest with 1 inch of rain.

Climatic Design Priorities*

1st Keep hot temperatures out during the summer. (VIII)
2nd Protect from the summer sun. (IV)
3rd Use evaporative cooling in the summer. (IX)
4th Use thermal mass to flatten day-to-night temperature swings during the summer. (VII)

Lower priority

5th Keep the heat in and the cool temperatures out during the winter. (I)
6th Let the winter sun in. (III)
7th Use natural ventilation to cool in the spring and fall. (VI)

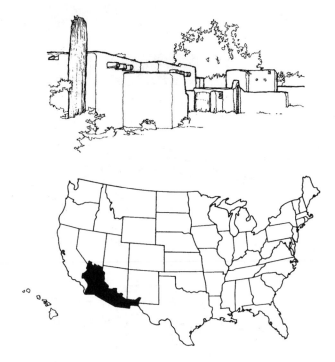

*These climatic design priorities are for envelope-dominated building types only. See Section 5.7 for a specific list of design strategies appropriate for achieving each of these priorities. Use the Roman numerals to find the relevant lists.

Note: For an explanation of these Climatic Data Tables, see Section 5.6. Much of the material in the Climate Data Tables comes from the book *Regional Guidelines for Building Passive Energy Conserving Homes* by the AIA Research Corporation.

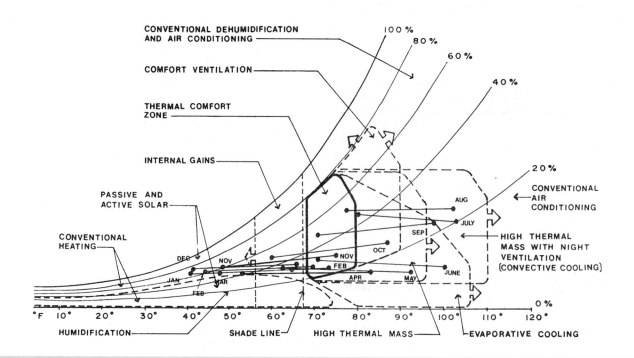

JAN. FEB. MAR. APR. MAY. JUNE JULY AUG. SEPT. OCT. NOV. DEC.

BASIC CLIMATIC CONDITION

		% of year
	Comfortable period	15
	too hot	37
	too cold	48

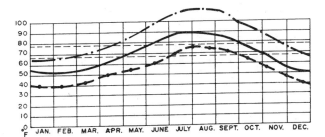

TEMPERATURE

– – – – – range of comfortable temp.
–·–·–·– afternoon maximum temperature
———— average daily temperature
—•—•— morning minimum temperature

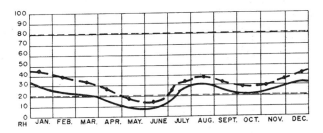

RELATIVE HUMIDITY

———— average morning humidity
—•—•— average afternoon humidity
– – – – range of comfortable humidity

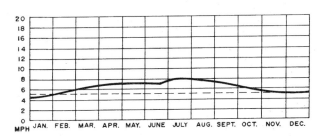

WIND SPEED

———— mean daily wind speed
– – – – wind speed for effective
natural ventilation

for wind direction see the
wind roses on pages 42–45.

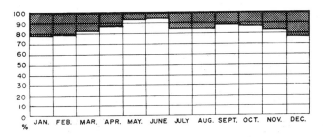

SUNSHINE
average % of daylight hours

annual sunshine = 85 %

peak solar radiation in January
horiz. sq. ft. = 1200 btu/day
vert. sq. ft. = 1600 btu/day

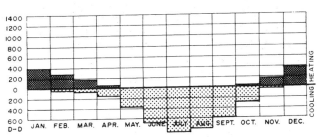

DEGREE-DAYS

		annual
	heating degree-days	1,442
	cooling degree-days	3,746

CLIMATE REGION 12

Reference City: Midland, TX

The Climate

This area of west Texas and southeast New Mexico is an arid climate of hot summers and cool winters. Plentiful sunshine, more than 60 percent in the winter, can supply ample solar heating. The low humidity in summer facilitates the effective use of evaporative cooling. Thus, in this region, climatic design can have a very beneficial impact on thermal comfort.

The annual precipitation is about 14 inches, and although it occurs throughout the year, most of it falls during the summer months.

*Climatic Design Priorities**

1st   Use evaporative cooling in summer. (IX)
2nd   Let the winter sun in. (III)
3rd   Protect from the summer sun. (IV)
4th   Keep the heat in and the cool temperatures out during the winter. (I)
5th   Keep hot temperatures out during the summer. (VIII)
6th   Protect from the cold winter winds. (II)
7th   Use natural ventilation for summer cooling. (V)
8th   Use thermal mass to flatten day-to-night temperature swings during the summer. (VII)

*These climatic design priorities are for envelope-dominated building types only. See Section 5.7 for a specific list of design strategies appropriate for achieving each of these priorities. Use the Roman numerals to find the relevant lists.

Note: For an explanation of these Climatic Data Tables, see Section 5.6. Much of the material in the Climate Data Tables comes from the book *Regional Guidelines for Building Passive Energy Conserving Homes* by the AIA Research Corporation.

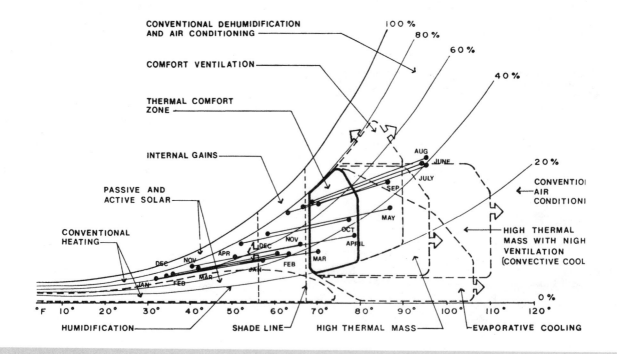

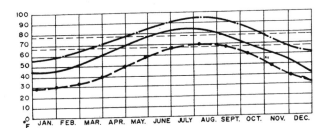

JAN. FEB. MAR. APR. MAY. JUNE JULY AUG. SEPT. OCT. NOV. DEC.

BASIC CLIMATIC CONDITION

		% of year
▨	Comfortable period	19
■	too hot	26
☐	too cold	55

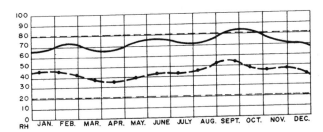

TEMPERATURE

Line	
– – – –	range of comfortable temp.
– · – · –	afternoon maximum temperature
———	average daily temperature
–•–•–	morning minimum temperature

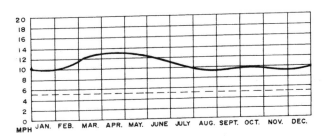

RELATIVE HUMIDITY

Line	
———	average morning humidity
–•–•–	average afternoon humidity
– – – –	range of comfortable humidity

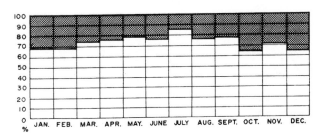

WIND SPEED

Line	
———	mean daily wind speed
– – – –	wind speed for effective natural ventilation

for wind direction see the wind roses on pages **42–45**.

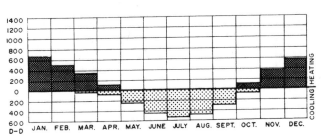

SUNSHINE
average % of daylight hours

annual sunshine = 74 %

peak solar radiation in January
horiz. sq. ft. = 1100 btu/day
vert. sq. ft. = 1450 btu/day

DEGREE-DAYS

		annual
■	heating degree-days	2658
▨	cooling degree-days	2,126

CLIMATE REGION 13 ▬

Reference City: Fort Worth, Texas

The Climate

This area of Oklahoma and north Texas has cold winters and hot summers. Cold winds come from the north and northeast. There is a significant amount of sunshine available for winter solar heating.

During part of the summer, the high temperatures and fairly high humidities combine to create uncomfortable conditions. During other times in the summer, the humidity drops sufficiently to enable evaporative cooling to work. There are also ample summer winds for natural ventilation.

During the drier parts of the summer and especially during spring and fall, the diurnal temperature range is large enough to encourage the use of thermal mass. The annual precipitation is about 29 inches, and it occurs fairly uniformly throughout the year.

Climatic Design Priorities*

1st Use natural ventilation for cooling in spring and fall. (V)
2nd Let the winter sun in. (III)
3rd Protect from the summer sun. (IV)
4th Protect from the cold winter winds. (II)

Lower priority

5th Use thermal mass to flatten day-to-night temperature swings during the summer. (VII)

*These climatic design priorities are for envelope-dominated building types only. See Section 5.7 for a specific list of design

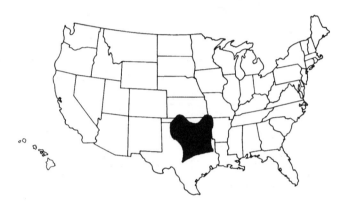

strategies appropriate for achieving each of these priorities. Use the Roman numerals to find the relevant lists.

Note: For an explanation of these Climatic Data Tables, see Section 5.6. Much of the material in the Climate Data Tables comes from the book *Regional Guidelines for Building Passive Energy Conserving Homes* by the AIA Research Corporation.

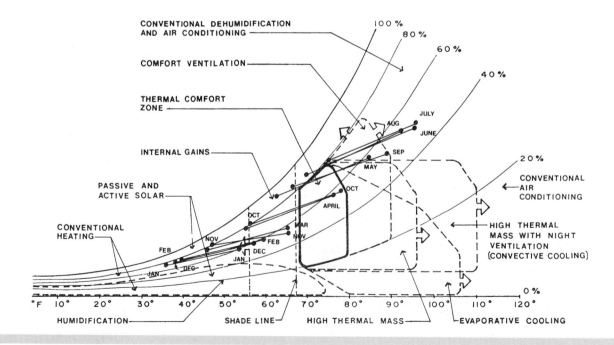

BASIC CLIMATIC CONDITION

		% of year
▦	Comfortable period	_14_
▪	too hot	_39_
☐	too cold	_47_

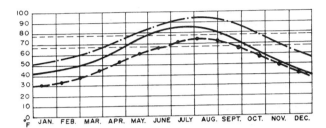

JAN. FEB. MAR. APR. MAY. JUNE JULY AUG. SEPT. OCT. NOV. DEC.

TEMPERATURE

– – – – – range of comfortable temp.
– · – · – afternoon maximum temperature
————— average daily temperature
—•—•— morning minimum temperature

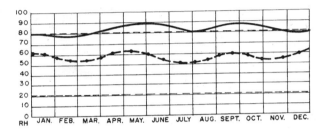

RELATIVE HUMIDITY

————— average morning humidity
—•—•— average afternoon humidity
– – – – – range of comfortable humidity

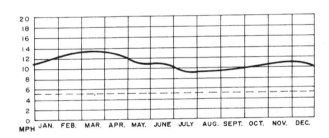

WIND SPEED

————— mean daily wind speed
– – – – – wind speed for effective natural ventilation

for wind direction see the wind roses on pages 42–45.

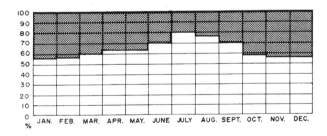

SUNSHINE
average % of daylight hours

annual sunshine = _64_ %

peak solar radiation in January
horiz. sq. ft. = _900_ btu/day
vert. sq. ft. = _1200_ btu/day

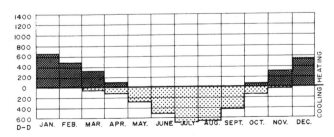

DEGREE-DAYS

		annual
▪	heating degree-days	_2,407_
▦	cooling degree-days	_2,809_

CLIMATE REGION 14

Reference City: New Orleans, Louisiana

The Climate

This Gulf Coast region has cool but short winters. Summers, on the other hand, are hot, very humid, and long. The flat damp ground and frequent rains create a very humid climate. Besides creating thermal discomfort, the high humidity also causes mildew problems. Much of the region has reliable sea breezes, which are strongest during the day, weaker at night, and nonexistent during the morning and evening when the wind reverses direction.

The annual precipitation is quite high at 60 inches, and it occurs fairly uniformly throughout the year.

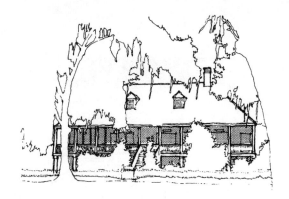

*Climatic Design Priorities**

1st   Allow natural ventilation to both cool and remove excess moisture in the summer. (VI)
2nd   Protect from the summer sun. (IV)
3rd   Avoid creating additional humidity during the summer. (X)

Lower priority

4th   Let the winter sun in. (III)
5th   Protect from the cold winter winds. (II)

*These climatic design priorities are for envelope-dominated building types only. See Section 5.7 for a specific list of design strategies appropriate for achieving each of these priorities. Use the Roman numerals to find the relevant lists.

Note: For an explanation of these Climatic Data Tables, see Section 5.6. Much of the material in the Climate Data Tables comes from the book *Regional Guidelines for Building Passive Energy Conserving Homes* by the AIA Research Corporation.

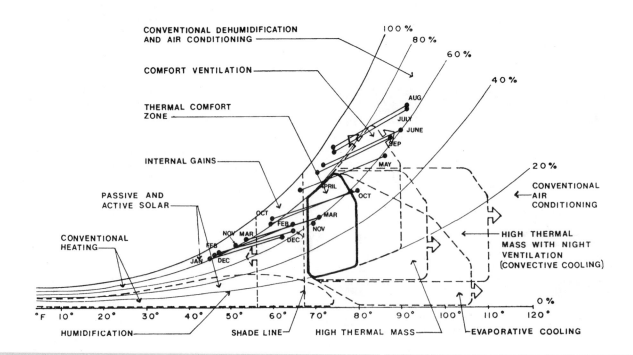

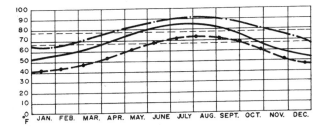

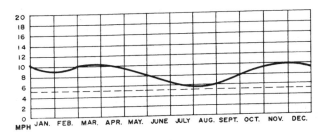

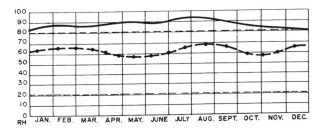

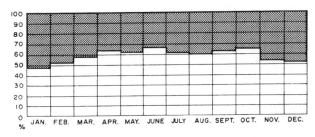

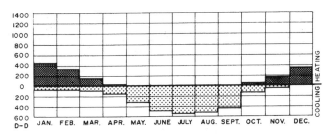

BASIC CLIMATIC CONDITION

		% of year
▨	Comfortable period	12
▦	too hot	52
☐	too cold	36

TEMPERATURE

– – – – – range of comfortable temp.
–·–·–·– afternoon maximum temperature
——— average daily temperature
–•–•–•– morning minimum temperature

RELATIVE HUMIDITY

——— average morning humidity
–•–•–•– average afternoon humidity
– – – – – range of comfortable humidity

WIND SPEED

——— mean daily wind speed
– – – – – wind speed for effective
　　　　natural ventilation

for wind direction see the
wind roses on pages 42–45.

SUNSHINE
average % of daylight hours

annual sunshine = 59 %

peak solar radiation in January
horiz. sq. ft. = 800 btu/day
vert. sq. ft. = 1250 btu/day

DEGREE-DAYS

		annual
▨	heating degree-days	1,490
▨	cooling degree-days	2,686

CLIMATE REGION 15 ▬▬▬▬▬▬▬▬▬

Reference City: Houston, Texas

The Climate

This part of the Gulf Coast is similar to region 14 except that the summers are more severe. Very high temperatures and humidity levels make this a very uncomfortable summer climate. The high humidity and clouds prevent the temperature from dropping much at night. Thus, the diurnal temperature range is quite small. Fortunately, frequent coastal breezes exist in the summer.

Winters are short and mild. Ample sunshine can supply most of the winter heating demands, but the main concern for the designer is summer overheating.

The annual precipitation is about 45 inches and occurs fairly uniformly throughout the year.

Climatic Design Priorities*

1st Keep hot temperatures out during the summer. (VIII)
2nd Allow natural ventilation to both cool and remove excess moisture in the summer. (VI)
3rd Protect from the summer sun. (IV)
4th Avoid creating additional humidity during the summer. (X)

Lower priority

5th Protect from the cold winter winds. (II)
6th Let the winter sun in. (III)
7th Keep the heat in and the cool temperatures out during the winter. (I)

* These climatic design priorities are for envelope-dominated building types only. See Section 5.7 for a specific list of design

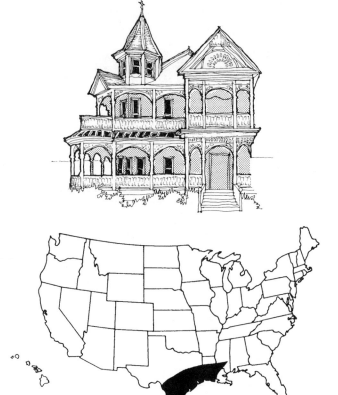

strategies appropriate for achieving each of these priorities. Use the Roman numerals to find the relevant lists.

Note: For an explanation of these Climatic Data Tables, see Section 5.6. Much of the material in the Climate Data Tables comes from the book *Regional Guidelines for Building Passive Energy Conserving Homes* by the AIA Research Corporation.

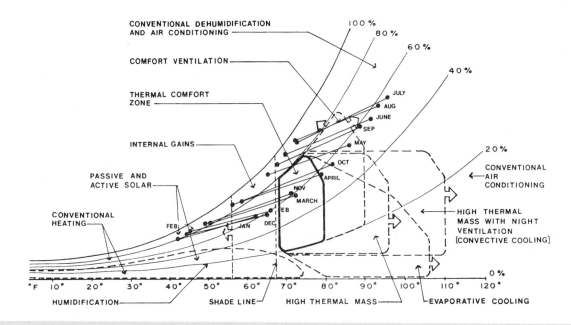

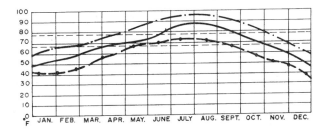

BASIC CLIMATIC CONDITION

		% of year
░	Comfortable period	11
▓	too hot	54
☐	too cold	35

TEMPERATURE

- – – – – – range of comfortable temp.
- – · – · – afternoon maximum temperature
- ——— average daily temperature
- —•—•— morning minimum temperature

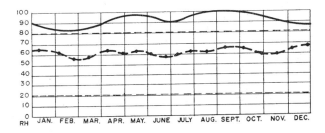

RELATIVE HUMIDITY

- ——— average morning humidity
- —•—•— average afternoon humidity
- – – – – – range of comfortable humidity

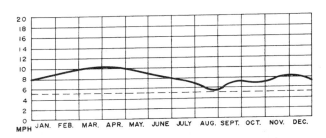

WIND SPEED

- ——— mean daily wind speed
- – – – – – wind speed for effective natural ventilation

for wind direction see the wind roses on pages 42–45.

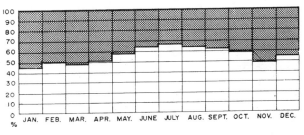

SUNSHINE
average % of daylight hours

annual sunshine = 56 %

peak solar radiation in January
horiz. sq. ft. = 1050 btu/day
vert. sq. ft. = 1300 btu/day

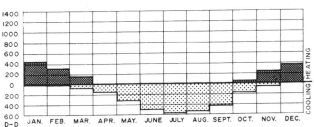

DEGREE-DAYS

		annual
▓	heating degree-days	1,549
░	cooling degree-days	2,761

CLIMATE REGION 16 ▬▬▬▬

Reference City: Miami, Florida

The Climate

The climate of southern Florida has long hot summers and no winters. When the slightly high temperatures are combined with high humidities, uncomfortable summers are the result. However, in spring, fall, and winter, the climate is quite pleasant. Ocean winds add significantly to year-round comfort.

The annual precipitation is quite high at about 58 inches, and much of the rain falls during the summer months.

Climatic Design Priorities*

1st Open the building to the outdoors since temperatures are comfortable much of the year. (XI)

2nd Protect from the summer sun. (IV)

3rd Allow natural ventilation to both cool and remove excess moisture most of the year. (VI)

4th Avoid creating additional humidity. (X)

Lower priority

5th Keep the hot temperatures out during the summer. (VIII)

6th Keep the heat in and the cool temperatures out during the winter. (I)

*These climatic design priorities are for envelope-dominated building types only. See Section 5.7 for a specific list of design strategies appropriate for achieving each of these priorities. Use the Roman numerals to find the relevant lists.

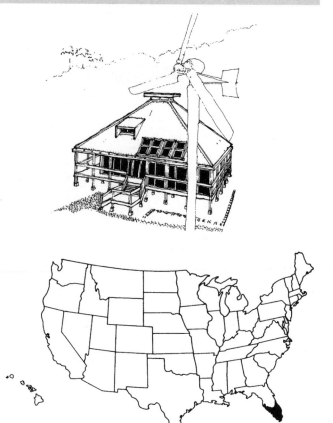

Note: For an explanation of these Climatic Data Tables, see Section 5.6. Much of the material in the Climate Data Tables comes from the book *Regional Guidelines for Building Passive Energy Conserving Homes* by the AIA Research Corporation.

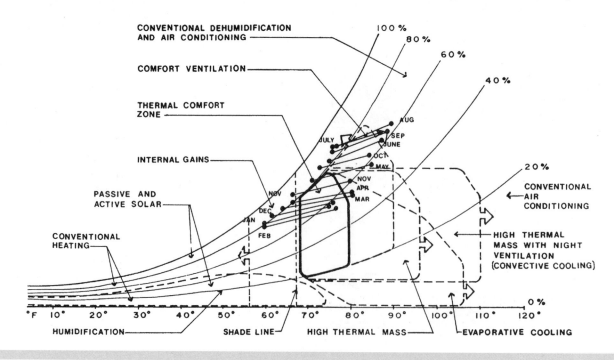

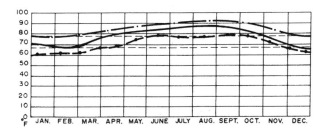

BASIC CLIMATIC CONDITION

		% of year
	Comfortable period	20
	too hot	69
	too cold	11

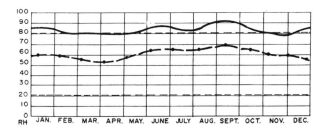

TEMPERATURE

- - - - - - range of comfortable temp.
— · — · — afternoon maximum temperature
———— average daily temperature
— • — morning minimum temperature

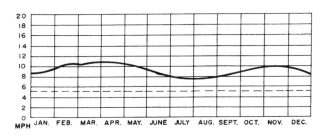

RELATIVE HUMIDITY

———— average morning humidity
— • — average afternoon humidity
- - - - - - range of comfortable humidity

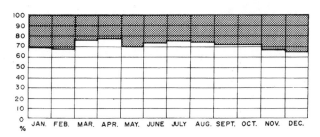

WIND SPEED

———— mean daily wind speed
- - - - - - wind speed for effective
natural ventilation

for wind direction see the
wind roses on pages 42–45.

SUNSHINE
average % of daylight hours

annual sunshine = 72 %

peak solar radiation in January
horiz. sq. ft. = 1300 btu/day
vert. sq. ft. = 1450 btu/day

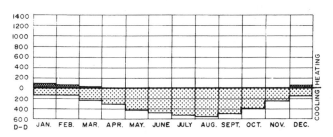

DEGREE-DAYS

		annual
	heating degree-days	199
	cooling degree-days	4.095

CLIMATE REGION 17
Reference City: Los Angeles, CA

The Climate

The semiarid climate of Southern California is very mild because of the almost constant cool winds from the ocean. Although these onshore winds bring high humidity, comfort is maintained because of the low temperatures .

Occasionally when the wind reverses, hot desert air enters the region. Because this air is dry, comfort is still maintained. There is a sharp increase in temperature and decrease in humidity as one leaves the coast. Thus, a large variation in the local microclimates exists.

Winter temperatures are very moderate, and little heating is required. Although the annual precipitation of about 15 inches is not very low, the rain falls mainly in the winter. Since there is almost no rain during the summer, few plants can grow year-round without irrigation. Since sunshine is plentiful all year, solar heating, especially for hot water, is very advantageous.

Climatic Design Priorities*

1st Open the building to the outdoors since temperatures are comfortable most of the year. (XI)
2nd Protect from the summer sun. (IV)
3rd Let the winter sun in. (III)
4th Use natural ventilation for summer cooling. (V)
5th Use thermal mass to flatten day-to-night temperature swings in the summer. (VII)

* These climatic design priorities are for envelope-dominated building types only. See Section 5.7 for a specific list of design strategies appropriate for achieving each of these priorities. Use the Roman numerals to find the relevant lists.

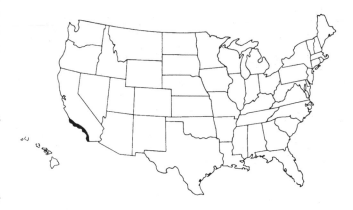

Note: For an explanation of these Climatic Data Tables, see Section 5.6. Much of the material in the Climate Data Tables comes from the book *Regional Guidelines for Building Passive Energy Conserving Homes* by the AIA Research Corporation.

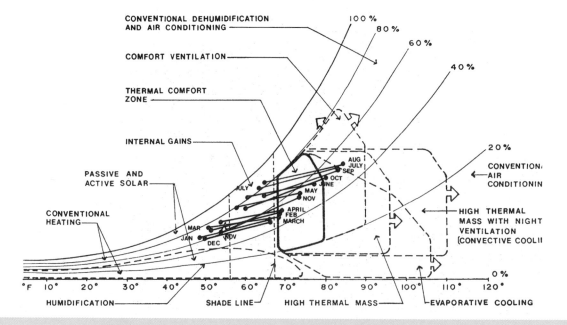

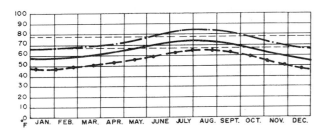

BASIC CLIMATIC CONDITION

		% of year
░░░	Comfortable period	64
▓▓▓	too hot	8
☐	too cold	28

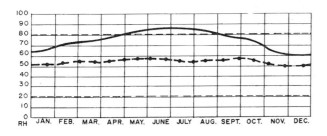

TEMPERATURE

– – – – – range of comfortable temp.
–·–·–·– afternoon maximum temperature
——— average daily temperature
—•—•— morning minimum temperature

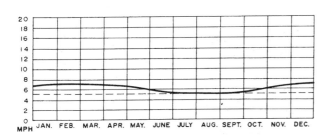

RELATIVE HUMIDITY

——— average morning humidity
—•—•— average afternoon humidity
– – – – – range of comfortable humidity

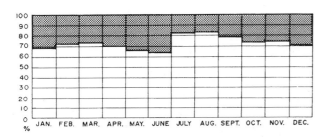

WIND SPEED

——— mean daily wind speed
– – – – – wind speed for effective
natural ventilation

for wind direction see the
wind roses on pages **42–45.**

SUNSHINE
average % of daylight hours

annual sunshine = _73_ %

peak solar radiation in January
horiz. sq. ft. = _900_ btu/day
vert. sq. ft. = _1200_ btu/day

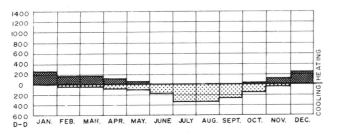

DEGREE-DAYS

		annual
▓▓▓	heating degree-days	1,204
░░░	cooling degree-days	1,339

5.7 DESIGN STRATEGIES

The following climate-related design strategies are appropriate ways of achieving the design priorities listed in the Climatic Data Tables (above). More detailed information is found in the chapters shown in parentheses.

Winter

I. *Keep the heat in and the cold temperatures out during the winter.*

   a. Avoid building on cold northern slopes. (Chapter 11)

   b. Build on the middle of slopes to avoid both the pools of cold air at the bottom and the high winds at the top of hills. (Chapter 11)

   c. Use a compact design with minimum surface-area-to-volume ratio. For example, use two- instead of one-story buildings. (Chapter 15)

   d. Build attached or clustered buildings to minimize the number of exposed walls. (Chapter 15)

   e. Use earth sheltering in the form of underground or bermed structures. (Chapter 15)

   f. Place buffer spaces that have lower temperature requirements (closets, storage rooms, stairs, garages, gymnasiums, heavy work areas, etc.) along the north wall. Place a sun-space buffer room on the south wall. (Chapters 7 and 15)

   g. Use temperature zoning by both space and time since some spaces can be kept cooler than others at all times or at certain times. For example, bedrooms can be kept cooler during the day, and living rooms can be kept cooler at night when everyone is asleep. (Chapter 16)

   h. Minimize window area on all orientations except south. (Chapters 7 and 15)

   i. Use double or triple glazing, low-e coatings, and movable insulation on windows. (Chapter 15)

   j. Use plentiful insulation in walls, roofs, under floors, over crawl spaces, on foundation walls, and around slab edges. (Chapter 15)

   k. Insulation should be a continuous envelope to prevent heat bridges. Avoid structural elements that are exposed on the exterior, since they pierce the insulation. Avoid fireplaces and other masonry elements that penetrate the insulation layer. (Chapter 15)

   l. Place doors on fireplaces to prevent heated room air from escaping through the chimney. Supply fireplaces and stoves with outdoor combustion air. (Chapter 16)

II. *Protect from the cold winter winds.*

   a. Avoid windy locations, such as hill tops. (Chapter 11)

   b. Use evergreen vegetation to create wind breaks. (Chapter 11)

   c. Use garden walls to protect the building and especially entrances from cold winds. (Chapter 11)

   d. In very windy areas, keep buildings close to the ground (one story).

   e. Use compact designs to minimize surface area exposed to the wind. (Chapter 15)

   f. Use streamlined shapes with rounded corners to both deflect the wind and minimize the surface-area-to-volume ratio.

Use attached buildings to reduce exposed wall area. Use compact building forms and two-story plans. Use triple glazing or double glazing with movable insulation. (Drawings from *Regional Guidelines for Building Passive Energy Conserving Homes* by the AIA Research Corporation.)

 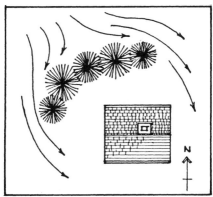 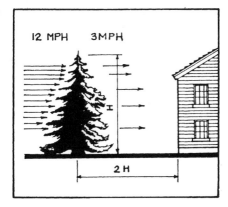

Build in wind-protected areas such as the side of a hill. Plant or build barriers against the wind. Evergreen trees are effective wind barriers. (Drawings from *Regional Guidelines for Building Passive Energy Conserving Homes* by the AIA Research Corporation.)

g. Cluster buildings for mutual wind protection. (Chapter 11)

h. Use long sloping roofs, as in the New England "saltbox" houses, to deflect the wind over the building and to create sheltered zones on the sunny side.

i. Place garages and other utility spaces on the winter windward side. This is usually the north, northwest, and sometimes the northeast side of the building.

j. Use sun spaces and glazed-in porches as windbreakers.

k. Use earth sheltering by building below the surface winds. Also, the wind can be deflected by earth berms built against the wall, or by constructing protective earth banks a short distance from the building. (Chapters 11 and 15)

l. Minimize openings, especially on the side facing the winter winds, and place the main entry on the leeward side. (Chapter 15)

m. Use storm windows, storm doors, air locks (vestibules), and revolving doors to minimize infiltration. (Chapter 15)

n. Close all attic and crawl-space vents. (*See* Chapter 15 for precautions for the hazards of water vapor and radon gas.)

o. Use tight construction, caulking, and weatherstripping to minimize infiltration. Use high-quality operable windows and doors. (Chapter 15)

p. Place outdoor courtyards on the south side of the building. (Chapter 11)

q. In winter, even windows in free-standing garden walls should be closed to protect the enclosure from cold winds.

r. In snow country, use snow fences and wind screens to keep snow from blocking entries and south-facing windows.

III. *Let the winter sun in (covered in Chapter 7 unless noted otherwise).*

a. Build on south, southeast, or southwest slopes. (Chapter 11)

b. Check for solar access that might be blocked by land forms, vegetation, and manmade structures. (Chapter 11)

c. Avoid trees on the south side of the building. (Chapter 11)

d. Use deciduous trees on southeast and southwest sides.

e. Also use deciduous trees on east and west sides if winter is very long.

f. Long axis of building should run east–west.

g. Most windows should face south.

h. Use south-facing clerestories instead of skylights.

i. Place spaces that benefit most from solar heating along the south wall. Spaces that benefit the least should be along the north wall (e.g., storage rooms, garages, etc.). (Chapter 15)

j. Use open floor plan to enable sun and sun-warmed air to penetrate throughout the building.

k. Use direct-gain, Trombe walls, and sun spaces for effective passive solar heating. (Chapter 7)

l. Use thermal mass on the interior to absorb and store solar radiation.

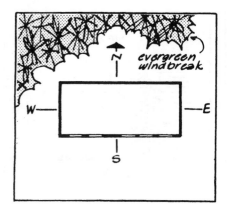  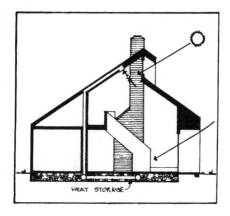

Orient building with long side facing south. Avoid trees or other structures on the south. Place most windows on the south facade. Use mainly vertical glazing. Use south-facing clerestory windows to bring the sun farther into the interior. (Drawings from *Regional Guidelines for Building Passive Energy Conserving Homes* by the AIA Research Corporation.)

m. Use light-colored patios, pavements, or land surfaces to reflect additional light through windows.

n. Use specular reflectors (polished aluminum) to reflect additional sunlight into windows.

o. Use active solar collectors for domestic hot water, swimming-pool heating, space heating, and process heat for industry. (Chapter 18)

p. If there is little or no summer overheating, use dark colors on exterior walls.

q. Create sunny but wind-protected outdoor spaces on the south side of the building. (Chapter 11)

Summer

IV. *Protect from the summer sun (covered in Chapter 9 unless noted otherwise).*

a. Avoid building on east and especially west slopes. North slopes are best if solar heating is not required in the winter, while south slopes are best if solar heating is desirable in the winter. (Chapter 11)

b. Use plants for shading. Evergreen trees can be used on the east, west, and north sides of a building. Deciduous plants are most appropriate for shading the southeast, the southwest, and the roof. Unless carefully placed, deciduous plants on the south side of a building might do more harm in the winter than good in the summer. The exception would be a very hot climate with a very mild winter. (Chapter 11)

c. Avoid light-colored ground covers around the building to minimize reflected light entering windows unless daylighting is a important strategy. Living ground covers are best because they do not heat the air while they absorb solar radiation.

d. Have neighboring buildings shade each other. Tall buildings with narrow alleys between them work best. (Chapter 11)

e. Avoid reflections from adjacent structures that have white walls and reflective glazing.

f. Build attached houses or clusters to minimize the number of exposed walls. (Chapter 15)

g. Use free-standing or wing walls to shade the east, west, and north walls.

h. Use the form of the building to shade itself (e.g., cantilever floors, balconies, courtyards).

i. Avoid east and especially west windows if at all possible. Minimize the size and number of any east and west windows that are necessary. Project windows on east and west facades so that they face in a northernly or southernly direction.

j. Use only vertical glazing. Any horizontal or sloped glazing (skylights) should be shaded on the outside during the summer. Only skylights on steep northern roofs do not require exterior shading.

k. Use exterior shading devices on all windows except north windows in cool climates.

l. Shade not only windows but also east and especially west walls. In very hot climates also shade the south wall.

m. Use a double or second roof (ice house roof) with the space between well ventilated.

n. Use shaded outdoor spaces, such as porches and carports, to protect the south, east, and especially the west facades.

o. Use open rather than solid shading devices to prevent trapping hot air next to the windows.

Orient short side of building to east and west and avoid windows on these facades if possible. Use overhangs, balconies, and porches to shade both windows and walls. Use large overhanging roofs and porticoes to shade both windows and walls. (Drawings from *Regional Guidelines for Building Passive Energy Conserving Homes* by the AIA Research Corporation.)

p. Use vines on trellises for shading. (Chapters 9 and 11)

q. Use movable shading devices that can retract to allow full winter sun penetration.

r. Use highly reflective building surfaces (white is best). The roof and west wall are the most critical.

s. Use interior shading devices if exterior shading is either not available or not sufficient.

t. Place outdoor courtyards, which are intended for summer use, on the north side of the building. The east side is the next best choice. (Chapter 11)

V. *Use natural ventilation for summer cooling (covered in Chapter 10 unless noted otherwise).*

a. Night ventilation that is used to cool the building in preparation for the next day is called night flush cooling and is described under priority VII below.

b. Natural ventilation that cools people by passing air over their skin is called comfort ventilation.

c. Site and orient building to capture the prevailing winds. (Chapters 10 and 11)

d. Direct and channel winds toward building by means of landscaping and landforms. (Chapter 11)

e. Keep buildings far enough apart to allow full access to the desirable winds. (Chapter 11)

f. In mild climates where winters are not very cold and summer temperatures are not extremely high, use a noncompact shape for maximum cross-ventilation.

g. Elevate the main living space since wind velocity increases with the height above ground.

h. Use high ceilings, two-story spaces, and open stairwells for vertical air movement and for the benefits of stratification.

Provide many large but shaded windows for ventilation. Provide openings at the ceiling level of high spaces. Provide large openings to vent attic spaces. (Drawings from *Regional Guidelines for Building Passive Energy Conserving Homes* by the AIA Research Corporation.)

i. Provide cross-ventilation by using large windows on both the windward and leeward sides of the building.

j. Use fin walls to direct air through the windows.

k. Use a combination of high and low openings to take advantage of the stack effect.

l. Use roof openings to vent both the attic and the whole building. Use openings, such as monitors, cupolas, dormers, roof turrets, ridge vents, and gable vents.

m. Use porches to create cool outdoor spaces and to protect open windows from sun and rain.

n. Use a double or parasol roof with sufficient clearance to allow the wind to ventilate the hot air collecting between the two roofs. (Chapter 9)

o. Use high-quality windows with good seals to allow summer ventilation while preventing winter infiltration. (Chapter 15)

p. Use open floor plan for maximum air flow. Minimize the use of partitions.

q. Keep transoms and doors open between rooms.

r. Use a solar chimney to move air vertically through a building on calm days.

s. Use operable windows or movable panels in garden walls to maximize the summer ventilation of a site while allowing protection against the winter winds.

VI. *Allow natural ventilation to both cool and remove excess moisture in the summer (covered in Chapter 10 unless otherwise noted).*

a. All the strategies from priority V above also apply here.

b. Elevate the main living floor above the high humidity found near the ground.

c. Use plants sparsely. Minimize trees, shrubbery, and ground covers to enable air to circulate through the site to remove moisture. Use only trees that have a high canopy. (Chapter 11)

d. Avoid deep basements that cannot be ventilated well.

VII. *Use thermal mass to flatten day-to-night temperature swings in the summer (covered in Chapter 10 unless noted otherwise).*

a. This cooling strategy is also known as "night flush cooling" because the thermal mass is usually cooled with night ventilation. See Chapter 10 for a description of this strategy.

b. Use massive construction materials since they have a high heat capacity. Use materials such as brick, concrete, stone, and adobe. (Chapter 15)

c. Place insulation on the outside of the thermal mass. (Chapter 15)

d. If massive materials are also to be used on the outside, sandwich the insulation between the inside and outside walls. (Chapter 15)

e. Use earth or rock in direct contact with the uninsulated walls. (Chapters 10 and 15)

f. Keep daytime hot air out by closing all openings.

g. Open the building at night to allow cool air to enter. Use the strategies of natural ventilation, listed above in priority V, to maximize the night cooling of the thermal mass.

h. Use water as a thermal mass because of its very high heat capacity. Use containers that maximize the heat transfer into and out of the water. (Chapter 7)

 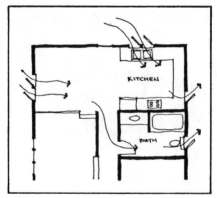 

Raise building above the moisture at ground level and ventilate under the building. Allow natural ventilation to carry away moisture from kitchens, baths, and laundry rooms. Avoid dense landscaping near ground level. A high canopy of trees is good, however. (Drawings from *Regional Guidelines for Building Passive Energy Conserving Homes* by the AIA Research Corporation.)

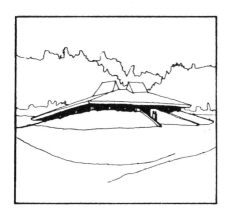

Use thermal mass to reduce the impact of high temperatures. Use the thermal mass of the earth. Use berms or sloping sites for earth sheltered buildings. (Drawings from *Regional Guidelines for Building Passive Energy Conserving Homes* by the AIA Research Corporation.)

i. Use radiant or evaporative cooling for additional temperature drop in the thermal mass at night.

j. Use earth tubes or a ground to air heat pump to tap the coolness of the ground. (Chapters 10 and 16)

VIII. *Keep hot temperatures out during the summer.*

a. Use compact designs to minimize surface-area-to-volume ratio. (Chapter 15)

b. Build attached houses to minimize the number of exposed walls. (Chapters 11 and 15)

c. Use vegetation and shade structures to maintain cool ambient air around the building, and to prevent reflecting sunlight into the windows. (Chapter 11)

d. Use earth sheltering in the form of underground or bermed structures. (Chapter 15)

e. Use ample insulation in the building envelope. (Chapter 15)

f. Use few and small windows to keep heat out.

g. Use exterior or interior window shutters. In hot climates, use double glazing, and in very hot climates also use movable insulation over windows to be used during the day when a space is unoccupied. (Chapter 15)

h. Isolate sources of heat in a separate room, wing, or building.

l. Zone building so that certain spaces are cooled only while occupied. (Chapter 16)

j. Use light-colored roofs and walls to reflect the sun's heat.

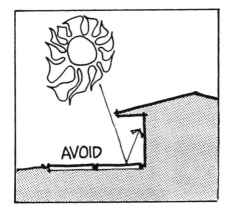

Use compact, well-insulated, and white painted buildings. Use attached housing units to minimize exposed wall area. Have buildings shade each other. Avoid reflecting sun into windows.

  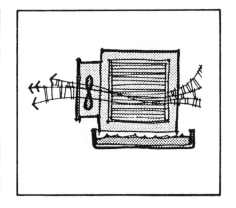

Use fountains, pools, and plants for evaporative cooling. Use courtyards to prevent cooled air from blowing away. Use energy-conserving evaporative coolers. (Drawings from *Regional Guidelines for Building Passive Energy Conserving Homes* by the AIA Research Corporation.)

IX. *Use evaporative cooling in the summer (covered in Chapter 10 unless otherwise noted).*

   a. Locate pools or fountains in the building, in a courtyard, or in the path of incoming winds.

   b. Use transpiration by plants to cool the air both indoors and outdoors.

   c. Spray roof, walls, and patios to cool these surfaces.

   d. Pass incoming air through a curtain of water or a wet fabric.

   e. Use a roof pond or other "indirect evaporative cooling" system.

   f. Use an "evaporative cooler." This simple and inexpensive mechanical device uses very little electrical energy.

X. *Avoid creating additional humidity during the summer.*

   a. Do not use evaporative cooling strategies in climates that specify priority IX.

   b. Use underground or drip rather than spray irrigation.

   c. Avoid pools and fountains.

   d. Keep area around the building dry by providing the proper drainage of land. Channel runoff water from the roof and paved areas away from the site.

   e. Use permeable paving materials to prevent puddles on the surface.

   f. Minimize plants, especially indoors. Use plants that add little water to the air by transpiration. Such plants are usually native to dry climates .

   g. Shade plants and pools of water both indoors and out because the heat of the sun greatly increases the rate of transpiration and evaporation.

   h. Use exhaust fans in kitchens, bathrooms, laundry rooms, etc., to remove excess moisture.

XI. *Open the building to the outdoors since temperatures are comfortable much of the year.*

   a. Create outdoor spaces with different orientations for use at different times of the year. For example, use outdoor spaces on the south side in the winter and on the north side in the summer.

   b. Create outdoor living areas that are sheltered from the hot summer sun and cool winter winds.

   c. Use noncompact building designs for maximum contact with the outdoors. Use an articulated building with many extensions or wings to create outdoor spaces.

   d. Use large areas of operable windows, doors, and even movable walls to increase contact with the outdoors.

   e. Create pavilion-like buildings that have few interior partitions and minimal exterior walls.

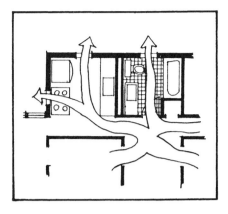

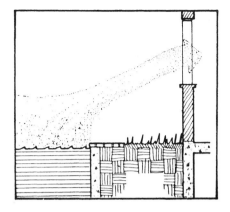

Use exhaust fans to remove excess moisture from kitchens, baths, and laundry rooms. Minimize indoor plants, and keep them out of direct sunlight to reduce evaporation. Avoid pools, fountains, and plants in the landscape. Minimize interior partitions and provide many openings in the exterior walls.

Use operable and movable wall panels. Create sheltered outdoor spaces with various orientations for use at different times of day and year. Drawings from *Regional Guidelines for Building Passive Energy Conserving Homes* by the AIA Research Corporation.)

KEY IDEAS OF CHAPTER 5

1. Because of water vapor and clouds in hot and humid climates, the daytime temperatures are lower and nighttime temperatures are higher. Thus, the diurnal temperature range is small.

2. Because of the lack of water in the air, hot and dry climates have high daytime temperatures and low nighttime temperatures. Thus, the diurnal temperature range is large.

3. Because of such features as elevation, form of land, large bodies of water, soil types, vegetation, and manmade objects, the microclimate can be significantly different from the regional climate.

4. The direction of the wind is given by wind roses.

5. The psychrometric-bioclimatic chart is an excellent tool for understanding a climate in terms of temperatures and coincident relative humidities.

6. The sunshine chart indicates the relative importance of direct sun-shine for solar heating and daylighting for each climate region.

7. The degree-days chart is an excellent tool for determining the depth and severity (heating and cooling loads) of winter and summer.

8. Design priorities are given for each of the seventeen climate regions detailed in this chapter. Design strategies or techniques for addressing the priorities are given at the end of the chapter.

Acknowledgment

Much of the material of this chapter was taken from the book *Regional Guidelines for Building Passive Energy Conserving Homes* by the AIA Research Corporation.

Resources

FURTHER READING

(See Bibliograqphy in back of book for full citations. This list includes valuable out-of-print books.)

AlA Research Corp. *Regional Guidelines for Building Passive Energy Conserving Homes.*

ASHRAE Handbook. Fundamentals.

Climatic Atlas of the United States.

Compatative Climatic Data for the United States.

Fitch, J. M., with W. Bobenhausen. *American Building: The Environmental Forces That Shape It.*

National Oceanic and Atmospheric Administration (NOAA). *Climate Atlas of U.S.*

Olgyay, V. *Design with Climate.*

Ruffner, J. A., and F. E. Bair. *The Weather Almanac.*

Stein, B., and J. Reynolds. *Mechanical and Electrical Equipment for Buildings,* Ninth Edition.

Watson, D., and K. Labs. *Climatic Design: Energy-Efficient Building Principles and Practices.*

SOLAR GEOMETRY

"It is the mission of modern architecture to concern itself with the sun."

Le Corbusier
from a letter to Sert

Figure 6.1 Part of the year the sun is our friend, and part of the year it is our enemy. (Drawing by Le Corbusier from Le Corbusier: *Oevre Complete, 1938–1944,* Vol. 4. by W. Boesiger, 7th ed. Verlag fuer Architektur Artemis © 1977.)

Please note that figure numbers are keyed to sections. Gaps in figure numbering result from sections without figures.

6.1 INTRODUCTION

People used to worship the sun as a god; they understood how much life depended on sunshine. However, with the rapid growth of science and technology, mankind came to believe that all problems could be solved by high technology and that it was no longer necessary to live in harmony with nature. An architectural example of this attitude is the construction in the desert of buildings with very large areas of glazing, which can be kept habitable only by means of huge energy-guzzling air-conditioning plants.

Crises and disappointments have persuaded us to reconsider our relationships with nature and technology. There is a deepening conviction that progress will come mainly from technology that is in harmony with nature. The growing interest in sustainable or green design illustrates this shift in attitude. In architecture, this point of view is represented by buildings that let the sun shine in during the winter and are shaded from the sun in the summer (Fig. 6.1).

This approach to architecture requires that the designer have a good understanding of the natural world. Central to this understanding is the relationship of the sun to the earth. This chapter discusses solar radiation and its effect on the seasons.

6.2 THE SUN

The sun is a huge fusion reactor in which light atoms are fused into heavier atoms, and in the process energy is released. This reaction can occur only in the interior of the sun, where the necessary temperature of 25,000,000°F exists. The solar radiation reaching earth, however, is emitted from the sun's surface, which is much cooler (Fig. 6.2a). Solar radiation is, therefore, the kind of radiation that a body having a temperature of about 12,500°F emits. The amount and composition of solar radiation reaching the outer edge of the earth's atmosphere are quite unvarying and is called the **solar constant.** The amount and composition of the radiation reaching the earth's surface, however, vary widely with sun angles and the composition of the atmosphere (Fig. 6.2b).

6.3 ELLIPTICAL ORBIT

The orbit of the earth is not a circle but an ellipse, so that the distance between the earth and sun varies as the earth revolves around the sun (Fig. 6.3). The distance varies about 3.3 percent and this results in a small annual variation in the intensity of solar radiation.

Does this explain why it is cooler in January than in July? No, because we are actually closer to the sun in January than July. In fact, this variation in distance from the sun slightly reduces the severity of winters in the northern hemisphere. What then is the cause of the seasons?

Since the sun is very far away and since it lies in the plane of the earth's orbit, solar radiation striking the earth is always parallel to this plane (Fig. 6.3). While the earth revolves around the sun, it also spins around its own north–south axis. Since this axis is not perpendicular to the orbital plane but is tilted 23.5 degrees off the normal to this plane, and since the orientation in space of this axis of rotation remains fixed as the earth revolves around the sun, the angle at which the sun's rays hit the earth continuously changes

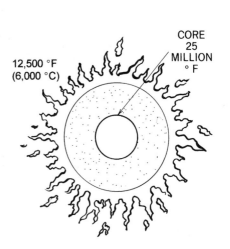

Figure 6.2a The surface temperature of the sun determines the type of radiation emitted.

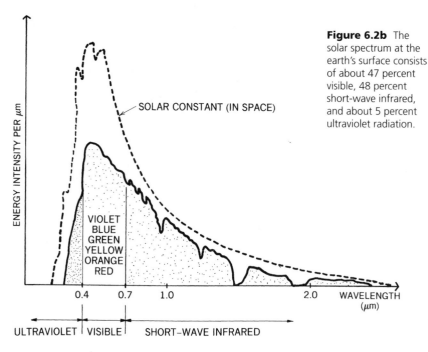

Figure 6.2b The solar spectrum at the earth's surface consists of about 47 percent visible, 48 percent short-wave infrared, and about 5 percent ultraviolet radiation.

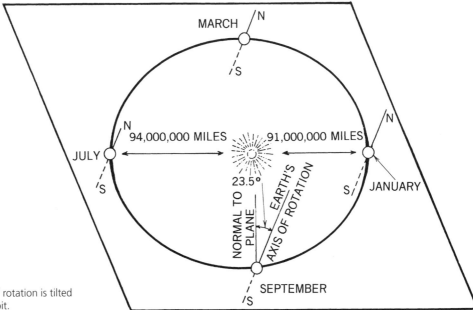

Figure 6.3 The earth's axis of rotation is tilted to the plane of the elliptical orbit.

throughout the year. This tilt of 23.5 degrees is the cause of the seasons and has major implications for solar energy.

6.4 TILT OF THE EARTH'S AXIS

Because the tilt of the earth's axis is fixed, the northern hemisphere faces the sun in June and the southern hemisphere faces the sun in December (Fig. 6.4a). The extreme conditions occur on June 21 when the north pole is pointing most nearly toward the sun and on December 21 when the north pole is pointing farthest away from the sun.

Notice that on June 21, all of the earth north of the Arctic Circle will have twenty-four hours of sunlight (Fig. 6.4b). This is the longest day in the northern hemisphere, and is called the **summer solstice.** Also on that day, the sun's rays will be perpendicular to the earth's surface along the Tropic of Cancer, which is, not by coincidence, at latitude 23.5 degrees north. No part of the earth north of

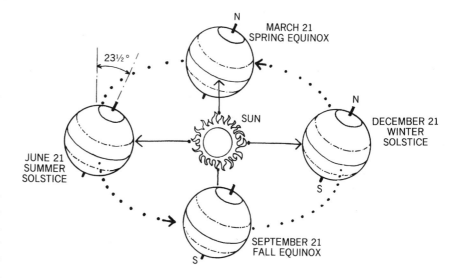

Figure 6.4a The seasons are a consequence of the tilt of the earth's axis of rotation. [From *Solar Dwelling Design Concepts* by AIA Research Corporation. U.S. Dept. Housing and Urban Development, 1976. HUD-PDR-154(4).]

the Tropic of Cancer ever has the sun directly overhead.

Six months later on December 21, at the opposite end of the earth's orbit around the sun, the north pole points so far away from the sun that now all of the earth above the Arctic Circle experiences twenty-four hours of darkness (Fig. 6.4c). In the northern hemisphere, this is the day with the longest

night and is also known as the **winter solstice.** On this day, the sun is perpendicular to the southern hemisphere along the Tropic of Capricorn, which, of course, is at latitude 23.5 degrees south. Meanwhile, the sun's rays that do fall on the northern hemisphere do so at much lower sun angles (altitude angles — see below) than those striking the southern hemisphere.

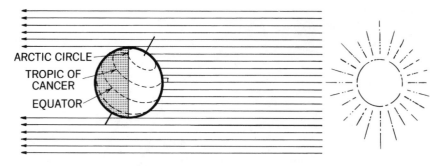

Figure 6.4b During the summer solstice (June 21), the sun is directly overhead on the Tropic of Cancer.

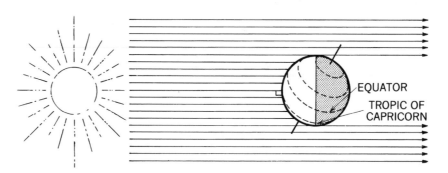

Figure 6.4c During the winter solstice (December 21), the sun is directly overhead on the Tropic of Capricorn.

the **cosine law** (Fig. 6.5c). This law says that a given beam of sunlight will illuminate a larger area as the sun gets lower in the sky. As the given sunbeam is spread over larger areas, the sunlight on each square foot of land is naturally getting weaker. The amount of sunlight that a surface receives changes with the cosine of the angle between the sun's rays and the normal to the surface.

6.6 WINTER

Now we can understand what causes winter. The temperature of the air, as well as that of the land, is mainly a result of the amount of solar radiation absorbed by the land. The air is main-ly heated or cooled by its contact with the earth. The reasons for less radia-tion falling on the ground in the win-ter are the following.

Most important is the fact that there are far fewer hours of daylight in the winter. The exact number is a function of latitude. As was men-tioned earlier, there is no sunlight above the Arctic Circle on December 21, and at 40 degrees latitude, for example, there are six fewer hours of daylight on December 21 than June 21.

The second reason for reduced heating of the earth is the cosine law. On December 21, the solar radiation falling on a square foot of land is sig-nificantly less than that on June 21.

Lastly, the lower sun angles increase the amount of atmosphere the sun must pass through and, there-fore, there is again less radiation reaching each square foot of land.

6.7 THE SUN REVOLVES AROUND THE EARTH

Despite threats of torture and death, Galileo and Copernicus spoke up and convinced the world that the earth revolves around the sun. Never-theless, I would like to suggest, for nonreligious reasons, that we again

Halfway between the longest and shortest day of the year is the day of equal nighttime and daytime hours. This situation occurs twice a year, on March and September 21, and is known as the spring and fall **equinox** (Fig. 6.4a). On these days the sun is directly overhead on the equator.

6.5 CONSEQUENCES OF THE ALTITUDE ANGLE

The vertical angle at which the sun's rays strike the earth is called the **alti-tude angle** and is a function of the geographic latitude, time of year, and time of day. In Fig. 6.5a we see how the altitude angle is derived from these three factors. The simplest situ-ation occurs at 12 noon on the equi-nox, when the sun's rays are perpendicular to the earth at the equa-

tor (Fig. 6.5a). To find the altitude angle of the sun at any latitude, draw the ground plane tangent to the earth at that latitude. By simple geometric principles, it can be shown that the altitude angle is equal to 90 degrees minus the latitude. There are two important consequences of this alti-tude angle on climate and the seasons.

The first effect of the altitude angle is illustrated by Fig. 6.5b, which indi-cates that at low angles the sun's rays pass through more of the atmosphere. Consequently, the radiation reaching the surface will be weaker and more modified in composition. The extreme case occurs at sunset when the radia-tion is red and weak enough to be look-ed at. This is because of the selective absorption, reflection, and refraction of solar radiation in the atmosphere.

The second effect of the altitude angle is illustrated in the diagram of

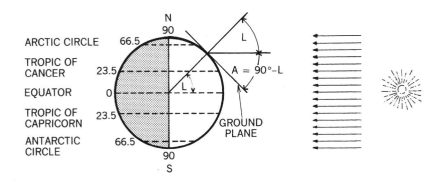

Figure 6.5a On the equinox, the sun's altitude (A) at solar noon at any place on earth is equal to 90 degrees minus the latitude (L).

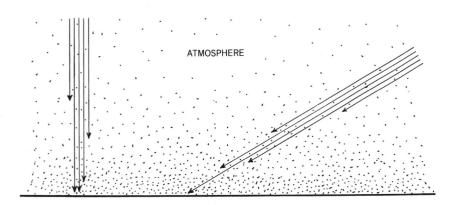

Figure 6.5b The altitude angle determines how much of the solar radiation will be absorbed by the atmosphere.

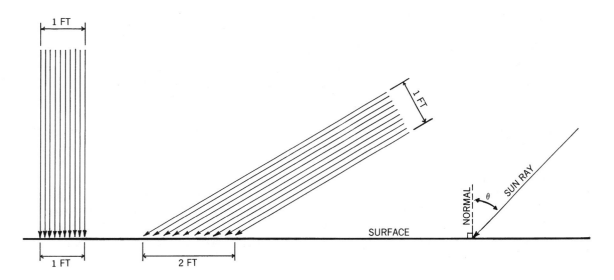

Figure 6.5c The cosine law states that the amount of radiation received by a surface decreases as the angle with the normal increases.

assume that the sun revolves around the earth or at least that the sun revolves around the building in question. For the moment this assumption makes it infinitely more convenient to understand sun angles. To make things even more convenient let us also assume a sky dome (see below), where a large clear plastic hemisphere is placed over the building site in question.

6.8 SKY DOME

In Fig. 6.8a we see an imaginary sky dome placed over the building site. Every hour the point at which the sun's rays penetrate the sky dome is marked. When all the points for one day are connected, we get a line called the **sun path** for that day. Figure 6.8a shows the highest sun path of the year (summer solstice), the lowest sun path (winter solstice), and the middle sun path (equinox). Note that the sun enters the sky dome only between the sun paths of the summer and winter solstices. Since the solar radiation is quite weak in the early and late hours of the day, the part of the sky dome

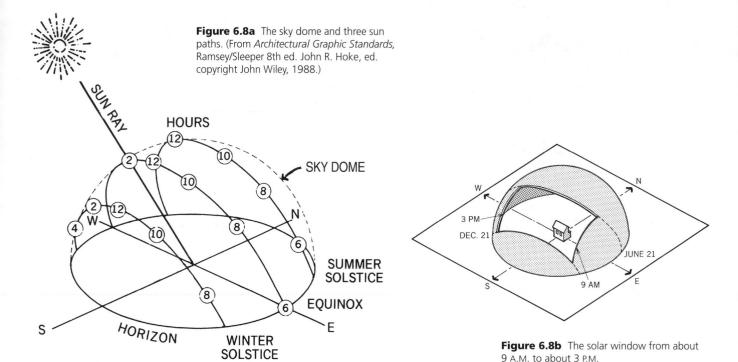

Figure 6.8a The sky dome and three sun paths. (From *Architectural Graphic Standards,* Ramsey/Sleeper 8th ed. John R. Hoke, ed. copyright John Wiley, 1988.)

Figure 6.8b The solar window from about 9 A.M. to about 3 P.M.

Figure 6.8c An east elevation of the sky dome. [From *Solar Dwelling Design Concepts* by AIA Research Corporation. U.S. Dept. Housing and Urban Development, 1976. HUD-PDR-154(4).]

through which the most useful of the sun's rays enter is called the **solar window.** Figure 6.8b shows the conventional solar window, which is assumed to begin at 9 A.M. and end at 3 P.M. Ideally, no trees, buildings, or other obstacles should block the sun's rays entering through the solar window during those months when solar energy is desired. Space heating requires solar access only during the winter months (lower portion of solar

window), while domestic hot-water heating requires solar access for the whole year (whole solar window).

An east elevation of the sky dome is illustrated in Fig. 6.8c. The sun paths for the summer solstice (June 21), equinoxes (March 21 and September 21), and winter solstice (December 21) are shown in edge view. The afternoon part of the sun path is directly behind the morning part of the path. The mark for 3 P.M.

is, therefore, directly behind the 9 A.M. mark labeled in the diagram. The sun's motion is completely symmetrical about a north–south axis. Notice in the diagram that the sun moves 23.5 degrees each side of the equinoxes because of the tilt of the earth's axis of rotation. The total vertical travel between winter and summer is, therefore, 47 degrees. The actual altitude, however, depends on the latitude.

6.9 DETERMINING ALTITUDE AND AZIMUTH ANGLES

By far the easiest way to work with the compound angle of the sun's rays is to use component angles. The most useful components are the altitude angle, which is measured in a vertical plane, and the **azimuth angle,** which is measured in a horizontal plane.

In Fig. 6.9a we see a sun ray enter the sky dome at 2 P.M. on the equinox. The horizontal projection of this sun ray lies in the ground plane. The vertical angle from this projection to the sun ray is called the **altitude.** It tells us how high the sun is in the sky. The horizontal angle, which is measured from a north-south line, is called the **azimuth.**

It is important to understand that the above discussions on sun angles refer only to direct radiation. Water and dust particles scatter the solar radiation (Fig. 6.9b) so that on cloudy, humid, or dusty days the diffuse radiation becomes the dominant form of solar energy.

6.10 SOLAR TIME

At 12 noon **solar time,** the sun is always due south. However, the sun will not be due south at 12 noon **clock time** because solar time varies from clock time. There are three reasons for this. The first is the common use of **daylight saving time.** The second is the deviation in longitude of the building site from the standard longitude of the time zone. The third reason is a consequence of the fact that the earth's speed in its orbit around the sun changes during the year. The amount of correction, therefore, depends on the time of year. Changing solar time to clock time or vice versa is quite complicated, and the conversion is not usually necessary since our goal is simply to collect the sun rays when it is too cold and to reject the sun rays when

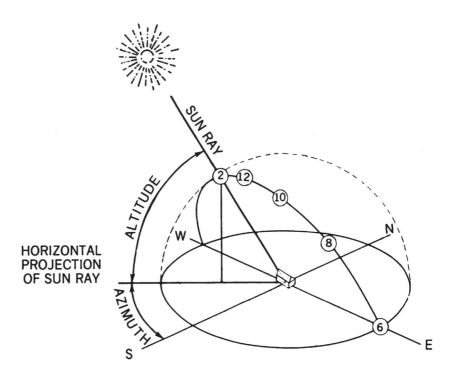

Figure 6.9a Definition of altitude and azimuth angles. (From *Architectural Graphic Standards,* Ramsey/Sleeper 8th ed. John R. Hoke, ed. copyright John Wiley, 1988.)

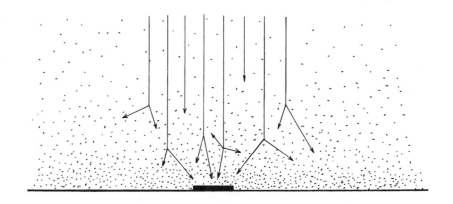

Figure 6.9b Diffuse radiation.

it is too hot. Therefore, the conversion is not explained in this book, and all references to time are in solar time. The author can think of only one situation where knowing the clock time of certain sun angles is important in architecture. The situation, very rare today, is designing a temple to the sun where a beam of sunlight hits the altar at a particular clock time.

6.11 HORIZONTAL SUN-PATH DIAGRAMS

Although altitude and azimuth angles can be readily obtained from tables, it is often more convenient and informative to obtain the information from sun-path diagrams. In Fig. 6.11a we again see the sky dome, but this time it has a grid of altitude and azimuth lines drawn on it just as a globe of the

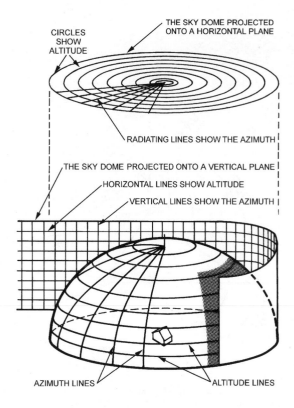

CIRCLES SHOW ALTITUDE

THE SKY DOME PROJECTED ONTO A HORIZONTAL PLANE

RADIATING LINES SHOW THE AZIMUTH

THE SKY DOME PROJECTED ONTO A VERTICAL PLANE

HORIZONTAL LINES SHOW ALTITUDE

VERTICAL LINES SHOW THE AZIMUTH

AZIMUTH LINES   ALTITUDE LINES

Figure 6.11a Derivation of the horizontal and vertical sun path diagrams.

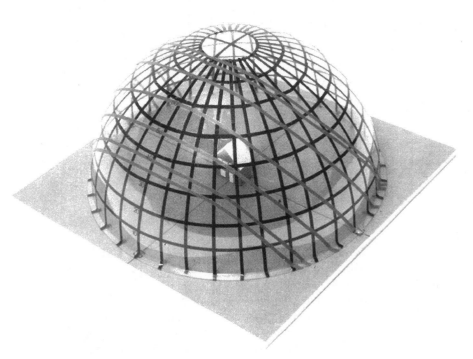

Figure 6.11b A model of the sky dome. The sun paths for the 21st day of each month are shown. Only seven paths are needed for twelve months because of symmetry (i.e., May 21 is the same path as July 21.

earth has latitude and longitude lines. Just as there are maps of the world that are usually either Mercator or polar projections, so there are vertical or horizontal projections of the sky dome (Fig. 6.lla). Notice how the grids project on the horizontal and vertical planes.

The sky dome shown in Fig. 6.11b has an azimuth grid, an altitude grid, and the sun paths for each month of the year for 32°N latitude. When the sun paths are plotted on a horizontal projection of the sky dome, we get a sun-path diagram such as shown in Fig. 6.11c. In these diagrams, the sun path of day 21 of each month is labeled by a Roman numeral (e.g., XII = December). The hours of the day are labeled along the sun path of June (VI). The concentric rings describe the altitude, and the radial lines define azimuth. The sun-path diagram for 36 degrees north latitude is shown in Fig. 6.11c. Additional sun-path diagrams, at 4° latitude intervals, are found in Appendix A.

Example: Find the altitude and azimuth of the sun in Memphis, Tennessee, on February 21 at 9 A.M.

Step 1. From a map of the United States, find the latitude of Memphis. Since it is at about 35 degrees latitude, use the sun-path diagram for 36 degrees north latitude (found in Appendix A and Fig. 6.11c).

Step 2. On this sun-path diagram, find the intersection of the sun path for February 21 (curve II) and the 9 A.M. line. This represents the location of the sun. The intersection is circled in Fig. 6.11c.

Step 3. From the concentric circles, the altitude is found to be about 27 degrees.

Step 4. From the radial lines, the azimuth is found to be about 51 degrees east of south.

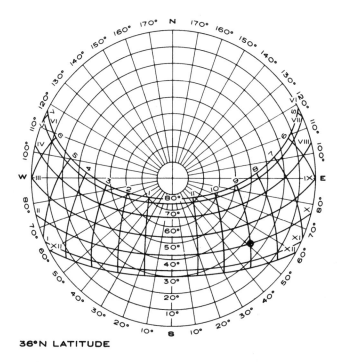

36°N LATITUDE

Figure 6.11c Horizontal sun path diagram. A complete set of these diagrams is found in Appendix A. (From *Architectural Graphic Standards.* Ramsey/Sleeper 8th ed. John R. Hoke, ed. copyright John Wiley, 1988.

6.12 VERTICAL SUN PATH DIAGRAMS

The illustration in Fig. 6.11a also shows how a vertical projection of the sky dome is developed. Notice, however, that the apex point of the sky dome is projected as a line. Consequently, severe distortions occur at high altitudes. In Fig. 6.12a, we see a vertical sun-path diagram for 36°N latitude.

Altitude and azimuth angles are found in a manner similar to that used in the horizontal sun-path diagrams. Appendix B has vertical sun-path diagrams from 28°N to 56°N latitude in 4-degree intervals. Below 26°N and above 58°N, refer to *Sun Angles for Design,* by Robert Bennett.

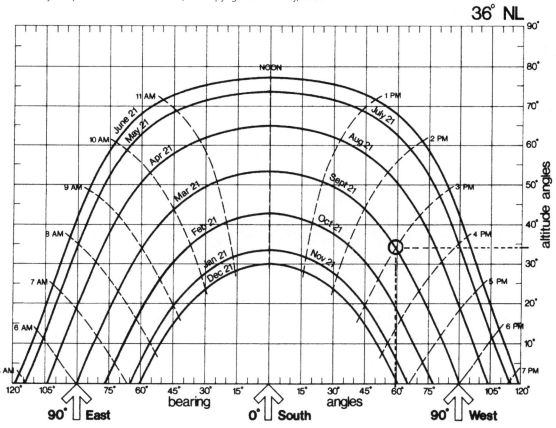

Figure 6.12a Vertical sun path diagram. A complete set of these diagrams is found in Appendix B. (Reprinted from *The Passive Solar Energy Book,* copyright E. Mazria, 1979, by permission.)

Example: Find the altitude and azimuth of a sun ray in Albuquerque, New Mexico, on March 21 at 3 P.M.

Step 1. From Appendix B, choose the sun-path diagram that is within 2 degrees of the place in question. Since Albuquerque is at 35°N latitude, use the sun path for 36°N.

Step 2. Find the intersection of the curves for March 21 and 3 P.M. (see circle in Fig. 6.12a).

Step 3. From the horizontal scale, the azimuth is found to be about 59° *west* of south.

Step 4. From the vertical scale, the altitude is found to be about 34° above the horizontal.

Besides being a source of sun-angle data, these diagrams are very helpful in creating a mental model of the sun's motion across the sky. The diagram can also be used for visualizing and documenting the solar window and any obstacles that might be blocking it. The finely shaded area of Fig. 6.12b is the winter solar window from 9 A.M. to 3 P.M. The roughly shaded area along the bottom represents the silhouette of trees and buildings surrounding a particular site. Notice that one building and one tree are partially blocking the solar window during the critical winter months. The easiest and quickest way to generate such a horizon profile is to use the site-evaluation tool described in Section 6.14 below.

6.13 SUN-PATH MODELS

Three-dimensional models of the sun-path diagrams are especially helpful in understanding the complex geometry of sun angles (Fig. 6.13). For simplicity only the sun paths for June 21, March/September 21, and December 21 are shown. These models can help a designer better visualize how the sun will relate to a building located at the center of the sun-path model.

The various models illustrate how sun paths vary with latitude. Models are shown for the special latitudes of the Equator at 0°, the Tropic of Cancer at 23.5° (the model is for 24 percent), the Arctic Circle at 66.5° (the model is for 64°), and the North

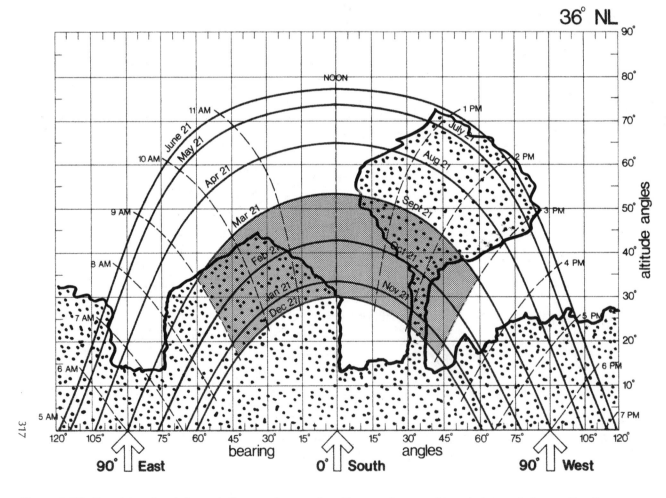

Figure 6.12b The winter solar window and silhouette of surrounding objects are shown on this vertical sun path diagram. The silhouette of a specific location was hand-drawn by means of a site-evaluator tool described in Section 6.14. (Sun path diagram from *The Passive Solar Energy Book,* copyright E. Mazria, 1979, reprinted by permission.)

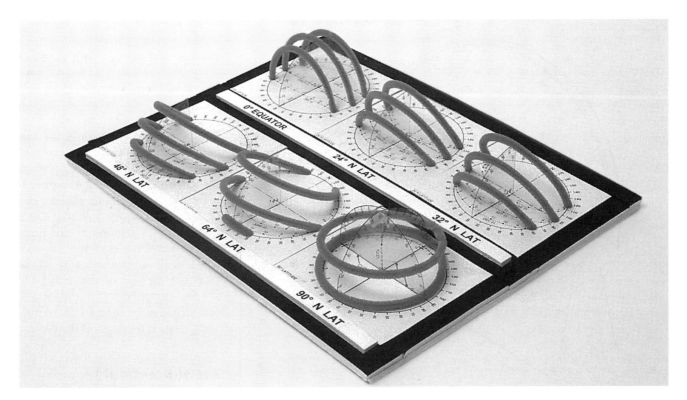

Figure 6.13 A comparison of various sun-path models. Note especially the sun paths for the Equator, Tropic of Cancer, Arctic Circle, and North Pole.

Pole at 90°. Appendix F presents complete instructions and a set of charts required to create a sun path model for any of the following latitudes: 0°, 24°, 28°, 32°, 36°, 40°, 44°, 48°, 64°, and 90°. It is very worthwhile to spend the fifteen minutes required to make one of these sun-path models. The model can be placed on the corner of the designer's table to be a reminder of where the sun is at different times of the day and year.

6.14 SOLAR SITE-EVALUATION TOOL

A solar building on a site that does not have access to the sun is a total disaster. Fortunately, there are good tools available for analyzing a site in regard to solar access. Appendix H presents information on how to build and use your own low-cost **site-evaluation tool** similar to the one shown in Fig. 6.14. The author strongly recommends the use of this tool.

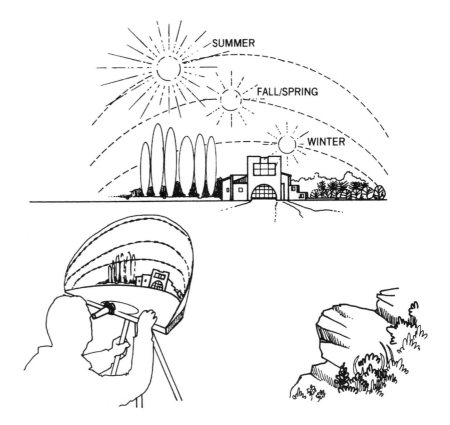

Figure 6.14 The sun-path diagram used as part of a solar site-evaluation tool.

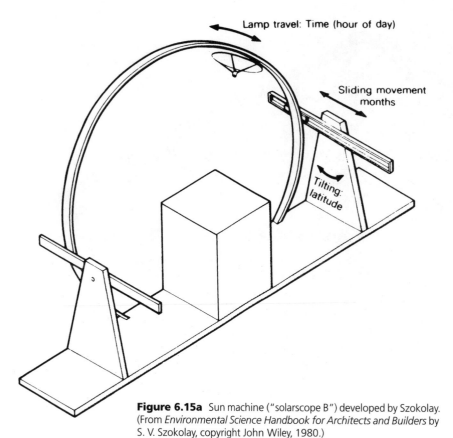

Figure 6.15a Sun machine ("solarscope B") developed by Szokolay. (From *Environmental Science Handbook for Architects and Builders* by S. V. Szokolay, copyright John Wiley, 1980.)

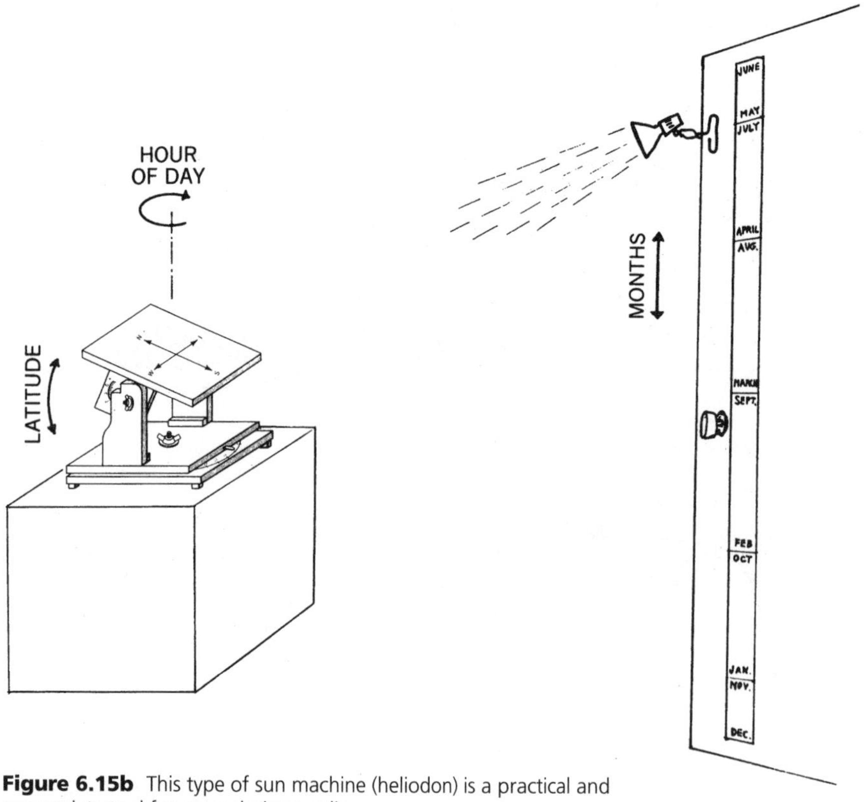

Figure 6.15b This type of sun machine (heliodon) is a practical and appropriate tool for every design studio.

As Fig. 6.14 illustrates, the site is viewed through the device in such a manner that the sun-path diagrams are superimposed on an image of the site. It is, then, immediately clear to what extent the solar window is blocked.

One serious drawback of the site-evaluation tool is that it indicates the solar access only for the spot where the tool is used. It cannot easily determine the solar access for the roof of a proposed multistory building. There is, however, a solution to this problem. A scale model of the site analyzed with a sun machine is an excellent method of evaluating the site for solar access. The scale model can then also be used for the design and presentation stages of the project.

6.15 SUN MACHINES

To simulate shade, shadows, sun penetration, and solar access on a scale model, a device called a **sun machine** or **heliodon** is used. The sun machine simulates the relationship between the sun and a building. The three variables that affect this relationship are latitude, time of year, and time of day. Every sun machine has a light source, an artificial ground plane, and three adjustments so that the light will strike the ground plane at the proper angle corresponding to the latitude, time of year, and time of day desired. In the sun machine shown in Fig. 6.15a, we can see the light moving on a circular track to simulate time of day. The track slides forward and back to simulate time of year and it is rotated to simulate the latitude. Although this kind of sun machine is very easy to use and understand, it is both expensive and difficult to construct.

The sun machine of Fig. 6.15b consists of a model stand, which rests on a table, and a clip-on lamp, which is supported by the edge of an ordinary door. The adjustment for time of year is made by moving the light up or down along the door edge. The model stand is tilted for the latitude adjustments and rotated about a vertical axis for the time of day adjustment. This simple but effective sun machine can be used at several stages during the design process:

1. Site analysis for solar access
2. Design (e.g., try different sized overhangs until the desired shadow is achieved)
3. Compare alternative designs
4. Live presentations or making photographs.

While this kind of sun machine is not as easy to use or understand as the first one mentioned, it is very easy and inexpensive to construct (about 20 dollars). Another virtue of this sun machine is that even though it can accommodate large models, it is lightweight and compact, making it easy to store or to carry. Because of the many virtues of this type of sun machine, complete instructions in its use and construction are included in Appendix C.

The author feels very strongly that the time spent on making this sun machine is very quickly repaid in better architectural designs. Although graphic techniques are available, they are difficult to learn and apply in any but the simplest cases. Many computer programs allow for the presentation of shadows and beams of sunlight entering into spaces. Although these computer images are great tools, the author believes that physical modeling is still the best way to gain understanding for designing solar-responsive architecture. Physical modeling is quite easy to understand, infinitely flexible, and inexpensive to use. It requires only the initial investment of acquiring a sun machine, and the investment can be quite modest.

6.16 SUNDIALS FOR MODEL TESTING

The least expensive way to test models for shading, solar access, and daylighting is to use a sundial (Fig. 6.16). Instead of using a sundial the conventional way to determine time from the position of the sun, the sundial is rotated and tilted until the desired testing time and date is achieved from the particular position of the sun at the moment. Thus, a sundial would be mounted on a model so that its south and that of the model align. The model along with the sundial is then rotated and tilted until the shadow of the gnomon points to the time and day to be tested. Instructions for making sundials can be found in Appendix E. The above-mentioned sun machine can be used to hold the model at the appropriate orientation and tilt. See the "Alternate Mode of Use of the Sun Machine" in Section C.4 of Appendix C for instructions on how to use the sundial in conjunction with the sun machine.

Sundials have important advantages and disadvantages in regard to testing physical models. When one uses the sun as a source of light, great accuracy can be achieved in modeling shadows and sun beams. However, this mode of testing is limited to daytime on sunny days, which are not common in some climates and some times of the year. A less accurate but sometimes more practical use of the sundial is in conjunction with an arti-

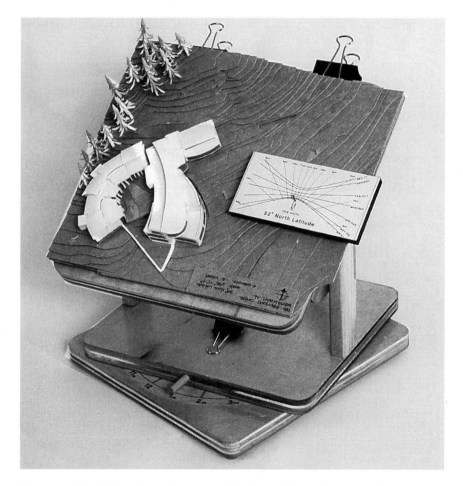

Figure 6.16 Sundials can be used to test models either under sunlight or an artificial light source.

ficial light source, such as a slide projector at the end of a corridor.

The author believes that sundials are great for making accurate photographs of finished designs and for studying daylight models outdoors, while sun machines are better during the design process for understanding solar access and shading.

6.17 INTEGRATING SUN MACHINE AND SUN EMULATOR

A new type of sun machine has been developed by the author at Auburn University, School of Architecture (Fig. 6.17a). Even before it is turned on, it acts as an extremely good educational tool because it is a three-dimensional model of the solar window. With it, the sun–earth relationship is easy to understand. A separate switch for each light facilitates easy simulation of the sun angles for any month or hour. Latitude adjustments are made by tilting the model table. The dynamic nature of the sun–earth relationship is simulated by an automatic sequencing of the lights.

It is called the "Integrating Sun Machine" not only because it simulates the instantaneous sun angles as do the machines mentioned before, but also because it can sum up the effect of a whole season. For example, when one checks the winter performance of a passive solar design, all the lights representing the winter season are turned on at once. Photocells inside the model then measure the combined effect of the sun during that whole season. This allows objective comparison of alternate design schemes.

After twenty years of use, the Integrating Sun Machine has proven to be a very effective tool for teaching solar geometry and design. Because of the expense and space requirement, this sun machine has not been duplicated to the author's knowledge. As a consequence, the author designed another sun machine with most of the advantages and few of the disadvantages of the Integrating Sun Machine. The new Sun Emulator (Fig. 6.17b) maintains the conceptual clarity of the previous integrating sun machine by keeping the model horizontal and making the lights revolve around the model. The Sun Emulator is small enough to be manufactured and shipped, and the author hopes that one day such a sun machine will be in every architecture school.

Figure 6.17a This "Integegrating Sun Machine" was developed by the author at Auburn University, Alabama.

Figure 6.17b The Sun Emulator is the latest sun machine developed by the author. Model included for scale.

6.18 SUMMARY

The concepts presented in this chapter on the relationship between the sun and the earth are fundamental for an understanding of much of this book. The chapters on passive solar energy, shading, passive cooling, and daylighting depend heavily on the information presented here.

KEY IDEAS OF CHAPTER 6

1. Solar radiation reaching the earth's surface consists of about 47 percent visible, 48 percent short-wave infrared (heat), and about 5 percent ultraviolet radiation.
2. Winter is the result of a shorter number of daylight hours, the filtering effect of lower sun angles, and the cosine law.
3. The sun is 47 degrees higher in the sky in the summer than in the winter.
4. Sun angles are defined by altitude and azimuth angles. The altitude is measured from the horizontal and the azimuth from the south.
5. The solar window is that part of the sky dome through which the sun shines.
6. Sun-path diagrams present both the pattern of the sun's motion across the sky and specific sun-angle data.
7. A sun machine is a powerful tool for achieving solar-responsive architecture.
8. Sun-path models and sundials are simple tools for achieving solar-responsive architecture.

Resources

SOURCE FOR A COMMERCIAL SUN
MACHINE

A sun machine called "SOLUX" is
commercially available for about 900
dollars from:

> Robert A. Little, FAJA
> Design & Architecture
> 5 Peper Ridge Road
> Cleveland, OH 44124
> (216) 292-4858

FURTHER READING

(See Bibliography in back of book for
full citations. This list includes valuable
out-of-print book.)

Anderson, B. *Solar Energy.*

Bennet, R. *Sun Angles for Design.*

Mazria, E. *The Passive Solar Energy
Book.*

Stein, B., and J. S. Reynolds.
*Mechanical and Electrical
Equipment for Buildings.*

Papers

Knowles, R. L. "Rhythm and Ritual,"
www-rcf.usc.edu/ ~ rknowles

Knowles, R. L. "On Being the Right
Size," www-rcf.usc.edu/ ~ rknowles

Knowles, R. L. "The Rituals of Place,"
www-rcf.usc.edu/ ~ rknowles

PASSIVE SOLAR

.

"The useful practice of the 'Ancients' should be employed on the site so that
loggias should be filled with winter sun, but shaded in the summer."

Leone Battista Alberti
from his treatise, De Re Aedificatoria, 1452,
the first modern work on architecture,
which influenced the development of
the Renaissance architectural style.

Please note that figure numbers are keyed to sections. Gaps in figure numbering result from sections without figures.

7.1 HISTORY

Although the ancient Greeks used the sun to heat their homes, the benefits were modest because much of the captured heat escaped again through the open windows. The efficient and practical Romans first solved this problem by using glass in their windows sometime around 50 A.D. The glass created an efficient heat trap by what we now call the **greenhouse effect.** The idea worked so well that the Romans found a variety of uses for it.

The upper classes often added a sun room (heliocaminus) to their villas. Greenhouses produced fruits and vegetables year-round. The later, more ''modern'' version of the Roman baths usually faced the winter sunset (southwest), when the solar heat was most needed. Solar heating was important enough that Roman architects, such as Vitruvius, wrote about it in their books.

With the fall of Rome, the use of solar energy declined, and Europe entered the "Dark" Ages. However, during the Renaissance, architects, such as Palladio, read and appreciated the advice of Vitruvius. Palladio utilized such classical principles as placing summer rooms on the north side and winter rooms on the south side of a building. Unfortunately, northern Europe copied only the style and not the principles that guided Palladio.

The seventeenth century in northern Europe saw a revival of solar heating, but not for people. Exotic plants from newly discovered lands, and the appetite for oranges and other warm climate fruits of a sizable upper class created a need for greenhouses (Fig. 7.1a). With the invention of better glass-making techniques, the eighteenth century became known as the "Age of the Greenhouse." Eventually, those greenhouses that were attached to the main building became known as **conservatories** (see Fig. 7.1b). These, like our modern **sun spaces,** were used for growing plants, added space to the living area, and helped heat the main house in the winter.

This use of the sun, however, was reserved for the rich.

The idea of solar heating for everyone did not start in Europe until the 1920s. In Germany, housing projects were designed to take advantage of the sun. Walter Gropius of the Bauhaus was a leading supporter of this new movement. The research and accumulated experience with solar design then slowly made its way across the Atlantic with men like Gropius and Marcel Breuer.

7.2 SOLAR IN AMERICA

Passive solar design also has Native American roots. Many of the early Native American settlements in the Southwest show a remarkable understanding of passive solar principles. One of the most interesting is Pueblo Bonito (Fig. 7.2a), where the housing in the south-facing semicircular village stepped up to give each home full access to the sun, and the massive construction stored the heat for nighttime use.

Figure 7.1a The orangery on the grounds of the royal palace in Prague, Czech Republic, has an all-glass south facade, which is typical of the greenhouses that became popular in the eighteenth century.

Figure 7.1b Conservatories supplied plants, heat, and extra living space for the upper classes in nineteenth-century Europe. Conservatory of Princess Mathilde Bonaparte, Paris, about 1869. (From *Uber Land und Meer, Allgemeine Illustrierte Zeitung,* 1868.)

Figure 7.2a Pueblo Bonito, Chaco Canyon, NM, built about 1000 AD, is an example of an indigenous American solar village. (From *Houses and House-Life of the American Aborigines* by Lewis Morgan, (contributions to *North American Ethnology,* Vol. 4) U.S. Department of the Interior/US. G.P.O., 1881.)

Figure 7.2b The New England saltbox faced the sun and turned its back to the cold northern winds. (From *Regional Guidelines for Building Passive Energy Conserving Homes,* by A.I.A. Research Corporation, U.S. G.P.O., 1980.)

Some of the colonial buildings in New England also show an appreciation of good orientation. The "saltbox," as shown in Fig. 7.2b, had a two-story wall with numerous windows facing south to catch the winter sun. The one-story north wall had few windows and a long roof to deflect the cold winter winds.

Aside from these early examples, the heating of homes with the sun made slow progress until the 1930s when a number of different American architects started to explore the poten-

tial of solar heating. One of the leaders was George F. Keck, who built many successful solar homes (Fig. 7.2c). The pioneering work of these American architects, the influence of the immigrant Europeans, and the memory of the wartime fuel shortages made solar heating very popular during the initial housing boom at the end of the war. But the slightly higher initial cost of solar homes and the continually falling price of fuels resulted in public indifference to solar heating by the late 1950s.

7.3 SOLAR HEMICYCLE

One of the most interesting solar homes built during this time was the Jacobs II House (Fig. 7.3a), designed by Frank Lloyd Wright. Figure 7.3b shows a floor plan of this house, which Wright called a **solar hemicycle.** As usual, Wright was ahead of his time because this building would in many ways make a fine passive home by present-day standards. For example, most of the glazing faces the winter sun but is well shaded from the summer sun by a 6-foot overhang (Fig. 7.3c). Plenty of thermal mass, in the form of stone walls and a concrete floor slab, stores heat for the night and prevents overheating during the day (Fig. 7.3d). The building is insulated to reduce heat loss, and an earth berm protects the northern side. The exposed stone walls are cavity walls filled with vermiculite insulation. Windows on opposite sides of the building allow cross-ventilation during the summer.

Like most of Wright's work, the design of this house is very well integrated. For example, the curved walls not only create a sheltered patio, but also very effectively resist the pressure of the earth berm, just as a curved dam resists the pressure of the water behind it. The abundant irregularly laid stone walls supply the ther-

Figure 7.2c One of the first modern solar houses in America. Architect, George Fred Keck, Chicago, 1940s. (Courtesy of Libby-Owens-Ford Co.)

Figure 7.3a The Jacobs II House, Architect, Frank Lloyd Wright, Madison, WI circa 1948. (Photograph by Ezra Stoller © Esto.)

Figure 7.3d Interior view of Jacobs II House. (Photograph by Ezra Stoller © Esto.)

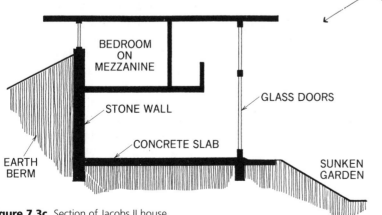

Figure 7.3b Plan of Jacobs II house.

Figure 7.3c Section of Jacobs II house.

mal mass while relating the interior with the natural environment of the building site. Successfully integrating the psychological and functional demands seems to produce the best architecture. This is what the truly great architects have in common.

7.4 LATEST REDISCOVERY OF PASSIVE SOLAR

From the late 1950s until the mid-1970s, it was widely assumed that active solar systems had the greatest potential for harnessing the sun. Slowly, however, it was realized that using active collectors for space heating would add significantly to the first costs of a home, while passive solar could be achieved with little or no additional first costs. It also became apparent that passive solar systems had lower maintenance and higher reliability.

Possibly the greatest advantage of passive solar is that it usually results in a more pleasant indoor environment while active collectors only supply heat. The Human Services Field Office in Taos, New Mexico, is

145

Figure 7.4a The Human Services Field Office, Taos, NM, (1979) has all of its windows facing 20 degrees east of south to take advantage of the morning sun. The clerestory windows, which cover the whole roof, supply both daylight and solar heat.

Figure 7.4b Integrated Passive and Hybrid Solar Multiple Housing, Berlin, 1988. (Courtesy of and copyright Institut fur Bau-, Umwelt- und Solar Forschung.)

a pleasant place to work because of the abundance of sunlight that enters, especially in the winter (Fig. 7.4a). A sawtooth arrangement on the east and west enables the windows on those facades to face south. There are also continuous clerestory windows across the whole roof so that even interior rooms have access to the sun. Black-painted water drums just inside the clerestory windows store heat for nighttime use, while insulated shutters reduce the heat loss.

Much of the renewed interest occurred in New Mexico not only because of the plentiful sun, but also because of the presence of a community of people who were willing to experiment with a different lifestyle. An example is the idealistic developer, Wayne Nichols, who built many solar houses, including the well-known Balcomb house, described later. As so often happens, successful experiments in alternate lifestyles are later adopted by the mainstream culture. Passive solar is now being accepted by the established culture because it has proved to be a very good idea.

Passive solar heating is also gaining popularity in other countries. Successful passive solar houses are even being built in climates with almost constant clouds and gloomy weather, as is found in northern Germany at a latitude of 54 degrees. This is the same latitude as that of southern Alaska (Fig. 7.4b). The success of passive buildings in so many different climates is a good indication of the validity of this approach to design.

7.5 PASSIVE SOLAR

"Passive solar" refers to a system that collects, stores, and redistributes solar energy without the use of fans, pumps, or complex controllers. It functions by relying on the integrated approach to building design, where the basic building elements, such as windows, walls, and floors, have as many different functions as possible. For example, the walls not only hold up the roof and keep out the weather but also act as heat-storage and heat-radiating elements. In this way, the various components of a building simultaneously satisfy architectural, structural, and energy requirements. Every passive solar heating system will have at least two elements: a collector consisting of south-facing glazing and an energy-storage element that usually consists of thermal mass, such as rock or water.

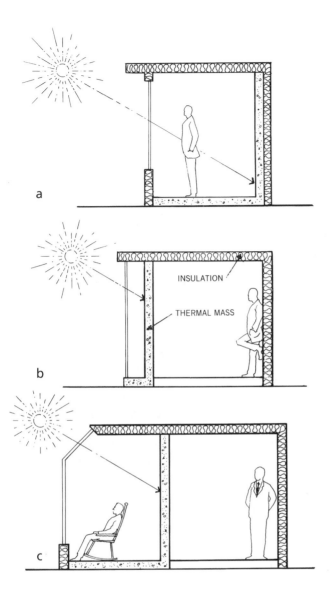

Figure 7.5a The three main types of passive solar space-heating systems. (a) direct gain, (b) Trombe wall, and (c) sun space.

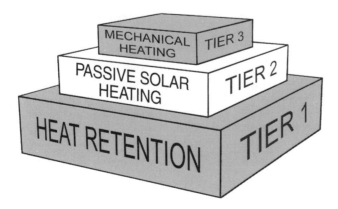

Figure 7.5b Passive solar heating is the second tier of the more efficient and sustainable three-tier design approach. The first tier is heat retention.

Depending on the relationship of these two elements, there are several possible types of passive solar systems. Figure 7.5a illustrates the three main concepts:

a. Direct Gain
b. Trombe Wall
c. Sun space

Each of these popular space-heating concepts will be discussed in more detail. The chapter will conclude with a discussion of a few less common passive space-heating systems.

Passive solar is part of the rational design accomplished through the three-tier design approach (Fig. 7.5b). The first tier consists of minimizing the heat loss through the building envelope by proper insulation, orientation, and surface-area-to-volume ratios. The better the architect designs the thermal barrier, the less heating will be required. The second tier, which consists of harvesting the sun's energy by passive means, is explained in this chapter. The mechanical equipment and fossil energy of the third tier is needed only to supply the small amount of heating not accomplished by tiers one and two.

7.6 DIRECT-GAIN SYSTEMS

Every south-facing window creates a **direct-gain system**, while windows at any other orientation lose more heat than they gain in the winter. The greenhouse effect, described in Chapter 3, acts as a one-way heat valve. It lets the short-wave solar energy enter but blocks the heat from escaping (Fig. 7.6a). The thermal mass inside the building then absorbs this heat both to prevent daytime overheating and to store it for nighttime use (Fig. 7.6b). The proper ratio of mass to south-facing glazing is critical.

The graph in Fig. 7.6c shows the effect of the south glazing in trapping heat in a building with the conventional heating turned off. Curve "A" is

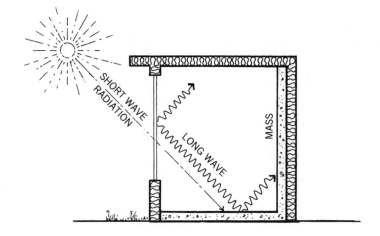

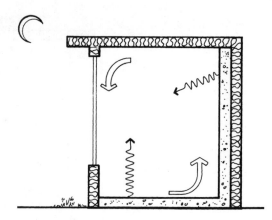

Figure 7.6a The greenhouse effect collects and traps solar radiation during the day.

Figure 7.6b The thermal mass stores the heat for nighttime use.

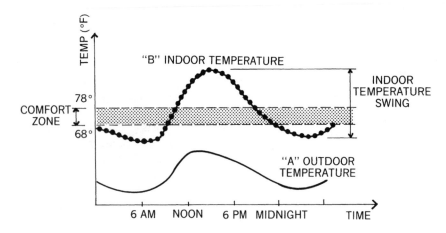

Figure 7.6c A low-mass passive solar building will experience a large indoor temperature swing during a twenty-four-hour period of a winter day.

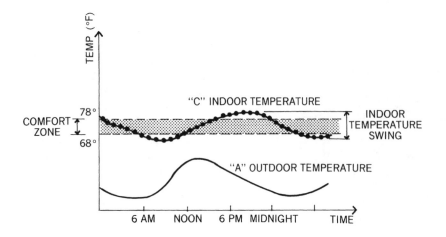

Figure 7.6d A high-mass passive solar building will experience only a small indoor temperature swing during a winter day.

the outdoor temperature during a typical cold but sunny day. Curve "B" describes the indoor temperature in a direct-gain system with little mass. Notice the large indoor temperature swing from day to night. In the early afternoon, the temperature will be much above the comfort zone, while late at night it will be below the comfort zone. Increasing the area of south glazing will not only raise the curve, but also increase the temperature swing. The overheating in the afternoon would then be even worse.

In the graph of Fig. 7.6d, we see the benefits of thermal mass. Notice that the indoor temperature (curve "C") is almost entirely within the comfort zone. The thermal mass has reduced the amplitude of the temperature swing so that little overheating occurs in the afternoon and little overcooling at night. Thus, the designer's goal is to get the right mix of south glazing area and thermal mass so that the indoor temperature fluctuates within the comfort zone.

Since in direct gain the building is the collector, all contents, such as the drywall, furniture, and books, act as thermal mass. However, the contents are usually not sufficient, and additional thermal mass must be added. The thermal mass can be masonry, water, or a phase-change material. These alternatives will be discussed later in this chapter.

Although solar heat can be supplied by convection to the rooms on the north side of a building, it is much better to supply solar radiation directly by means of south-facing clerestory windows as shown in Fig. 7.6e. Skylights, although not as good as clerestories, can still be used if a reflector is included, as shown in Fig. 7.6f. The same reflector can also shade some of the summer sun, if it is moved to the position shown in Fig. 7.6g. Such a reflector is commercially available.

Frank L. Wright's "Solar Hemicycle," described above, is a good example of the direct-gain approach. Of all the passive systems, direct gain is the most efficient when energy collection and first costs are the main concerns.

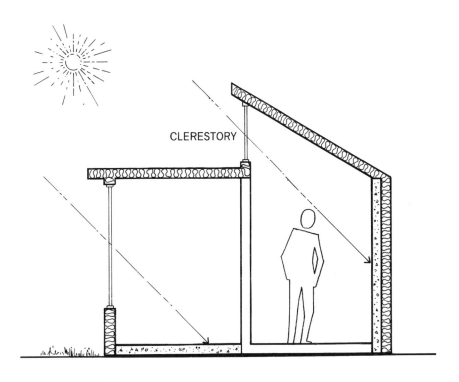

Figure 7.6e Use clerestory windows to bring the solar radiation directly to interior or north-facing rooms.

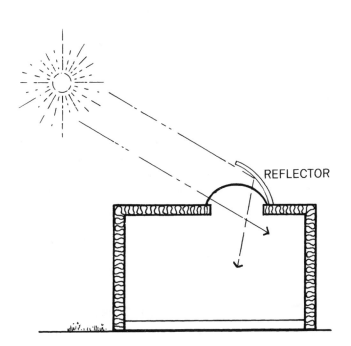

Figure 7.6f Skylights should use a reflector to make them more effective in the winter.

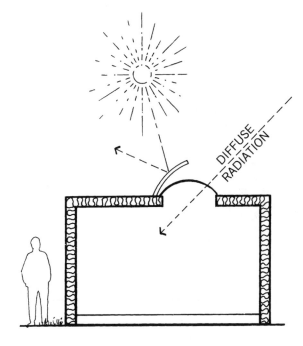

Figure 7.6g The same reflector can be used to block excessive summer sun.

TABLE 7.7A RULES FOR ESTIMATING AREAS OF SOUTH-FACING GLAZING FOR DIRECT-GAIN AND TROMBE WALLS

Climate Region (see Chapter 5)	Reference City	South Glazing Area as a Percentage of Floor Area[a]	Heating Load Contributed by Solar (%)	
			No Night Insulation	With Night Insulation
1	Hartford, CT	35	19	64
2	Madison, WI	40	17	74
3	Indianapolis, IN	28	21	60
4	Salt Lake City, UT	26	39	72
5	Ely, NE	23	41	77
6	Medford, OR	24	32	60
7	Fresno, CA	17	46	65
8	Charleston, SC	14	41	59
9	Little Rock, AK	19	38	62
10	Knoxville, TN	18	33	56
11	Phoenix, AZ	12	60	75
12	Midland, TX	18	52	72
13	Fort Worth, TX	17	44	64
14	New Orleans, LA	11	46	61
15	Houston, TX	11	43	59
16	Miami, FL	2	48	54
17	Los Angeles, CA	9	58	72

[a]Use the floor area of those parts of the building that will receive benefits from solar heating either by direct radiation, or by convection from the solar-heated parts of the building.

TABLE 7.7B RULES FOR ESTIMATING REQUIRED THERMAL MASS IN DIRECT-GAIN SYSTEMS

Thermal Mass	Thickness (inches)	Surface Area per Square Foot of Glazing (ft^2)
Masonry or concrete exposed to direct solar radiation (Figure 7.7a)	4 to 6	3
Masonry or concrete exposed to reflected solar radiation (Figure 7.7b)	2 to 4	6
Water	About 6	About ½

7.7 DESIGN GUIDELINES FOR DIRECT-GAIN SYSTEMS

Area of South Glazing

Use Table 7.7A as a guideline for initial sizing of south-facing glazing. The table is based on the seventeen climate regions described in Chapter 5. Column 5 shows how much more effective passive heating systems are when night insulation is used over the windows or when high-efficiency windows are used.

Notes on Table 7.7A

1. The table presents optimum south-glazing areas.
2. Smaller south-glazing areas than those shown in the table will still supply a significant amount of heat.
3. Larger glazing areas can be used if some daytime overheating is permissible, but they nevertheless tend to become less cost effective.
4. Adequate thermal storage must be supplied (see Table 7.7B).
5. Windows should be double-glazed except in very mild climates.
6. High-efficiency windows using low-e coatings can be used instead of night insulation (see Chapter 15).
7. Building must be well insulated.
8. Unless large amounts of light are desired for daylighting, sunbathing, etc., direct-gain glazing areas should not exceed about 20 percent of the floor area. In those cases where Table 7.7A recommends more than 20 percent glazing, use either Trombe walls or sun spaces to supply the additional glazing area.

Thermal-Mass Sizing

Use Table 7.7B as a guideline for sizing thermal mass for direct-gain systems. Keep in mind that slabs and walls of concrete, brick, or rock should be between 4 and 6 inches thick. From a thermal point of view, anything over 6 inches thick is only slightly helpful in a direct-gain system. Also, any thermal mass not receiving either direct or reflected solar radiation is mostly ineffective.

Notes on Table 7.7B

1. A mixture of mass directly and indirectly exposed to the sun is quite common.
2. The table specifies minimum mass areas. Additional mass will increase thermal comfort by reducing temperature extremes.
3. Keep mass as close as possible to the floor for structural as well as thermal reasons.
4. The thermal mass should be medium to dark in color while surfaces of nonmassive materials should be very light in color to reflect the solar radiation to the darker mass materials (Fig. 7.7b).
5. If the mass is widely distributed in the space, then diffusing glazing or diffusing elements should be used (Fig. 7.7c).
6. For more information on thermal mass, see Section 7.17.

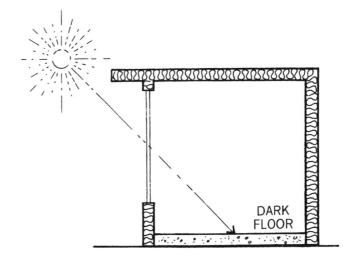

Figure 7.7a Massive floors should be medium to dark in color.

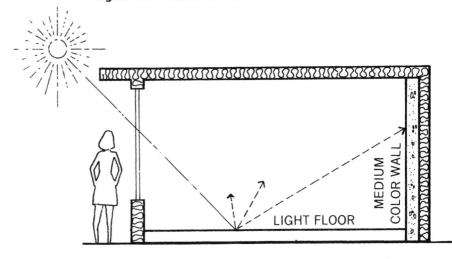

Figure 7.7b The surface finish of nonmassive materials should consist of very light colors to reflect the sun to the darker massive material.

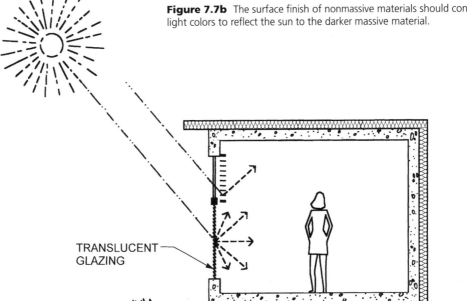

Figure 7.7c Diffused radiation will distribute the heat more evenly in the space. It is especially useful where ceilings are massive.

7.8 EXAMPLE

Design a direct-gain system for a 1000 ft² building in Little Rock, Arkansas, as shown in Fig. 7.8.

Procedure:

1. Table 7.7A tells us that if the area of south-facing glazing is 19 percent of the floor area, then we can expect solar energy to supply 62 percent of the winter heating load (if high-efficiency windows or night insulation is used). Use this recommendation unless there are special reasons to use larger or smaller glazing areas.
2. Thus, the area of south-facing glazing should be about 19% × 1000 square feet = 190 square feet.
3. Table 7.7B tells us that we will need 3 square feet of mass for each square foot of glazing if the mass is directly exposed to the sun. Thus, 190 × 3 = 570 square feet is required. If we use a concrete slab, then we have a slab area of 1,000 square feet, of which only 570 square feet is required for storing heat. The remaining 430 square feet can be covered by carpet if so desired.

7.9 TROMBE WALL SYSTEMS

The Trombe wall was named after professor Felix Trombe, who developed this technique in France in 1966. In this passive system, the thermal mass consists of a wall just inside the south-facing glazing (Fig. 7.9a). As before, the greenhouse effect traps the solar radiation. Because the surface of the wall facing the sun is either covered with a selective coating (black foil) or painted a dark color, it gets quite hot during the day, causing heat to flow into the wall. Since the Trombe wall is quite thick — often about 12 inches thick — and the time lag is quite long, the heat does not reach the interior surface until evening. This time-lag effect of mass was explained in Section 3.18. If there is enough mass, the wall can act as a radiant heater all night long (Fig. 7.9b).

When only the sun's heat and not its light is desired, the Trombe wall is the system of choice. Because this is a rare occurrence, the Trombe wall is ordinarily used in combination with direct gain. The direct gain part of the system delivers heat early in the day, functional light, views, and the delight of winter sunshine, while the Trombe wall stores heat for nighttime use. The combination of systems prevents the need for excessive light levels, which can cause glare and fading of colors. When one carefully chooses the right combination of systems, great thermal and visual comfort is possible.

Although the Trombe wall is usually made of solid materials, such as concrete, brick, stone, or adobe, it can also be made of containers of water. The classic example was designed by Steve Baer, a solar entrepreneur, for his own residence in New Mexico. He used 55-gallon drums that are stacked, as shown in Fig. 7.9c. The side of the drums facing the glazing are painted black, while the side facing the interior is painted white. (Any color except a polished metallic finish is a good emitter of radiant heat.) An exterior insulating shutter

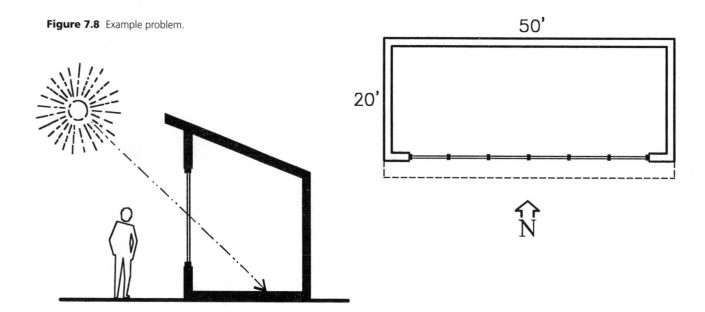

Figure 7.8 Example problem.

keeps the heat in during a winter night or out during a summer day. This shutter, when it is down on the ground, also acts as a reflector to increase the total amount of solar radiation collected. The shutter can be raised or lowered from indoors by means of a cable. Also, on a mild winter afternoon, when heating is not yet required, a curtain can be drawn across the interior side of the drums to delay the transfer of heat until it is needed.

Most water walls consist of vertical tubes. If steel tubes are used, they can be painted a dark color on the glazing side and a light color on the room side. Often, however, the tubes are made of translucent or transparent plastic to allow some light to pass through (Fig. 7.9d). The water can be left clear or tinted any color. Transparent tubes are especially beautiful because of the way they refract the light. Tests have shown that clear water is almost as efficient as tinted water or opaque containers when it comes to storing heat. See the water

Figure 7.9a The Trombe wall passive-solar heating system collects heat without having light enter the space.

Figure 7.9b Because of the wall's 8- to 12-hour "time lag," most of the heat is released at night.

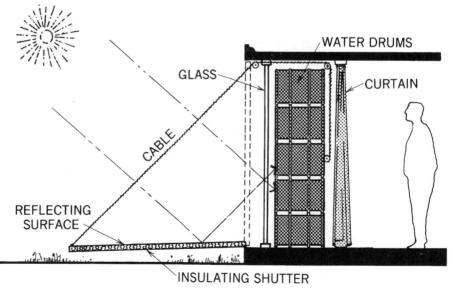

Figure 7.9c A section through the Baer residence shows the insulating shutter, which is covered with a reflective surface to increase the performance of the Trombe wall.

WATER DRUMS

GLASS

CURTAIN

CABLE

REFLECTING SURFACE

INSULATING SHUTTER

Figure 7.9d A Trombe wall can consist of vertical tubes filled with water. The tubes can be opaque, translucent, or transparent. (Courtesy of and © Solar Components Corporation.)

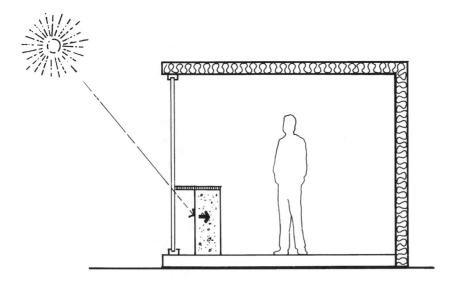

Figure 7.9e A half-height wall allows controlled direct gain for daytime heating and daylighting while also storing heat for the night.

tubes on each side of the main entrance of the Hood College Resource Management Center presented as a case study in Section 17.5.

As mentioned before, a mix of Trombe walls and direct gain usually yields the best design solution. Although most Trombe walls are full height with punched windows, sometimes they are built as a parapet wall (Fig.7.9e). An arrangement, as shown in Fig. 7.16b, can increase morning pickup, prevent afternoon overheating, and provide adequate storage through the night.

The Shelly Ridge Girl Scout Center near Philadelphia, Pennsylvania, is a wonderful example of mixed Trombe wall and direct-gain systems (Fig. 7.9f). Since most scout activities happen during the day and early evening, the Trombe wall was made of brick only 4 inches thick so that the resulting short time lag delivered the heat during the afternoon and evening. There was plenty of direct gain for early heating and daylighting (Fig. 7.9g). Because the 4-inch-thick Trombe wall was much too high to be stable, a timber grid was used to support the brick and the glazing (Fig. 7.9h). Crank-out awnings were extended in the summer to shade most of the direct-gain windows. Summer evenings, however, would be cooler if the Trombe wall were also shaded. A screen hung in front of the glazing during the summer would be especially effective since it would shade the Trombe wall from direct, diffuse, and reflected radiation (Fig. 7.9i).

The Girl Scouts learned the methods and benefits of passive solar firsthand from using the building. They learned more experiential lessons from the direct-gain heated lobby (Area #4 in Figure 7.9j), where a stained-glass gnomon and hour marks imbedded in the floor create a sundial.

Some early Trombe walls had indoor vents to supply daytime heat in winter and outdoor vents to prevent summer overheating. It is now clear that these vents do not work

Figure 7.9f The Shelley Ridge Girl Scout Center near Philadelphia, PA, utilizes both direct-gain and Trombe wall passive systems. Glass is used for the direct gain, and translucent Fiberglas panels are used for the Trombe wall. Awnings are extended in the summer to shade the direct-gain windows. (Photograph © Otto Baitz/ESTO.)

Figure 7.9g The direct-gain windows provide daylighting, views, and heat early in the day, while the Trombe wall provides heat in the early evening. (Photograph © Otto Baitz/ESTO.)

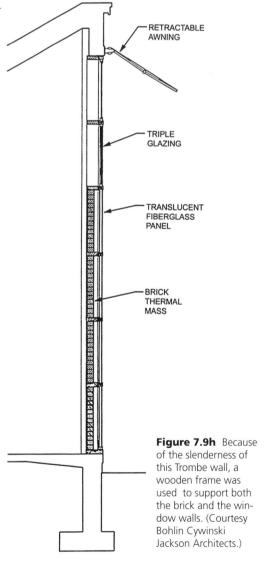

RETRACTABLE AWNING

TRIPLE GLAZING

TRANSLUCENT FIBERGLASS PANEL

BRICK THERMAL MASS

Figure 7.9h Because of the slenderness of this Trombe wall, a wooden frame was used to support both the brick and the window walls. (Courtesy Bohlin Cywinski Jackson Architects.)

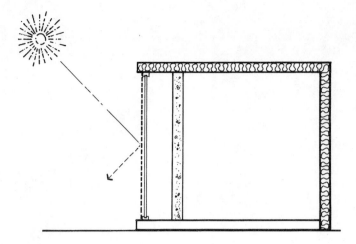

Figure 7.9i In hot climates, a shade screen should be draped over the Trombe wall glazing during the summer.

well in either summer or winter. Instead, direct gain should supply daytime heat in the winter, and an outside shading device should prevent heat collection in the summer.

From the outside, Trombe walls are sometimes indistinguishable from windows. Under certain lighting conditions, however, the dark wall is visible. If this is undesirable, textured glass can be used both to hide the dark wall and to serve as an aesthetic expression to differentiate the Trombe wall from ordinary windows.

Since dark colors are almost as effective as black, some Trombe walls use clear glass to show the dark brick, dark natural stones, water tubes, or another attractive thermal-mass system.

Not all applications are for new buildings, however, since the Trombe wall is well suited for some renovation work. Old masonry buildings can benefit in several ways by having a glass-curtain wall built over their south facade, which then becomes a Trombe wall (Fig. 7.9k). Other benefits include a more weather-resistant skin, greater thermal insulation, and, sometimes, a more attractive facade for the building.

The Trombe wall system is also known as the **thermal-storage wall system**. This author prefers the term "Trombe wall" because not all thermal walls are part of this system — only those that are just inside the glazing.

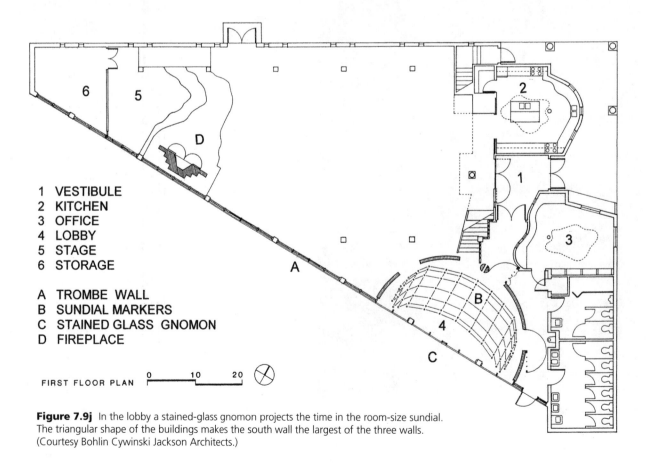

1 VESTIBULE
2 KITCHEN
3 OFFICE
4 LOBBY
5 STAGE
6 STORAGE

A TROMBE WALL
B SUNDIAL MARKERS
C STAINED GLASS GNOMON
D FIREPLACE

FIRST FLOOR PLAN   0   10   20

Figure 7.9j In the lobby a stained-glass gnomon projects the time in the room-size sundial. The triangular shape of the buildings makes the south wall the largest of the three walls. (Courtesy Bohlin Cywinski Jackson Architects.)

Figure 7.9k The addition of glazing can turn an existing wall into a solar collector, as in this Boston rowhouse. (Cover photo of Solar Age, August 1981. ~ Solar Vision Inc., 1981.)

TABLE 7.10 RULES FOR ESTIMATING THE REQUIRED THICKNESS OF A TROMBE WALL

Thermal Mass	Thickness (inches)	Surface Area per Square Foot of Glazing (ft²)
Adobe (dry earth)	6 to 10	1
Concrete or brick	10 to 16	1
Water[a]	8 or more	1

[a] If tubes are used, they should be at least 10 inches in diameter. Source: *Solar Age,* May, 1979, p. 64.

7.10 DESIGN GUIDELINES FOR TROMBE WALL SYSTEMS

Area of South Glazing

Table 7.7A is used for Trombe wall, as well as direct-gain systems. The total area of south glazing can be divided between the two systems as the designer wishes.

Thermal-Mass Sizing

Each square foot of south-facing glazing should be matched by 1 square foot of thermal mass. However, the mass must be much thicker than that in direct-gain systems. The thickness for various materials is shown in Table 7.10. For best results, the mass should be at least 1 inch from the glazing. The surface facing the glazing should be covered with a black high-efficiency, "selective" coating, while the surface facing the living space can be any color, including white. If a "selective" coating is not used, then double glazing is recommended.

7.11 EXAMPLE

Redesign the building of Example 7.8 to be half Trombe wall and half direct gain.

Procedure:

1. Total required south-facing glazing is again obtained from Table 7.7A: 19 percent × 1000 = 190 square feet.
2. Since half of the glazing will be for direct gain, the Trombe wall will require 50 percent × 190 = 95 square feet.
3. If we use a brick Trombe wall, it will have an area of 95 square feet and a thickness of at least 10 inches (from Table 7.10). The slab for direct gain will have an area of 95 square feet × 3 = 285 square feet (Table 7.7B).
4. Consider using a 3-foot-high and 32-foot-long wall (96 square feet) as shown in Fig. 7.9e. Do not block the

inside of the wall with furniture since it must act as a radiator at night.

5. The window above the wall would also be 3 feet high and 32 feet long to supply the 95 square feet of glazing required for the direct-gain system.

7.12 SUN SPACES

A sun space is a room designed to collect heat for the main part of a building, as well as to serve as a secondary living area. This concept is derived from the "conservatories" popular in the eighteenth and nineteenth centuries. Until recently, this design element was usually called an "attached greenhouse," but that was a misleading name because growing plants is a minor function. More appropriate

terms were "solarium" or "sun room," but the term **"sun space"** seems to have become most common. Sun spaces are one of the most popular passive systems, not only because of their heating efficiency, but even more so because of the amenities that they offer. Most people find the semioutdoor aspect of sun spaces extremely attractive. Almost everyone finds it pleasurable to be in a warm sunny space on a cold winter day.

Sun spaces are considered adjunct living spaces because the temperature is typically allowed to swing from a high of 90°F during a sunny day to a low of 50°F during a winter night. Here we have an efficient solar collector that can also be used as an attractive living space much, but not all, of the time. Consequently, a sun space must be designed as a separate ther-

mal zone that can be isolated from the rest of the building. Figure 7.12a shows the three ways a sun space can relate physically to the main building.

In Fig. 7.12b we see a sun space collecting solar heat during the day. Much of the heat is carried into the main building through doors, windows, and vents. The rest of the captured solar heat is absorbed in the sun space's thermal mass such as the floor slab and the masonry common wall. At night, as seen in Fig. 7.12c, the doors, windows, and vents are closed to keep the main part of the building warm. The heat in the massive common wall keeps the house comfortable and prevents the sun-space from freezing.

Because sun spaces are generally poorly insulated and shaded, they should not be heated or cooled. Mechanical heating and cooling

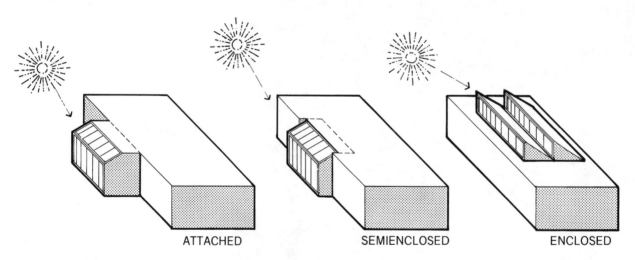

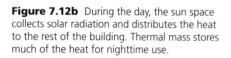

Figure 7.12a Possible relationships of a sun space to the main building.

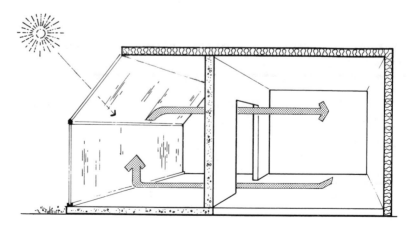

Figure 7.12b During the day, the sun space collects solar radiation and distributes the heat to the rest of the building. Thermal mass stores much of the heat for nighttime use.

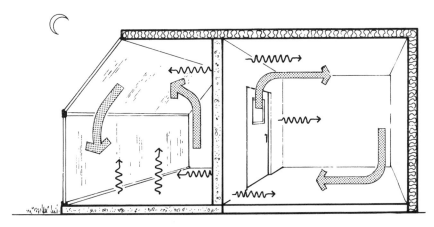

Figure 7.12c At night, the sun space must be sealed from the main building to keep it from becoming an energy drain on the main building.

require so much energy that a sun space would become a net loser rather than a gainer of energy. Large temperature swings should be expected, and when the temperatures get extreme, the space should be temporarily abandoned. However, a well-designed sun space is a delight most of the year.

7.13 BALCOMB HOUSE

One of the best known sun-space houses (Fig. 7.13a) belongs to J. Douglas Balcomb, a foremost researcher in passive solar systems. Because it is located in historic Santa Fe, New Mexico, adobe is used for the common wall (Fig. 7.13b). Double doors enable convection currents to heat the house during the day, but seal off the sun space at night (Fig. 7.13c). Also at night the common adobe wall heats both the house and the sun space. The sun space not only contributes about 90 percent of the heating, but it is a delightful place to spend an afternoon.

Another solar heating strategy used here actually makes this a **hybrid** solar building. It is partly **active solar** because fans force the hot air, collecting at the ceiling of the sun space, to pass through a rock bed below the first floor slab. This strategy does not contribute much heat, but it does increase the comfort level within the building.

On the roof is a large vent to allow hot air to escape during the overheated periods of summer and fall (Fig. 7.13b). Unfortunately, venting is usually not enough to prevent overheating, and in most climates shading of the glass is of critical importance. Since shading inclined glass is more complicated than shading vertical glass, the slope of the glazing will be a major consideration in the sun-space design guidelines.

The graph in Fig. 7.13d illustrates the performance of the Balcomb House during three winter days. Such performance is typical of a well-designed sun-space system. Note that the outdoor temperatures are quite cold, with a low of about 18°F. The sun-space temperature swing was quite wide, with a low of 58°F and a high of 88°F. The house, on the other hand, was fairly comfortable, with a

Figure 7.13a One of the first and most interesting sun space houses is the Balcomb residence in Santa Fe, NM.

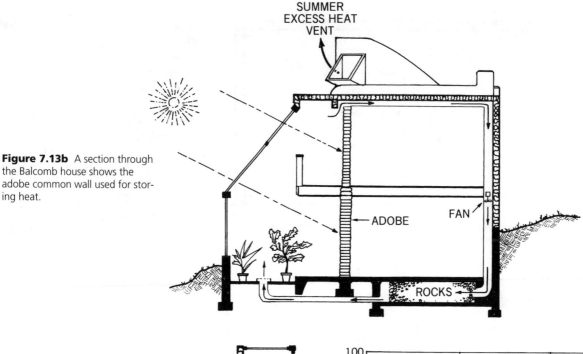

SUMMER
EXCESS HEAT
VENT

ADOBE

FAN

ROCKS

Figure 7.13b A section through the Balcomb house shows the adobe common wall used for storing heat.

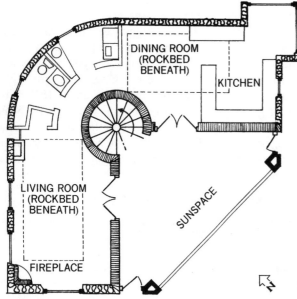

DINING ROOM
(ROCKBED
BENEATH)

KITCHEN

LIVING ROOM
(ROCKBED
BENEATH)

SUNSPACE

FIREPLACE

Figure 7.13c This plan of the Balcomb house shows how the building surrounds the sun space.

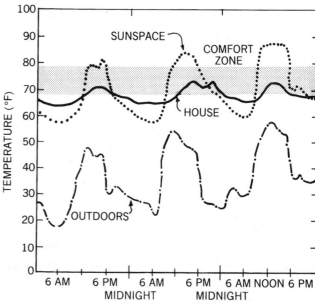

Figure 7.13d The performance of both the house and sun space are shown for three sunny but cold winter days.

modest swing of temperatures from a low of about 65°F to a high of 74°F. The graph shows the different character of the two thermal zones. Although the addition of a small amount of auxiliary heating can create complete comfort in the house, the sun space is always allowed its large temperature swing.

7.14 SUN-SPACE DESIGN GUIDELINES

Slope of Glazing

To maximize solar heating in the continental United States, the slope of the glazing should be between 50 and 60

degrees, as shown in Fig. 7.14a. However, from the point of view of safety, water leakage, and, most important, sun shading, vertical glazing is best (Fig. 7.14b). A compromise, as shown in Fig. 7.14c, will often work quite well. In hot climates, where shading to prevent overheating is critical, vertical glazing is usually

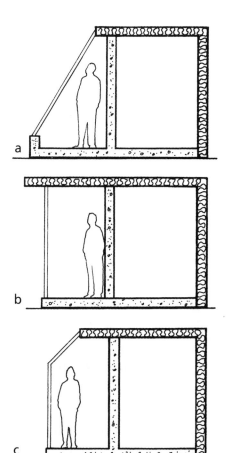

Figure 7.14a-c Variations on slope of sun-space glazing: (a) 50 to 60 degrees; (b) vertical; (c) combined. In most cases, vertical glazing gives the best year-round performance.

best. Shading strategies will be explained in detail in Chapter 9.

Very little if any glazing should be used on the end walls facing east and west. Such glazing is a thermal liability in both winter and summer.

Area of Glazing

Use Table 7.7A to determine the minimum glazing area desired for the total south facade. Since sun spaces are often sized on other than thermal considerations, the total glazing area might end up much larger than suggested in Table 7.7A This is acceptable because overheating of sun spaces will have much less of an impact on the main house than overheating of direct-gain or Trombe wall systems.

Vent Sizing

To prevent overheating, especially in summer and fall, venting of the sun space to the outside is required, as shown in Fig. 7.14d. The low inlet vent area should be about 5 percent of the south-facing glazing area and the upper exhaust vent should be another 5 percent. Smaller openings will suffice if an exhaust fan is used.

To heat the house in the winter, openings in the form of doors, windows, or vents are required in the common wall between the house and the sun space. The total area of any combination of these openings must add up to a minimum of 10 percent of the glazing area. Larger openings are better.

Thermal-Mass Sizing

The size of the mass depends on the function of the sun space. If it is primarily a solar collector, then there should be little mass so that most of the heat ends up in the house. On the other hand, if the sun space is to be a useful space with a modest temperature swing, then it should have much more mass.

A good solution for temperate climates is a common thermal-storage

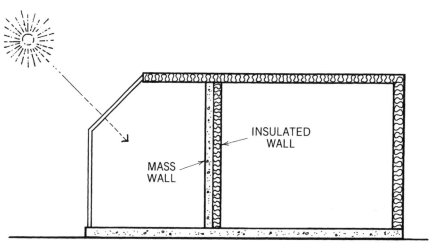

Figure 7.14d To prevent overheating in the summer, the sun space must be vented to the outdoors. Inside vents are only used in the winter and then have the same purpose as doors or windows.

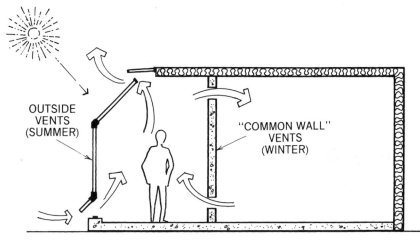

Figure 7.14e In extreme climates, the sun space should be completely isolated from the main building by an insulated wall.

TABLE 7.14 RULES FOR ESTIMATING THE REQUIRED THERMAL MASS IN SUN SPACE SYSTEMS

Thermal Mass	Thickness (inches)	Surface Area per Square Foot of Glazing (ft²)
Masonry common wall (noninsulated)	8 to 12	1
Masonry common wall (insulated)[a]	4 to 6	2
Water[b]	About 12[b]	About ½

[a] Since this mass is exclusively for the sun space, some additional mass will be required for the main building.

[b] Use about 2 gallons of water for each square foot of glazing. Source: *Solar Age,* June, 1984, p. 32.

TABLE 7.15 COMPARISON OF PASSIVE SOLAR HEATING SYSTEMS

System	Advantages	Disadvantages
Direct gain	Promotes the use of large "picture" windows Least expensive Most efficient Can effectively use clerestories and skylights Daylighting and heating can be combined, which makes this system very appropriate for schools, small offices, etc. Very flexible and best when total glazing area is small	Too much light, which can cause glare and fading of colors Thermal-storage floors must not be covered by carpets Only few and small paintings can be hung on thermal-mass walls Overheating can occur if precautions are not taken Fairly large temperature swings must be tolerated (about 10°F)
Trombe wall	Gives high level of thermal comfort Good in conjunction with direct gain to limit lighting levels Easy to retrofit on existing walls Medium cost Good for large heating loads	More expensive than direct gain Less of glazing will be available for views and daylighting Not good for very cloudy climates
Sun spaces	A very attractive amenity Extra living space Can function as a greenhouse Most appropriate for residential units or public spaces such as atriums, lobbies, restaurants, etc.	Most expensive system Least efficient

wall, as shown in Fig. 7.12b. In extremely hot or cold climates, it might be desirable to completely isolate the house from the sun space. In this case, an insulated, less massive masonry wall might be used, as shown in Fig. 7.14e. When heat is desired, the doors, windows, or vents in the common wall are opened. When the sun space needs to be isolated from the main building, the openings are closed and the insulated wall acts as a thermal barrier. With either type of wall, water or a phase-change material can be efficiently used instead of masonry for thermal storage. For rules in sizing the mass in a sun space, see Table 7.14.

7.15 COMPARISON OF THE THREE MAIN PASSIVE HEATING SYSTEMS

Table 7.15 compares the three main passive solar heating systems by listing the main advantages and disadvantages of each approach.

7.16 GENERAL CONSIDERATIONS FOR PASSIVE SOLAR SYSTEMS

The following comments refer to all of the above passive systems.

Orientation

"Orientation is 80 percent of passive solar design," says Doug Balcomb, the foremost solar scientist. Usually solar glazing should be oriented to the south. In most cases, this orientation gives the best results for both winter heating and summer shading. The graph in Fig. 7.16a illustrates how the solar radiation transmitted through a vertical south window is maximal in the winter and minimal in the summer. This ideal situation is not true for any other orientation. Note how the curves for horizontal, east–west, and north windows indicate minimal heat collec-

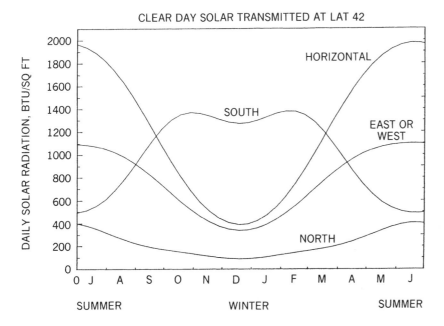

CLEAR DAY SOLAR TRANSMITTED AT LAT 42

Figure 7.16a Vertical south glazing is usually the best choice because it transmits the maximum solar radiation in the winter and the minimum in the summer. (From *Workbook for Workshop on Advanced Passive Solar Design,* by J. Douglas Balcomb and Robert Jones, © J. Douglas Balcomb, 1987.)

Even at 45 degrees to true south passive solar still works quite well.

There are special conditions, however, when true south is not best. Consider the following examples:

1. Schools, which need heating early in the morning and little heating in late afternoon or night, should be oriented about 30 degrees to the east.
2. It is sometimes desirable, as in schools, to use a combination of systems, each of which has a different orientation. For example, the solution shown in Fig. 7.16b will give quick heating by direct gain in the morning. In the afternoon, overheating is prevented by having the solar radiation charge the Trombe walls for nighttime use.
3. Areas with morning fog or cloudiness should be oriented west of south.
4. Buildings that are used mainly at night (e.g., some residences where no one is home during the day) should be oriented about 10 degrees west of south.
5. To avoid shading from neighboring buildings, trees, etc., reorient either to the east or west as needed.

tion in the winter and maximal in the summer. Who would want that? The south orientation is not just better—it is significantly better. In winter, south glazing collects about three times the solar radiation that east or west glazing collects, and in summer, south glazing collects only about one-third the radiation that east or west collects.

Since in the real world a due-south orientation is not always possible, it is useful to know that solar glazing will still work well if oriented up to 20 degrees east or west of true south.

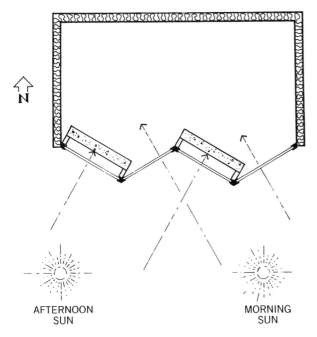

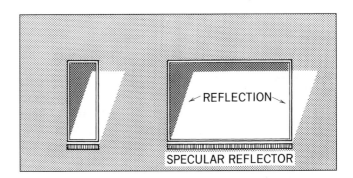

Figure 7.16c Specular reflectors are much less efficient on narrow than on wide windows. The diagram shows how on a south orientation the afternoon sun is reflected toward the east side of the window.

Figure 7.16b Plan view of a combined system of direct gain and Trombe walls to get quick morning heating and to prevent afternoon overheating.

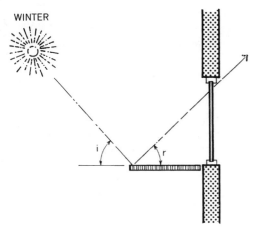

WINTER

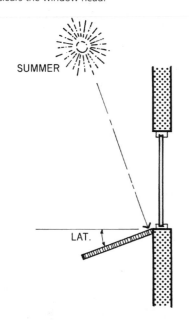

Figure 7.16d The length of a specular reflector is determined by the sun ray that just clears the window head.

SUMMER

LAT.

Figure 7.16e When summer sun is not desired, specular reflectors that are not removed should be tilted at an angle roughly equal to the latitude.

Plan

The plan should be designed to take advantage of the daily cycle of the sun (Fig. 15.5f). Waking up to the morning sun and eating breakfast on a table flooded with sunlight is both physiologically and psychologically satisfying. Later in the day, relaxing in a sun-filled living or family room on the south or southwest side of the building is a great amenity. There is growing scientific evidence that supporting our natural circadian rhythms can promote well-being.

Distribute south-facing glazing throughout the building proportionally to the heat loss of each space. Use clerestories to bring a southern exposure to northern rooms (Fig. 7.6e).

Slope of Glazing

Although the optimum collection slope for glazing in the United States is about 55 degrees, vertical glazing is almost always preferred. Vertical glazing is less expensive, safer, easier to shade on both the exterior and interior, and easier to fit with night insulation, and actually collects more heat when snow acts as a reflector.

Shading

Passive solar heating systems can become a liability during the overheated periods of the year if they are not properly shaded. Not only should direct sun rays be rejected but reflected and diffuse radiation should also be blocked. The problem of reflected heat is most acute in hot and dry areas, while that of diffuse radiation is most critical in humid regions. A full discussion on shading strategies is presented in Chapter 9.

Reflectors

Exterior **specular** (mirror-like) reflectors can increase the solar collection without some of the drawbacks of using larger glazing areas. Both winter heat loss and summer heat gain can be minimized by using reflectors rather than larger window sizes to increase the solar collection. Specular reflectors can also be very beneficial in daylighting designs, which are discussed in Chapter 13. However, specular reflectors are not inexpensive, and they are quite inefficient when used on narrow windows (Fig. 7.16c).

Since a specular surface reflects light so that the angle of incidence equals the angle of reflectance, the length of the reflector is determined by the ray of sunshine that just clears the head of the window (Fig. 7.16d). For the angle of incidence use the altitude angle of the sun on December 21 at 12 noon. These angles are determined from the sun-path diagrams found in Appendix A and B.

To prevent unwanted collection, one should remove the specular reflector or rotate it out of the way in the summer. If the reflector cannot be removed, then it should be at least rotated so that the summer sun is not reflected into the window. The angle of tilt should be about equal to the latitude (Fig. 7.16e).

Diffusing reflectors (white) are also beneficial, but they must be much larger than specular reflectors because only a small percentage of the sun is reflected in the direction of the window. Where it exists snow is ideal because of its large area, seasonal existance, and "low" cost. Although neither light-colored concrete nor gravel has a very high reflectance factor, they can still be beneficial if used in large areas.

Conservation

High-performance windows or night insulation over the solar glazing is recommended highly. In most parts of the country, these strategies can significantly improve the performance of passive heating systems in winter. Night insulation can also be used to reject the sun during summer days. Furthermore, it can offer privacy control and eliminate the "black hole" effect of bare glazing at night. Night insulation is most appropriate for direct-gain systems. It is less critical but still useful in sun spaces and Trombe walls, where high-performance windows may be the better choice.

Night insulation can take several forms. Drapes with thermal liners have many advantages but are limited in their thermal resistance. Rigid panels can have high R-value but are complicated to install and use.

High-efficiency windows are increasingly popular because they have no special needs in regard to installation and require no user involvement. However, they do not eliminate the need for shading, privacy, or the elimination of the "black-hole" effect at night. These strategies are discussed further in Chapter 15.

Even if movable insulation is used over windows, double glazing with low-e is recommended in all but the mildest climates, such as those found along the coast of Southern California or southern Florida. As with night insulation, efficient glazing is most appropriate for direct-gain systems.

7.17 HEAT-STORAGE MATERIALS

The success of passive solar heating and some passive cooling systems depends largely on the proper use of heat-storage materials. When comparing various materials for storing heat in buildings, the architectural designer is mainly interested in the heat capacity in terms of the BTU per volume rather than the BTU per pound. Table 7.17A shows the large variation in **volumetric heat capacity** among materials.

Air, for example, is almost completely worthless as a heat-storage material because has so little mass. Insulation, too, can store only insignificant amounts of heat because it consists mostly of air. Water, on the other hand, is one of the best heat-storage materials, and steel is almost as good. Except for wood, it seems that the heavy materials are good and the light materials are bad for storing heat. Although wood has a high heat capacity because of its water content, it is not very suitable for heat storage because it has a low conductance of heat. This low conductance or high resistance prevents the center of a mass of wood from participating efficiently in the storage of heat. Thus, for a material to be a good heat-stor-

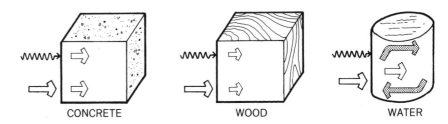

Figure 7.17a Because the conduction of heat into the interior of a material is critical for heat storage, wood is not good for storing heat, while water is excellent.

age medium, it must have both a high heat capacity and high conductance. For this reason water, steel, brick, and concrete (stone) are some of the best choices.

Water is an excellent heat-storage material not only because it has the highest heat capacity of any material, but also because it has a very high heat-absorption rate. In water natural convection currents as well as conduction help to move the heat to the interior of the mass (Fig. 7.17a). Because of the somewhat slow conductance of heat in concrete, brick, stone, etc. the thermally effective thickness of these solids is limited. Although concrete, stone, brick, and adobe are not as efficient as water when it comes to storing heat, they have a number of advantages. They don't leak or freeze, and they usually also serve as the structure. Note that Tables 7.7B and 7.10, which are used to estimate the required thermal mass, give the optimum thickness of the mass. When more

heat storage is required, the surface area and not the thickness should be increased.

There are potentially even more efficient materials for storing heat. These are called **phase-change materials** (PCM). They store the energy in the form of latent heat, while the previously mentioned materials store the energy as sensible heat. Since for passive heating the phase change must occur near room temperature, the salt hydrates (calcium chloride hexahy-

TABLE 7.17A HEAT CAPACITY OF MATERIALS BY VOLUME

Material	Heat Capacity per Volume (BTU/Ft²°F)
Water	62.4
Steel	59
Wood	26
Brick	25
Concrete (stone)	22
Foam insulation	1
Air	0.02

TABLE 7.17B COMPARISON OF VARIOUS HEAT-STORAGE MATERIALS

Material	Advantages	Disadvantages
Water	Quite compact Free	A storage container is required and can be expensive Leakage is possible
Concrete (stone)	Very stable Can also serve as wall, floor, etc.	Expensive to buy and install because of weight
Phase-change material (PCM)	Most compact Can fit into ordinary wood-frame construction	Most expensive Long-term reliability is not yet proven

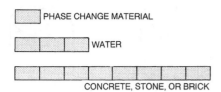

Figure 7.17b Relative volumes required for equal heat storage.

drate, Glauber's salt, etc.) are the most promising PCM materials.

Fig. 7.17b shows that to store equal amounts of heat, water requires only one-third the volume of concrete. Although phase-change materials are even more efficient when it comes to storing heat, they are not yet a practical alternative. See Table 7.17B for a comparison off the various heat-storage mediums.

7.18 OTHER PASSIVE HEATING SYSTEMS

Convective-Loop System (Thermosiphon)

In Fig. 7.18a, we see the basic elements of a convective-loop system. The collector generates a hot fluid (air or water) that rises to the storage area. Meanwhile, the cooler fluid sinks from storage and flows into the collector. This flow by natural convection is also called **thermosiphoning.** At night, the convection currents would reverse if the storage container were not located higher than the collector. The key to success in this system, therefore, is to place the thermal storage at a higher elevation than the collector. Unfortunately, this is usually difficult to do because of the weight of the water or rocks that are the typical storage mediums. Placing such a mass in an elevated position can be quite a problem unless you are lucky enough to have a building site that slopes down steeply to the south.

The Paul Davis house is located on a steep slope so that a hot-air collector can heat a rock bed under the house (Fig. 7.18b). At night, the heat from the rock bed is allowed to flow into the house, while cool air from the house returns to the rock bed. This is also a convective loop and is controlled by a damper. The house is heated during the day by direct gain.

Although strictly speaking this is a passive system, it has more in common with active than with passive solar systems. It is not an integrated approach since both collector and rock storage have only one function. This raises the cost significantly and, therefore, it is not a very popular passive system. A further discussion of related collector and heat storage systems can be found in Chapter 8, under active solar systems.

Roof Ponds

This concept is similar to the Trombe wall except that in this case we have a thermal-storage roof (Fig. 7.18c). In this roof-pond system, water is stored in black plastic bags on a metal deck roof, and during a winter day the sun heats the water bags (Fig. 7.18d). The heat is quickly conducted down and radiated from the ceiling into the living space. At night movable insulation covers the water to keep the heat from being lost to the night sky (Fig. 7.18e).

In theory, the roof pond is an excellent system because it not only heats passively in winter but can also give effective passive cooling in the summer. During the overheated part of the year, the insulation covers the house in the day and is removed at night. This passive cooling strategy is explained further in Section 10.10.

Unfortunately, this concept has some serious practical problems. The main difficulty is that no one seems to have been able to develop a workable, movable insulation system for the roof. Poor seals along the edge of the movable insulation result in major heat leaks. Another problem is the weight of the water. In the United States, lightweight construction is the norm, and heavy roofs cost significantly more. Water leakage is also a concern. Because the idea has potential for high efficiency and thermal comfort, it is worthwhile to investigate how the practical problems might be overcome.

There is one additional problem. Because of the cosine law, flat roofs receive less solar radiation than sloped or vertical surfaces in the winter. The higher the latitude, the worse

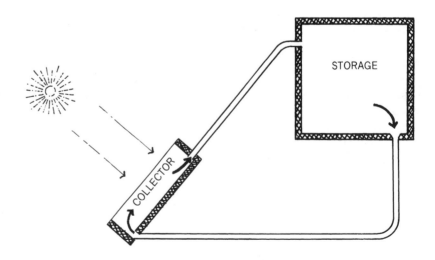

Figure 7.18a The convective loop (thermosiphon) system requires the storage to be above the collector.

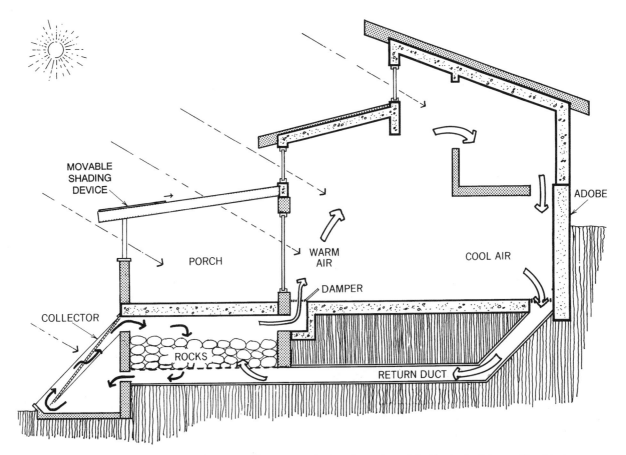

Figure 7.18b A convective loop heats the rock bed in the Davis house, New Mexico, designed by Steve Baer.

Figure 7.18c The Harold Hay house in Atascadero, CA, 1967, utilizes the roof pond concept. [From *Solar Dwellings Designs Concepts* by A.I.A. Research Corporation, U.S. G.P.O., 1976. (HUD-PDR-1564).]

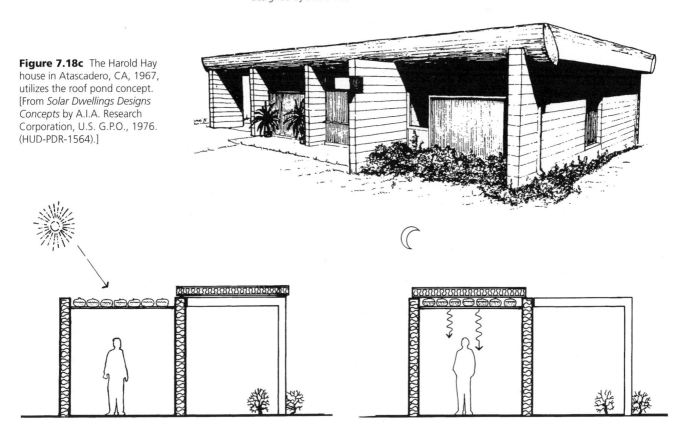

Figure 7.18d During the winter day, the black plastic bags of water are exposed to the sun, in the roof-pond system.

Figure 7.18e During the winter night, a rigid insulation panel is slid over the water.

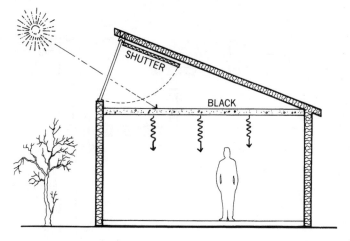

Figure 7.18f The roof radiation trap system developed by Givoni in Israel.

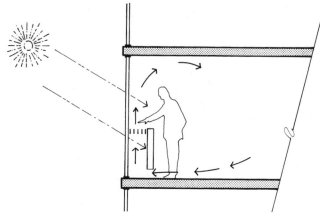

Figure 7.18g A lightweight collecting wall can supply additional daytime heating by natural convention and radiation without the introduction of excessive light.

this problem becomes. Therefore, the concept as shown above is only for the southern parts of the United States. For the northern latitudes, Harold Hay proposed a different solution, which is similar to the **Roof Radiation Trap** developed by B. Givoni (Fig. 7.18f).

Roof Radiation Trap

To overcome some of the serious difficulties with roof ponds, B. Givoni developed the Roof Radiation Trap system. As shown in Fig. 7.18f, the glazing on the roof is tilted to maximize winter collection at any latitude (tilt = latitude + 15 degrees). After passing through the glazing, the solar radiation is absorbed by the black-painted concrete ceiling slab. The building is, thus, heated by radiation from the ceiling. The sloped roof is well insulated, and a movable shutter can reduce heat loss through the glass at night. Since this shutter is on the inside of the glazing, the edge seals are not critical. This system can also be adapted for summer passive cooling and is described further in Section 10.11.

Lightweight Collecting Walls

The lightweight collecting wall shown in Fig. 7.18g is useful in very cold climates for those types of buildings in which extra heating is required during the day and where little heat is required at night. This is typical of many schools, office buildings, and factories. Since there is no special storage mass in this system, all of the solar radiation falling on the collector is used to heat the interior air while the sun is shining.

The Modern Horticultural Greenhouse

The conventional horticultural greenhouse, as seen in Fig. 7.18h, has several serious faults. It experiences such severe heat loss during winter nights that much fuel must be burnt to keep the plants from freezing. Also, only a part of the glazing sees the sun during a winter day, while the rest is losing heat. Even during the summer there is a problem. Since large areas of glazing face the summer sun, overheating is a major problem. It is common to have large fans and sometimes cooling systems running continuously during a

summer day to prevent the plants from dying.

This approach to horticultural greenhouse design can be traced to its historical roots in such countries as England and Holland. Both these countries have very mild and cloudy climates. Thus, neither heat loss in winter nor overheating from hot sunny days is much of a problem. In addition, large glazing areas are required in cloudy climates to supply the plants with sufficient light.

Most of the United States, however, has much harsher climates than England or Holland, and so we need a different greenhouse design. In Fig. 7.18i, we see an example of a modern horticultural greenhouse appropriate for most of the climate zones in the United States. The north wall and roof are well insulated. The south wall and roof are constructed with double glazing. Thermal mass of the earth and flooring are supplemented with drums of water. The summer heat load is limited not only by the reduced glazing area but also by shading from an overhang. Further cooling comes from natural ventilation, which is permitted by the large movable vents at the highest and lowest points in the greenhouse.

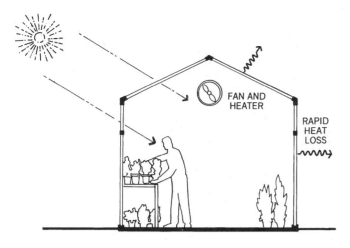

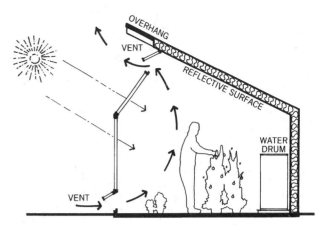

Figure 7.18h A conventional horticultural greenhouse is a heavy energy user because it typically utilizes a fan to prevent overheating, and a heater to prevent freezing.

Figure 7.18i A modern passive horticultural greenhouse utilizes shading and venting to prevent overheating. Insulation, direct gain, and mass are used to prevent freezing.

To minimize the tendency of plants growing toward the source of light (south in this case), the north ceiling is covered with a specular surface so that the plants also receive some light on their northern side.

7.19 SUMMARY

Many factors determine the best choice of a passive heating system.

Climate, building type, user preference, and cost are some of the major considerations. Often it is best to use a combination of systems to satisfy the demands of a particular problem. Other times a variation of only one system will prove best. It is also likely that some good ideas have not yet been developed. It is certain, however, that most buildings that require heating can benefit from some type of passive heating system.

Many of the buildings included as case studies in Chapter 17 use one or more passive solar techniques. See especially the Colorado Mountain College and Hood College.

This chapter started with a look at historical examples of passive buildings. One such example was the New England saltbox type. The chapter ends with a building in western Maryland that alludes to this historical prototype (Fig. 7.19).

Figure 7.19 This country home, designed by the author for western Maryland, expresses its roots in the "saltbox" of New England and "pent" roof of Pennsylvania. Most glazing faces south, and the roof monitor is used for summer cooling.

KEY IDEAS OF CHAPTER 7

1. Maximize south-facing glazing because south windows:
   a) Collect much more sun during the day than they lose at night
   b) Collect much more sun in the winter than in the summer.

2. Passive solar heating consists essentially of south-facing glazing and thermal mass.

3. In a direct-gain system, the more mass receiving direct or reflected solar radiation, the better. The mass should have a large surface area rather than depth.

4. The three main passive solar systems are:
   a. Direct gain
   b. Trombe wall
   c. Sun space

5. Direct-gain systems are the simplest and most economical but result in high light levels.

6. Trombe walls supply heat without light.

7. Sun spaces are delightful living spaces and solar heaters.

8. To keep a sun space from being an energy liability, one must design it to be a semi-temperate living space (i.e., neither mechanically heated or cooled).

9. Passive systems must be shaded in summer to prevent an asset from becoming a liability.

10. Passive solar is 80 percent orientation.

Resources

FURTHER READING
(See Bibliography in back of book for full citations. This list includes valuable out-of-print books.)

Anderson, B. *Solar Energy: Fundamentals in Building Design.*

Anderson, B., and M. Wells. *Passive Solar Energy.*

ASHRAE. *Passive Solar Heating Analysis: A Design Manual.*

Balcomb, J. D. *Passive Solar Buildings.*

Brown, G. Z., and M. DeKay. *Sun, Wind, and Light: Architectural Design Strategies.*

Buckley, S. *Sun Up to Sun Down.*

Clegg, P., and D. Watkins. *Sunspaces: New Vistas for Living and Growing.*

Cook, J. *Award Winning Passive Solar House Designs.*

Daniels, K. *The Technology of Ecological Building.*

Givoni, B. *Climate Considerations in Building and Urban Design.*

Heinz, T. A. "Wright's Jacobs II House."

Hestnes, A. G., R. Hastings, and B. Saxhof, eds. *Solar Energy Houses: Strategies, Technologies, Examples.*

Jones, R. W., and R. D. McFarland. *The Sunspace Primer: A Guide to Passive Solar Heating.*

Kachadorian, J. *The Passive Solar House: Using Solar Design to Heat and Cool Your Home.*

Kohlmaier, G., and B. von Sartory. "Das Glashaus: Ein Bautypus des 19. Jahrhunderts."

Los Alamos National Lab. "Passive Solar Heating Analysis: A Design Manual."

Mazria, E. *The Passive Solar Energy Book.*

Miller, B. *Buildings for a Sustainable America Case Studies.*

Panchyk, K. *Solar Interior: Energy Efficient Spaces Designed for Comfort.*

Potts, M. *The New Independent Home: People and Houses that Harvest the Sun.*

Singh, M. *The Timeless Energy of the Sun for Life and Peace with Nature.* Preface by Federico Mayor.

Boonstra, C., ed. *Solar Energy in Building Renovation.*

Stein, B., and J. Reynolds. *Mechanical and Electrical Equipment for Buildings.*

Steven Winter Associates. *The Passive Solar Design and Construction Handbook.*

Watson, D. *Designing and Building a Solar House: Your Place in the Sun.*

PERIODICALS
Solar Today

ORGANIZATIONS
(See Appendix J at end of book for full citations.)

American Solar Energy Society
http://www.ases.org
National group for renewable energy education.

Caddet Center for Renewable Energy
www.caddet.co.uk
Summaries of renewable-energy projects around the world.

Center for Renewable Energy and Sustainable Technology (CREST)
www.crest.org
Comprehensive educational source for renewables.

Florida Solar Energy Center (FSEC)

National Center for Appropriate Technology (NCAT)

National Renewable Energy Laboratory (NREL)
www.nrel.gov
Good source for detailed information on renewables.

Sustainable Buildings Industry Council (SBIC)
www.sbic.org

PHOTOVOLTAICS AND ACTIVE SOLAR

"The Earth belongs to each generation during its course, fully and in its own right; no generation can contract debts greater than may be paid during the course of its own existence."

Thomas Jefferson

Please note that figure numbers are keyed to sections. Gaps in figure numbering result from sections without figures.

8.1 INTRODUCTION

Although from a distance, photovoltaic and active solar collectors appear quite similar, they produce distinctly different forms of energy. **Photovoltaic (PV) panels** produce the very high-grade energy of electricity while **active solar panels** produce the low-grade thermal energy of low-temperature heat. Electricity is called a **high-grade energy source** because it can be used to do all kinds of work (i.e., generate light, move elevators, etc.), while low-temperature heat can do little more than heat water or a building.

These two systems are discussed in the same chapter because they are often mounted side-by-side on a building, they look similar, and their needs for orientation and tilt are similar. Photovoltaics are discussed first.

8.2 THE ALMOST-IDEAL ENERGY SOURCE

As was mentioned in Chapter 2, the conventional energy sources all have serious drawbacks. What, then, would be the almost-ideal energy source? What are the characteristics of the ideal energy source?

Characteristics of the Ideal Energy Source

1. Sustainable (renewable).
2. Non-polluting.
3. Not dangerous to people or the planet.
4. High-grade energy useful for any purpose.
5. Silent.
6. Supplies power where it is needed (no need to transport energy).
7. Most available at peak demand time, which is frequently a hot, sunny summer day.
8. Has additional benefit of creating the building envelope (i.e., displaces conventional building materials).
9. High reliability.
10. No moving parts.
11. No maintenance required.
12. Modular (can come in any size required).
13. Low operating cost.
14. Low initial cost.
15. Supplies energy all the time.

Photovoltaics (PV) is the only energy source that comes close to meeting these characteristics. PV meets or exceeds all except the last two characteristics: "low initial costs" and "supplies energy all the time." The fact that PV doesn't generate electricity at night is not a big problem because batteries or the power grid can be used to store the electricity. During the day, clouds are not as severe a problem as one might guess because PV can use diffused light quite well.

Cloudiness and PV-Output Rules of Thumb (from *Tapping into the Sun*)

1. About 80 percent output on partly cloudy days.
2. About 50 percent output on hazy/humid days.
3. About 30 percent output on extremely overcast days.

PV is still more expensive than conventional electricity in most cases. In places where conventional electricity is either very expensive or unavailable, as on islands or areas remote from the power grid, PV is the economical choice.

As more PV is used, its cost will continue to decline through rapid technological improvements and the economics of scale (Fig. 8.2). PV has the potential of becoming very inexpensive because of its inherent simplicity and the very small amount of material that it requires. At the same time that PV becomes less expensive, it will become more attractive as more people recognize the need for a sustainable, clean energy source that does not produce any greenhouse gases.

8.3 HISTORY OF PHOTO-VOLTAICS (PV)

Becquerel discovered the photoelectric effect in 1839, and Bell Laboratories developed the first crystalline silicon cell in 1954. Little practical progress was made until 1958, when the space program needed an extremely light and reliable source of electricity for its satellites. Although

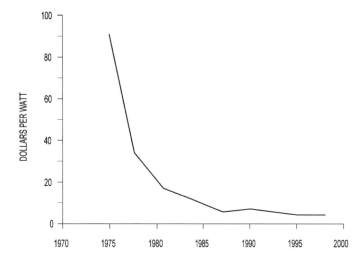

Figure 8.2 The dramatic decline in the cost of photovoltaic (PV) systems (source: Maycock, 1998).

Figure 8.3a The Lord House in Maine is a grid-connected building with half of the roof generating electricity and the other half, hot water. During the day, electricity is sold to the power company, and at night it is bought back. The house is also passively heated and superinsulated. (Courtesy Solar Design Associates, Inc.)

Figure 8.3b The Intercultural Center at Georgetown University is a fine example of building integrated photovoltaics (BIPV). The south-facing roof is covered with a 300-kilowatt PV array that provides about 50 percent of the building's electrical needs. (Courtesy and © BP Solarex.)

Figure 8.3c Several power companies have set up PV farms to harvest the sun. This energy is most available on hot, sunny summer days, precisely when it is needed to supply electricity for air conditioning. Shown is the 2-megawatt, single-axis-tracking system of the Sacramento Municipal Utility District at Rancho Seco, California. The nuclear-power plant, seen in the background, is shut down. (Courtesy Sacramento Municipal Utility District.)

photovoltaics (PV) proved very reliable, the cost was initially too high for earthbound applications. Continuing research brought the price down to make PV useful for powering telephone relay stations, buoys, railroad signals, and other electrical installations remote from the grid.

Currently, PV is inexpensive enough to power stand-alone houses and to supply power to the grid during times of peak demand. PV is also widely used in developing countries where villages are often far from the power grid. In the developed world, many individuals, communities, and countries use PV to address the issues of global warming, sustainability, and energy independence (see Figs. 8.3a, 8.3b, & 8.3c).

8.4 THE PHOTOVOLTAIC CELL

Photovoltaic cells, sometimes also known as **solar cells,** are made of wafers that convert light directly into electricity. Like transistors, the wafers are usually made of thin layers of a semiconductor material like silicon with small amounts of certain impurities added to create an excess of electrons in one layer and a lack of electrons in the other layer. Photons of light create free electrons in one layer, and a conducting strip enables the electrons to flow through an external circuit to reach the layer that lacks electrons (Fig. 8.4a).

Single-crystal silicon cells are the most efficient but also the most expensive. To reduce the cost, polycrystalline and thin-film PV cells have been developed. Thin-film PV is made of amorphous silicon, copper indium diselenide, or cadmium telluride. Although these cells convert sunlight into electricity at half the efficiency (about 8 percent) of the single-crystal silicon cells (about 15 percent), their lower cost more than compensates whenever collector area is not limited. Continuing research is raising the efficiency of all cells, with 33 percent the highest reached in a laboratory so far.

Because PV cells are small and fragile, and produce only a small amount of power, they are encased to form modules. Modules come in many sizes, but to make handling easy, they are rarely over 3 feet wide and 5 feet long. Some modules are combined to form panels, which are further combined to form an array (Fig. 8.4b).

When two cells or modules are connected in series, the voltage doubles; when they are connected in parallel, their current doubles. Thus, a sufficient number of cells in the right combination can produce any combination of voltage and current. Since electrical power is the product of voltage and current, any amount of power can be produced with enough cells (see Sidebox 8.4).

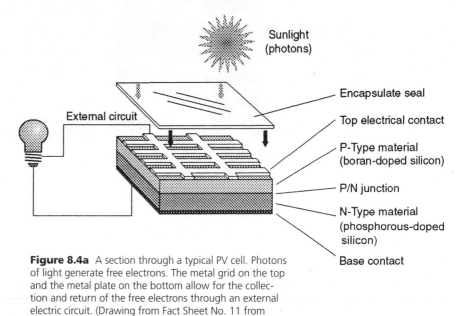

Figure 8.4a A section through a typical PV cell. Photons of light generate free electrons. The metal grid on the top and the metal plate on the bottom allow for the collection and return of the free electrons through an external electric circuit. (Drawing from Fact Sheet No. 11 from State Energy Conservation Office of Texas.)

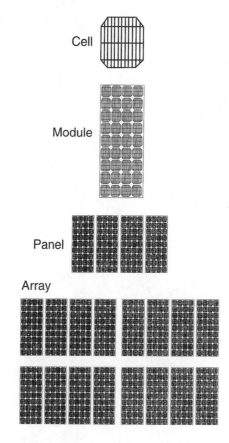

Figure 8.4b Cells combine to form modules, modules form panels, and panels combine to form an array. (From Federal Technology Alert: Photovoltaics.)

SIDEBOX 8.4

The electrical power produced is equal to the product of the current and voltage.
$$W = I \cdot V$$
where:
W = power in watts
I = current in Amperes
V = voltage in volts

The electrical energy produced is equal to the power times the amount of time that the power is produced.
$$E = W \cdot H$$
where
E = Electrical energy in watt-hours
W = Power in watts
H = time in hours

Since the number of watt-hours is usually a very large number, electrical energy is usually measured in kilo-watt-hours.
$$E = K \cdot W \cdot H$$
where
1 KW = 1,000 W

8.5 TYPES OF PHOTOVOLTAIC SYSTEMS

Two basic types of photovoltaic (PV) systems exist for buildings: stand-alone and grid-connected. When connection to a power grid is not possible or not wanted, a stand-alone system is required. In such cases, batteries are needed to supply power at nighttime, on overcast days, and when peak power is required. The PV arrays are sized to handle both normal daytime loads and battery charging. The batteries add significantly to the cost and maintenance of the systems.

When PV is used where the power grid exists, batteries are not needed. During a sunny day, excess PV power is sold to the utility, and at night, power is drawn from the grid. In effect, the grid acts as a giant storage battery. This can be an advantage to both the PV owners and many power companies because the greatest demand on their grid is often on hot, sunny summer days, while at night the power companies have excess capacity that they are eager to sell.

Right now most people pay the average cost over a day and year for electricity, but with the deregulation of the power industry time-of-day pricing is expected to become common. Then the price at peak hours will be so high that PV will be able to compete. Grid connection with PV systems is already possible because the Public Utilities Regulatory Policy Act of 1978 states that utilities are required to buy power from owners of PV systems. **Net-metering laws** in many states require the power companies to buy PV power from individuals at the same rate they charge.

In a grid-connected system, an inverter is required to change the direct current from the PV array to the alternating current (AC) at the correct voltage of the grid (Fig. 8.5a). All appliances in the building are then ordinary AC, 120v (or whatever the baseline voltage is). Note again that batteries are not needed, which is a

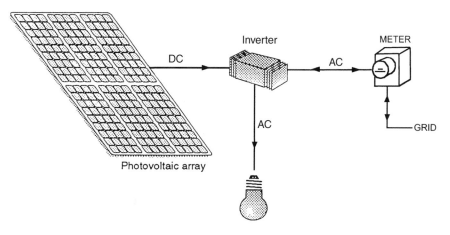

Figure 8.5a A typical grid-connected PV system. On a sunny day, the excess power will flow into the grid and the meter will run backwards.

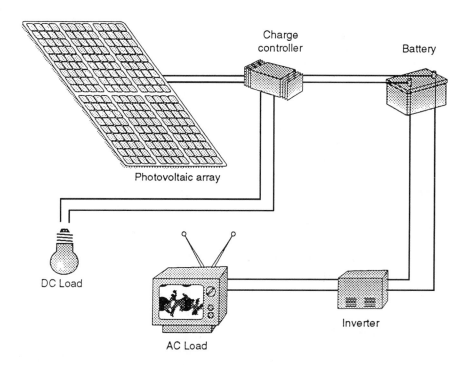

Figure 8.5b A stand-alone system needs batteries to store the electricity for nighttime and an inverter to change DC to AC.

significant saving of both money and maintenance.

In stand-alone systems, the excess electricity produced during the day is stored in batteries for nighttime and dark, cloudy days (Fig. 8.5b). Since inverters are expensive and can consume as much as 20 percent of the power the PV produces, buildings must use as many low-voltage direct-current (DC) appliances in place of standard AC appliances as possible. A small, less

expensive inverter can then supply the AC appliances. Also, since PV cells and batteries are expensive, backup power is preferable over an extra-large PV system for lengthy overcast periods or storms. A hybrid system using wind power is often the ideal complement to PV because not only can the wind blow at night, but also it is usually extra windy during bad weather. Furthermore, in the winter, when there is less solar energy to harvest, it is usually

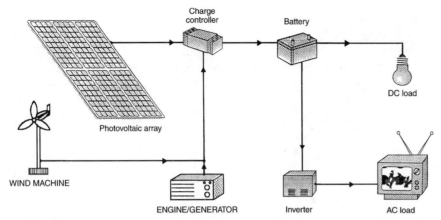

Figure 8.5c Hybrid systems give the most reliable power at the least cost for stand-alone installations.

windier than it is in the summer (Fig. 8.5c). Not all regions, however, are suitable for wind power. See the discussion of windpower in Chapter 2.

When a reliable wind source is not available, an engine generator unit should be used to back up the PV system. It is reasonable to ask: why bother with the PV system at all? Why not just use the engine generator? Many remote installations that previously used only an engine generator are switching to the PV hybrid system because it is more reliable, requires much less maintenance, makes no noise, and needs little fuel, which is always a burden to bring to a remote site. In a well-designed hybrid system, the engine generator will operate only a few times a year, during unusually long cloudy periods.

8.6 BALANCE OF SYSTEM EQUIP-MENT

A photovoltaic (PV) installation consists of the PV array and the **balance of systems (BOS) equipment.** Typically, the BOS equipment comprises a charge controller, an inverter (for alternating current), switches, fuses, wires, etc. For stand-alone systems, the BOS equipment also includes the batteries. A small wall panel is usually sufficient for most of the BOS equipment except for the batteries.

In stand-alone systems, the excess electricity produced when the sun is shining is stored in batteries. Although several different types of batteries are available, many systems use a deep-cycle version of the lead-acid batteries found in cars. For safety and long life, batteries should be stored in well-ventilated, cool, and dry chambers. Batteries need to be vented outdoors because of the hydrogen that is produced during charging. Consequently, they should be stored either indoors next to an outdoor wall or in an insulated shed against the outside wall. A separate structure will work only if the batteries can be kept from getting too warm or too cold; 77°F is the optimum.

In the near future, an alternative to batteries might be the use of non-polluting **fuel cells.** During the day, the sun will provide energy for the production of hydrogen via the electrolysis of water. At night, the fuel cell will generate electricity, and with hydrogen as the fuel the only byproduct will be water.

Commercial fuel cells are just coming on the market as this book is being written. Many experts are convinced that hydrogen is the fuel of the future, but, of course, hydrogen is not a source of energy: it must be produced, and PV is one sustainable way to produce it.

8.7 BUILDING-INTEGRATED PHOTOVOLTAICS (BIPV)

Photovoltaic (PV) systems can power buildings in a variety of ways, ranging from remote PV farms to being part of the building fabric. Some utilities are augmenting their electrical capacity through large centralized PV farms (Fig. 8.3c), while other power companies are setting up smaller PV fields closer to the electrical users. PV arrays can also be set up on the land adjacent to the building (Fig. 8.7a), placed on the roof (Fig. 8.7b), or integrated into the building envelope (Fig. 8.7c), in which case the phrase **Building-Integrated Photovoltaics (BIPV)** is used. Such BIPV systems can replace the roofing, siding, curtain wall, glazing, or special elements such as overhangs and canopies.

There are a number of important benefits to using BIPV:

1. The elimination of the cost of transporting electricity to the building, which can be more than 50 percent of the cost of the electricity.
2. The elimination of the energy waste of transporting electricity, which can be as much as 25 percent (Fig. 8.7d).
3. The avoidance of using valuable open space to mount the PV array.
4. The avoidance of part of the cost of the building envelope by using PV modules instead.
5. The avoidance of a support structure since the building structure exists anyway.
6. The aesthetic potential of using a new type of cladding material.
7. The benefit of generating all or at least a significant portion of the required electricity in an environmentally friendly way.

Although at this time PV modules are still expensive, they are no more expensive than some premium architectural cladding. PV is not more expensive, for example, than granite facing. Thus, one can save either all

Figure 8.7a An on-site PV array requires a support structure and some open land for solar access. A sun-tracking system is shown. (Courtesy Ecological Design Institute ©.)

Figure 8.7b On existing sloped roofs, the PV is usually placed on the roofing. (Courtesy Sacramento Municipal Utility District.)

Figure 8.7c The BIPV standing seam roofing on this Maryland townhouse is almost indistinguishable from conventional, standard seam roofing on the building to the left. (Courtesy United Solar Systems Corp., UNI-solar® Photovoltaic Roofing.)

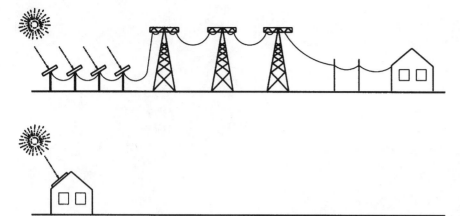

Figure 8.7d Besides displacing part of a building's weathering skin, BIPV also eliminates the electrical-transmission losses, which can be as high as 25 percent with the existing power grid.

can be used as glazing. Cells can be round, semicircular, octagonal, square, or rectangular, and custom modules and panels can be produced for large projects.

There are four major parts of the building envelope into which PV can be integrated: walls, roofs, glazing, and ancillary structures such as overhangs, entrance canopies, and shading structures for parking areas. Each of the four parts will be discussed in some detail after a short discussion of PV characteristics important in building integration.

Since the PV modules are fairly dark in color and *must not* be shaded, much heat is produced, which not only degrades the performance of the PV cells but also may heat the building. Thus, cooling the cells is an important concern and will be discussed later. Also important are the orientation and tilt of the PV array which will be discussed next.

or part of the cost of the PV array by eliminating the cost of the building element being replaced.

PV modules come in many sizes, finishes, and colors. Polycrystalline silicon cells are a beautiful blue color, with the crystalline structure forming an attractive pattern (Fig. 8.7e). Most thin-film PV modules are dark brown in color, and some are flexible so that they can be used on curved surfaces (Fig 8.7f). Cells are now being developed that are gold, violet, or green in color. A variety of semitransparent PV modules

Figure 8.7e PV modules can be very attractive, as shown in this blue pattern of polycrystalline silicon cells.

Figure 8.7f Because thin-film modules are flexible, they are easily integrated into curved architecture.

8.8 ORIENTATION AND TILT

The maximum collection of solar radiation occurs when the collector is perpendicular to the direct beam radiation. Since the sun moves both daily and annually, only a two-axis tracking collector can maximize the collection over a year. Tracking collectors, however, only excel in dry climates that have mostly direct beam radiation, and even there, 10 to 20 percent of the solar radiation is diffuse. In most sunny and humid climates only about one-half of the solar radiation is direct, while in cloudy climates 80 percent or more of the radiation is diffuse. When PV cells were very expensive it made good sense to invest in a tracking device in most climates (Fig. 8.7a). Now tracking collectors should be considered only for sunny and dry climates, and even there the advantages of building-integration may be greater than the benefits of tracking.

Even with building integration, orientation and tilt should still be considered. The best tilt for a PV array is primarily a function of the time of year that maximum power is required. Hot climates need the most electricity in the summer for air conditioning, while cold climates need maximum electricity in the winter for lighting- and heating-system pumps and blowers. Use the design guidelines in Section 8.16 for choosing the optimum tilt of south-facing PV.

Usually, the optimal orientation is due south, but there is little loss up to 20° east or 20° west of south. The daily-load profile, however, can influence the orientation. For example, elementary schools that start early in the morning and end in the middle of the afternoon should have the array facing about 30° east of south or, in a climate with a morning fog, a southwest orientation would be appropriate.

As PV modules get less expensive, the optimal orientation and tilt will become less important. Already, it makes sense to cover the roof and

sometimes the south facade. East and west facades are not far behind because they can produce up to 60 percent of the optimal south output.

8.9 ROOFS CLAD WITH PHOTOVOLTAICS

Ideally, the roof should have the slope (tilt) described above with the PV replacing the roofing (Fig. 8.7c). On flat roofs, a support structure can offer the ideal tilt, but, of course, the

benefits of building integration are lost (Fig. 8.9a). A sawtooth clerestory is superior to a flat roof because the north-facing slope can be glazed for daylighting, while the south slope supports the building-integrated photovoltaics (BIPV) (Fig. 8.9b). The south-facing PV can also be of the semitransparent kind, so that the clerestories can both collect south daylight and generate electricity. Also, sloped roofs are easier to waterproof. A flat slope is not desirable for the PV because it is too far from the

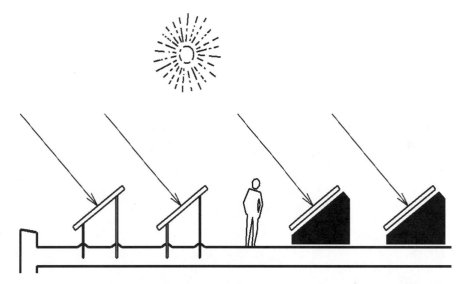

Figure 8.9a Support structures can provide the appropriate tilt for the PV on flat roofs, but they do not provide the benefits of being building-integrated. The use of heavy concrete or gravel-filled support structures makes it unnecessary to penetrate the roof membrane.

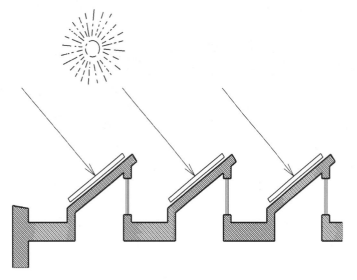

Figure 8.9b Sawtooth clerestories can provide both daylighting and the proper tilt for BIPV.

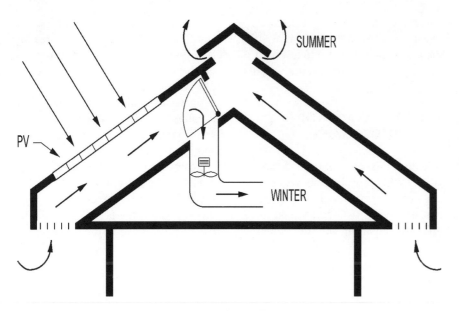

Figure 8.9c Ventilate the underside of the PV in summer to keep the cells from overheating. In winter, this warm air can be used to heat the building.

Figure 8.9d PV shingles are designed to "blend" in with standard shingles.(Courtesy United States Solar Systems Corp., manufacturing solar roofing under the brand name of UNI-solar®.)

8.10 FACADES CLAD WITH PHOTOVOLTAICS

Not only south but also east and west facades can be clad with PV and still generate a significant amount of electricity. If mullions are used on the exterior they should be as shallow as possible to prevent shading of the PV (Fig. 8.10a). Only use the PV on the upper facade if the lower areas are shaded, as is common in dense urban areas or sites with many trees (Fig. 8.10b). As with the roof, it is best to leave an airspace behind the PV to cool the panels, and in winter this warm air can be collected to heat the building (Fig. 8.10c).

8.11 GLAZING AND PHOTOVOLTAICS

There are two types of PV glazing systems. One is semitransparent, much like tinted glazing (Fig. 8.11a). The other consists of opaque cells mounted on clear glazing, with the spacing of the cells determining the ratio of clear to opaque. This is much like ordinary glazing painted with frets (Fig. 8.11b).

With either type of PV glazing, it is possible to control the amount of light being transmitted. Of course, the more light that is transmitted the lower the power production. Even with very transparent PV systems, large amounts of power can be produced because of the large amount of glazing in most modern buildings. The PV glazing is especially appropriate in clerestories or skylights since these are not designed for view. The PV glazing can also be incorporated into high-performance windows to gain good insulating qualities.

Under development is a transparent PV glazing that utilizes only the solar infrared radiation to generate electricity. Thus, both cool daylight and electricity are produced by the glazing—just what most office buildings in the South need.

ideal tilt and because of dirt and snow accumulation. A steep angle will allow the snow to slide off the smooth PV array.

If the PV is integrated into the roof, the underside needs to be ventilated to cool it. In winter this waste heat can be collected to heat the building (Fig. 8.9c). For more traditional roofs, PV shingles, slates, and tiles are available. They are used in the conventional manner except that an electrical connection must be made with every unit (Fig. 8.9d).

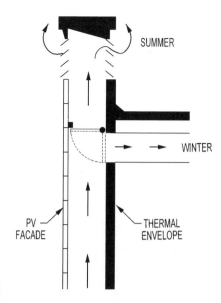

Figure 8.10c This double-wall design enables the air behind the PV to be vented in the summer to cool both the PV and the building. In winter, the hot air can be used to heat the north side of the building.

Figure 8.10a The APS Building in Fairfield, California, uses thin-film PV integrated into the curtain wall, skylight, and awning. (Courtesy Kiss & Cathcart, architects. © Richard Barnes, photographer.)

Figure 8.10b The New York City skyscraper, 4 Times Square, uses BIPV spandrel panels only on floors 35 to 48, because the lower floors are shaded too much. (Courtesy Kiss & Cathcart, architects.)

Figure 8.11a Semitransparent PV glazing is used as skylight glazing in the APS factory in California. (Courtesy Kiss & Cathcart, architects. © Richard Barnes, photographer.)

Figure 8.11b Opaque PV cells are mounted on clear glass. The spacing between cells determines the degree of shading. As seen from below, the entranceway canopy of the Aquatic Center at Georgia Institute of Technology is roofed with a 4.5-kilowatt array of Solarex PowerWall™ MSX-240/AC laminates. With their integrated inverters, these laminates provide grid-synchronized AC power. They use an optional clear Tedlar® backing material that emphasizes the precision of solar cell placement and provides soft, natural lighting under the canopy. (Courtesy and © BP Solarex.)

8.12 PHOTOVOLTAIC SHADING DEVICES

Shading devices are a very good application for PV because they can be designed to be tilted at the optimum angle (Fig. 8.12a and b). Shading devices either can be opaque or can use PV glazing with a wide range of transparency).

PV can also be integrated into entrance canopies or free-standing shading structures (Fig. 8.12c). Although at present, the shading of automobiles in hot parking lots is usually too expensive, as a combination PV generator and shading structure, its cost becomes more reasonable. As electric cars become more common, these structures will serve as ideal charging stations (Fig. 8.12d).

8.13 PHOTOVOLTAICS: PART OF THE SECOND TIER

It is important to understand that electrical generation by PV is part of the second-tier of the three-tier approach to environmental design (Fig. 8.13). The first tier consists of utilizing efficient electrical appliances and lighting systems to minimize the electrical load. The second

Figure 8.12a These shading devices are tilted at the optimum collection angle on the Center for Environmental Sciences and Technology Management (CESTM), State University of New York (SUNY), Albany. (Courtesy Kawneer Company, Inc., © Gordon H. Schenck, Jr., 1996, photographer.)

Figure 8.12b This close-up of the CESTM buiding clearly shows the tilt of the PV shading devices. (Courtesy Kawneer Company, Inc., © Gordon H. Schenck, Jr., 1996, photographer.)

Figure 8.12c The barrel-vaulted entrance canopy to the Georgia Tech Natatorium (aquatic center) is made entirely of PV modules. See the underside of this canopy in Fig. 8.11b (Courtesy and © BP Solarex.)

Figure 8.12d Automobile shading structures covered with PV are also ideal electric car-charging stations. (Courtesy Sacramento Municipal Utility District.)

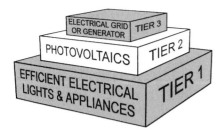

Figure 8.13 As always, the three-tier approach is the most logical and sustainable strategy for environmental design.

tier consists of using PV to generate clean sustainable electricity. The third tier, then, consists of utilizing a generator or an electrical grid to supply the small amount of electrical power and energy still required. This is not only the most sustainable approach, but also usually the most economical.

8.14 SIZING A PHOTOVOLTAIC SYSTEM

Grid-connected systems are sized differently from stand-alone systems, and their design will be discussed later. For stand-alone systems, sizing is critical because too large a system is a very expensive waste. As mentioned above, the three-tier approach to design should be utilized to minimize the electrical load. Furthermore, the two very large users of low-grade energy, which are space heating and domestic hot water, should not be supplied by the photovoltaics (PV). Passive solar is the first choice for space heating, and active solar, the best choice for domestic hot water. Backup energy for both space heating and domestic hot water could come from gas or wood.

There should be backup power from an engine-generator or a reliable source of wind power, otherwise the PV system would have to be significantly larger in order to store power for extended periods of bad weather. The resultant cost would be unnecessarily high.

Sizing of a Stand-Alone System

The following guidelines are for stand-alone systems with a backup power source occasionally used. Use the rules of thumb presented in Table 8.14 for a first approximation and the calculations in the next section for a more precise estimate.

It is critical to consider winter conditions when designing a PV system in the North because the lights are on for the most hours, and the PV array sees the sun for the fewest hours. In the South, winter design is less critical because there are about two more hours of daylight. Because summer air conditioning is almost mandatory in the Southeast, peak electrical loads usually occur in the summer. In high elevation areas of the Southwest where buildings can be designed to use little if any air conditioning, winter conditions are usually critical.

Sizing a Grid-Connected PV System

There is no limit to the desired array area since the grid can use all the power produced at peak times in most areas of the United States. The cost of the array is the main determinant of size. Because the present cost of PV is high, only the orientations that can generate the most power should be used. As PV prices decline, more and more building surfaces will be clad in PV. In homes, the roof area might be all that is needed, but in commercial, institutional, and industrial buildings, the walls, roofs, and glazing will also be put to work generating electricity. Section 8.16 will discuss the efficiency of different orientations.

If the local power company shows little interest in promoting PV, one will have to size the grid-connected PV system as a stand-alone system but without batteries. This approach will minimize the need to buy power from the utility.

Deregulation of the electric-power industry will encourage time-of-day pricing, which will make electricity very expensive at peak times. Since much of the United States electrical demand peaks on hot, sunny summer days, most grid-connected PV systems should also peak at that time.

One likely possibility is that the power companies will rent roof space from the user in order to generate the power where it is needed. This will enable the utility to avoid the great expense of upgrading the distribution

TABLE 8.14 APPROXIMATE METHOD FOR SIZING STAND-ALONE SYSTEMS*

Building Type	Climate*	Array Size** (ft²)	Battery storage*** (ft²)
Small residence	Mild and sunny	50	10
	Cold and cloudy	100	20
	Hot and humid	100	
Average residence	Mild and sunny	100	20
	Cold and cloudy	500	100
	Hot and humid	500	100
Large residence	Mild and sunny	500	100
	Cold and cloudy	1,000	200
	Hot and humid	1,000	200

* Hot and humid climates need a large array because of the air-conditioner load. In cold and cloudy climates, the large array is due to the combined effect of lighting long winter nights and short cloudy days for generating power.

** The size of the array is very approximate in part because of the large variance in cell efficiency.

*** A space with this floor area will also contain the balance of system (BOS), which includes: controller, inverter, switches, and circuit breakers.

system. This is already the case with the Sacramento Municipal Utility District (SMUD) in California. About ninety electric utilities have formed the Utility Photovoltaics Group to promote and sponsor PV installations on or around buildings.

To prevent the future problem of collectors being added at unattractive odd angles, new buildings should be oriented so that their roofs face in a southerly direction close to the optimum tilt. *From now on, all buildings should use or anticipate PV cladding.*

8.15 FINDING PHOTOVOLTAIC ARRAY SIZE FOR A STAND-ALONE BUILDING BY THE SHORT CALCULATION METHOD

The size of the photovoltaic (PV) array depends on the following factors:

1. Amount of electrical energy needed per day (KWH/day).
2. System efficiency—losses in inverters, controllers, etc. (50 percent is typical).

3. Amount of solar radiation available per day (KWH/m^2/day).
4. Power produced by PV module (KW/m2).

Steps for Sizing a PV Array

1. Use Tables 8.15A and 8.15B to determine the electrical load per day in watt-hours per day.
   WH/day = (___)
2. Find the adjusted load by multiplying the WH/day (Step 1) by 1.5 to account for system losses.
   (WH/day) × 1.5 = (___)
3. Determine the solar energy available at the site for one day in terms of "sun-hours" from the map of Fig. 8.15a for winter-peaking loads and Fig. 8.15b for summer-peaking loads.
   "sun-hours" = (___)
4. To find the required peak-watts (W$_P$) divide the adjusted load (Step 2) by the sun-hours (Step 3).
   $W_P = \dfrac{\text{adjusted (WH/day)}}{\text{"sun-hours"}} = (\_\_\_)$
5. Find array size by dividing W$_P$ (Step 4) by:

12 W/ft^2 for single-crystal silicon cells
or 10 W/ft^2 for polycrystalline silicon cells
or 5 W/ft^2 for amorphous silicon or thin-film cells
or 5 W/ft.^{2} for PV standing seam roof
or 2.5 W/ft.^{2} for PV shingles.
   $A = \dfrac{W_P}{W/ft^2} = (\_\_\_) \text{ ft}^2$

As might be expected, the more efficient cells cost more. The less efficient cells sometimes give more watts/dollar, and the roof area of most residences is more than adequate to use low-efficiency cells and still generate all the power needed, especially if the roof faces south at the appropriate tilt angle.

Example:
Find the array size for a stand-alone residence located in the middle of Pennsylvania (40° latitude). The total load in watt-hours per day is 3,000 WH/day, and the roof is made of a standing seam PV material facing south at a tilt angle of 55° (latitude plus 15 degrees for winter peaking).

TABLE 8.15A WORKSHEET FOR DETERMINING THE ELECTRICAL LOAD

Appliances	Watts*	Number of Hours Used Per Day	Watt-Hours Per Day
Lights			
Refrigerator**			
Washing machine			
Furnace blower			
Air conditioning			
Exhaust fans			
Television			
Computer			
Miscellaneous plug loads			
Other			
		Total Watt-Hours Per Day	_____

* Find actual watts from appliances to be used, or refer to Table 8.15B.
** Because refrigerators cycle on and off, they typically run about 6 hours per day.

TABLE 8.15B TYPICAL APPLIANCE WATTAGE

Device	Wattage
Incandescent lights* (wattage of lamps)	—
Fluorescent lamps (wattage of lamps plus 10% for ballast)	—
Coffee pot	200
Microwave	1,000
Dishwasher	1,300
Washing machine	500
Vacuum cleaner	500
Clothes dryer (uses gas)	350
Furnace blower	500
Air conditioner (central)	3,000
Ceiling fan	30
Computer & printer	200
Television and videocassette recorder	200
Stereo	20
Refrigerator—conventional	500
Refrigerator—high efficiency	200

* Minimize use of incandescent and halogen lamps (see Chapter 14).

Step 1   Use the given load
$$3,000 \text{ WH/day}$$

Step 2   Find the adjusted load by multiplying by 1.5
$$3,000 \ (1.5) = 4,500 \text{ WH/day}$$

Step 3   Determine the sun-hours by using Fig. 8.15a for this northern location the Sun-hours = 2H/day

Step 4   Find the peak-watts (W_p) by dividing the adjusted load by the sun-hours $4,500/2 = 2,250 \ W_p$

Step 5   Since a standing seam roof is specified, divide W_p by 5 W/ft^2 to find the area of the array.
$$2,250/5 = 450 \text{ ft}^2$$

8.16   DESIGN GUIDELINES

Use the following guidelines to get the full benefits of the photovoltaic (PV) system sized with the methods described above:

1. Use building-integrated photovoltaics (BIPV) to save money and improve the aesthetics.

2. Use the following orientations and slopes in descending order of efficiency (most efficient first).

    a. Southerly orientation tilted at (see Fig. 8.16):
       • latitude for maximum annual energy production
       • latitude -15 degrees for summer peaking (the Southeast)
       • latitude +15 degrees for winter peaking (the North)
    b. south wall
    c. west wall
    d. east wall

3. Make sure the array is shaded as little as possible. Avoid having even a small area of array shaded because of the "Christmas-lights-in-series-syndrome," where if one light goes out, all the lights go out.

    Fortunately, solar performance is not degraded too much with some deviation from the optimum tilt and orientation. For example, in most of the United States, there will

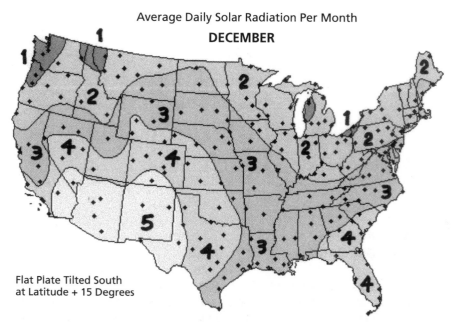

Average Daily Solar Radiation Per Month
DECEMBER

Flat Plate Tilted South
at Latitude + 15 Degrees

Figure 8.15a The average December solar radiation in sun-hours for a surface with a tilt equal to latitude plus 15 degrees. Because December has the least sunshine, it is usually used for sizing PV systems. (From the National Renewable Energy Laboratory-NREL.

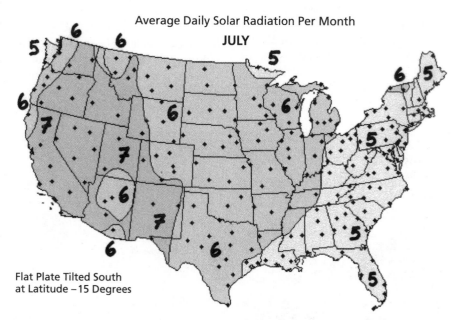

Average Daily Solar Radiation Per Month
JULY

Flat Plate Tilted South
at Latitude – 15 Degrees

Figure 8.15b The average July solar radiation in sun-hours for a surface with a tilt equal to latitude minus 15 degrees. (From the National Renewable Energy Laboratory-NREL.)

be only a 10 percent loss if the tilt is anywhere from 15 degrees to 60 degrees and the orientation is within 45 degrees of true south (*Environmental Building News,* 1999).

4. Keep modules cool by venting their backs (cool cells generate more electricity than hot cells), and use this heat in winter.

5. Avoid horizontal arrays since they will collect dirt and snow.

6. Mount modules at a steep tilt in snow country so that snow will quickly slide off.

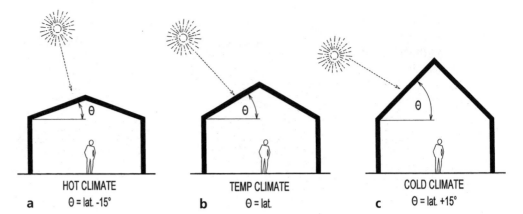

Figure 8.16 The recommended tilt of a PV array for: (a) maximum summer generation; (b) maximum annual generation; and (c) maximum winter generation.

HOT CLIMATE
Θ = lat. -15°
a

TEMP CLIMATE
Θ = lat.
b

COLD CLIMATE
Θ = lat. +15°
c

8.17 THE PROMISE OF PHOTOVOLTAICS

In architecture, an elegant solution is one that has many benefits besides solving the immediate problem. Thus, a roof would be much more elegant if, besides its traditional function, it also produced all the energy the building needed. If photovoltaics (PV) replaced all of the roofing, the amount of power produced would be much more than any residence could need. Consequently, each home would be an exporter of energy, and instead of being a burden on the environment, many buildings could become environmental assets.

I remember looking out at the roofscape of New York City from a high building and thinking what a waste these thousands of empty, flat roofs were. Very few roofs were used either for gardens or for cooling-tower supports. Consider the enormous value of this resource, which is about equal to the built land area of the city. If these roofs were covered by PV, the energy would be produced right where it is needed with minimal transmission losses (Fig. 8.17). With the deregulation of the electric-power industry, an entrepreneur could offer a building owner a free, no-maintenance, high-quality roof for the privilege of generating power. What a

gigantic amount of clean, renewable power would be generated right where it is needed.

If all roofs and most south walls were covered with PV, most towns and small cities would produce all the electricity they needed. Although large cities, especially with dense clusters of high-rise buildings, could not be energy-independent, they would need to import much less power than at present. All new construction should either have building-integrated photovoltaics (PV) or be designed to have the PV added when the cost has declined further. In such cases, a temporary low-cost weathering skin could be used until building-integrated materials are cost-competitive.

PV applications are not limited just to the Sunbelt but are appropriate in almost all climates. In Norway, which is as far north as Alaska, more than 50,000 vacation houses are powered solely by PV systems.

8.18 THE COST-EFFECTIVENESS OF ACTIVE SOLAR APPLICATIONS

The main competition to photovoltaics (PV) for roofspace are solar collectors designed to harvest the sun in order to produce hot air or water. Hot air is used primarily for space heating, while hot water can be used for a number of different purposes: domestic hot water, space heating, space cooling, swimming-pool heat-

Figure 8.17 A 360-kilowatt array using Solarex's MSX-120 power modules mounted to the roof of the aquatic center. The installation is located on the campus of Georgia Tech University, Atlanta, GA, the site of the 1996 Summer Olympics. (Courtesy and © BP Solarex.)

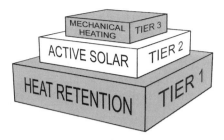

Figure 8.18a A building's hot-air and water needs are best provided by the three tier design approach.

TABLE 8.18 THE COST-EFFECTIVENESS OF ACTIVE SOLAR APPLICATIONS

Application	Economical?
Swimming pool-heating	Yes, in most cases (two- to three-year payback)
Domestic hot water	Frequently
Solar space heating	Sometimes (rarely in the South, frequently in cold northern climates)
Solar space cooling	Never*
Dehumidification	Possibly with desiccants
Process hot water	Usually

* This does not apply to air conditioners running on PV, which is not included under the strict definition of "active solar."

ing, and commercial hot water. Regarding the present and near future costs of PV, it is much more economical to create hot air or water directly rather than to use PV electricity. Also, as was explained in Chapter 3, it is not sensible to convert a high-grade energy, such as electricity, directly into a low-grade energy, such as heat.

The most sustainable way to meet a building's need for hot air and water is achieved by the three-tier design approach (Fig. 8.18a). Tier one consists of minimizing the needs through efficiency. In the second tier, active collectors harvest the sun. Only in the third tier will mechanical systems using nonrenewable energies be called upon to supply the small load that remains after tiers one and two.

The term **active solar** is used to designate mechanical devices whose sole purpose is to collect solar energy in the form of heat and then store it for later use. The working fluid is sometimes air, but usually it is water pumped from the collector to the stor-

age tank. This system is referred to as active solar because of the pump and, more important, because it does just one thing: collect heat. On the other hand, a passive system, as discussed in the previous chapter, uses only the basic building fabric to collect the heat (Fig. 8.18b). Since the building fabric is there anyway, passive solar is free or nearly so. Furthermore, since there are no mechanical parts to break or wear out, passive solar is more reliable and requires little or no maintenance. Thus, it is generally agreed that passive solar is the best choice for space heating when it will work. It won't work, for example, if only the roof of a multistory building has access to the winter sun. In that case, active solar space heating might be appropriate. Solar access is often a problem in urban areas, in wooded sites (especially if evergreen), and in high latitudes where the very low winter sun is easily blocked. Thus, active solar is sometimes used for space heating, but mostly it is used

for heating water for other purposes.

Five different applications exist for hot water in buildings: swimming-pool heating, domestic hot water, commercial/institutional hot water, space heating, and solar cooling. Unfortunately, active solar cooling, although technologically feasible, is not economical, and, therefore, will not be discussed here. PV is a much better way to air-condition with solar energy. Commercial/institutional hot water is much like domestic hot water. Buildings, such as hospitals, apartment buildings, schools, jails, car washes, nursing homes, health clubs, restaurants, and hotels, all use large quantities of hot water that can be produced economically by an active solar system. See Table 8.18 for the relative cost effectiveness of these different applications.

Why is it so appropriate to use active solar for heating swimming pools? One reason is that pools are heated only in the spring, summer, and fall to extend the swimming season when much more solar energy is available than in the winter. Another reason is based on the Laws of Thermodynamics: all solar collectors have the highest efficiency at their lowest operating temperature. Since pool-water temperatures are rather low (about 80°F), the efficiency of the solar collectors is very high (Fig. 8.18c). This low-temperature efficiency also applies to two forms of space heating. A radiant floor heating system can make good use of water heated to only 90°F, and a heat pump can **upgrade** the heat solar col-

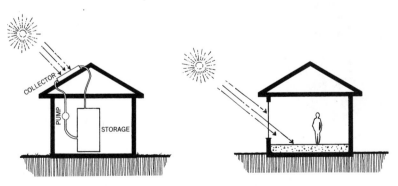

Figure 8.18b Active solar requires specialized mechanical equipment to make it work, while passive solar relies only on the building itself. Both use the natural energy of the sun.

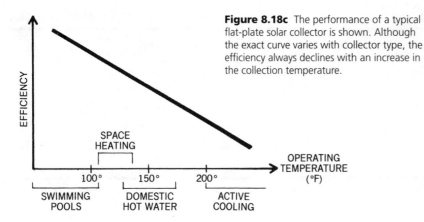

Figure 8.18c The performance of a typical flat-plate solar collector is shown. Although the exact curve varies with collector type, the efficiency always declines with an increase in the collection temperature.

lectors gather. Heat pumps will be explained in Chapter 16.

Domestic hot water is also a good application for active solar but for a different reason. Since domestic hot water is required year-round, the equipment is never idle. This, unfortunately, is not true for space heating. Not only is an active solar space heating system idle for much of the year, but when it does work in the winter, the supply of solar energy is at its lowest point. Consequently, active systems for space heating are less efficient than they are for domestic hot water or swimming-pool heating. However, an important exception exists. Preheating ventilation air in cold climates can be accomplished very economically with a system

called "Solarwall™," which will be discussed in Section 8.24.

Limitations on active solar systems are more economic than technical. Where cheap, alternative energy sources are not available, solar energy is popular. Active solar systems are common is many countries — millions of systems are now operating in Europe, Japan, and Israel. As a matter of fact, active solar systems were already sold in the United States at the turn of the century and became quite popular in Florida and Southern California (Fig 8.18d). By 1941, about 60,000 solar hot-water systems existed in the United States. Solar energy declined after that not because it did not work but because there were cheap alternatives, and it was no

longer fashionable. Because only the poor kept their solar rooftop systems, the public developed a negative image of solar energy.

Because solar swimming-pool heaters are the most cost-effective and simplest of the active solar systems, they are described first.

8.19 ACTIVE SOLAR SWIMMING POOL HEATING

As mentioned before, solar swimming-pool heaters are very cost-effective because they collect the sun during the part of the year when solar energy is plentiful and because they collect heat at rather low temperatures (70°F to 90°F). At these relatively low temperatures, the collectors are not only very efficient but they can also be rather simple and inexpensive to manufacture (Fig. 8.19a). The rest of the system is also inexpensive because the existing filtration equipment can often be used to circulate the water (Fig. 8.19b). And unlike other active solar heaters, freeze protection is not needed because the systems are not used in the winter.

As with buildings, the first step in heating a pool is to reduce heat loss. Since most of the heat is lost by evaporation, a pool cover is a must. The above discussion refers to outdoor pools. For indoor pools that can be used year-round, active solar systems similar to the kinds used for domestic hot water described below are required.

Rules of Thumb for Sizing Swimming-Pool Collectors

1. For the hot southern United States, with good collector orientation and tilt,* use a collector area equal to 50 percent of the pool area.
2. For the cold northern United States and good orientation and tilt,* use a collector area equal to the pool area.

Figure 8.18d This advertisement for a solar hot-water heater appeared in 1892. (From Special Collections, Romaine Collection, University of California, Santa Barbara; cited in *A Golden Thread* by Ken Butti and John Perlin.)

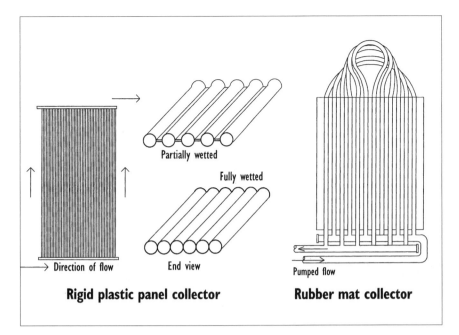

Figure 8.19a Solar swimming-pool collectors are simple and inexpensive because they need neither glass covers nor insulation. Some are made of flexible extruded plastic that can be shipped in a roll. (Courtesy Dan Cuoshi © *Home Energy Magazine.*)

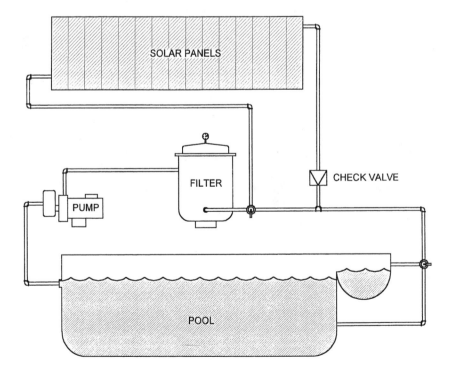

Figure 8.19b A typical swimming-pool heating system that uses the existing filtration equipment to circulate the water. (After Sun Trapper Solar System Inc.)

3. For states in the middle, use a collector area equal to 75 percent of the pool area.

* Good orientation is within 20 degrees of south and good tilt is near (latitude -15 degrees). For fair orientation and tilt, add 25 percent, and for poor orientation and tilt, add 50 percent additional collector area.

8.20 SOLAR HOT-WATER SYSTEMS

Since most buildings need domestic hot water, most active systems use water rather than air as a heat-transfer and -storage medium. Such water-based systems can also be easily used for space heating. Each system must have a collector, heat-transfer fluid, and a storage device. To protect against freezing and boiling, some water systems use a mixture of water and antifreeze. An insulated tank located indoors stores the hot water. The collector must be more sophisticated than the swimming-pool kind because it has to generate medium high temperatures (120°F to 140°F) even on very cold winter days.

The most common kind of collector used for producing domestic hot water is called a **flat plate collector,** which essentially consists of a metal plate coated with a black **selective surface** to reduce heat loss by re-radiation (see Section 3.11). A glass cover creates the greenhouse effect to maximize the energy collected, and insulation is used to reduce heat loss from the back and sides of the collector (Fig. 8.20a). Water is pumped through pipes attached to the hot collector plate, and the heated water is then stored in tanks located inside the building. To prevent contamination of the potable domestic hot water, one ordinarily uses a heat exchanger.

When the controller senses that the water is warmer in the collector than in the storage tank, the pump is activated, and when the sun sets and the collector is cooler than the storage tank, the controller shuts off the pump. In a particular arrangement called a **drain-back system,** the water completely drains into the indoor tank whenever the pump does not operate (Fig. 8.20b). If the pump fails to operate, this safety measure will prevent freezing or boiling in the collector without the need for antifreeze. Other arrangements are used depending on climate and the particular company that makes the system.

It is not practical to try to design a 100 percent solar system because the supply of sunshine is irregular. It would take a very large solar collector and storage system to supply hot water after a week of cloudy cold weather. Since such a large system would be completely overdesigned for a week of consistent sunshine, the overall efficiency would be low. Thus, a 100 percent solar heating system is not economical; the optimum percentage varies with climate but is usually between 60 percent and 80 percent.

When very hot water is required, special collectors are used. Either by concentrating the sun (Fig. 8.20c) or by reducing the heat loss (Fig. 8.20d), water can be heated above the boiling point.

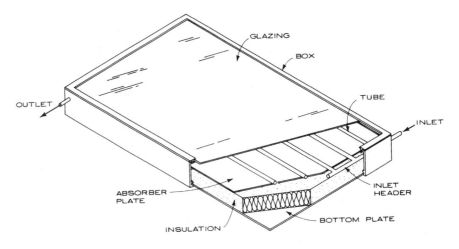

Figure 8.20a A typical flat-plate collector designed to heat a liquid. (From *Architectural Graphic Standards, Ramsey/Sleeper,* 8th ed., John R. Hoke, editor © John Wiley & Sons, Inc., 1988.)

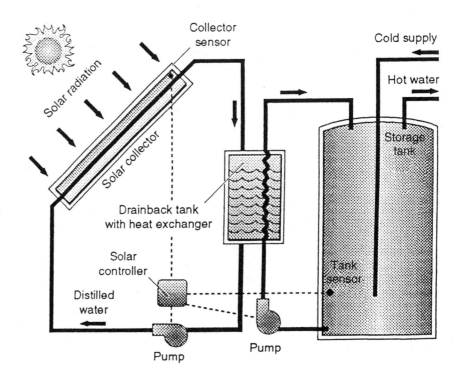

Figure 8.20b The drain-back, solar hot-water system. To prevent contamination of the domestic hot water, a double-walled heat exchanger is submerged in the storage tank. A space-heating system can also utilize the hot water in the tank. (From the Texas State Energy Conservation Office, Fact Sheet #10.)

Figure 8.20c A concentrating collector uses a parabolic mirror to achieve high temperatures.

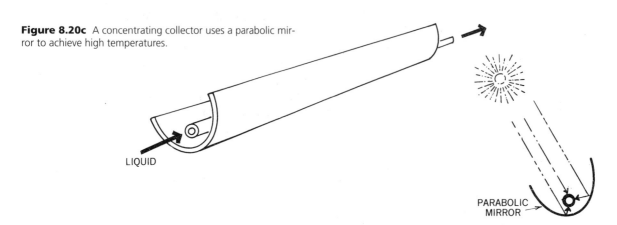

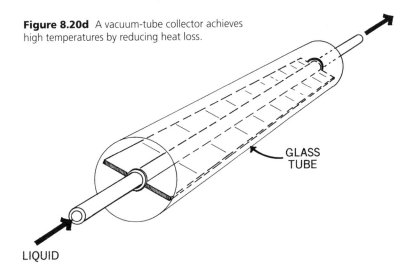

Figure 8.20d A vacuum-tube collector achieves high temperatures by reducing heat loss.

GLASS
TUBE

LIQUID

8.21 SOLAR HOT-AIR COLLECTORS

Hot-air collectors are used primarily for space heating. The main disadvantages of air as a collecting fluid are: the collectors and ducts are bulky, it is hard to prevent air leakage, and it is awkward to heat domestic hot water with the hot air. The advantages are: air doesn't freeze or boil, leaks don't cause damage, and the warm air can be used directly to heat the building. Early systems usually stored the heat in a rock bin (Fig. 8.21a). It might be more economical instead to use the mass of the building, as in the popular Japanese system described in Figs. 8.21b through 18.21f.

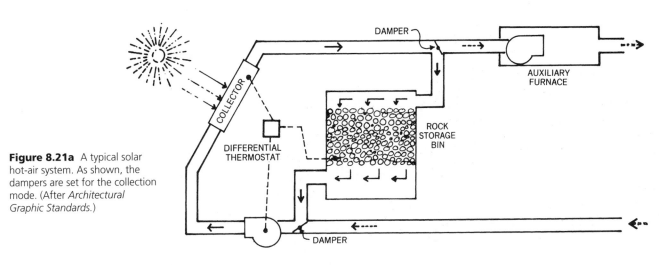

Figure 8.21a A typical solar hot-air system. As shown, the dampers are set for the collection mode. (After *Architectural Graphic Standards.*)

DAMPER

AUXILIARY
FURNACE

COLLECTOR

ROCK
STORAGE
BIN

DIFFERENTIAL
THERMOSTAT

DAMPER

Figure 8.21b One popular active solar system in Japan uses special hot-air collectors to gather the heat and concrete floor slabs to store the heat. The collectors cover the whole roof, but for economic reasons only the upper third are covered with glass. This system is used in homes, schools, and small community buildings. (After Oku Mura of the OM Solar Association/OM Institute.)

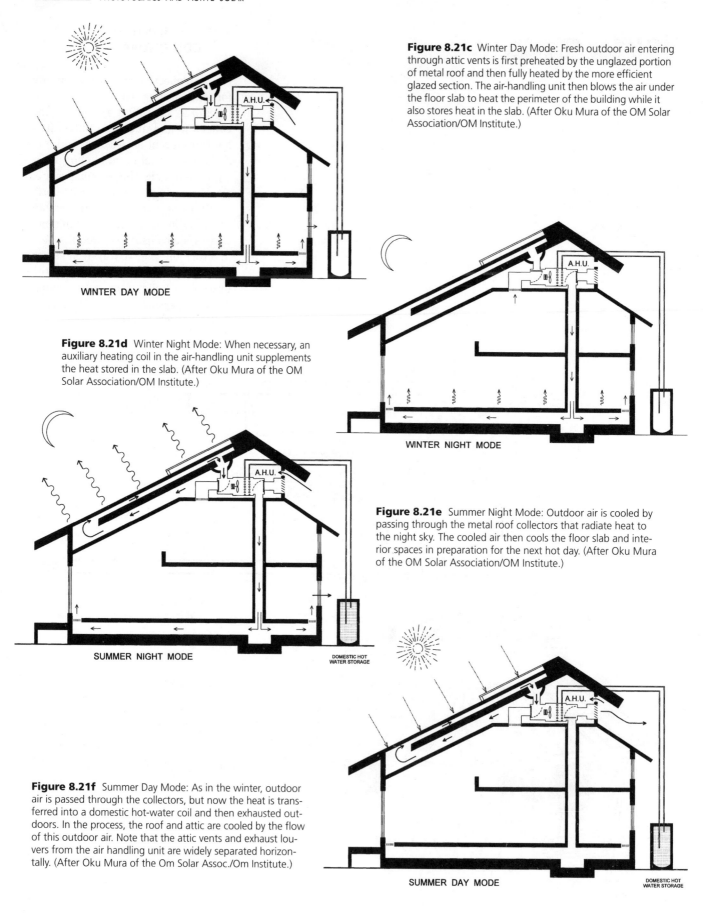

Figure 8.21c Winter Day Mode: Fresh outdoor air entering through attic vents is first preheated by the unglazed portion of metal roof and then fully heated by the more efficient glazed section. The air-handling unit then blows the air under the floor slab to heat the perimeter of the building while it also stores heat in the slab. (After Oku Mura of the OM Solar Association/OM Institute.)

WINTER DAY MODE

Figure 8.21d Winter Night Mode: When necessary, an auxiliary heating coil in the air-handling unit supplements the heat stored in the slab. (After Oku Mura of the OM Solar Association/OM Institute.)

WINTER NIGHT MODE

Figure 8.21e Summer Night Mode: Outdoor air is cooled by passing through the metal roof collectors that radiate heat to the night sky. The cooled air then cools the floor slab and interior spaces in preparation for the next hot day. (After Oku Mura of the OM Solar Association/OM Institute.)

SUMMER NIGHT MODE

DOMESTIC HOT WATER STORAGE

Figure 8.21f Summer Day Mode: As in the winter, outdoor air is passed through the collectors, but now the heat is transferred into a domestic hot-water coil and then exhausted outdoors. In the process, the roof and attic are cooled by the flow of this outdoor air. Note that the attic vents and exhaust louvers from the air handling unit are widely separated horizontally. (After Oku Mura of the Om Solar Assoc./Om Institute.)

SUMMER DAY MODE

DOMESTIC HOT WATER STORAGE

8.22 DESIGNING AN ACTIVE SOLAR SYSTEM

Because of the expense of solar equipment, the system must be designed to be as efficient as possible. It is most important to maximize the solar exposure by aiming the collectors at the sun and by minimizing the shading of a collector while it gathers energy from about 9 A.M. to 3 P.M. In most cases, rooftops have the best solar access (Figs. 8.22a and b). Rooftop mounting also saves land and minimizes the potential for damage that exists with ground-mounted collectors. A model study on a sun machine is the most effective way to check for solar access. See Chapter 11 for a discussion on solar access.

Collector Orientation

Usually it is best to orient solar collectors toward true south. Variations up to 20° east or 20° west are acceptable (Fig. 8.22c). For special conditions,

Figure 8.22a This add-on to an existing building consists of two flat-plate collectors for domestic hot water. There is also a special low-temperature, swimming-pool collector made of flexible plastic that is draped over the roof tiles (upper right.)

Figure 8.22b The solar collectors are an integral part of this roof design. (Courtesy and © Chromagen-Solar Energy Systems, Israel.)

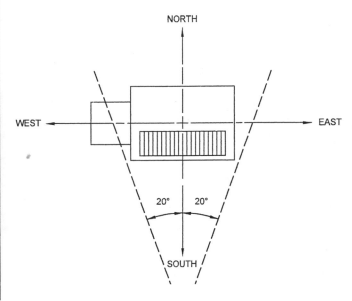

Figure 8.22c A collector orientation and allowable deviation from south.

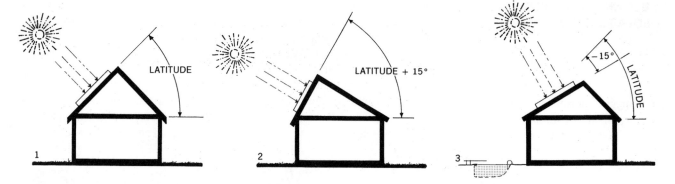

Figure 8.22d A collector tilt: (1) a collector tilt for domestic hot water; (2) a collector tilt for space heating and for combination space heating and domestic hot water; (3) a collector tilt for heating swimming pools.

such as a need for morning heat or the prevalence of morning fog, a twenty-to-thirty-degree shift to the east or west can even be beneficial.

Collector Tilt

The optimum **tilt** of the collectors is a function of latitude and the purpose of the solar collectors. Fig. 8.22d illustrates the tilt angle for different heating applications as a function of latitude. Collectors are most efficient when they are perpendicular to the sun rays. However, with the daily and seasonal motions of the sun, that is possible only with tracking collectors which, unfortunately, are too complicated in most circumstances. The tilt angles given in Fig. 8.22d are the optimum slopes for fixed collectors.

Collector Size

The collector size depends on a number of factors: type of heating (pool water, domestic water, or space heating), amount of heat required, climate, and efficiency of the collector system. Table 8.22A gives the approximate collector areas and storage-tank sizes for domestic hot water heating, while Table 8.22B gives the approximate sizes of collectors and storage tanks for a combined space-heating and domestic hot-water system. For sizing swimming-pool heating sys-

TABLE 8.22A APPROXIMATE SIZING OF A SOLAR DOMESTIC HOT-WATER SYSTEM*

Number of People Per Household	Approximate Collector Size (Ft²)** in Regions				Approximate Tank Size Gallons	Ft³
	A	B	C	D		
1–2	30	40	60	80	60	8
3	40	53	80	107	80	11
4	50	67	100	133	100	13
5	60	80	120	160	120	16
6	70	93	140	187	140	19

* Based on rules of thumb and map from AAA Solar Service and Supply, Inc., Albuquerque, NM.

** The collector area is also a function of its efficiency. High-efficiency collectors, such as the vacuum-tube type, will require a smaller area but cost more, and vice versa.

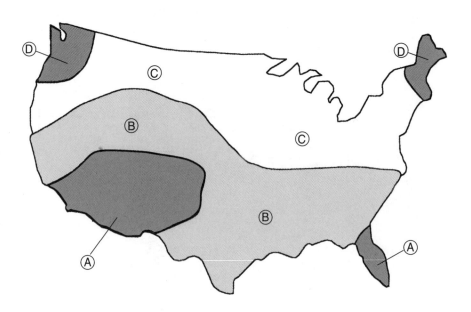

TABLE 8.22B APPROXIMATE SIZING OF A COMBINED SPACE-HEATING AND DOMESTIC HOT-WATER SYSTEM FOR A 1,500-SQUARE-FOOT HOME

Climate Region	Reference City	Approximate Collector Area (ft²)	Approximate Storage Size Water (ft³)	Rock Bin* (ft³)
1	Hartford, CT	800	200	600
2	Madison, WI	750	200	600
3	Indianapolis, IN	800	200	600
4	Salt Lake City, UT	750	200	600
5	Ely, NE	750	200	600
6	Medford, OR	500	100	300
7	Fresno, CA	300	70	210
8	Charleston, SC	500	100	300
9	Little Rock, AK	500	100	300
10	Knoxville, TN	500	100	300
11	Phoenix, AZ	300	70	210
12	Midland, TX	200	40	120
13	Fort Worth, TX	200	40	120
14	New Orleans, LA	200	40	120
15	Houston, TX	200	40	120
16	Miami, FL	50	10	30
17	Los Angeles, CA	50	0	30

Sizes are approximate and vary with actual microclimate and efficiency of specific equipment.
*Rock bin is used with a hot air system.

tems, see Section 8.19. In all cases, the collector area should be increased to compensate for a poor tilt angle, a poor orientation, or a partial shading of collectors.

8.23 ACTIVE/PASSIVE SOLAR SYSTEMS

The end of Chapter 7 mentioned that although the convective-loop (thermosiphon) system (Fig. 7.18a) is a passive system because it uses no pumps, it is more closely related to active systems. The first active solar systems at the turn of the century used this natural-convection technique (Fig. 8.18d). Because of their simplicity and low cost, two different thermosiphon systems are becoming popular for domestic hot water. In one system, called a **batch heater,** the storage tank is also the collector (Fig. 8.23a). The other popular thermosiphon system is called the **integral collector storage (ICS) system** because the storage tank and collector are combined in one unit (Fig. 8.23b). Because these systems have no mov-

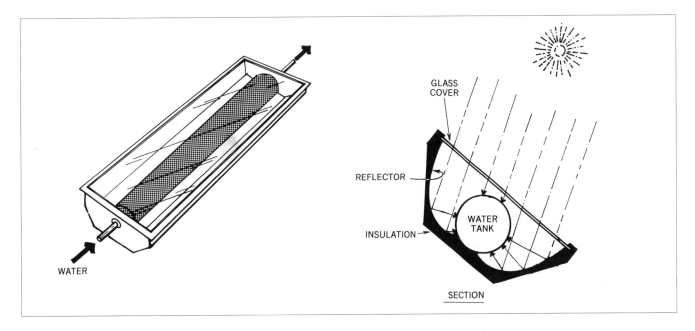

Figure 8.23a In a batch-type hot-water heater, the collector and storage are one and the same.

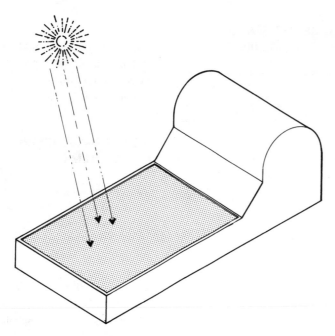

Figure 8.23b In an integral collector storage (ICS) system, the collector and storage tank are supplied in one package.

ing parts and a minimum of plumbing, they are very cost-effective and popular. They are most appropriate in mild to temperate climates because the storage, although well insulated, is outdoors. For thermosiphons to work, the storage tank must be above the collector. Their major disadvantage is aesthetic since they cannot be easily integrated into the roof, while flat-plate collectors can look a great deal like skylights.

8.24 PREHEATING OF VENTILATION AIR

As mentioned earlier, active solar swimming-pool heating is very cost-effective partly because simple inexpensive collectors that don't use glass covers are sufficient for the low-temperature heat that is collected. Similarly, simple, inexpensive solar collectors are sufficient to preheat ventilation air because even a small rise above the winter outdoor-air temperature is a benefit.

A simple ventilation preheater can be made by mounting dark metal

cladding a few inches in front of a south wall. A ventilation fan moves cold outdoor air across the back of the metal cladding before the air enters the building. Unfortunately, this strategy does not capture the warm air

film that forms on the front of the dark metal collector. Through the use of perforated cladding, however, heat is collected from both the front and the back of the collector, yielding efficiencies as high as 75 percent (Fig. 8.24a). Such transpired solar collectors even save energy at night because the heat lost through the south wall is brought back in by the incoming ventilation air. In summer, the ventilation system bypasses the solar collector, while the air being heated inside the collector rises by the stack effect and exits through the top perforations. Thus, minimal heat gain results in the summer from the solar collector.

Since the collector is the facade, the collector's appearance is very important. Fortunately, any dark color will work almost as well as black. On sunny days, the air can be heated anywhere from 30°F to 50°F (17°C to 30°C) above the ambient temperature. Even on cloudy days, the system is still about 25 percent efficient.

All buildings need to bring in outdoor air for health reasons. Because we use both toxic materials and materials that give off toxic components, **indoor-**

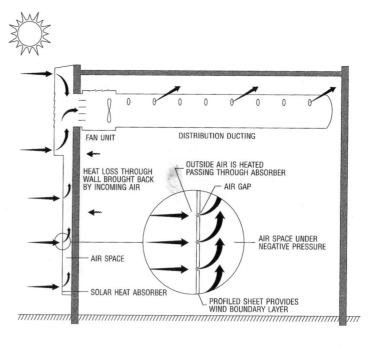

Figure 8.24a Much greater efficiencies are achieved when metal collectors are perforated. (Courtesy Conserval Systems Inc., makers of the Solarwall™ system.)

Figure 8.24b This Windsor, Ontario, apartment building uses the world's tallest solar collector to preheat ventilation air. (Courtesy Conserval Systems Inc., makers of the Solarwall™ system.)

apartment building that uses an active solar ventilation preheater made of perforated cladding panels. Preheating ventilation air is also appropriate for existing buildings. It is especially useful for buildings with badly deteriorated south facades, for which new cladding might be needed anyway.

8.25 THE FUTURE OF ACTIVE SOLAR

There is no reason why 100 percent of outdoor swimming pools should not be heated with active solar since the cost of solar is easily competitive with fossil energy. The choice is not as clear-cut with domestic hot water because the payback is fairly long. For commercial/institutional situations, for which large amounts of hot water are needed, the economics becomes very attractive. Active solar space heating should be used where limited solar access to walls makes passive solar impractical. In cold climates, preheating of ventilation air is very appropriate.

Despite active solar's fairly long payback, if it were used wherever it is appropriate and somewhat economical, the resulting mass production would lower the price and make active solar practical for almost everyone. Thus, active solar has a great potential not only to reduce our dependency on nuclear and fossil energy, but also to help the environment and reduce global warming.

8.26 CONCLUSION

Instead of importing energy, buildings could become net exporters of energy. Also, instead of being a burden on the environment, buildings could be assets of a healthy environment. This new role for buildings is completely possible today, at least in regard to energy, by simply implementing the techniques described in this book and outlined below.

air quality (IAQ) has become an important issue. Small buildings, such as residences, have traditionally relied on infiltration to supply the needed fresh air, while large buildings have relied on a designed ventilation system. Because energy-efficient buildings have a tight envelope, all buildings now need a carefully designed ventilation system, and in winter preheating this fresh air will save a great deal of energy. Fig. 8.24b shows a high-rise

Figure 8.26a South-facing roofs could consist of a mix of passive solar, active solar, and PV panels, as in The Oxford Solar House, Oxford, United Kingdom. The skylight supplies daylight and direct-gain heat, the active solar supplies the domestic hot water, and the PV supplies the electricity. (Photo from Caddet Technical Brochure #84.)

Figure 8.26b When active solar is not considered in the original design, retrofit installations can prove to be very awkward.

First, use efficiency to reduce the need for energy. Second, use the passive techniques to further reduce the energy load. Finally, use active solar to supply most of the low-grade heat requirements for domestic hot water and space heating, and use PV to generate the high-grade energy of electricity to operate all the lights and appliances.

Because of a similarity in appearance and solar-access requirements, active solar and PV can be integrated side-by-side in buildings. In residences, the entire south-facing roof can be used for a combination of active solar and PV panels (Figs. 8.3a and 8.26a). In commercial, institutional, and industrial buildings, the south wall should also be used.

It is most important that all new buildings either utilize these technologies or at the very least allow for them in the future. Thus, buildings should face south and have south-facing roofs at the appropriate tilt angle. Roads should be laid out to accommodate building orientation (see Chapter 11). If these steps are not taken, many buildings in the future will have solar collectors at odd angles to the architecture (Fig. 8.26b). Rather, the use of passive solar, active solar, and PV should generate a new aesthetic. There will be opportunities to create new forms appropriate for a sustainable world (see again Fig. 8.3b).

KEY IDEAS OF CHAPTER 7

---

1. Photovoltaic (PV)-generated electricity is an almost-ideal energy source. PV converts sunlight directly into electricity.
2. Most of the electricity in the twenty-first century will probably come from PV.
3. PV should be building-integrated (i.e., PV modules should displace the weathering skin of a building).
4. PV is appropriate not only in the Sunbelt but in every climate.
5. Stand-alone PV systems are best for buildings some distance from the power grid. Such systems require batteries and backup power.
6. Grid-connected PV systems are best for buildings on or near the power grid, which replaces the need for batteries or backup power.
7. Wind power is sometimes a good complement to PV power.
8. A southern orientation and a tilt equal to the latitude maximizes the PV output.
9. To match energy demand to supply, the optimum orientation and tilt can be off 15 degrees to 30 degrees from that of the maximum output.
10. PV has the potential to be very inexpensive because little material is used and the assembly is very simple.
11. As PV costs decline because of research and mass production, the optimum tilt and orientation will become less important. Someday, all south-facing roofs, south walls, west walls, and east walls will be clad with PV.
12. Transparent PV modules can be used as glazing.
13. Energy-efficient appliances and lighting, as well as passive systems, should be used in conjunction with PV.
14. Use PV to generate the necessary high-grade energy of electricity, and use active solar to generate the low-grade thermal energy needed in buildings.
15. Solar swimming-pool heating is the best and most effective way to extend the swimming season in outdoor pools.
16. The area of most solar collectors for pool heating is approximately equal to the pool area in the North and about half the pool area in the South.
17. Wherever possible, active solar should be used to heat domestic hot water.
18. Active solar space heating should be considered if passive solar is not possible.
19. Use radiant floor heating or a heat pump in conjunction with the active solar space-heating system.
20. All south-facing roofs should be covered with a mix of active solar and PV in order to protect the environment, control global warming, and lead society to a sustainable economy.
21. Green solar architecture will produce an exciting new aesthetic in harmony with the world

References

1. *Environmental Building News,* July/August, 1999, p. 13
2. Maycock, P. "1997 World Cell/Module Shipments," *PV News,* February, 1998.
3. *Tapping into the Sun: Today's Applications of Photovoltaic Technology,* U.S. Department of Energy, Rev. April, 1995.

Resources

FURTHER READING

(See Bibliography in back of book for full citations. This list includes valuable out-of-print books.)

GENERAL

Boonstra, C. *Solar Energy in Building Renovation.*
Hestnes, A. G., R. Hastings, and B. Saxhof. *Solar Energy Houses.*
Kachadorian, J. *The Passive Solar House.*
Miller, B. *Buildings for a Sustainable America Case Studies.*
Potts, M. *The New Independent Home.*
Røstvik, H. *The Sunshine Revolution.*
Singh, M. *The Timeless Energy of the Sun for Life and Peace with Nature.*
Stein, B., and J. Reynolds. *Mechanical and Electrical Equipment for Buildings.*

PHOTOVOLTAICS

Cathcart, K., Anders Architects. "Building Integrated Photovoltaics."
Davidson, J. *The New Solar Electric Home.*
Humm, O., and P. Toggweiler. *Photovoltaics in Architecture: The Integration of Photovoltaic Cells in Building Envelopes.*
Komp, R. *Practical Photovoltaics- Electricity from Solar Cells.*
Perlin, K. *From Space to Earth: The Story of Solar Electricity.*
Sick, F. *Photovoltaics in Buildings: A Design Handbook for Architects and Engineers.*
Strong, S. J. *The Solar Electric House.*
Zweibel, K. *Harnessing Solar Power: The Photovoltaics Challenge.*

ACTIVE SOLAR

Butti, K., and J. Perlin. *A Golden Thread.*
Hestnes, A. G., R. Hastings, and B. Saxhof. *Solar Energy Houses.*

PERIODICALS

Solar Today.

ORGANIZATIONS

(See Appendix J for full citations.)

American Solar Energy Society (ASES)
www.ases.org

Caddet Center for Renewable Energy
http://www.caddet.co.uk
Summaries of renewable energy projects around the world.

Department of Energy's (DOE) Energy Efficiency and Renewable Energy Clearinghouse (EREC)
www.eren.doe.gov/erec/factsheets
EREC provides free general and technical information to the public on the many topics and technologies pertaining to energy efficiency and renewable energy, including PV systems, solar energy, and solar-radiation data.

Florida Solar Energy Center (FSEC)
www.fsec.ucf.edu/

Interstate Renewable Energy Council (IREC)

International Solar Energy Association (ISES)
www.ises.org

National Renewable Energy Laboratory (NREL)
http://www.nrel.gov
Good source for detailed information on renewables.

Sandia National Laboratories
www.sandia.gov

Solar Energy Industries Association (SEIA)
http://www.seia.org
SEIA is the national trade organization of PV and thermal manufacturers and component suppliers.

Utility Photo Voltaic Group (UPVG)
www.ttcorp.com/upvg

VIDEOS

(See Appendix J for full citation.)
Living on the Sun. 30 minutes.

9

SHADING

"The sun control device has to be on the outside of the building, an element of the facade, an element of architecture. And because this device is so important a part of our open architecture, it may develop into as characteristic a form as the Doric column."

Marcel Breuer
from Sun and Shadow,
Dodd Mead & Co., 1955

9.1 HISTORY OF SHADING

The benefits of shading are so great and obvious that we see its application throughout history and across cultures. We see its effect on classical architecture as well as on unrefined vernacular buildings ("architecture without architects").

Many of the larger shading elements had the dual purpose of shading both the building and an outdoor living space. The portico and colonnades of the ancient Greek and Roman buildings certainly had this as part of their function (see Fig. 9.1a). Greek Revival architecture was so successful in the American South because it offered the much needed shading, as well as symbolic and aesthetic benefits. In hot and humid regions, large windows are required to maximize natural ventilation, but at the same time any sunlight that enters through these large windows increases the discomfort. Large overhangs that are supported by columns can resolve this conflict (see Fig. 9.1b). The white color of Greek Revival architecture is also very appropriate for hot climates.

In any good architecture, building elements are usually multifunctional. The fact that the Greek portico also protects against the rain does not negate its importance for solar control. It just makes the concept of a portico all the more valuable in hot and humid regions where rain is common and the sun is oppressive.

While we need not be as literal as revival architecture, borrowing from the past can be very useful when there are functional as well as aesthetic benefits (Fig. 9.1c). There is a rich supply of historical examples from which to draw. Traditional design features from around the world, while appearing different, often developed in response to the same needs. The Greek portico, mentioned above, is closely related to the porch, verandah (from India), balcony, loggia, gallery, arcade, colonnade, and engawa (from Japan). See Figs. 9.1a–h).

Figure 9.1a Ancient Greek architecture made full use of colonnades and porticoes for protection against the elements. Stoa of Attalos II Athens.

Figure 9.1b Greek Revival architecture was especially popular in the South where it contributed greatly to thermal comfort. The Hermitage, Andrew Jackson's home near Nashville, TN.

Figure 9.1c Postmodernism, with its allusion to classical architecture, can draw on time-tested ideas for thermal comfort, as in the public library for San Juan Capistrano, designed by Michael Graves.

Figure 9.1d Loggias supported on arcades and colonnades shielded the large windows necessary for natural ventilation in the hot and humid climate of Venice. Sometimes an open-walled extra floor was added above the top floor to ventilate the heat collecting under the roof.

Figure 9.1e Victorian architecture made much use of the porch, veranda, and balcony to shade the building and create cool outdoor spaces. Eufaula, AL.

Chinese and Japanese architecture is dominated by the use of large overhangs (Fig. 9.1f). The Japanese made much use of a veranda-like element called the engawa. This large overhang protected the sliding wall panels that could be opened to maximize access to ventilation, light, and view. When the panels are closed, light enters through a continuous translucent strip window above. Also note how rainwater is led down to a drain by means of a hanging chain (Fig. 9.1g). In the early years of the twentieth century, the Green brothers developed a style appropriate for California by using concepts derived from Japanese architecture (Fig. 9.1h).

Many great architects have understood the importance of shading and used it to create powerful visual statements. Frank Lloyd Wright used shading strategies in most of his buildings. Early in his career, he used large overhangs both to create thermal comfort and to make an aesthetic statement about building on the prairie. In his Robie House (Fig. 9.1i), Wright used large areas of operable glazing to maximize natural ventilation during the hot and humid Chicago summers. He understood, however, that this would do more harm than good unless he shaded the glazing from the sun. The very long canti-levered overhangs not only supply the much needed shade, but they also create strong horizontal lines that reflect the nature of the region. See Fig. 10.8b for a plan view of the Robie House.

Of all architects, Le Corbusier is most closely linked with an aesthetic based on sun shading. It is interesting to note how this came about. In 1932, Le Corbusier designed a multistory building in Paris known as the Cité de Refuge. It was designed with an all-glass south facade so that a maximum

Figure 9.1f Much Asian architecture is dominated by large overhangs. Golden Pavilion, Kyoto, Japan. (Courtesy of Japan National Tourist Organization.)

Figure 9.1g The sliding wall panels can be opened for maximum access to ventilation, light, and view. The engawa or porch is clearly visible in this building in the Japanese Garden of Portland, OR.

Figure 9.1h The Gamble House in Pasadena, CA, 1908, by Green and Green shows strong influence from Japanese architecture. Note especially the large roof overhangs. (Model by Gary Kamemoto and Robert Takei, University of Southern California.)

Figure 9.1i Large overhangs dominate the design of the Robie House, Chicago, 1909, by Frank Lloyd Wright.

Figure 9.1j Sunshades known as brise-soleil were retrofitted on the Cité de Refuge, Paris, which Le Corbusier designed in 1932 without sunshades. (Photograph by Alan Cook.)

Figure 9.1k The brise-soleil and parasol roof shade the High Court at Chandigarh. Evaporation from the reflecting pool helps cool the air.

Figure 9.1m The Maharajah's Palace at Mysore illustrates the extensive shading techniques used in some Indian architecture. (Courtesy of Government of India Tourist Office.)

of sunlight could warm and cheer the residents. In December it worked wonderfully, but in June the building became unbearably hot. As a result of this mistake, Le Corbusier invented the fixed structural sunshade now known as **brise-soleil** (sun-breaker). In Fig. 9.1j, we see the building after it was retrofitted with a brise-soleil.

Le Corbusier realized the dual nature of the sun—our friend in the winter and our enemy in the summer. His own artwork says it best (Fig. 6.1). After this realization, shading became a central part of his architecture. For him, the aesthetic opportunities were as important as the protection from the sun. Thus, many of his buildings use sun shading as a strong visual element. Some of the best examples come from the Indian city of Chandigarh, where Le Corbusier designed many of the government buildings. The brise-soleil and parasol roof create powerful visual statements in the High Court Building (Fig. 9.1k). The Maharaja's Palace at Mysore has some similarity with the High Court, and it is, therefore, tempting to speculate on how much Le Corbusier was influenced by native Indian Architecture (Fig. 9.1m). For another example of the parasol roof see Fig. l.3h, and for another example of the brise-soleil see Fig. 10.6y and z.

Traditional buildings often provided shading even when it was not a conscious goal. Because windows were usually set back into deep bearing walls or even thick masonry curtain walls, the effect was that of a shallow brise-soleil (Fig. 9.1n).

9.2 SHADING

Shading is a key strategy of achieving thermal comfort in the summer. Shading, as part of heat avoidance, is tier one of the three-tier design approach to cooling a building (Fig. 9.2a). The second tier consists of passive cooling, and the third uses

Figure 9.1n An often-overlooked benefit of the traditional, thick masonry wall was the shading produced from both the vertical and horizontal elements. (Old office building in lower Manhattan.)

mechanical equipment to cool whatever the architectural strategies of tiers one and two could not accomplish.

Although shading of the whole building is beneficial, shading of the windows is crucial. Consequently, most of the following discussion refers to the shading of windows.

Which window orientations need the most shading in the summer? The graph in Fig. 9.2c shows that on June 21, a skylight (horizontal glazing) collects about four times more solar radiation than a south window. Clearly then, skylights need very effective shading or better yet should be avoid-

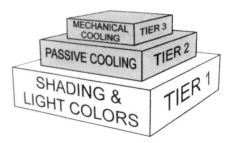

Figure 9.2a The three-tier approach to design is the most logical and sustainable method for achieving thermal comfort in the summer.

Figure 9.2b In the Conoco Inc. Headquarters complex in Houston, Kevin Roche was inspired by the local Texas plantation style with its large overhangs and column-supported porches. The awning-like translucent overhangs are 13 feet deep because they shade two floors, face in all directions of the compass, protect verandas, and block the sky and its strong diffused radiation in this humid climate. Trellises covered with jasmine and fig ivy protect the first floor, as well as the courtyards where the second story verandas leave off. In this very hot and wet climate, even the on grade parking is protected by the awning-like Fiberglas sunshades (not shown).

ed. Figure 9.2c also shows that east or west glazing collects almost three times the solar radiation of south windows. Thus, the shading of east and west windows is also more important than the shading of south windows.

If the graph in Fig. 9.2c looks familiar, it is not surprising since a variation of it appeared before in the Passive Solar chapter, as Fig. 7.16a which was used to show how much more solar radiation south windows collected than any other orientation in the winter. Thus, south windows are very desirable from both a shading and passive solar heating point of view. Skylights should be avoided because they collect a great amount of solar radiation in the summer and little in the winter. Similarly, east and west windows are not desirable from a heating and cooling point of view.

The total solar load consists of three components: direct, diffuse, and reflected radiation. To prevent passive solar heating when it is not wanted, one must always shade a window from the direct solar component and often also from the diffuse sky and reflected components. In sunny humid regions, like the Southeast, the diffuse-sky radiation can be very significant (Fig. 9.2b). Sunny areas with much dust or pollution can also create much diffuse radiation (Fig. 9.2d). Reflected radiation, on the other hand, can be a large problem in such areas as the Southwest, where intense sunlight and high-reflectance surfaces often coexist. The problem also occurs in urban areas where highly reflective surfaces can be quite common. Concrete paving, white walls, and reflective glazing can all reflect intense solar radiation into a window. There are cases where the north facade of a building experiences the solar load of a south orientation because a large building with reflective glazing was built toward the north (Fig. 9.2e).

The type, size, and location of a shading device will, therefore, depend in part on the size of the direct, diffuse, and reflected components of the total solar load. The reflected component is usually best controlled by reducing the reflectivity of the offending

Figure 9.2c All orientations except south receive maximum solar radiation in summer. A skylight receives about four times the solar heating that south windows receive on June 21. (After the *Workbook for Workshop on Advanced Passive Solar Design* by J. Douglas Balcomb and Robert Jones, © J. Douglas Balcomb, 1987.)

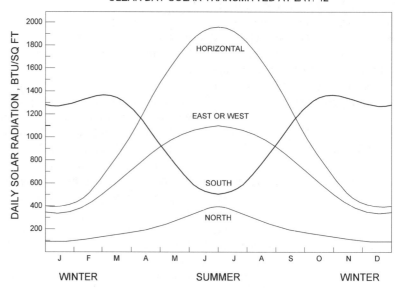

CLEAR DAY SOLAR TRANSMITTED AT LAT. 42

surfaces. This is often best accomplished by the use of plants. The diffuse-sky component is, however, a much harder problem because radiation comes from a large exposure angle. It is, therefore, usually controlled by additional indoor shading devices or shading within the glazing. The direct solar component is effectively controlled by exterior shading devices.

The need for shading might seem to conflict with the demand for daylighting. Fortunately, when solar energy is brought into a building in a very controlled manner, it can supply high-quality lighting as well as reduce the heat gain. This is accomplished by allowing just enough light to enter so that the electric lights can be turned off. A more detailed discussion of daylighting versus shading is found in Section 15.8.

When it is not used for daylighting, solar radiation should be blocked during the overheated period of the year. A residence in the north would experience an overheated period that was only a few months long. That same residence in the south or a large office building in the north could experience overheated periods that are two to three times as long. Thus, the required shading period for any building depends on both the climate and the nature of the building. Shading periods are discussed below.

The ideal shading device will block a maximum of solar radiation while still permitting views and breezes to enter a window. Table 9.2 shows some of the most common fixed external shading devices. They are all variations of either the horizontal overhang, the vertical fin, or the eggcrate, which is a combination of the first two. The louvers and fins can be angled for additional solar control. Almost an infinite number of variations are possible, as can be seen by looking at the work of such architects as Le Corbusier, Oscar Niemeyer, Richard Neutra, Paul Rudolph, and E. D. Stone. For examples of the work of

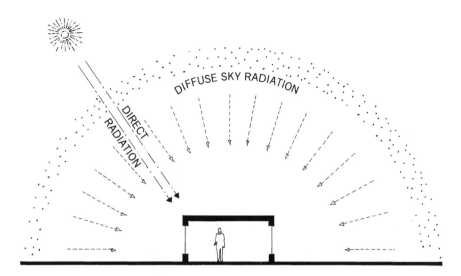

Figure 9.2d In humid and dusty regions, the diffuse-sky component is a large part of the total solar load.

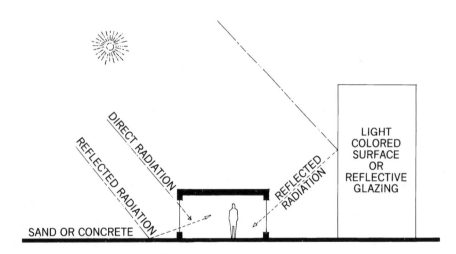

Figure 9.2e In dry regions, the solar load consists mainly of the direct and reflected components. However, reflective glazing can be a problem in all climates.

these and many other architects, the author highly recommends the book *Solar Control and Shading Devices* by Olgyay. Although this chapter is concerned with the shading the building itself produces, there is often significant shading from the surrounding environment. Neighboring buildings, trees, and landforms can all produce substantial shade. These shading con-

ditions are covered in Chapter 11, Site and Community Planning.

External shading devices are discussed first and in most detail because they are the most effective barrier against the sun and have the most pronounced effect on the aesthetics of a building.

TABLE 9.2 EXAMPLES OF FIXED SHADING DEVICES

		Descriptive Name	Best Orientation	Comments
I		Overhang Horizontal panel	South, east, west	Traps hot air Can be loaded by snow and wind
II		Overhang Horizontal louvers in horizontal plane	South, east, west	Free air movement Snow or wind load is small Small scale **Best buy!**
III		Overhang Horizontal louvers in vertical plane	South, east, west	Reduces length of overhang View restricted Also available with miniature louvers
IV		Overhang Vertical panel	South, east, west	Free air movement No snow load View restricted
V		Vertical fin	East, west, north	Restricts view For north facades in hot climates only
VI		Vertical fin slanted	East, West	Slant toward north Restricts view significantly
VII		Eggcrate	East, west	For very hot climates View very restricted Traps hot air
VIII		Eggcrate with slanted fins	East, west	Slant toward north View very restricted Traps hot air For very hot climates

From *Architectural Graphic Standards,* 8th ed. John R. Hoke, ed. Wiley, 1988.

9.3 ORIENTATION OF SHADING DEVICES

The horizontal overhang on south-facing windows is very effective during the summer because the sun is then high in the sky. Although less effective, the horizontal overhang is also the best on the east, southeast, southwest, and west orientations. In hot climates, north windows also need to be shaded because during the summer the sun rises north of east and sets north of west. Since the sun is low in the sky at these times, the horizontal overhang is not very effective and small vertical fins work best on the north facade (Fig. 9.3a).

East- and west-facing windows pose a difficult problem because of the low-altitude angle of the sun in the morning and afternoon. The best solution by far is to avoid using east and especially west windows as much as possible. The next best solution is to have the windows on the east and west facades face north or south as is shown in the plans of Fig. 9.3b. If that is also not possible, then horizontal overhangs and/or vertical fins should be used, but it must be understood that if they are to be very effective, they will severely restrict the view. Even movable devices described below, although better, still severely limit the view at certain times of day.

For more effective fixed shading devices, a combination of vertical and horizontal elements should be used, as shown in Fig. 9.3c. When these vertical and horizontal elements are closely spaced, the system is called an **eggcrate.** This device is most appropriate on east and west facades in *hot* climates and on the southeast and southwest facades in *extremely* hot climates.

Since the problem of shading is one of blocking the sun at certain angles, many small devices can have the same effect as a few large ones, as shown in Fig. 9.3d. In each case, the ratio of length of overhang to the vertical portion of window shaded is the

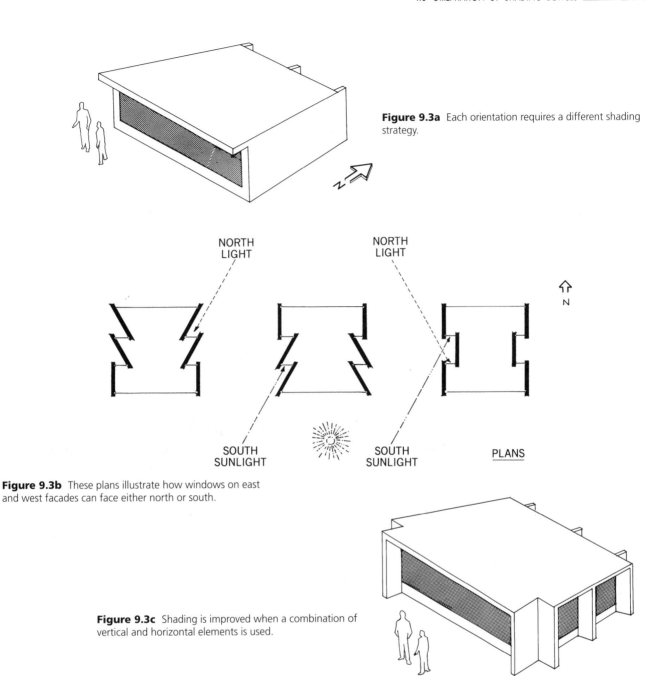

Figure 9.3a Each orientation requires a different shading strategy.

Figure 9.3b These plans illustrate how windows on east and west facades can face either north or south.

Figure 9.3c Shading is improved when a combination of vertical and horizontal elements is used.

same. There are screens available that consist of miniature louvers (about ten per inch) that are very effective in blocking the sun and yet are almost as transparent as insect screens.

Remember, view is the highest priority for most windows. Obstructing the view to shade a window is, therefore, counterproductive. For this reason, the horizontal overhang is usually the best choice. Although it obstructs the high sky, the more important hori-

zontal view is unimpeded.

Skylights (horizontal glazing systems) create a difficult shading problem because they face the sun most directly during the worst part of the year, summer at noon (Fig. 9.3e). Therefore skylights, like east and west windows, should be avoided. A much better solution for letting daylight and winter sun enter through the roof is the use of clerestory windows (Fig. 9.3f). The vertical glazing in the

clerestory can then be shaded by the window techniques explained in this chapter. If domed-type skylights are to be used, consider using shade/reflectors as illustrated in Fig. 7.6g.

Fixed, rather than movable shading devices are often used because of their simplicity, low cost, and low maintenance. Their effectiveness is limited, however, for several significant reasons, and movable shading devices should be seriously considered.

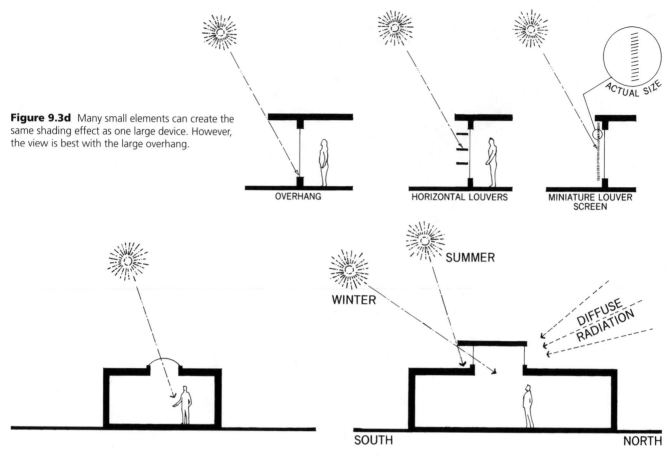

Figure 9.3d Many small elements can create the same shading effect as one large device. However, the view is best with the large overhang.

OVERHANG

HORIZONTAL LOUVERS

MINIATURE LOUVER SCREEN

ACTUAL SIZE

SUMMER

WINTER

DIFFUSE RADIATION

SOUTH

NORTH

Figure 9.3e Skylights (horizontal glazing) should usually be avoided because they face the summer sun.

Figure 9.3f Clerestory windows should be used instead of skylights because they allow the sun to enter in a controlled manner.

9.4 MOVABLE SHADING DEVICES

It is not surprising that movable shading devices respond better to the dynamic nature of weather than do static devices. Since we need shade during the overheated periods and sun during the underheated periods, a shading device must be in phase with the thermal conditions. With a fixed shading device, the period of solar exposure to the window is not a function of temperature but rather of sun position (Fig. 9.4a). Unfortunately, sun angles and temperature are not completely in phase. For one, daily weather patterns vary widely, especially in spring and fall when one day might be too hot while the next might be too cold. A fixed device that is wide enough to block the sun in late

April cannot make adjustments for a cold April day.

The other more important reason for the discrepancy between sun angles and temperature is that the solar year and thermal year are out of phase. Because of its great mass, the earth heats up slowly in spring and does not reach its maximum summer temperature until one or two months after the summer solstice (June 21). Similarly in the winter, there is a one- or two-month time lag in the cooling of the earth. The minimum heating effect from the sun comes on December 21, while the coldest days are in January or February. Figure 9.4b shows the overheated and underheated periods of the year for one of the U.S. climate regions. Note that the overheated period is not symmetrical

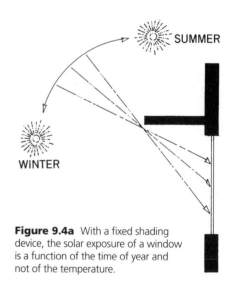

SUMMER

WINTER

Figure 9.4a With a fixed shading device, the solar exposure of a window is a function of the time of year and not of the temperature.

about June 21. A fixed shading device will shade for equal time periods before and after June 21. For example, April 21 and August 21 will receive the same shade, even though August is significantly hotter.

To get full shading, we might try a fixed shading device (Fig 9.4b), which is sized to provide shade through the end of the overheated period. Although we now have shade throughout the entire overheated period, we also shade the windows during part of the underheated period. Only a movable shading device, as shown in Fig. 9.4c, can overcome this problem as well as the problem of daily variations. The exception is in those buildings in which passive solar heating is not required. Here, a fixed shading device would be more appropriate.

The movement of shading devices can be very simple or very complex. An adjustment twice a year can be quite effective and yet simple. Late in spring, at the beginning of the overheated period, the shading device would be manually extended. At the end of the overheated period in late fall, the device would be retracted for full solar exposure (Fig. 9.4d).

Before air conditioning became available, awnings were used to effectively shade windows in the summer. Awnings were used on many buildings but were particularly common on luxury buildings, such as major hotels (Fig. 9.4e). In the winter, the awnings were removed to let more sun and light enter the building. Modern awnings are excellent shading devices. They can be durable, attractive, and easily adjustable to meet requirements on a daily and even hourly basis. Movable shading devices, which adjust to the sun on a daily basis, are often automated, while those that need to be adjusted only twice a year are usually manually operated. Table 9.4 presents a variety of movable shading devices.

In many ways, the best shading devices are the deciduous plants, most of which are in phase with the

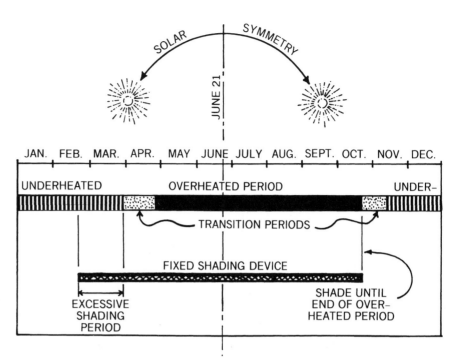

Figure 9.4b For a fixed shading device, the shading period is symmetrical about June 21. Since, however, the thermal year is not symmetrical about June 21, excessive shading occurs in late winter.

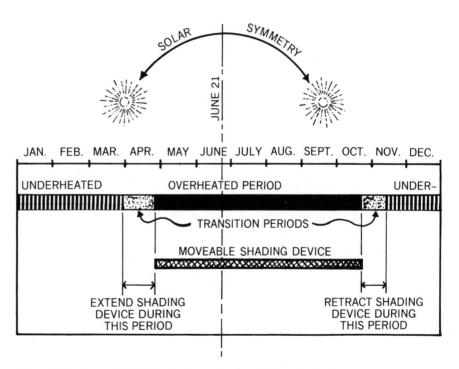

Figure 9.4c A movable shading device enables the shading to be in phase with the thermal year.

thermal year because they gain and lose their leaves in response to temperature changes. Other advantages of deciduous plants include low cost, aesthetically pleasing quality, ability to reduce glare, visual privacy, and ability to cool the air by evaporation from the leaves.

The main disadvantage of using plants is the fact that leafless plants still create some shade with some types much more than others (Fig. 9.4f). Other disadvantages include slow growth, limited height, and the possibility of disease destroying the plant. However, vines growing on a trellis or hanging from a planter can overcome many of these problems (Fig. 9.4g and h). Given enough time, vines will grow to great heights (Fig 9.4i). In hot climates, there is great benefit in shading not only windows but also walls. The darker the wall, the greater the benefit. A study in Miami of a moderately sparse vine (3 inches thick) on a west wall resulted in a drop of the wall temperature by 8°F in the morning and 14°F in the afternoon. For examples of vines and trees for shading see Chapter 11. In general, the east and west orientations are the best locations for deciduous plants.

Another very effective movable shading device is the exterior roller shade. The Bateson Office building makes very effective use of exterior fabric roller shades (see Fig. 17.7c). A roller shade made of rigid slats is very popular in Europe and is now available here (Fig. 9.4j). It offers security as well as very effective shading. These kinds of shading devices are especially appropriate on those difficult east and west exposures, where for half a day almost no shading is necessary and for the other half almost full shading is required.

There is a general conviction that since a building should be as low maintenance as possible, movable shading devices are unacceptable. This is a little like saying that because an automobile should be low mainte-nance, the wheels should be fixed and should not turn. The author believes that the use of existing technology and careful detailing can produce trouble-free, low-maintenance movable shading devices.

TABLE 9.4 EXAMPLES OF MOVABLE SHADING DEVICES

		Descriptive Name	Best Orientation	Comments
IX		Overhang Awning	South, east, west	Fully adjustable for annual, daily, or storm conditions; Traps hot air; Good for view; **Best buy!**
X		Overhang Rotating horizontal louvers	South, east, west	Will block some view and winter sun
XI		Fin Rotating fins	East, west	Much more effective than fixed; Less restricted view than slanted fixed fins
XII		Eggcrate Rotating horizontal louvers	East, west	View very obstructed but less than fixed eggcrate; For very hot climates only
XIII		Deciduous plants Trees Vines	East, west, southeast, southwest	View restricted but attractive for low-canopy trees; Air cooled
XIV		Exterior roller shade	East, west, southeast, southwest	Very flexible from completely open to completely closed; View is restricted when shield is used

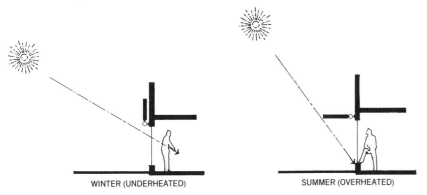

WINTER (UNDERHEATED)   SUMMER (OVERHEATED)

Figure 9.4d A movable shading device with just two simple adjustments per year can function extremely well.

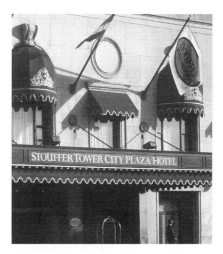

Figure 9.4e Awnings were a common element on many buildings during the first half of the twentieth century. After giving effective shade in the summer, they are removed in order to allow more sun and light to enter the building in the winter.

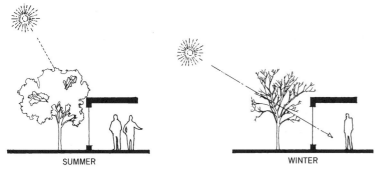

SUMMER   WINTER

Figure 9.4f The shading from trees depends on the species, pruning, and maturity of the plants. Transmission can be as low as 20 percent in the summer and as high as 70 percent in the winter. Unfortunately, with some trees the winter transmission can be as low as 40 percent.

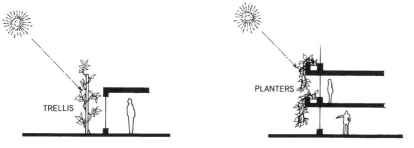

TRELLIS   PLANTERS

Figure 9.4g Vines can be very effective sun shading devices. Some vines grow as much as 30 feet in one year.

Figure 9.4h Since trees grow too slowly to help much on multistory buildings, planters can bring the shade plants to each level almost immediately.

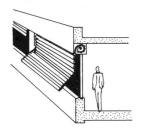

Figure 9.4i Medium-to-dark colored walls in hot climates benefit greatly from a vine cover.

Figure 9.4j Exterior roller shades made of rigid slats not only move in a vertical plane but can also project out like an awning.

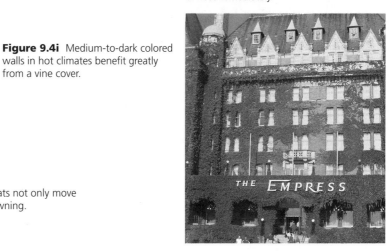

9.5 SHADING PERIODS OF THE YEAR

Windows need shading during the overheated period of the year, which is both a function of climate and building type. From an energy point of view, buildings can be divided into two main types: **envelope-dominated** and **internally dominated.**

The envelope-dominated building is very much affected by the climate because it has a large surface-area-to-volume ratio and because it has only modest internal heat sources. The internally dominated building, on the other hand, tends to have a small surface-area-to-volume ratio and large internal heat gains from such sources as machines, lights, and people. See Table 9.5A for a comparison of the two types of buildings.

A more precise way to define buildings than by the above two types is the concept of **balance point temperature** (BPT). Buildings do not need heating when the outdoor temperature is only slightly below the comfort zone because there are internal heat sources (lights, people, machines, etc.) and because the skin of the building slows the loss of heat. Thus, the greater the internal heat sources and the more effectively the building skin can retain heat, the lower will be the outdoor temperature before heating will be required. The BPT is that outdoor temperature below which heating is required. It is a consequence of building design and function and not climate. The BPT for a typical internally dominated buildings is about 50°F; for a typical envelope-dominated buildings it is about 60°F.

Since the comfort zone has a range of about 10°F wide (68°F to 78°F), the overheated period of the year starts at about 10°F above the BPT of any building. For example, for an internally dominated building (BPT = 50°F) the overheated period would start when the average daily outdoor temperature reached about 60°F. Consequently, the lower the BPT of a particular building, the shorter will be the underheated period (heating season), and the longer will be its overheated period (cooling season) during which time shading is required.

Table 9.5B shows the over- and underheated periods of the year for internally dominated (BPT = 50°F) buildings in each of seventeen climate regions, while Table 9.5C gives the same information for envelope-dominated buildings (BPT = 60°F). Note how much shorter the overheated periods are in Table 9.5C as compared to Table 9.5B. Also, it is very important to note that the overheated periods are not symmetrical about June 21 in every case. As mentioned earlier, the thermal year is always out of phase with the solar year.

TABLE 9.5A COMPARISON OF ENVELOPE AND INTERNALLY DOMINATED BUILDING TYPES

Characteristic	Envelope Dominated	Internally Dominated
Balance point temperature[a]	60°F	50°F or less
Building form	Spread out	Compact
Surface-area-to-volume ratio	High	Low
Internal heat gain	Low	High
Internal rooms	Very few	Many
Number of exterior walls of typical room	2 to 3	0 to 1
Use of passive solar heating	Yes, except in very hot climates	No, except in very cold climates
Typical examples	Residences, small office buildings, some small schools	Large office and school buildings, auditoriums, theaters, factories

[a]Superinsulated buildings tend to have a balance point temperature of about 50°F even though the other characteristics are those of an envelope-dominated building.

TABLE 9.5B OVERHEATED AND UNDERHEATED PERIODS FOR INTERNALLY DOMINATED BUILDINGS[a,b,c]

Climate Region	Reference City	JAN.	FEB.	MAR.	APR.	MAY	JUNE	JULY	AUG.	SEPT.	OCT.	NOV.	DEC.
1	Hartford, CT												
2	Madison, WI												
3	Indianapolis, IN												
4	Salt Lake City, UT												
5	Ely, NE												
6	Medford, OR												
7	Fresno, CA												
8	Charleston, SC												
9	Little Rock, AK												
10	Knoxville, TN												
11	Phoenix, AZ												
12	Midland, TX												
13	Fort Worth, TX												
14	New Orleans, LA												
15	Houston, TX												
16	Miami, FL												
17	Los Angeles, CA												

JUNE 21

■ OVERHEATED PERIOD
▨ UNDERHEATED PERIOD
▧ TRANSITION PERIOD

[a] Table is for well-constructed, modern internally dominated buildings (BPT=50°F).
[b] Overheated period—when average daily outdoor temperature is greater than 60°F.
[c] Underheated period—when average daily outdoor temperature is under 50°F.

TABLE 9.5C OVERHEATED AND UNDERHEATED PERIODS FOR ENVELOPE-DOMINATED BUILDINGS[a,b,c]

Climate Region	Reference City	JAN.	FEB.	MAR.	APR.	MAY	JUNE	JULY	AUG.	SEPT.	OCT.	NOV.	DEC.
1	Hartford, CT												
2	Madison, WI												
3	Indianapolis, IN												
4	Salt Lake City, UT												
5	Ely, NE												
6	Medford, OR												
7	Fresno, CA												
8	Charleston, SC												
9	Little Rock, AK												
10	Knoxville, TN												
11	Phoenix, AZ												
12	Midland, TX												
13	Fort Worth, TX												
14	New Orleans, LA												
15	Houston,TX												
16	Miami, FL												
17	Los Angeles, CA												

JUNE 21

■ OVERHEATED PERIOD
▨ UNDERHEATED PERIOD
▦ TRANSITION PERIOD

[a]Table is for well-constructed, modern envelope dominated buildings (BPT=60°F).
[b]Overheated period—when average daily outdoor temperature is greater than 70°F.
[c]Underheated period—when average daily outdoor temperature is under 60°F.

9.6 HORIZONTAL OVERHANGS

All shading devices consist of either horizontal overhangs, vertical fins, or a combination of the two. The horizontal overhang and its many variations are the best choice for the south facade. Because they are directionally selective, the can allow the low winter sun to enter while fully shading the high summer sun with minimum obstruction of the view. They are usually also the best choice for east, southeast, southwest, and west orientations.

Horizontal louvers have a number of advantages over solid overhangs. Horizontal louvers in a horizontal plane reduce structural loads by allowing wind and snow to pass right through. In the summer, they also minimize the collection of hot air next to the windows under the overhang (Fig. 9.6a). Horizontal louvers in a vertical plane (Diagram III in Table 9.2) are appropriate when the projecting distance from the wall must be limited. This could be important if a building is on or near the property line. Louvers can also be useful when

the architecture calls for small-scale elements and a rich texture.

When designing an overhang for the south facade, one must remember that the sun comes from the southeast before noon and from the southwest after noon. Therefore, the sun will outflank an overhang the same width as a window. Narrow windows need either a very wide overhang or vertical fins in addition to the overhang (Fig. 9.6b). Wide strip windows are affected less by this problem as can be seen in Fig. 9.6c.

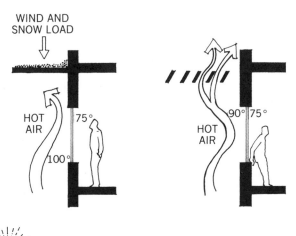

Figure 9.6a Horizontal louvered overhangs both vent hot air and minimize snow and wind loads.

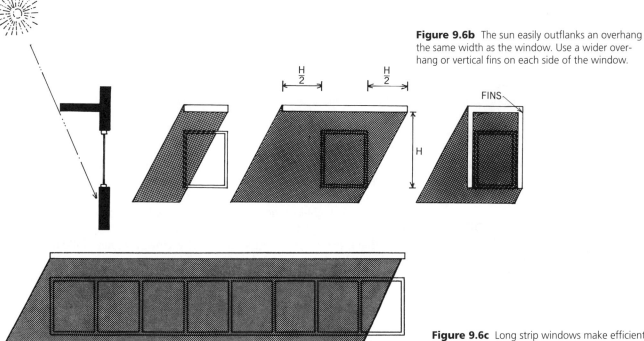

Figure 9.6b The sun easily outflanks an overhang the same width as the window. Use a wider overhang or vertical fins on each side of the window.

Figure 9.6c Long strip windows make efficient use of the horizontal overhang.

9.7 SHADING DESIGN FOR SOUTH WINDOWS

Since vertical fins alone are not appropriate on the south facade, the first step is to decide on either a fixed or movable horizontal overhang. Use the following rules for this purpose.

Rules for Selecting a South-Shading Strategy

1. If shading is the main concern and passive heating is not required, then a fixed overhang may be used.
2. If *both* passive heating and shading are important (long over- and underheated periods), then a movable overhang should be used.

The next step is to choose or design a particular kind of horizontal overhang. Refer to Tables 9.2 and 9.4 for examples of the generic types.

The size, angle, and location of the shading device can be determined by several different methods. The most powerful, flexible, and informative is the use of physical models. This method will be explained in detail in Section 9.17. There are also graphic methods, which are explained in other books (see recommended reading at the end of the chapter). Finally, there are rules and design guidelines. Because this last method is the quickest and easiest, it is presented here in some detail. It must be noted, however, that this method is always limited in flexibility. The author, therefore, strongly recommends the use of physical models in conjunction with the design guidelines described below.

9.8 DESIGN GUIDELINES FOR FIXED SOUTH OVERHANGS

As stated in the rules above, a fixed horizontal overhang is most appropriate when passive solar heating is not desired. The goal, then, is to find the length of overhang that will shade the

TABLE 9.8A SIZING SOUTH OVERHANGS ON INTERNALLY DOMINATED BUILDINGS[a,b,c]

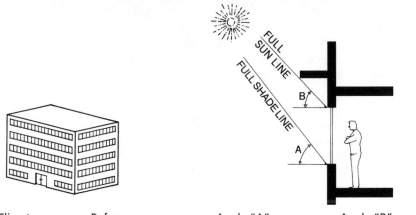

Climate Region	Reference City	Angle "A" (Full Shade)	Angle "B" (Full Sun)
1	Hartford, CT	59	54
2	Madison, WI	58	47
3	Indianapolis, IN	53	47
4	Salt Lake City, UT	60	49
5	Ely, NE	69	59
6	Medford, OR	59	45
7	Fresno, CA	55	33
8	Charleston, SC	54	36
9	Little Rock, AK	54	43
10	Knoxville, TN	53	41
11	Phoenix, AZ	48	d
12	Midland, TX	52	40
13	Forth Worth, TX	54	41
14	New Orleans, LA	49	d
15	Houston, TX	49	d
16	Miami, FL	40	d
17	Los Angeles, CA	33	d

[a]This table is for south-facing windows or windows oriented up to 20 degrees off south.
[b]An overhang reaching the "full shade line" will shade a window for most of the overheated period.
[c]An overhang not projecting beyond the "full sun line" will allow full solar exposure of a window for most of the underheated period.
[d]Use a fixed overhang projecting to the full shade line because passive solar heating is not required.

south windows during most of the overheated period.

Figure 9.8a shows the sun angle at the end of the overheated period. Since the sun is higher in the sky during the rest of the overheated period, any overhang that extends to the line shown will fully shade the window for the whole overheated period. This **full shade line** is defined by angle "A" and is drawn from the windowsill. This angle is given for each climate region in Table 9.8A for internally dominated

buildings and in Table 9.8B for envelope-dominated buildings.

Overhangs that are higher on the wall and that extend to the "full shade line" will still block the direct radiation and yet give a larger view of the sky. However, this would not be desirable in regions with significant diffuse radiation since both increased overheating and visual glare will result from the increased exposure to the bright sky (Fig. 9.8b). Even the overhang shown in Fig. 9.8a might not be

TABLE 9.8B SIZING SOUTH OVERHANGS ON ENVELOPE-DOMINATED BUILDINGS[a,b,c]

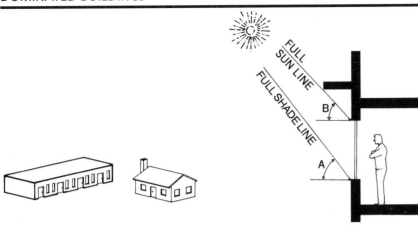

Climate Region	Reference City	Angle "A" (Full Shade)	Angle "B" (Full Sun)
1	Hartford, CT	65	59
2	Madison, WI	64	55
3	Indianapolis, IN	63	55
4	Salt Lake City, UT	65	60
5	Ely, NE	72	69
6	Medford, OR	71	61
7	Fresno, CA	64	45
8	Charleston, SC	65	49
9	Little Rock, AK	63	52
10	Knoxville, TN	62	51
11	Phoenix, AZ	56	49
12	Midland, TX	63	50
13	Forth Worth, TX	61	54
14	New Orleans, LA	63	44
15	Houston, TX	60	42
16	Miami, FL	50	d
17	Los Angeles, CA	61	43

[a]This table is for south-facing windows or windows oriented up to 20 degrees off south.

[b]An overhang reaching the "full shade line" will shade a window for most of the overheated period.

[c]An overhang not projecting beyond the "full sun line" will allow full solar exposure of a window for most of the underheated period.

[d]Use a fixed overhang projecting to the full shade line because passive solar heating is not required.

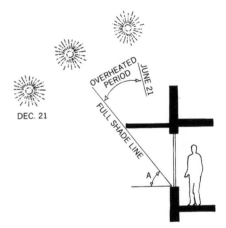

Figure 9.8a The "full shade line" determines the length of overhang required for shade during the overheated period.

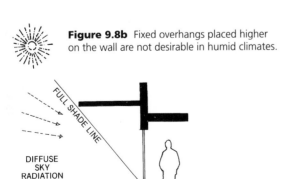

Figure 9.8b Fixed overhangs placed higher on the wall are not desirable in humid climates.

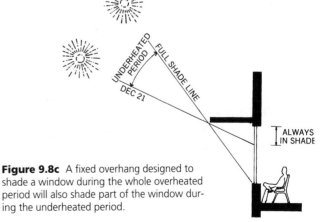

Figure 9.8c A fixed overhang designed to shade a window during the whole overheated period will also shade part of the window during the underheated period.

sufficient in very humid regions where over 50 percent of the total radiation can come from the diffuse sky. Rather than increasing the length of the overhang, it might be desirable to use other devices, such as curtains or plants, to block the diffuse radiation from the low sky.

As the sun dips below the "full shade line" later in the year, the window will gradually receive some solar radiation. However, the upper portion of the window will be in shade even at the winter solstice (Fig. 9.8c). Remember that *any fixed overhang that is very effective in the summer will also block some of the passive heating in the winter.*

Furthermore, an overhang extending to the full shade line can result in a quite dark interior. If daylighting is desired, as is often the case, then the strategies of Chapter 13 should be followed. The techniques described there allow ample light to enter a building while minimizing the overheating effect.

Procedure for Designing Fixed South Overhangs

1. Determine the climate region of the building from Fig. 5.5.
2. Determine angle "A" from Table 9.8A for internally dominated and from Table 9.8B for envelope-dominated buildings.
3. On a section of the window draw the "full shade line" from the windowsill.
4. Any overhang that extends to this line will give full shade during most of the overheated period of the year.
5. Shorter overhangs would still be useful, even though they would shade less of the overheated period.

9.9 DESIGN GUIDELINES FOR MOVABLE SOUTH OVERHANGS

The design of movable overhangs is the same as for fixed overhangs for the overheated period of the year. However, to make effective use of passive solar heating, the overhang must retract to avoid shading the window during the underheated period.

To ensure full-sun exposure of a window during the underheated period (winter), two points must be addressed. The first is to determine at which times of year the overhang must be retracted, and the second is to determine how far it must be retracted.

The simplest and most practical approach to the first question is to extend and retract the shading device during the spring and fall transition periods. These periods are described in Tables 9.5B and 9.5C. Making the twice-annual changeover could be no more complicated than washing the windows and could be done at the same time.

The sun angle at the *end* of the underheated period (winter) determines the **"full sun line"** (Fig. 9.9a). Since the sun is lower than this position during the rest of winter, any overhang short of this line will not block the sun when it is needed. This "full sun line" is defined by angle "B" and is drawn from the window's head. The appropriate angle is given for each climate region in Tables 9.8A and 9.8B.

Procedure for Designing Movable South Overhangs

1. Determine the climate region of the building from Fig. 5.5.
2. Determine angles "A" and "B" from Table 9.8A for internally dominated buildings and from Table 9.8B for envelope-dominated buildings.
3. On a section of the window, draw the "full shade line" (angle "A") from the windowsill, and draw the "full sun line" (angle "B") from the window head (Fig. 9.9b).
4. A movable overhang will have to extend to the "full shade line" during the overheated portion of the year and be retracted past the "full sun line" during the underheated period of the year. See Fig. 9.9c for some typical solutions.
5. The overhang should be extended during the spring transition period and retracted during the fall transition period. The dates for these transition periods can be determined from Table 9.5B for internally dominated and Table 9.5C for envelope-dominated buildings.

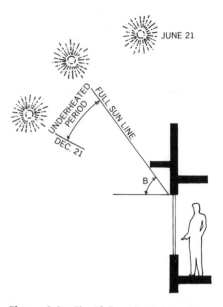

Figure 9.9a The "full sun line" determines the maximum allowable projection of an overhang during the winter period.

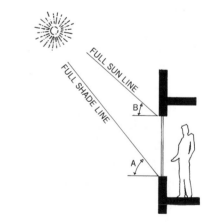

Figure 9.9b A fixed overhang, unlike a movable overhang, will not work well because it cannot both meet the "full shade line" and stay behind the "full sun line."

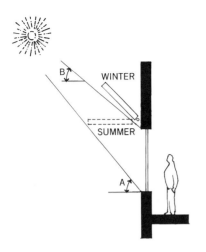

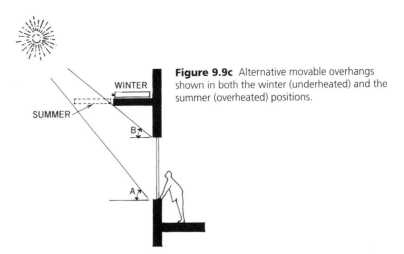

Figure 9.9c Alternative movable overhangs shown in both the winter (underheated) and the summer (overheated) positions.

9.10 SHADING FOR EAST AND WEST WINDOWS

On the east and west orientations, unlike the south, it is not possible to fully shade the summer sun with a fixed overhang. Fig. 9.10a shows how futile it would be to try to completely shade east or west windows with a horizontal overhang. Even though the direct sun rays cannot be shaded for the whole overheated period, it is nevertheless worthwhile to shade the windows part of the time.

Since little winter heating can be expected from east and west windows, shading devices on those orientations can be designed purely on the basis of the summer requirement.

No shading device will fully shade the east or west windows and allow a good view because the low sun will be part of the view. Since the view is a very high priority for windows, it is the author's opinion that a horizontal overhang offers the best combination of view and shade on the east and west. These overhangs must be much longer than the one on the south and should be backed up by another device, such as Venetian blinds.

Vertical fins are often presented as the shading device of choice for east and west. In fact, they shade no better

Figure 9.10a The 33-foot overhang needed to shade a 4-foot window on August 21 at 6 P.M. at 36N latitude illustrates the futility of trying to fully shade east and west windows with horizontal overhangs.

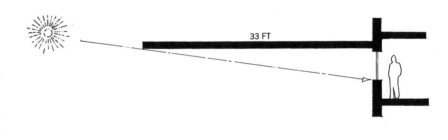

Figure 9.10b This plan view illustrates the sweep of the sun's azimuth angle at different times of the year from sunrise to sunset.

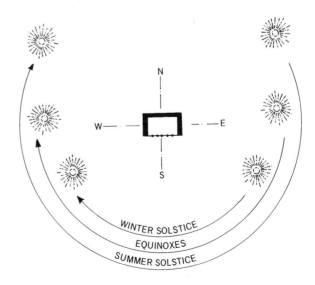

than the horizontal overhang, and they obstruct the view much more. Fig. 9.10b illustrates the fact that there is a time every morning and afternoon when the sun shines directly at the east and west facades of a building during the summer six months of the year (March 21 to September 21). Therefore, vertical fins that face directly east or west will allow some sun penetration every day during the worst six months of the year. To minimize this solar penetration, we need to minimize the "exposure angle (Fig. 9.10c). We can accomplish this by decreasing the spacing of the fins, by making the fins deeper, or some of both. To be highly effective, the fins must be so deep and so closely spaced that a view through them becomes almost impossible.

Vertical fins can be appropriate either when there is a desire to control the direction of view (e.g., slant fins to the northeast to block the view to the west and southwest) or when the view is not important. In that case, the fins could be slanted either to the south for more winter sun or to the north for more cool daylight (Fig. 9.12) or both if the fins are movable.

By moving in response to the daily cycle of the sun, movable fins allow somewhat unobstructed views for most of the day and yet block the sun when necessary. For example, movable fins on a west window would be held in the perpendicular position until the afternoon when the sun threatened to outflank them (Fig. 9.10d, top). Either at once or gradually, they would then rotate to the position shown in Fig. 9.10d, bottom. Movable fins on the east windows would, of course, work similarly. Thus, if both effective shading and views to the east and west are desirable, then movable rather than fixed vertical fins should be considered.

A note of caution is in order. Just as the sun outflanked a horizontal overhang the same width as the win-

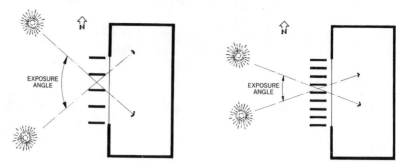

Figure 9.10c A plan view of vertical fins on a west (east) facade illustrates how solar penetration is reduced by moving fins closer together, by making them deeper, or both.

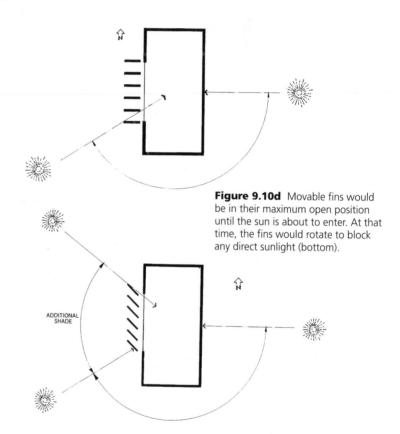

Figure 9.10d Movable fins would be in their maximum open position until the sun is about to enter. At that time, the fins would rotate to block any direct sunlight (bottom).

dow (Fig. 9.6b), so it will peak over the top of vertical fins that reach only the head of the window. Either extend the fins above the top of the window or cap the fins with an overhang the same width as the fins (Fig. 9.3c).

The advantages of both overhangs and vertical fins are such that the combination of the two devices is quite complementary. Consequently, on east or west orientations, a frame (Fig. 9.3c) or eggcrate system (Fig. 9.14a–d) is an effective shading system.

Rules for East and West Windows

1. Use as few east and especially west windows as possible.
2. Have windows on east or west facades face north or south, as shown in Fig. 9.3b.
3. If views of the ground and horizon are important, use a horizontal overhang with backup indoor shading.
4. Use trees, plant-covered trellises, or hanging plants (Figs. 9.4f–h).
5. When some obstructions to view

are acceptable, vertical fins are an alternative. Slant the fins toward the northwest if shading is required most of the year. Slant the fins toward the southwest if winter sun is desired.

6. Use movable shading devices for both better shading and better views.

7. When shading is critical and views are not, use an eggcrate system.

9.11 DESIGN OF EAST AND WEST HORIZONTAL OVERHANGS

When views to the east and west are desirable, it is especially worthwhile to consider using horizontal overhangs. A long overhang can be reasonably effective and yet give a better view of the landscape than do vertical fins. Although there is a fair amount of time when the sun peeks under an east and west overhang, there are many hours of useful shade. On a year-round basis, a horizontal overhang is a good alternative to vertical fins for east and west orientations.

When the sun peeks under overhangs or around vertical fins, it is important to have some additional protection in the form of Venetian blinds, roller shades, drapes, etc.

Table 9.11 allows the designer to determine the length of overhang needed to shade east and west windows from 8 A.M. to 4 P.M. (solar time) during most of the overheated period. The length of overhang thus determined should be a guide rather than a rigid requirement. Shorter overhangs, although less effective, are still worthwhile.

The absurdly long overhangs required in hot climates, as shown by Table 9.11 indicate the problem with east and west windows. It is, therefore, worth repeating once more that east and west windows should be avoided if at all possible in hot climates.

Procedure for Designing East and West Fixed Overhangs

1. Determine climate region from Fig. 5.5.
2. Determine angle "C" from Table 9.11.
3. On a section of the east or west window draw the "shade line" from the window sill.
4. Any overhang that projects to this line will shade east and west windows from 8 A.M. to 4 P.M. during

most of the overheated period. Of course, shorter overhangs would still be useful and longer ones would be even better.

TABLE 9.11 SIZING EAST AND WEST HORIZONTAL OVERHANGS*

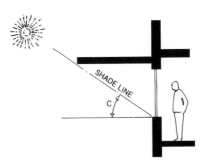

| Climate Region | Reference City | Angle "C" | |
		Internally Dominated	Envelope-Dominated
1	Hartford, CT	30	34
2	Madison, WI	30	34
3	Indianapolis, IN	25	32
4	Salt Lake City, UT	30	33
5	Ely, NE	34	36
6	Medford, OR	30	37
7	Fresno, CA	25	30
8	Charleston, SC	23	29
9	Little Rock, AK	24	29
10	Knoxville, TN	23	29
11	Phoenix, AZ	19	24
12	Midland, TX	22	28
13	Fort Worth, TX	23	28
14	New Orleans, LA	19	27
15	Houston, TX	19	25
16	Miami, FL	14	19
17	Los Angeles, CA	9	28

*Any overhang that extends to the "shade line" defined by angle "C" will shade east and west windows from 8 am to 4 P.M. during most of the overheated period. Choose the column for angle "C" according to the building type (internally or envelope dominated).

*The extremely long overhang required in hot climates indicates the problem of shading east and west windows.

9.12 DESIGN OF SLANTED VERTICAL FINS

The following procedure will yield a slanted vertical fin system that will shade east and west windows from direct sun for the whole year between the hours of 7 A.M. and 5 P.M. (solar time). Since the sun is fairly low in the sky after 5 P.M., neighboring trees and buildings often provide additional shade for windows on the ground floor.

Procedure for Designing Slanted Vertical Fins

1. Find the latitude of the building site from Fig. 5.6d.
2. From Table 9.12 determine the "shade line" angle "D."
3. On a plan of the east or west window, draw the "shade line" at angle "D" from an east–west line (Fig. 9.12, left).
4. Draw slanted vertical fins so that the head of one fin and the tail of the adjacent one both touch the "shade line" (Fig. 9.12, left). Different solutions are possible by varying the size, spacing, and slant of the fins (Fig. 9.12, right).

TABLE 9.12
SHADE LINE ANGLE FOR SLANTED VERTICAL FINS *

Latitude	Angle "D"
24	18
28	15
32	12
36	10
40	9
44	8
48	7

* This table is for vertical fins slanted toward the north on east or west windows. Designs based on this table will provide shade from direct sun for the whole year between the hours of 7 A.M. and 5 P.M. (solar time). This table can also be used to design vertical fins on north windows for the same time period.

9.13 DESIGN OF FINS ON NORTH WINDOWS

Buildings with long overheated periods may also require shading of north windows. Because of the sun angles involved, small vertical fins are sufficient to give full shade from 7 A.M. to 5 P.M. (solar time) (See Fig. 17.10b and c). Figure 9.13 illustrates how the fins are determined by the same angle "D" used for sizing slanted fins on the east and west.

Procedure for Designing North Fins

1. Find the latitude of the building site from Fig. 5.6d.
2. From Table 9.12 determine the appropriate angle "D."
3. On a plan of the north window, draw the "shade line" at angle "D" from an east–west line (Fig. 9.13 left).
4. Draw vertical fins to meet this "shade line," and note that if intermediate fins are used, all fins will be shorter (Fig. 9.13 right).
5. Remember that fins are required on both the east and west sides of north windows.

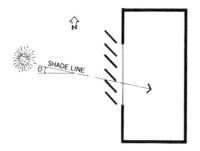

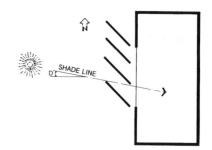

Figure 9.12 The "shade line" at angle "D" determines the combination of fin spacing, fin depth, and fin slant on east and west windows. An alternative solution is also shown.

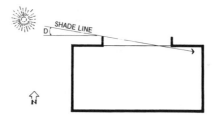

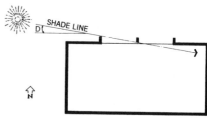

Figure 9.13 The "shade line" at angle "D" also determines the vertical-fin design on north windows. An alternative solution is also shown.

9.14 DESIGN GUIDELINES FOR EGGCRATE SHADING DEVICES

Eggcrate shading devices are mainly for east and west windows in *hot* climates and for the additional southeast and southwest orientations in *very hot* climates. An eggcrate is a combination of horizontal overhangs (louvers) and vertical fins. By controlling sun penetration by both the altitude and azimuth angle of the sun, very effective shading of windows can be achieved. The view, however, is usually very obstructed.

The "brise-soleil," developed by Le Corbusier, is an eggcrate system with dimensions frequently at the scale of rooms (Figs. 9.1j and k). Since shading is a geometric problem, many small devices are equivalent to a few large ones (see again Fig. 9.3d). Therefore, eggcrates can also be made at the scale of a fine screen. In India these screens were often cut from a single piece of marble (Fig. 9.14a). Today, these screens are most often made of metal (Fig. 9.14b) or masonry units (Fig. 9.14c). The shading effect of eggcrates at different scales is identical, but the view from the inside and the aesthetic appearance from the outside vary greatly.

The designer should first decide on the general appearance of the eggcrate system. The required dimensions of each unit (Fig. 9.14d) are best determined experimentally by means of a sun machine. As far as sun penetration is concerned, the scale of the eggcrate can be changed at any time as long as the ratios of h/d and w/d are kept constant. The use of the sun machine for this purpose will be explained in detail below.

Figure 9.14a This marble screen, carved from a single piece of stone, is actually a miniature eggcrate shading device. (Photograph by Suresh Choudhary.)

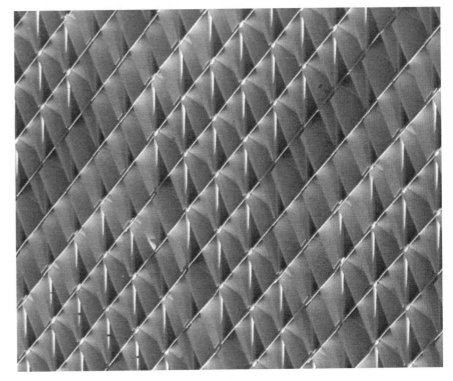

Figure 9.14b An eggcrate shading device made of metal. (Courtesy of Construction Specialties, Inc.)

Figure 9.14c Eggcrate shading device made of masonry units.

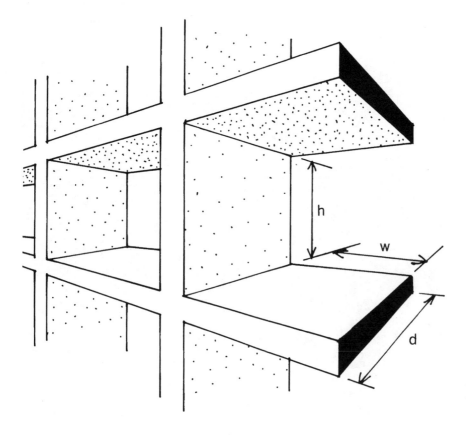

9.15 SPECIAL SHADING STRATEGIES

Most external shading devices are variations of the horizontal overhang, vertical fin, or eggcrate. However, some interesting exceptions exist.

The geodesic dome designed by Buckminster Fuller for the U.S. Pavillion, Expo '67, created an artificial climate for the structures within (Fig. 4.2b). To prevent overheating inside the clear plastic dome, the upper panels had vents and special roller shades. Each glazed hexagonal structural unit had six triangular roller shades operated by a servomotor. Figure 9.15a shows these shades in various positions.

A completely different approach is to rotate the building with the changing azimuth of the sun. A solid wall, which could be covered with solar collectors, would always face the sun (Fig. 9.15b). If this sounds far-fetched, consider the fact that rotating buildings have already been built (Fig. 9.15c). To enjoy the beautiful panoramic view of his Connecticut property, Richard Foster built a revolving house. Although he did not build a blank wall facing the sun, he did include a wide peripheral porch for shade (Figs. 9.15d and e). As the building revolves, the large area of glazing and concrete floor slab allow passive solar to uniformly heat the building even on the coldest sunny day.

A similar but simpler approach would be to have the building stand still but a shade move around the building. For example, a barn door hanging from a curved track could follow the sun around a building (Fig. 9.15f). If the barn door were covered with photovoltaic cells, you would have a tracking solar collector as well.

Figure 9.14d The shading effect is a function of the ratios h/d and w/d. It is not a function of actual size.

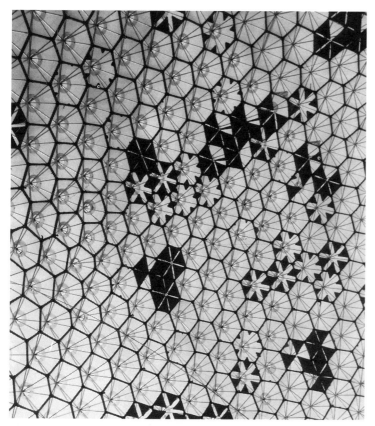

Figure 9.15a The U.S. Pavilion, Expo '67, Montreal, Canada, was designed by Buckminster Fuller. This view of the dome from the inside shows the vent holes (upper left panels) and triangular roller shades that prevent overheating in the summer.

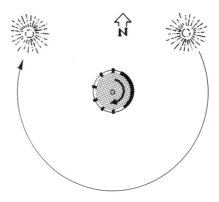

Figure 9.15b Plan view of a building that rotates during the day so that an opaque wall or solar collector always faces the sun.

Figure 9.15c The residence that architect Richard Foster built for himself in Wilton, CT in 1967 is round and rotates 360 degrees to take full advantage of the panoramic view and passive solar heating. (Courtesy of Richard Foster, architect.)

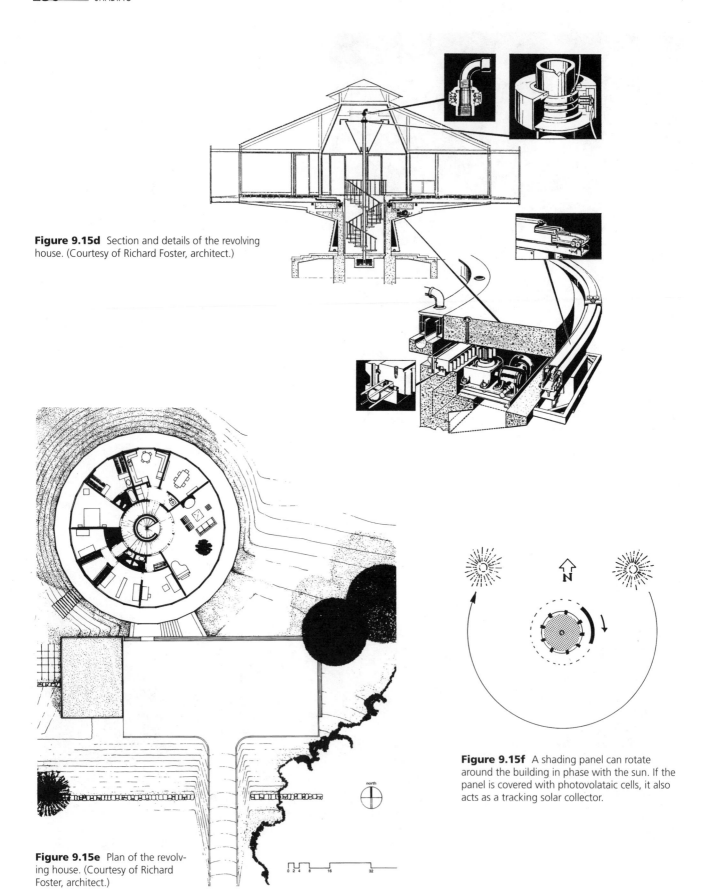

Figure 9.15d Section and details of the revolving house. (Courtesy of Richard Foster, architect.)

Figure 9.15e Plan of the revolving house. (Courtesy of Richard Foster, architect.)

north

0 2 4 8   16   32

Figure 9.15f A shading panel can rotate around the building in phase with the sun. If the panel is covered with photovolataic cells, it also acts as a tracking solar collector.

Figure 9.16a A membrane tension structure for shading outdoor seating, Snowmass, CO.

9.16 SHADING OUTDOOR SPACES

Shading of outdoor spaces can be just as critical as shading buildings. Open-air amphitheaters and stadiums are a special problem because of their size and need for unobstructed sightlines. The most popular solution is the use of membrane tension structures since they can span large distances at relatively low cost. Most often waterproof membranes are used because they also protect against rain and snow (Fig. 9.16a), but in dry climates, an open weave fabric might be appropriate. This is not a new idea, however, for the Romans covered not only their theaters but even their gigantic Colosseum with an awning (Fig. 9.16b). Modern small-scale versions of removable awnings called toldos are shown in Figs. 9.16c and d.

Many traditional shading structures were designed to create shade while letting air and rain pass right through. The **pergola, trellis,** and **arbor** are examples of such structures (Fig. 9.16f). Interesting nontraditional shading structures are shown in Fig. 9.16g and Fig. 9.16h.

When the outdoor shading system is fixed and permanent, it should be designed to let the winter sun enter while rejecting the summer sun (Fig. 9.16e). The author has seen many shading structures that created more shade in the winter than in the summer. Designing a successful shading structure is not as simple as it might appear. The best way to design a shading system for an outdoor space as well as for a building is to use physical models on a sun machine. This technique will now be explained.

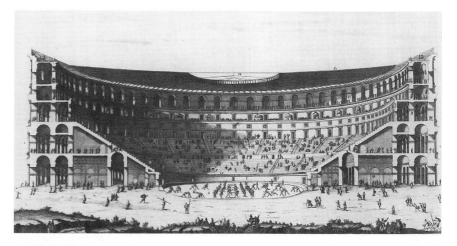

Figure 9.16b The Roman Colosseum, which was built about 80 AD and which seated about 50,000 spectators, was covered with a giant awning for sun protection. (From *Lanfiteatro Flavio Descritto e Deliniato* by Carlo Fontana, Vaillant, 1725.)

Figure 9.16c This type of retractable awning suspended from cables is called a toldo. Disneyland, CA.

Figure 9.16d This winter garden in Washington, DC, is protected from the summer sun by folding awnings also known as toldos.

Figure 9.16e Fixed, outdoor shading systems should allow hot air to escape and winter sun to enter as in this structure in San Jose, California.

Figure 9.16f Traditional outdoor shading structures. (Top) Trellis around outdoor reading areas of the public library in San Juan Capistrano, California, designed by Michael Graves; (middle) pergola; (bottom) arbor in the garden of the Governor's Palace, Colonial Williamsburg, Virginia. (Courtesy of Richard Kenworthy.)

Figure 9.16g Perforated screen shading structure. (Courtesy of *ARAMCO World* magazine.)

Figure 9.16h Antione Predock used a trellis of steel bars to shade outdoor walkways and sculpture gardens at the Fine Arts Center at Arizona State University in Tempe.

9.17 USING PHYSICAL MODELS FOR SHADING DESIGN

Sun machines were introduced earlier and the author suggested that building a sun machine is an excellent investment for an architectural office or school. Appendix C gives detailed instructions for making and using a sun machine. One of its main applications is the design of shading devices. Testing a model of a shading device not only gives feedback on the performance of the device but also teaches the designer much about the whole question of sun shading. Since this method of design is conceptually simple, it is easy to learn and to remember. The step-by-step procedure for designing shading devices by means of physical models is followed by an illustrative example.

Procedure for Shading Design by Means of Physical Models

1. Build a scale model of the building or at least a typical portion of the building facade.

2. Set up the sun machine and adjust it for the correct latitude (see Appendix C).

3. Place the model on center of the sun machine tilt table. Be sure to orient the model properly (e.g., south window should face south, as shown in Fig. 9.17a).

4. Determine the end of the over- and underheated periods from Tables 9.5B and 9.5C for internally and envelope-dominated buildings, respectively.

5. Set the sun-machine light to match the date of the end of over-

heated period, check the shading on the model, and adjust the overhang as necessary.

6. Rotate the model stand to simulate the changing shadows at different hours on that date.

7. Make more adjustments on model to achieve the desired shading.

8. Set the sun-machine lamp to match the date of the end of underheated period, and check the sun penetration.

9. Make changes on the model if the sun penetration is not sufficient.

10. Rotate the model stand to simulate the changing shadows on this date.

11. Repeat steps 5 to 10 until an acceptable design has been developed.

Illustrative Example

Problem: A horizontal overhang is required for a small office building (envelope-dominated) in Indianapolis, Indiana. Daylighting will not be considered in this example. The overhang is for a 5-foot-wide and 4-foot-high window on a wall facing south.

Solution

1. Build a model of the window with some of the surrounding wall. For convenience, the model should be about 6 inches on a side (Fig. 9.17a). Use a clear plastic film, such as acetate, for the glazing.
2. Appendix C explains how to set up and use the sun machine. Adjust the tilt table for the latitude of Indianapolis, which is 40°N latitude.
3. With pushpins or double-stick tape, tack the model to the center of the tilt table and orient it south (Fig. 9.17a).
4. From Fig. 5.5, determine that Indianapolis is in climate region 3. Since it is given that the building is envelope dominated, use Table 9.5C to determine that the over- heated period ends about September 15 and the underheated period ends about May 7.
5. Move the light on the sun machine to correspond with September 15. Cut and attach an overhang of such length that the shadow just reaches the windowsill (Fig. 9.17a).
6. When one rotates the model stand, the shadows for different times of the day can be investigated. Note how the sun outflanks the window at 4 P.M., because the overhang was not wide enough (Fig. 9.17b).
7. The overhang is made wider (Fig. 9.17c).
8. Move the lamp to the position corresponding to the end of the underheated period, which we determined above to be about May 7. At this time, the window should still be in sun and not shaded (Fig. 9.17d). Since a shorter overhang would decrease the summer shading, use a movable overhang instead.
9. Swing the overhang up until the window is fully exposed to the winter sun (Fig. 9.17e).
10. Rotate the model stand to see how shade changes during the day.
11. The solution in this case is for an overhang that is moved twice a year. During the summer, the overhang is as shown in Fig. 9.17d and during the winter it is up as in Fig. 9.17e.

Model testing can reveal many surprises. For example, little sun penetrates the glazing at acute angles. The glazing acts almost like a mirror at these angles (Fig. 9.17f). This phenomenon is explained later in this chapter.

Even complicated shading problems are easy to solve by physical modeling. For example, the analysis of a shading system for a complex building with odd angles and round features (Fig. 9.17g) is no more difficult than the analysis for a conventional building.

Since this tool can also easily simulate the shading from trees, neighboring buildings, and landforms, its use is also very appropriate for site planning (Chapter 11).

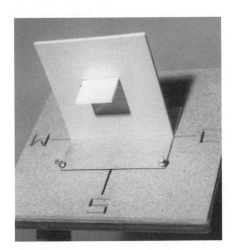

Figure 9.17a The construction and placement of the model on the sun machine. The shadow exactly covers the window and correspond to 12 noon at the end of the overheated period (September 15) for this design.

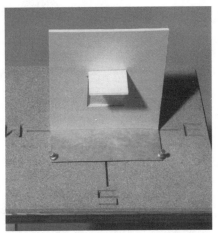

Figure 9.17b The shadow now corresponds to 4 P.M. on September 15. Note how the sun is outflanking the overhang.

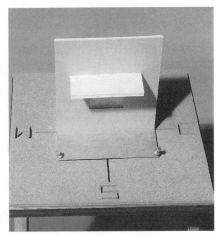

Figure 9.17c The overhang is redesigned by making it wider.

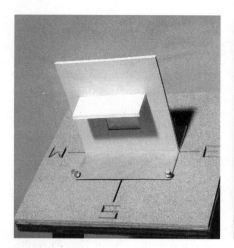

Figure 9.17d The light is readjusted to simulate the shading just before the end of the underheated period (May 7 in this case) at which time sun is still desired. Instead, the window is in shade.

Figure 9.17e The overhang is rotated up until the window is fully exposed to the winter sun. This determines the position for the overhang during the whole underheated period.

Figure 9.17f The model shows that at very large angles of incidence the sun is mostly reflected off the glazing. Note the reflections onto the ground below the window.

Figure 9.17g No matter how complicated the shading problem is, physical modeling can help the designer.

9.18 GLAZING AS THE SHADING ELEMENT

Even the clearest and thinnest glass does not transmit 100 percent of the incident solar radiation. The radiation that is not transmitted is either absorbed or reflected off the surface (Fig. 9.18a). The amount that is absorbed depends on the type of, additives to, and thickness of the glazing. The amount that is reflected depends on the nature of the surface and the angle of incidence of the radiation. Each of these factors will be explained below, starting with absorption.

Most types of **clear** glass and plastic vary only slightly in the amount of radiation they absorb. The thickness is not very critical either. Absorption is mainly a function of additives that give the glazing a tint or shade of gray. Although tinted glazing reduces the light transmission, it usually does not decrease the heat gain by much because the absorbed radiation is then reradiated indoors (Fig. 9.18b). One type of tinted glazing is called **heat absorbing** because it absorbs the short-wave infrared part of solar radiation much more than the visible part. But even this type of glazing reduces the solar heat gain by only a small amount.

Tinted glass was very popular in the 1960s because it reduced, even if only slightly, the solar load through glass-curtain walls. It also provided color to what was otherwise often stark architecture. It was originally available only in greens, grays, and browns, but recently it has become available also in blue and its popularity is again on the rise.

Glazing also blocks solar radiation by reflection. The graph in Fig. 9.18d shows how transmittance is a function of the angle of incidence. It also shows how the angle of incidence is always measured from the normal to the surface. Notice that the transmittance is almost constant for an angle of incidence from 0 degrees to about 45 degrees. Above 70 degrees, however, there is a pronounced reduction in the transmittance of solar radiation through glazing. Several architects have used this phenomenon as a shading strategy. One of the most dramatic examples is the Tempe, Arizona, City Hall (Fig. 9.18e).

The amount of solar radiation that is reflected from glazing can be increased significantly by adding a reflective coating. One surface of the glazing is covered with a metallic coating thin enough that some solar radiation still penetrates. The percentage reflectance depends on the thickness of this coating, and a mirror is nothing more than a coating that is thick enough so that no light is transmitted. **Reflective glazing** can be extremely effective in blocking solar radiation while still allowing a view (Fig. 9.18c). It is most appropriate on the east- and west-facing windows.

When reflective glazing became available in the 1970s, it quickly became popular for several reasons. It blocked solar radiation better than heat-absorbing glass, and did it without any color distortion. Compare the total solar transmittance in Figs. 9.18a, 9.18b, and 9.18c. Reflective glazing also mirrored dramatic images of other buildings, clouds, etc.

Although tinted and reflective glazing systems can be effective shading devices, they are very undiscerning. They do not differentiate between light from the sun and light from the view. They filter out light whether daylighting is desired or not. And they block the desirable winter sun as much as the undesirable summer sun. Thus, tinted or reflective glazing is not appropriate where either daylighting or solar heating is desired. It is

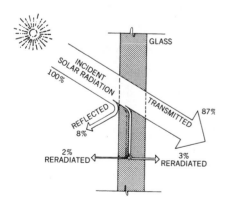

Figure 9.18a The total heat gain from the incident solar radiation consists of both the transmitted and reradiated components. For clear glazing, about 90 percent of the incident solar radiation, ends up as heat gain.

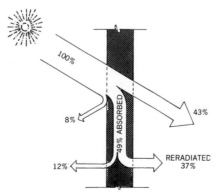

Figure 9.18b Since with heat-absorbing glass a large proportion of the absorbed solar radiation is reradiated indoors, the total heat gain is quite high (80 percent).

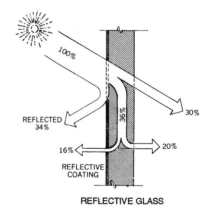

Figure 9.18c Reflective glazing effectively blocks solar radiation without color distortion. Reflective glass is available in a variety of reflectances—50 percent is shown.

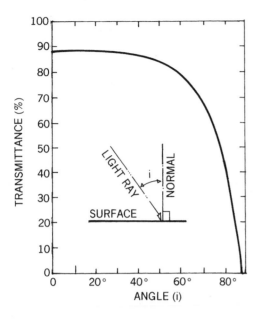

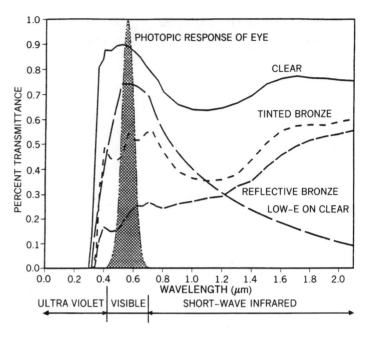

Figure 9.18d The transmittance of solar radiation through glazing is a function of the angle of incidence, which is always measured from the normal to the surface.

Figure 9.18f Spectally selective low-e glazing transmits cooler daylight because it reflects the short-wave infrared much more than the visible radiation. (From "Effects of Low Emissivity Glazings on Energy Use Patterns," *ASHRAE Transactions,* Vol. 93, Pt. I 1989.)

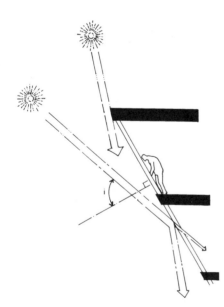

Figure 9.18e The City Hall of Tempe, Arizona, is an inverted pyramid as a consequence of a shading concept. When the sun is high in the sky, the building shades itself. At low sun angles, much of the solar radiation is reflected off the glazing because of the very large angle of incidence.

also not appropriate when only the sun should be excluded, but not the view. When the glazing is expected to do all the shading, it has to be of a very low transmittance type. The view through this kind of glazing makes even the sunniest day look dark and gloomy. Thus, external overhangs, fins, etc., which are more discerning, are usually still the best shading devices. Tinted or reflective glazing is excellent, however, for blocking diffuse sky radiation in very humid regions, and for glare control, which will be discussed in Chapter 13.

The directionally selective control possible with external shading can also be achieved by the glazing itself in certain special circumstances. The opaque mortar joints in glass-block construction can act as an eggcrate shading system. A new type of glass incorporates photoetched slats that can be ordered at any preset angle. The resultant effect will be similar to the horizontal louvers illustrated in Fig. 9.3d.

When daylighting is desired and solar heating is not, having the visible component of solar radiation pass through while heat radiation is blocked would be advantageous. Certain "spectrally selective" glazing systems can do that to a limited extent. As Fig. 9.18f illustrates, spectrally selective low-e glazing transmits cooler daylight than other glazing materials because it transmits a much higher ratio of visible-to-infrared radiation.

In the near future, there might be even better glazing systems than the "selective" types mentioned above. These are known as **responsive glazing** systems because they change in response to light, heat, and electricity. The sunglasses that darken when exposed to sunlight are an example of this type of glass.

Responsive glazing can be either the passive or the active kind. Passive glazing responds directly to environmental conditions, such as light level or temperature (photochromics or thermochromics, respectively). The active system can be controlled as needed and can include such devices as liquid crystal, dispersed particle, and electrochromics.

Photochromics: These materials change their transparency in response to light intensity. They are ideal for automatically controlling the quantity of daylight allowed into a building. The goal is to let in just enough light to eliminate the need for electric lighting, but not so much that the cooling load would increase.

Thermochromics: These materials change transparency in response to temperature. They are transparent when cold and reflective white when hot. They can be used in skylights, where the loss of transparency on a hot day is not a problem as it would be in a view window. These materials could also be used to prevent passive systems from overheating in the summer.

Liquid-Crystal Glazing: When electric power is applied, the transparent liquid crystals align and become translucent. Thus, liquid-crystal glazing has some application for shading, but its real potential is in privacy control.

Dispersed-Particle Glazing: Although similar to liquid-crystal glazing, this material is more promising for solar control because the applied power can change the transmittance of the material in a range between clear and dark states, thereby preserving the view.

Electrochromic Glazing: This is the most promising material because it can change transparency—not translucency—continuously over a wide range (about 10 percent to 70 percent) and can be easily controlled. Consequently, either a computer, a photocell, a thermostat, or the occupant can adjust the transparency as the local conditions require.

Rules for Glazing Selection

1. Use clear glazing when solar heating is a major concern (especially on a south facade).
2. Use reflective glazing when solar heat gain must be minimized and the use of external shading devices is not possible (especially on east and west facades).
3. Use the new selective low-e glazing when solar heat gain must be minimized but daylighting is still desired and the use of external shading devices is not available.
4. Blue-green heat-absorbing glazing or other tinted glazing materials are a less efficient alternative to reflective glazing.

9.19 INTERIOR SHADING DEVICES

From an energy-rejection point of view, the external shading devices are by far the most effective. But for a number of practical reasons, the interior devices, such as curtains, roller shades, Venetian blinds, and shutters, are also very important (Fig. 9.19a). Interior devices are often less expensive than external shading devices, since they do not have to resist the elements. They are also very adjustable and movable, which enables them to easily respond to changing requirements. Besides shading, these devices provide numerous other benefits, such as privacy, glare control, insulation, and interior aesthetics. At night, they also prevent the "black hole" effect created by exposed windows.

Since internal devices are usually included whether or not external devices are supplied, we should use them to our advantage. They should be used to stop the sun when it outflanks the exterior shading devices. They are also useful for those exceptionally hot days during the transition or underheated periods of the year, when exterior shading is not designed to work. In the form of Venetian blinds or light shelves (Fig. 9.19b), they can also produce fine daylighting.

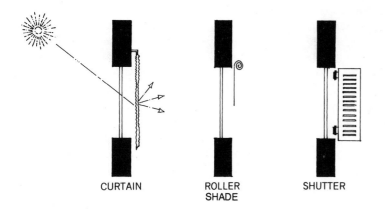

Figure 9.19a Interior shading devices for solar control.

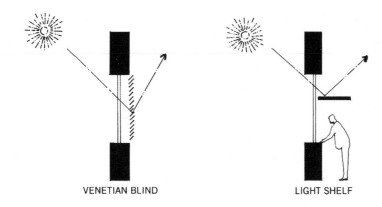

Figure 9.19b Interior shading devices that contribute to quality daylighting.

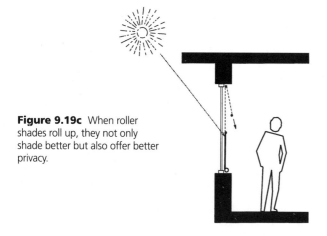

Figure 9.19c When roller shades roll up, they not only shade better but also offer better privacy.

One of the main drawbacks of interior devices is that they are not always discerning. They cannot block the sun while admitting the view, something that can be effectively done with an external overhang. Since they block the solar radiation on the inside of the glazing, much of the heat remains indoors. The side of the shade facing the glass should be as light as possible (white) in order to reflect solar radiation back out through the glass before it is converted to heat.

When indoor shades are used in conjunction with overhangs, the shades should move up from the window sill instead of down from the window head (Fig. 9.19c). The lower portion of a window always needs more shade than the upper. Thus, some view, privacy, and daylighting can be maintained while still shading the sun.

There are, however, a few very sophisticated interior shading systems. The Hooker Chemical Corporate Office Building in Niagara Falls, New York, is a good example. The window wall consists of horizontal louvers between two layers of glass that are about 4 feet apart (see Fig. 13.11i). The heat that builds up in this space during the summer is allowed to vent out through the roof. Shading is only one of the functions of the louvers. They also reflect daylight into the interior, and they can close up to insulate the building at night.

One of the best inventions of all time is the Venetian blind. It is extremely flexible and can be used in a very directionally selective way (i.e., it can block the sun but not the view or can send the light to the ceiling for quality daylighting). Recently Venetian blinds have become even better. A new type, with tiny perforations, allows some light and view to enter even when the shades are completely closed. Although the Venetian blinds come in many colors, the ones with white or mirrored finishes are most functional for heating, cooling, and daylighting.

9.20 SHADING COEFFICIENT AND SOLAR HEAT-GAIN COEFFICIENT

The various shading devices mentioned in this chapter can be compared in a quantitative way by the concept of the **shading coefficient (SC)**. Although this coefficient was developed for use in analytical work, we will use it here as a rough way to

TABLE 9.20 SHADING COEFFICIENTS (SC) AND SOLAR HEAT GAIN COEFFICIENTS (SHGC) FOR VARIOUS SHADING DEVICES

Device	SC	SHGC
Single glazing		
Clear glass, 1/8-inch thick	1.0	0.86
Clear glass, _-inch thick	0.94	0.81
Heat absorbing or tinted	0.6–0.8	0.5–0.7
Reflective	0.2–0.5	0.2–0.4
Double glazing		
Clear	0.84	0.73
Bronze	0.5–0.7	0.4–0.6
Low-e clear	0.6–0.8	0.5–0.7
Spectrally selective	0.4–0.5	0.3–0.4
Triple-clear	0.7–0.8	0.6–0.7
Glass Block	0.1–0.7	
Interior shading		
Venetian blinds	0.4–0.7	
Roller shades	0.2–0.6	
Curtains	0.4–0.8	
External shading		
Eggcrate	0.1–0.3	
Horizontal overhang	0.1–0.6	
Vertical fins	0.1–0.6	
Trees	0.2–0.6	

NOTES:
1. The smaller the number, the less solar radiation enters through a window. A value of zero indicates that the window allows no solar radiation to enter either directly or reradiated after being absorbed.
2. Ranges are given either because of the large variety of glazing types available (e.g., slightly or heavily tinted) or because of the varying geometry due to differences in orientation, sun angle, size and type of shading device, and variations in window size.
3. Source: ASHRAE Fundamentals Handbook 1997, and Egan, 1975.

compare various shading devices. The author uses the word "rough" because the SC does not account for the fact that often we want to block the high summer sun and not the low winter sun, or that we want to block the sun but not the view, as is possible with directionally selective shading systems, such as a horizontal overhang. Because the effectiveness of external shading devices depends on the specific design and sun angles, it is not possible to assign a precise number to generic types. Nevertheless, Table 9.20 can give us some idea of the relative effectiveness of the various shading devices.

The SC is most precise for traditional, simple glazing systems. Because it is not precise for the new complex glazing systems of multiple air spaces and coatings, a new coefficient has been introduced. The new **solar heat-gain coefficient (SHGC)** is similar to the SC as a measure of how much solar radiation enters a window. They are both dimensionless numbers ranging from zero to one with zero indicating that no solar radiation is passing through. However, the solar radiation passing through clear, single glazing results in an SC of one and an SHGC of 0.86. Table 9.20 also gives the SHGC for the glazing systems.

Figure 9.21a The south and east facades of the Biological Sciences Building at the University of California at Davis illustrate how unity can be maintained while each orientation responds to its unique conditions.

Figure 9.21b The north and west facades of the same building shown in Fig. 9.21a

Figure 9.21c The office slab of the United Nations Headquarters in New York City, 1950, became the prototype for many office buildings. Le Corbusier was very upset when he discovered that the blank walls faced mostly north and south, while the glass facades faced mostly east and west and were in no way protected by sunshades. (Courtesy of New York Convention and Visitors' Bureau, Inc.)

9.21 CONCLUSION

The ideal building from a shading point of view will have windows only on the north and south facades, with some type of horizontal overhang protecting the south-facing windows. The size and kind of overhang will depend on the type of building, climate, and latitude of the building site.

Even if there are windows on all sides, as there often will be, a building should not look the same from each direction. Each orientation faces a very different environment. James M. Fitch in his *American Building: 2. The Environmental Forces That Shape It* pointed out that moving from the south side of a building to the north side is similar to traveling from Florida to Maine. A building can still have unity without the various facades being identical. Even the east and west facades, which are symmetrical from a solar point of view, should rarely be identical. They differ because afternoon temperatures are much higher than morning temperatures, and because site conditions are rarely the same (e.g., trees toward the east but not the west). One of the buildings at the University of California at Davis exemplifies very well this unity with diversity (Figs. 9.21a and b).

If Le Corbusier had his way with the United Nations Building in New York City, our cities might have a very different appearance today. Le Corbusier wanted to use a brise-soleil to shield the exposed glazing. Not only were there no shading devices, but the building is so oriented that the glass facades face mostly east and west, while the solid stone facades face mostly north and south. Just rotating the plan could have greatly improved the performance of the building. Instead of a symbol of energy-conscious design the building became the prototype for the glass-slab office tower that could be made habitable only by use of energy-

Figure 9.21d The Price Tower, Bartlesville, OK, designed by Frank Lloyd Wright, uses sunshading as a major design concept. (Photograph by James Bradley.)

guzzling mechanical equipment (Fig. 9.21c).

Frank Lloyd Wright also had a different image of the high-rise building. His Price Tower makes full use of shading devices (Fig. 9.21d). He, like many other great architects, realized that shading devices were central to the practice of architecture because they

not only solve an important functional problem but also make a very strong aesthetic statement. This powerful potential for aesthetic expression has been largely ignored in recent years. The energy crisis of 1973 has renewed our interest in this very important and visible part of architecture.

KEY IDEAS OF CHAPTER 9

1. Shading is an integral part of architecture that has traditionally offered great opportunity for aesthetic expression (e.g., the Greek portico).

2. Every orientation should have a different shading strategy.

3. **Maximize south glazing** because it is the only orientation that can be effectively shaded in the summer while preserving the view. It is also the orientation with the maximum harvest of solar radiation in the winter.

4. **Minimize east and west glazing** because it is impossible to effectively shade those orientations while maintaining the view.

5. Glazing on the east and west facades should face north or south.

6. Movable shading devices are much better than fixed devices because the thermal year and the solar year are out of phase, and because of the variability of the weather.

7. Plants can be excellent shading devices. Deciduous plants can act as movable shading devices.

8. On the east and west facades, the horizontal overhang is usually preferred over vertical fins because the overhang preserves the view better while the solar energy blocked is about the same.

9. **Exterior shading is superior** to both interior shading and the shading from the glazing itself.

10. Use indoor shading devices to back up the outdoor shading

11. Light transmission through glazing is a function of the angle of incidence.

12. Outdoor spaces should be carefully shaded to make them attractive in the summer.

Resources

FURTHER READING
(See Bibliography in back of book for full citations. This list includes valuable out-of-print books.)

ASHRAE. *ASHRAE Fundamentals Handbook.*

Brown, G. Z., and M. DeKay. *Sun, Wind, and Light,* 2nd edition.

Carmody, J., S. Selkovwitz, and L. Heschong. *Residential Windows: A Guide to New Technologies and Energy Performance.*

Egan, M. D. *Concepts in Thermal Comfort.*

Fitch, J. M., with W. Bobenhausen. *American Building: The Environmental Forces That Shape It,* revided 1999 edition.

Franta, G., K. Anstead, and G. D. Ander. *Glazing Design Handbook for Energy Efficiency.*

Olgyay, A., and V. Olgyay. *Solar Control and Shading Devices.*

Ramsey Sleeper Architectural Graphic Standards, 9th edition. CD-ROM version is also available.

Stein, B., and J. S. Reynolds. *Mechanical and Electrical Equipment for Buildings,* 9th edition.

PASSIVE COOLING

"True regional character cannot be found through a sentimental or imitative approach by incorporating either old emblems or the newest local fashions which disappear as fast as they appear. But if you take, for instance, the basic difference imposed on architectural design by the climatic conditions of California, say, as against Massachusetts, you will realize what diversity of expression can result from this fact alone...."

Walter Gropius
Scope of Total Architecture © Harper & Row, 1955

10.1 INTRODUCTION TO COOLING

To achieve thermal comfort in the summer in a more sustainable way, one should use the three-tier design approach. (Fig. 10.1). The first tier consists of **heat avoidance.** At this level the designer does everything possible to minimize heat gain in the building. Strategies at this level include the appropriate use of shading, orientation, color, vegetation, insulation, daylight, and the control of internal heat sources. These and other heat-avoidance strategies are described throughout this book.

Since heat avoidance is usually not sufficient by itself to keep temperatures low enough all summer, the second tier of response, called **passive cooling,** is used. With some passive cooling systems, temperatures are actually lowered and not just minimized as is the case with heat avoidance. Passive cooling also includes the use of ventilation to shift the comfort zone to higher temperatures. The major passive cooling strategies will be discussed in this chapter.

In many climates, there will be times when the combined effort of heat avoidance and passive cooling is still not sufficient to maintain thermal comfort. For this reason, the third tier of mechanical equipment is usually required. In a rational design process, as described here, this equipment must cool only what heat avoidance and passive cooling could not accom-

Figure 10.2a Hot and dry climates typically have buildings with small windows, light colors, and massive construction. Thera, Santorini, Greece. (From *Proceedings of the International Passive and Hybrid Cooling Conference,* Miami Beach, FL, Nov. 6–16, © American Solar Energy Society, 1981.)

plish, and, consequently, the mechanical equipment will be quite small and use only modest amounts of energy.

10.2 HISTORICAL AND INDIGENOUS USE OF PASSIVE COOLING

Examples are sometime better than definitions in explaining concepts. The following examples of historical and indigenous buildings will illustrate what is meant by passive cooling.

Passive cooling is much more dependent on climate than passive heating. Thus, the passive cooling strategies for hot and dry climates are very different from those for hot and humid climates.

In hot and dry climates, one usually finds buildings with few and small windows, light surface colors, and massive construction, such as adobe, brick, or stone (Fig. 10.2a). The massive materials not only retard and delay the progress of heat through the walls, but also act as a heat sink during the day. Since hot and dry climates have high diurnal temperature ranges, the nights tend to be cool. Thus, the mass cools at night and then acts as a heat sink the next

day. To prevent the heat sink from being overwhelmed, small windows and light colors are used to minimize the heat gain. Closed shutters further reduce the daytime heat gain, while still allowing good night ventilation when they are open.

In urban settings and other places with little wind, wind scoops are sometimes used to maximize ventilation. Wind scoops were already used several thousand years ago in Egypt (Fig. 10.2b), and they are still found in the Middle East today. When there is a strong prevailing wind direction as in Hyderabad, Pakistan, the scoops are all aimed in the same direction (Fig. 10.2c). In other areas, where there is no prevailing wind direction, such as Dubai on the Persian Gulf, wind towers with many openings are used. These rectangular towers are divided by diagonal walls, which create four separate airwells facing four different directions (Fig. 10.2d).

Wind towers have shutters to keep out unwanted ventilation. In dry climates, they also have a means of evaporating water to cool the incoming air. Some wind towers have porous jugs of water at their base, while others use fountains or trickling water (Fig. 10.2e).

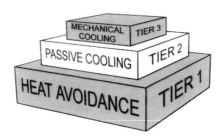

Figure 10.1 Cooling needs of buildings are best and most sustainably achieved by the three-tier design approach, and this chapter covers tier two.

Figure 10.2b Ancient Egyptian houses used wind scoops to maximize ventilation. (Replica of wall painting in the Tomb of Nebamun, circa 1300 BC, courtesy of the Metropolitan Museum of Art, New York City, #30.4.57.)

Figure 10.2c The wind towers in Hyderabad, Pakistan, all face the prevailing wind.

Figure 10.2d The wind towers in Dubai, United Arab Emirates, are designed to catch the wind from any direction. (Photograph by Mostafa Howeedy.)

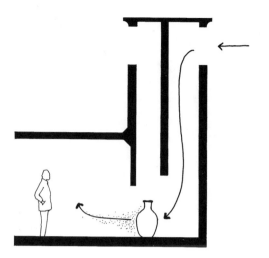

Figure 10.2e Some wind towers in hot and dry areas cool the incoming air by evaporation.

Figure 10.2f A mashrabiya is a screened bay window popular in the Arabic Middle East. It shades, ventilates, and provides evaporative cooling. Cairo, Egypt. (Photograph by Mostafa Howeedy.)

Figure 10.2g In the evening, the orchestra platform provided entertainment while the water cooled the air at the Panch Mahal Palace at Fathepur, India. (Photograph by Lena Choudhary.)

The mashrabiya is another popular wind-catching feature in the Arabic Middle East (Fig. 10.2f). These bay windows were comfortable places to sit and sleep since the delicate wood screens kept most of the sun out yet allowed the breezes to blow through. Evaporation from porous jugs of water placed in the mashrabiya cooled not only drinking water but the houses as well. Of course, the mashrabiya also satisfied the cultural need to give women an inconspicious place to view the activity of the outside world.

Wherever the humidity is low, evaporative cooling will be very effective. Fountains, pools, water trickling down walls, and transpiration from plants can all be used for evaporative cooling (Fig. 10.2g). The results are best if the evaporation occurs indoors or in the incoming air stream. In India, it was quite common to hang wetted mats at openings to cool the incoming air. Indian palaces had pools, streams, and waterfalls brought indoors to make the cooling more effective.

Evaporative cooling of courtyards is especially efective when the courtyard is the main source of air for the building (Fig. 10.2h). Small and deep courtyards or atriums are beneficial in hot and dry climates also because they are self-shading most of the day. Despite the privacy and security that they offer, courtyards are not much used in hot and humid climates because of the obstruction to the necessary cross-ventilation.

Massive domed structures are successful in hot and dry regions. Besides the thermal benefit of their mass, their form yields two different benefits. During the day, the sun sees little more than the horizontal footprint of the dome, while at night almost a full hemisphere sees the night sky. Thus, radiant heating is minimized while radiant cooling is maximized. Domes also have high spaces where stratification will enable the occupants to inhabit the cooler lower levels (Fig. 10.2i). Sometimes vents are located at the top to allow the hottest air to escape (Fig. 10.2j). The most dramatic example of this kind of dome is the Pantheon, in Rome. Its "oculus" allows light to enter while the hot air escapes. The same concept was used

Figure 10.2h A small shaded courtyard was cooled by a series of fountains. Court of the Lions, Alhambra, Spain. (Courtesy of *ARAMCO World* magazine.)

Figure 10.2i The trulli are conical stone houses in Apulia, Italy. Their large mass and high ceilings, with the resultant stratification of air, make these houses fairly comfortable. (From *Proceedings of the International Passive and Hybrid Cooling Conference*, Miami Beach, FL, Nov. 6–16, © American Solar Energy Society, 1981.)

Figure 10.2j Small domes made of sun-dried mud bricks work well in very hot and dry climates, such as those found in Egypt. Small vents allow the hot air to escape and a small amount of light to enter. Narrow alleys enable buildings to shade each other. Small courtyards provide outdoor sleeping areas at night. (Courtesy of the Egyptian Tourist Authority.)

Figure 10.2k Dwellings and churches are carved from the volcanic tuffa cones in Cappadocia, Turkey. (Photograph by Tarik Orgen.)

in the U.S. Pavilion, Expo '67, in Montreal. The upper panels of the geodesic dome had round openings to vent the hot air (Fig. 9.l5a).

A large quantity of earth or rock is an effective barrier to the extreme temperatures in hot and dry climates. The deep earth is usually near the mean annual temperature of a region, which in many cases is cool enough for the soil to act as a heat sink during summer days. Earth sheltering is discussed in more detail in Chapter 15.

In Cappadocia, Turkey, thousands of dwellings and churches have been excavated from the volcanic tuffa cones over the last 2,000 years (Fig. 10.2k). Many of these spaces are still inhabited today in part because they provide effective protection from extreme heat and cold.

A structure need not be completely earth-sheltered to benefit from earth contact cooling. The dwellings leaning against the cliffs at Mesa Verde, Colorado, make use of the heat-sink capacity of both the rock cliff and the massive stone walls. The overhanging, south-facing cliffs also offer much shade during summer days (Fig. 10.2m). In areas where rock was not

available, thick earthen walls were used. The Navajo of the dry Southwest built **hogans** for the insulating effect of their thick earthen walls and roofs (Fig. 10.2n). The Spanish settlers used adobe for the same purpose (Fig. 10.2o).

In hot and humid climates, we find a different kind of building, one in which the emphasis is on natural ventilation. In very humid climates, mass is a liability and very lightweight structures are best. Although the sun is not as strong as in dry climates, the humidity is so uncomfortable that any additional heating from the sun is very objectionable. Thus, in very humid regions we find buildings with- large windows, large overhangs, and low

Figure 10.2m The cliff dwellings at Mesa Verde, CO, benefit from the heat-sink capacity of the stone walls and rock cliff.

Figure 10.2n The Navajo **hogan** with its thick earthen walls provides comfort in the hot and dry Southwest.

Figure 10.2o Spanish missionaries and settlers used thick adobe walls for thermal comfort.

Figure 10.2p These "chickees, "built by the Indians of southern Florida, respond to the hot and humid climate by maximizing ventilation and shade while minimizing thermal mass. The diagonal poles successfully resist hurricane winds.

mass (Fig. 10.2p). These buildings are often set on stilts to catch more wind and to get above the humidity near the ground. High ceilings allow the air to stratify, and vents at the gable or ridge allow the hottest air to escape (Fig. 1.2c).

Much of Japan has very hot and humid summers. To maximize natural ventilation, the traditional Japanese house uses post-and-beam construction, which allows the lightweight paper wall panels to be moved out of the way in the summer (Fig. 10.2q). Large overhanging roofs protect these panels and also create an outdoor space called an **engawa**. Large gable vents further increase the ventilation through the building (Fig. 10.2r).

Gulf Coast houses and their elaborate version, the French Louisiana plantation houses, were well adapted to the very humid climate (Fig. 10.2s). In that region, a typical house had its main living space, built of a light wood frame, raised off the damp and muggy ground on a brick structure. Higher up there was more wind and less humidity. The main living spaces had many tall openings to maximize ventilation. The ceiling was very high (sometimes as high as 14 feet) to permit the air to stratify and the people to occupy the lower, cooler layers. Vents in the ceiling and high attic allowed the stack effect to exhaust the hottest air from the building. Deep verandas shaded the walls and created cool out-

Figure 10.2q Ventilation is maximized by movable wall panels in traditional Japanese houses. (Courtesy of Japan National Tourist Organization.)

Figure 10.2r The movable wall panels open onto the engawa (veranda), which is protected by a large overhang. Also note the large gable vent. Japanese Garden, Portland, Oregon.

door areas. The open central hallway was derived from the **dog trot** houses of the early pioneers of that region, who would build one roof over two log cabins spaced about 10 feet apart (Fig. 10.2t). In the summer, this shady, breezy outdoor space became a desirable hangout for dogs and people alike.

Many of these same concepts were incorporated in the Classical Revival architecture that was so popular in the South during the nineteenth century. As was mentioned in Chapter 9, the classical portico was a very suitable way to build the large overhangs needed to shade the high doors and windows. These openings were often as high as 12 feet and the windows were frequently triple hung so that two-thirds of the window could be opened. Louvered shutters allowed ventilation when sun shading and privacy were desired (Fig. 10.2u). The classical image of white buildings was also very appropriate for the hot climate.

The Waverly plantation is a good example of the classical idiom adapted to the climate (Fig. 10.2v). It has a large, many-windowed belvedere, which offers a panoramic view, light, and a strong stack effect through the two-story stair hall. Since every door has operable transoms, all rooms have cross-ventilation from three sides (Fig. 10.2w). The author visited this non-air-conditioned building on a hot, humid summer day and found it to be comfortably cool inside.

Often the "temperate" climate is the hardest to design for. This is partly true because many so called "temperate" climates actually have very hot summers and very cold winters. Buildings in such climates cannot be designed to respond to either hot or cold conditions alone. Rather, they must be designed for both summer and winter, which often make opposing demands on the architect. The Governor's Mansion in Williamsburg, Virginia, is located in such a climate. The building is compact and the windows are of a medium size (Fig.

Figure 10.2s This Gulf Coast house incorporated many cooling concepts appropriate for hot and humid climates. Note the large shaded porch, ventilating dormers, large windows, and ventilation under the house.

Figure 10.2t The breezy passage of the dog trot house was a favorite for both man and beast during the hot and humid summers.

Figure 10.2u Shutters with adjustable louvers were almost universally used in the Old South.

10.2x). The brick construction allows passive cooling during much of the summer when the humidity is not too high. The massive fireplaces can act as additional heat sinks in the summer, as well as heat storage mediums in the winter. Every room has openings on all four walls for maximum cross-ventilation. The little tower on the roof can go by several different names, depending on its main function. It is a belvedere if the panoramic view is most important, it is a lantern if it acts as a skylight, it is a cupola if it has a small dome on it and is mainly for decoration or image, and it is a monitor if ventilation is most important. In this case, the tower's main purpose was to create the image of a governmental building. They usually had several of the above functions and sometimes all.

10.3 PASSIVE COOLING SYSTEMS

The passive cooling systems described here include not only the well-known traditional techniques mentioned above, but also the more sophisticated techniques that are still somewhat experimental but very promising. As much as possible passive cooling uses natural forces, energies, and heat sinks. When some fans and pumps are used, the systems are sometimes called **hybrid**. Because the systems described here use only small, low-energy pumps and fans, if any, these systems are also included under the passive category.

Since the goal is to create thermal comfort during the summer (the overheated period), we can either cool the building or raise the comfort zone sufficiently to include the high indoor temperature. In the first case, we have to remove heat from the building by finding a heat sink for it. In the second case, we increase the air velocity so that the comfort zone shifts to higher temperatures. In this second case, people will feel more comfortable even though the building is not

Figure 10.2v The Waverly plantation near Columbus, MS, has a large belvedere for view, light, and ventilation. (Photograph by Paul B. Watkins. Courtesy of the Mississippi Department of Economic Development. Division of Tourism.)

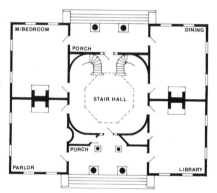

Figure 10.2w A strong stack effect is created by the octagonal belvedere over the open stair hall. All interior doors have transoms. [From *Mississippi Houses: Yesterday Toward Tomorrow*, by Robert Ford (copyright).]

Figure 10.2x The Governor's Mansion in Colonial Williamsburg, VA, is well suited for a "temperate" climate. (Courtesy of the Colonial Williamsburg Foundation.)

actually being cooled.

There are five methods of passive cooling:

Types of Passive Cooling Systems

I. Cooling with Ventilation
   A. *Comfort ventilation:* Ventilation during the day and night to increase evaporation from the skin and thereby increasing thermal comfort.
   B. *Night flush cooling:* Ventilation to precool the building for the next day.

II. Radiant Cooling
   A. *Direct radiant cooling:* A building's roof structure cools by radiation to the night sky.
   B. *Indirect radiant cooling:* Radiation to the night sky cools a heat-transfer fluid, which then cools the building.

III. Evaporative Cooling
   A. *Direct evaporation:* Water is sprayed into the air entering a building. This lowers the air's temperature but raises its humidity.
   B. *Indirect evaporative cooling:* Evaporation cools the incoming air or the building without raising the indoor humidity.

IV. Earth Cooling
   A. *Direct coupling:* An earth-sheltered building loses heat directly to the earth.
   B. *Indirect coupling:* Air enters the building by way of earth tubes.

V. Dehumidification with a Desiccant: Removal of latent heat.

A combination of these techniques is necessary in some cases. For example, in the South the earth might be too warm for cooling unless its temperature is first lowered by evaporation. Each of these techniques will now be discussed in more detail.

10.4 COMFORT VENTILATION VERSUS NIGHT FLUSH COOLING

Until recently, ventilation has been the major cooling technique throughout the world. It is very important to note that not only are there two very different ventilation techniques, but also that they are mutually exclusive. **Comfort ventilation** brings in outdoor air, especially during the daytime when temperatures are at their highest. The air is then passed directly over people to increase evaporative cooling on the skin. Although thermal comfort might be achieved, the warm air is actually heating the building.

Night-flush cooling is quite different. With this technique, cool night air is introduced to flush out the heat of the building, while during the day very little outside air is brought indoors so that heat gain to the building can be minimized. Meanwhile, the mass of the relatively cool structure acts as a heat sink for the people inside. Before these techniques can be explained in more detail, some basic principles of air flow and their applications in buildings must be discussed.

10.5 BASIC PRINCIPLES OF AIR FLOW

To design successfully for ventilation in the summer or for wind protection in the winter, the following principles of air flow should be understood.

1. *Reason for the flow of air:* Air flows either because of natural convection currents, caused by differences in temperature, or because of differences in pressure (Fig. 10.5a).
2. *Types of air flow:* There are four basic types of air flow: laminar, separated, turbulent, and eddy currents. Figure 10.5b illustrates the four types by means of lines representing air streams. These diagrams are similar to what one would see in a wind-tunnel test using smoke streams. Air flow

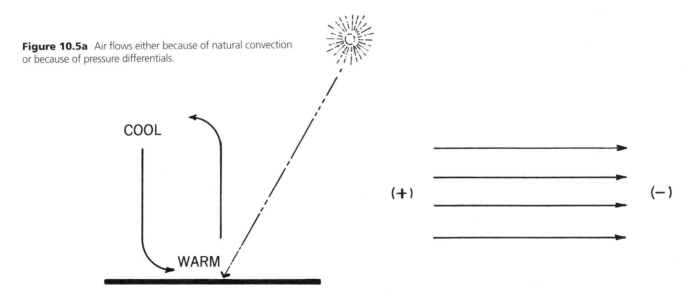

Figure 10.5a Air flows either because of natural convection or because of pressure differentials.

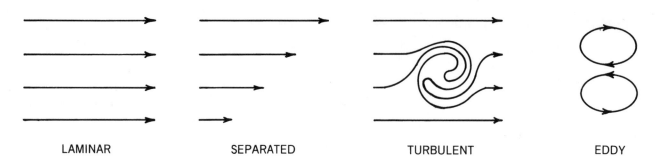

LAMINAR　　　SEPARATED　　　TURBULENT　　　EDDY

Figure 10.5b The four different kinds of air flow. (After Art Bowen, 1981.)

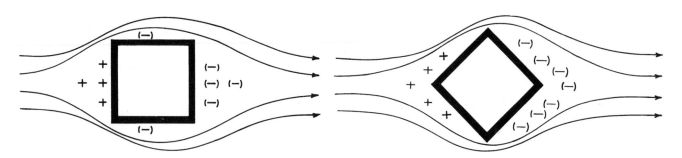

Figure 10.5c Air flowing around a building will cause uneven positive and negative pressure areas to develop. (After Art Bowen, 1981.)

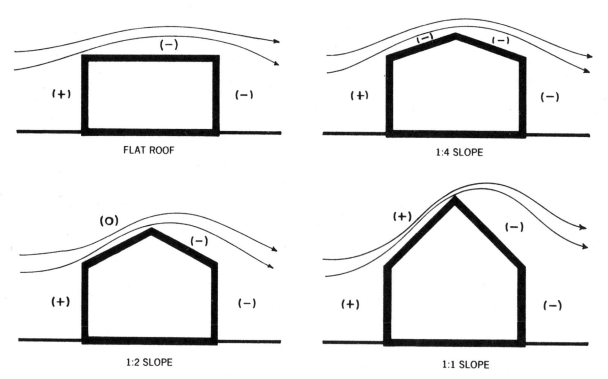

FLAT ROOF

1:4 SLOPE

1:2 SLOPE

1:1 SLOPE

Figure 10.5d The pressure on the leeward side of a roof is always negative (–), but on the windward side it depends on the slope of the roof. (After Art Bowen, 1981.)

changes from laminar to turbulent when it encounters sharp obstructions, such as buildings. Eddy currents are circular air flows induced by laminar or turbulent air flows (Fig. 10.5e).

3. *Inertia:* Since air has some mass, moving air will tend to go in a straight line. When forced to change direction, air streams will follow curves and never right angles.

4. *Conservation of air:* Since air is neither created nor destroyed at the building site, the air approaching a building must equal the air leaving the building. Thus, lines representing air streams should be drawn as continuous.

5. *High and low pressure areas:* As air hits the windward side of a building, it compresses and creates a positive pressure (+). At the same time, air is sucked away from the leeward side, thus creating a negative pressure (–).

Air deflected around the sides will generally also create a negative pressure. Note that these pressures are not uniformly distributed (Fig. 10.5c). The type of pressure created over the roof depends on the slope of the roof (Fig. 10.5d). These pressure areas around the building determine how air flows through the building.

It should also be noted that these high- and low-pressure areas are not necessarily places of calm but also of air flow in the form of turbulence and eddy currents (Fig. 10.5e). Note how these currents reverse the air flow in certain locations. For simplicity's sake, turbulence and eddy currents, although usually present, are not shown on all diagrams.

6. *Bernoulli effect:* The Bernoulli effect states that an increase in the velocity of a fluid decreases its static pressure. Because of this phenomenon, there is a negative pressure at the constriction of a venturi tube (Fig. 10.5f). A cross section of an airplane wing is like half a venturi tube (Fig. 10.5g).

A gabled roof is like a half a venturi tube. Thus, air will be sucked out of any opening near the ridge (Fig. 10.5h). The effect can be even stronger by designing the roof to be like a full venturi tube (Fig. 10.5i).

There is another phenomenon at work here. The velocity of air increases rapidly with height above ground. Thus, the pressure at the ridge of a roof will be lower than that of windows at the ground level. Consequently, even without the help of the geometry of a venturi tube, the Bernoulli effect will exhaust air through roof openings (Fig. 10.5j).

7. *Stack effect:* The stack effect can exhaust air from a building by the action of natural convection. The stack effect will exhaust air only if the indoor-temperature difference between two vertical openings is greater than the outdoor-tempera-

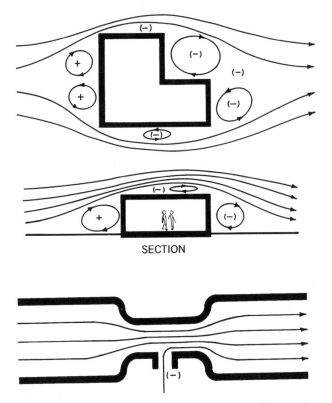

SECTION

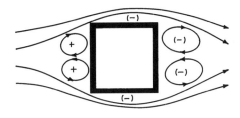

Figure 10.5e Turbulence and eddy currents occur in the high- and low-pressure areas around a building. (After Art Bowen, 1981.)

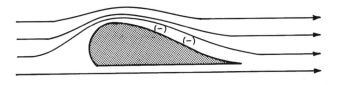

Figure 10.5f The venturi tube illustrates the Bernoulli effect: As the velocity of air increases, its static pressure decreases.

Figure 10.5g An airplane wing is like half of a venturi tube. In this case, the negative pressure is also called lift.

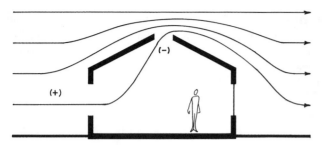

Figure 10.5h The venturi effect causes air to be exhausted through roof openings, and at or near the ridge.

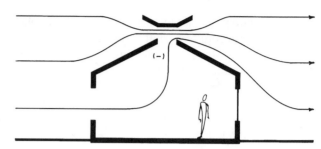

Figure 10.5i A venturi tube used as a roof ventilator.

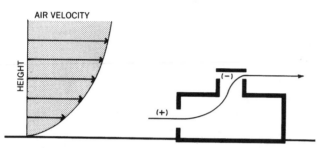

Figure 10.5j Because the air velocity increases rapidly with height above grade, the air has less static pressure at the roof than on the ground (the Bernoulli effect).

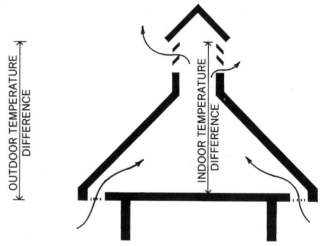

Figure 10.5k The stack effect will exhaust hot air only if the indoor-temperature difference is greater than the outdoor-temperature difference between the vertical openings.

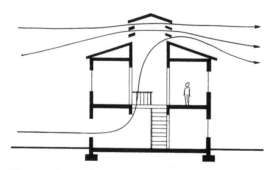

Figure 10.5m The central stair and geometry of this design allow effective vertical ventilation by the combined action of stratification, the stack effect, and both the Bernoulli and Venturi effects.

ture difference between the same two openings (Fig. 10.5k). To maximize this basically weak effect, the openings should be as large and as far apart vertically as possible. The air should be able to flow freely from the lower to the higher opening (i.e., minimize obstructions).

The advantage of the stack effect over the Bernoulli effect is that it does not depend on wind. The disadvantage is that it is a very weak force and cannot move air quickly. It will, however, combine with the Bernoulli and venturi effects mentioned above to create particularly good vertical ventilation on many hot summer days. Figure 10.5m illustrates how strati-

fication, the stack effect, the shape of the roof (the venturi effect), and the increased wind velocity at the roof (the Bernoulli effect) can all combine to naturally ventilate a building. Roof monitors and ventilators high on the roof are especially helpful: because of stratification, the hottest indoor air is exhausted first.

An interesting variation on the stack effect is the **solar chimney.** Since the stack effect is a function of temperature differences, heating the indoor air increases the air flow. But, of course, that would conflict with our goal of cooling the indoor air. Therefore, the solar chimney heats the air after it leaves the building (Figs. 10.5n and 10.5o). Thus, the stack effect is increased but without additional heating of the building.

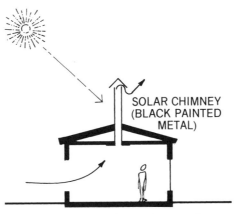

Figure 10.5n A solar chimney increases the stack effect without heating the indoors

Figure 10.5o This outhouse is ventilated with a solar chimney. The author can verify that even on windless days no odor was present when the sun was shining.

10.6 AIR FLOW THROUGH BUILDINGS

The factors that determine the pattern of air flow through a building are the following: pressure distribution around the building; direction of air entering windows; size, location, and details of windows; and interior partitioning details. Each of these factors will be considered in more detail.

Site Conditions

Adjacent buildings, walls, and vegetation on the site will greatly affect the air flow through a building. These site conditions will be discussed in Chapter 11.

Window Orientation and Wind Direction

Winds exert maximum pressure when they are perpendicular to a surface, and the pressure is reduced about 50 percent when the wind is at an oblique angle of about 45 degrees. However, the indoor ventilation is often better with the oblique winds because they generate greater turbulence indoors (Fig 10.6a). Consequently, a fairly large range of wind directions will work for most designs. This is fortunate because it is a rare site that has winds blowing mainly from one direction. Even where there are strong prevailing directions, it might not be possible to

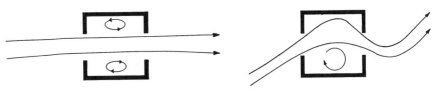

Figure 10.6a Usually indoor ventilation is better from oblique winds than from head-on winds because the oblique air stream covers more of the room.

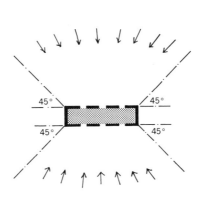

Figure 10.6b Acceptable wind directions for the orientation that is best for summer shade and winter sun.

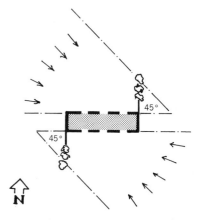

Figure 10.6c Deflecting walls and vegetation can be used to change air-flow direction so that the optimum solar orientation can be maintained.

face the building into the wind.

In most climates, the need for summer shade and winter sun calls for an east–west orientation of a building, and Fig. 10.6b shows the range of wind directions that works

well with that orientation. Even when winds are east–west, the solar orientation usually has priority because winds can be rerouted more easily than the sun (Fig. 10.6c).

Window Locations

Cross-ventilation is so effective because air flows from strong positive-pressure to strong negative-pressure areas located on opposite walls (Fig. 10.6d). Ventilation from windows on adjacent walls can be either good or bad, depending on the pressure distribution, which varies with wind direction (Fig. 10.6e).

Ventilation from windows on one side of a building can vary from fair to poor depending on the location of windows. Since the pressure is greater at the center of the windward wall than at the edges, there is some pressure difference in the asymmetric placement of windows, while there is no pressure difference in the symmetric scheme (Fig. 10.6f).

Fin Walls

Fin walls can greatly increase the ventilation through windows on the same side of a building by changing the pressure distribution (Fig. 10.6g). Note, however, that each window must have only a single fin. Furthermore, fin walls will not work if they are placed on the same side of each window (Fig. 10.6h). Fin walls work

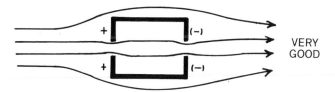

Figure 10.6d Cross-ventilation between windows on opposite walls is the ideal condition.

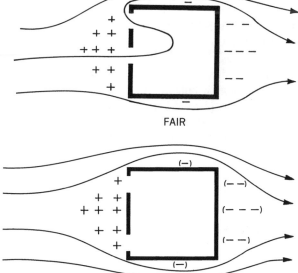

FAIR

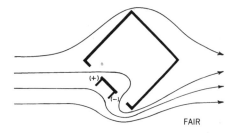

Figure 10.6e Ventilation from adjacent windows can be poor or good, depending on the wind direction.

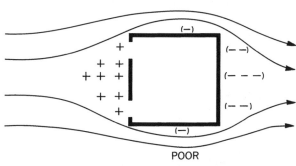

POOR

Figure 10.6f Some ventilation is possible in the asymmetric placement of windows because the relative pressure is greater at the center than at the sides of the windward wall. (After Art Bowen, 1981.)

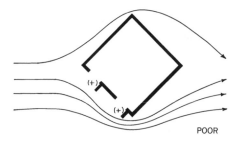

FAIR

Figure 10.6g Fin walls can significantly increase ventilation through windows on the same wall. (After Art Bowen, 1981.)

POOR

Figure 10.6h Poor ventilation results from fin walls placed on the same side of each window or when two fins are used on each window.

best for winds at 45 degrees to the window wall. Casement windows can act as fin walls at no extra cost.

The placement of windows on a wall determines not only the quantity but also the initial direction of the incoming air. The off-center placement of the window gives the airstream an initial deflection because the positive pressure is greater on one side of the window (Fig. 10.6i). To better ventilate the room, one should deflect the airstream in the opposite direction. A fin wall can be used to change the pressure balance and thus, the direction of the air stream (Fig. 10.6j).

Horizontal Overhangs and Air Flow

A horizontal overhang just above the window will cause the airstream to deflect up to the ceiling because the solid overhang prevents the positive pressure above it from balancing the positive pressure below the window (Fig. 10.6k). However, a louvered overhang or gap of 6 inches or more in the overhang will allow the positive pressure above it to affect the direction of the air flow (Fig. 10.6l). Placement of the overhang higher on the wall can also direct the airstream down to the occupants (Fig. 10.6m).

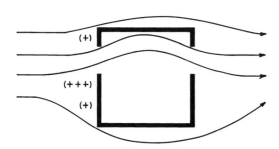

Figure 10.6i The greater positive pressure on one side of the window deflects the airstream in the wrong direction. Much of the room remains unventilated.

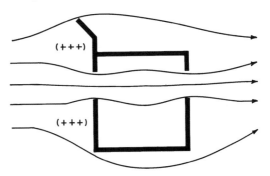

Figure 10.6j A fin wall can be used to direct the airstream through the center of the room.

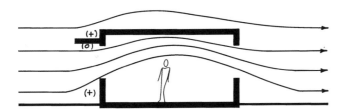

Figure 10.6k The solid horizontal overhang causes the air to deflect upward. (After Art Bowen, 1981.)

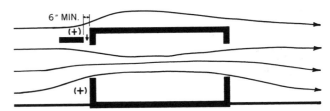

Figure 10.6l A louvered overhang or at least a gap in the overhang will permit the airstream to straighten out.

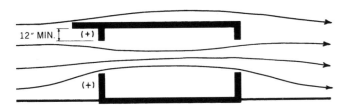

Figure 10.6m A solid horizontal overhang placed high above the window will also straighten out the airstream. (After Art Bowen, 1981.)

Window Types

The type and design of windows have a great effect on both the quantity and direction of the air flow. Although double-hung and sliding windows do not change the direction of the airstream, they do block at least 50 percent of the air flow. On the other hand, casement windows allow almost full air flow, but they can deflect the airstream (Fig. 10.6n). They also act as fin walls, as described above.

For the vertical deflection of the airstream, use hopper, awning, or jalousie windows (Fig. 10.6n). These types also deflect the rain while still admitting air, which is very important in hot and humid climates. Unfortunately, with this kind of inclination, they deflect the wind upward over people's heads, which is undesirable for comfort ventilation.

Movable opaque louvers, frequently used on shutters, are like jalousie windows except that they also block the sun and view. The large amount of crack and resultant infiltration makes these types of windows and louvers inappropriate in climates with cold winters.

Horizontal strip or ribbon windows are often the best choice when good ventilation is required over large areas of a room.

Vertical Placement of Windows

The purpose of the air flow will determine the vertical placement and height of windows. For comfort ventilation, the windows should be low, at the level of the people in the room. That places the windowsill between 1 and 2 feet above the floor for people seated or reclining. A low windowsill is especially important when hopper or jalousie windows are used because of their tendency to deflect air upward. Additional high windows or ceiling vents should be considered for exhausting the hot air that collects near the ceiling (Fig. 10.6o). High

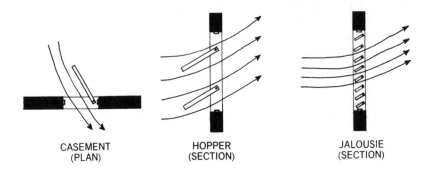

Figure 10.6n All but double-hung and sliding windows have a strong effect on the direction of the airstream.

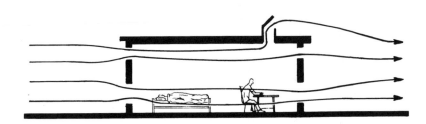

Figure 10.6o For comfort ventilation, openings should be at the level of the occupants. High openings vent the hot air collecting near the ceiling and are most useful for night flush cooling. (After Art Bowen, 1981.)

openings are also important for night-flush cooling where air must pass over the structure of the building.

Traditional Southern buildings had windows that were almost the full height of the room. Thomas Jefferson's home in Monticello, Virginia, had triple-hung windows that went from the floor to the ceiling. By being triple-hung, the windows' upper and lower sashes could be opened for maximum ventilation. Jefferson could also open the lower two sashes to create a door to the porch.

It is often advantageous to place windows too high to reach for direct manual operation. Mechanical devices are readily available for both manual and automatic operation. Some work with mechanical linkage (Fig. 10.6p) and others with electric motors (Fig. 10.6q).

Inlet and Outlet Sizes and Locations

Generally, the inlet and outlet size should be about the same since the amount of ventilation is mainly a function of the smaller opening. However, if one opening is smaller, it should usually be the inlet because that maximizes the velocity of the indoor airstream, and it is the velocity that has the greatest effect on comfort. Although velocities higher than the wind can be achieved indoors by concentrating the air flow, the area served is, of course, decreased (Fig. 10.6r). The inlet opening not only determines the velocity, but also determines the air-flow pattern in the room. The location of the outlet, on the other hand, has little effect on the air-velocity and flow-pattern.

Insect Screens

Air flow is decreased about 50 percent by an insect screen. The actual resistance is a function of the angle at which the wind strikes the screen, with the lowest resistance for a head-on wind. To compensate for the effect of the screen, larger openings are required. A screened-in porch is especially effective because of the very large screen area that it provides (Fig. 10.6s).

Roof Vents

Passive roof ventilators are typically used to lower attic temperatures. If, however, local winds are high enough, and the ventilator is large enough or high enough on the roof, these devices can also be used to ventilate habitable spaces. The common wind turbine enhances ventilation about 30 percent over an open stack. Research has shown that other designs can enhance the air flow as much as 120 percent (Fig. 10.6t).

Although cupolas, monitors, and roof vents are often a part of traditional architecture (see Figs. 10.2v and x at the front of this chapter), they can also be integrated very successfully into modern architecture (Fig. 10.6u). Some kind of shutter or trap door is required to prevent unwanted ventilation, especially in the winter.

Fans

In most climates, wind is not always present in sufficient quantity when needed, and usually there is less wind at night than during the day. Thus, fans are usually required to augment the wind.

There are three quite different purposes for fans. The first is to exhaust hot, humid, and polluted air. This is part of the heat-avoidance strategy and is discussed in Chapter 15. The second is to bring in outdoor air to either cool the people (comfort ventilation) or cool the building at night

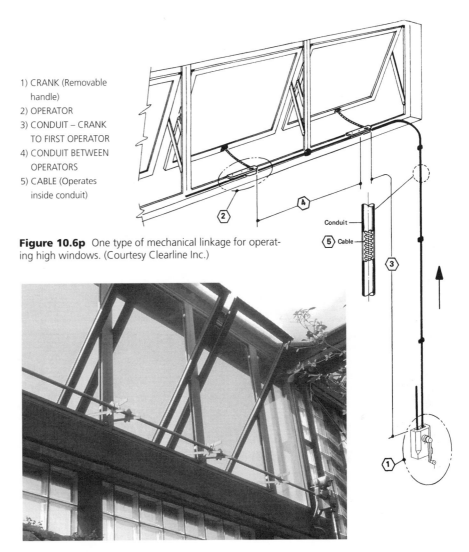

1) CRANK (Removable handle)
2) OPERATOR
3) CONDUIT – CRANK TO FIRST OPERATOR
4) CONDUIT BETWEEN OPERATORS
5) CABLE (Operates inside conduit)

Conduit

Cable

Figure 10.6p One type of mechanical linkage for operating high windows. (Courtesy Clearline Inc.)

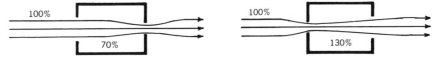

Figure 10.6q Each motor opens a bank of pivoting windows high above the floor in a greenhouse at Callaway Gardens, Georgia.

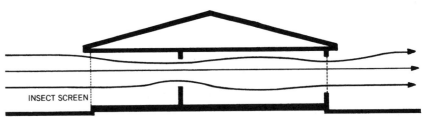

100%   70%   100%   130%

Figure 10.6r Inlets and outlets should be the same size. If they cannot be the same size, the inlet should be smaller to maximize the velocity. (After Art Bowen, 1981.)

INSECT SCREEN

Figure 10.6s The resistance to air flow by insect screens can be largely overcome by means of larger openings or screened-in porches.

OPEN STACK
100%

TURBINE
130%

DEFLECTOR
220%

Figure 10.6t The design of a roof ventilator has a great effect on its performance. Percentages show relative effectiveness. (After Shubert and Hahn, 1983.)

Figure 10.6u Monitors on the roof of the bathing pavilions at Callaway Gardens, GA.

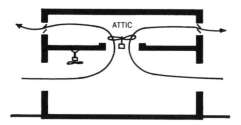

Figure 10.6v Whole-house or window fans are used to bring in outdoor air for either comfort ventilation or night-flush cooling. Ceiling or table fans are mainly used when the air temperature and humidity is lower indoors than outdoors.

(night-flush cooling). The third purpose is to circulate indoor air at those times when the indoor air is cooler than the outdoor air.

Separate fans are required for each purpose. Window or **whole house fans** are used for comfort ventilation or night-flush cooling (Fig. 10.6v). Ceiling or table fans are used whenever the indoor air is cooler and or less humid than the outdoor air.

Partitions and Interior Planning

"Open plans" are preferable because partitions increase the resistance to air flow, thereby decreasing total ventilation.

When partitions are used but are in one apartment or one tenant area, cross-ventilation is often possible by leaving open doors between rooms. Cross-ventilation is almost never pos-

sible, however, when a public double-loaded corridor plan is used. Before air conditioning became available, **transoms** (windows above doors) allowed for some cross-ventilation. An alternative to the double-loaded corridor is the open single-loaded corridor since it permits full cross-ventilation (Fig. 10.6w). Single-story buildings can improve on the double-loaded corridor by using clerestory windows instead of transoms (Fig. 10.6x).

Le Corbusier came up with an ingenious solution for cross-ventilation in his Unite d'Habitation at Marseilles (Fig. 10.6y). He has a corridor on only every third floor, and each apartment is a duplex with an opening to the corridor as well as the opposite sides of the building (Fig. 10.6z). The balconies have perforated parapets to further encourage ventilation, and they form a giant brise-soleil for sun shading.

Le Corbusier opened the area under the building to the wind by resting the building on columns that he called **pilotis.** In a hot climate, such an area becomes a cool breezy place in summer, but in cold climates the same area becomes very unpleasant in the winter. The wind patterns around buildings will be discussed further in Chapter 11.

10.7 EXAMPLE OF VENTILATION DESIGN

Ventilation design is greatly aided by the use of **air-flow diagrams.** These diagrams are based on the general principles and rules mentioned above and not on precise calculations. They are largely the product of a trial-and-error process. The following steps are a guide to making these air-flow diagrams.

Air-Flow Diagrams

1. Determine the prevailing summer-wind direction from local weather data or from the wind roses given in Fig. 5.6f.
2. On an overlay of the plan and site, draw a series of arrows parallel to the prevailing wind direction on both the upwind and downwind sides of the building (Fig. 10.7a). These arrows should be spaced about the width of the smallest window.
3. By inspection, determine the positive- (+) and negative- (–) pressure areas around the building

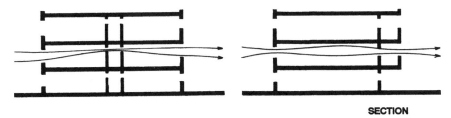

SECTION

Figure 10.6w In regard to natural ventilation, single-loaded corridor plans *(right)* are far superior to double-loaded plans *(left)*.

SECTION

Figure 10.6x In single-story buildings, a double-loaded corridor plan can use clerestory windows instead of transoms.

Figure 10.6y The Unite d'Habitation outside of Marseilles was designed by Le Corbusier to provide cross-ventilation for each apartment. (Photograph by Alan Cook.)

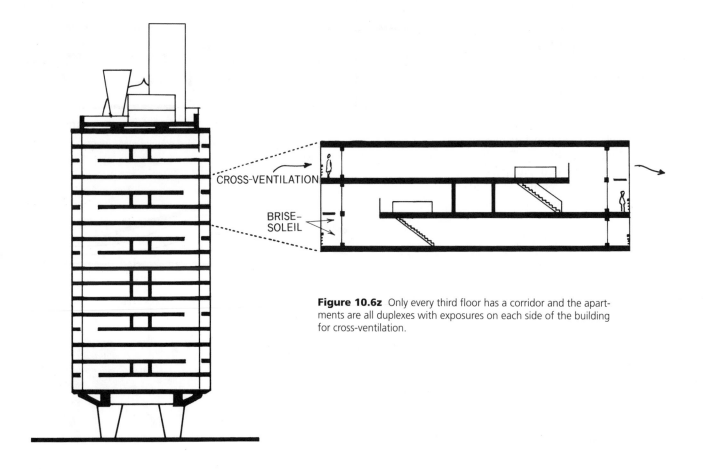

Figure 10.6z Only every third floor has a corridor and the apartments are all duplexes with exposures on each side of the building for cross-ventilation.

and record these on the overlay (Fig. 10.7a).

4. By means of a trial-and-error process, trace each windward arrow through or around the building to meet its downwind counterpart. Lines should not cross, end, or make sharp turns. Air flow through the building should go from positive- to negative-pressure areas (Fig. 10.7b).

5. When the airstream is forced to flow vertically to another floor plan, show the point where it leaves any plan by a circle with a dot and the return point by a circle with a cross (see Fig. 10.7b). Also show the vertical movement in a section of the building (Fig. 10.7c).

6. Since spaces that are not crossed by air-flow lines might not receive enough ventilation, relocate win-

dows, add fins, etc. to change the air-flow pattern as necessary.

7. Repeat steps 2 to 5 until a good air-flow pattern has been achieved.

This technique is based on work by Murray Milne, UCLA.

10.8 COMFORT VENTILATION

Air passing over the skin creates a physiological cooling effect by evaporating moisture from the surface of the skin. The term **comfort ventilation** is used for this technique of using air motion across the skin to promote thermal comfort. This passive cooling technique is useful for certain periods in most climates, but it is especially appropriate in hot and humid climates, where it is typical for

air temperatures to be only moderately hot and ventilation is required to control indoor humidity. See Fig. 4.11 for the conditions under which comfort ventilation is appropriate.

Comfort ventilation can rarely be completely passive because in most climates winds are not always sufficient to create the necessary indoor air velocities. Window or whole-house attic fans are usually needed to supplement the wind. See Table 10.8 for the effect on comfort due to various air velocities. For comfort ventilation, the air-flow techniques mentioned above should be used to maximize the air flow *across the occupants* of the building.

If the climate is extremely humid and little or no heating is required, lightweight construction is appropriate. In such climates, any thermal mass

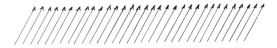

Figure 10.7a Initial setup for drawing an air-flow diagram.

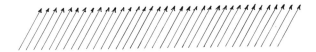

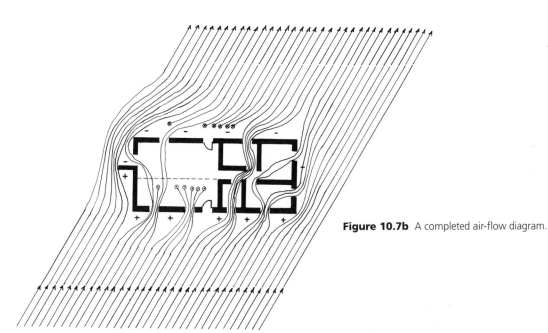

Figure 10.7b A completed air-flow diagram.

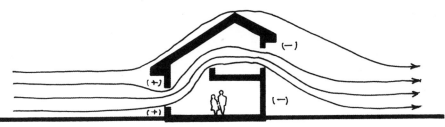

Figure 10.7c Air flow should also be checked in section. (This technique is based on work by Murray Milne, Prof., UCLA.)

TABLE 10.8 AIR VELOCITIES AND THERMAL COMFORT

Air Velocity		Equivalent Temperature Reduction (°F)[a]	Effect on Comfort
FPM	Approximate MPH		
10	0.1	0	Stagnant air slightly uncomfortable
40	0.5	2	Barely noticeable but comfortable
80	1	3.5	Noticeable and comfortable
160	2	5	Very noticeable but acceptable in certain high-activity areas if air is warm
200	2.3	6	Upper limit for air-conditioned spaces
			Good air velocity for natural ventilation in hot and dry climates
400	4.5	7	Good air velocity for natural ventilation in hot and humid climates
900	10	9	Considered a "gentle breeze" when felt outdoors

[a] The values in this column are number of degrees Fahrenheit that the temperature would have to drop to create the same cooling effect as the given air velocity.

Figure 10.8a The Mayan Indians of the hot and humid Yucatan Peninsula build lightweight, porous buildings for maximum comfort. Note that although mud and rocks are available, experience has led the Mayans to the most comfortable construction method.

used will only store up the heat of the day to make the nights less comfortable (Fig. 10.8a). In the United States only southern Florida and Hawaii (climate region 13) fit in this category. In these climates, a moderate amount of insulation is still required to keep the indoor surfaces from getting too hot due to the action of the sun on the roof and walls. The insulation keeps the mean radiant temperature (MRT) from rising far above the indoor air temperature since that would decrease thermal comfort. Insulation is also required when the building is air conditioned.

Some control is also possible over the temperature of the incoming air. For example, tests have shown that when the air temperature above unshaded asphalt was 110°F, it was only 90°F over an adjacent shaded lawn. This 20°F difference in temperature will have a great effect on the heat load of the building. Thus, comfort ventilation will be much more effective if a building is surrounded by vegetation rather than asphalt.

For comfort ventilation, the operable window area should be about 20 percent of the floor area with the openings split about equally between windward and leeward walls. The windows should also be well shaded on the exterior as explained in Chapter 9. One of the examples presented there was Frank Lloyd Wright's Robie House (Fig. 9.1i). It has very large roof overhangs to shade walls made entirely of glass doors and windows that can be opened for ventilation (Fig. 10.8b). Since Chicago has very hot and humid summers, plentiful ventilation and full shade were the major cooling strategies before air conditioning became available.

Comfort ventilation is most appropriate when the indoor temperature and humidity are above the outdoor level. This is often the case because of internal heat sources and the heating effect of the sun. However, when it is hotter outdoors than indoors, the win-

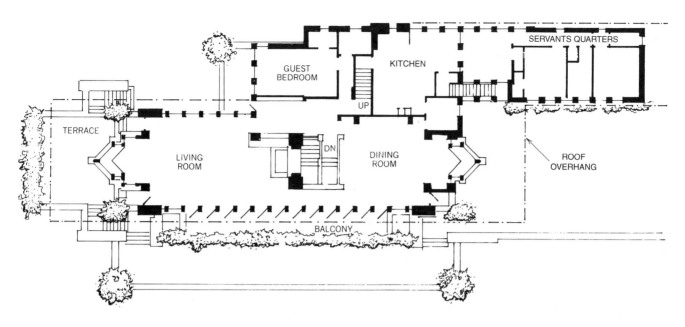

Figure 10.8b Frank Lloyd Wright's Robie House (1909) in Chicago had whole walls of doors and windows that opened for natural ventilation.

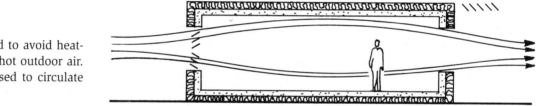

Figure 10.9a With "night-flush cooling," night ventilation cools the mass of the building.

dows should be closed to avoid heating the building with hot outdoor air. Ceiling fans can be used to circulate the cooler indoor air.

Rules for Comfort Ventilation in Hot and Very Humid Climates

1. See Fig. 4.11 for the climatic conditions for which comfort ventilation is appropriate.
2. Use fans to supplement the wind.
3. Maximize the air flow across the occupants.
4. Lightweight construction is appropriate only in climates that are very humid and that do not require passive solar heating.
5. Use at least a moderate amount of insulation to keep the MRT near the air temperature.
6. Operable window area should be about 20 percent of the floor area split about equally between windward and leeward walls.
7. Windows should be open both during the day and during the night.

10.9 NIGHT-FLUSH COOLING

In all but the most humid climates, the night air is significantly cooler than the daytime air. This cool night air can be used to flush out the heat from a building's mass. The precooled mass can then act as a heat sink during the following day by absorbing heat. Since the ventilation removes the heat from the mass of the building at night, this time-tested passive technique is called **night-flush cooling.**

This cooling strategy works best in hot and dry climates because of the large diurnal (daily) temperature ranges found there, above 30°F. A large range implies cool nighttime temperatures of about 70°F even

though daytime temperatures are quite high—about 100°F. However, good results are also possible in somewhat humid climates, which have only modest diurnal temperature ranges—about 20°F. The map of the United States in Fig. 5.6c illustrates that most of the country, including the humid East, has diurnal temperature ranges of more than 20°F. The daily ranges are smaller only very close to the coast.

Night-flush cooling works in two stages. At night, natural ventilation or fans bring cool outdoor air in contact with the indoor mass, thereby cooling it (Fig. 10.9a). The next morning, the windows are closed to prevent heating the building with outdoor air (Fig.

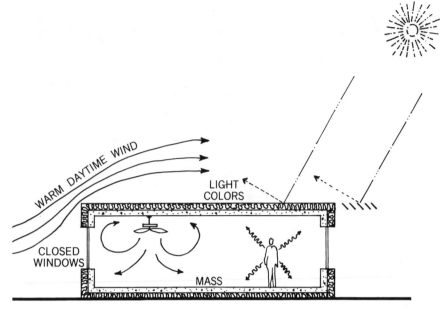

Figure 10.9b During the day, the night-flush cooled mass acts as a heat sink. Light colors, insulation, shading, and closed windows keep the heat gain to a minimum. Interior circulating fans can be used for additional comfort.

10.9b). The mass now acts as a heat sink and, thus, keeps the indoor air temperature from rising as fast as it would otherwise. However, when the indoor air temperature has risen above the comfort zone, internal circulating fans are required to maintain comfort for additional hours. As with passive heating, a significant temperature range indoors will result. Although more thermal mass will reduce the swing, it is advantageous to allow night flushing to cool the building below the comfort zone in preparation for the hot day to follow.

The thermal mass is critical because without it there is no heat sink to cool the building during the day. The requirements for the mass are similar to those for passive solar heating, and, of course, the mass can serve both purposes. Ideally, the mass should equal 80 pounds per square foot of floor area (concrete weighs about 150 pounds per cubic foot). The surface area of the mass should be more than two times the floor area.

Minimize the heat gain to minimize the amount of mass required. Use heat-avoidance techniques, such as well-shaded windows, a heavily insulated envelope, and light colors. These and many other techniques are mentioned throughout the book.

To flush out the heat at night, the operable window area should be about 10 percent to 15 percent of the floor area. When natural ventilation is not sufficient, exhaust fans should be used. With night-flush cooling, the air flow should be directed *over the mass* and not the occupants.

In normal buildings, it is difficult to completely flush the mass of its heat at night. A more sophisticated version of convective cooling passes the night air through channels in the structural mass. The Emerald PUD Building in Oregon uses night flush cooling as a major design strategy. Cool night air is passed through a hollow-core, concrete-plank roof. This excellent design also incorporates many other energy-conscious design concepts and it is more fully explained as a case study in Section 17.4.

The Bateson State Office Building in Sacramento, California, uses two different night-flush cooling techniques. Night air cools both the exposed interior concrete structure for direct cooling and a rock bed for indi-

rect cooling. This building is also discussed more in Section 17.7.

Rules for Night Flush Cooling

1. Night-flush cooling works best in hot and dry climates with a daily temperature range that exceeds 30°F, but is still effective in somewhat humid regions as long as the daily range is above 20°F.
2. Except for areas with consistent night winds, window or whole-house fans should be used. Ceiling or other circulating indoor fans should be used during the day when the windows are closed.
3. Ideally, there should be about 80 pounds of mass for each square foot of floor area, and the surface area of this mass should be more than two times the floor area.
4. The air flow at night must be directed over the mass to ensure good heat transfer.
5. Window area should be between 10 and 15 percent of the floor area.
6. Windows should be open at night and closed during the day.

10.10 RADIANT COOLING

As was explained in Chapter 3, all objects emit and absorb radiant energy, and an object will cool by radiation if the net flow is outward. At night the long-wave infrared radiation from a clear sky is much less than the long-wave infrared radiation emitted from a building, and thus there is a net flow to the sky (Fig. 10.10a).

Since the roof has the greatest exposure to the sky, it is the best location for a long-wave radiator. Since only shiny metal surfaces are poor emitters, any other surface will be a good choice for a long-wave radiator. Painted metal (any color) is especially good because the metal conducts heat quickly to the painted surface, which then readily emits the energy. Such a radiator on a clear night will cool as much as 12°F below the cool night air. On humid nights, the radiant cooling is less efficient but a temperature depression of about 7°F is still possible. Clouds, on the other hand, almost completely block the radiant cooling effect (Fig. 10.10b).

Direct Radiant Cooling

Potentially the most efficient approach to radiant cooling is to make the roof itself the radiator. For example, an exposed-concrete roof will rapidly lose heat by radiating to the night sky. The next day, the cool mass of concrete can effectively cool a building by acting as a heat sink. The roof, however, must then be protected from the heat of the sun and hot air. Consequently, insulation must be added to the roof every morning and removed every evening.

Harold Hay has designed and built several buildings using this concept, except that he used plastic bags filled with water rather than concrete for the heat-sink material. At night, the water bags are exposed to the night sky by removing the insulation that covered them during the day (Fig. 10.10c). When the sun rises the next day, the water bags are covered by the movable insulation. During the day, the water bags, which are supported by a metal deck, cool the indoors by acting as a heat sink (Fig. 10.10d). This system is especially attractive because it can also work in a passive heating mode during the winter (Figs. 7.18c, d, and e). Although this concept has been tested and shown to be very effective, an inexpensive, reliable, and convenient movable insulation system has still not been achieved.

Another direct-cooling strategy uses a lightweight radiator with movable insulation on the inside. This eliminates two of the problems associated with the above concept: heavy roof structure and a movable insulation system exposed to the weather. With this system, a painted sheet-metal radiator, which is also the roof, covers movable insulation (Fig. 10.10e). At night, this insulation is in the open position so that heat from the building can migrate up and be emitted from the radiator. For the cooling effect to be useful during the day, sufficient mass must be present in the building to act as a heat sink. Also, during the day, the insulation is moved into the closed position to block the heat gain from the roof (Fig. 10.10f).

Indirect Radiant Cooling

The difficulty with movable insulation suggests the use of specialized radiators that use a heat-transfer fluid. This approach is much like active solar

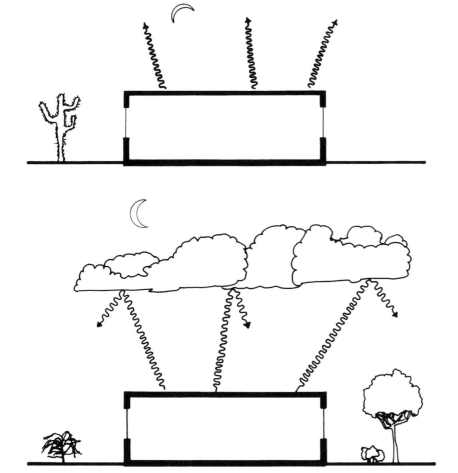

Figure 10.10a On clear nights with little humidity, there is strong radiant cooling.

Figure 10.10b Humidity reduces radiant cooling, and clouds practically stop it.

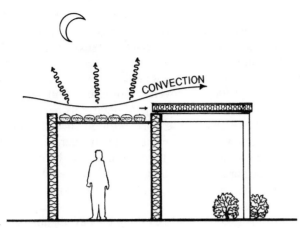

Figure 10.10c During a summer night, the insulation is removed and the water is allowed to give up its heat by radiant cooling.

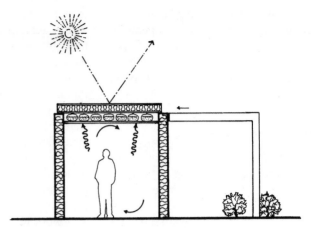

Figure 10.10d During a summer day, the water is insulated from the sun and hot outdoor air, while it acts as a heat sink for the space below.

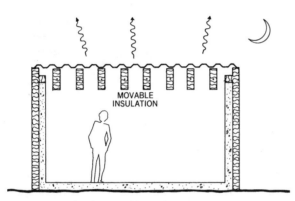

Figure 10.10e At night, the movable insulation is in the "open" position so that the building's heat can be radiated away. This is an example of direct radiant cooling.

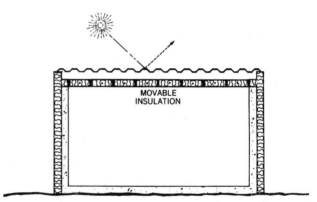

Figure 10.10f During the day, the insulation is in the "closed" position to keep the heat out. The cool interior mass now acts as a heat sink.

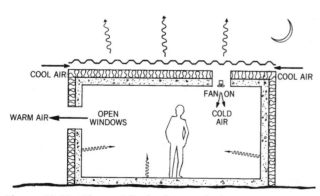

Figure 10.10g The specialized radiator cools air, which is then blown into the building to cool the mass. This is an example of indirect radiant cooling.

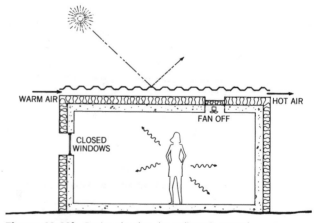

Figure 10.10h During the day, the radiator is vented outdoors, while the building is sealed and the cooled mass acts as a heat sink.

heating in reverse. In Fig. 10.l0g the painted metal radiator cools air at night, which is then blown into the building to cool the indoor mass. The next morning the fan is turned off, and the building is sealed. The cooled indoor mass now acts as a heat sink. The radiator is vented during the day to reduce the heat load to the building (Fig. 10.l0h). Unless the radiator is also used for passive heating, it should be painted white since that color is a good emitter of long-wave radiation and a poor absorber of short-wave (solar) radiation.

If there is not enough exposed mass in the building, a rock bed can be used. At night, the cooled air is blown through the rock bed to flush out the heat. During the day, indoor air blown across the rock bed is cooled by giving up its heat to the rocks. This is similar to one of the passive cooling techniques used by the Bateson State Office building (see Section 17.7).

Rules for Radiant Cooling
1. Radiant cooling will not work well in very cloudy regions. It performs best under clear skies and low humidity, but will still work at lower efficiency in humid regions.
2. This cooling concept applies mainly to one-story buildings.
3. Unless the radiator is also used for passive heating, the radiator should be painted white.
4. Since the cooling effect is small, the whole roof area should be used.

10.11 EVAPORATIVE COOLING

When water evaporates, it draws a large amount of sensible heat from its surroundings and converts this type of heat into latent heat in the form of water vapor. As sensible heat is converted to latent heat, the temperature drops. This phenomenon is used to cool buildings in two very different ways. If the water evaporates in the building or in the fresh-air intake, the air will be not only cooled, but also humidified. This method is called direct evaporative cooling. If, however, the building or indoor air are cooled by evaporation without humidifying the indoor air, the method is called indirect evaporative cooling.

Direct Evaporative Cooling

When water evaporates in the indoor air, the temperature drops but the humidity goes up. In hot and dry climates, the increase in humidity actually improves comfort. However, **direct evaporative cooling** is not appropriate in humid climates because the cooling effect is low and the humidity is already too high. See Fig. 4.11 for applicability of direct evaporative cooling. Horticultural greenhouses are an exception because most plants thrive on high humidity but not high temperature.

The most popular form of direct evaporative cooling is accomplished with commercially available **evaporative coolers** (swamp coolers). Although they look like active mechanical devices from the outside, they are actually quite simple and use little energy (Fig. 10.11a and b). A fan is used to bring outdoor air into the building by way of a wet screen. A modest amount of water is required to keep the screen wet. To maintain comfort, a high rate of ventilation is required during the day (about twenty air changes per hour).

Misting the air has become quite a popular, direct evaporative-cooling strategy. Water under high pressure is atomized into tiny droplets, which then readily evaporate to cool the air. Misting is mainly used to cool outdoor spaces. Unfortunately, if the area is too sunny or too windy, the benefit of misting will be minimal. However, the cooling effect can be significant in sheltered outdoor spaces and greenhouses. Misting is often used more for the "atmosphere" it creates than for its cooling benefits.

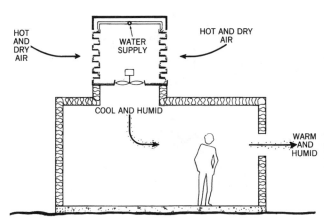

Figure 10.11a Evaporative coolers (swamp coolers) look a great deal like central Air Conditioning units, but their cooling mechanism is very simple and inexpensive. They are appropriate only in dry climates.

Figure 10.11b Evaporative coolers are widely used in hot and dry regions. This is an example of a direct evaporative cooler on the roof of a house.

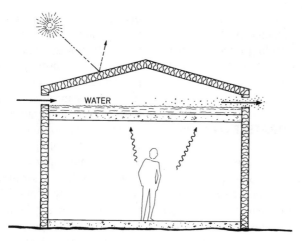

Figure 10.11c This indirect evaporative-cooling system uses a roof pond. Note that no humidity is added to the indoors.

Indirect Evaporative Cooling

The cooling effect from evaporation can also be used to cool the roof of a building, which then becomes a heat sink to cool the interior. This technique is an example of **indirect evaporative cooling,** and its main advantage is that the indoor air is cooled without increasing its humidity.

A critical aspect of evaporative cooling is that the heat of vaporization must come from what is to be cooled. Thus, spraying a sunlit roof is not especially good because most of the water will be evaporated by the heat of the sun. On the other hand, the heat to evaporate water from a shaded roof pond comes mainly from the building itself.

Figure 10.11c illustrates the basic features of roof-pond cooling. An insulated roof shades the pond from the sun. Openings in the roof enable air currents to pass over the pond during the summer. As water evaporates, the pond will become cooler and together with the ceiling structure will act as a heat sink for the interior of the building. During the winter, the pond is drained and the roof openings are closed. The main disadvantage of this system is the cost of the double-roof structure and waterproofing.

A clever alternative to the above roof pond is the "roof pond with floating insulation."" A double-roof structure is now not needed because the insulation floats on the roof pond

(Fig. 10.11d). At night, a pump sprays the water over the top of the insulation, and it cools by both evaporation and radiation. When the sun rises, the pump stops and the water remains under the insulation, where it is protected from the heat of the day. Meanwhile, the water together with the roof structure act as a heat sink for the interior. Although the cooling occurs only at night, it is very effective because of the combined action of evaporation and radiation.

"Indirect evaporative coolers" are now commercially available as packaged units. They are similar to the "evaporative coolers" mentioned above except that they do not humidify the indoor air (Fig. 10.11e). Outdoor air is used to evaporate water off the surface of tubes. The necessary heat of vaporization is drawn in part from these tubes through which indoor air is flowing. Thus, indoor air is cooled but not humidified. These units are sometimes used in series with "evaporative coolers" for extra cooling.

Rules for Evaporative Cooling

1. Direct evaporative cooling is appropriate only in dry climates.
2. Indirect evaporative cooling works best in dry climates but can also be used in somewhat humid climates because it does not add to the indoor humidity.

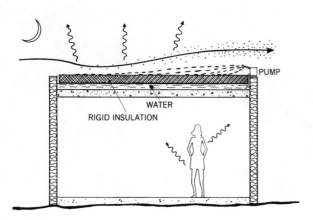

Figure 10.11d This indirect evaporative cooling system uses floating insulation instead of a second roof to protect the water from the sun and heat of the day.

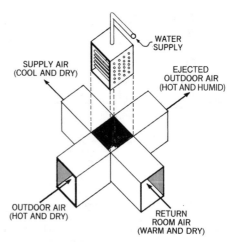

Figure 10.11e Indirect evaporative coolers reduce the indoor air temperature without increasing its humidity.

10.12 EARTH COOLING

Before earth-cooling techniques can be discussed, the thermal properties of soil must be considered. Since the temperature of soil near the surface is close to the air temperature, it fluctuates widely from summer to winter. However, due to the large time lag of earth, the soil temperature fluctuates less and less as the soil depth increases. At about 20 feet in depth, the summer/winter fluctuations have almost disappeared and a steady state temperature exists year-round, which is equal to the average annual air temperature.

The graph of Fig. 10.12a shows the earth temperatures as a function of depth. One curve represents the maximum summer temperatures, and the other represents the minimum winter temperatures of the soil. The ground temperatures at any depth fluctuate left and right between these curves.

Notice that the ground temperature is always below the maximum air temperature, and the difference increases with depth. Thus, the earth can always be used as a heat sink in the summer. However, unless the temperature difference is great enough, earth cooling might not be practical. The map of Fig. 10.12b shows that the deep soil temperatures are low enough for earth cooling (approx. 60°F or less) in much of the country. Even if the soil is too warm for actual cooling, it will, nevertheless, be much cooler than the outdoor air. Earth sheltering, which is described in Section 15.11, can be very advantageous in some climates.

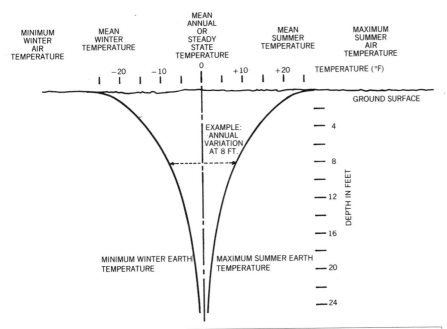

Figure 10.12a Soil temperature varies with time of year and depth below grade. To find the maximum or minimum soil temperature, first find the mean annual steady-state temperature from Fig. 10.12b, and then according to depth add (summer) or subtract (winter) the deviation from the centerline.

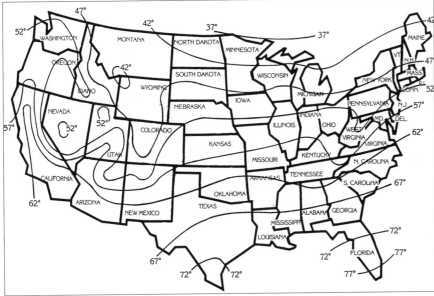

Figure 10.12b Deep earth temperatures are approximately equal to these well-water temperatures. (Reprinted with permission of National Water Well Association.)

Cooling the Earth

Since the sun heats the soil, shading the surface significantly reduces the maximum earth temperature. Water evaporating directly from the surface will also cool the soil. Both techniques together can reduce soil surface temperatures as much as 18°F. It must be noted, however, that transpiration from plants (especially trees and bushes) is not very helpful because the evaporation occurs in the air high above the soil.

A canopy of trees, an elevated patio deck, and even a building over a crawl space are all possibilities for shading soil while letting air motion cause evaporation from the surface (Fig. 10.12c). When rain is not sufficient, a sprinkler should keep the soil moist. In dry climates, a light-colored gravel bed about 4 inches deep can effectively shade the soil while still allowing evaporation from the earth's surface below the gravel (Fig. 10.12d). However, when sprinklers are used they should operate only at night, otherwise sun-warmed water will percolate into the soil.

Direct Earth Coupling

When earth-sheltered buildings have their walls in direct contact with the ground (i.e., there is little or no insulation in the walls), we say that there is **direct earth coupling.** In regions where the mean annual temperature is below 60°F, direct coupling will be a significant source of cooling. This asset becomes a liability, however, in the winter when excess heat will be lost to the cold ground. One solution is to insulate the earth around the building from the cold winter air but not from the building (Fig. 10. 12e). This horizontal insulation buried in the ground will not change the steady-state temperature but will reduce the maximum and minimum ground temperatures.

When the mean annual temperature is above 60°F, heat loss in winter is not a big problem, but the earth might not be cool enough in the summer. Unless other reasons exist for choosing an earth-sheltered building, earth cooling might be better achieved by using an **indirect earth coupling** strategy described below.

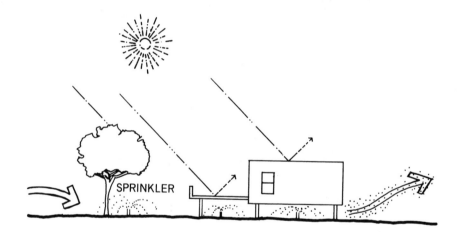

Figure 10.12c Soil can be cooled significantly below its natural temperature by shading it and by keeping it wet for evaporative cooling.

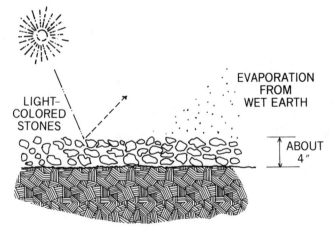

Figure 10.12d In dry climates, soil can be cooled with a gravel bed, which shades the soil while it allows evaporation to occur.

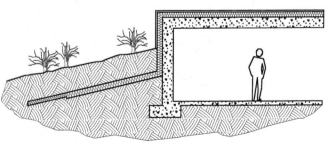

Figure 10.12e In earth-sheltered buildings in cold climates, the earth should be insulated from the cold winter air.

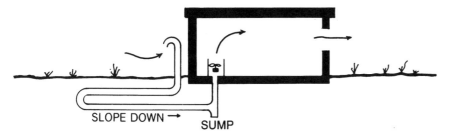

Figure 10.12f Indirect earth cooling is possible by means of tubes buried in the ground. Sloped tubes and a sump are required to catch condensation. An open-loop system is shown, while a closed-loop system would return the air from indoors.

Indirect Earth Coupling

A building can be indirectly coupled to the earth by means of earth tubes. When cooling is desired, air is blown through the tubes into the building (Fig. 10.12f). The earth acts as a heat sink to cool the air.

The tubes should be buried as deeply as possible to take advantage of that constant, deep earth temperature, which is the coolest available in the summer. See again Fig. 10.12b for deep-ground temperatures in the United States.

Use an open-loop system if large amounts of fresh air are required. On the other hand, a closed-loop system offers greater efficiency and reduced amounts of condensation.

The greatest problem with earth tubes is condensation, which occurs mainly in humid climates where the earth temperature is frequently below the dew point, or saturation temperature, of the air. The tubes, therefore, must be sloped to drain into a sump. The consequences of biological activity (e.g., mold growth) in the moist tubes are unknown at this time, and caution is advised. Condensation is not a big problem in dry climates. Tubes must be absolutely tight to prevent radon gas or water from entering. (Radon is discussed further in Section 16.17.)

Rules for Earth Cooling

1. The steady-state deep earth temperature is similar to the mean annual temperature at any location (Fig. 10.12b).

2. Directly coupled earth cooling works well when the steady-state earth temperature is a little below 60°F. If the earth is much colder, the building must be insulated from the ground.

4. Earth tubes are very effective in dry climates with cold winters. The condensation in humid climates causes a health risk.

5. In humid climates, the condensation on walls or in earth tubes might cause biological activity, which may be a problem.

6. Directly coupled (underground) structures have reduced access to natural ventilation, which has top priority in hot and humid climates. Thus, earth cooling works best in dry climates and can be problematic in humid climates.

10.13 DEHUMIDIFICATION WITH A DESICCANT

In humid regions, dehumidifying the air in summer is very desirable for thermal comfort and control of mildew. Two fundamental ways to remove moisture from the air exist. With the first method, the air is cooled below the dew point temperature. Water will then condense out of the air. Conventional air conditioning and dehumidification use this principle. Some of the passive cooling techniques mentioned above will also dehumidify in the same way. For example, in humid climates, water will often condense in earth tubes.

The second method (Fig. 10.13) involves the use of a **desiccant** (drying agent). A number of chemicals, such as silica gel, natural zeolite, activated alumina, and calcium chloride, will absorb large amounts of water vapor from the air. However, there are two serious difficulties with the use of these materials. First, when water vapor is absorbed and turned into liquid water, heat is given off. This is the same heat that was required to vaporize the water in the first place (heat of vaporization). Thus, if a desiccant is placed in a room, it will heat the air as

Figure 10.13 Dehumidification for thermal comfort by means of a desiccant is very expensive and complicated as this schematic diagram indicates. (From the *Passive Cooling Handbook* by Lawrence Berkeley Laboratory, CA, 1980. DOE Pub-375.)

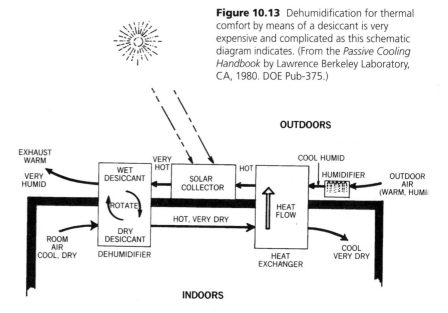

it de-humidifies it (i.e., the desiccant converts latent heat into sensible heat). Thermal comfort will, therefore, require another cooling stage to lower the temperature of the air.

The second problem with the use of a desiccant is that the material soon becomes saturated with water and stops dehumidifying. The desiccant must then be **regenerated** by boiling off the water. Although desiccant dehumidification works in theory, no one has been able to build an inexpensive system due to its inherent complexity.

10.14 CONCLUSION

Passive cooling strategies have the greatest potential in hot and dry climates. Just about every cooling technique will work there. On the other hand, in *very* humid regions only comfort ventilation will be very helpful. Many regions, however, are not too humid for night-flush cooling and night-radiation cooling to be beneficial. Most of the eastern United States has this moderately humid climate, where passive cooling can replace or reduce the need for air conditioning much of the summer. However, in every climate, the first and best strategy for summer comfort is heat avoidance.

KEY IDEAS OF CHAPTER 10

1. Make full use of **heat avoidance** before applying any passive cooling strategies.
2. **Comfort ventilation** is used day and night to cool people and to keep the indoor temperature close to the outdoor temperature. This type of ventilation is used mainly in very humid climates and in temperate climates when the humidity is high.
3. **Night flush cooling** uses night ventilation to cool the mass of the building. During the day, the windows are closed, and the mass acts as a heat sink. This passive cooling strategy is used in both dry climates and temperate climates whenever the humidity is low.
4. Air flows from positive- to negative-pressure areas.
5. There is a positive pressure on the windward side, and a negative pressure on both the leeward side and the sides of the building parallel to the wind.
6. Hot air can be exhausted from the top of a building by stratification, the stack effect, the shape of the roof (the venturi effect), and the increased wind velocity found at higher elevations (the Bernoulli effect).
7. Use cross-ventilation whenever possible.
8. Have air flow across people for comfort ventilation and across the building mass for night-flush cooling.
9. **Radiant cooling** from the roof works well in climates in which the humidity is low and clouds are few.
10. Use **direct evaporative cooling** when the relative humidity is low.
11. Use **indirect evaporative cooling** when the relative humidity is high.
12. Use **earth cooling** mainly in the North and West. Condensation can be a problem in the humid Southeast.

Resources

FURTHER READING
(See Bibliography in back of book for full citations. This list includes valuable out-of-print books.)

Abrams, D. W. *Low-Energy Cooling.*
Akbari, H., et al. *Cooling Our Communities: A Guidebook on Tree Planting and Light-Colored Surfacing.*
Allard, F., ed. *Natural Ventilation in Buildings.*
Boutet, T. S. *Controlling Air Movement: A Manual for Architects and Builders.*
Brown, G. Z., and M. DeKay. *Sun, Wind, and Light: Architectural Design Strategies.*
Cook, J. "Cooling as the Absence of Heat."
Cook, J., ed. *Passive Cooling.*
Daniels, K., *The Technology of Ecological Building: Basic Principles and Measures, Examples and Ideas.*
Environmental Protection Agency. "Cooling Our Communities."
Givoni, B. *Climate Considerations in Building and Urban Design.*
Givoni, B. *Man, Climate and Architecture.*
Givoni, B. *Passive and Low Energy Cooling of Buildings.*
Golany, G. *Housing in Arid Lands— Design and Planning.*
Konya, A. *Design Primer for Hot Climates.*
Olgyay, V. *Design with Climate.*
Santamouris, M., and D. Asimakopoulos, eds. *Passive Cooling of Buildings.*
Solar Energy Research Inst. *Cooling with Ventilation.*
Stein, B., and J. S. Reynolds. *Mechanical and Electrical Equipment for Buildings,* 9th ed.
Underground Space Ctr. Unv. Minn. *Earth Sheltered Housing Design: Guidelines, Examples, and References.*
Watson, D., and K. Labs. *Climatic Design: Energy Efficient Building Principles and Practices.*

Reference
Schubert, R. P., and P. Hahn. From *Progress in Passive Solar Energy* by American Solar Energy Society, Boulder, CO, 1983.

SITE DESIGN AND COMMUNITY PLANNING

"The sun is fundamental to all life. It is the source of our vision, our warmth, our energy, and the rhythm of our lives. Its movements inform our perceptions of time and space and our scale in the universe....Assured access to the sun is thus important to the quality of our lives."

Ralph L. Knowles
from **Sun Rhythm Form** © *Ralph L. Knowles, 1981*

11.1 INTRODUCTION

The heating, cooling, and lighting of a building are very much affected by the site and community in which the building is located. Although many aspects of a site have an impact a building, only those that affect solar access and wind penetration will be discussed here. All of the strategies discussed fall under tiers one and two of the three-tier design approach (Fig. 11.1a).

The ancient Greeks realized the importance of site and community planning for the heating and cooling of buildings. Since they wanted their buildings to face the winter sun and reject the summer sun, their new towns were built on southern slopes and the streets ran east–west whenever possible. The ancient Greek city of Olynthus was planned so that most buildings could front on east–west streets (Fig. 11.1b). See Fig. 2.15 for how the buildings were designed to take advantage of this street layout. The ancient Greeks considered their solar design of buildings and cities to be "modern" and "civilized."

The Romans were also convinced of the value of solar heating, so much so that they protected solar access by law. The great Justinian Code of the sixth century states that sunshine may not be blocked from reaching a heliocaminus (sun room).

While winter heating was critical to the ancient Greeks and Romans, summer shade was also very desirable. By building continuous rows of

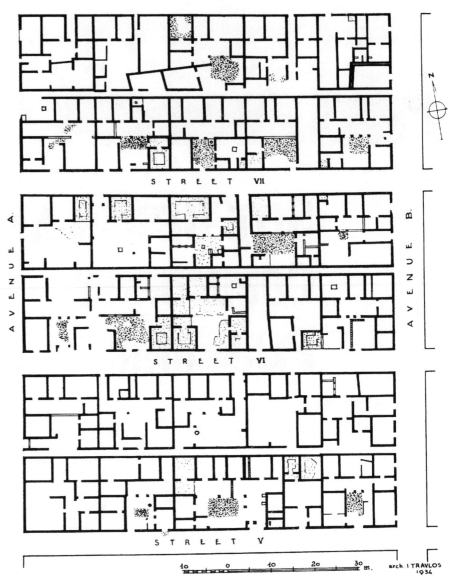

Figure 11.1b The ancient Greek city of Olynthus was oriented toward the sun. (From *Excavations at Olynthus. Part 8, The Hellenic House* © Johns Hopkins University Press, 1938.)

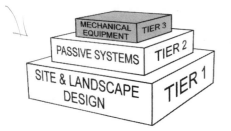

Figure 11.1a The heating, cooling, and lighting needs of a building are best and most sustainably achieved by the three-tier design approach, and this chapter covers both tiers one and two.

homes along east–west streets, only the end units would be exposed to the low morning and afternoon summer sun.

In climates with very hot summers and mild winters, shade is more desirable than solar access. Often multistory buildings are built on narrow streets to create shade both for the street and for the buildings (Fig. 11.1c). When buildings are not tall enough to cast much shade, pedestrian streets can have their own shading system (Fig. 11.1d). And when protection from rain as well as

sun is desirable, arcades and colonnades are frequently used (Fig. 11.1e).

Wind is also an important factor in vernacular design. When there is too much wind and the temperature is cool, windbreaks are common, and windbreaks of dense vegetation are most common (Fig. 11.1f). On the other hand, when the climate is warm and humid, cross-ventilation is very desirable. Many native communities maximize the benefit of natural ventilation by building far apart and by eliminating low vegetation that would block the cooling breezes (Fig. 11.1g).

Figure 11.1c Multistory buildings facing narrow streets create desirable shade in very hot climates, as here in Jidda, Saudi Arabia. (Photograph by Richard Millman.)

Figure 11.1d The shading structure over this Moroccan street blocks much of the sun but still allows air and daylight to filter through. (Courtesy of Moroccan National Tourist Office.)

Figure 11.1e This colonnade in Santa Fe, NM, protects pedestrians from rain as well as sun.

Figure 11.1f Farms in the Shimane Prefecture of Japan use "L"-shaped windbreaks for protection from the cold wind. (From *Sun, Wind, and Light: Architectural Design Strategies* by G. Z. Brown, © John Wiley, 1985.)

Figure 11.1g In hot and humid climates, such as Tocamacho, Honduras, buildings are set far apart to maximize the cooling breezes. (From *Sun, Wind, and Light: Architectural Design Strategies* by G. Z. Brown, © John Wiley, 1985.)

11.2 SITE SELECTION

When the United States was first settled and land was still plentiful, farms were almost exclusively built on south slopes. It was well known that a south slope is warmer and has the longest growing season. When a choice of sites is available, a south slope is still the best for most building types.

In the winter, the south slope is the warmest land for two reasons. The south slope receives the most solar energy on each square foot of land because it most directly faces the winter sun (Fig. 11.2a). This phenomenon, called the cosine law, was discussed in depth in Chapter 6. The south slope will also experience the least shading because objects cast their shortest shadows on south slopes (Fig. 11.2b).

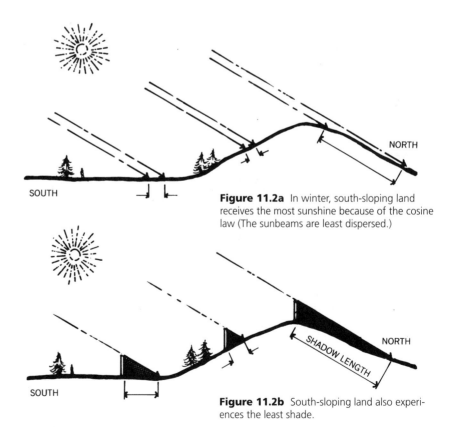

Figure 11.2a In winter, south-sloping land receives the most sunshine because of the cosine law (The sunbeams are least dispersed.)

Figure 11.2b South-sloping land also experiences the least shade.

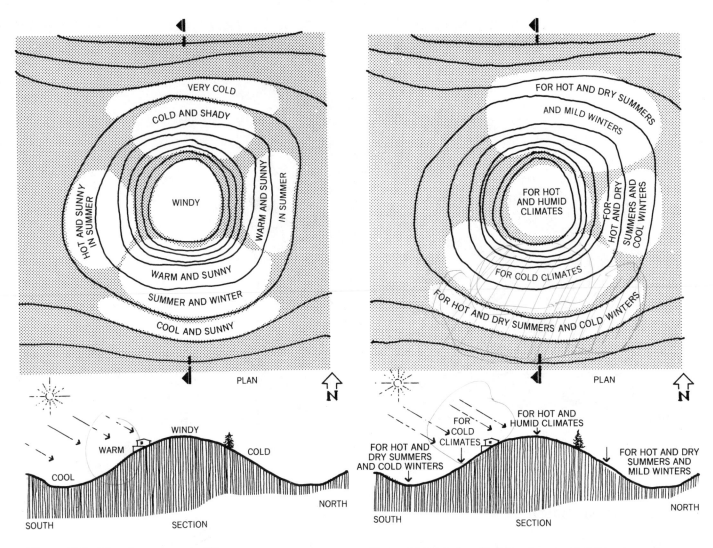

Figure 11.2c Microclimates around a hill.

Figure 11.2d Preferred building sites around a hill in response to climate for envelope-dominated buildings.

Figure 11.2c illustrates the variation in microclimate with different slope orientations. The south slope gets the most sun and is the warmest in the winter while the west slope is the hottest in the summer. The north slope is the shadiest and coldest, while the hilltop is the windiest location. Low areas tend to be cooler than slopes because cold air drains into them and collects there.

The best site for a building on hilly land depends on both climate and building type. For envelope-dominated buildings, such as residences and small office buildings, the climate would suggest the sites shown in Fig. 11.2d.

For example, in

Cold climates: South slopes maximize solar collection and are shielded from cold northern winds. Avoid the windy hilltops and low-lying areas that collect pools of cold air.

Hot and dry climates: Build in low-lying areas that collect cool air. If winters are very cold, build on bottom of south slope. If winters are mild, build on north or east slopes, but in all cases avoid the west slopes.

Hot and humid climates: Maximize natural ventilation by building on hilltops but avoid the west side of hilltops because of the hot afternoon sun.

For internally dominated buildings, such as large office buildings that require little if any solar heating, the north and northeast slopes are best. Also appropriate are the cool low-lying areas especially to the north of hills.

11.3 SOLAR ACCESS

Nothing is as certain and consistent as the sun's motion across the sky. It is, therefore, possible to design for solar access with great accuracy, barring the possibility that future construction on neighboring property will block the sun. If neighbors are sufficiently faraway or if restrictions exist as to what can be built next door, solar access can be assured.

Although laws protecting solar access are rare, they do exist in the United States and they have existed for centuries in England. These legal aspects of solar access will be considered later. A discussion of the physical principles of solar access must come first.

In Chapter 6, the sun's motion was explained by means of a sky dome.

That part of the sky vault through which *useful* solar energy passes was called the **solar window** (Fig. 6.8b). The bottom of this solar window is defined by the sun path of the winter solstice (December 21). The sides of the window are usually set at 9 A.M. and 3 P.M. The window thus includes the time period during which more than 80 percent of the winter solar radiation is available. Of course if sunlight is available before 9 A.M. or after 3 P.M., it should be used.

The conical surface generated by the sun's rays on the winter solstice from 9 A.M. to 3 P.M. is called the **solar-access boundary.** Any object that projects through this surface will obstruct the winter sun (Fig. 11.3a). A north–south section through this solar-access boundary is shown in Fig. 11.3b.

Because the conical surface of the solar-access boundary is difficult to deal with, a simplified surface is used instead (Fig. 11.3c). In plan, this simplified surface can be defined by contour lines, which not only define the elevation of the solar-access boundary but at the same time define the maximum height an object can have without penetrating the boundary and obstructing the sun. The solar contour lines for a simple rectangular building are given in Fig. 11.3d. Since the slope of the boundary is a function of latitude, use Table 11.3A to determine the location of the contour lines for the latitude in question.

To draw sections of the solar-access boundary, use Table 11.3B, which presents the solar altitude angles when the sun is due south (12 noon) and when the azimuth is 45 degrees (about 9 A.M. and 3 P.M.).

The contour lines shown in Fig. 11.3d are for a building on level ground. Any rise or fall of the land will have a positive or negative effect on how high objects can be before they block the winter sun. To account

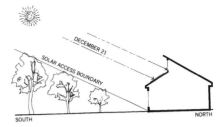

Figure 11.3b The solar-access boundary determines how high objects can be before they obstruct the sun.

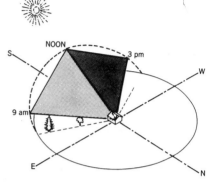

Figure 11.3c To simplify construction of the solar-access boundary, two inclined planes replace the conical surface.

for the effect of sloping land, superimpose the contour lines for the solar-access boundary and the contour lines for the slope of the land as in Fig. 11.3e. Where the land is lower than the building, add the drop in elevation to the height of the solar-access boundary. For example, at point "X" an object can be 35 feet high without blocking the sun because the total height from grade level to the solar-access boundary has increased (Fig. 11.3f). However, at point "Y" an object can be only 15 feet high because the rise of the land has decreased the distance from grade level to the solar-access boundary (Fig. 11.3g).

For passive solar heating through south-facing windows, the solar-access boundary should start from the base of the building as was discussed above. However, if solar access is required only for the roof or clerestory windows, the solar-access boundary should be raised as shown in Fig. 11.3h.

Figure 11.3a The solar-access boundary is a conical surface generated by sunrays on December 21 from 9 A.M. to 3 P.M. Any trees or buildings projecting through this surface will obstruct solar access to the site.

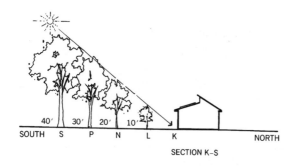

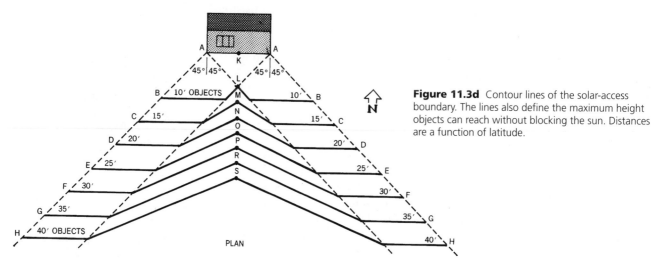

Figure 11.3d Contour lines of the solar-access boundary. The lines also define the maximum height objects can reach without blocking the sun. Distances are a function of latitude.

Figure 11.3e Sloping land will affect how high objects can be before they block the winter sun.

TABLE 11.3A DISTANCES IN FEET FROM A BUILDING TO CONTOUR LINES OF SOLAR-ACCESS BOUNDARY

	Latitude					
Segment	28	32	36	40	44	48
AB	24	29	37	47	71	95
AC	35	44	56	71	107	143
AD	47	58	75	94	142	190
AE	59	73	93	118	178	238
AF	71	87	112	141	214	285
AG	83	102	131	165	249	333
AH	94	116	149	188	285	380
KL	13	15	17	20	25	29
KM	19	22	26	29	37	44
KN	26	30	35	39	50	58
KO	32	37	43	49	62	73
KP	38	45	52	59	74	87
KR	45	52	61	69	87	102
KS	51	59	69	79	99	116

TABLE 11.3B ALTITUDE ANGLES OF SOLAR-ACCESS BOUNDARY

	Latitude					
	28	32	36	40	44	48
At noon	38	34	30	27	22	19
At 45° azimuth	23	19	15	12	8	6

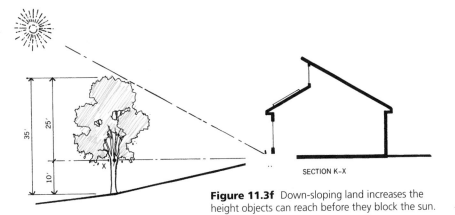

Figure 11.3f Down-sloping land increases the height objects can reach before they block the sun.

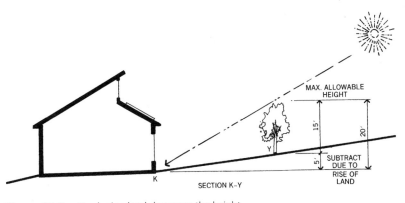

Figure 11.3g Up-sloping land decreases the height objects can reach before they block the sun.

The solar-access boundary reaches 45 degrees east and west of south even if a building is not facing due south. The distances from Table 11.3A can still be used as before. However, the contour lines will look a little different as can be seen in Fig. 11.3i.

Some books suggest using deciduous trees on the south side, as shown in Fig. 11.3j, to produce summer shade while still permitting access to the winter sun. Unfortunately, deciduous trees without their leaves still block a significant amount of sunlight. Most deciduous trees block somewhere between 30 percent and 60 percent of sunlight (Figs. 11.3k and 11.9b). Also, if the roof has collectors for domestic hot water, pool heating, or photovoltaics, it should *not* be shaded in the summer. Thus, on the south side, trees should usually be kept below the solar-access boundary (Fig. 11.3b).

If large trees already exist on the south side of a building, it is rarely appropriate to cut them down to improve the solar access. Especially in hot climates, the summer shade from mature trees might be more valuable than the winter sun. Instead, it is often possible to prune the lower branches so that both summer shade and winter south-wall access are possible (Fig. 11.3l). Of course, roof collectors are then not appropriate, and the use of clerestory windows is questionable.

For additional information about solar-access boundaries, see Chapter 7 of *Energy-Conserving Site Design* by McPherson. Since the above-mentioned graphic method can become very complicated except for the simplest situations, the author highly recommends the use of physical models in conjunction with a sun machine, which will be described in Section 11.7.

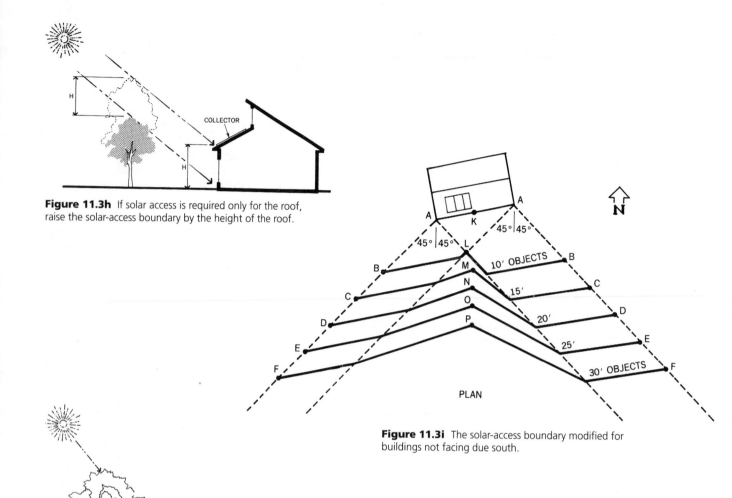

Figure 11.3h If solar access is required only for the roof, raise the solar-access boundary by the height of the roof.

PLAN

Figure 11.3i The solar-access boundary modified for buildings not facing due south.

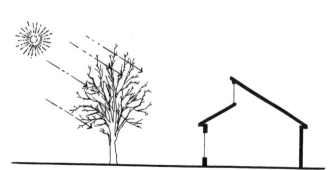

Figure 11.3j Trees on the south side of a building can shade a domestic hot-water solar collector as well as windows during the summer.

Figure 11.3k Even without leaves, deciduous trees still block from 30 percent to 60 percent of sunshine.

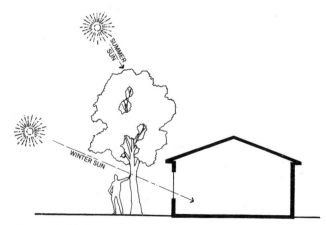

Figure 11.3l If large trees exist on the south side, trim the lower branches to form a high canopy.

11.4 SHADOW PATTERNS

When we are designing only one building on a site, solar access is best achieved by working with the solar-access boundary. When we are designing a complex of buildings or a whole development, **shadow patterns** are more useful for achieving solar access to all the buildings. By drawing the shadow pattern for each building and tree, it is possible to quickly determine conflicts in solar access (Fig. 11.4a). The main difficulty with this technique is the generation of accurate shadow patterns.

A shadow pattern is a composite of all shadows cast during winter hours when access to the sun is most valuable. It is generally agreed that solar access should be maintained, if possible, from about 9 A.M. to 3 P.M. during the winter months. During those six hours more than 80 percent of a winter's day total solar radiation will fall on a building.

The easiest way to understand shadow patterns is by examining the shadows cast by a vertical pole. Figure 11.4b illustrates the shadows cast by a pole on December 21. The same pole is shown in plan in Fig. 11.4c with the shadows cast at each hour between 9 A.M. and 3 P.M. at 36° north latitude. The shaded area represents the land where solar access is blocked by the pole at some time during these hours and is called the shadow pattern of the pole. Of course, the taller the pole, the longer the shadow pattern. The shadow pattern would be longer at higher latitudes and shorter at lower latitudes.

To make the construction of shadow patterns easier, we can determine the shadow lengths only for the hours of 9 A.M., 12 noon, and 3 P.M. and then connect them with straight lines (Fig. 11.4d). The construction of a shadow pattern is further simplified by drawing the ends of the shadow pattern at 45 degrees, which corresponds closely with 9 A.M. and 3 P.M. (actual azimuth at these times varies

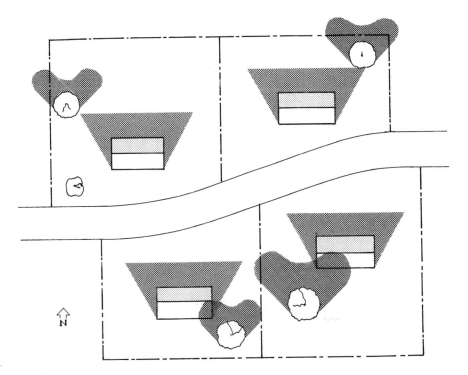

Figure 11.4a Shadow patterns demonstrate conflicts in solar access. Notice that the trees shade the lower two buildings during the winter.

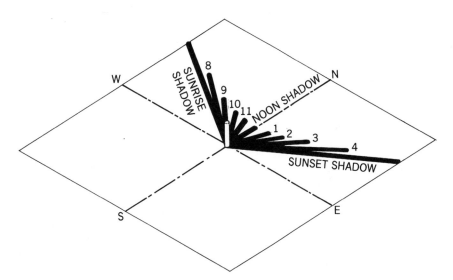

Figure 11.4b Shadows cast by a vertical pole from sunrise to sunset on December 21 (shadows vary with latitude).

with latitude). The length of the shadow is determined by drawing a section through the pole at 12 noon and 9 A.M./3 P.M. (45-degree azimuth) (Fig. 11.4e). Use Table 11.3B, Appendix A, or Appendix B for the altitude angles of the sun's rays at those times.

To construct the shadow pattern for a building, we assume that the building consists of a series of poles (Fig. 11.4f). The morning, noon, and afternoon shadows are then constructed from these poles (Fig. 11.4g). The composite of these shadows creates the shadow pattern (Fig. 11.4h).

Additional poles would be required for more complex buildings.

The shading pattern, like the solar-access boundary, is affected by the slope of the land (Fig. 11.4i). Also, just because the footprint of a building is shaded, it cannot be assumed that the roof is also shaded. Because of the limitations and complexity of this graphic method, it is often easier to use physical models for generating accurate shadow patterns and for determining how a building is actually shaded. The use of physical models for this purpose will be explained below.

However, there is a quick graphic method for creating approximate shadow patterns for simple buildings on flat land. Figure 11.4j illustrates this quick method for a gabled house, while Fig. 11.4k illustrates the quick method for creating shadow patterns for trees.

It is important to remember that some solar access is better than none. Even if solar access cannot be provided from 9 A.M. to 3 P.M., assured access from 10 A.M. to 2 P.M. would still make available over 60 percent of the total daily solar radiation in the winter.

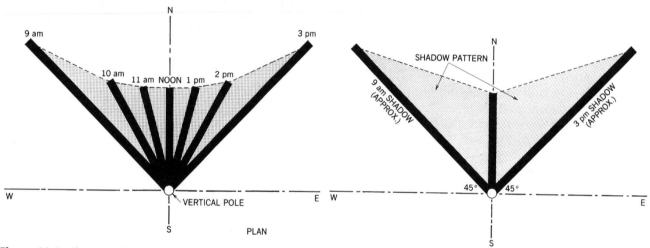

Figure 11.4c Plan view of shadows cast by a pole at 36°N. latitude on December 21.

Figure 11.4d Simplified shadow pattern of a pole on December 21.

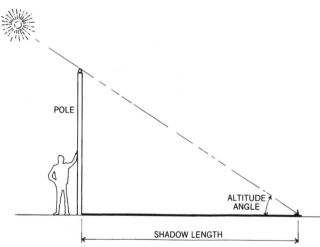

Figure 11.4e Determination of shadow length.

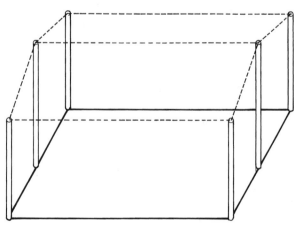

Figure 11.4f To generate the shadow pattern of a building, assume that the building consists of a series of poles.

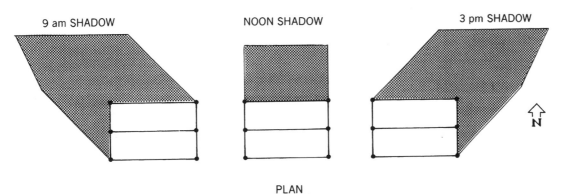

9 am SHADOW NOON SHADOW 3 pm SHADOW

N

PLAN

Figure 11.4g The morning, noon, and afternoon shadows are constructed by assuming six poles. (After *Solar Energy Planning* by Tabb.)

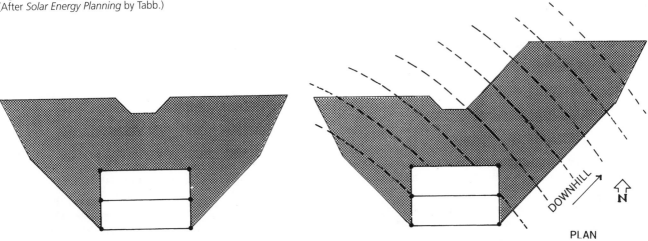

Figure 11.4h The shadow pattern is the composite of the morning, noon, and afternoon shadows. (After *Solar Energy Planning* by Tabb.)

DOWNHILL

N

PLAN

Figure 11.4i Sloping land changes the length of the shadow pattern. (After *Solar Energy Planning* by Tabb.)

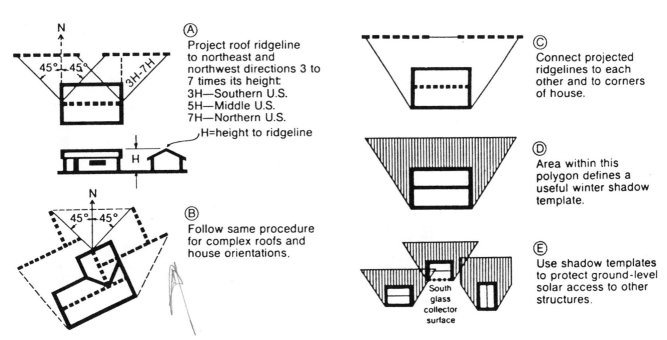

N

45° 45° 3H-7H

Ⓐ Project roof ridgeline to northeast and northwest directions 3 to 7 times its height
3H—Southern U.S.
5H—Middle U.S.
7H—Northern U.S.

H=height to ridgeline

H

N

45° 45°

Ⓑ Follow same procedure for complex roofs and house orientations.

Ⓒ Connect projected ridgelines to each other and to corners of house.

Ⓓ Area within this polygon defines a useful winter shadow template.

Ⓔ Use shadow templates to protect ground-level solar access to other structures.

South glass collector surface

Figure 11.4j Quick method for constructing shadow patterns (templates). (Reprinted with permission from *Energy Conserving Site Design* by E. Gregory McPherson, © 1984, The American Society of Landscape Architects.)

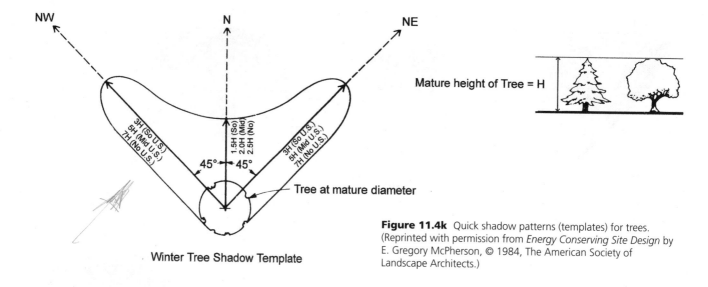

Mature height of Tree = H

Winter Tree Shadow Template

Figure 11.4k Quick shadow patterns (templates) for trees. (Reprinted with permission from *Energy Conserving Site Design* by E. Gregory McPherson, © 1984, The American Society of Landscape Architects.)

11.5 SITE PLANNING

Access to winter sun and avoidance of the summer sun are greatly affected by building orientation. Because it is common in suburban areas for buildings and especially homes to present their long facade to the street, building orientation is determined largely by road design.

Fortunately, there is a road orientation that is ideal for both winter-heating and summer-cooling needs. Streets that run east–west not only maximize winter solar access from the south but also maximize shade from the low morning and afternoon summer sun (Fig. 11.5a). On the other hand, with north–south streets there is little if any winter solar access, and the east and west facades are exposed to the summer sun (Fig. 11.5b).

It is usually possible to design new developments to maximize the lots fronting on east–west streets. A good example of this, the Village Homes

subdivisions in Davis, California, is described at the end of the chapter (Fig. 11.11b and c). The performance of buildings on north–south streets can be improved significantly by a number of different methods. Orienting the short facade to the street is the most obvious technique (Fig. 11.5c). The use of **flag lots,** interior lots with only driveway access to the street, is less common but a very effective technique for achieving a good orientation for each building, as

well as a quiet off-the-street location for some buildings (Fig. 11.5d). The increasing popularity of duplexes makes the technique shown in Fig. 11.5e very promising.

Good orientation can also be achieved on diagonal streets if the buildings are rotated to face south. Although the practice of orienting facades parallel to streets is a widely held convention, there are a number of benefits in the alternate arrangement shown in Fig. 11.5f. Besides the

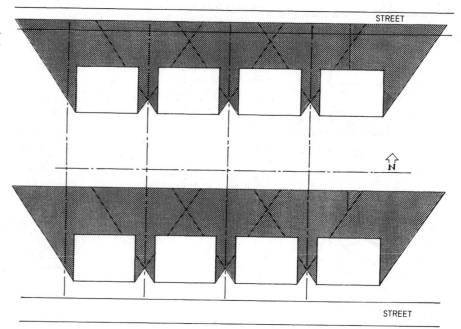

Figure 11.5a East–west streets are ideal for both winter solar access from the south and summer shading from the low east and west sun.

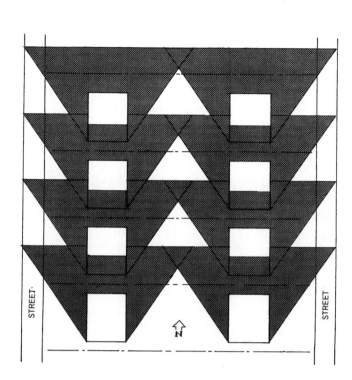

Figure 11.5b With conventional development, north–south streets promote neither solar access in winter nor shading on east or west facades in summer.

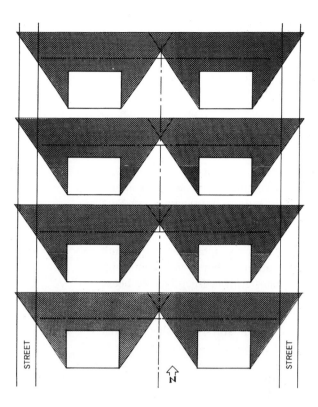

Figure 11.5c Buildings on north–south streets should have their narrow facade face the street.

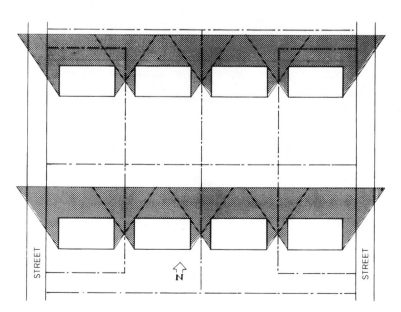

Figure 11.5d Flag lots (driveway looks like a flagpole) can achieve a good orientation for each building on a north–south street.

better solar orientations, this arrangement also yields much greater privacy since windows do not face each other. There are also aesthetic possibilities in this nonconventional design.

Although lots facing east–west streets have the greatest potential for good solar access and shading, these benefits are not guaranteed. For example, uneven setbacks can significantly reduce both winter sun and summer shading (Figs. 11.5g and 11.5h). Also, for buildings two or more stories high, the north–south separation between buildings becomes critical. Deep lots are better than shallow lots on east–west streets (Fig. 11.5i). However, when depth of lots is not sufficient, the designer must adjust setback requirements to benefit solar access.

Adjust the size, shape, and location of buildings to maximize solar access. Since streets are in most cases fairly wide, it is usually best to have the higher buildings and trees on the south side of east–west streets (Fig. 11.5j). If sufficient spacing is not possible, collect the sun at the roof level with south facing clerestory windows and rooftop collectors (Fig. 11.5k).

The foregoing discussion on solar access concerned itself with the most challenging demand: solar space heating in winter. Solar access for domestic hot water can be easier to achieve because much of the solar collection occurs during the higher sun angles of spring and summer. Access for daylighting is also less demanding because of the year-round use of the sun and because both diffuse-sky radiation and reflected sunlight are useful. Thus, daylighthng can be achieved with considerably less solar access than was described here for space heating. See Chapter 13 for guidelines on daylighting design.

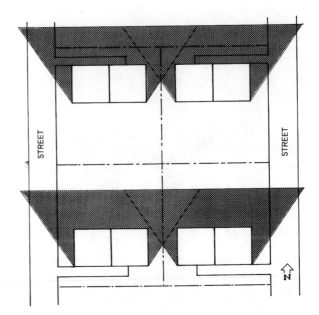

Figure 11.5e Duplexes can achieve good solar access even on a north–south street with this arrangement.

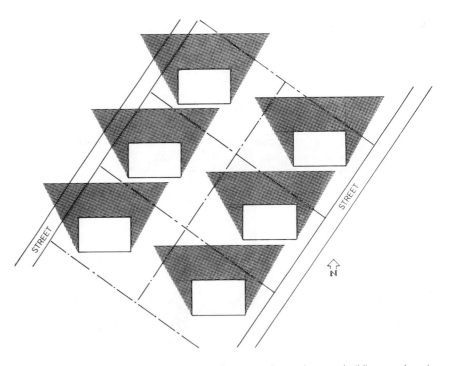

Figure 11.5f Even on diagonal streets, buildings can be oriented toward the south.

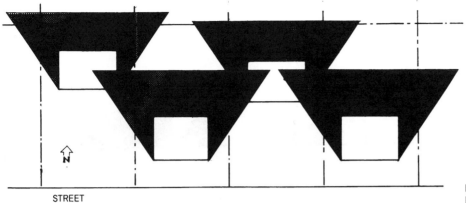

STREET

Figure 11.5g Uneven setbacks cause both winter and summer problems.

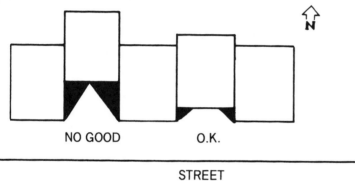

NO GOOD   O.K.

STREET

Figure 11.5h Very small setbacks, as those sometimes used in row housing, can be acceptable.

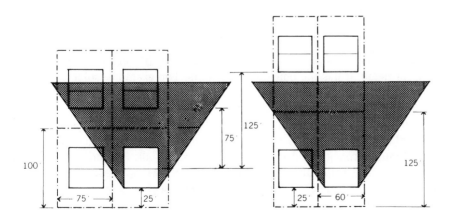

Figure 11.5i On east–west streets, deep lots are better than wide lots for solar access.

11.6 SOLAR ZONING

As the above discussion shows, solar access is very much dependent on what occurs on neighboring properties in all but the largest sites. Thus, solar-access laws are very important. However, since, the United States does not have a doctrine of "ancient light" like England and since there is no constitutional right to solar access, state and local laws must be passed to ensure solar access. Although a few states, such as California and New Mexico, have passed solar-rights acts, most protection comes from zoning codes. Because of the many variables involved, this type of zoning must be very local in nature. Climate, latitude, terrain, population density, and local tradition all play an important part.

The amount of solar access to be achieved is another variable in the creation of zoning laws. If only the roofs of buildings require solar access, the zoning will be least restrictive and easiest to achieve (Fig. 11.6a). Although south-wall access for passive solar is more difficult to achieve, it is important to have, if possible. Providing south-lot access is less critical than it might seem because the sun is normally required there only in the spring and fall when the sun is high in the sky. At least in northern climates, outdoor spaces are usually too cold in winter even if sunlight were available.

"Solar zoning" controls the shadows cast by a building on neighboring properties by defining the buildable volume on a site. Conventional zoning usually defines a rectangnlar solid (Fig. 11.6b), while solar zoning defines a sloped volume. Zoning can define this sloped buildable volume several ways. In the **bulk-plane method,** a plane slopes up from the north property line (Fig. 11.6c). The **solar-envelope method** is more sophisticated but also more complicated (Fig. 11.6d). Professor Ralph Knowles at the University of Southern

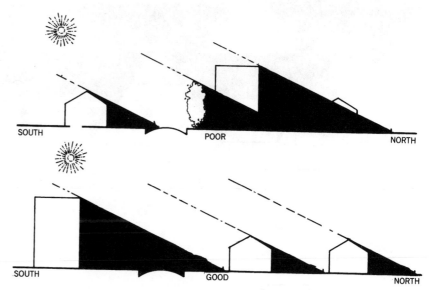

Figure 11.5j Place taller buildings and trees on the south side of east–west streets to take advantage of the wide right of way of the streets.

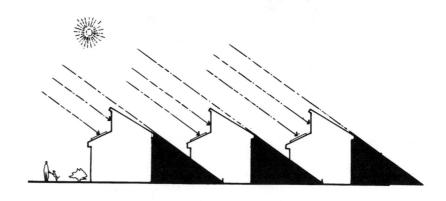

Figure 11.5k When solar access is not possible for the whole building, clerestories and active roof collectors should be used.

California has developed this method in great detail. It has the potential of not only ensuring high-quality solar access, but also of generating attractive architecture (Fig. 11.6e and f). The **solar-fence method** is the third solar zoning strategy, and it utilizes an imaginary wall of prescribed height over which no shadow can be cast (Fig. 11.6g).

In addition to zoning, several other possible ways exist to ensure solar access. Legal agreements or con-

tracts can be set up between neighbors. For example, solar easements can be placed on neighboring properties. This is most effective in new developments where solar covenants and restrictions can guarantee solar access for all. For further discussion on the legal ways of ensuring solar access; see *Solar Energy Planning* by Phillip Tabb and *Protecting Solar Access for Residential Development*, U.S. Department of Housing and Urban Development.

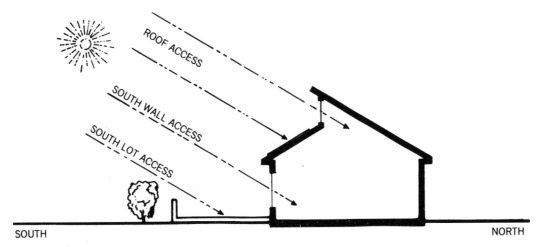

Figure 11.6a The three levels of solar access.

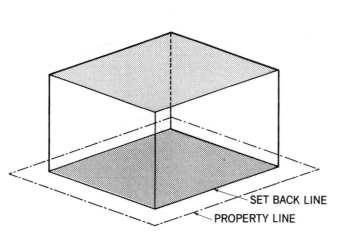

Figure 11.6b Conventional zoning.

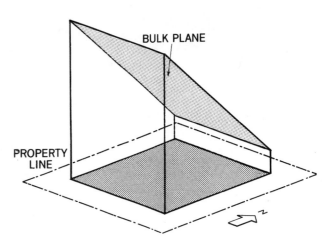

Figure 11.6c Bulk-plane zoning.

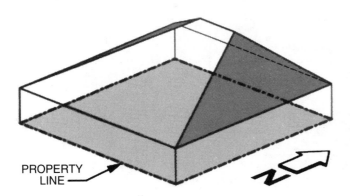

Figure 11.6d Solar-envelope zoning.

Figure 11.6e Example of architectural form encouraged by solar-envelope zoning. The model was created in Ralph Knowles's studio at the University of Southern California. (Photo courtesy Ralph Knowles ©.)

Figure 11.6g Solar-fence zoning.

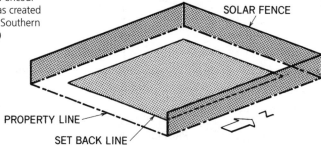

Figure 11.6f This development in downtown Denver, CO, demonstrates the kind of architecture that solar-envelope zoning would encourage.

11.7 PHYSICAL MODELS

As was mentioned several times above, the use of a physical model in conjunction with a sun machine is a very powerful design tool. With a physical model of the site, solar access can be accurately determined no matter how complex the situation. The shading due to any number of buildings, trees, and the lay of the land is, therefore, easy to analyze for any latitude, time of year, and time of day (Fig. 11.7a).

Physical models can also be used to easily generate shadow patterns no matter how complex the building or site. The following procedure is illustrated with a building on the side of a hill sloping down to the northeast at 32°N latitude.

Procedure for Creating Shadow Patterns

1. Place the model of the building with site on the sun machine (see Appendix C for details on how to use the sun machine).
2. Illuminate model to simulate shadows on December 21 at 9 A.M. (Fig. 11.7b).
3. Outline shadows. (If drawing the shadow pattern directly on the model is not desirable, then first tape down a sheet of paper.)
4. Repeat steps 2 and 3 but for 12 noon (Fig. 11.7c).
5. Repeat steps 2 and 3 again but for 3 P.M. (Fig. 11.7d).
6. The composite of the above three shadows is a rough **shadow pattern** (Fig. 11.7e). A more refined shadow pattern can be obtained by also drawing the shadows for 10 A.M., 11 A.M., 1 A.M., and 2 P.M.

Figure 11.7a Physical modeling is an excellent design tool for providing each site with access to the sun. Note how an east–west street (left to right in model) promotes solar access, while the north–south street inhibits it.

Figure 11.7b Shadow for December 21 at 9 A.M.

Figure 11.7c The shadow for December 21 at noon. Note the dashed outline for the 9 A.M. shadow.

Figure 11.7d The shadow for December 21 at 3 P.M. Note the outlines for both 9 A.M. and noon shadows.

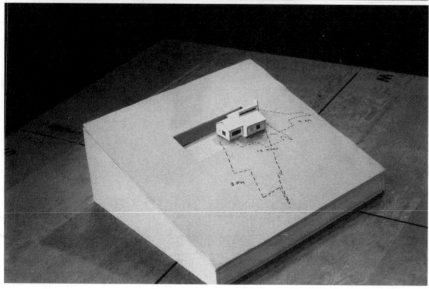

Figure 11.7e The composite of the three shadows form the shadow pattern.

11.8 WIND AND SITE DESIGN

Since in most climates the wind is an asset in the summer and a liability in the winter, a different wind strategy is required for summer and winter. Fortunately, this is made easier by a significant change in wind direction from summer to winter in most parts of the United States (see wind roses in Figs. 5.6d, e, f, and g).

Even if a particular region does not have a strong prevailing wind direction in winter, it is safe to say that the northerly winds will be the colder ones, and it is from these that the primary protection is required. Thus, it seems possible to design a site that diminishes the cooling effect of the northerly winds in the winter while it still encourages the more southerly summer winds. Much easier to design for are those few climates that are either so cold or so hot that only winter or summer winds need to be considered.

The design implications of wind on the building itself are discussed in Chapter 10 (summer condition) and Chapter 15 (winter condition). In this chapter, the impact of the wind on site design will be investigated.

In winter, the main purpose for blocking the wind is to reduce the heat losses infiltration causes. Although infiltration is normally responsible for about one-third of the total heat loss in homes, on a windy day on an open site, infiltration can account for more than 50 percent of the total heat loss. Since infiltration is approximately proportional to the square of the wind velocity, a small reduction in wind speed will have a large effect on heat loss. For example, if the wind speed is cut in half, the infiltration heat loss will be only one-fourth as large (Fig. 11.8a).

It is also worthwhile to block the winter wind for several other reasons. Heat transmission through the building envelope is also affected by wind speed. The heat loss from door operations is greatly reduced if the entrance is protected from the wind. Finally, outdoor spaces are usable in winter only if they are protected from the cold winds.

Windscreens can effectively reduce the wind velocity by three methods: they deflect air to higher levels, they create turbulence, and they absorb energy by frictional drag. Solid windbreaks, such as buildings, tend to use the first two methods the most, while porous windbreaks, such as trees, rely mainly on the third method. Fig. 11.8b illustrates the effect of porosity and height on the performance of a windbreak. Since the depth of wind protection is proportional to the height of the windbreak, the horizontal axis of the graph depicts multiples of the height of the windbreak. Notice that the densest windbreak results in the greatest reduction of air velocity but also has the smallest downwind coverage. Thus, dense windbreaks should be used on small lots or whenever the building is close to the windbreak, while medium-dense windbreaks would be better for protection at distances greater than four times the height of the windbreak.

Since windbreaks are never continuous, the end condition must be considered. At gaps or ends of windbreaks, the air velocity is actually greater than the free wind (Fig. 11.8c). This phenomenon can be an asset in the summer but certainly is not in the winter. The same undesirable situation arises in cities where buildings channel the wind along streets. A similar situation also occurs at passages through buildings, such as dogtrots (Fig. 10.2t). Buildings raised on columns (e.g., Le Corbusier's pilotis) create very windy conditions at grade level and are recommended only for climates without cold winters (Fig. 11.8d). Even without underpasses, high-rise buildings often create such high winds at grade level that entrance doors are prevented from closing. This occurs because much of the wind hitting the facade of the building is deflected downward, as shown in Fig. 11.8e. By the simple addition of a lower extension or large canopy, this downward wind can be deflected before reaching pedestrians at grade level (Fig. 11.8f).

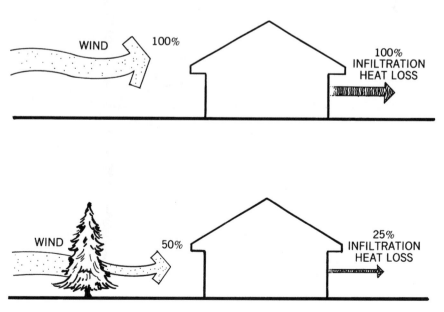

Figure 11.8a A small reduction in wind velocity results in a high reduction in heat loss.

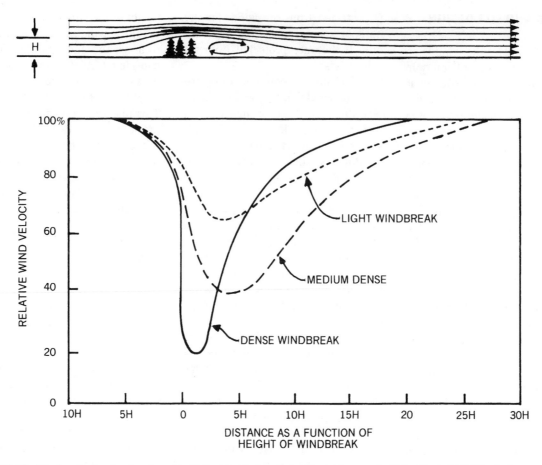

Figure 11.8b Wind protection is a function of both the height of a windbreak and its porosity. [After Naegeli (1946), cited in J.M. Caborn (1957). *Shelterbelts and Microclimate,* Edinburgh, Scotland: H. M. Stationery Office.]

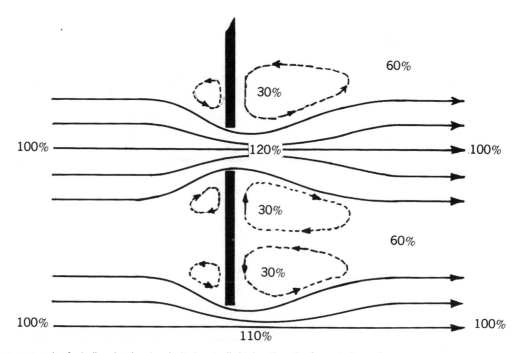

Figure 11.8c In gaps or at ends of windbreaks, the air velocity is actually higher than the free-wind speed.

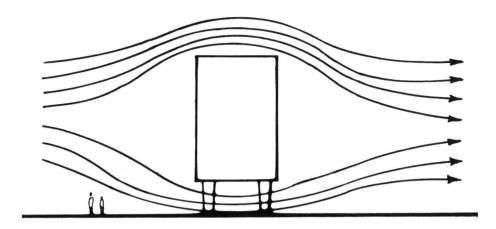

Figure 11.8d Buildings on columns (pilotis) experience very high wind speeds at ground level.

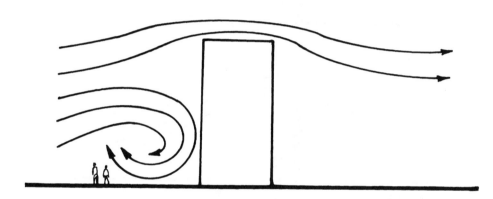

Figure 11.8e Tall buildings often generate severely windy conditions at ground level.

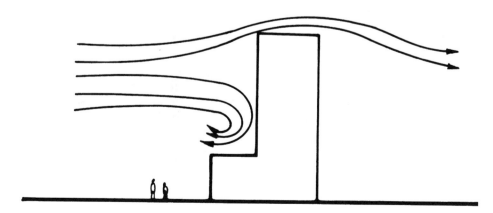

Figure 11.8f A building extension deflects winds away from ground-level areas.

When one designs a community or even a small development, it would be advantageous to place the higher buildings on the northern end. Not only will this arrangement block some of the cold northern wind, it also will provide better access to the winter sun (Fig. 11.8g).

Windbreaks can also be used to control snow, dust, or sand, which are carried by the wind and will settle out in the "windstill" area behind a windbreak. Thus, snowfences are used in snow country, and garden walls surround buildings in areas prone to dust storms or sandstorms. For lightweight dust, the protective walls need to be as high as the building, but for relatively heavy sand even 6-foot walls will reduce the wind velocity enough for the sand particles to settle out.

Guidelines for Windbreak Design

1. The higher the windbreak, the longer will be the wind shadow (Fig. 11.8h).
2. To get full benefit of hehght, the width of the windbreak should be at least ten times the height (Fig. 11.8i).
3. The porosity of the windbreak determines both the length of the wind shadow and the reduction of wind velocity (Fig. 11.8b).

In summer or in hot climates, breezes are welcome. Instead of using trees for windbreaks, they can be used to funnel more of the wind into the building (Fig. 11.8j). Even if the trees do not create a funnel, they can still increase ventilation by preventing the wind from easily spilling around the sides of the building (Fig. 11.8k). This concept works best when the wind comes predominantly from the south because it is then easier to maintain winter solar access and winter wind protection. When there is no dominant summer wind direction, shade trees with a high canopy are desired (Fig. 11.8l). If bushes are used, they should be placed away from the building as shown in Fig. 11.8m. If, instead, they are placed

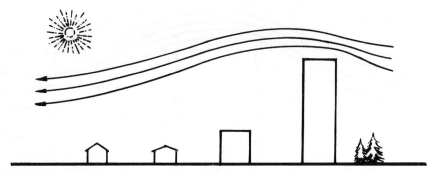

Figure 11.8g Tall buildings placed toward the north not only protect from the cold winter winds, but also permit good solar access.

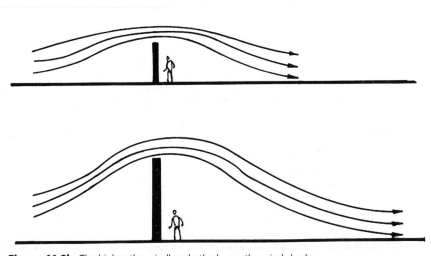

Figure 11.8h The higher the windbreak, the larger the wind shadow.

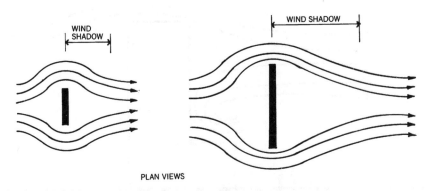

PLAN VIEWS

Figure 11.8i Up to a point, the width of a windbreak also affects the length of a wind shadow. Both windbreaks are of the same height.

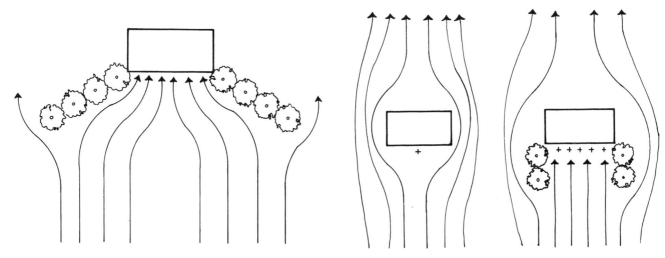

Figure 11.8j Trees and bushes can funnel breezes through buildings.

Figure 11.8k By preventing the wind from spilling around the sides of a building, a few trees or bushes can significantly increase natural ventilation.

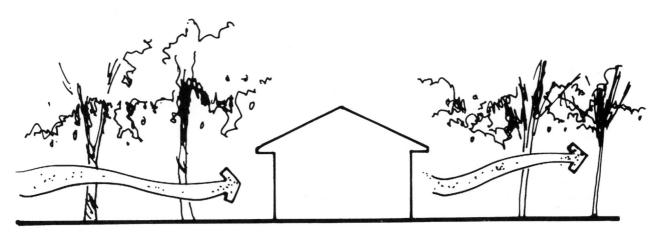

Figure 11.8l To maximize summer winds, use trees with high canopies.

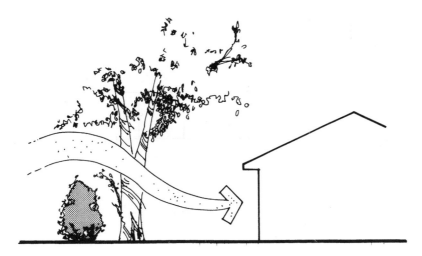

Figure 11.8m To maximize summer ventilation, place bushes away from the building and trees, as shown.

between the trees and buildings, the wind will be deflected over the building (Fig. 11.8n). This is the appropriate way to place bushes on the north side for winter wind protection.

Through the staggering of the location of buildings, cooling breezes can be maximized (Fig. 11.8o). Since buildings cannot be moved for the winter, this strategy is appropriate only for hot and humid climates with mild winters. In cold climates where the priority is protection from the cold winter winds, row or cluster housing is most appropriate (Fig. 11.8p).

Through the placement of a pool of water upwind or by building downwind from an existing lake, the air can be cooled by evaporation before entering a building. This strategy works best in hot and dry climates but can also be used in moderately humid areas. However, it is definitely counterproductive in very humid areas where additional humidity is to be avoided. Pools and fountains were popular among the Romans in part for their cooling benefits (Figs. 11.8q and r). Frank Lloyd Wright also recognized the advantages of fountains in hot and dry climates. At Taliesin West, he used several pools and fountains, at least in part, to cool the desert air (Fig. 11.8s).

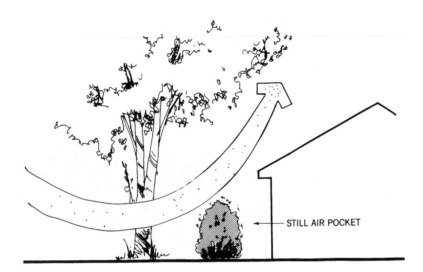

STILL AIR POCKET

Figure 11.8n For winter wind protection, place bushes between the building and trees.

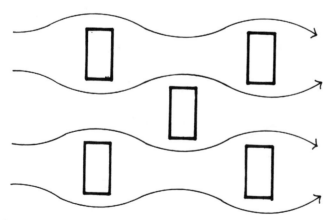

Figure 11.8o In hot and humid climates, buildings should be staggered to promote natural ventilation.

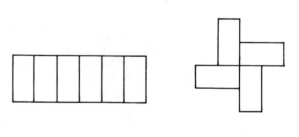

Figure 11.8p Use row or cluster housing for protection against wind in cold climates.

Figure 11.8q Large pools of water frequently helped cool Roman villas. The Getty Museum in California is a careful replica of a Roman villa. (Courtesy of the John Paul Getty Museum, Malibu, CA, Julius Schulman, photographer.)

Figure 11.8r This dining terrace in the House of Loreio Tiburtino in Pompeii, Italy, is cooled by an indoor canal. A grapevine-covered pergola provides shade. The terrace is oriented to the south for winter heating. (Courtesy Richard Kenworthy, photographer.)

Figure 11.8s In Taliesin West, near Pheniz, AZ, Frank Lloyd Wright used pools and fountains to help cool the desert air.

11.9 PLANTS AND VEGETATION

Plants are immensely useful in the heating, cooling, and lighting of buildings. Although plants are very popular, they are usually used for their aesthetic and psychological benefits. The famous biologist E. O. Wilson holds the theory that people experience **biophilia,** an innate need for contact with a wide variety of species (Kellert and Wilson, 1993).

Ideally , along with their aesthetic and psychological function they could act as windbreaks in the winter, as shading devices and evaporative coolers in the summer, and as light filters all year long (Fig. 11.9a). Vegetation can reduce erosion and attract wildlife. It can also reduce noise, dust, and other air pollution. It is sometimes most useful in blocking visual pollution or in creating privacy. Of

course, just like in the design elements of the building itself, the more functions vegetation has, the better.

Before we discuss specific design techniques, some general comments about plants are in order. Perennial plants are usually better than annuals because they do not have to start from the beginning each year. Deciduous plants can be very useful for solar access, but some cautions are in order. It must be understood that even without leaves, the branches create significant shade (30 percent to 60 percent). A few trees shade even more of the winter sun. For example, some oaks hold on to their dead leaves and, thus, shade up to 80 percent of the winter sun (Fig. 11.9b). The best trees are those that have a dense summer canopy and an almost branchless, open winter canopy. Certain species will be deciduous only if the tempera-

ture gets cold enough. Thus, the same plant might be deciduous in the north and evergreen in the south. The time of defoliation in fall and time of foliation in spring vary with species. Some deciduous plants respond to length of daylight rather than temperature and, thus, might defoliate at the wrong time. This is especially true when bright outdoor lighting confuses the biological timing of the plants.

The size and shape of a fully grown tree or shrub vary not only with species, but also with local growing conditions. See Table 11.9 for a sample of tree sizes, growth rates, percentage of winter and summer sun blockage, and time of fall defoliation and spring foliation. The location of the hardiness zones listed in the table are shown on the map of Fig. 11.9c. Since a design is often based on the more mature size of a tree or bush,

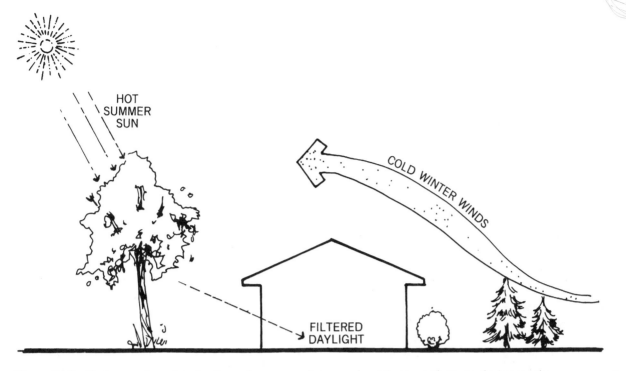

Figure 11.9a Plants can reduce winter heating and summer cooling as much as 50 percent. Plants can also improve the quality of daylight by filtering the light.

Figure 11.9b Deciduous trees vary greatly in the amount of sunlight they block in the winter (30 to 60 percent). A few deciduous trees like the particular oak at left do not even lose their dead leaves until spring.

Name	Shape	Mature Size FT (M) HT x Spread	GROWTH RATE	Shade Provided		Time in Leaf		USDA Hardiness Zone	Site Features
				Winter	Summer	Fall Defoliates	Spring in Leaf		
Acacia greggii (Catclaw Acacia)	spreading	15(8)x10(3)	moderate	light	light	late	early	7–11	sun/dry soils
Acer platanoides (Norway Maple)	round	50(15)x40(12)	moderate	light	dense	late	early	3–8	sun/well-drained soils
Acer rubrum (Red Maple)	oval/round	60(18)x40(12)	fast	light	moderate	average	average	3–9	sun-partial shade/moist soils
Acer saccharinum (Silver Maple)	oval/round	75(23)x40(12)	fast	moderate	dense	average	average	3–9	sun/wide variety of soils
Acer saccharum (Sugar Maple)	oval/spreading	75(23)x40(12)	moderate/fast	moderate	dense	average	average	3–8	sun/moist well-drained, acid soils
Betula nigra (River Birch)	oval	30(9)x20(6)	fast	light	moderate	early	early	4–9	sun/moist, sandy, acid soil
Cercidium floridum (Blue Palo Verde)	spreading	30(9)x30(9)	moderate	light	light	late	late	8–11	sun/dry soils
Cercis canadensis (Eastern Redbud)	spreading	30(9)x35(11)	moderate	light	light	late	average	4–9	sun-shade/moist-dry soils
Chilopsis linearis (Desert Willow)	vase	20(6)x15(8)	fast	light	light	average	average	8–10	sun/dry, alkalinesoils
Cornus florida (Flowering Dogwood) select disease-resistant hybrids	round	35(11)x35(11)	slow	light	medium	early	late	6–9	partial shade/moist, well-drained soils
Cornus kousa (Kousa Dogwood)	round	25(8)x20(6)	moderate	moderate	moderate	early	late	4–8	sun-part shade
Fagus sylvatica (European Beech)	oval/round	100(30)x70(21)	slow	moderate	dense	late	late	4–8	sun/moist, well-drained soils
Fraxinus pennsylvanica (Green Ash)	oval/round	50(15)x30(9)	moderate	light	moderate	average	average	3–9	sun/wide variety of soils
Fraxinun veluntina (Arizona Ash)	pyramid	35(11)x25(8)	fast	light	moderate	average	late	7–9	sun/tolerates dry alkaline soil
Ginkgo biloba (Ginkgo)	pyramid	70(21)x40(12)	moderate/slow	light	dense	average	late	4–9	sun/wide variety of soils
Gleditsia triacanthos var. inermis (Honey Locust)	oval/spreading	60(18)x20(6)	fast	light	moderate	early	late	4–9	sun-part shade/dry-wet soils
Gymnocladus dioica (Kentucky Coffeetree)	oval	60(18)x40(12)	slow	light	light	average	average	4–8	sun/wide variety of soils
Lagerstroemia indica (Crepe Myrtle)	vase	20(6)x15(4)	moderate	light	moderate	late	late	7–10	sun/moist soil
Liquidambar styraciflua (Sweet Gum) select fruitless cultivar	pyramid	80(24)x40(12)	moderate/fast	moderate	moderate	late	average	5–10	sun/rich, wet, acid soils
Liriodendron tulipifera (Tulip Poplar)	columnar/oval	80(24)x40(12)	moderate	moderate	dense	average	average	4–9	sun/moist, fertile soil
Magnolia acuminata (Cucumber Magnolia)	pyramid/spreading	80(24)x70(21)	moderate	light	medium	average	average	4–8	sun/moist, slightly acid soils
Pistacia chinensis (Chinese Pistache)	oval/round	40(12)x40(12)	slow	medium	dense	average	average	8–10	sun/well-drained soil
Platanus x acerifolia (London Planetree)	pyramid/spreading	90(27)x70(21)	moderate/fast	light	moderate	late	late	4–8	sun/wide variety of soils
Platanus occidentalis (American Sycamore)	round	90(27)x90(27)	moderate/fast	light	moderate	late	late	4–9	sun/moist soils
Populus deltoides (Eastern Cottonwood) select seedless cultivars	vase	75(23)x50(15)	fast	light	light	early	early	2–9	sun/moist soil/invasive roots
Prosopis glandulosa (Honey Mesquite)	spreading	25(8)x15(4)	fast	light	light	early	late	8–9	sun/dry, alkaline soils
Quercus palustris (Pin Oak)	pyramid/columnar	75(23)x40(12)	moderate	moderate	moderate	early	late	5–10	sun/acid soils
Quercus phellos (Willow Oak)	round	60(18)x40(12)	moderate	medium	dense	late	late	6–9	sun/moist, well-drained soils
Robinia pseudoacacia (Black Locust)	columnar	50(15)x30(9)	moderate/fast	moderate	moderate	early	late	3–8	sun/wide variety of soils
Tilia cordata (Littleleaf Linden)	pyramid	50(15)x40(12)	moderate/slow	moderate	dense	early	late	3–7	sun/wide variety of soils
Zelkova serrata (Japanese Zelkova)	vase	70(21)x20(6)	moderate	light	moderate	average	late	5–8	sun/moist soils

TABLE 11.9 USEFUL TREES

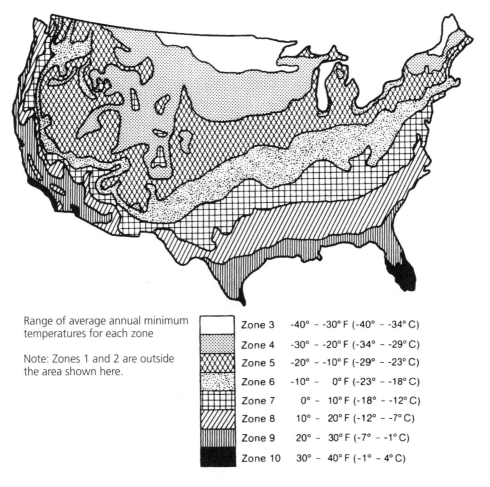

Range of average annual minimum temperatures for each zone

Note: Zones 1 and 2 are outside the area shown here.

	Zone 3	-40° – -30° F (-40° – -34° C)
	Zone 4	-30° – -20° F (-34° – -29° C)
	Zone 5	-20° – -10° F (-29° – -23° C)
	Zone 6	-10° – 0° F (-23° – -18° C)
	Zone 7	0° – 10° F (-18° – -12° C)
	Zone 8	10° – 20° F (-12° – -7° C)
	Zone 9	20° – 30° F (-7° – -1° C)
	Zone 10	30° – 40° F (-1° – 4° C)

Figure 11.9c The zones of plant hardiness as listed in Table 11.9. (From Richard Montgomery, *Passive Solar Journal,* Vol. 4 (1), p. 91. Courtesy of the American Solar Energy Society.)

the growth rate is very important. Choosing a fast-growing tree or bush (2 or more feet per year) is not always a good choice because most fast-growing trees are weak-wooded (have poor strength). However, a vine can be the ideal fast-growing plant. Since some vines grow as much as 40 feet per year, a vine can create as much shade in 3 years as it can take a tree 50 years to achieve. Physical strength is not required since a vine can be supported by a man-made structure such as a wall, a trellis, or a cable network. Unlike a tree the growth of a vine can be directed exactly where it is needed (Fig. 11.10g).

The growth rate of any tree, shrub, or vine can be accelerated by supplying ample nutrients and a steady source of water. A drip-irrigation system is excellent for this purpose. And, of course, starting with a large plant will further shorten the time required to reach maturity.

Sometimes it is desirable to stop the growth of a plant when it has reached the desired size. That this is possible is proven by the existence of *bonsai plants.* The usual methods for creating bonsai include limiting the supply of nutrients, limiting the space for root growth, pruning, and wire chokes to constrict the flow of nutri-

ents. However, for the health of the plant, sufficient water must always be supplied.

Plants help in heating primarily by reducing infiltration and partly by creating still-air spaces next to buildings, which act as extra insulation. Use dense evergreen trees and shrubs for breaking the wind.

Summer cooling is more complicated, with most of the benefit derived from the shade that the plants provide. The shade from a tree is better than the shade from a manmade canopy because the tree does not heat up and reradiate down (Fig. 11.9d). This is the case because of the multiple layers that are ventilated and because the leaves stay cool by the transpiration (evaporation) of water from the leaves. Since only a small amount of the sun's energy is used in photosynthesis, the chemical energy created has little effect on reducing the temperature.

Transpiration not only cools the plant, but also the air in contact with the vegetation (Fig. 11.9e). Thus, the cooling load on a building surrounded by lawns will be smaller than on a building surrounded by asphalt or concrete. Trees are even more effective than grass in providing comfort. They provide shade, and unlike grass, their evaporative cooling occurs high above the ground and, therefore, does not raise the humidity level at the ground.

Using large amounts of water to achieve cooling is not the answer in the dry Southwest, nor even in the Southeast, which is also experiencing water shortages. Georgia, Alabama, and Florida are in the courts fighting over the water in the Chattahoochee River. It is increasingly important to use a **Xeriscape** design, which is landscaping that saves water and energy. Incidentally, the word "Xeriscape" uses the same Greek root as Xerox, which refers to a "dry" process.

It is not unrealistic to expect a reduction of up to 50 percent in the cooling load when an envelope-dominated building is effectively shaded by plants. A 60 percent reduction in the use of electricity was realized in a Florida school when plants shaded the walls and windows. In another experiment, the temperature inside a mobile home was reduced 20°F when it was well shaded by plants.

At night, trees work against natural cooling by blocking long-wave radiation. There will be more radiant cooling in an open field than under a canopy of trees (Fig. 11.9f). Consequently, the diurnal temperature range is much smaller under trees than in an open field. In hot and humid climates this is a liability, but in cold climates it is an asset.

Plants can also improve the quality of daylight entering through windows. Direct sunlight can be scattered and reduced in intensity, while the glare from the bright sky can be moderated by plants (Fig. 11.9h). Vines across the windows or trees farther away can have the same beneficial effect. Since light reflected off the ground penetrates deeper into a building than direct light, it is sometimes desirable not to have vegetation right outside a window on the ground (Fig. 11.9g). See Chapter 13 for a detailed discussion on daylighting.

There is no doubt that the proper choice and positioning of plants can greatly improve the microclimate of a site. However, choosing a specific plant can be difficult not only because of the tremendous variety that exists, but also because of the specific needs of plants, such as minimum safe temperatures, rainfall, exposure to sun, and soil type. For these reasons, advice should be obtained from such sources as local nurseries, agricultural extension agents, state foresters, and landscape architects. **Rule of thumb: Choose plants suited for the local environment.**

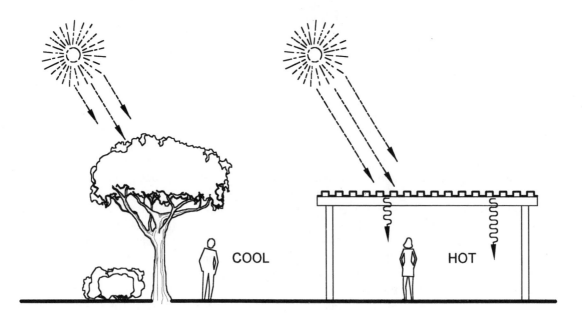

Figure 11.9d Shade from trees is so effective because trees do not get hot and reradiate heat (long-wave infrared) as do most manmade shade structures.

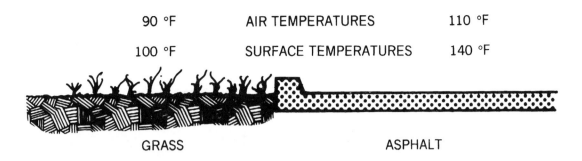

Figure 11.9e Since air is heated by contact with the ground, the air over asphalt is much hotter than the air over grass.

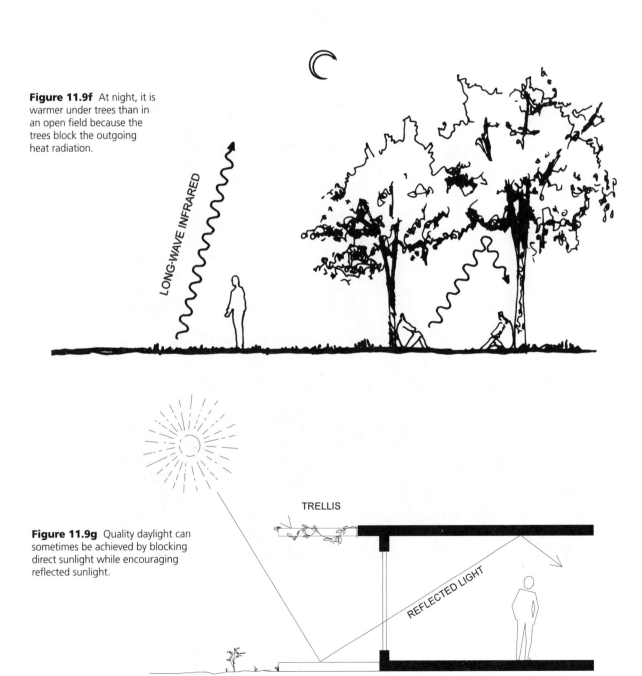

Figure 11.9f At night, it is warmer under trees than in an open field because the trees block the outgoing heat radiation.

Figure 11.9g Quality daylight can sometimes be achieved by blocking direct sunlight while encouraging reflected sunlight.

Figure 11.9h Plants can soften and diffuse daylight and reduce the glare from the bright sky and light-colored surfaces. Apartment complex in Vienna designed by F. Hundertwasser.

11.10 LANDSCAPING

The above concepts can now be combined into landscaping techniques that promote the heating, cooling, and lighting of buildings. Fig. 11.10a illustrates the general tree-planting logic for most of the country, while Figs. 11.10b to 11.10e present landscaping techniques appropriate for four different climates (temperate, very cold, hot and dry, and hot and humid).

When trees are not available to shade the east, west, and north windows, high bushes or a vine-covered trellis can be used. Bushes can shade windows just like vertical fins, which were discussed in Chapter 9. And, like fins, shrubs should extend above the windows and be fairly deep (Fig. 11.10f). A vertical vine-covered trellis is very effective on east and west facades, while a horizontal trellis can be used on any orientation (Fig. 11.10g).

Outdoor shading structures, such as trellises, pergolas, and arbors, are described in Fig. 9.16f. Other functional landscaping elements include allees, pleached allees, and hedgerows (Fig. 11.10h). **Allees** are garden walks bordered with shrubs and trees; they primarily control sight lines but can also be used to control air movement. In **pleached allees,** closely spaced trees or tall shrubs are intertwined and pruned to form a tunnel-like structure. Not only do these features effectively frame views, they also create cool, shady walkways. The term **hedgerow** refers to a row of bushes, shrubs, or trees forming a

hedge. Depending on the orientation, such hedges can be used for shading, wind protection, or wind funneling.

As mentioned before, pools of water and especially fountains can be used to cool the air. However, if these features are used in hot and humid climates, they should be placed downwind to avoid adding more humidity to the air. On the other hand, in very dry climates, the water should be placed in a sheltered courtyard. Unless the water is chlorinated, a healthy ecosystem should be created. As waterfalls or fountains cool the air, they also oxygenate the water for fish and snails, which are required to control mosquitoes and algae growth. In very hot climates, the water should be shaded to prevent overheating and excessive algae growth (Fig. 11.10i). A natural black dye added to the water will also prevent algae growth. The water should also circulate and not become stagnant.

A private sunny garden is considered a basic amenity in much of the world. It is, therefore, surprising how little use is made of roof gardens in the cities of the western world. Some guidelines can help make the design of roof gardens more of a success. Use lightweight materials for both structural and thermal reasons. Heavy elements, such as trees or pools, should be placed over columns or bearing walls. Use light-colored materials to prevent heat buildup that could make its way through the roof. A vine-covered trellis can be a lightweight element for creating shade. Trees and bushes can be planted in lightweight soils made up of materials, such as perlite and vermiculite.

Of course, the design of roof gardens will vary with climate. A design of a roof garden for a northern climate might include high parapet walls on the north, east, and west facades to deflect the cold winds. An open railing on the south side would allow the winter sun to enter (Fig. 11.10j). A roof garden in a southern climate, on the other hand, would be open to cooling breezes and be well shaded by trees, trellises, etc. (Fig. 11.10k).

Figure 11.10a General logic for tree planting around a building.

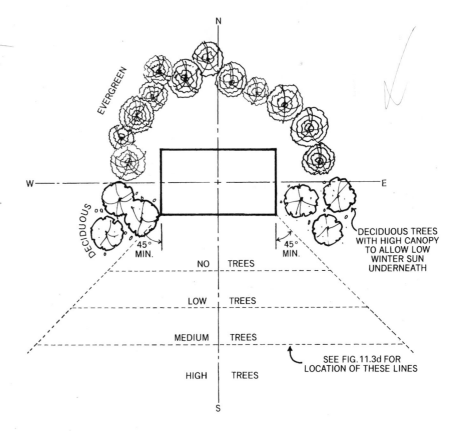

Figure 11.10b Landscaping techniques for a temperate climate. The windbreak on the north side of the building should be no farther away than four times its height.

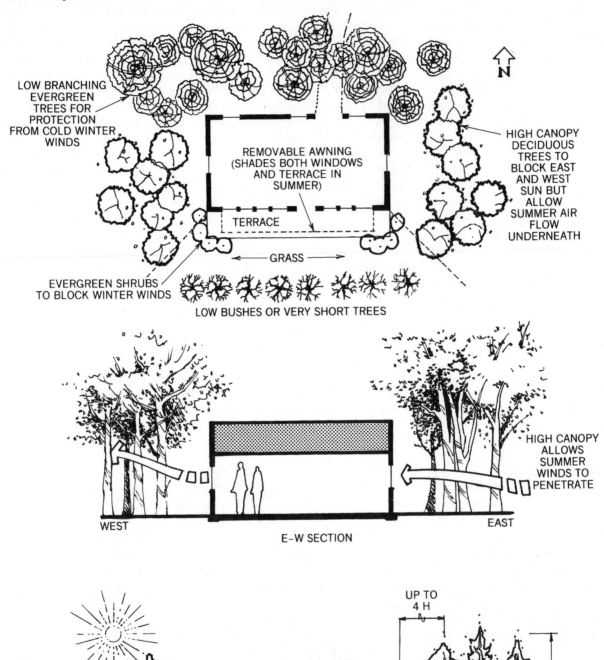

LOW BRANCHING EVERGREEN TREES FOR PROTECTION FROM COLD WINTER WINDS

N

HIGH CANOPY DECIDUOUS TREES TO BLOCK EAST AND WEST SUN BUT ALLOW SUMMER AIR FLOW UNDERNEATH

REMOVABLE AWNING (SHADES BOTH WINDOWS AND TERRACE IN SUMMER)

TERRACE

GRASS

EVERGREEN SHRUBS TO BLOCK WINTER WINDS

LOW BUSHES OR VERY SHORT TREES

WEST

E–W SECTION

EAST

HIGH CANOPY ALLOWS SUMMER WINDS TO PENETRATE

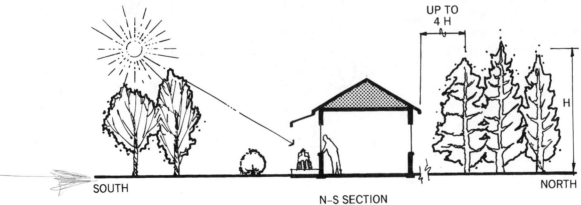

UP TO 4 H

H

SOUTH

N–S SECTION

NORTH

TEMPERATE CLIMATE

Figure 11.10c Landscaping techniques for very cold climates.

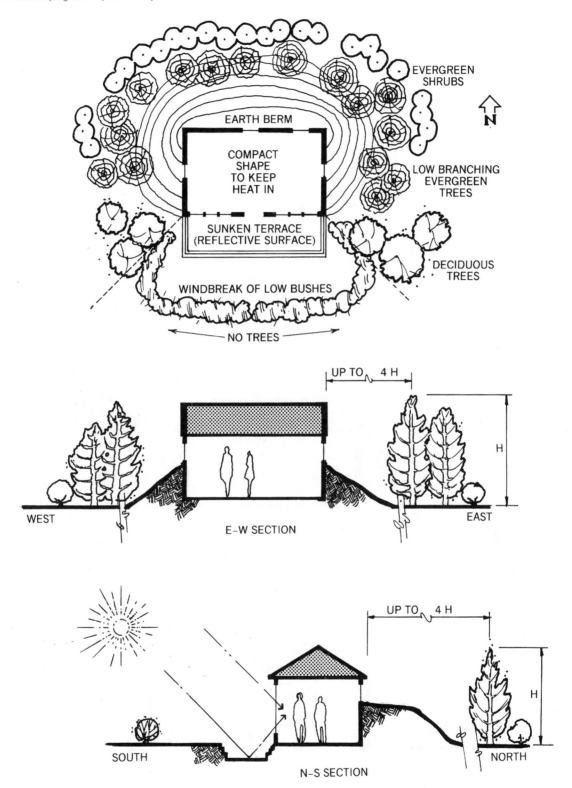

COLD CLIMATE

Figure 11.10d Landscaping techniques for hot and dry climates.

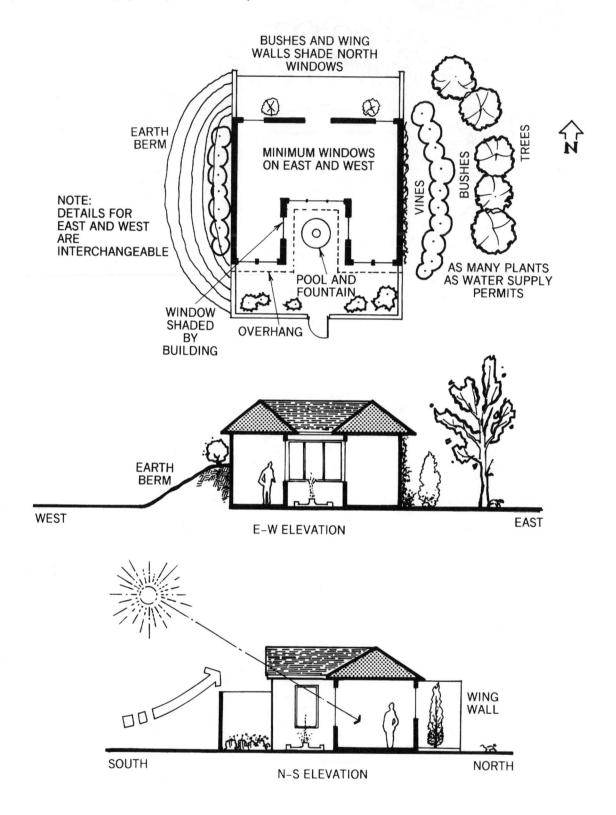

BUSHES AND WING
WALLS SHADE NORTH
WINDOWS

EARTH
BERM

MINIMUM WINDOWS
ON EAST AND WEST

VINES

BUSHES

TREES

N

NOTE:
DETAILS FOR
EAST AND WEST
ARE
INTERCHANGEABLE

AS MANY PLANTS
AS WATER SUPPLY
PERMITS

POOL AND
FOUNTAIN

WINDOW
SHADED
BY
BUILDING

OVERHANG

EARTH
BERM

WEST

E–W ELEVATION

EAST

WING
WALL

SOUTH

N–S ELEVATION

NORTH

HOT AND DRY CLIMATE

Figure 11.10e Landscaping techniques for hot and humid climates.

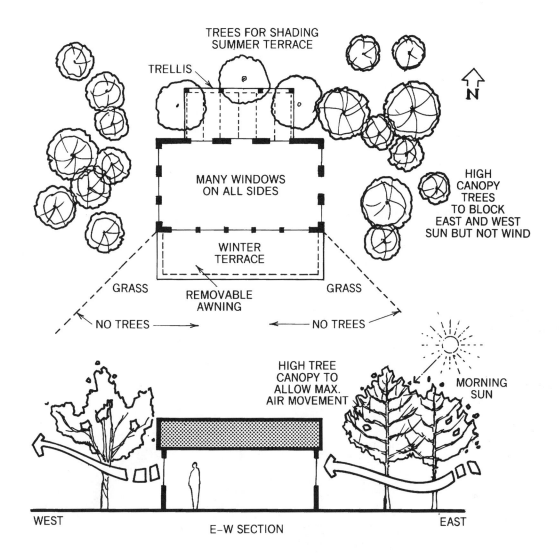

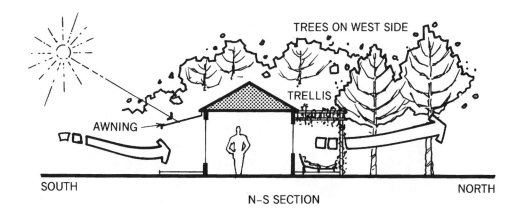

HOT AND HUMID CLIMATE

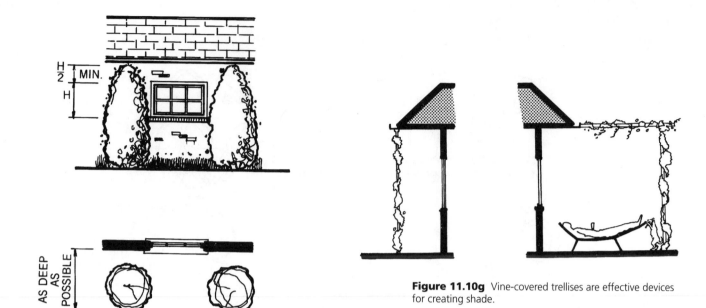

PLAN

Figure 11.10f Bushes can act as vertical fins to block the low sun from the east and west.

Figure 11.10g Vine-covered trellises are effective devices for creating shade.

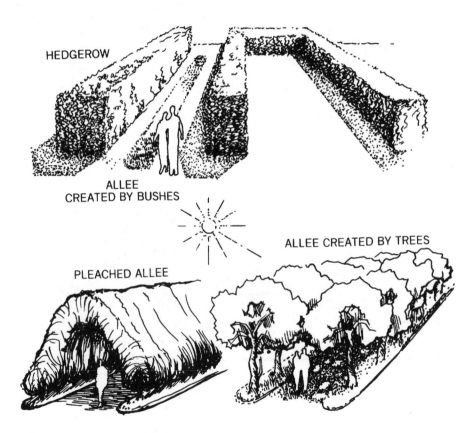

Figure 11.10h Landscaping elements for creating shade or controlling air movement.

Figure 11.10i Waterfalls, fountains, and pools can cool the air in all but very humid climates. If the climate is very hot, the water should be shaded by trees, bushes, lily pads, etc.

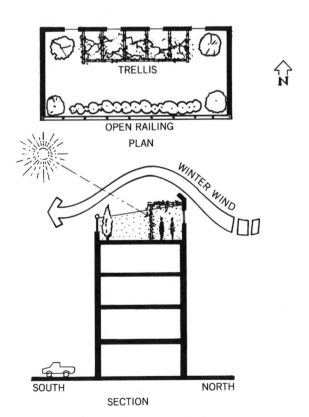

Figure 11.10j Roof-garden design for a cold climate.

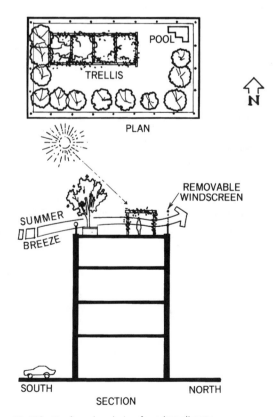

Figure 11.10k Roof-garden design for a hot climate.

11.11 COMMUNITY DESIGN

Community planning can either promote or hinder the design for each lot. Outside of Phoenix, Arizonza, is a place called Sun City. Although the name might suggest a place in harmony with the sun, in fact, the street layout shows a total disregard for sun angles (Fig. 11.11a). On circular streets, every building has a different orientation. Only two buildings on each street have the ideal east-west orientation, where the small facades face and west and where shading from neighbors is at a maximum.

A quite different approach is illustrated in the street plan of Village Homes in Davis, California (Fig. 11.11b). Although the site runs north–south, the streets run mostly east–west. Cluster housing is used to save both land and energy (Fig. 11.11c). Bicycle and pedestrian paths are included in part to reduce the use of automobiles. Studies have shown that houses in the community of Village Homes use on average about half of the energy required by comparable nonsolar buildings in the same area. In addition, these houses tend to be more comfortable, desirable, and economical.

As much as possible, Village Homes encourages employment opportunities within the community. Tremendous amounts of energy and time could be saved if people did not have to commute so much. Although the heavy use of automobiles is a major part of America's national energy drain, that aspect of sustainability is unfortunately beyond the scope of this book.

The existence of communities such as Village Homes is in part due to the zoning technique called **PUD (Planned Unit Development).** PUD provisions allow for modification of lot size, shape, and placement for increasing siting flexibility. Thus, solar access and community open land can be maximized.

11.12 COOLING OUR COMMUNITIES*

Urban areas are much hotter than the rural areas around them. Fig. 5.3e shows the **heat-island effect** caused mainly by the excessive absorption of solar energy and the lack of cooling that trees can produce. Significantly cooler communities are possible by using light-colored materials for roofs, walls, and especially paved surfaces. Use materials that have a very high albedo (a factor that measures the reflectivity of solar radiation). A surface with an albedo of one reflects all radiation, while a surface with an albedo of zero absorbs all radiation. Black roofs are especially bad because they heat both the buildings and the urban area.

Certain paved areas, such as driveways and parking areas, not only can be made of light colored materials, such as concrete, but also can be made of special blocks that allow grass to grow in the openings. All plants cool by transpiration. Trees have the double benefit of creating shade while the transpiration occurs high above the ground where it will blow away faster. Significant comfort and energy savings are possible from an official policy of using light colors and planting trees.

*This section is based on the excellent book, *Cooling Our Communities: A Guidebook on Tree Planting and Light Colored Surfacing*, U.S. Environmental Protection Agency, 1992.

11.13 CONCLUSION

Site and community planning can have a tremendous effect on energy consumption. For example, the annual per-capita energy use in New York City is about one-half of the United States average. While in Davis, California, good planning and building design has resulted in some houses that have achieved 100 percent natural cooling and 80 percent solar heating.

Planning decisions made today will be with us for decades if not centuries. It *will* be very unfortunate if future interest in solar energy is frustrated by the poor planning decisions made today. It is almost certain that photovoltaic electricity, the almost ideal energy source, will become economical in the near future (see Section 8.2). When that day arrives, some buildings will be able to make much better use of it than others. Solar access and proper orientation will be critical. Good planning decisions not only reward us now, but also create a decent legacy for the future.

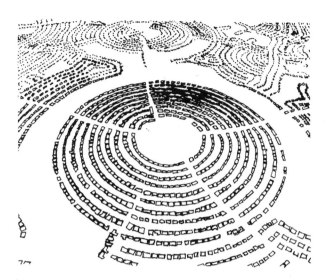

Figure 11.11a With a circular street layout, every building has a different orientation.

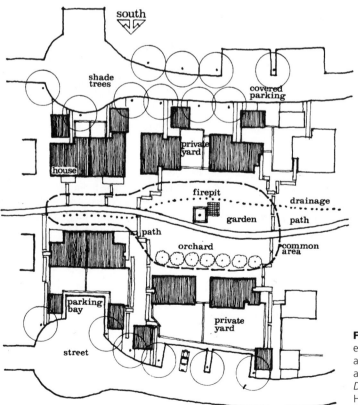

Figure 11.11b In the community of "Village Homes" in Davis, CA, most streets run east–west for winter solar access and summer shading. (From *Village Homes' Solar House Design* by David Bainbridge, Judy Corbett, John Hofacre. Rodale Press, 1979. © Michael N. Corbett.)

Figure 11.11c Cluster housing saves land and energy. The community land is used for pedestrian/bicycle paths, recreation, vegetable gardens, and orchards. (From *Village Homes' Solar House Design* by David Bainbridge, Judy Corbett, John Hofacre. Rodale Press, 1979 © John C. Hofacre.)

KEY IDEAS OF CHAPTER 11

1. East-west streets offer the best access to the winter sun and the best shading from the summer sun.

2. The solar-access boundary can be used on plans and sections to ensure winter solar access.

3. Shadow patterns can be used on site plans to ensure winter solar access to all buildings.

4. Solar zoning ensures solar access through ordinances.

5. Use physical models as a design tool for ensuring winter solar access and summer shading.

6. Use site features to block the cold winter wind and to funnel in the summer wind.

7. Use plants to block the winter wind and the summer sun.

8. Landscaping that supports the heating, cooling, and lighting of buildings varies with the climate.

9. Design communities for sustainability. Emphasize east-west streets, and de-emphasize the need for automobiles.

References

(See Bibliography in back of book for full citations.)

Kellert, S. R., and E. O. Wilson, eds. *The Biophilia Hypothesis.*

Resources

FURTHER READING

(See Bibliography in back of book for full citations. This list includes valusable out-of-print books.)

Akbari, H., et al. *Cooling Our Communities: A Guidebook on Tree Planting and Light- Colored Surfacing.*

Boutet, T. S. *Controlling Air Movement: A Manual for Architects and Builders.*

Brown, G. Z., and M. DeKay. *Sun, Wind, and Light: Architectural Design Strategies.*

Brown, R., and T. Gillespie. *Microclimatic Climate Design.*

Druse, K., and M. Roach. *The Natural Habitat Garden.*

Foster. R. S. *Homeowner's Guide to Landscaping That Saves Energy Dollars.* *

Givoni, B. *Climate Considerations in Building and Urban Design.*

Groesbeck, W., and J. Stiefel. "The Resource Guide to Sustainable Landscapes and Gardens."

Hightshoe, G. L. *Native Trees, Shrubs, and Vines for Urban and Rural America.*

Hottes, A. C. *Climbers and Ground Covers: Including a Vast Array of Hardy and Subtropical Vines Which Climb or Creep.* *

Jaffe, M. S., and E. Duncan. "Protecting Solar Access for Residential Development: A Guidebook for Planning Officials."

Knowles, R. L. *Energy and Form: An Ecological Approach to Urban Growth*

Knowles, R. L. *Sun Rhythm Form*

Lyle, J. T. *Design for Human Ecosystems: Landscape, Land Use, and Natural Resources.*

McHarg, I. L. *Design with Nature.*

McPherson, , E. G., ed. *Energy Conserving Site Design.*

Moffat, A., and M. Schiler. *Energy- Efficient and Environmental Landscaping.*

Moffat, A., and M. Schiler. *Landscape Design that Saves Energy.* *

Norwood, K., and K. Smith. *Rebuilding Community in America: Housing for Ecological Living, Personal Empowerment, and the New Extended Family.*

Ottsen, C. *The Native Plant Primer: Trees, Shrubs, and Wildflowers for Natural Gardens.*

Petit, J., D. Bassert, and C. Kollin. *Building Greener Neighborhoods: Trees as Part of the Plan.*

Robinette, G. O. *Energy Efficient Site Design.*

Robinette, G. O. *Landscape Planning for Energy Conservation.*

Rocky Mountain Institute Staff. *Green Development: Integrating Ecology and Real Estate*

Tabb, P. *Solar Energy Planning.*

Van der Ryn, S., and P. Calthorpe. *Sustainable Communities: A New Design Synthesis for Cities, Suburbs, and Towns.*

Vickery, R. L. *Sharing Architecture.*

Watson , D., and K. Labs. *Climatic Design: Energy Efficient Building Principles and Practices.*

*These books have extensive lists of plants that are useful for energy conscious landscape design.

PAPERS

Knowles, R. L. "The Solar Envelope." At http://www.rcf.usc.edu/ ~ rknowles

12

LIGHTING

"More and more, it seems to me, light is the beautifier of the building."

Frank Lloyd Wright from **The Natural House**
©Frank Lloyd Wright Foundation, 1958

"The design of human environments is, in effect, the design of human sensory experience; all visual design is de facto also lighting design, . . ."

William M. C. Lam, **Perception and Lighting as Formgivers for Architecture** *(p. 13), ©Wm. Lam Associates, 1977.*

12.1 INTRODUCTION

The form of matter is known to us primarily by the way it reflects light. Sensitive designers have always understood that what we see is a consequence of both the quality of the physical design and the quality of light falling on it. The ancient Egyptians found that shallow, negative relief created powerful patterns under the very clear, bright, and direct sunlight of Egypt (Fig. 12.1a). The Greeks found that relief sculpture and mouldings were well modeled under the somewhat less bright sun of Greece (Fig. l2.1b). The designers of the Gothic cathedrals had to create powerful state-ments in the cloudy and diffused light of northern Europe. Here, sculpture in the round could be placed in niches and portals and still be seen because of the softness of the shadows (Fig. l2.1c). Most of the sculpture of a Gothic cathedral would disappear in the dark shadows of an Egyptian sun. The quality of light and the quality of architecture are inextricably intertwined.

Sometimes the architect must accept the light as it is and design the form in response to it. Other times both the form and the light source are under the architect's control. This is true not only for interiors, but also the exterior at night. Thus, the architect creates the visual environment by both molding the material and by controlling the lighting.

The following three chapters on lighting present the information required by the designer to create a quality lighting environment. Such an environment includes the lighting necessary for satisfying aesthetic and biological needs, as well as the lighting required to perform certain tasks. Since a quality lighting environment is not achieved just by supplying large quantities of light, the emphasis in this book is not on the quantification of light. This chapter explains the basic concepts required for a quality lighting environment, which is achieved primarily through the geo-

Figure 12.1a Low sunken relief is ideal for the very bright and direct sun of Egypt. (Courtesy of the Egyptian Tourist Authority.)

Figure 12.1b High relief is modeled well by the direct sun of Greece.

Figure 12.1c The cloudy and subdued lighting of northern Europe allows highly sculptured forms. Even when the sun does come out, as in this photograph, it is not so intense that details are lost in dark shadows. (Photograph by Nicholas Davis.)

metric manipulation of light and the color of finishes. This aspect of lighting is tier one of the three-tier design approach (Fig. 12.1d). Tier two consists of the natural energy of daylighting, and tier three consists of electric lighting. To design quality lighting, we must start with an understanding of light, vision, and perception.

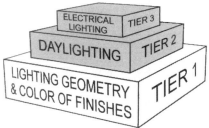

Figure 12.1d High-quality and more sustainable lighting is best achieved via the three-tier design approach. This chapter covers tier one.

327

12.2 LIGHT

Light is defined as that portion of the electromagnetic spectrum to which our eyes are visually sensitive. In Fig. 6.2b, we see the intensity of the solar radiation reaching the earth as a function of wavelength. It is no accident that our eyes have evolved to make use of that portion of the solar radiation that was most available.

Not all animals are limited to the visible spectrum. Rattlesnakes can see the infrared radiation emitted by a warm-blooded animal. Many insects can see ultraviolet radiation. We can get a glimpse of how the world looks to them when we see certain materials illuminated by black light, which is ultraviolet light just beyond violet. Certain materials glow (fluoresce) when exposed to radiation at this wavelength. Ultraviolet radiation of a somewhat shorter wavelength causes our skin to tan or burn. Even shorter ultraviolet is so destructive that it is germicidal and can be used in sterilization (Fig. 12.2a). Although we cannot see beyond the visible spectrum, we can feel infrared radiation on our skin as heat.

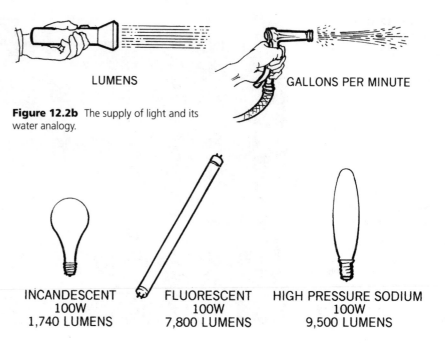

Figure 12.2b The supply of light and its water analogy.

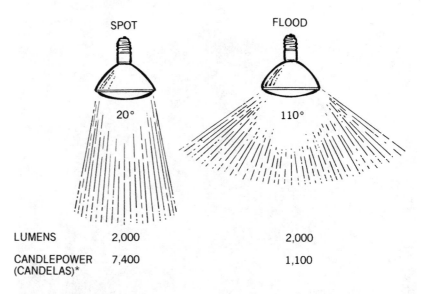

Figure 12.2c The power or rate at which lamps emit light is measured in lumens. Because of large differences in efficiency, lamps of equal wattage can emit very different amounts of light.

Figure 12.2d Candlepower describes the intensity of a light source. Although both lamps emit the same amount of light (lumens), the intensity and width of the beams are very different. *Average value of the central 10-degree cone.

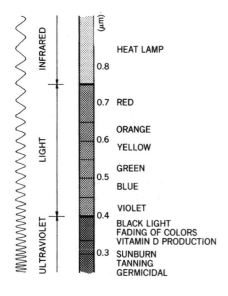

Figure 12.2a Light is only a small part of the electromagnetic spectrum. Light is the radiation to which our eyes are visually sensitive.

Lumen

The rate at which a light source emits light energy is analogous to the rate at which water sprays out of a garden hose (Fig. 12.2b). The power with which light is emitted from a lamp is measured in lumens. We can say that the quantity of light a lamp emits in all directions is indicated by the lumen value (Fig. 12.2c).

Candlepower

Lumens, however, do not reveal how the emitted light is distributed. In Fig. 12.2d, we see two reflector lamps that give off equal amounts of light

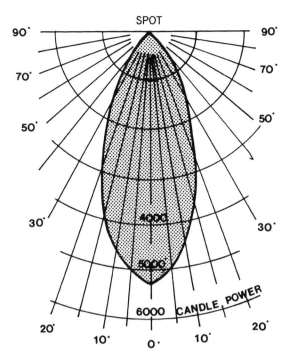

SPOT

6000 CANDLE POWER

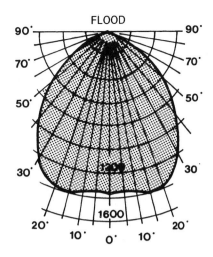

FLOOD

1600

Figure 12.2e Candlepower distribution curves illustrate how light is emitted from lamps and lighting fixtures. In this vertical section, the distance from the center determines the intensity of the light in that direction.

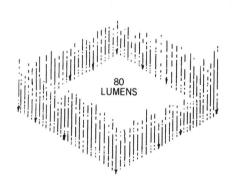

80 LUMENS

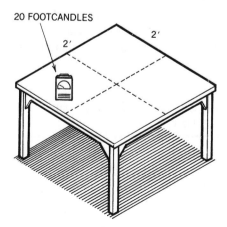

20 FOOTCANDLES

2′   2′

Figure 12.2f Illumination (footcandles) is the amount of light (lumens) falling on 1 square foot. illumination can be measured with a photometer (light meter).

(lumens) but with very different distribution patterns. The spot lamp has an intense narrow beam, while the flood lamp has a much wider beam with less intensity. Candlepower, measured in **candelas,** describes the intensity of the beam in any direction. Manufacturers supply candlepower-distribution graphs for each of their lighting fixtures (Fig. 12.2e).

Illuminance

The lumens from a light source will illuminate a surface. A meaningful comparison of various illumination schemes is possible only when we compare the light falling on equal areas. **Illuminance** is, therefore, equal to the number of lumens falling on each square foot of a surface. The unit of illumination is the **footcandle.** For example, when the light of 80 lumens falls uniformly on a 4-square-foot table, the illumination of that table is 20 lumens per square foot, or 20 footcandles (see Fig. 12.2f and Sidebox 12.2a). Illumination is measured with footcandle meters, which are also known as **illuminance meters** or **photometers.** Such instruments are available in a wide range of prices.

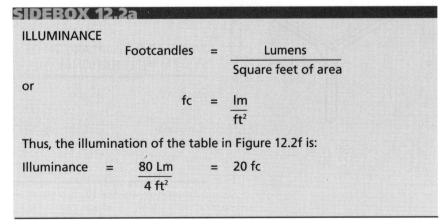

SIDEBOX 12.2a

ILLUMINANCE

$$\text{Footcandles} = \frac{\text{Lumens}}{\text{Square feet of area}}$$

or

$$\text{fc} = \frac{\text{lm}}{\text{ft}^2}$$

Thus, the illumination of the table in Figure 12.2f is:

$$\text{Illuminance} = \frac{80 \text{ Lm}}{4 \text{ ft}^2} = 20 \text{ fc}$$

Brightness/Luminance

The words **brightness** and **luminance** are closely related. The brightness of an object refers to the perception of a human observer, while the object's luminance refers to the objective measurement of a light meter. The perception of brightness is a function of the object's actual luminance, the adaptation of the eye, and the brightness of adjacent objects. Although the words are interchangeable much of the time, under certain conditions a significant discrepancy exists between what we see (brightness) and what a light meter reads (luminance).

Luminance is the amount of light that is reflected off an object's surface and reaches the eye. The luminance of an object is a function of: the illumination; the geometry of the viewer in relation to the light source; the **specularity,** or mirror-like reflection, of the object; and the color, or reflectance, of the object (Fig. 12.2g). Light emitted from glowing or translucent objects is also called **luminance.** Thus, we can talk of the luminance of a table, an electric lamp, or a translucent window.

It is usually more important to consider the perceived brightness than the objective luminance. Lights will generally appear much brighter at night than during the day, but they will register the same luminance on a light meter at any time. In this book, the more familiar word "brightness" is used except when a serious error would result.

Conversions Between System International and American System

In lighting, the switch to the System International (SI) from the American System (AS) will be rather painless. Both systems use the unit of the lumen, and the unit of candlepower is equal to the SI candela. Lux is the SI unit for illumination and is approximately equal to one-tenth of a footcandle or

1 footcandle = approximately 10 lux (see Table 12.2).

12.3 REFLECTANCE/ TRANSMITTANCE

Light falling on an object can be transmitted, absorbed, or reflected. The **reflectance factor (RF)** indicates how much of the light falling on a surface is reflected. To determine RF of a surface, divide the reflected light by the incident light. Since the reflected light (brightness) is always less than the incident light (illumination), the RF is always less than one, and since a little light is always reflected, the RF is never zero. A white surface has an RF of about 0.85, while a black surface has an RF of only 0.05. The RF does not predict how the light will be reflected, only how much. Very smooth polished surfaces, such as mirrors, produce specular reflections where the angle of incidence is equal to the angle of reflection. Very flat or matte surfaces scatter the light to produce diffuse reflections. Most materials reflect light in both a specular and a diffuse manner (Fig. 12.3a).

Similarly the **transmittance factor** describes the amount of light that is transmitted as compared to the incident light. To determine the transmittance factor of a surface, divide the transmitted light by the incident light. A clear, transparent material transmits the image of the light sources, while a diffusing translucent material like frosted glass scatters the light, thereby obscuring the image of the light sources (Fig. 12.3b). In general, diffusion does not affect the quantity of

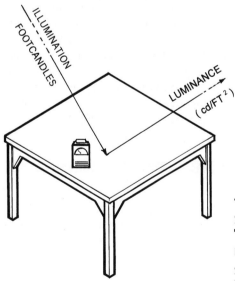

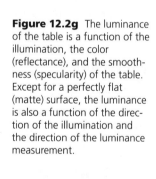

Figure 12.2g The luminance of the table is a function of the illumination, the color (reflectance), and the smoothness (specularity) of the table. Except for a perfectly flat (matte) surface, the luminance is also a function of the direction of the illumination and the direction of the luminance measurement.

TABLE 12.2 COMPARISON OF AMERICAN STANDARD (AS) AND SYSTEM INTERNATIONAL (SI) LIGHTING UNITS

PROPERTY	(AS)	(SI)	Conversion Factor
Supply of light	Lumen (lm)	Lumen (lm)	1
Illuminance	Footcandle (fc)	Lux (lx)	1 fc ≈ 10 lx*
Luminous intensity (candlepower)	Candela (cd)	Candela (cd)	1
Luminance	cd/ft²	cd/m²	1 cd/ft² = 0.09 cd/m²

*The approximation of 10 lux per footcandle is more than sufficient for most purposes (actually 1 fc = 10.764 lux).

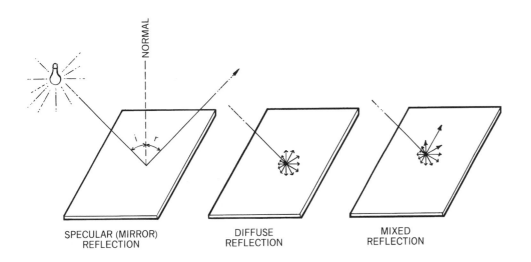

Figure 12.3a The characteristics of a surface determine not only how much but also in what way light is reflected. Most real materials tend to give mixed reflections.

Figure 12.3b Transparent materials do not distort the transmitted image, while translucent (frosted or milky) materials diffuse the light and destroy any image.

light transmitted (both clear and frosted glass transmit about 85 percent of the incident light).

There are three aspects to the reflection of light: the quantity (the RF), its manner (specular versus diffuse), and spectral selection (color). We have discussed the first two and will now discuss the third.

12.4 COLOR

White light is a mixture of various wavelengths of visible light. Figure 12.4a illustrates the composition of daylight on a clear day in June at noon. The horizontal axis describes the colors (wavelengths in millionths of a meter) and the vertical axis the amount of light (relative energy) at the various wavelengths. This kind of graph is the best way to describe the color composition of any light, and it is known as an **SED (Spectral Energy Distribution)** diagram. The mostly horizontal curve reflects the even mixture of the various colors that make up daylight. Only violet light is present in less quantity. North light, which is often considered the ideal white light for painters, is not such an even mixture. It has more light in the blue end than the red end of the spec-

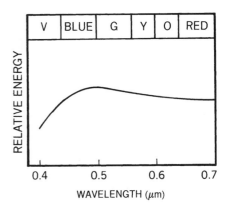

Figure 12.4a The spectral-energy distribution of average daylight at noon on a clear day in June. Notice the almost even distribution of the various colors. (See also Color Plate 4).

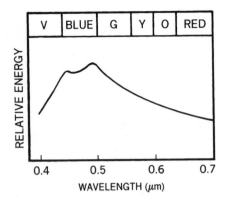

Figure 12.4b The spectral-energy distribution of daylight from a north-facing window. Note that there is more energy in the blue end than the red end of the spectrum. (See also Color Plate 4).

trum, as can be seen in the SED diagram of north light (Fig. 12.4b and color plate 4). Artist studios used to face north not because north light was considered to have the best color balance but because until recently it was the most consistent source of white light. The main advantage of north glazing is the constancy of the light. Light from windows of other orientation varies greatly throughout the day and year. Late-afternoon daylight has much more energy in the red end of the spectrum and less in the blue end. Although all of the above varieties of daylight and many artificial light sources supply "white" light, there is a great difference in the composition of these sources.

The color of a surface is due not only to its spectrally selective reflectance characteristics but also to the spectral composition of the illumination. A completely saturated (pure) red paint that is illuminated by monochromatic (pure) red light will appear bright red because most of the light is reflected (Fig. 12.4c). However, if this same red paint is illuminated with monochromatic blue light, it will appear black because the color red absorbs all colors except red (Fig. 12.4d). Unless the red paint is illuminated with light that contains red, it will not appear red.

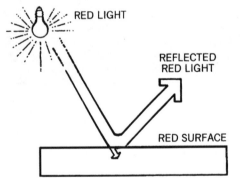

Figure 12.4c A red-colored surface reflects most red light and absorbs most of the light of the other colors.

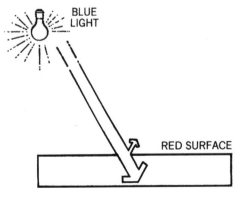

Figure 12.4d Under pure blue light, a pure red color will appear black since almost no light is reflected.

PLATE 1 ▶
Light is the small portion of the electromagnetic spectrum that the human eye is sensitive to. Waves can be described either by their frequency, as is typical for radio waves, or by their wavelength. Here the wavelength of the visible spectrum is described by nanometers (1 billionth of a meter). (Courtesy General Electric Co.)

PLATE 2 ▶
The perceived color of an object is a function of both the reflectance characteristics (color) of an object and the spectral composition of the light source. In this example, the center graph describes the reflectance characteristics of a reddish complexion, note the high reflectance in the red end of the visible spectrum. Nevertheless, a person will appear pale if illuminated by a red-poor light source, such as a cool white fluorescent lamp. (Courtesy General Electric Co.)

PLATE 3 ▶
The same complexion (note the identical reflectance graph) will appear to have a healthy reddish color when illuminated by a source rich in red light, such as a deluxe cool white fluorescent lamp. (Courtesy General Electric Co.)

PLATE 4 ▶
These graphs describe the **spectral-energy distribution (SED)** of various light sources. (Courtesy General Electric Co.)

Daylight: The SED for daylight varies greatly. Note how the light from the north sky is rich in blues and poor in reds (the bottom curve on the red side). Sunlight, on the other hand, has almost equal amounts of all wavelengths (the upper curve on the red side).

Incandescent (Halogen): Although most energy is in the red end of the spectrum, all colors are rendered well because all wavelengths are present. A halogen lamp emits a "whiter" light than the incandescent lamp because its filament is hotter.

Fluorescent (SPX 35™): Modern high-quality and high-efficacy fluorescent lamps emit energy at carefully selected wavelengths. See Plates 2 and 3 above for the SED diagrams of the traditional cool white and the much warmer deluxe cool white lamps.

Mercury: The light emitted from a clear mercury lamp is almost completely in the blue, green, and yellow parts of the spectrum, and, thus, reds are rendered poorly. Green plants, however, look lush under the decorative landscape lighting of clear mercury lamps.

Metal-Halide (MULTI-VAPOR™): Because these lamps emit light in all parts of the spectrum, they render colors well.

High-Pressure Sodium (LUCALOX ™): Although the efficacy is very high, the color rendering quality of these lamps is only fair. Most of the energy is emitted in the yellow portion of the spectrum where our eyes are most sensitive.

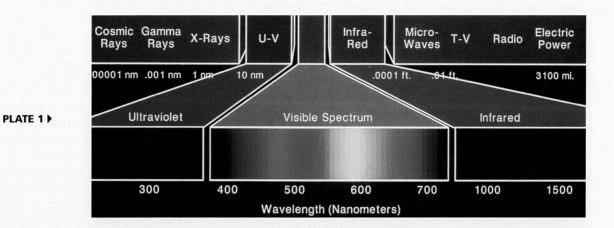

PLATE 1 ▶

| Cosmic Rays | Gamma Rays | X-Rays | U-V | | Infra-Red | Micro-Waves | T-V | Radio | Electric Power |

00001 nm   .001 nm   1 nm   10 nm   .0001 ft.   .01 ft.   3100 mi.

Ultraviolet   Visible Spectrum   Infrared

300   400   500   600   700   1000   1500

Wavelength (Nanometers)

PLATE 2 ▶

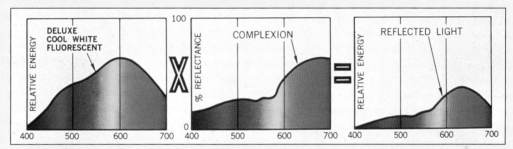

RELATIVE ENERGY — COOL WHITE FLUORESCENT — X — % REFLECTANCE — COMPLEXION — = — RELATIVE ENERGY — REFLECTED LIGHT

400 500 600 700   400 500 600 700   400 500 600 700

WAVELENGTH (Nanometers)

PLATE 3 ▶

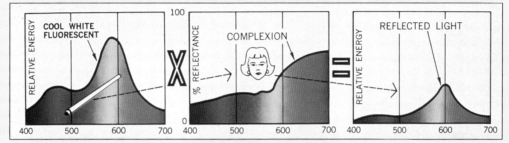

RELATIVE ENERGY — DELUXE COOL WHITE FLUORESCENT — X — % REFLECTANCE — COMPLEXION — = — RELATIVE ENERGY — REFLECTED LIGHT

400 500 600 700   400 500 600 700   400 500 600 700

WAVELENGTH (Nanometers)

PLATE 4 ▼

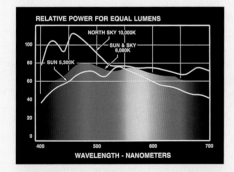

RELATIVE POWER FOR EQUAL LUMENS

NORTH SKY 10,000K
SUN & SKY 6,000K
SUN 5,300K

400   500   600   700

WAVELENGTH - NANOMETERS

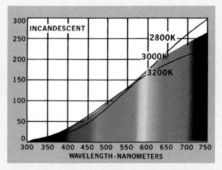

INCANDESCENT

2800K
3000K
3200K

300 350 400 450 500 550 600 650 700 750

WAVELENGTH - NANOMETERS

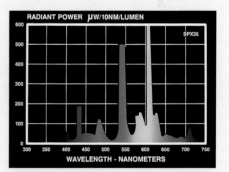

RADIANT POWER   μW/10NM/LUMEN

SPX35

300 350 400 450 500 550 600 650 700 750

WAVELENGTH - NANOMETERS

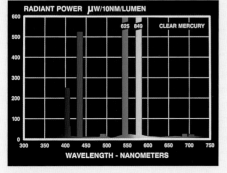

RADIANT POWER   μW/10NM/LUMEN

625 849   CLEAR MERCURY

300 350 400 450 500 550 600 650 700 750

WAVELENGTH - NANOMETERS

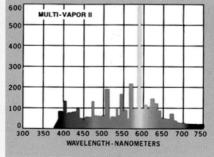

MULTI-VAPOR II

300 350 400 450 500 550 600 650 700 750

WAVELENGTH - NANOMETERS

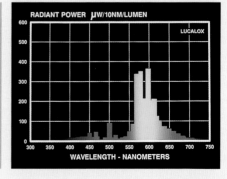

RADIANT POWER   μW/10NM/LUMEN

LUCALOX

300 350 400 450 500 550 600 650 700 750

WAVELENGTH - NANOMETERS

PLATE 5 ▼

PLATE 6 ▼

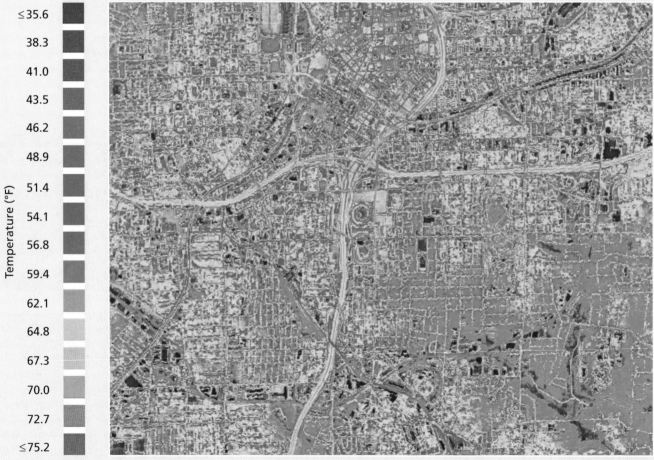

Temperature (°F)

≤35.6
38.3
41.0
43.5
46.2
48.9
51.4
54.1
56.8
59.4
62.1
64.8
67.3
70.0
72.7
≤75.2

Source: NASA / EPA

In the real world, where the colors are not completely saturated (pure), the situation is more complicated. Ordinary colors, such as red, reflect not only most of the red light, but also small amounts of the other colors. This can create problems when the illumination does not have a good mixture of the various colors. A bright red car reflects plenty of red light when it is illuminated by daylight (Fig. 12.4e). However, when this same bright red car is parked at night under a clear mercury street light, it will appear to be brown (Fig. 12.4f). Since clear mercury lamps emit mostly blue and green light, there is little red light that the car can reflect. Although much of the light of the other colors was absorbed, enough blue and green was reflected to overwhelm the red light.

When an object's color is very important as, for example, in the displaying of meat or tomatoes, the selection of the light source is critical. A full-spectrum, white light source will accurately render the red colors of these items. To make them look even fresher and more appetizing, use a light source rich in red.

The transmission of light through colored glass or plastic is a selective

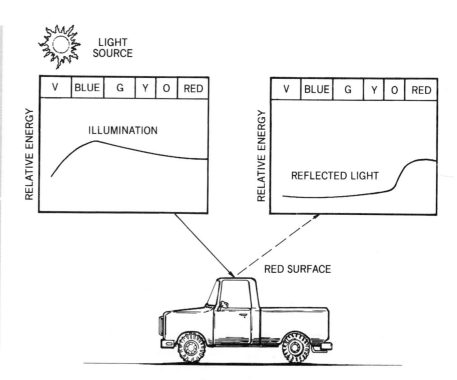

Figure 12.4e A red car will appear red if it is illuminated by a full-spectrum white light source, such as the sun. See also Color Plate 2.

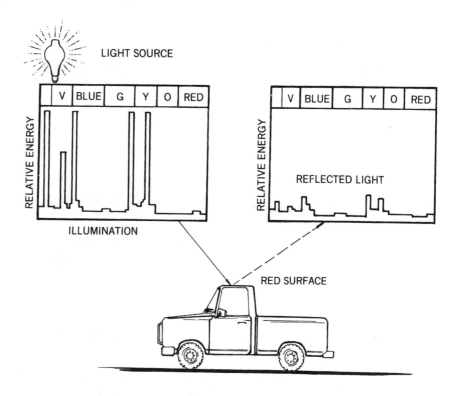

Figure 12.4f A red car will appear brown under light sources such as clear mercury lamps, because these lamps emit only small amounts of red light. Thus, only a very small amount of red light is reflected from the car. See also Color Plate 3.

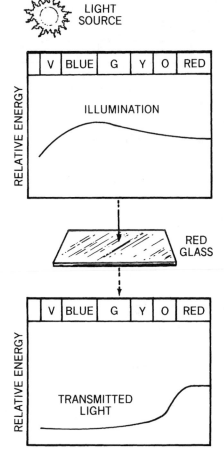

Figure 12.4g Red glass transmits most of the red light but only very little of the other colors.

process similar to reflection. A white light viewed through red glass will appear red because red light is mostly transmitted, and the light of the other colors is mostly absorbed (Fig. 12.4g).

Color Temperature

To completely describe the color content of a light source, the amount of light at each wavelength must be defined as in the SED diagrams of Color Plate 4. Because this method is quite cumbersome, the concept of **color temperature (CT)** is often used. It was noticed that as many materials were heated, they would first glow red, then white, and finally blue. Thus, there is a relation between temperature and color. A color-temperature

scale was developed that describes the color of a light source in degrees Kelvin (Fig. 12.4h).

CT is mostly used to describe the warmness or coolness of a light source. Low CT, or warm, light sources tend to render red colors well, while high CT, or cool, light sources tend to render blue colors well. It must be noted, however, that this scale can give only a very crude description of the color-rendering ability of light sources. Another attempt to simplify the description of light sources was the development of the **color-rendering index (CRI),** but it, too, has limitations and must be used with care. The CRI compares light sources to a standard source of white light. A perfect match would yield a CRI of 100. A CRI of 90 is considered quite good, and a CRI of 70 is usually still acceptable.

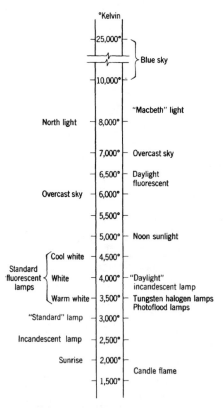

Figure 12.4h The correlated color-temperature scale gives a rough indication of the color balance (spectral-energy distribution) of various sources of "white" light. (From Mechanical and Electrical Equipment for Buildings, by B. Stein et al., 9th ed., © John Wiley & Sons, Inc., 2000.)

Because of the simplicity of the color-rendering index (CRI), it is tempting to rely on it too much. The CRI can be used only to compare light sources of the same color temperatures. Even then there is no guarantee that any specific color will appear natural.

Color selection or matching is best accomplished by actual tests. If colors are to be matched or selected, they must be examined with the type of light source by which they will be illuminated. Many designers have been shocked when they saw their carefully chosen colors under a different light source. If two light sources are to be compared, then critical color samples must be compared side by side, with only one light source illuminating each. Fig. 12.4i shows a device built by the author to compare the color-rendering qualities of different light sources. The effect of light sources on color appearance is called **color rendition** and will be discussed some more in Chapter 14.

Rule: Full-spectrum white light is required for the accurate judgment of color.

Figure 12.4i This device is used for comparing the color-rendering effect of two light sources.

12.5 VISION

Vision is the ability to gain information through light entering the eyes. Rather than compare the eye to a photographic camera, which is the usual analogy, let's compare it to the video camera of a robot (Fig. 12.5a). The light rays that enter the video camera are changed into electrical signals, and the robot's computer then processes these signals for their information content. The meaning of the signals is determined by both the hardware and software of the robot. Similarly, our eyes convert light into electrical signals that the brain then processes (Fig. 12.5b). Here also, the meaning of the visual information is a consequence of the hardware (eye and brain) and the software (associations, memory, and intelligence). The brain's interpretation of what the eyes see is called perception. Although a lighting design must ultimately be based on an understanding of perception, we must start by understanding vision.

Light enters the eye through an opening called the **pupil** and is focused onto the light sensitive lining at the back of the eye called the **retina.** The retina consists of **cone** cells that are sensitive to colors and **rod** cells that respond to motion and dim lighting conditions.

To accommodate the large range of brightness levels in our environment, the eye adapts by varying the size of the pupil with the muscle called the **iris,** as well as by a change in the sensitivity of the retina. The eye is able to effectively see in a range of brightness of 1,000 to 1 and to partially see in a range of over 100,000,000 to 1. However, it takes about an hour for the eye to make a full adaptation, and, in the meantime, vision is not at its best.

Table 12.5 lists commonly experienced brightness levels and shows how these relate to vision. Notice that each item listed is ten times brighter than the previous item. This illus-

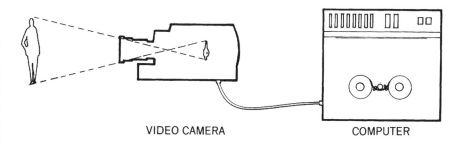

VIDEO CAMERA   COMPUTER

Figure 12.5a In robots that can "see," the computer brain interprets the electrical signals that come from the video camera.

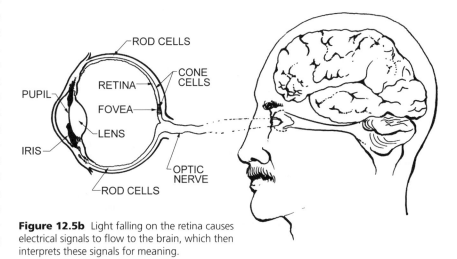

Figure 12.5b Light falling on the retina causes electrical signals to flow to the brain, which then interprets these signals for meaning.

trates the nonlinear sensitivity of the eyes and the consequence that it takes large increases in light for the eyes to notice a small increase in brightness.

Although rapid and extreme changes in brightness cause stress and fatigue, the eye is very well adapted to the gradual changes of brightness that are associated with daylighting. Because of the eye's adaptation, we perceive only small changes in light levels when clouds move across the sky. A gradual change in brightness is not a liability and might even be an asset because changes are more stimulating than static conditions.

A very small area of the retina surrounding the center of vision called the **fovea** consists mainly of "cone" cells. Here, the eye receives most of the information on detail and color (Fig. 12.5b). The foveal (sharp) vision occurs in a 2-degree cone around the

TABLE 12.5 COMMONLY EXPERIENCED BRIGHTNESS LEVELS

	Brightness (cd/sq. ft.)*	
Sidewalk on a dark night	0.0003	▲
Sidewalk in moonlight	0.003	Poor vision
Sidewalk under a dim streetlight	0.03	
Book illuminated by a candle	0.3	
Wall in an office	3	Normal indoor brightness
Well-illuminated drafting table	30	
Sidewalk on a cloudy day	300	Normal outdoor brightness
Fresh snow on a sunny day	3,000	
500-watt incandescent lamp	30,000	▼ Blinding glare

*For S.I., (cd/sq. m.) ≈ (cd/sq. ft.) × 11

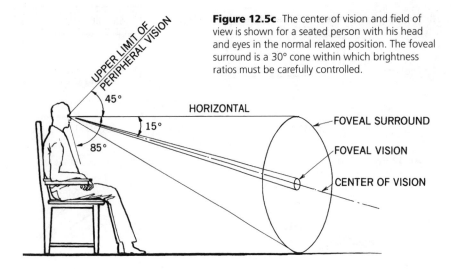

Figure 12.5c The center of vision and field of view is shown for a seated person with his head and eyes in the normal relaxed position. The foveal surround is a 30° cone within which brightness ratios must be carefully controlled.

UPPER LIMIT OF PERIPHERAL VISION

45°

HORIZONTAL

15°

85°

FOVEAL SURROUND

FOVEAL VISION

CENTER OF VISION

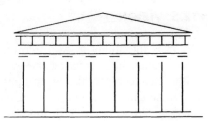

Figure 12.6a Greek temples appear to be built with straight lines, square corners, and uniform spacing of the repeating elements.

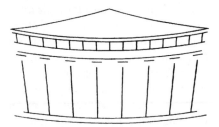

Figure 12.6b When a temple was actually built as shown in Fig. 12.6a, it was perceived as distorted in this manner (optical illusions).

Figure 12.6c The Parthenon was actually built in this distorted way so that it would be perceived as shown in Fig. 12.6a.

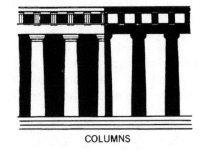

COLUMNS

Figure 12.6d Some columns appear bright because of the shaded wall behind them. Corner columns seen agains the bright sky seem dark in comparison. Because the darker corner columns appear smaller and weaker than the brighter columns, the Greeks made the end columns stouter than the central columns. Figures 12.6a through 12.6d are from Banister Fletcher's *A History of Architecture, 19th ed., edited by John Musgrove,* © Royal Institute of British Architects, 1987.)

center of vision, which moves as the eye scans. Focus, color, and awareness in the field of view decrease with distance from the central 2-degree cone of vision, because the density of cone cells decreases while the density of "rod" cells increases. Rod cells respond to low light levels and movement but not to color or detail. Awareness is still quite high in the **foveal surround,** which is within a 30-degree cone around the center of vision. The cheeks and eyebrows are the limiting factors for peripheral vision, and the total field is, therefore, generally about 130 degrees in the vertical direction and about 180 degrees in the horizontal direction.

For a seated person whose head and eyes are at rest, the center of vision is about 15 degrees below horizontal (Fig. 12.5c). The location and brightness of objects in the field of view will have a major impact on the quality of the lighting environment (this will be discussed in more detail later.)

12.6 PERCEPTION

The ancient Greeks realized that we do not perceive the world as it actually is. They found that when they built their early temples with straight lines, right angles, and uniform spacing of columns, the results were perceived not as they built them (Fig. 12.6a) but distorted, as shown in Fig. 12.6b. Consequently, the Greeks built later temples, like the Parthenon, in a very cleverly distorted manner (Fig. 12.6c) so that they would be perceived as correct (Fig. 12.6a).

In the Parthenon, the columns are all inclined inward to preclude the illusion that they are falling outward. The columns have a slight bulge (entasis) to counteract the illusion of concavity that characterizes columns with straight sides. The column spacing and thickness vary because of the effect of high brightness ratios. Fig. 12.6d illustrates how bright columns on a dark background look sturdier than dark columns on a bright background. This is important because the Parthenon's central columns are seen against the dark, shaded building wall, while the end columns are seen against the bright sky. Therefore, the ancient Greeks made the end columns thicker than the central columns.

This example of temple design was not included to suggest that we should proportion our buildings as subtly as the ancient Greeks did, but to suggest how much perception can vary from what we might expect. To create a successful lighting system, the designer must understand human perception, some of the more important aspects of which are described below.

Relativity of Brightness

The absolute value of brightness as measured by a **photometer (light meter)** is called **luminance.** A human being, however, judges the brightness of an object relative to the brightness of the immediate surroundings. Since the Renaissance, painters have used this principle to create the illusion of bright sunshine. The puddle of light on the table in the painting in Fig. 12.6e will appear as bright sunshine no matter how little light illuminates the painting. The painter was able to "highlight" objects by creating a dark setting rather than by high illumination levels. Fig. 12.6f shows this same principle in an abstract diagram. The middle gray triangles are identical in every way, including their reflectance factor. Their luminance as measured by a photometer will be the same but their perceived brightness will depend on the brightness of the surrounding area.

Because of the importance of this aspect of perception, one more example is in order. Car headlamps seem very bright at night but are just noticeable during the day. Although a meter would show the luminance to be the same, the brightness we perceive depends on the relative brightness of the headlamps to the overall lighting condition.

This is partly due to the fact that at night, our wide-open pupils allow much of the light of the headlights to enter, while during the day, our small pupils shield us from the excess light of the headlights.

Figure 12.6e This Italian painter of the nineteenth century fully understood the concept of relativity of brightness. He could simulate bright sunshine by creating dark surroundings. The mind visualizes bright sunshine no matter how little light is falling on this painting. *The 26th of April 1859,* painted by Odoardo Borrani in 1861. (Courtesy of the Guiliano Matteucci-Studio d'Arte Matteucci, Rome, Italy.)

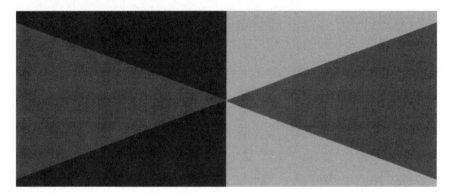

Figure 12.6f The two triangles are exactly the same, yet they appear to have different reflectance factors because of the phenomenon of the relativity of brightness. To see the triangles as equal, cover the dark areas around the left triangle with two pieces of white paper.

Brightness Constancy

To make sense of the visual environment, the brain has to make adjustments to what the eyes see. For example, in a room with windows on one end, the ceiling plane will appear of constant brightness although a photometer would clearly show greater luminance near the windows. The brain knows that the reflectance factor is constant and that it is the illumination level that is varying. Consequently, the brain interprets the ceiling as having uniform brightness. This ability of the brain to ignore differences in luminance under certain conditions is called **brightness constancy.**

Color Constancy

It is common to have the experience of photographing a white building at sunset and then being surprised when the photograph comes back showing a pink building. The photograph tells the truth. Our perception fooled us into seeing a white building when we took the picture. Our brain "filtered" out much of the red light from the setting sun, something a camera can do also—but only when the lens is covered with a color filter. This ability of the brain to eliminate some of the differences in color due to differences in illumination is called **color constancy.** This ability has very important survival implications because without it we would never recognize our own home if we returned at a different time of day.

Color constancy is not possible, however, if more than one type of light source is used simultaneously. Fig. 12.6g illustrates what happens if an object is illuminated by north light from one side and an incandescent lamp from the other. One shadow will appear bluish and the other reddish.

Because the brain cannot adjust to the color balance of each source simultaneously. A lighting design with different light sources must take this into account. Often, the best solution is not to mix light sources that are very different. The placement of clear window glazing adjacent to tinted glazing should also be avoided.

Other Color Perception Phenomena

The warm colors (red, orange, and yellow) appear to advance toward the eye, while the cool colors (blue, green and dark gray) appear to recede. The choice of wall colors can make a space seem larger or smaller than it actually is.

Prolonged concentration on any color will result in an afterimage of the complementary color. A surgeon staring at a bright red organ will see a cyan (blue-green) organ as an afterimage when he moves his eyes to look elsewhere. To minimize this upsetting phenomenon, hospitals now use green sheets and wall surfaces in their

Figure 12.6h Venetian blinds are often disturbing, because of the figure/background confusion.

operating rooms. A cyan afterimage superimposed on a green sheet is much less noticeable than when it is superimposed on a white sheet.

Figure/Background Effect

The brain is always trying to sort out the visual signal from the visual noise. When this becomes difficult or impossible, the image becomes disturbing. Figure 12.6h illustrates a view interrupted by venetian blinds. Either miniature blinds or larger overhangs would change the **figure/background** effect and therefore create a more comfortable view.

Gestalt Theory

The purpose of seeing is to gather information. The brain is always looking for meaningful patterns. In Fig. 12.6i, we see only a small circle and a long rectangle. But in Fig. 12.6j, the first thing we see is the exclamation mark. In Fig. 12.6k, we see a disturbing arrangement because it reminds us of something, but it is not quite right. The brain's search for greater meaning than the parts themselves would suggest is called **gestalt theory.** A particular lighting scheme will,

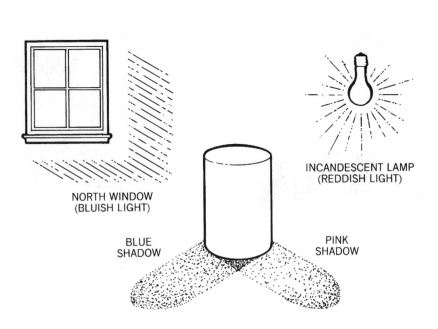

NORTH WINDOW
(BLUISH LIGHT)

INCANDESCENT LAMP
(REDDISH LIGHT)

BLUE
SHADOW

PINK
SHADOW

Figure 12.6g When more than one type of white light is used, color constancy cannot operate and the shadows appear colored. However, if either source is eliminated, the remaining shadow will appear normal because of color constancy.

Figure 12.6i The brain perceives only a bar and a circle.

Figure 12.6j The brain perceives an exclamation mark rather than a bar and a circle. This perception of greater meaning is explained by gestalt theory.

Figure 12.6k Since the brain is not sure if there is greater meaning, this pattern is disturbing.

therefore, be successful not so much if all the parts are well designed but more so if the whole composition is meaningful and not disturbing or distracting.

Other Perception Phenomena

Bright ceilings and upper walls make a room look larger and friendly, while dark ceilings and upper walls make a room seem smaller and less inviting.

Dramatic lighting is achieved by utilizing large brightness ratios in the field of view.

Romantic light, as from a candle on a dining table in a dark room, has several benefits. It creates an intimate space, emits a very warm light whose reds complement skin complexion, and the nearly horizontal light tends to minimize wrinkles and shading all over the face.

12.7 PERFORMANCE OF A VISUAL TASK

Many factors affect the performance of a visual task (i.e., a task where visibility is important). Some of these factors are inherent in the task, some describe the lighting conditions, and the remainder reflect the condition of the observer. Most of the important factors can be easily understood by examining the common but critical seeing task of reading an interstate-highway sign (Fig. 12.7). Since the time of exposure is very limited, the signs are made large, bright, of high contrast, and of a consistent design. They are well illuminated at night but are often obscured by the glare of oncoming cars. The health and alertness of the driver are also factors. Thus, we can see that the basic factors that affect the performance of a visual task can be categorized as:

A. The Task
 1. Size/proximity
 2. Exposure time
 3. Brightness
 4. Contrast
 5. Familiarity

B. The Lighting Condition
 1. Illumination level
 2. Brightness ratios
 3. Glare

C. The Observer
 1. Condition of eyes
 2. Adaptation
 3. Fatigue level

Most of these factors will now be discussed in more detail.

Figure 12.7 Since exposure time is limited, the other factors of visual performance are used to their maximum effect: size/proximity, brightness (night illumination), contrast, and familiarity (always white on green).

12.8 CHARACTERISTICS OF THE VISUAL TASK

Size/Proximity

The most important characteristic of a visual task is the **exposure angle,** a function of the viewed object's size and proximity (Fig. 12.8a). The exposure angle will increase when the object is either enlarged or brought closer (Fig. 12.8a). In most cases, the brain then determines the difference by its familiarity with the real world and by means of binocular vision.

Whenever possible, the designer should increase the size of the task because a small increase in size is equivalent to a very large increase in illumination level. For example, a 25 percent increase in lettering size on a blackboard increases the visual performance as much as a change in illumination from 10 to 1,000 footcandles.

Exposure Time

Other factors of visual performance can offset a short exposure time, but, as with size, very high increases of illumination are required to offset small decreases in the exposure time. Thus, exposure time should not be cut short if at all possible.

Brightness

Fig. 12.8b illustrates how an increase in task brightness at first results in significant improvements in visual performance, but additional increases yield smaller and smaller benefits. The "law of diminishing returns" is in effect because of the nonlinear relationship between brightness and visual performance. For example, raising the illumination from 0 to 50 footcandles, the brightness also increases, and the visual performance improves to about 85 percent, while another increase of 50 footcandles improves the visual performance only by 5 percent. Since large increases in brightness are made possible only by large increases in illumination, high brightness is a very expensive route to visual performance.

The discussion so far has been about absolute brightness (luminance), but, as we saw earlier, we perceive brightness in relative terms. Therefore, it is often possible to increase performance by reducing the background brightness, thereby increasing the relative brightness of the task. The reduction of the background brightness increases the eye's sensitivity to light, making the task easier to see.

This concept is used to its fullest by museums exhibiting artifacts that light can damage. Wood, paper, cloth, and natural pigments all fade significantly when exposed to light. Such damage can be minimized by keeping the light level as low as possible. The museum shown in Fig. 12.8c manages to highlight its fragile objects with fewer than 4 footcandles of illumination simply by having an even darker background illumination.

Although ultraviolet radiation causes the most damage to delicate objects, visible light also causes some fading. Short-wavelength radiation, such as that of violet, is worse than the long-wavelength radiation of red.

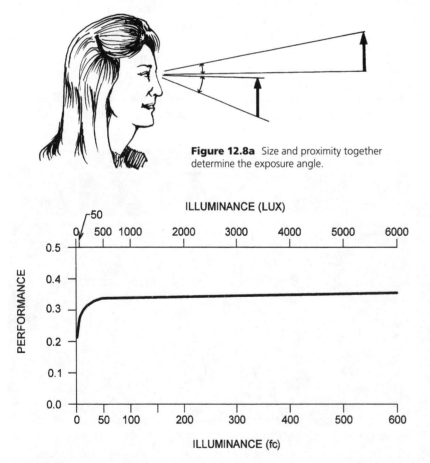

Figure 12.8a Size and proximity together determine the exposure angle.

Figure 12.8b At first, visual performance improves rapidly with increases of illumination, but soon the "law of diminishing returns" governs. Above about 100 fc (1000 lux) there is little benefit with even large increases of illumination. Graph shown is for small tasks of high contrast like black print on white paper. Results vary somewhat with contrast and task size. (After Fig. 3.42 of the IESNA *Lighting Handbook,* 1993.)

Contrast

The difference in brightness between a detail and its immediate background is called **contrast.** Most critical visual tasks will benefit when the contrast

Figure 12.8c Fragile artifacts from ancient Egypt are brightly illuminated by only 4 footcandles of illumination because of the very dark background. (Courtesy of Memphis State Photo Services, Memphis State University, TN.)

HIGH CONTRAST	LOW CONTRAST
ILLUMINATION OF ONE FOOT CANDLE IS SUFFICIENT	ILLUMINATION OF OVER 100 FOOT CANDLES IS REQUIRED

Figure 12.8d Contrast is an extremely important factor for visual performance. High levels of illumination are required to compensate for poor contrast.

12.9 ILLUMINATION LEVEL

Since brightness is directly proportional to illumination, the previous discussion on brightness is directly relevant to illumination. The graph in Fig. 12.8b also describes the relationship between visual performance and illumination. As the light level increases to about 50 footcandles, there is a significant improvement in visual performance. Above 100 footcandles, however, the law of diminishing returns begins to govern and large increases in illumination result in only minor improvements in visual performance. The reason for this is that the pupil gets smaller as the illumination increases. Thus, the amount of light reaching the retina increases only slightly.

It is, therefore, usually appropriate to keep the general area illumination below 30 footcandles, and to supply higher light levels only if specific tasks require it. The additional light should be localized to the tasks that

between the task and its immediate surroundings is maximized. Writing, for example, is most easily seen when the contrast between ink and paper is at a maximum. When contrast decreases, the other factors of visual performance can be adjusted to compensate. Again, however, very large increases of illumination are required to offset poor contrast (Fig. 12.8d).

It is important to note that the concept of contrast refers to detailed visual tasks (foveal vision), such as the print on a piece of paper. It does not refer to the brightness relationship of the paper to the desk, or the desk to surrounding area. The brightness differences in these peripheral areas will have different effects on vision and will be discussed later.

TABLE 12.9   GUIDELINES FOR ILLUMINATION LEVELS

Approximate Type of Activity	Footcandles*
1. General lighting throughout space	
a. Public spaces with dark surroundings	3
b. Simple orientation for short, temporary visits	8
c. Working spaces where visual tasks are only occasionally performed	15
2. Illumination on task	
a. Performance of visual tasks of high contrast or large size	30
b. Performance of visual tasks of medium contrast or small size	75
c. Performance of visual tasks of low contrast and very small size over a prolonged period	150

*Because of the variability of actual conditions, the final design illumination values will often be 50 percent larger or smaller than these guideline values. Precise values are not appropriate because of the large tolerance of human vision, and because the quality of the light determines whether more or less light is required. These values can be reduced by 25 percent if the quality of the lighting is very high and they should be increased 35 percent if the average age is over forty. This table is adapted from IESNA tables for recommended illumination levels.

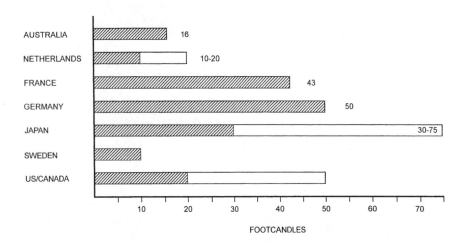

Figure 12.9 A comparison of recommended lighting levels for general office work in horizontal footcandles. Note that some countries recommend ranges instead of a specific value. (After Mills and Borg. "Rethinking Light Levels." IAEEL Newsletter, 7 (20), 4-7 (1998).)

The ASHRAE Standard 90-75, which has been widely accepted as an energy code, makes the following recommendations about lighting:

1. Task lighting should be consistent with the IESNA recommendations.
2. General area lighting should be one-third of task lighting.
3. Noncritical circulation lighting should be one-third of general area lighting.

For example, in an office the task lighting might be 75 footcandles; the general area lighting, 25 footcandles; and the corridor, 8 footcandles.

There is nothing absolute about these recommended light levels, as can be seen by the threefold increase in recommended light levels in the last seventy years. Also, consider the very large discrepancy in what is recommended in various industrialized countries (Fig. 12.9). More important than quantity is the quality of the light. The following discussion will explain some of the critical aspects of the quality of light.

12.10 BRIGHTNESS RATIOS

Although the eye can adapt to large variations in brightness, it cannot adapt to two very different brightness levels simultaneously. This problem can be easily visualized by looking at photographs of a building entrance. In Fig. 12.10a, the camera was set to correctly expose the exterior, and consequently the view of the interior is too dark to see. On the other hand, in Fig. 12.10b, we see a picture where the camera was set to correctly expose the interior, and the result is that the outdoors is too bright to see. There is no way that the camera itself can overcome the problem that the brightness ratio between indoors and outdoors is too great. Professional photographers usually wait until early evening when the indoor and outdoor brightnesses are equal. Another

require it. This nonuniform approach to lighting is called **task lighting.**

The Illuminating Engineering Society of North America (IESNA) publishes recommended illumination levels for various activities. These values are based on such factors as task activity, occupant age, required speed, required accuracy, and room-surface reflection factors (dark finishes require more light). At the schematic design stage, however, only a very rough approximation of illumination levels is required for determining lighting strategies and for model studies. Table 12.9 gives some very approximate guidelines for illumination levels appropriate for various activities. Unless otherwise specified, illumination levels are always given for horizontal work surfaces, and most tasks are performed on tables or desks that are about 2½ feet high.

option is to greatly increase the indoor illumination.

Although the eye can minimize this problem by concentrating on one brightness area at a time, all brightness areas in the field of view have some impact. The result of too high a brightness ratio is visual stress. If the eye keeps switching back and forth between areas of very different brightness, the additional stress of constant readaptation will also be present.

Lighting designers can avoid these sources of visual stress by controlling the brightness ratios in the field of view. Designers can accomplish this by adjusting reflectance factors as well as the illumination of surfaces, since brightness is a function of both. The eye is most sensitive to brightness ratios near the center of vision, and least sensitive at the edge of peripheral vision. Consequently, the acceptable brightness ratios depend on the part of the field of view that is affected. For good visual performance such as that required in an office, the brightness ratios should be kept within the limits shown in Table 12.10.

The first step in designing brightness ratios is to choose the reflectance factors of all large surfaces. In work areas, such as offices, the following minimum reflectances are recommended: ceiling, 70 percent; vertical surfaces, such as walls, 40 percent; and floors, 20 percent (Fig. 12.11h). Dark walls, especially, should be avoided. A small sample of dark wood panelling can be quite attractive, but a whole wall of it is likely to be oppressive. Additional control of brightness ratios is then achieved by selective illumination. Although the illumination on the work surface is more than adequate, the walls in Fig. 12.10c are not bright enough. Figure 12.10d shows the same room with additional illumination on the vertical surfaces.

Figure 12.10a In this photograph, the camera was adjusted to correctly expose for the high brightness of the exterior. We cannot see indoors because the brightness there is too low compared to the outdoors.

Figure 12.10b In this photograph, the camera was adjusted to correctly expose for the interior. Consequently, we cannot clearly see the outdoor view because it is too bright compared to the interior.

TABLE 12.10 RECOMMENDED BRIGHTNESS RATIOS FOR INDOOR LIGHTING FOR MAXIMUM PRODUCTIVITY*

Ratio	Areas	Example
3:1	Task to immediate surroundings	Book to desk top
5:1	Task to general surroundings	Book to nearby partitions
10:1	Task to remote surroundings	Book to remote wall
20:1	Light source to large adjacent area	Window to adjacent wall

*For high visual performance in a normal work area, these brightness ratios should not be greatly exceeded. However, uniform brightness is not desirable either. The task should be slightly brighter than the immediate surroundings to avoid distraction. This table does not apply in situations in which dramatic highlighting, mood lighting, or aesthetic concerns should be dominant.

343

Figure 12.10c Although this room has more than enough illumination on the horizontal work surface, it appears dark because of the low brightness of the vertical surfaces. (Photograph by James Benya.)

Figure 12.10d Additional illumination on the vertical surfaces makes this room appear as well illuminated as the table actually is. (Photograph by James Benya.)

12.11 GLARE

Glare is "visual noise" that interferes with visual performance. Two kinds of glare exist, direct and reflected, and each can have very detrimental effects on the ability to see.

Direct Glare

Direct glare is caused by a light source sufficiently bright to cause annoyance, discomfort, or loss in visual performance. It is called **discomfort glare** when it produces physical discomfort, and **disability**

glare when it reduces visual performance and visibility. The severity of the glare that a light source causes is in large part due to its brightness. Both absolute brightness and apparent brightness produce glare. High-beam headlights can be blinding at night, but hardly noticeable during the day. Similarly, a bare lamp against a black ceiling causes much more glare than the same lamp seen against a white ceiling. This is one of several reasons why ceilings should usually be white.

Direct glare is also a consequence of geometry. The closer an offending light source is to the center of vision, the worse the glare. For this reason, windows are often a serious source of glare (Fig. 12.11a). Of all the lights in Figure 12.11a, lamp "C" is closest to the center of vision and, therefore, a serious source of glare, while lamp "A" is not a source of glare at all because it is completely outside the field of view for the observer at the location shown. Geometry also affects the exposure angle, which is a function of both size and proximity. Large exposure angles from large or close light sources also result in more glare than small exposure angles from small or more distant light sources. Thus, glare changes greatly from point to point in a room.

Since ceiling-mounted lighting fixtures are a likely source of direct glare, much research has been conducted to quantify and reduce the glare these fixtures cause. Eggcrates, parabolic louvers, lenses, and diffusers are commonly used to minimize glare from lighting fixtures (Fig. 12.11b). These optical controllers eliminate or reduce the light emitted in the direct-glare zone (see Fig. 12.11c). Note how the direct-glare zone of the light source (45 degrees below the horizontal) corresponds to the direct-glare zone of the viewer (45 degrees above the horizontal). Because indirect lighting uses the ceiling as a large-area, low-brightness

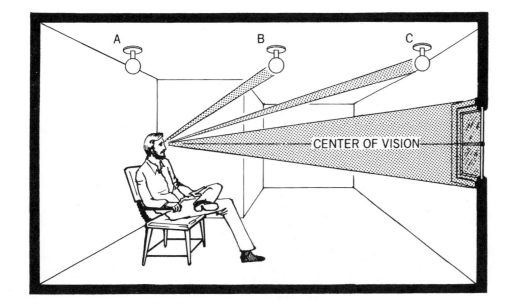

Figure 12.11a Light sources near the center of vision cause more direct glare than those at the edge of the field of view. For the person seated as shown, light "A" causes no glare at all.

reflector, it creates almost no glare at all. The impact of lighting-fixture design and the benefits of indirect light are explained in more detail in Chapter 14.

Since lighting fixtures vary greatly in the amount of direct glare that they produce, the concept of **visual comfort probability (VCP)** was developed to compare fixtures' potential to cause direct glare. The VCP factor predicts the percentage of people who will find a specific lighting system acceptable with regard to visual comfort. Indirect lighting fixtures, because of their low brightness, come close to the maximum of 100 percent.

We must not forget that lighting design is not just a problem in physics, but also a problem in human perception. The same light source that creates glare in an office might create sparkle in a nightclub. What is noise in one situation can be an information signal in another.

Indirect Glare and Veiling Reflections

Reflections of light sources on glossy table tops or polished floors cause a problem similar to direct glare (Fig. 12.10b). This **indirect glare** is often

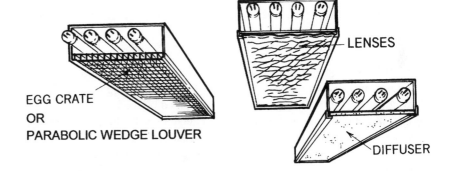

Figure 12.11b Eggcrates, parabolic louvers, baffles, and lenses limit direct glare by controlling the direction of the emitted light. Diffusers limit glare by reducing the brightness of the light source.

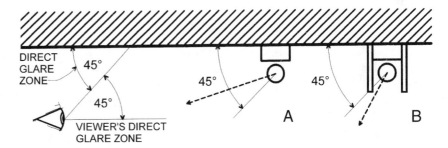

Figure 12.11c Lighting fixture "A" causes glare because light is emitted within the direct-glare zone, while the shielded fixture "B" does not cause direct glare. Note how the fixture's and viewer's direct-glare zones correspond.

best avoided by specifying flat or matte finishes. However, when the task has a glossy surface, the lighting system has to be designed to avoid producing this reflected glare.

The reflections of bright light sources on such tasks as glossy printed pages are known as **veiling reflections** because they reduce the contrast necessary for good visual performance

Figure 12.11d Veiling reflections impair the visual performance of a task by reducing contrast.

(Fig. 12.11d). Veiling reflections are specular, or mirror-like, reflections that are most severe on very smooth materials, but exist to a lesser degree also on semigloss and even matte surfaces. Pencil marks and some inks quickly disappear under veiling reflections because of their glossy finish.

Veiling reflections are at a maximum when the angle of incidence, established by the light source, equals the angle of reflection, established by the location of the eye (Fig. 12.11e). Most people seated at a desk will do their reading and writing in a zone from 25 degrees to 40 degrees mea-

sured from the vertical. Any glossy material in this zone will reflect light from a corresponding zone in the ceiling (Fig. 12.11f). Any light source in this **offending zone** of the ceiling will be a cause of veiling reflections. In an existing lighting design, this offending zone is easy to spot simply by substituting a mirror for the visual task (Fig. 12.11g).

Veiling reflections are the most serious problem that the lighting designer faces. It is a problem not only for people working behind a desk, but also for others, such as workers handling smooth parts, people viewing artwork through glass, etc. The problem is getting even more serious with the growing use of computers or video display terminals (VDT). The avoidance of glare and veiling reflections is the top priority in many lighting designs. Figure 12.11h illustrates the sources of veiling reflections and direct glare. The diagram also shows common reflectances that will produce acceptable brightness ratios in the field of view. Much of the material presented in the next two chapters on lighting explains how to manage the serious problems of glare and veiling reflections.

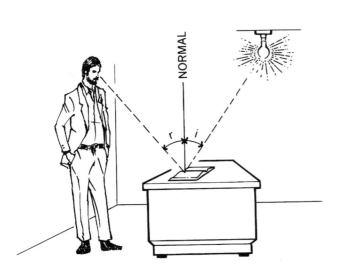

Figure 12.11e Veiling reflections are at a maximum when the angle of incidence (i) equals the angle of reflection (r).

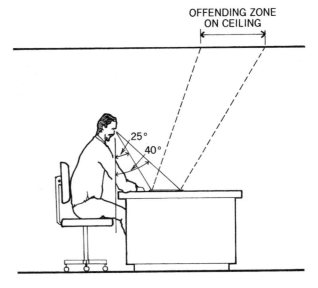

Figure 12.11f Any light source in the offending zone can create severe veiling reflections for a person working at a table or desk. The offending zone will shift if the task or table is tilted (e.g., drafting table).

Figure 12.11g By replacing the task with a mirror, any bright light source in the offending zone will be visible. Hold the mirror vertically if the task is in a vertical plane (e.g., a painting or computer monitor).

Figure 12.11h Sources of direct glare and veiling reflections are shown. Common ranges for reflectance factors are also shown. (Courtesy of General Electric Lighting.)

12.12 EQUIVALENT SPHERICAL ILLUMINATION

Veiling reflections are so detrimental to visual performance that increased lighting at the wrong angle can actually reduce our ability to see. Clearly then, the quality of the light is just as important as the quantity. Ordinary "raw" footcandles of illumination as measured by a photometer can be quite meaningless if the geometry of the lighting is not included. To correct this serious deficiency, the concept of **equivalent spherical illumination (ESI)** was developed.

Sphere illumination is a standard reference condition with which the actual illumination can be compared. In sphere illumination, the task receives light from a uniformly illuminated hemisphere (Fig. l0.12a). Since the task is illuminated from all directions, only a small amount of the total light will cause veiling reflections. Although spherical illumination is of high quality, it could be improved by eliminating that portion of the light causing veiling reflections. Sphere illumination is a valuable concept not because it represents the best possible lighting, but because it is a very good reproducible standard with which any actual lighting system can be compared.

An actual lighting system that supplied an illumination of 250 foot-candles might be no better than an equivalent spherical illumination of 50 ESI footcandles. This means that the quality of the actual system is so poor that 200 out of 250 footcandles are noneffective. Therefore, the ESI footcandles can tell use how effective the "raw" footcandles are. Equivalent spherical illumination enables us to describe the quality as well as the quantity of the illumination.

In the lighting layout plan shown in Fig. 12.12b, we can see that the quality of the lighting varies greatly with location and that "raw" footcandles are not a good indication of visual performance. Notice that if only "raw" footcandles were considered, location "C"

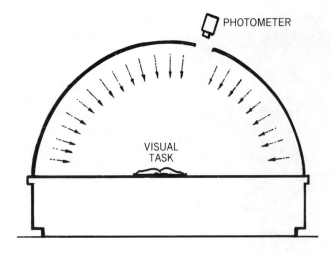

Figure 12.12a Test chamber for measuring equivalent spherical illumination (ESI).

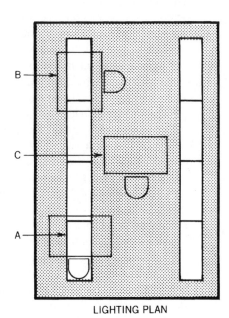

LIGHTING PLAN

	DESK		
	A	B	C
RAW FOOTCANDLES	100	120	90
ESI FOOTCANDLES	17	30	90

Figure 12.12b A comparison of "raw" and ESI footcandles for three different locations clearly shows that desk "C" has the best lighting. Although desk "B" has the highest illumination level, the veiling reflections neutralize most of the footcandles. Desk "A" has the worst lighting of all.

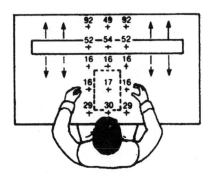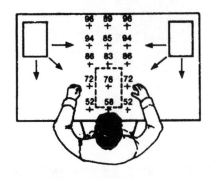

Figure 12.13a The ESI footcandles are quite low for the lighting fixture in front of the task, because of the veiling reflections. With fixtures on each side, there will be almost no veiling reflections, and consequently the ESI footcandles are much higher although the total wattage is lower (Left) One 40-W lamp; (Right) two 14-W lamps. (Courtesy of Cooper Lighting.)

would be the worst choice for the desk because it had the lowest illumination level. In fact, location "C" is by far the best, as the very high ESI footcandle level indicates. Locations "A" and "B" will both experience serious veiling reflections from the overhead lighting fixtures. This situation can be visualized via the mirror test (Fig. 12.11g). Imagine a mirror placed on each desk in place of the task. The person occupying desk "A" will see the most lighting fixture in the offending zone, while the person at desk "C" will see only the ceiling in the offending zone.

Since quality compensates for quantity, the lighting levels that the IESNA recommends can be reduced by 25 percent if veiling reflections are largely avoided. Recommended light levels are still given in raw" footcandles because at this time it is still very difficult and expensive to measure ESI footcandles.

12.13 ACTIVITY NEEDS

The requirements for good visual performance, mentioned so far, apply to most visual tasks. However, additional requirements vary with the specific visual task.

1. *Reading and writing:* The avoidance of veiling reflections has the highest priority for the activities of reading and writing. Light should, therefore, come from the sides or from behind but never from in front of the observer. Notice in Fig. 12.13a how much higher the ESI footcandles are on a desk when the light comes from the sides rather than from in front. The light should come from at least two sources to prevent workers from casting shadows on their own task.

2. *Drafting:* Because of the somewhat glossy finish of drafting films, pencil lines and ink, veiling reflections are a major problem. Shadows from drafting instruments also obscure the work. A very diffuse lighting mainly from the sides and back is

appropriate on both accounts.

3. *Observing sculpture:* Shades and shadows are necessary in understanding the three-dimensional form of an object. The appropriate lighting should, therefore, have a strong directional component (Fig. l0.13b). Unless there is some diffused light, however, the shadows will be so dark that many details will be obscured (Fig. l0.13c), but a completely diffused lighting is not appropriate either because it makes objects appear flat and three-dimensional details will tend to disappear (Fig. 12.13d). Usually, the directional light should come from above and slightly to one side because that is the way the sun generally illuminates objects, and we are used to that kind of modeling. Seeing familiar objects like the human face lit from below can be a very eerie experience.

4. *Seeing texture:* The appearance of texture depends on the pattern created by shades and shadows. A texture is, therefore, made most visible by glancing light that maximizes the shades and shadows (Fig. 12.13e). The same material seen under diffused or straight-on lighting will appear to have almost no texture (Fig. 12.13f). Glancing light can also be used to investigate surface imperfections, and, con-

Figure 12.13b The best modeling occurs with strong directional light along with some diffused light to soften the shades and shadows.

Figure 12.13c If all the light comes from one direction, strong shadows and shade will obscure much of the object.

Figure 12.13d An object will appear flat under completely diffused light.

Figure 12.13e Texture is most visible under glancing light. Note the long shadow of the pushpin in the upper right corner.

Figure 12.13f The same texture seen under straight-on or diffused light. Note the lack of shadow from the pushpin.

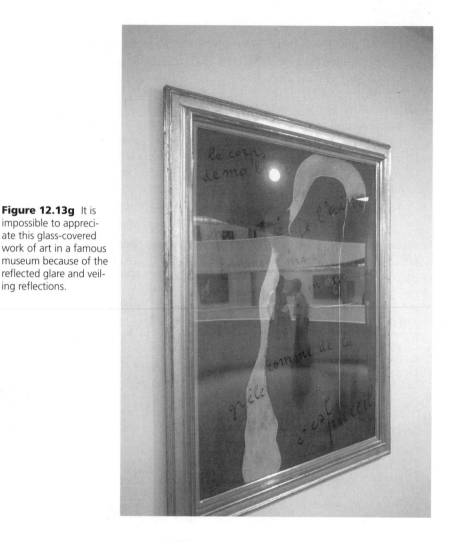

Figure 12.13g It is impossible to appreciate this glass-covered work of art in a famous museum because of the reflected glare and veiling reflections.

versely, glancing light should be avoided if surface imperfections are to be hidden.

5. *Looking at paintings:* When one highlights paintings or graphics that have a glossy or a semigloss finish, the challenge is to prevent specular reflections of the light sources into the viewers' eyes. Many a fine print protected by glass has become invisible because of veiling reflections (Fig. 12.13g). The accent light must be placed in front of the offending zone so that people of various heights and different locations will not see the specular reflection of the light source (Fig. 12.13h). However, if the light is too close to the wall above the painting, the top of the picture frame will cast a shadow and the texture of the painting will be emphasized. In most cases, a 60-degree aiming angle will be a good compromise (Fig. 12.13i).

When the artwork is covered with glass, special anti-reflectivity glass, which makes the lighting much less critical, can be specified (Fig. 12.13j).

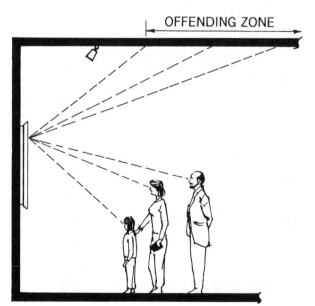

Figure 12.13h Accent lighting must be placed in front of the offending zone.

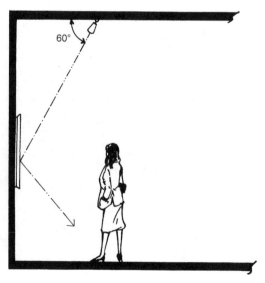

Figure 12.13i Under normal conditions, a 60-degree fixture aiming angle is quite satisfactory.

Figure 12.13j When veiling reflections on glazing cannot be controlled (right glazing), special glass with an anti-reflective coating is available (left glazing). (Courtesy Schott Co.)

6. *Computer Monitors:* The tasks of looking at computer monitors and VDTs is becoming critical. The glossy surface and vertical format of the screen makes veiling reflections a major problem. Avoid bright light sources or bright surfaces behind the operator (Fig. 12.12k). If it is not possible to eliminate these offending light sources, place a partition behind the operator. Indirect lighting from large areas of ceiling and walls will work quite well as does direct, almost vertical lighting from the ceiling. Such lighting will have a Visual Comfort Probability (VCP) close to 100 percent. Computer monitors can be purchased with low-reflectance glass or retrofitted with a low-reflectance screen. Either approach makes the lighting less critical.

12.14 BIOLOGICAL NEEDS

A good lighting design must address not only the previously mentioned requirements for visual performance, but also the biological needs of all

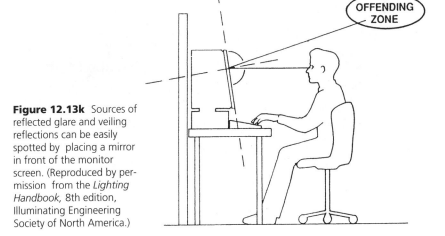

Figure 12.13k Sources of reflected glare and veiling reflections can be easily spotted by placing a mirror in front of the monitor screen. (Reproduced by permission from the *Lighting Handbook,* 8th edition, Illuminating Engineering Society of North America.)

human beings, which are independent of culture and style. These needs relate to the biological requirements of orientation, stimulation, sustenance, defense, and survival. The following list of biological nees is largely based on the book *Perceptions and Lighting* by William Lam:

1. *The need for spacial orientation:* The lighting system must help define slopes and changes of level. It must also help people know where they are and where to go. For example, an elevator

lobby or reception area might be brighter than the corridor leading to it. (Fig. 12.14a). Windows are very helpful in relating one's position inside a building to the outside world.

2. *The need for time orientation:* Jet lag is a result of internal clocks being out of synchronization with what the eyes see. The internal clock might expect darkness and the time for sleep, while the eyes experience bright sunshine. Melatonin might be produced or

Figure 12.14a An elevator lobby or reception area can become the focus for direction by making it brighter than the corridor leading to it. (Photograph courtesy of Hubbell/Lighting Division.)

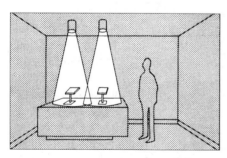

Figure 12.14b Important areas can be highlighted, while less important areas can receive subdued illumination. (From *Architectural Graphic Standards,* Ramsey/Sleeper, 8th ed. John R. Hoke, editor, © John Wiley & Sons. Inc., 1988.)

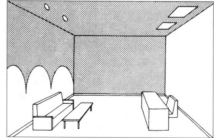

Figure 12.14c Changing light levels can help in defining personal space. (From Architectural Graphic Standards, Ramsey/Sleeper, 8th ed. John R. Hoke, editor, © John Wiley & Sons, Inc., 1988.)

suppressed at the wrong times during a person's circadian rhythm. The least stress occurs when the eyes see what the internal clocks expect. For example, views of the exterior through clear glazing give people the feedback on the progress of time that their internal clocks seem to need.

3. *The need to understand structural form:* The need to understand the physical world is frustrated by lighting that contradicts the physical reality, by excessive darkness or by excessively diffuse lighting. Directional light gives form to objects, while diffuse light tends to flatten their appearance. The sculpture in Fig. 10.13b is well modeled by the mostly directional light, while it loses its three-dimensional quality when illuminated by completely diffuse lighting (Fig. 12.13d). Fog and luminous ceilings both create this excessively diffuse type of lighting.

4. *The need to focus on activities:* To prevent information overload, the brain has to focus its attention on the most important aspects of its environment and largely ignore the rest. The lighting can help by creating order and by highlighting the areas and activities that are most relevant. Low illumination for the less important areas is just as important as highlighting (Fig. 12.14b).

5. *The need for personal space:* Light and dark areas in a large room can help define the personal space of each individual (Fig. 12.14c). Uniform lighting tends to reduce individuality, while local or furniture-integrated lighting emphasizes personal territory. People appreciate the ability to control their own environment. Personal lighting fixtures that can be adjusted are an easy way to satisfy this need for control.

6. *The need for cheerful spaces:* Dark walls and ceilings create a cave-like atmosphere (see Fig. 12.10c).

This gloom can be caused by specifying dark surfaces, low illumination levels, or excessively vertical light. However, a dark restaurant with candlelight is not gloomy because we expect it to be dark. A space is, therefore, gloomy only if we expect it to be bright and it is not. Small patches of sunlight can be especially welcome in the winter. Direct sunlight, however, must be kept off the task to prevent excessive brightness ratios at the center of vision. Gloom can also be created without dark surfaces. Most people find the lighting from an overcast rainy day to be quite gloomy. An all-indirect-lighting scheme, as shown in Fig. 12.14d, can create this same dull, dreary appearance. Instead, a combination of direct, indirect, and accent lights creates a most interesting and cheerful design (Fig. 12.14e).

7. *The need for interesting visual input:* Dull spaces are not made interesting just by increasing the light levels. A very barren space might be interesting for a short period, when it is first perceived, but it will not remain interesting long. Furthermore, there is a need to occasionally look up from one's work and to scan the environment. Interesting objects, such as windows, people, paintings, sculpture, and plants, can act as visual rest centers. Viewing distant objects allows the eye muscles to relax.

8. *The need for order in the visual environment:* When order is expected but not present, we perceive chaos. For example, when the lighting fixtures in the ceiling have no relationship with the structure, we find the design disturbing (Fig. 12.14f).

9. *The need for security:* Darkness is a lack of visual information. In a situation in which we expect danger, this lack of information causes fear. Dark alleys, dark corners,

Figure 12.14d An all indirect lighting scheme creates a feeling of gloom. (Photograph by James Benya.)

Figure 12.14e A cheerful and interesting lighting design is achieved by the combination of direct, indirect, and accent lights. (Photograph by James Benya.)

Figure 12.14f When the lighting fixture pattern is not in harmony with the structure, the need for order is frustrated. (From *Architectural Graphic Standards,* Ramsey/Sleeper, 8th ed. John R. Hoke, editor, © John Wiley & Sons, Inc., 1988.)

Figure 12.14g Light-colored buildings can be a source of gentle, diffused area lighting at night. (Courtesy Spaulding Lighting, Inc.)

and shadows from trees are best eliminated by numerous closely spaced street lights and not by a few very bright lights. Light-colored buildings help greatly by reflecting diffuse light into dark corners (Fig. 12.14g).

Most lighting systems that satisfy these biological needs automatically also satisfy the needs of the visual tasks mentioned before.

12.15 LIGHT AND HEALTH

In northern latitudes where winter days are short, depression is more common in the winter than in the summer. Dr. Alfred J. Lewy discovered that light therapy could help some patients who become depressed during the short winter days. This illness is now called **Seasonal Affective Disorder (SAD).** Recent research has shown that bright light (more than 150 footcandles) through the eyes will cause the pineal gland in the brain to stop making melatonin, which is produced whenever people are in the dark. High melatonin levels cause drowsiness, while low levels produce

alertness; thus, melatonin plays a critical part in our circadian cycles.

Many industrial accidents occur during the night shift. It is now believed that many of these accidents, as well as jet lag, result from drowsiness caused by activity at a time when our internal clocks tell us to be sleeping. Research in light therapy shows promising developments in several areas: fighting SAD depression, making people more alert at night, regulating sleep cycles for older people, and alleviating the problem of jet lag. Research also suggests that periods of high light levels are needed in the morning to suppress the previous night's melatonin production. The easiest way for architects to supply this light is with views of the bright outdoors and with the introduction of high levels of daylight (about 250 footcandles).

Other photobiological effects of light exist besides melatonin suppression. Excessive ultraviolet (UV) radiation can cause serious burns to skin and eyes, blindness, and, in some cases, skin cancer. For this reason, most lamps sold in the United States have glass covers that absorb most of

the damaging UV. However, excessive visible light and especially blue and violet light can also cause serious eye and skin burns. Light also plays a role in a variety of body functions, and it is used in the treatment of such diseases as hyperbilirubinemia, psoriasis, and Vitamin D deficiency (rickets).

A few people seem to get headaches from the 120 flashes per second that fluorescent lights operating with magnetic ballasts create (see Chapter 14). This problem is easily solved by switching to electronic ballasts, which have the additional benefit of increased efficiency.

12.16 THE POETRY OF LIGHT

The previous objective discussion of lighting principles was both necessary and useful. A full understanding of lighting also requires a poetic perspective. Richard Kelly, who was one of the foremost lighting designers, fully understood the role of poetry in design conceptualization. His own words say it best:

> In dealing with our visual environment, the psychological sensations can be broken down into three elements of visual design. They are **focal glow, ambient luminescence** and the **play of brilliants.**
>
> **Focal glow** is the campfire of all time...the welcoming gleam of the open door...the sunburst through the clouds....The attraction of the focal glow commands attention and creates interest. It fixes gaze, concentrates the mind, and tells people what to look at. It separates the important from the unimportant....
>
> **Ambient luminescence** is a snowy morning in open country. It is twilight haze on a mountaintop...a cloudy day on the ocean...a white tent at high noon....It fills people with a sense of freedom of space and can suggest infinity....

The background of ambient luminescence is created at night by fixtures that throw light to walls, curtains, screens, ceilings and over floors for indirect reflection from these surfaces.

Play of brilliants is the aurora borealis...Play of brilliants is Times Square at night....It is sunlight on a tumbling brook....It is a birch tree interlaced by a motor car's headlights. Play of brilliants is the magic of the Christmas tree...the fantasy excitement of carnival lights and restrained gaiety of Japanese lanterns....A play of brilliants excites the optic nerves, stimulates the body and spirit, and charms the senses...

12.17 RULES FOR LIGHTING DESIGN

The following rules are for general lighting principles. Specific rules for electric lighting and daylighting will be given in the next two chapters.

Figure 12.17a Direct glare from bright lighting fixtures is not a problem with an indirect lighting system. The whole ceiling becomes a low brightness source. (Courtesy, © Peerless Lighting Corporation.)

1. First, establish the lighting program by fully determining what the seeing task is in each space. For example, is the illumination mainly for vertical or horizontal surfaces? Are colors very important? Does the task consist of very fine print? Will day-lighting be used to reduce the need for electric lighting?

2. Illuminate those things that we want or need to see. Since this usually includes the walls and some furnishings, the light reflected from these surfaces can supply much of the required illumination. Except for decorative light fixtures, such as chandeliers, we usually want to see objects and not light sources.

3. Quality lighting is largely a problem of geometry. Direct glare and veiling reflections are avoided mainly by manipulating the geometry between the viewer and the light source. The main light source should never be in front of the viewer. Glare can also be prevented by baffling the light sources from normal viewing angles. Baffles can be louvers, eggcrates, or parts of a building. With indirect lighting, the ceiling becomes a large-area, low-brightness source with minimal glare and veiling reflections (Fig. 12.17a).

4. In most situations, the best lighting consists of a combination of direct and diffuse light. The resulting soft shadows and shading enable us to fully understand the three-dimensional quality of our world.

5. "Darkness is as important as light: it is the counterpoint of light—each complements the other" (Hopkinson). Avoid, however, very large brightness ratios that force the eye to readapt continually.

6. An object or area can be highlighted by either increasing its brightness or by reducing the brightness of the immediate surroundings. The absolute brightness matters little. What does matter, however, is that the brightness ratio should be about 10 to 1 (Fig. 12.17b) for the purpose of highlighting.

7. Paint is one of the most powerful lighting tools. In most cases, light colors are desirable, and with indirect lighting, the most reflective white is almost mandatory. This is one of the most economical lighting tools; usually it costs nothing because paint or some finish is specified anyway. Dark colors should be considered only when drama rather than performance of visual tasks is the goal for the lighting design. Important examples of places where drama is the goal are museums and the-

Figure 12.17b This artwork is highlighted by reducing the background brightness. MIT Chapel by Eero Saarinen.

9. Flexibility and quality are more important than the quantity of light. Generally, illumination greater than 30 footcandles can be justified only for small areas where difficult visual tasks are performed (task lighting) or for the elderly.

12.18 CAREER POSSIBILITIES

Because lighting is so important and complex, the new profession of lighting designers has emerged. Since the emphasis is on both design and aesthetics, architects have a good appreciation and background for this field. With some additional education and experience, an architect can become a lighting consultant, working on a wide variety of projects with some of the world's most prominent architects and interior designers. For more information, contact the International Association of Lighting Designers (IALD) (see Appendix J). For a Master's degree in lighting, consider the Lighting Research Center at Rensselaer Polytechnic Institute (see Appendix I).

Since 1998, the National Council for the Qualifications for the Lighting Professions (NCQLP) has administered a national certification test for lighting designers. Architects qualify for taking the test that leads to the title "Lighting Certified" (LC). See www.ncqlp.org for information about the test.

aters; here the highlighting of objects or the stage, respectively, draws the viewer's attention. Dark paint is often used to hide the clutter of pipes, ducts, and beams where a suspended ceiling is not possible or desirable. In such a case, direct lighting that does not depend on reflections off the ceiling should be used. Remember, you cannot make a room with dark finishes look light.

8. Use daylighting wherever possible. Most people prefer the quality and variety of daylight. They especially desire and need the views that often accompanies daylighting. Eye muscles are released only when we look at a distant object. There is evidence that both health and productivity benefit from the use of daylight in buildings. Recent research has shown that children performed better in daylit rather than electrically lit classrooms.

12.19 CONCLUSION

The importance of quality over quantity cannot be overemphasized because for too long the general opinion has been "more is better." It is a bit strange that this attitude was so widely held because we do not hold this view in regard to the other senses. We do not appreciate sound according to its loudness. The difference between noise and music is certainly not its amplitude. We do not appreciate touch by its hardness. And we do not appreciate smell or taste by its strength. In each case, a minimum level is required, but above that it is quality and not quantity that counts. Our sense of sight is no different in this regard.

Ignoring quality has always been at the expense of visual performance. Often, the detrimental effects are not even recognized. There is, however, a very dramatic example in which the impairment to visual performance could not be ignored. The Houston Astrodome was built with translucent plastic bubbles over a very interesting steel structure, as seen in Fig. 12.18a. The illumination level was high enough indoors to allow grass to grow on the playing field. However, high-flying balls could not be seen against the visual "noise" of the structure. The problem was solved by painting over the skylights and using electric lights even during the day. Since the grass then died, it was replaced by "Astro-turf." This disaster arose because lighting was considered only as a problem in quantity and not as a problem of quality that must be integrated with the architecture. A pneumatic structure, on the other hand,

Figure 12.18a The structure and skylights of the Houston Astrodome had created an interesting visual pattern, which, unfortunately, became very strong visual "noise" when people tried to see a high-flying ball.

solves the lighting problem in a well-integrated manner (Fig. 12.18b). It creates a neutral background for the ball, it allows soft daylight to enter, and at night it reflects low-brightness light from electric light sources.

To best satisfy all the biological and activities needs, one must usually integrate daylighting and electric

lighting. With the basic concepts of this chapter and the more specific information of the next two chapters on daylighting and electric lighting, designers should be able to design a high-quality-lighting environment that will satisfy the environmental, biological, psychological, and activity needs of the occupants.

Figure 12.18b Pneumatic structures with translucent membranes are an example of well-integrated designs. These structures provide daylight without visual "noise," and at night work well with indirect lighting. (Courtesy of Tensar Structures, Inc.)

KEY IDEAS OF CHAPTER 12

1. The rate at which light is emitted from a source is measured in **lumens,** while the intensity in any direction is called **candlepower** and is measured in **candelas.**

2. Illumination (illuminance) is measured in **footcandles** and is equal to lumens per square foot (SI = lux).

3. **Luminance** is measured brightness, and its units are candelas per square foot. (SI = candelas per square meter).

4. The reflectance factor (reflectance) describes how much of the light falling on an object is reflected.

5. Light can be reflected in a specular, diffuse, or mixed fashion.

6. The color of an opaque object is the result of both the spectrally selective reflections of the object and the spectral composition of the light source.

7. The consistency of white light can be described by color temperature, the color-rendering index (CRI), or by a spectral-energy distribution diagram.

8. The iris controls the size of the pupil to adjust the amount of light falling on the retina.

9. Lighting should be designed for what people perceive and not what meters measure.

10. The performance of a visual task is a nonlinear function of brightness (i.e., a little light is very helpful but a lot of light is only slightly better.

11. Use IESNA (Illuminating Engineering Society of North America) recommendations for task lighting, one-third of those levels for general illumination, and only one-third of those levels for circulation areas (i.e. one-ninth of task levels). Use somewhat higher levels for daylighting in summer, and use much higher levels for daylighting in winter.

12. Rules for quality lighting
    a. Control brightness ratios
    b. Avoid glare
    c. Avoid veiling reflections.

13. Don't forget the poetry of light
    a. Focal glow
    b. Ambient luminescence
    c. Play of brilliants.

14. Rules for lighting design
    a. Determine the visual task
    b. Illuminate the things that we want or need to see
    c. Quality lighting is largely a problem of geometry
    d. Darkness is as important as light
    e. Use light-color finishes when ever possible
    f. Use efficient electric lighting
    g. Use daylighting wherever possible
    h. Flexibility and quality are much more important than the quantity of light above the minimum task-illumination level.

Resources

FURTHER READING

(See Bibliography in back of book for full citations. The list includes valuable out-of-print books.)

Brown, G.Z., and M. DeKay. *Sun, Wind, and Light: Architectural Design Strategies.*

Egan, D. *Concepts in Architectural Lighting.*

Flynn, J. E. *Architectural Interior Systems: Lighting, Acoustics, and Air-Conditioning.*

Futagawa, Y., ed. *Light and Space: Modern Architecture.*

Gordon, G., and J. L. Nuckolls. *Interior Lighting for Designers.*

Grosslight, J. *Light, Light, Light: Effective Use of Daylight and Electric Lighting in Residential and Commercial Spaces.*

IESNA. *Lighting Handbook.*

Lam, W. H. *Perception and Lighting as Formgivers for Architecture.*

Millet, M. S. *Light Revealing Architecture.*

Moore, F. *Environmental Control Systems.*

Schiler, M. *Simplified Design of Building Lighting.*

Steffy, G. R. *Architectural Lighting Design.*

Steffy, G.R. *Time-Saver Standards for Architectural Lighting.*

Stein, B., and J. S. Reynolds. *Mechanical and Electrical Equipment for Buildings.* A basic resource.

ORGANIZATIONS

(See Appendix J for full citations.)

IESNA Illuminating Engineering Society of North America

International Association of Lighting Designers (IALD). www.iald.org

LBNL (Lawrence Berkeley National Laboratory) http://eande.lbl.gov/BTP/BTP.html

Lighting Design Lab www.northwestlighting.com

Lighting Research Center–Rensselaer Polytechnic Institute www.erc.rpi.edu

Information is available on light and health, lighting publications, and free online lighting-technology publications.

DAYLIGHTING

"We were born of light. The seasons are felt through light. We only know the world as it is evoked by light.... To me natural light is the only light, because it has mood — it provides a ground of common agreement for man — it puts us in touch with the eternal. Natural light is the only light that makes architecture architecture."

Louis I. Kahn

13.1 HISTORY OF DAYLIGHTING

Until the second half of the twentieth century when fluorescent lighting and cheap electricity became available, the history of daylighting and the history of architecture were one. From the Roman groin vault (Fig. 1.3c), to the Crystal Palace of the nineteenth century, the major structural changes in buildings reflected the goal of increasing the amount of light that entered. Because artificial lighting had been both poor and expensive until then, buildings had to make full use of daylight.

Gothic architecture was primarily a result of the quest for maximum window area. Only small windows

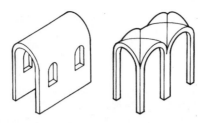

Figure 13.1a Few windows were possible in the massive bearing walls required to support the Romanesque barrel vault. (From *Architectural Lighting*, 1987. © Architectural Lighting. Reprinted Courtesy of Cassandra Publishing Corporation, Eugene, OR.)

were possible when a barrel vault rested on a bearing wall. The Roman groin vault supplanted the barrel vault partly because it allowed large windows in the vaulted spaces (Fig. 13.1a). Gothic groin vaulting with flying buttresses provided a skeleton construction that permitted the use of very large windows (Fig. 13.1b).

Large and numerous windows were a dominant characteristic of Renaissance architecture. Windows dominated the facade, especially in regions with cloudy climates such as England. The increase in window size was so striking that one English manor was immortalized in rhyme: "Hardwick Hall, more window than wall" (Fig. 13.1c). Bay windows, too, became very popular (see Fig. 16.2d). Although the facades of such Renaissance palaces were designed to give the impression of great massive structures, their "E"- and "H"-shaped floor plans provided for their ventilation and daylight requirements. As a matter of fact, such shapes were typical of floor plans for most large buildings until the twentieth century (Fig. 13.1d). These buildings or their wings were rarely more than 60 feet deep so that no point would be more than 30 feet from a window. High ceilings with high windows allowed daylight to reach about 30 feet in from the exterior walls while today's lower ceiling allow daylight to reach only about 15 feet.

During the nineteenth century, all glass buildings became possible because of the increased availability of glass combined with the new ways of using iron for structures. The Crystal Palace by Paxton is the most famous example (Fig. 4.2d).

More modest amounts of glass and iron could be found in many buildings of the day. The Bradbury Building, designed around a glass-covered atrium, is a precursor for many of today's office buildings (Fig. 13.1e).

In older neighborhoods of many cities, such as New York, it is still possible to find sidewalks paved with glass blocks that allow daylight to enter basements. New York City enacted zoning codes to ensure minimum levels of daylighting. In England, laws that tried to ensure access to daylight date back as far as the year 1189.

Figure 13.1b Groin vaulting and flying buttresses allowed Gothic cathedrals to have windows where there had been walls.

Figure 13.1c Hardwick Hall, Derbyshire, England, 1597. (From *Mansions of England in Olden Times,* by Joseph Nash, Henry Sotheran & Co., 1871.)

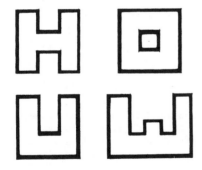

Figure 13.1d These were the common floor plans for large buildings prior to the twentieth century because of the need for light and ventilation.

Figure 13.1e The Bradbury Building, Los Angeles, 1893, has a glass-covered atrium as the circulation core. Delicate ironwork allows light to filter down to the ground level. The building was cooled by natural ventilation. Air entered exterior windows, passed through transoms and interior windows facing the atrium, and then left through hopper windows just below the skylight.

361

The masters of twentieth-century architecture have continued to use daylight for both functional and dramatic purposes. In New York's Guggenheim Museum, Frank Lloyd Wright used daylight to illuminate the artwork both with indirect light from windows and with light from an atrium covered by a glass dome (Figs. 13.1f and 13.1g). In the Johnson Wax Building in Racine, Wisconsin, he created a space with no apparent upper boundaries by letting daylight enter continuously along the upper walls and the edge of the roof. Daylight also enters through skylights around the mushroom columns (Figs. 13.1h and 13.1i).

Le Corbusier created very dramatic effects with the splayed windows and light towers of the Chapel at Ronchamp (Figs. 13.1j, 13.1k and 13.1l). Eero Saarinen used a most fascinating form of daylight in the MIT chapel. A skylight over the altar was fitted with a black eggcrate so that only vertical light could enter the chapel. This vertical light was then reflected into the room by a sculpture consisting of small brass reflectors like leaves on a tree (see Fig. 12.17b).

This short history demonstrates what an important roll daylight has had in architecture. We will look at more examples after we discuss some of the basic daylighting concepts.

Figure 13.1f The Guggenheim Museum, New York City, 1959, by Frank Lloyd Wright uses a glass-domed atrium for diffused daylighting.

Figure 13.1g Continuous strip windows bring additional daylight to the gallery space. (Courtesy of New York City Convention and Visitors Bureau, Inc.)

Figure 13.1h The Johnson Wax Administration Building, Racine, WI 1939, by Frank Lloyd Wright. Note the skylights between the mushroom columns, as well as the glazing at the junction of roof and walls. The two circular shafts (center left) are fresh-air intakes ('nostrils" as Wright called them). (Courtesy of SC Johnson Wax.)

Figure 13.1i Glazing dematerialized the upper walls and ceiling of the Johnson Wax Administration Building. (Courtesy of SC Johnson Wax.)

Figure 13.1j In Notre Dame du Haut at Ronchamp, 1955, Le Corbusier used thick walls with splayed windows, colored glass, and light scoops to bring carefully controlled light into the interior. (Photograph by William Gwin.)

Figure 13.1k Interior of Notre Dame du Haut. (Photograph-by William Gwin.)

Figure 13.1l Slit openings in the light scoops are seen in this rear view of a model built by Simon Piltzer at the University of Southern California.

13.2 WHY DAYLIGHTING?

Daylighting became a minor architectural issue as we entered the second half of the twentieth century because of the availability of efficient electric light sources; cheap, abundant electricity; and the perceived superiority of electric lighting. Perhaps the most important advantage of electric lighting was — and still is — the ease and flexibility it permitted in floor-plan design by enabling designers to ignore window locations.

Supplying adequate daylight to work areas can be quite a challenge because of the great variability in available daylight. Electric lighting is so much simpler. It offers consistent lighting that can be easily quantified. But it also has some drawbacks.

The energy crisis of the mid-1970s led to a reexamination of the potential for daylighting. At first, only the energy implications were emphasized, but now daylighting is also valued for its aesthetic possibilities and its ability to satisfy biological needs.

For most climates and many building types, daylighting can save energy. For example, a typical office building in Southern California can reduce its energy consumption 20 percent by using daylighting. Buildings such as offices, schools, and industrial facilities often devote 40 percent of their energy usage to lighting (Fig. 13.2a). In these kinds of buildings, most of the work is performed during daylight hours, which make up most of the 2,500 day-shift hours per year. Consequently, most of the lighting energy load could be fully provided by daylighting.

There is another energy-related factor in using daylight, and it is usually the more important factor. Figure 13.2b shows the rate at which energy is used in a typical office building during a sunny summer day. The horizontal axis represents time and the vertical axis describes the rate at which electricity is used. The greatest annual electrical demand usually

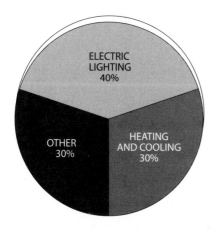

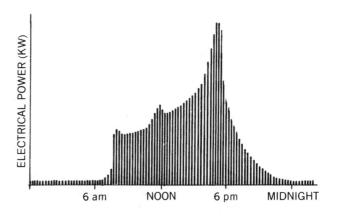

Figure 13.2a Typical distribution of energy use for buildings such as offices, schools, and many industrial facilities.

Figure 13.2b In most parts of the country, the maximum demand for electricity in an office building occurs on a hot summer afternoon when daylighting is plentiful.

occurs during sunny summer afternoons, when the air conditioning is working at full capacity. Since the sun creates this maximum cooling load, it simultaneously also supplies a maximum of daylight. Consequently, some or most of the electric lights, which consume about 40 percent of the total energy, are then not needed. The maximum demand for electrical power can, therefore, be reduced up to 40 percent by the proper utilization of daylighting. Electric power plants are, and must be, built and sized not for the total energy used but for the maximum demand. Heavy consumers of electricity are, therefore, charged not only for the total energy they use, but also for the maximum demand they make. For such users (e.g., large office buildings, schools, and factories) daylighting can significantly reduce the cost of electricity because of both the reduced energy use and the reduced demand charge. Much of the extra cost for the daylighting design can be offset by these savings. Society also benefits if the demand for electricity can be reduced because fewer electric power plants will have to be built in the future.

The dynamic nature of daylight is now seen as a virtue rather than a liability. It satisfies the biological need to respond to the natural rhythms of the day. The generally slow but occasionally dramatic changes in the quality and intensity of the natural light can be stimulating.

Even when daylighting was completely ignored, architects continued to use plenty of windows for the enjoyment of views, for visual relief, and for satisfying biological needs. The irony is that the all-glass-curtain walls were most popular in the 1960s, a time when daylighting was not utilized. Consequently, daylighting design does not require adding windows to otherwise windowless buildings. In most cases, it does not even require increasing the window area. Daylighting design does, however, require the careful design of the fenestration for the proper distribution and quality of daylighting.

13.3 THE NATURE OF DAYLIGHT

The daylight that enters a window can have several sources: direct sunlight, clear sky, clouds, or reflections from the ground and nearby buildings (Fig. 13.3a). The light from each source varies not only in quantity and heat content, but also in such qualities as color, diffuseness, and efficacy.

Although sky conditions can be infinitely variable, it is useful to understand the daylight from two extreme conditions: overcast sky and clear sky with sunlight. A daylighting design that works under both of these conditions will also work under most other sky conditions.

The brightness distribution of an overcast sky is typically three times

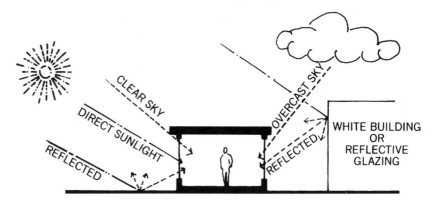

Figure 13.3a The various sources of daylight.

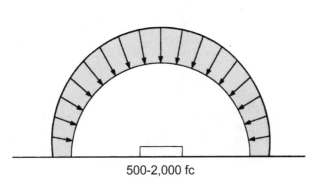

500-2,000 fc

Figure 13.3b The brightness distribution on an overcast day is typically about three times greater at the zenith than the horizon. (After *Architectural Lighting*, 1987. © *Architectural Lighting*. Reprinted courtesy of Cassandra Publishing Corporation, Eugene, OR.)

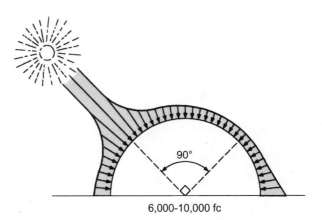

6,000-10,000 fc

Figure 13.3c The brightness distribution on a clear day is typically about ten times greater near the sun than at the darkest part of the sky. (After *Architectural Lighting*, 1987. © *Architectural Lighting*. Reprinted courtesy of Cassandra Publishing Corporation, Eugene, OR.)

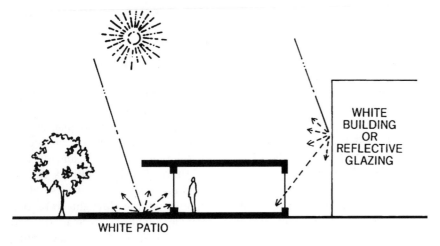

WHITE BUILDING OR REFLECTIVE GLAZING

WHITE PATIO

Figure 13.3d Sometimes reflected light is the major source of daylight. Under certain circumstances, a north window can receive as much light as a south window.

greater at the zenith than at the horizon (Fig. 13.3b). Although the illumination from an overcast day is quite low (500–2,000 footcandles), it is still ten to fifty times greater than what is needed indoors.

On a clear day, the brightest part of the sky, which is in the direction of the sun, is about ten times brighter than the darkest part of the sky, which is found at about 90 degrees to the sun (Fig. 13.3c). Under a clear sky, the illumination is quite high (6,000–10,000 footcandles) or 100 to 200 times greater than the requirements for good indoor illumination. Under such conditions, windows and skylights can be quite small. The

main difficulty with the clear sky is the challenge of the direct sunlight, which is not only extremely bright but is continually changing direction. Consequently, to understand clear-day illumination, it is necessary to also understand the daily and seasonal movements of the sun as explained in Chapter 6.

In most climates, there are enough days of each sky condition to make it necessary to design for both conditions. The main exceptions are parts of the Pacific Northwest, where overcast skies predominate, and the Southwest, where clear skies predominate. Under overcast skies, the main challenge for the designer is one of

quantity, while for clear-sky conditions the challenge is one of quality.

The daylight from clear skies consists of the two components of skylight and direct sunlight. The light from the blue sky is diffuse and of low brightness, while the direct sunlight is very directional and extremely bright. Because of the potential for glare, excessive brightness ratios, and overheating, it is sometimes assumed that direct sunlight should be excluded

TABLE 13.3 TYPICAL REFLECTANCE FACTORS

Material	Reflectance (Percent)
Aluminum, polished	70—85
Asphalt	10
Brick, red	25—45
Concrete	30—50
Glass	
Clear or tinted	7
Reflective	20—40
Grass	
Dark green	10
Dry	35
Mirror (glass)	80—90
Paint	
Black	4
White	70—90
Porcelain enamel (white)	60—90
Snow	60—75
Stone	5—50
Vegetation, average	25
Wood	5—40

from a building. It is sometimes erroneously believed that direct sunshine is appropriate only for solar heating. Figure 13.6a illustrates the efficacy (lumens/watt) of various light sources. Although direct-beam sunlight has a lower efficacy than skylight, its efficacy is comparable to the best electric sources, while its color-rendering ability is superior. Since direct sunlight is an abundant and free source of light, it should be utilized in any daylight design. With the proper design it can supply high quality as well as high quantity daylight.

The light from clear skies, especially the light from the northern sky, is rich in the blue end of the spectrum. While the color-rendering quality of such light is excellent, it is slightly on the cool side (Color Plate 5).

Reflected light from the ground and neighboring structures is often a significant source of daylight (Fig. 13.3d). The reflectance factor of the reflecting surface is critical in this regard. A white painted building will frequently reflect about 80 percent of the incident light, while lush green grass will reflect about 10 percent and mostly green light. Table 13.3 gives the reflectance factors in percent for some common surfaces.

13.4 CONCEPTUAL MODEL*

Direct-beam light can be nicely modeled with arrows, but a diffused source cannot. For one to understand and to predict the effect of a diffuse light source, a different kind of visual model is required. The illumination due to a diffused light source is similar to the concept of mean radiant temperature (MRT) described in Section 3.12. The illuminating effect of a diffused source on a point is a function of both the brightness of the source and the apparent size of the source, which is defined

*This section is based on material from the book *Concepts and Practice of Architectural Daylighting* by Fuller Moore, 1985.

by the exposure angle. Fig. 13.4a illustrates the fact that illumination increases with the brightness of the source. Fig. 13.4b illustrates how the apparent size is a consequence of actual size, proximity, and tilt. For example, the apparent size (angle theta) decreases if the source is moved further away, if the actual size is decreased, or if the source is tilted. If the source is tilted 90 degrees, the apparent size is zero.

Let us now use this model to help visualize how the table in Fig. 13.4c is illuminated by the window. If we move the table from position "A" to

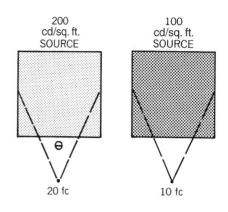

Figure 13.4a The illumination of a point by a diffuse source is partly a function of the brightness of the source. (After Moore, 1985.)

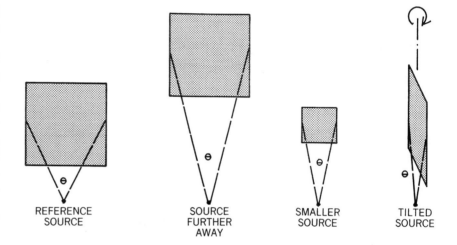

Figure 13.4b The illumination of a point by a diffuse source is also a function of the apparent size (exposure angle), which, in turn, is a function of distance, size, and tilt of the source. (After Moore, 1985.)

Figure 13.4c When a table is moved from point "A" to point "B," the illumination on it is decreased not only because the distance is increased, but also because of the relative tilt of the window. (After Moore, 1985.)

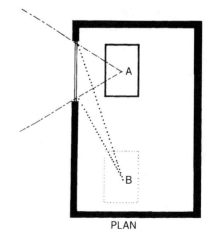

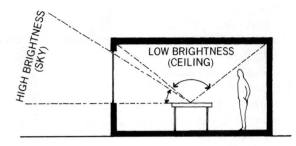

Figure 13.4d To determine the total illumination on the table, one must consider the brightness and exposure angle of each source. (After Moore, 1985.)

position "B," we decrease the illumination for two reasons: we change both the proximity and tilt of the window in relation to a point on the table. Everything else remains constant.

In Fig. 13.4d, we see a section of the same room. The two main sources of daylight for a point on the table are the sky and the ceiling. Some of the daylight entering the window is reflected off the ceiling, which, in turn, becomes a low brightness source for the table. The illumination from the ceiling is significant, even though the brightness is low, because of the large apparent size of this source. The sky, despite its smaller apparent size, is the major source of light because it is much brighter than the ceiling. If the walls are of a light color, they will also reflect some light on the table. For simplicity, the contribution of the walls is not shown in Fig. 13.4d.

13.5 ILLUMINATION AND THE DAYLIGHT FACTOR

One of the best ways for the architect to determine both the quantity and the quality of a daylighting design is through the use of physical models. Although most daylighting model tests are conducted under the real sky, the actual measured illumination is of limited usefulness. Unless the model can be tested under the worst daylight conditions, the illumination inside the model will not indicate the lowest illumination level to be expected. Fortunately, it is not necessary to test

the model under the worst conditions because of a concept called the **daylight factor.** The daylight factor is the ratio of the illumination indoors to outdoors on an overcast day (Fig. 13.5), which is an indication of the effectiveness of a design in bringing daylight indoors. For example:

A daylight factor of 5 percent means that on an overcast day when the illumination is 2,000 footcandles outdoors, the illumination indoors will be 100 footcandles (2,000 fc × .05 = 100 fc).

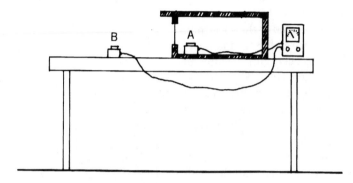

Figure 13.5 The daylight factor is determined by the ratio of indoor to outdoor illumination on an overcast day. D.F. = A/B.

TABLE 13.5A   TYPICAL MINIMUM DAYLIGHT FACTORS

Type of Space	Daylight Factor (%)
Art studios, galleries	4–6
Factories, laboratories	3–5
Offices, classrooms, gymnasiums, kitchens	2
Lobbies, lounges, living rooms, churches	1
Corridors, bedrooms	0.5

TABLE 13.5B   AVERAGE ILLUMINATION FROM OVERCAST SKIES[a]

North Latitude (degrees)[b]	Illumination (footcandles)
46	700
42	750
38	800
34	850
30	900

[a] The illumination values are typical for overcast-sky conditions, available about 85 percent of the day from 8 A.M. to 4 P.M. For more detailed information, see *IES Lighting Handbook* (1981 Reference Volume, pages 7-5 to 7-7).

[b] See map of Fig. 5.6d.

Although winter overcast skies are usually the worst design condition, the model can be tested under an overcast sky at any time of year or day. Table 13.5A presents typical daylight factors for different kinds of spaces. If the measured daylight factor is greater than that of Table 13.5A, there will be more than enough daylight for most of the year. By multiplying the daylight factor by the average minimum daylight of Table 13.5B, one can also determine the average minimum indoor illumination.

Remember, however, that absolute illumination is not a good indicator of visibility because of the eye's great power of adaptation. The relative brightness between the interior and the window is, however, a critical consideration in daylight design, and the daylight factor is a good indicator of this relationship. The higher the factor, the less extreme are the brightness differences.

If a design excludes direct sunlight, clear days will behave similarly to the overcast conditions explained above. If direct sunlight is included, as it generally should be, then the model will have to be tested with a sun machine to simulate the various sun angles throughout the year. Model testing will be explained later in this chapter.

Often, models are used to compare alternative designs. Since actual outdoor lighting varies greatly from hour to hour and day to day, footcandle measurements made at different times cannot be compared, but the daylight factor can. As the outdoor illumination changes, the indoor illumination changes proportionally and the daylight factor remains constant for any particular design.

13.6 LIGHT WITHOUT HEAT?

All light, whether electric or natural, is radiant energy that is eventually absorbed and turned into heat. Along with the light energy, there is usually also a large amount of infrared (heat) radiation. Electric light sources also produce large amounts of sensible heat as a byproduct. In the winter, all this heat is beneficial. During the cheap energy years, whole buildings were heated by the electric lighting. Unfortunately, in the summer this light energy increases the overheating of the building. Thus, from a heating and cooling point of view, we want maximum amounts of light in the winter and minimum amounts in the summer.

As was explained in Chapter 10, there is a very strong "law of diminishing returns" for visibility due to illumination. A certain minimum is required, but above that the benefits are small compared to the cost. It is, therefore, not wise to bring more light than necessary into a building in the summer.

If the direct sunlight that enters a building is well distributed and does not create excessively high illumination levels, the heat load from direct sunlight will be less than from electric lights. Fig. 13.6a shows the efficacy of various light sources. Note that light from the sky has the highest light-to-heat ratio. Although sunlight introduces more heat per unit of light than light from the sky, it is still better than any white, electric light source. For the same light level, incandescent lamps introduce about six times more heat than daylighting, and fluorescent lamps introduce about two times more heat than daylighting (60 lumens/watt for fluorescent versus 120 lumens/watt for daylighting).

In the winter, there is no limit to the amount of sunlight that might be introduced as long as glare and excessive-brightness ratios are controlled. After all, the daylighting indoors can never be brighter than the daylighting outdoors, for which the eyes evolved.

The seasonal problem of daylighting is most acute with skylights since horizontal openings receive much

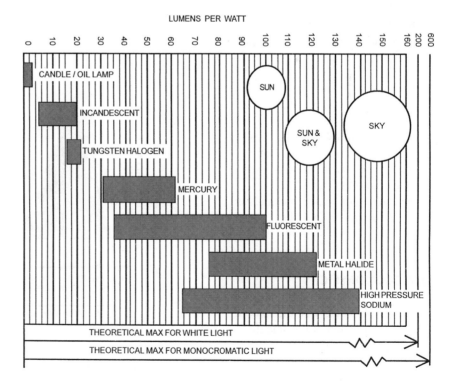

Figure 13.6a The efficacy (lumens/watt) of various light sources is compared.

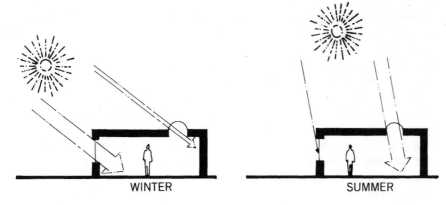

Figure 13.6b South-facing vertical glazing is more in phase with sunshine demand than is horizontal (skylight) glazing. Although the amount of daylight collected should be just adequate in the summer, there is almost no limit to the amount that is desirable in the winter.

more sunlight in the summer than in the winter. South-facing vertical glazing is much better in this regard because it captures more sunlight in the winter than in the summer (Fig. 13.6b).

Rules

1. During the summer, introduce only as much daylight as can be effectively used. The sunlight must be well distributed and it must not raise the illumination levels much above what is required.
2. During the winter, introduce as much sunlight as possible in most envelope-dominated and many internally dominated buildings. There is no upper limit as long as it does not create glare or excessive brightness ratios.
3. Internally dominated buildings in mild climates should obey rule #1 in winter as well as summer.

13.7 COOL DAYLIGHT

In addition to the fact that the amount of heat is a function of the amount of light, the amount of heat is also a function of the quality of the light. Figure 13.7 (lower graph) illustrates the fact that about 50 percent of solar

radiation is in the infrared part of the electromagnetic spectrum. This radiation enters a building through glazing just as visible light does, but it contributes nothing to daylighting. Light reflected off clouds or from the blue sky has a smaller proportion of this infrared radiation and, therefore, it has a higher efficacy (lumens/watt) (Fig. 13.6a).

Because most materials have the same reflectance for short-wave infrared as for visible light, there is no selective advantage to using reflected light. For example, the light reflected into a clerestory from a white roof has

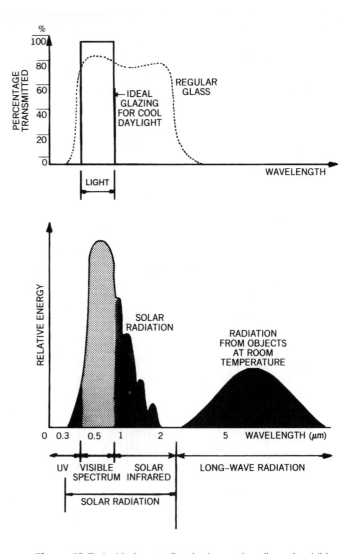

Figure 13.7 An ideal spectrally selective coating allows the visible but not the infrared part of the solar radiation to pass through the glazing. Compare this behavior to that of glass.

TABLE 13.7 LIGHT-TO-SOLAR-GAIN RATIOS (LSG) FOR VARIOUS GLAZING SYSTEMS

Glass Type	Visible Transmittance (VT)	Solar Heat-Gain Coefficient (SHGC)	Light-to-Solar-Gain Ratio (LSG)
Double glazing			
Clear	0.82	0.75	1.20
Bronze	0.62	0.60	1.03
Reflective	0.20	0.16	1.25
Spectrally selective	0.70	0.46	1.52

SIDEBOX 13.7

Light-to-solar-gain ratio = $\dfrac{\text{Visible Transmittance}}{\text{Solar Heat-Gain Coefficient}}$

or $\text{LSG} = \dfrac{\text{VT}}{\text{SHGC}}$

the same ratio of visible to short-wave infrared as the original sunlight did. However, reflectors are still useful because they can increase the total radiation collected, as well as control its direction.

Heat-absorbing glass was developed to deal with this unwanted infrared radiation. Although it absorbs a slightly larger proportion of infrared than visible light, much of the absorbed radiation is reradiated inside and makes the glass uncomfortably hot. Its green tint also affects the color of the view. What is really needed for cool daylight is a **selective glass** that reflects the infrared but not the visible portion of daylight (see upper graph of Fig. 13.7). There are now some new glazing materials commercially available that begin to do this, and they are generally known as **spectrally selective low-e glazing.** See also Fig. 9.18f for a graph comparing the transmission characteristics of various types of glazing.

The ideal glazing type shown in Fig. 13.7 (top) also filters out the ultraviolet (UV) radiation that fades colored materials, such as carpets, fabrics, artwork, paints, and wood. Coatings on glass can filter out about 75 percent of UV radiation, and special plastic films within the glazing system can reduce UV radiation by 99

percent. It is important to note, however, that visible light (blue especially) fades colors at one-third the rate of UV radiation. Valuable and delicate artifacts need to be shielded from both UV radiation and high-level light (i.e., above 15 footcandles).

Visible transmittance (VT) is the factor that quantifies the amount of visible light that passes through glazing. It varies from 0.9 for very clear glass to less than 0.1 for highly reflective or tinted glass. For cool daylight, the VT should be high while the transmission of the solar infrared should be low.

As mentioned in Section 9.20, the **solar heat-gain coefficient (SHGC)** is a factor that quantifies the total solar radiation (visible and solar infrared) that passes through glazing. When one compares the VT to the SHGC, one can predict the coolness of the transmitted light. The ratio of the VT to the SHGC is called the **light-to-solar-gain (LSG)** ratio (see Sidebox 13.7). The higher the ratio, the cooler the light. See Table 13.7 for the LSG ratio for various types of glazing. Note that tinted glazing has a low LSG ratio because it blocks visible light and solar heat about equally, which is certainly not desirable for daylighting purposes. On the other hand, spectrally selective glazing has a high LSG

ratio because solar heat is blocked much more than light.

The same logic applies if you replace the SHGC with the still used but older shading coefficient (SC). Thus, the following recommendation:

Rule of thumb: The visible-transmittance (VT) factor should be much larger than the solar heat- gain coefficient (SHGC) or the shading coefficient (SC) when one is choosing glazing for cool daylighting purposes.

13.8 GOALS OF DAYLIGHTING

The general goal for daylighting is the same as that for electric lighting: to supply sufficient light of high quality while minimizing direct glare, veiling reflections, and excessive-brightness ratios.

Because of the limitations of window location and the variability of daylight, some specific goals refer only to daylighting. The diagram in Fig. 13.8a shows that typically there is too little light at the back of the room and more than enough right inside the window. Thus, the first goal is to get more light deeper into the building to both raise the illumination level there and to reduce the illumination gradient across the room (Fig. 13.8b).

Figure 13.8a The light from windows creates an excessive illumination gradient across the room (too dark near the back wall compared to the area near the window).

Figure 13.8b One goal of daylighting design is to create a more acceptable illumination gradient.

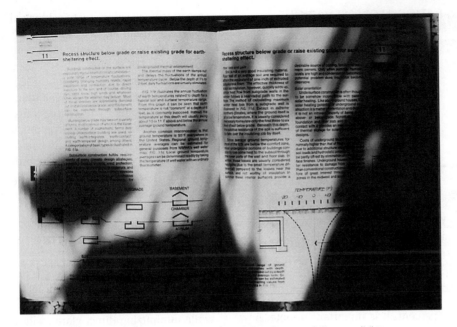

Figure 13.8c Excessive brightness ratios can result from beams of direct sunlight.

Figure 13.8d Veiling reflections are a common problem from any overhead lighting.

The second goal is to reduce or prevent the severe direct glare of unprotected windows and skylights. This glare is aggravated if the walls adjacent to the windows are not illuminated and, therefore, appear quite dark (see Fig. 13.10f).

If a beam of sunlight creates a puddle of light over part of the work area, severe and unacceptable brightness ratios will exist. Thus, the third goal is to prevent excessive-brightness ratios (especially those caused by direct sunlight) (Fig. 13.8c).

Although the low-angle light from windows is usually not a source of veiling reflections, the light from overhead openings generally is (Fig. 13.8d). Thus, the fourth goal is to prevent or minimize veiling reflections (especially from skylights and clerestory windows).

In most situations, lighting should not be too directional because of the dark shadows that result. The fifth goal, therefore, is to diffuse the light by means of multiple reflections off the ceiling and walls.

In areas where there are no critical visual tasks, the drama and excitement of direct sunlight can be a positive design element. Therefore, the sixth goal, which is limited to those spaces in which there are few if any critical visual tasks, is to use the full aesthetic potential of daylighting and sunlight. In all spaces, however, the dynamic nature of daylight should be seen as an asset rather than a liability. The ever-changing nature of daylight needs only to be limited — not eliminated.

The remainder of this chapter discusses techniques and strategies for achieving the above-mentioned goals.

Both the orientation and form of the building are critical to a successful daylighting scheme. We must consider not only the external form, but also the shape of internal spaces. Internal partitions, unless made of glass, will stop the penetration of daylight. Only after these basic issues have been determined can fenestration design proceed.

Normally, the selection of finishes is one of the last steps in the design process, but for effective daylighting it must be considered early.

13.9 BASIC DAYLIGHTING STRATEGIES

Although daylighting is primarily a second-tier strategy (Fig. 13.9a), it is important to utilize as many of the tier-one strategies listed below as possible to enhance and prepare for the daylighting design.

1. *Orientation*: Because of the usefulness of direct sunlight, the south orientation is usually best for daylighting. The south side of a building gets sunlight most consistently throughout the day and the year. This extra sunlight is especially welcome in the winter, when its heating effect is usually desirable. Sun-control devices are also most effective on this orientation.

   The second best orientation for daylighting is north because of the constancy of the light. Although the quantity of north light is rather low, the quality is high, if a cool white light is acceptable (see Color Plate 5). There is also little problem with glare from the direct sun. In very hot climates, the north orientation might even be preferable to the south orientation. North might also be preferable because it does

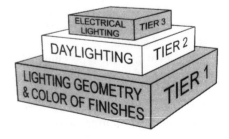

Figure 13.9a The best and most sustainable lighting is achieved by the three-tier design approach, and this chapter covers primarily tier two (daylighting).

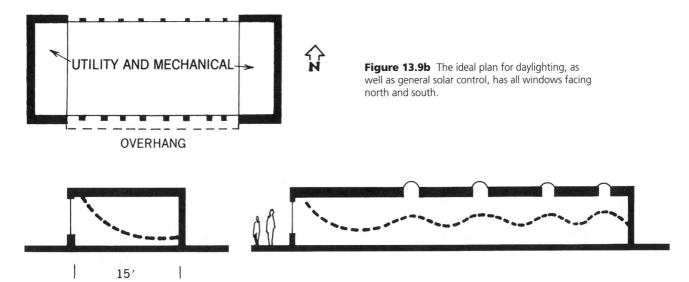

Figure 13.9b The ideal plan for daylighting, as well as general solar control, has all windows facing north and south.

Figure 13.9c While daylighting from windows is limited to the area about 15 feet from the outside walls, roof openings can yield fairly uniform lighting over unlimited areas.

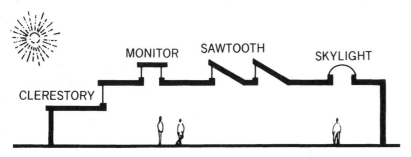

Figure 13.9d The various possibilities for overhead openings for daylighting.

not require movable sun-control devices, which must be maintained to remain effective.

The worst orientations are east and west. Not only do these orientations receive sunlight for only half of each day, but the sunlight is at a maximum during summer instead of winter. The worst problem, however, is that the east or west sun is low in the sky and, therefore, creates very difficult glare and shading problems. Figure 13.9b illustrates an ideal floor plan in regard to building orientation.

Rules for Orientation

1. (a) For daylighting when winter heat is desirable, use south-facing glazing.

(b) For daylighting when winter heat is not desirable, use north facing glazing.

(c) For daylighting without summer overheating or severe glare, *avoid* east and west glazing.

2. *Lighting through the roof:* Except for the use of light wells, only one story or the top floor of multistory buildings can use overhead openings. When applicable, horizontal openings offer two important advantages. First, they allow fairly uniform illumination over very large interior areas, while daylighting from windows is limited to about a 15-foot depth (Fig. 13.9c). Second, horizontal openings also receive much more light

than vertical openings. Unfortunately, a number of important problems are associated with this orientation. The intensity of light is greater in the summer than the winter. It is also difficult to shade horizontal glazing. For these reasons, it is often appropriate to use vertical glazing on the roof in the form of clerestory windows, monitors, or sawtooth arrangements (Fig. 13.9d).

3. *Form:* The form of the building not only determines the mix of vertical and horizontal openings that is possible, but also how much of the floor area will have access to daylighting. Generally, in multistory buildings a 15-foot perimeter zone can be fully daylit and another 15 feet beyond that can be partially daylit. All three floor plans in Fig. 13.9e have the same area (10,000 square feet). In the square plan, 16 percent is not daylit at all, and another 33 percent can be only partially daylit. The rectangular plan can eliminate the core area that receives no daylight, but it still has a large area that is only partially daylit, while the atrium scheme is able to have all of its area daylit. Of course, the actual percentage of core versus perimeter zones depends on the actual area. Larger buildings will have larger cores and less surface area.

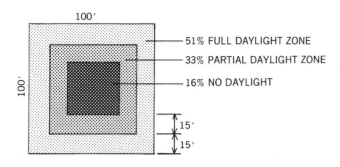

51% FULL DAYLIGHT ZONE

33% PARTIAL DAYLIGHT ZONE

16% NO DAYLIGHT

100'

100'

15'

15'

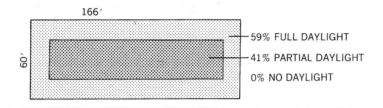

166'

60'

59% FULL DAYLIGHT

41% PARTIAL DAYLIGHT

0% NO DAYLIGHT

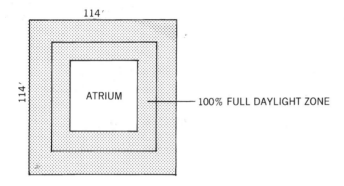

114'

114'

ATRIUM

100% FULL DAYLIGHT ZONE

Figure 13.9e These plans of a multistory office building illustrate the effect of massing on the availability of daylight.

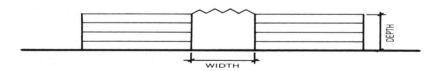

DEPTH

WIDTH

Figure 13.9f It is not the actual depth or width but their ratio that determines how much daylight will be available at the base of an atrium.

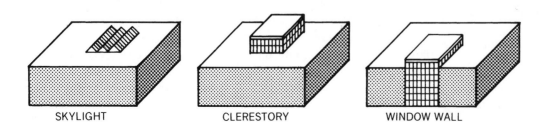

SKYLIGHT   CLERESTORY   WINDOW WALL

Figure 13.9g Generic types of daylit atriums.

The modern atrium is typically an enclosed space whose temperature is maintained close to the indoor conditions. Buildings with atriums are, therefore, compact from a thermal point of view and yet have a large exposure to daylight. The amount of light available at the base of the atrium depends on a number of factors: the translucency of the atrium roof, the reflectance of atrium walls, and the geometry of the space (depth versus width), as shown in Fig. 13.9f. Physical models are the best way to determine the amount of daylight that can be expected at the bottom of an atrium. When atriums get too small to be useful spaces, they are known instead as light wells. Atriums can be illuminated by skylights, clerestories, or window walls (Fig. 13.9g). The advantage of each approach will be explained below in the discussion of windows, skylights, and clerestories.

One of the most sophisticated modern atrium buildings, the Bateson State Office Building in Sacramento, California, is discussed further as a case study in Section 17.7.

4. *Space Planning:* Open space planning is very advantageous for bringing light to the interior. Glass partitions can furnish acoustical privacy without blocking the light. If or when visual privacy is also needed, curtains or Venetian blinds can cover the glass, or translucent materials could be

Figure 13.9h Full- or partial-height glass partitions can enable borrowed light to enter interior spaces.

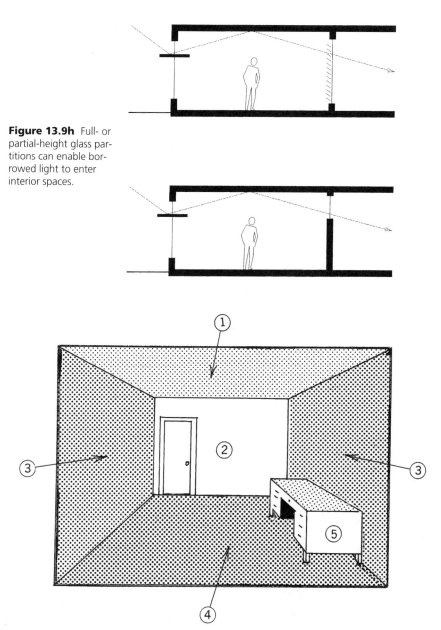

Figure 13.9i For good distribution and penetration of light, the order of importance for high reflectance factors is shown (e.g., 1 should have the highest reflectance factor).

reduce dark shadows, glare, and excessive-brightness ratios. The ceiling should have the highest reflectance factor possible. The floor and small pieces of furniture are the least critical reflectors and, therefore, *might* have fairly low reflectance factors (dark finishes). The descending order of importance for reflecting surfaces is: ceiling, back wall, side walls, floor, and small pieces of furniture (Fig. 13.9i).

6. *Use separate openings for view and daylighting.* Use high windows, clerestories, or skylights for excellent daylighting, and use low windows for view. High glazing should be clear or spectrally selective, while low glazing could be tinted or reflective for glare control.

13.10 BASIC WINDOW STRATEGIES

To understand window daylighting strategies, one will find it worthwhile to first examine the lighting from an ordinary window. As mentioned earlier, the illumination is greatest just inside the window and rapidly drops off to inadequate levels for most visual tasks (Fig. 13.10a left). The view of the sky is often a source of direct glare, and direct sunlight entering the window creates excessive-brightness ratios (puddles of sunlight), as well as overheating during the summer. To overcome these negative characteristics of ordinary windows, designers should keep in mind the following strategies:

1. *Windows should be high on the wall, widely distributed, and of optimum area.* Daylight penetration into a space will increase with the mounting height of the window (Fig. 13.10a right). The useful depth of a daylit space is limited to about 1½ times the height of the top of the window.

used. Alternatively, the partitions could have glass above eye level only (Fig. 13.9h).

5. *Color:* Use *light* colors both indoors and outdoors to reflect more light into the building and farther into the interior, as well as to diffuse the light. Light-colored roofs can greatly increase the light that clerestories collect.

Windows adjacent or opposite to light-colored exterior walls will receive more daylight. Light-colored facades are especially important in urban areas to increase the availability of daylighting at the lower floors.

Light-colored interiors will not only reflect light farther into the building, but also diffuse it to

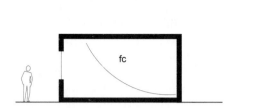

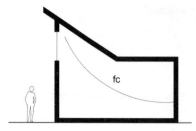

Figure 13.10a Daylight penetration increases with window height.

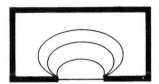

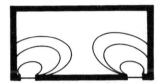

Figure 13.10b These plans, with contours of equal illumination, illustrate how light distribution is improved by admitting daylight from more than one point.

Figure 13.10c Strip or ribbon windows, as seen here in the Maison LaRoche by Le Corbusier, admit a uniform light, which is further improved by placing the windows high on the wall. (Note that photographic film exaggerates brightness ratios.) Photograph by William Gwin.

fusers, small-window areas can collect large amounts of daylight.

2. *If possible, place windows on more than one wall.* Whenever possible, avoid **unilateral lighting** (windows on one wall only), and use **bilateral lighting** (windows on two walls) for much better light distribution and reduced glare (Fig. 13.10d). Windows on adjacent walls are especially effective in reducing glare. The windows on each wall illuminate the adjacent wall and, therefore, reduce the contrast between each window and its surrounding wall.

3. *Place windows adjacent to interior walls.* Here, the interior walls adjacent to windows act as low-brightness reflectors to reduce the overly strong directionality of daylight (Fig. 13.10e). The glare of the window is also reduced because of the reduced brightness ratio between the window and its wall due to reflections back from the side wall (Fig. 13.10f).

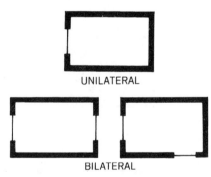

Figure 13.10d Bilateral lighting is usually preferable over unilateral lighting (plan view).

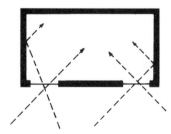

Figure 13.10e Light distribution and quality are improved by the reflection off sidewalls.

Whenever possible, ceiling heights should be increased so that windows can be mounted higher.

Daylight will be more uniformly distributed in a space if windows are horizontal rather than vertical and if they are spread out rather than concentrated (Fig. 13.10b). Architects, such as Le Corbusier, often used ribbon windows for this reason (Fig. 13.10c).

Window area as a percentage of floor area should rarely exceed 20 percent because of summer overheating and winter heat losses. By means of reflectors and dif-

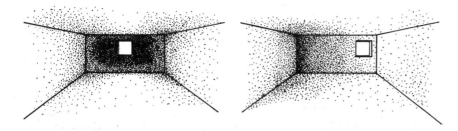

Figure 13.10f The glare from a window next to a sidewall is less severe than that from a window in the middle of a room.

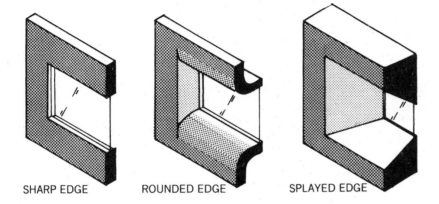

SHARP EDGE ROUNDED EDGE SPLAYED EDGE

Figure 13.10g The excessive contrast between a window and a wall can be reduced by splaying or rounding the inside edges. (After M. D. Egan, *Concepts in Architectural Lighting.*)

4. *Splay walls to reduce the contrast between windows and walls.* Windows create less glare when the adjacent walls are not dark in comparison to the window. Splayed or rounded edges create a transition of brightness that is more comfortable to the eye (Fig. 13.10g).

5. *Filter daylight.* Sunlight can be filtered and softened by trees or by such devices as trellises and screens (Fig. 13.10h). Translucent glazing or very light drapes, however, can make the direct-glare problem much worse. Although they diffuse direct sunlight, they often become excessively bright sources of light in the process (Fig 13.10i).

6. *Shade windows from excess sunlight in summer.* Ideally, only a small amount of sunlight should be admitted through the windows in the summer and a maximum amount in the winter. At all times, however, the light should be diffused by reflecting it off the ceil-

Figure 13.10h Trees, supported on a grid of wires, filter the light before it enters the Kimbell Art Museum. Louis I. Kahn, architect.

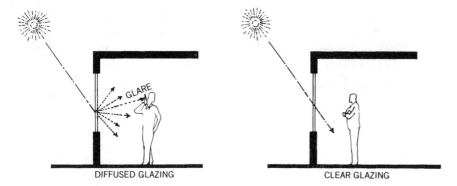

Figure 13.10i Translucent glazing can be a major source of glare because some of the sunlight is directed into the eyes of the observer.

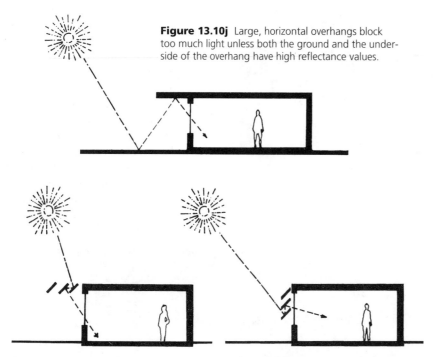

Figure 13.10j Large, horizontal overhangs block too much light unless both the ground and the underside of the overhang have high reflectance values.

Figure 13.10k Light-colored, horizontal louvers block direct sunlight but allow some diffused light to enter the windows.

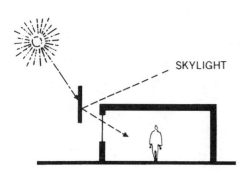

Figure 13.10m A vertical panel can block direct sunlight while it reflects diffuse skylight.

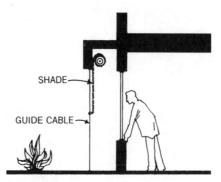

Figure 13.10n On the east and west facades the Bateson Building, Sacramento, CA, uses exterior roller shades that automatically respond to sun and wind conditions.

ing. If that is not possible, the light must be shaded before it enters. Overhangs on south windows can give us this ideal seasonal control. They can also eliminate puddles of sunlight, reduce glare, and even out the light gradient across the room. If a large, solid horizontal overhang is used, its underside should be painted white to reflect ground light (Fig. 13.10j). A light-colored overhang, especially with louvers, will also reduce the brightness ratio between itself and the sky.

Vertical or horizontal louvers painted a light color are beneficial because they block direct sunlight, yet reflect diffused sunlight (Fig. 13.10k). A vertical panel in front of the window can block direct sunlight, while reflecting diffuse skylight into the window (Fig. 13.10m).

Chapter 9 contains an extensive description of shading devices. Physical models can be used to determine how well these and other devices can admit quality daylight while shading direct sunlight. Of course, in certain spaces, such as lobbies, lounges, and living rooms, where visual tasks are not critical, some direct sunlight can be welcome for its visual and psychological benefits, especially in winter.

7. *Use movable shades.* A dynamic environment calls for a dynamic response. Variations in daylighting are especially pronounced on the east and west exposures, which receive diffused light for half a day and direct sunshine for the other half. Movable shades or curtains can respond to these extreme conditions (see Fig. 13.10n). To reduce heat gain, the interior shade or curtain should be highly reflective, while darker colors can be acceptable outdoors. Also, as mentioned earlier, shading is much more effective when placed outside the glazing.

13.11 ADVANCED WINDOW STRATEGIES

The challenges of getting daylight from windows farther into the building while still maintaining the quality of the lighting can best be met by reflecting daylight off the ceiling. In one-story buildings, walkways, roads, and light-colored patios can reflect a significant amount of light to the ceiling (Fig. 13.11a). In multistory buildings, parts of the structure can be

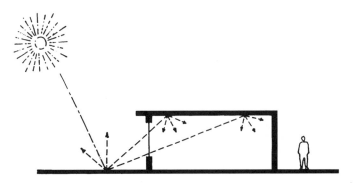

Figure 13.11a Light-colored pavement or gravel can reflect light deep into the interior.

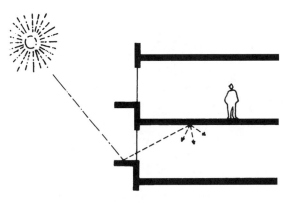

Figure 13.11b Wide windowsills can be used as light reflectors to send light deep into the interior.

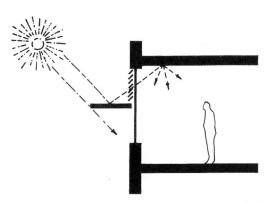

Figure 13.11c Light shelves are usually placed above eye level to prevent glare from the top of the shelf. In this position, they also act as overhangs for the view windows underneath. Louvers can be used to prevent glare from the clear glazing above the light shelf. Tinted glazing can be used for the view window.

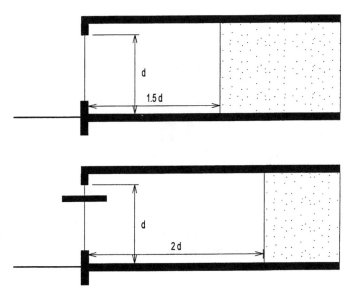

Figure 13.11d A rule of thumb for daylight penetration is 1½ times the height of a standard window and 2 times the height of a window with a light shelf for south-facing windows under direct sunlight. (After *Tips for Daylighting* by Jennifer O'Connor, © Regents of the University of California, 1997.)

used to reflect light indoors. Deep windowsills can be quite effective but are a potential source of glare (Fig. 13.11b). Light shelves prevent the glare problem when placed just above eye level (Fig. 13.11c). If glazing is used below the light shelf, it will be mainly for view. The light shelf acts as an overhang for this lower glazing to prevent direct sunlight from entering and creating puddles of sunlight. The overhang also reduces glare by blocking the view of the bright sky in the lower window. Glare from the upper window can be controlled by louvers (Fig. 13.11c) or by an additional light shelf on the inside. Light shelves will not only improve the quality of the daylighting, but also increase the depth of the daylighting zone (Fig. 13.11d).

Figure 13.11e The Ventura Coastal Corp. Administration Building, Ventura, CA, uses light shelves to daylight the building. Architect: Scott Ellinwood. (Courtesy of and © Mike Urbanek, 1211 Maricopa Highway, Ojai, CA 93023.)

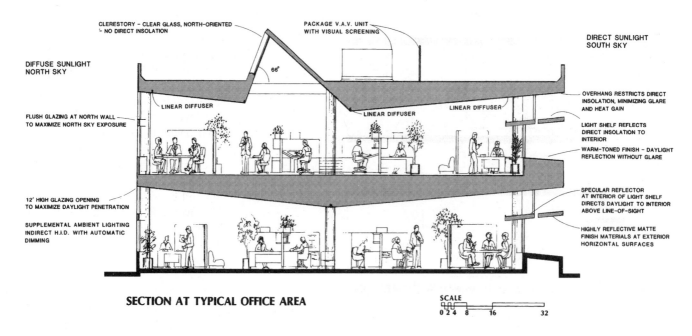

Figure 13.11f Light shelves, sloped ceilings, clerestories, north-facing windows, and open planning all help to illuminate the Ventura Coastal Building during the day. (Courtesy of and ©Mike Urbanek, 1211 Maricopa Highway, Ojai, CA 93023.)

Figure 13.11g Thin, metal light shelves are supported by cables. The top of the shelf is a high-reflectance white, while the rest is painted a bright yellow. The photo was taken before the indoor light shelves were installed.

The Ventura Coastal Corp. Administration Building is an excellent example of a building using light shelves (Fig. 13.11e). A sloped ceiling allows the windows to be large and high for extensive daylight penetration, and the necessary mechanical equipment to be concentrated near the center of the building where more room exists above the ceiling (Fig. 13.11f). North windows are also very large and high because of the sloping ceiling. In this very mild climate, the north-facing clerestory brings light into the center of the building so that the illumination from daylighting is more evenly distributed.

The very thin metal light shelves of the Florida Solar Energy Center are made hurricane-safe via cable stays (Fig. 13.11g). One of the most effective strategies for reflecting light onto the ceiling is the indoor Venetian blind or a similar outdoor louver system (Fig. 13.11h). The Venetian

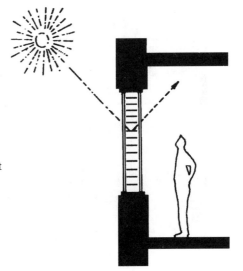

Figure 13.11h Venetian blinds are very effective in redirecting light up to the ceiling, controlling glare, and limiting the collection of light. They are especially appropriate on east and west facades because of their adjustability.

blind's main drawback, dirt accumulation, can largely be avoided by sandwiching the blind between two layers of glass. Miniature slats reduce the annoying figure/background effect described in Fig. 12.6h. Dynamic systems, such as Venetian blinds, are much more effective than

static systems because they can better respond to the varying conditions of daylight and sunlight.

The Occidental (Hooker) office building in Niagara Falls, New York (1981) is covered by two glass skins that create a 4-foot air space in which horizontal louvers have been placed

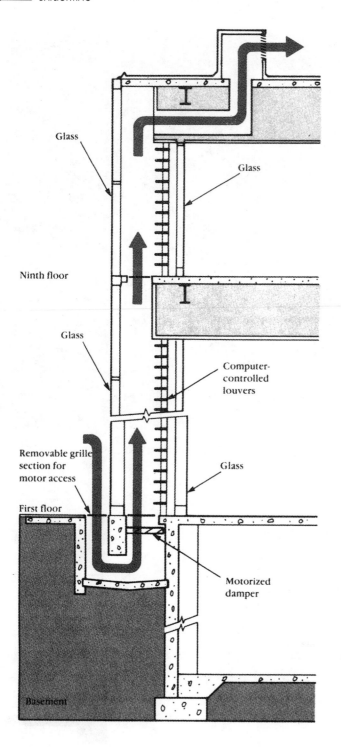

Glass

Glass

Ninth floor

Glass

Computer-
controlled
louvers

Removable grille
section for
motor access

Glass

First floor

Motorized
damper

Basement

(Fig. 13.11i). These white louvers are automatically rotated to control daylight entering the building. Direct sunlight is always intercepted and reflected to the ceiling. For insulating purposes, the louvers rotate to form an additional barrier at night.

In all cases, the ceiling should be a diffuse reflector, but the devices reflecting light onto the ceiling *could* have a specular finish in order to maximize the depth of sunlight penetration. Figures like 13.11b were drawn as if the reflectors had a specular finish for clarity. Unless they are specifically labeled, do not assume that the reflectors shown in the diagrams must be specular in nature. A disadvantage of specular reflectors is that they often cast excessively bright patches of sunlight on the ceiling. Curved specular reflectors minimize this problem by spreading the sunlight over a large part of the ceiling (Fig. 13.11j). Matte reflectors, on the other hand, create a very even distribution of light and are much less sensitive to sun angles. Model studies are a good way to determine whether specular or diffuse reflectors should be used.

Figure 13.11i The Occidental (Hooker) Office Building, Niagara Falls, NY, uses computer-controlled louvers to regulate the light and heat entering the building. (In summer, the hot air generated between the inner and outer glazing is vented outdoors. In winter, this hot air is sent to the north side of the building to provide heat. (From *Architectural Lighting*, 1987. © *Architectural Lighting*. Reprinted courtesy of Cassandra Publishing Corporation, Eugene, OR.)

Figure 13.11j Both concave and convex specular reflectors can be used to distribute daylight over a wide area of the ceiling.

13.12 WINDOW GLAZING MATERIALS

Choosing the right glazing material is critical to a successful daylighting design. Transparent glazing comes in a variety of types: clear, tinted, heat-absorbing, reflective, and spectrally selective.

The tinted, heat-absorbing, and reflective types are rarely appropriate for the collection of daylight because they reduce light transmittance. In daylighting, they are sometimes used to control the glare caused by the excessive-brightness ratios between windows and walls. These three types of glazing do not, however, solve the problem automatically because they might reduce the interior brightnesses as much as they reduce the brightness of the view. Thus, the brightness ratios remain the same—as does the glare. Tinted or reflective glazing can reduce window glare only if the interior is also illuminated by other sources, such as skylights or clerestory windows, and not by the view windows. In such cases, reducing the transmission of the view glazing improves the glare problem because the reduced brightness of the view window (Fig. 13.12a) is closer to the interior brightness. Of course, electric lights can also increase interior brightnesses, but using them to reduce the glare from sunlight defeats the whole idea of daylighting.

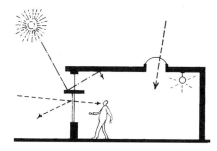

Figure 13.12a Tinted or reflective glazing will reduce glare only if the space is lighted by other sources, such as electric lights, skylights, etc.

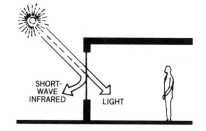

Figure 13.12b A selectively reflecting glazing blocks the sun's infrared radiation while it transmits the visible radiation.

As mentioned earlier in this chapter in the section on "cool daylight," one should use spectrally selective glazing when mainly light—and little heat—is required (Fig. 13.12b); one should use clear low-e glazing when light and winter heat are desired. Both of these glazing types (curves 2 and 3 of Fig. 13.12c) will create a cheerful interior. Avoid dark glazing because it can create a gloomy atmosphere that might reduce productivity.

Most glass blocks are not especially useful for daylighting because they afford little control over the direction or the quality of the light (Fig. 13.12d top). However, one type of glass block is made especially for daylighting design. It is called **light directing,** because of built-in prisms that refract the light up toward the ceiling for an even and deep penetration of daylight into a space (Fig. 13.12d bottom). These light-directing blocks were once very common but are now quite hard to obtain. At least one Japanese firm, however, still exports them to the United States.

Translucent glazing material with very high light transmittance and small area is not usually appropriate for window glazing for several reasons. A translucent material becomes an excessively bright source when

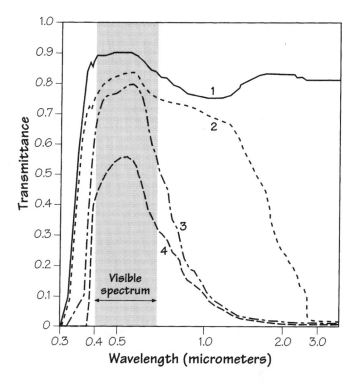

Figure 13.12c Curve 1 represents ordinary clear glazing. Use high-transmission low-e glazing (curve 2) when winter heat is a priority. Use high-transmission spectrally selective glazing (curve 3) when much cool daylight is desired, and use spectrally selective glazing (curve 4) for an east- or west-view window that has inadequate outdoor shading. (From *Residential Windows: A Guide to New Technologies and Energy Performance* by John Carmody, Stephen Selkowitz, and Lisa Heschong. © 1996 by John Carmody, Stephen Selkowitz, and Lisa Heschong. Reprinted by permission of W. W. Norton & Company, Inc.)

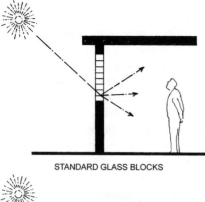

STANDARD GLASS BLOCKS

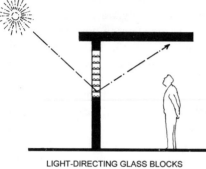

LIGHT-DIRECTING GLASS BLOCKS

Figure 13.12d "Light-directing" glass blocks refract the light up to the ceiling. Keep blocks high in the room to avoid reflected glare.

sunlight falls on it (see Fig. 13.10i). Because it diffuses light in all directions equally, it does not help much in improving the illumination gradient across the room. And, of course, translucent glazing does not allow for a view.

On the other hand, translucent glazing materials of relatively low light transmittance can be used suc-cessfully for daylighting when the glazing area is quite big. A large area of low- transmittance glazing, espe-cially overhead, creates a large, low-brightness source that will contribute a significant amount of light without glare. A discussion of translucent walls and roofs will follow later.

13.13 TOP LIGHTING

Skylights, monitors, and clerestories are all methods of **top lighting.** The main advantage of top lighting is the uniform and high illumination levels possible. Unfortunately, top lighting also has some serious drawbacks. It is not a workable strategy for multi-story buildings, and since it does not satisfy the need for view and orienta-tion, it should supplement, not replace, windows.

Top lighting also presents some potential glare problems. All over-head sources are a potential source of veiling reflections. These reflections are best avoided by keeping light sources out of the offending zones. This is possible, of course, only when the location of the visual task is fixed and the roof openings can be appro-priately placed (Fig. 13.13a). In many spaces, the best solution is to careful-ly diffuse the light so that there are no bright sources to cause veiling reflections. Either reflect the light off the ceiling (Fig. 13.14f), or use baf-fles to shield the light sources (Fig.

13.13b). Both of these strategies also solve the problem of direct glare and puddles of sunlight falling on the work surfaces. Because skylights behave differently from monitors and clerestories, they are discussed sepa-rately.

13.14 SKYLIGHT STRATEGIES

Skylights are horizontal or slightly sloped, glazed openings in the roof. As such, they see a large part of the unobstructed sky and, consequently, transmit very high levels of illumina-tion. Because beams of direct sun-light are not desirable for difficult visual tasks, the entering sunlight must be diffused in some manner. For skylights, unlike windows, translucent glazing can be appropri-ate since there is no view to block and direct glare can be largely avoid-ed.

A fundamental problem with all skylights is that they face the summer sun more than the winter sun. Thus, they collect much more light and heat in the summer than the winter, which is exactly the opposite of what is needed. As a result, clerestories should be used instead of skylight whenever possible.

The following are some common skylight strategies:

1. *Skylight spacing for uniform light-ing.* If there are no windows, the

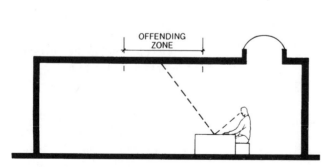

Figure 13.13a Veiling reflections are avoided when skylights are placed outside the offending zone.

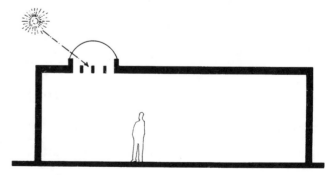

Figure 13.13b A system of baffles can control direct glare and to some extent veiling reflections.

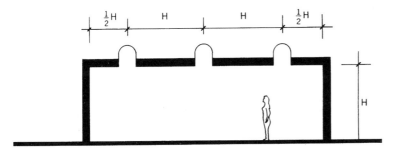

Figure 13.14a Recommended spacing for skylights without windows as a function of ceiling height.

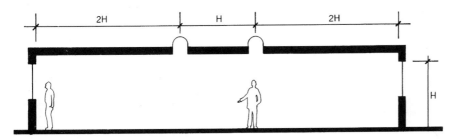

Figure 13.14b Recommended spacing for skylights with high windows as a function of ceiling height.

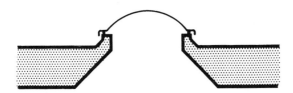

Figure 13.14c Splayed openings distribute light better and also cause less glare than square openings.

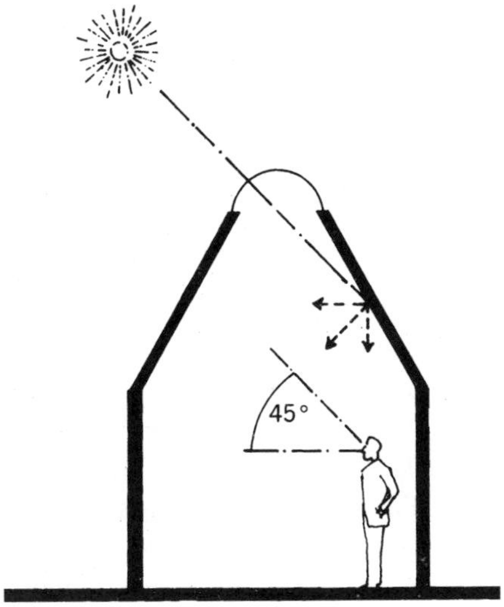

skylights should be spaced, as shown in Fig. 13.14a. With windows, the skylights can be farther from the perimeter, as shown in Fig. 13.14b.

2. *Use splayed openings to increase the apparent size of skylights.* Better light distribution and less glare result when the walls of the light well are sloped (Fig. 13.14c).

3. *Place the skylight high in a space.* A skylight mounted high above a space will enable the light to diffuse before it reaches the floor (Fig. 13.14d). Direct glare is largely prevented, because the bright skylight is at the edge of or beyond the observer's field of view.

4. *Place skylights near walls.* Any wall, and especially the north wall, can be used as a diffuse reflector for a skylight (Fig. 13.14e). The bright wall will make the space appear larger and more cheerful. The north wall will balance the illumination from the south window. Avoid puddles of sunlight on the lower parts of the walls.

5. *Use interior reflectors to diffuse the sunlight.* A skylight can deliver very uniform and diffused light when a reflector is suspended under the opening to bounce light up to the ceiling (Fig. 13.14f). Louis I. Kahn used this strategy

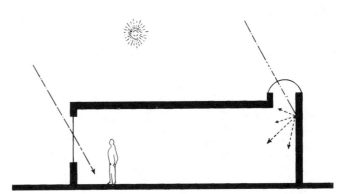

Figure 13.14d In high and narrow rooms, glare is minimal because the high light source is outside the field of view.

Figure 13.14e Place a skylight in front of a north wall for more uniform lighting and less glare.

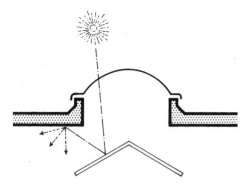

Figure 13.14f Use interior reflectors (daylight fixtures) to diffuse sunlight and reduce glare.

very successfully in the Kimbell Art Museum (Fig. 13.14g). The light entering a continuous skylight is reflected by a daylight fixture onto the underside of the concrete barrel vault. The result is extremely high-quality lighting. No direct glare results because the daylight fixture shields the skylight from view. Small perforations in the daylight fixture allow some light to filter through, so the fixture does not appear dark against the bright ceiling.

The Menil Collection in Houston, Texas, designed by architect Renzo Piano, uses skylights with diffusing baffles in all parts of the building except the two-story section (Fig. 13.14h). The skylight glazing is slightly sloped for drainage (Fig. 13.14i). The ferro-cement baffles were carefully designed to keep all direct sunlight out of the building. Baffles can also be made of light-colored and/or translucent cloth panels (Fig. 13.15i).

6. *Use exterior shades and reflectors to improve the summer/winter balance.* Shade the skylight from the summer sun, and use reflectors to increase the collection of winter skylight (Fig. 13.14j). Movable devices can be more effective (see Figs. 7.6f, 7.6g, and 17.6g).

7. *Use steeply sloped skylights to improve the summer/winter balance.* Since horizontal skylights collect more light and heat in summer than winter, skylights *steeply sloped toward the north or south* will supply light more uniformly throughout the year (Fig. 13.14k). As the slope is increased, the skylights eventually turn into monitors or clerestories, as described in the next section.

8. *Use sunlight for dramatic effect.* In lobbies, lounges, and other spaces without critical visual tasks, use sunlight and sun puddles to create delight. Splashes of sunlight moving slowly across surfaces can create dramatic effects and display the passage of time. To minimize summer overheating, use either small skylights or large skylights with reflective glazing, frets painted on the glass, or photovoltaic (PV) cells (Fig. 8.11b) to block a significant amount of the sun.

Perhaps no modern building uses daylighting as exuberantly as the Crystal Cathedral, Garden Grove, CA (Fig. 13.14m). The

Figure 13.14g In the Kimbell Art Museum, Fort Worth, TX, Louis I. Kahn very successfully used daylight fixtures to diffuse light and to eliminate direct glare.

Figure 13.14i The skylights in the Menil Collection are above the baffles which allow only soft, diffused daylight to enter.

Figure 13.14h All gallery spaces in the Menil Collection designed by Renzo Piano are daylit except for those under the second story. However, the color of the daylight varies for those areas in direct sunlight and those areas in shade that are receiving only skylight. (See color plate 5).

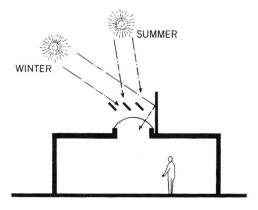

Figure 13.14j Shade the skylight from some of the summer sun, and use a reflector to increase the winter collection.

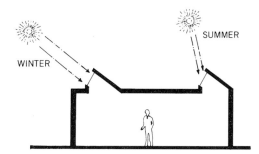

Figure 13.14k Steeply sloped skylights will perform better than horizontal ones because they collect more winter light and less summer light.

Figure 13.14m Highly reflective glazing and a gossamer space frame filter the light entering the Crystal Cathedral, Garden Grove, CA, by Johnson and Burgee.

387

Figure 13.14n Overheating problems in the Crystal Cathedral, Garden Grove, CA, are minimized by the highly reflective glazing and large sections of window-wall that can be opened. Only one of many operable panels is open (see left center).

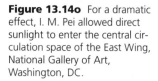

Figure 13.14o For a dramatic effect, I. M. Pei allowed direct sunlight to enter the central circulation space of the East Wing, National Gallery of Art, Washington, DC.

walls and roof are all glass which is supported by a gossamer space frame. The light is filtered first by the highly reflective glazing (8 percent transmittance) and then again by the white space frame. From the outside, the building is a large mirror mainly reflecting the blue sky (Fig. 13.14n). Overheating is minimized by the low- transmission glass, by large wall and roof panels that can open for natural ventilation, by stratification, and by the fact that the building is primarily used in either the morning or the evening in the rather mild climate found just south of Los Angeles.

In the East Wing of the National Gallery, Washington, D.C., I. M. Pei used skylights in the form of tetrahedrons to create dramatic daylighting in the atrium lobby (Fig. 13.14o). To limit the solar gain through the very large skylight, a fret of parallel white lines was applied to the glass.

13.15 CLERESTORIES, MONITORS, AND LIGHT SCOOPS

Clerestories, monitors, and light scoops are all portions of a large space that is raised above the main roof in order to bring light to the center of the space (Fig. 13.15a). The word "monitor" is ordinarily used when the windows face more than one direction and are operable (Fig. 13.15b), while the term "light scoops" is ordinarily used when the windows face one direction only and the opposite side is curved to reflect the light down (Fig. 13.15c). These devices have been used in architecture for at least 4,000 years to bring daylight into the central area of a large space. Egyptian hypostyle halls had taller columns in the center to raise the roof, thereby creating a clerestory for light and ventilation.

The vertical or near-vertical glazing of clerestories have the characteristics of windows rather than skylights. When they face south, they have the desirable effect of collecting more sunlight in the winter than in the summer. Vertical, south-facing openings can also be shaded easily from unwanted direct sunlight. North-facing openings deliver a low but constant source of light with little or no

Figure 13.15a These south-facing clerestories illuminate classrooms at the Durant Middle School, Raleigh, NC. The sawtooth arrangement keeps one clerestory from shading the next, and the sloped ceiling more efficiently directs the light down. (Innovative Design, Architect)

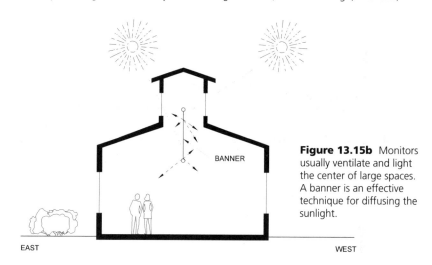

BANNER

EAST

WEST

Figure 13.15b Monitors usually ventilate and light the center of large spaces. A banner is an effective technique for diffusing the sunlight.

Figure 13.15c These light scoops on the roof of the Florida Solar Energy Center in the town of Cocoa face north because passive solar heating is not required for this building in that climate.

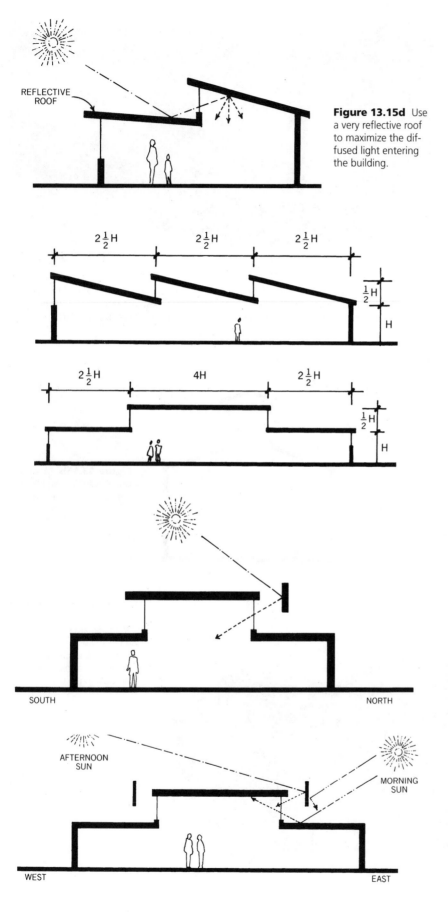

REFLECTIVE ROOF

Figure 13.15d Use a very reflective roof to maximize the diffused light entering the building.

$2\frac{1}{2}$H $2\frac{1}{2}$H $2\frac{1}{2}$H

$\frac{1}{2}$H

H

$2\frac{1}{2}$H 4H $2\frac{1}{2}$H

$\frac{1}{2}$H

H

Figure 13.15e Typical spacing of clerestories is shown as a function of ceiling height. It is usually best to have clerestories face either north or south depending on climate.

SOUTH   NORTH

Figure 13.15f A suncatcher baffle outside a north window can significantly increase daylighting on a sunny day. (After W. Lam, *Sunlighting as Formgiver for Architecture.*)

AFTERNOON SUN

MORNING SUN

WEST   EAST

Figure 13.15g Suncatcher baffles can greatly improve the performance of east and west clerestories. (After W. Lam, *Sunlighting as Formgiver for Architecture.*)

glare. East and west openings are usually avoided because of the difficulty in shading the low sun.

Another advantage of this type of top lighting is the diffused nature of the light, which results because much of the entering light is reflected off the ceiling (Fig. 13.15d). Since the light can be easily diffused once it is inside, the glazing can be transparent.

The main disadvantage of any vertical opening is that it sees less of the sky than a horizontal opening and, consequently, collects less light. As with skylights, direct glare and veiling reflections can be serious problems. Some of the more common strategies for clerestories, monitors, and light scoops follow.

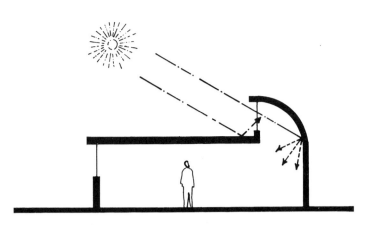

Figure 13.15h Reflect clerestory light off an interior wall. South-facing clerestories work best in this regard.

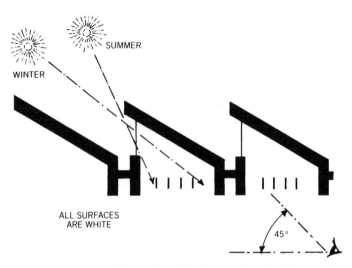

Figure 13.15j The baffles for the public library in Mt. Airy, NC, not only prevent direct sunlight from entering, but also prevent glare within the normal field of view.

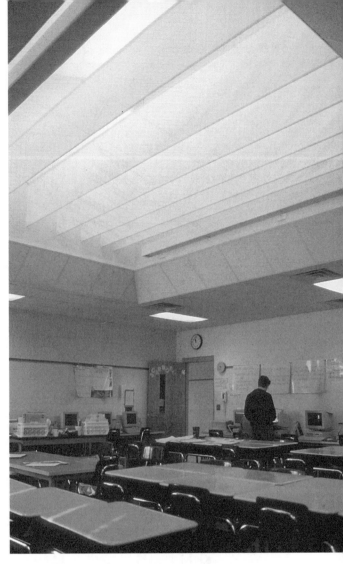

Figure 13.15i These cloth baffles prevent glare and diffuse the light entering from a south-facing clerestory at the Durant Middle School, Raleigh, NC. (Innovative Design, Architect)

1. *Orientation.* Face openings south to get the most constant year-round lighting, as well as winter solar heating. Design openings carefully to prevent problems associated with direct sunlight. In extremely hot climates with no winters, north clerestories are preferred, while in hot climates with short winters, a combination of north and south glazing might be best.
2. *Spacing:* Fig. 13.15e illustrates spacing for typical clerestories.
3. *Reflective Roof:* Use a white or very light-colored roof to reflect light into the clerestory where a matte white surface diffuses the light (Fig. 13.15d).
4. *Suncatcher Baffles:* Use suncatcher baffles outside of north clerestories to increase light collection on clear, sunny days (Fig. 13.15f).

   Although east and west clerestories are not usually recommended, their performance can be greatly improved via suncatcher baffles. Ordinarily, east clerestories receive too much morning sun and not enough afternoon light. A suncatcher can produce a more balanced light level by shading some of the morning sunlight while increasing the afternoon reflected light (Fig. 13.15g). Of course, the same is true for west clerestories.
5. *Reflect Light Off Interior Walls.* Walls can act as large, low-brightness diffusers. A well-lit wall will appear to recede, thereby making the room seem larger and more cheerful than it actually is. Furthermore, glare from a direct view of the sky or the sun can be completely avoided (Fig. 13.15h).
6. *Diffusing Baffles:* Use diffusing baffles to prevent puddles of sunlight on work surfaces, to spread the light more evenly over the work areas, and to eliminate glare from the clerestory (Fig. 13.15i). The baffle spacing must be designed both to prevent direct sunlight from entering the space and to prevent direct glare in the field of view below 45 degrees (Fig. 13.15j). The ceiling and baffles should have a matte, high-reflectance finish.

391

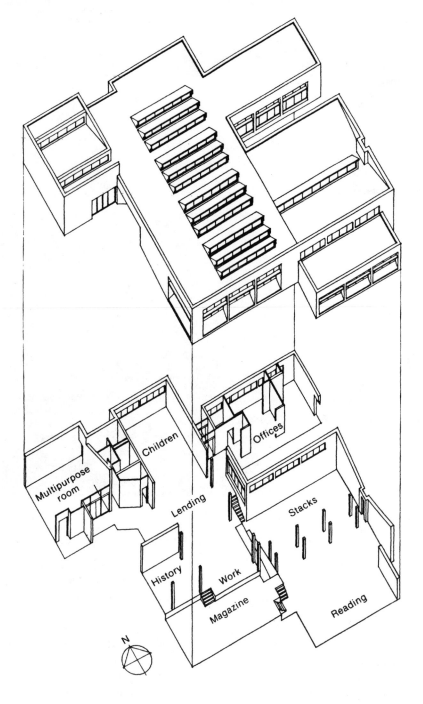

The Mt. Airy Public Library in North Carolina is an excellent example of a building largely daylit by clerestory windows. The axonometric view in Fig. 13.15k shows the location of the south-facing clerestories, and Fig. 13.15m illustrates how the direct sunlight is captured and diffused. Also, note how the electric lights and mechanical equipment are integrated into the daylighting system. This building also uses light shelves on the windows (Fig. 13.15n).

Alvar Aalto, the master of daylighting, made extensive use of clerestories and light baffles in the Parochial Church of Riola, Italy (Fig. 13.15o). He used the concrete structure to create both the light scoops and the baffles (Fig. 13.15p). Although his light scoops face north for cool, constant light, a good case could have been made for them facing south or especially east (Fig. 13.15q). East-facing light scoops might seem contradictory with everything said before, but it is not if one considers the building type, a church which is used primarily on Sunday mornings and rarely in the afternoon.

Figure 13.15k Axonometric view of the Mt. Airy Public Library, Mt. Airy, NC. Architects: Edward Mazria Assoc. and J. N. Pease Assoc. [From *Passive Solar Journal,* Vol 3 (4). © American Solar Energy Society.]

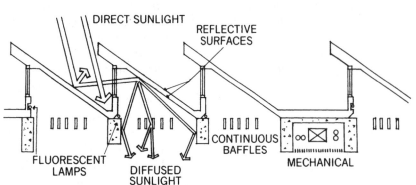

Figure 13.15m Section through the clerestory of the Mt. Airy Public Library, Mt. Airy, NC. [From *Passive Solar Journal,* Vol. 3 (4). © American Solar Energy Society.]

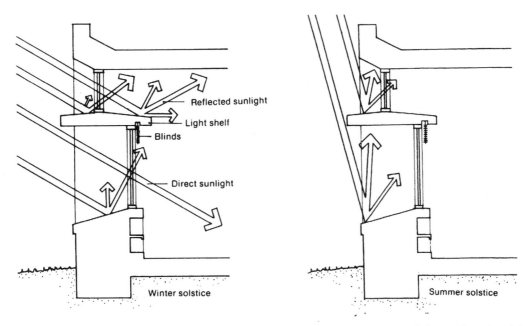

Figure 13.15n Sections through south windows of the Mt. Airy Public Library, Mt. Airy, NC. More reflected sunlight can enter in the winter than the summer. [From *Passive Solar Journal,* Vol. 3 (4). © American Solar Energy Society.]

Figure 13.15o Clerestories can also be used in the form of light scoops. The Parochial Church of Riola, Italy (1978), designed by Alvar Aalto, uses bent concrete frames to support the roof and to block the glare from the light scoops. (Photograph by William Gwin.)

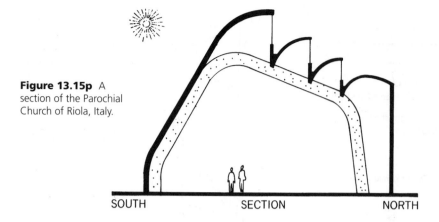

Figure 13.15p A section of the Parochial Church of Riola, Italy.

SOUTH   SECTION   NORTH

Figure 13.15q The light scoops of the Parochial School of Riola, Italy, collect constant and cool north light. (Photograph by Clark Lundell.)

13.16 SPECIAL DAYLIGHTING TECHNIQUES

The following mostly innovative daylighting strategies have the potential of being useful for special lighting problems:

1. *Light Wells or Shafts:* Light wells become more inefficient as the depth-to-width ratio increases (Fig. 13.9f) because much of the light is absorbed by the increasing number of reflections. If the well walls were very reflective, more light would be transmitted or the well could be made narrower for the same light transmission. With modern, very reflective, specular (mirrored) surfaces, which absorb less than 5 percent at each reflectance, it is possible to successfully transmit light one story with fairly small light wells. Moshe Safdie and associates used such light shafts in the National Gallery of Canada (Fig. 13.16a). Physical models were used to prove the viability of this strategy.

2. *Tubular Skylights:* Circular, duct-like tubes are commercially available with highly reflective, specular inner surfaces that transmit about 50 percent of the outdoor light through the attic. The amount of light depends on the diameter, and they are available in a range of sizes from 8 inches to 24 inches in diameter. Although they are an economic way to add light shafts to existing one-story, gabled, or flat-roofed buildings, the quality of the light is no better than that of a ceiling-mounted, circular, fluorescent lamp. Splaying the ceiling decreases the glare from this daylighting fixture (Fig. 13.16b).

3. *Beamed Daylighting.* A mirror mounted on a heliostat can track the sun and reflect a vertical beam of light through the roof regardless of the sun angle. Because the sunlight enters a building at a constant angle, it can be easily and effectively controlled.

   When large mirrors mounted on heliostats are used to light whole sections of a building, the technique is known as **beamed daylighting.** The Civil/Mineral Engineering Building at the University of Minnesota exemplifies this concept. Mirror and lenses are used to beam sunlight throughout the building (Fig. 13.16c). Where light is required, a diffusing element intercepts the beam and scatters the light.

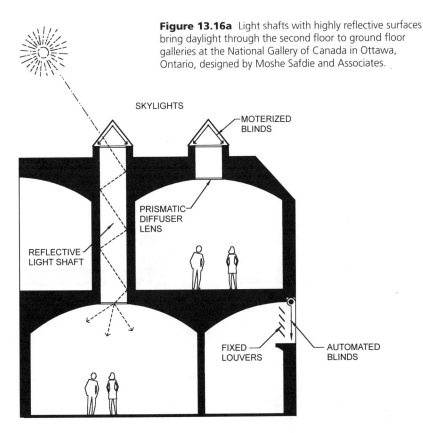

Figure 13.16a Light shafts with highly reflective surfaces bring daylight through the second floor to ground floor galleries at the National Gallery of Canada in Ottawa, Ontario, designed by Moshe Safdie and Associates.

SKYLIGHTS

MOTERIZED BLINDS

PRISMATIC DIFFUSER LENS

REFLECTIVE LIGHT SHAFT

FIXED LOUVERS

AUTOMATED BLINDS

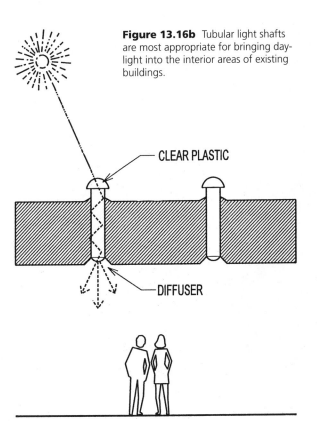

Figure 13.16b Tubular light shafts are most appropriate for bringing daylight into the interior areas of existing buildings.

CLEAR PLASTIC

DIFFUSER

Fig. 13.16c also shows how a similar type of optical system can be used to transmit views of the outdoors deep into the building or underground.

4. *Fiber Optics and Light Pipes:* Unlike the above-mentioned systems that use surface reflections to conduct light, fiber optics and light pipes use the much more efficient phenomenon of total internal reflection. These light

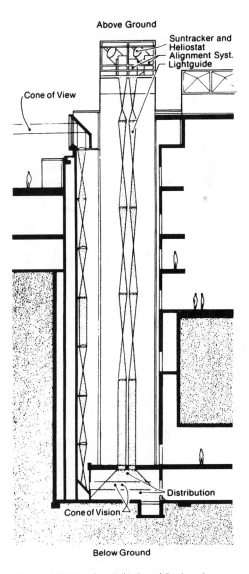

Above Ground

Suntracker and Heliostat Alignment Syst.

Lightguide

Cone of View

Distribution

Cone of Vision

Below Ground

Figure 13.16c The Civil/Mineral Engineering Building at the University of Minnesota uses beamed daylighting to light underground areas of the building. Images of the outdoors can also be beamed to the far interior or underground. (From *Building Control Systems* by V. Bradshaw, © John Wiley & Sons, Inc., 1985.)

Figure 13.16d This commercially available heliostat feeds sunlight into a fiber-optic bundle to illuminate a series of small displays indoors.

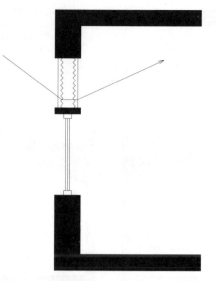

Figure 13.16e Prisms can refract light farther into the building, much like the way light shelves reflect the light.

Figure 13.16f Glass blocks illuminate this subterranean swimming pool while paving a terrace above. (Courtesy of Circle Redmont, Inc.)

guides are illuminated on one end with daylight or an electric light source, and can be designed to conduct the light either to the end or to lose it along its length to form a linear, diffused light source.

Light pipes are hollow, duct-like light guides made of prismatic, plastic film that transmit light by total internal reflection, unlike the tubular skylights mentioned above that use surface reflection. Since light pipes are essentially straight elements, mirrors are used to change direction. Fiber optics, on the other hand, consist of fine strands of plastic or glass that can easily be bent to conduct light around corners. Heliostats are used to channel the sunlight into the receiving end of the fiber bundles (Fig. 13.16d). Figs. 14.12a and 14.12b show how light pipes and fiber optics work with electric light sources.

5. *Prismatic Systems:* A popular daylighting technique in the early twentieht century, prismatic systems are being rediscovered. One of the main challenges for daylighting has always been the need

to get quality light deep into the interior of a building from the window wall. Glass or plastic prisms can be placed at the top of windows to refract light up to the ceiling, much like light shelves (Fig. 13.16e). See references, Willmert, 1999 and Baker, 1993, at the end of this chapter for more information on prismatic systems.

6. *Glass Floors:* In the nineteenth century, glass blocks were commonly used in order to enable light to reach basements. In some of the older commercial areas of New York City, one can still walk on glass-block-imbedded sidewalks that have withstood a century of trucks parking on them. Although glass blocks are making a comeback mostly for stylish reasons, they are still a wonderful way to bring light from one floor to the next (Fig. 13.16f).

Because of the brittle nature of glass, the blocks had to be small so that the failure of one

unit would not result in a serious overall failure. Today, laminated glass can accomplish the same end. As in automobile windshields and bulletproof glass, the failure of one or more glass laminations does not result in a catastrophic failure of the whole unit. Since glass has become a predictable material, engineers can calculate the thickness and number of laminations based on the size of the glass floor panel and its loading. Glass floors in block or panel form can be a powerful tool for transmitting daylight to the lower floors.

13.17 TRANSLUCENT WALLS AND ROOFS

Most translucent walls and roofs are made of either fabric membranes or composite panels. Membrane tension structures are most appropriate for large spaces with long spans. The

translucent membranes provide a very diffused, low brightness, and low-glare light source. Unfortunately, none of the available translucent-membrane materials has very good insulating value, and consequently, membrane roofs are appropriate only for special building types.

Many stadiums, tennis courts, and other similar facilities are now covered with these translucent membranes. They are usually made of a Teflon- or silicon-coated Fiberglas fabric. Even though the light transmittance of those fabrics is often less than 10 percent, abundant high-quality light is available inside because of the very large area of the translucent material. The Denver Airport is an excellent example of the high-quality daylight that filters through such membranes (Fig. 13.17a). Over-heating is avoided because the white membrane reflects about 90 percent of the solar radiation (Fig. 13.17b).

Sometimes double membranes are used to increase the insulating value

Figure 13.17a Although the roof membrane of the Denver Airport Terminal Building has very low transmittance, there is ample quality daylight because of the large-area, low-brightness translucent roof (Courtesy and © of Greg Stueber, photographer.)

Figure 13.17b Overheating is minimized because more than 90 percent of sunlight is reflected off the white membrane. The high ceiling enables the heat to rise, which is an advantage in the summer but a disadvantage in the winter.

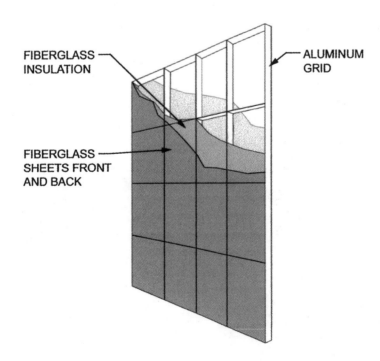

FIBERGLASS INSULATION

ALUMINUM GRID

FIBERGLASS SHEETS FRONT AND BACK

Figure 13.17c As the thermal resistance of translucent, composite sandwich panels is increased by the addition of fiberglass, the light transmittance is decreased. The panels can be used for walls and roofs of various shapes (e.g., barrel vaults)

of the skin to a point at which heating becomes feasible, but a double membrane with an R-value of about two is still a poor thermal envelope. For good thermal resistance with translucency, a composite sandwich panel system is appropriate (Fig. 13.17c). Panels with only air spaces have an R-value of about two, but if translucent fiberglass is added, the R-value can be raised as high as ten, which is about as good as a standard wall with windows. Unfortunately, as fiberglass is added to increase the thermal resistance, the transmittance of light decreases. For example, a panel with R-value of two can transmit 74 percent of the light, while a panel with an R-value of ten will transmit only 3 percent of the light. For comparison, see the R-value of various glazing systems in Figure 15.8b. A delightful byproduct of the translucency is the nighttime glow of the building walls and/or roof (Fig. 13.17d).

Figure 13.17d Translucent and insulated composite walls provide increased lighting by day and a spectacular luminescent architecture by night. (PA Technology, Princeton, NJ, by Richard Rogers, Kelbaugh & Lee Architects, photographs courtesy Kalwall Corporation.)

13.18 ELECTRIC LIGHTING AS A SUPPLEMENT TO DAYLIGHTING

Even if a building is designed to be fully daylit, an electric lighting system is still required for stormy weather and nighttime use. A daylit building can save a significant amount of energy and electrical demand only if the electric lights are turned off when sufficient daylight is available. Although people can be relied upon to turn on the lights, few will turn off the lights when they are no longer necessary. This is understandable because having both daylight and the electric lights on, thereby doubling the required illumination, is not visually objectionable and barely noticeable. The eye easily adapts to the higher illumination.

Consequently, automatic controls are necessary if daylighting is to save electricity. The automatic controls consist of a photocell placed in the ceiling of the work area and a control panel of either the on/off or the dimming type. The on/off type is less expensive, while the dimming type saves more energy and is less disturbing to the users. To take advantage of these automatic controls, the lighting fixtures must be arranged to complement the available daylight. Fig. 13.18a illustrates how the lighting gradient from part of the electric lighting can supplement the lighting gradient from daylighting. Fig. 13.18b illustrates how the fixtures are arranged in rows parallel to the windows so that

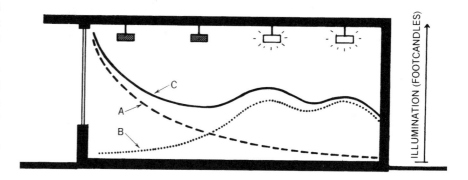

ILLUMINATION (FOOTCANDLES)

Figure 13.18a Daylighting (curve A) is supplemented by part of the electric lighting (curve B) to create a rather uniform light level (curve C).

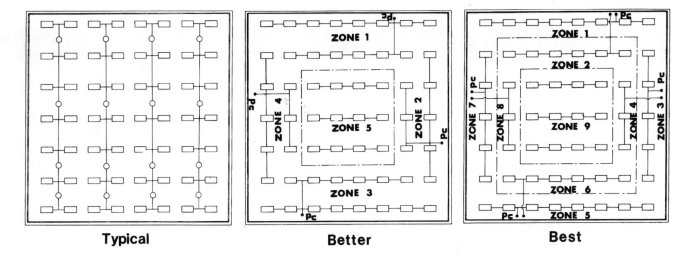

Figure 13.18b Lighting zones should consist of fixtures in rows parallel to the windows on each orientation. "Pc" stands for photocell light sensor. (From *Daylighting: Performance and Design,* by Gregg D. Ander. © Gregg D. Ander, AIA., Southern California Edison.)

any number of rows can be on or off as needed. An alternative approach is to use individually controlled fixtures; here, each fixture has its own light sensor and automatic switch.

Fluorescent lighting is the best choice for dimming and switching. The lamps can be dimmed to about 15 percent of their light output without changes in color, and they can be turned on and off almost instantaneously. Since most high intensity discharge sources (metal halide, mercury, and high-pressure sodium) change color as they are dimmed and have a long restrike time (five to ten minutes), they are not as suitable for dimming and switching strategies.

Not only do people leave electric lights on when there is more than enough daylight, they also leave them on when no one is in the room. **Occupancy sensors** are a very cost-effective solution to this problem. These sensors use either infrared radiation or ultrasonic vibrations to detect the presence of people. Combination occupancy- and light-sensing devices are also available for daylit spaces.

Daylighting should serve as the ambient part of a task/ambient lighting system. As described in the previous chapter, the ambient illumination is usually about one-third of the recommended task illumination. The user-controllable, electric task lighting then gives people the control they need to get abundant high-quality light for their tasks.

13.19 PHYSICAL MODELING

Physical-model simulation is by far the best tool for designing daylighting for a number of reasons:

1. Because of the physics of light, no error due to scale is introduced, and, therefore, the model can reproduce exactly the conditions of the actual building. Photographs made of a real space and of an accurate model show identical lighting patterns.
2. No matter how complicated the design, a model can accurately predict the result.
3. Physical models illustrate both the qualitative and the quantitative aspects of a lighting system (Figs. 13.19a and 13.19b). This is especially significant since glare, veiling reflections, and brightness ratios are often more important than illumination levels (Fig. 13.19c).

4. Simple hand calculations or basic software can produce erroneous conclusions because of the very complex nature of daylighting, and computer programs require a significant amount of time and experience. On the other hand, physical modeling requires little learning time and provide very reliable feedback and insight for the next project.
5. Physical modeling is a familiar, popular, and appropriate medium for architectural design.
6. Physical models are very effective in communicating with the client, as well as with the design team.
7. Although physical models are expensive to build and test, the resultant quality of design and improved skills of the designer make the investment worthwhile.
8. Excellent results are obtainable from even crude models as long as a few basic requirements are met.

Important Considerations for Physical Modeling

1. Architectural elements that affect the light entering a space must be carefully modeled (e.g., in terms of size, depth, and location of windows, overhangs, or baffles).

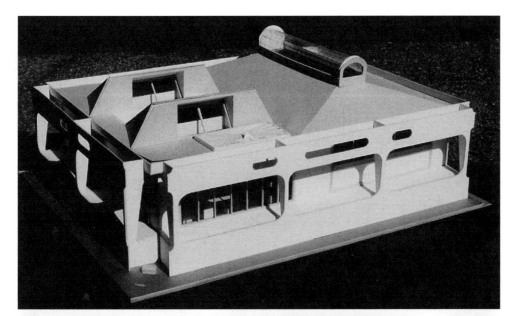

Figure 13.19a Daylighting model of a library built as a school project at Auburn University, School of Architecture, by S. Etemadi, T. Peters, and C. Scaglione.

Figure 13.19b A part of the library's roof is lifted off to show the interior.

Figure 13.19c This photograph through a view port shows the quality of the daylighting. Glare, excessive brightness ratios (puddles of sunlight), and the general lighting atmosphere are all easily and accurately determined.

Use plastic film to simulate glazing, especially if tinted, reflective, or translucent glazing is being modeled.

2. Reflectance factors should be reasonably close to the desired finishes. The best solution is to use actual finishes whenever possible.

3. External objects that reflect or block light entering the windows should be included in the model test. Include adjacent walls, trees, ground finishes, and anything else that will reflect light into or prevent light from entering the model.

4. Opaque walls must be modeled with opaque materials. Note that foamcore boards are translucent unless covered with opaque paper. All joints must be sealed with opaque material, such as black-cloth tape, black-duct tape, or aluminum tape. Vinyl (electrical) tape is not useful because it does not stick well to cardboard.

5. Test the model under the appropriate sky conditions as described below.

Helpful Hints for Constructing Models

1. Use a scale of at least ½ inch = 1 foot if possible, but for larger spaces or buildings this scale can result in difficult models to build and transport. A scale of ⅜ inch = 1 foot can still work quite well for large models.

2. Use modular construction so that alternative schemes can be easily tested. For example, the model might be constructed with interchangeable window walls.

3. Add view ports on the sides and back for observing or photographing the model. Make the ports large enough for a camera lens to get an unobstructed view and for the photometer probe to pass through. A 2-inch-square hole is usually sufficient. Windows on the model cannot be used as view ports because the observer's head would block a significant amount of light.

4. The quality of the model and the effort expended in its construction depend on the purpose of the model. If the model is not going to

be used for presentations to clients, even a crude model is sufficient for determining illumination levels and gross glare problems.

5. Since furniture can have an important effect on the lighting, especially if it is dark, large, and extensive, it should be included. Simple blocks painted with colors of the appropriate reflectance factors can act as furniture in crude models.

6. A photometer (light meter) is a very valuable instrument for measuring the illumination (footcandles or lux) inside the model. The less expensive meters are usually quite adequate, but make sure to get a meter with the sensor at the end of a wire lead.

Testing the Model

The climate of the site will determine which of the two critical sky conditions must be utilized in testing the model. In most parts of the United States, a model must be tested under both an overcast sky and a clear sky with sun. The overcast sky deter-

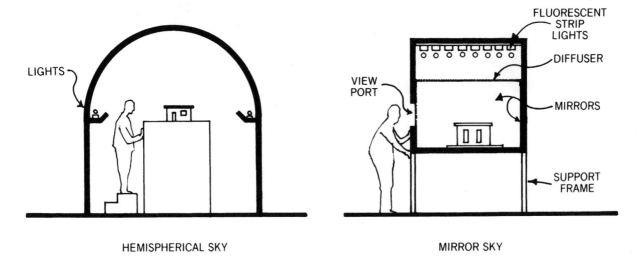

HEMISPHERICAL SKY

MIRROR SKY

Figure 13.19d Artificial skies for testing models usually consist of either a white dome or a mirror lined box.

Figure 13.19e This view of the inside of a mirror sky shows how a standard overcast sky is created, but it also shows the confusing images that multiple mirrors create.

mines whether minimum illumination levels will be met, and the clear sky with sun indicates possible problems with glare and excessive-brightness ratios.

For consistent results, artificial skies are used for the model tests. Unfortunately, artificial skies are not available to most designers. The hemispherical artificial skies are more accurate but are very expensive and bulky to build. The rectilinear mirror skies are smaller and less expensive than hemispherical artificial skies but still quite rare (Fig. 13.19d). See Fig. 13.19e for the confusing effect that opposing mirrors create. Also, most artificial skies simulate only "standard" overcast conditions, yet the consequences of direct sunlight are most critical. Thus, for a number of reasons, the real sky and sun are usually used to test daylighting models.

Avoid testing a model under partly cloudy conditions because the lighting will change too quickly to allow reliable observations. Although overcast and clear skies are quite consistent minute to minute, they vary greatly from day to day. For this reason, all quantitative comparisons between alternative schemes should be based on the daylight factor (DF) and not footcandles. The DF is a relative factor determined by measuring the horizontal indoor and outdoor illumination levels more or less at the same time. Under overcast skies, the DF remains constant even when the outdoor illumination changes. See Section 13.5 for a discussion of the daylight factor. **Rule of thumb: The design alternative with the highest daylight factor yields the highest indoor illumination.**

The human eye is the ideal tool for checking qualitative aspects of the design. The eye is also quite good at determining the adequacy of illumination levels. To get accurate results, look into the view ports for several minutes to allow your eyes to adapt to the lower light levels inside the model.

Outdoor Model-Testing Procedure for Overcast Skies

1. Place the model in the correct orientation on a table at the actual site, or at a site with similar sky access and ground reflectances. If neither of the above is possible, include the major site characteristics in the model and test it on a roof or other clear site.
2. Place the photometer sensor at the various critical points to be measured. Usually, the critical points to check are the center of a room and 3 feet from each corner.

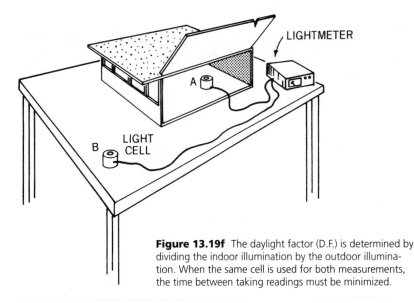

Figure 13.19f The daylight factor (D.F.) is determined by dividing the indoor illumination by the outdoor illumination. When the same cell is used for both measurements, the time between taking readings must be minimized.

The top of the sensor should be about 30 inches above the floor in the model.

3. Measure the horizontal outdoor illumination level by moving the sensor to point B (Fig. 13.19f), then calculate the DF. (D.F. = A/B)

4. Use the view ports on the sides and back of the model to visually check for glare, excessive-brightness ratios, and the general quality of the lighting. Use hands or a black cloth to prevent stray light from entering the view port during observations. Make photographs for a permanent record.

Outdoor Model-Testing Procedure for Clear Skies

The procedure is basically the same as that for overcast skies except that the model must be tilted to simulate the range of sun angles throughout the year. At the very minimum, the model should be tested for the conditions of June 21 at 8 A.M., noon, and 4 P.M.; and December 21 at 9 A.M., noon, and 3 P.M. It is very important to test the model under varying sun angles to prevent potentially serious glare and sun puddle problems. One can best accomplish this procedure by means of a sun machine and sundial, as described in Appendix C. Use the directions given under "Alternative Mode of Use of the Sun Machine," which is at the end of Appendix C.

Photographing the Model

Photographs greatly enhance the usefulness of physical models as a design tool. Photographs of model interiors facilitate the careful analysis and comparison of various lighting schemes. Photographs of well-constructed models also make for effective presentations to clients. Remember, though, that the camera does not see the way the human eye does. Brightness ratios always appear worse on film than they are in reality. The eye can also change focus as required, while the camera freezes one view. Either the near or the far image might be out of focus. Nevertheless, photography is a valuable adjunct to physical modeling. The following suggestions are for photographing the interiors of physical models.

1. Use wide-angle lenses for their large field of view, as well the increased depth of field:

2. Use a tripod and high-speed film in order to achieve maximum depth of field (e.g., film with a speed of 200 or 400 ISO.) Digital cameras seem to be less sensitive to brightness ratios than film cameras.

3. **Bracket** each photograph by taking shots at least one exposure setting higher and one setting lower than what the camera meter says.

4. Keep the center of lens at eye level of a standing scale figure in the model.

5. Avoid allowing light to leak into the model through the view port around the camera lens.

More Information On Physical Modeling

For more information on model building and testing, see *Simulating Daylight With Architectural Models* edited by Marc Schiler (see Bibliography in back of book for full citation).

13.20 CONCLUSION

As Louis Kahn suggests at the beginning of this chapter, daylighting brings meaning and richness to architecture. People want and need natural light for psychological, spiritual, and physiological reasons. Although more complicated to design than electric lighting, daylighting has profound aesthetic consequences for both the interior and exterior of buildings. The various daylighting strategies, such as light shelves and clerestories, change the appearance of buildings. Daylighting allows architects to economically justify additional visual elements that enrich the design.

The environment is also enriched. With daylighting, fewer fossil energy sources need to be extracted from the earth and less pollution is dumped into the environment. Thus, daylighting enriches life as well as architecture.

KEY IDEAS OF CHAPTER 13

1. Until the middle of the twentieth century, all buildings were daylit.
2. Daylighting is still appropriate because:
   a. People need and enjoy the qualities of daylight.
   b. It saves energy for a sustainable future and it can reduce electrical demand.
3. Daylight is a very plentiful resource. On an overcast day, the illumination on the roof is about 30 times what is required indoors, and on a sunny day it is about 160 times greater.
4. Daylighting designs should evenly distribute light throughout the space throughout the day.
5. South lighting is best because it is warm, plentiful, easy to control, and in tune with the seasons (a maximum in winter and a minimum in the summer).
6. North lighting is second best because it is constant and cool. However, it is not as plentiful or warm as south light.
7. Avoid east and west lighting if possible because of the glare from low sun angles and summer overheating.
8. Although all light turns into heat, electric light sources heat a building more than daylight.
9. In the summer, introduce only enough daylight to supply the required illumination levels; in winter, collect all the daylight possible (except internally dominated buildings in mild climates).
10. When winter heating is not desired, use spectrally selective glazing for cool daylight because it filters out the solar (short-wave) infrared radiation.
11. Strategies for a daylighting design:
    a. Use the form of the building to maximize daylit areas (e.g., long rectangle or atrium).
    b. Use open planning or glass partitions to allow light to penetrate to the interior.
    c. Use light colors on the exterior to reflect more light into openings; and use light colors on the interior to reflect light deeper into the building, to diffuse the light, and to reduce glare.
    d. Place windows high on walls.
    e. Use louvers or a light shelf to reflect light deep into building.
    f. Filter daylight to reduce glare.
    g. Use movable shades for flexibility.
    h. Use south clerestories with baffles for light that is free of glare and sun puddles.
    i. Use skylights with summer shading devices.
    j. Use translucent walls and roofs of large area and low transmittance.
    k. Use the smallest windows possible to collect the required amount of light.
12. Use electric lighting as a supplement to daylighting.
13. Use physical models to help design high-quality daylighting.
14. Use photocell controls to automatically dim or turn off lights when sufficient daylight is present.

References

Baker, N., A. Fanchiotti, and K. Steemers, eds. *Daylighting in Architecture.* London: James & James, 1993.

Moore, Fuller. *Concepts and Practice of Architectural Daylighting.* New York: Van Nostrand Reinhold, 1985.

Wilmert, Todd. "Prismatic daylighting systems, once commonly used, reemerge as a promising technology for the future." *Architectural Record,* 08 99, pp. 177–179.

Resources

FURTHER READING
(See Bibliography in back of book for full citations. The list includes valuable out-of-print books.)

Ander, G. D. *Daylighting Performance and Design.*

Brown, G. Z., and M. DeKay. *Sun, Wind, and Light: Architectural Design Strategies.*

Carmody, J., S. Selkowitz, and L. Heschong. *Residential Windows: A Guide to New Technologies and Energy Performance.*

Commission of the European Communities Directorate-General XII for Science Research and Development. *Daylighting in Architecture: A European Reference Book.*

Daniels, K. *The Technology of Ecological Building.*

Egan, M. D. *Concepts in Architectural Lighting.*

Evans, B. E. *Daylight in Architecture.*

Fitch, J. M., with W. Bobenhausen. *American Building: The Environmental Forces That Shape It.*

Flynn, J. E. *Architectural Interior Systems: Lighting, Acoustics, Air-Conditioning.*

Franta, G., K. Anstead, and G. D. Ander. "Glazing Design Handbook For Energy Efficiency."

Futagawa, Y., ed. *Light and Space: Modern Architecture.*

Guzowski, M. *Daylighting for Sustainable Design.*

Energy Design Resources. *Skylighting Guidelines.* www.energydesignresources.com

Hopkinson, R. G., P. Petherbridge, and J. Longmore. *Daylighting.*

Lighting Handbook.

Lam, W. M. C. *Sunlighting as Formgiver for Architecture.*

Millet, M. S. *Light Revealing Architecture.*

Moore, F. *Concepts and Practice of Architectural Daylighting.*

Robbins, C. *Daylighting: Design and Analysis.*

Schiler, M. *Simplified Design of Building Lighting.*

Schiler, M. *Simulating Daylighting with Architectural Models.*

Steffy, G. R. *Architectural Lighting Design.*

Steffy, G. R. *Time-Saver Standards for Architectural Lighting.*

Stein, B., and J. S. Reynolds *Mechanical and Electrical Equipment for Buildings.* A basic resource.

ORGANIZATIONS

(See Appendix J for full citations.)

IESNA Illuminating Engineering Society

International Association of Lighting Designers (IALD). www.iald.org

LBNL (Lawrence Berkeley National Laboratory) http://eande.lbl.gov/BTP/BTP.html

Lighting Design Lab. www.northwest-lighting.com

Lighting Research Center

ELECTRIC LIGHTING

"Light has always been recognized as one of the most powerful form-givers available to the designer.... Theoretically, the possibilities for imaginative lighting are limitless. And, theoretically, our ability to create great architecture should have increased in proportion to the availability of more, and more versatile, artificial [light] sources. Yet we have scarcely begun to scratch the surface of these 'limitless' possibilities."

William M. C. Lam,
Perception and Lighting as Formgivers for Architecture,
1977, p. 10.

14.1 HISTORY OF LIGHT SOURCES

Through most of human history, activities requiring good light were reserved for daylight hours. This was true not only because of the poor quality of the available light sources, but more so because of the expense. Oil lamps (Fig. 14.1a) and candles, the main sources of light, were so expensive that even the rich did not use more than a few at a time. For the poor, the choice was often light or food since lamps burnt cooking oil and most candles were made from animal fat (tallow). During the eighteenth and early nineteenth centuries, the whaling industry existed mainly for supplying oil and wax for lighting needs. The importance of whaling declined in the middle of the nineteenth century when kerosene, extracted from petroleum, became the oil of choice (Fig. 14.1b).

Coal gas was an important light source in the nineteenth century. At first, it was considered safe only for street lighting (Fig. 14.1c), but eventually it was accepted indoors as well. The light, however, was not much better than that from oil lamps until the invention of the mineral-impregnated mantle in the 1880s (Fig. 14.1d) which greatly improved both the quality and quantity of gas light. Since gas lighting, even with the mantle, generated high levels of heat and indoor air pollution, it was quickly replaced by electric lighting at the beginning of the twentieth century.

Thomas Edison did not invent the idea of the electric incandescent lamp, but he was the first to make it practical, around 1880. He also developed efficient electric generators and distribution systems without which the electric lamp was worthless. Although initial improvements made the incandescent lamp an excellent light source for general illumination, the development of electric-discharge lamps has largely rendered it obsolete. The first major lamp in this category was the fluorescent lamp, which was introduced in the late 1930s.

Until the invention of modern lighting, streets and public buildings were largely abandoned after dank. Now more and more facilities, such as offices, factories, stores, and even outdoor tennis courts, are available 24 hours a day. It has been suggested that the today's frontier is not space but "nighttime."

Neither during the day nor during the night is quality and more sustainable lighting achieved merely by supplying electric lights. As mentioned in

Figure 14.1a The history of the oil lamp is about as old as the history of mankind.

Figure 14.1b The kerosene lamp launched the petroleum age in the middle of the nineteenth century.

Figure 14.1c The mantles can be seen in this two-burner street light in old Mobile, AL. Lamplighters used to turn the gas on and off. Today, gaslights are wasteful because there is no economical way to turn them on and off every day.

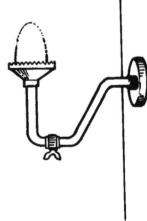

Figure 14.1d Not until the invention of the mantle did gas lamps significantly improve the quality of artificial lighting.

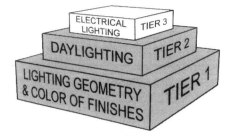

Figure 14.1e The best and most sustainable lighting design results when the three-tier design approach is applied, and this chapter covers tier three (electric lighting).

the previous two chapters, electric lighting is only the third tier of the three-tier design approach for lighting (Fig. 14.1e). The architect should make full use of geometry, color of finishes, and daylighting before the electric lighting system is designed.

14.2 LIGHT SOURCES

Fig. 13.6a shows the relative efficiency of various light sources by giving the number of lumens emitted for each watt of electricity used. This specific ratio of **lumens per watt** is called **effi-**cacy. The figure clearly shows that although the modern incandescent lamp is a great improvement over previous light sources, it, too, is inefficient when compared to the modern discharge lamps such, as fluorescent, metal-halide, and high-pressure sodium. The efficacy of each lamp type is shown as a range because efficacy is a function of several factors including wattage. High-wattage lamps have greater efficacy than low-wattage lamps. For example, a 100-watt lamp gives off much more light than the combined effect of two 50-watt lamps. The spectral distribution also influences the efficacy of lamps. Unfortunately, the lamps with the best quality white light do not have the highest efficacy.

The theoretical maximum efficacy is where 100 percent of the electrical energy is converted into light. For monochromatic yellow-green light, this would be about 680 lumens/watt, while for white light it is only about 200 lumens/watt. This difference exists because the human eye is not equally sensitive to all colors. Since the eye is most sensitive to yellow-green light, a lamp of that color will have the highest efficacy. The eye is not very sensitive to such colors as red and blue, and any light containing these colors, such as white, will have a lower efficacy than yellow-green monochromatic light. Therefore, whenever color rendition is important, we must accept the lower efficacy of white light.

The modern incandescent lamp turns only about 7 percent of the electricity into light; the other 93 percent is immediately turned into heat (Fig. 14.2). Although the fluorescent lamp is a great improvement, it still converts only about 22 percent of the electricity into light. Consequently, lighting, and especially incandescent lighting, uses large amounts of valuable electrical energy while contributing greatly to the air-conditioning load of a building.

As discussed in the previous chapter, daylighting has a higher efficacy than any white electric light source, and it is free. Thus, electric lighting should be supplemental to daylight whenever possible. Electric light sources will now be discussed in ascending order of efficacy.

Figure 14.2 These pie charts show how much of the electrical energy is converted into light and how much is converted directly into heat. Clearly, the incandescent lamp is a very hot and inefficient light source since only 7 percent of the electricity is converted into light. It is important to note that daylight is only the coolest light source if the *quantity* that is brought into the building is carefully controlled.

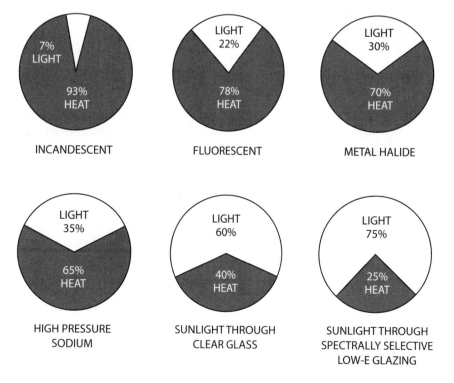

INCANDESCENT

FLUORESCENT

METAL HALIDE

HIGH PRESSURE SODIUM

SUNLIGHT THROUGH CLEAR GLASS

SUNLIGHT THROUGH SPECTRALLY SELECTIVE LOW-E GLAZING

14.3 INCANDESCENT LAMPS

Incandescent lighting has maintained its popularity for a number of reasons. Besides its low initial cost, it is also very flexible. Incandescent lamps come in a wide variety of sizes, types, and wattages (Fig. 14.3a). Since many of the lamps are interchangeable, adjustments in light pattern and intensity are easily accomplished at any time.

In an incandescent lamp, the light is emitted by electrically heating a tungsten filament (Fig. 14.3b). The

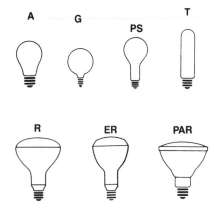

Figure 14.3a Some common shapes for incandescent lamps. Shapes of lamps are classified by uppercase letters (e.g., R, reflector; PAR, parabolic aluminized reflector; ER, ellipsoidal reflector). (Courtesy of Osram-Sylvania.)

Figure 14.3b The tungsten filaments of incandescent lamps are frequently coils of coils to concentrate the light source. (Courtesy of Osram-Sylvania.)

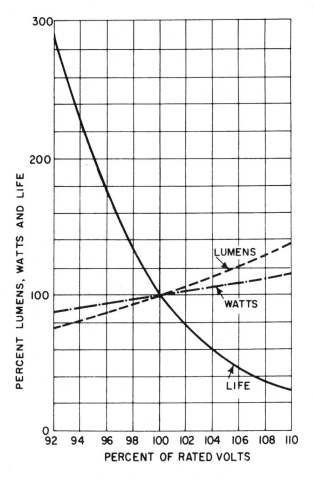

Figure 14.3c All incandescent lamps are sensitive to voltage variations. When the voltage is decreased, the life of the lamp greatly increases but the light output decreases. Thus, the lamp is much less efficient. (Courtesy of Osram-Sylvania)

hotter the filament glows, the more light is emitted and the higher the color temperature. Unfortunately, the life of the lamp is shortened. Fig. 14.3c shows the relationship between voltage, **lumens (light output),** and lamp life. Most long-life lamps are nothing more than lamps designed for a higher voltage than will be used. The operating voltage is, then, less than the design voltage of the lamp, and, consequently, the lamps will operate at a rather low temperature. The penalty, however, for the longer life is severely reduced light output and efficacy. Long-life incandescent lamps are, therefore, rarely economical. The average life of an ordinary incandescent lamp is about 1,000 hours, while a long-life lamp lasts about 3,000

hours. This life span is still very small compared to the 10,000-hour life span of fluorescents and the 20,000 hour life span of HID lamps.

Tungsten evaporation causes the blackening of the lamp and eventually lamp failure. This evaporation of the filament can be reduced by adding halogen elements to the inert gases inside the lamp. These types of incandescent lamps can, therefore, be operated at higher temperatures without shortening lamp life excessively. This variation of the incandescent lamp is known as the **tungsten halogen** or **quartz iodine lamp** (Fig. 14.3d). Because of their intense light and small size, they are very popular as automobile headlamps, projector lamps, and spotlights for accent lighting.

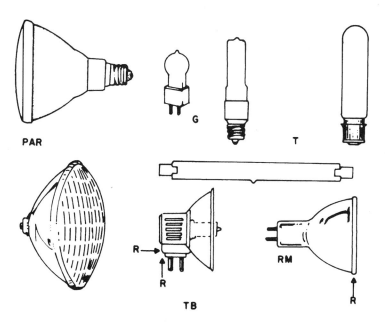

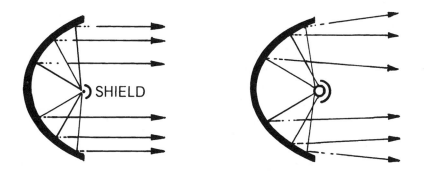

Figure 14.3d Common shapes of tungsten halogen lamps. (Courtesy of Osram-Sylvania.)

Figure 14.3e Parabolic reflectors will reflect light as a parallel beam if a point source is located at the focal point. Since all real sources are larger than a point, lamps cannot generate completely parallel beams of light.

One of the main advantages of the incandescent family of lamps is the optical control that is possible. A point source of light at the focal point of a parabolic reflector will produce a beam of parallel light (Fig. 14.3e). Although there is no point source of light available, incandescent lamps come closer than most other types of lamps. A tightly wound coil of a coil, as shown in Fig. 14.3b, when placed at the focal point of a parabolic reflector, will create a narrow but not parallel beam of light.

Incandescent downlights can be extremely wasteful if the wrong lamp is used. For example, if a standard "A"-type lamp is placed in a downlight designed for a reflector lamp, almost no light will exit the luminaire (Fig. 14.3f left). When the reflector (R) lamp is recessed to prevent direct glare, much light is still absorbed (Fig. 14.3f middle). An ellipsoidal reflector (ER) lamp is best because the beam is focused at the opening, and, therefore, most of the light exits the luminaire (Fig. 14.3f right). Of course, "A"-type lamps can be used if the luminaire is designed for that lamp, but a compact florescent would be much more efficient.

Low voltage (5.5-volt or 12-volt) lamps have smaller filaments than 120-volt lamps, and are, therefore, more of a point light source than regular lamps. They can yield beams

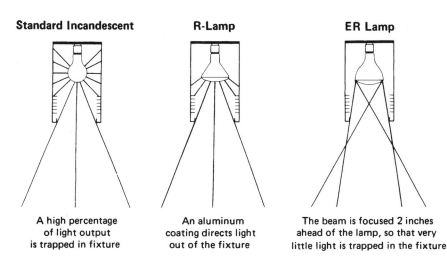

Standard Incandescent

A high percentage of light output is trapped in fixture

R-Lamp

An aluminum coating directs light out of the fixture

ER Lamp

The beam is focused 2 inches ahead of the lamp, so that very little light is trapped in the fixture

Figure 14.3f It is very important to use the correct lamp in a luminaire. Ellipsoidal reflector (ER) lamps are the most efficient choice for canned downlights with small openings. A 75-watt ER lamp is as effective as a 150-watt reflector lamp.

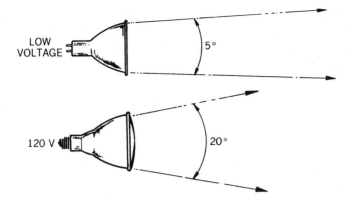

Figure 14.3g Low-voltage lamps can generate beams of light narrower than is possible with regular (120-volt) lamps.

as narrow as 5 degrees, while regular 120-volt lamps produce light beams 20 degrees or wider (Fig. 14.3g). This makes low-voltage lamps very appropriate for accent lighting. They can save energy as well because with the narrow beam more light is on target and less is spilled on adjacent areas.

The color-rendering quality of incandescent lamps is generally considered to be very good. Like daylight, the incandescent lamp emits a continuous spectrum, but unlike daylight, the color spectrum is dominated by the reds and oranges (Color Plate 4). The warm colors, including skin tones, are, therefore, complemented by this kind of lighting. Incandescent lamps are also popular because of their association with traditional surroundings and, consequently, still dominate in the home.

Because of the above-mentioned reasons of low first cost, beam control, and very good color rendition, incandescent lamps will continue to find specialty applications. These lamps are appropriate for accent lighting of small areas or objects, such as retail displays, sculpture, and paintings. Incandescent lamps are especially appropriate when sparkle and specular reflectances are desired as in chandeliers or in the display of glassware, silverware, or jewelry. Incandescent lamps are also appropriate where a *low light level* and warm

atmosphere are desirable, as in restaurants, lounges, and residences. Even here, however, compact fluorescent lamps are often more appropriate.

Incandescent lighting is *not* appropriate in situations where moderate or high light levels are required over large areas. The low efficacy of these lamps makes such applications extremely energy wasteful and often also very expensive. When equipment and cooling costs are also included they are even expensive on a first-cost basis.

14.4 DISCHARGE LAMPS

A major improvement in electric lighting came first with the development of the fluorescent lamp and then again with the development of high-intensity discharge lamps (mercury, metal-halide, high-pressure sodium).

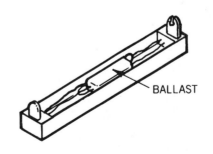

Figure 14.4 All discharge lamps require a ballast first to start the lamp and then to maintain the proper operating current.

All of these lamps are based on a phenomenon known as **discharge,** in which an ionized gas rather than a solid filament emits light.

All discharge lamps require an extra device known as a **ballast** (Fig. 14.4), which first ignites the lamp with a high voltage and then limits the electric current to the proper operating level. Traditional ballasts made of copper coils are being replaced by electronic ballasts. Ballasts are a large part of any discharge lighting system and can be a source of noise if poorly made or installed. They are rated for noisiness, with "A" types the most quiet and "F" types the noisiest.

The long life of the discharge lamps and their high efficacy are usually more than enough to offset the extra cost of the ballast and the higher cost of each lamp when compared to incandescent lamps. Since the various groups of discharge lamps have significant differences, each group will be discussed separately.

14.5 FLUORESCENT LAMPS

Although the fluorescent lamp was the first major discharge lamp, it is still very popular. It is available in a wide variety of sizes, colors, wattages, and shapes (Fig. 14.5a). Because of the concern with energy, compact fluorescent lamps have been developed that can directly replace the much less efficient incandescent lamp (Fig. 14.5b). Compact fluorescents contribute much less to global warming, and they are less expensive overall. Although their initial cost is high, their high efficiency and long life make them much less expensive on a life-cycle basis (Table 14.5).

In the fluorescent lamp, the radiation is emitted from a low-pressure mercury vapor that is ionized. Since much of the radiation is in the ultraviolet part of the spectrum, the inside surface of the glass tube is coated

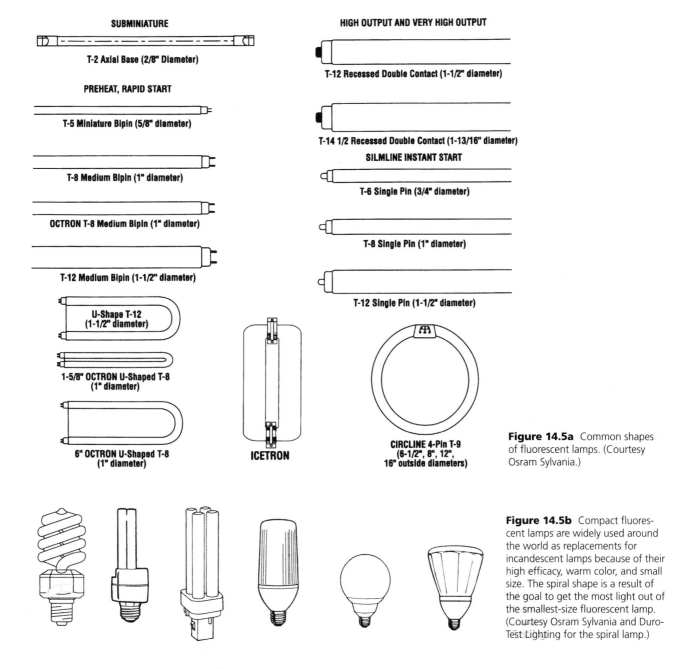

Figure 14.5a Common shapes of fluorescent lamps. (Courtesy Osram Sylvania.)

Figure 14.5b Compact fluorescent lamps are widely used around the world as replacements for incandescent lamps because of their high efficacy, warm color, and small size. The spiral shape is a result of the goal to get the most light out of the smallest-size fluorescent lamp. (Courtesy Osram Sylvania and Duro-Test Lighting for the spiral lamp.)

Bulb type Watts/Lumens*	Typical bulb cost dollars	Rated life in hours	Efficacy lumens per watt	Energy cost (@ 8¢/kwh) for 10,000 hrs.	Bulb(s) + Energy Cost for 10,000 hrs.
Compact Fluorescent 27 watt/1,750 Lumens	$20	10,000	64	$22	$42
Incandescent 100 watt/1,750 Lumens	$0.50	750	17	$80	$85

Table 14.5 As this chart shows, compact fluorescent lamps are much less expensive in real terms than incandescent lamps. (Taken from Energy Fact Sheet #13 prepared by Southface Energy Institute, a good source of sustainable building information—see Appendix J.)

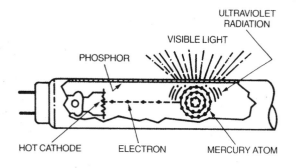

PHOSPHOR

ULTRAVIOLET RADIATION

VISIBLE LIGHT

HOT CATHODE   ELECTRON   MERCURY ATOM

Figure 14.5c The basic features of a fluorescent lamp are shown. The ultraviolet radiation is converted into visible light by the phosphor coating on the inside of the glass tube. (Courtesy of GTE Products Corporation, Sylvania Lighting Center.)

Figure 14.5d In standard fluorescent lamp designation, the "T" stands for tubular, and the number after the "T" stands for the diameter in one-eighths of an inch. Although the T12 was the traditional size for decades, the T8 is now considered the standard size. The T5 is also available in 4-foot lengths while the longest T2 is only 20 inches long.

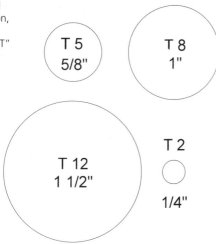

T 5
5/8"

T 8
1"

T 12
1 1/2"

T 2
1/4"

with phosphors to convert that invisible radiation into light (Fig. 14.5c). By using different kinds of phosphors, fluorescent lamps can be designed to emit various types of white light. For example, warm white lamps emit more energy in the red end of the spectrum, while cool white lamps emit more energy in the blue end of the spectrum (Color Plate 2). Special formulated fluorescent lamps are available that give excellent color rendition.

Because of its traditionally large physical size, the fluorescent lamp was only suitable as a large area source of light. This made it an excellent source for diffused lighting but an inappropriate source when beam control was required.

The large physical size was also a disadvantage when a small fixture

Figure 14.5e Neon lights help define the entrance way into this office building on John Street in New York City.

of the lamp. In normal use, a life of 15,000 hours is not unusual. It was once considered prudent to leave lamps on to maximize their life, but the high cost of energy and the need to protect the environment make it proper to turn lights off when they are not required.

Neon and Cold-Cathode Lamps

Neon and cold-cathode lamps are close relatives of fluorescent lamps. These lamps use such gases as neon, which gives off red light, and argon, which gives off blue light. Through the use of different combinations of gases, colored glass, and phosphors, a large variety of rich, colored light sources is possible.

The main advantage of these lamps is that they can be custom made to almost any desired shape. Neon, which uses about 0.5-inch-diameter glass tubes, can be bent into very complex shapes, while cold-cathode lamps, which use 1-inch-diameter glass tubes, can be bent only with more gentle curves. Neon lamps are hard-wired into place, while cold-cathode lamps usually fit into sockets. Both lamp types can be easily dimmed and have long lives of about 25,000 hours. Instead of a ballast, they require a transformer to generate the necessary high voltage.

Neon and cold-cathode lamps will not replace fluorescent lamps for general lighting because of their lower efficacy and lower light output. A cold-cathode lamp's light output is about half and the neon lamp's output is about a sixth of that of a fluorescent lamp of equal length. Neon and cold-cathode lamps are appropriate for applications that require special colors and special shapes. These lamps are most suitable when the shape of the lamp is closely integrated with the form of the architecture (Fig. 14.5e) or when the shape of the lamp is itself the design element (Fig. 14.5f).

Figure 14.5f Cold-cathode tubes used for both form generation and illumination in the Town Center, Boca Raton, FL. (Courtesy of National Cathode Corporation.)

size was desired. The availability of compact fluorescent lamps now allows manufacturers to make incandescent-like luminaires for them. Also, much more slender linear fluorescent lamps are now available (Fig. 14.5d). These new lamps are not only more efficient than standard lamps, but they also allow the creation of slender luminaires with a fair amount of beam control.

Long lamp life is another great virtue of the fluorescent lamp, but frequent starting cycles decrease the life

14.6 HIGH-INTENSITY DISCHARGE LAMPS (MERCURY, METAL-HALIDE, AND HIGH-PRESSURE SODIUM)

High-intensity discharge lamps are very efficient light sources that in size and shape are more like incandescent than fluorescent lamps (Fig. 14.6a), but like all discharge lamps they need a ballast to work. In all of the high-intensity discharge lamps, the light is emitted from a small arc tube that is inside a protective outer bulb (Fig. 14.6b). The relatively small size of this arc tube permits some optical control similar to that possible with a point source (see Fig. 14.3e). When increased color rendition is desired, metal halides are added to the mercury in the arc tube or phosphors are added to the inside of the outer bulb. However, the addition of phosphors greatly increases the size of the source, and some optical control is lost.

High-intensity discharge lamps have two other important characteristics in common. They all require a few minutes to reach maximum light output, and they will not restrike immediately when there is a temporary voltage interruption. The lamps must cool down about five minutes before the arc can restrike. Recently, a few instant-restrike lamps have come on the market. In public areas, a supplementary emergency light source (e.g., an incandescent or fluorescent lamp) should be part of the design.

Mercury Lamps

Besides lower efficacy compared to other discharge lamps, mercury lamps also have poor color rendition. They produce a very cool light, rich in blue and green and deficient in the red and orange parts of the spectrum (Color Plate 4). Because of their blue-green light, mercury lamps are still appropriate in landscape lighting, but otherwise they are largely obsolete.

Metal-Halide Lamps

The white light metal-halide lamps emit is moderately cool, but there is enough energy in each part of the spectrum to give very good color rendition (Color Plate 4). Metal-halide lamps are

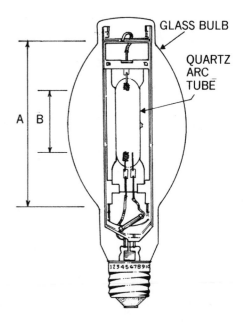

Figure 14.6b High-intensity discharge lamps generate the light in the arc tube. This relatively small source (dimension B) allows a fair amount of optical control. When a phosphor coat is used, however, the light source is much larger (dimension A) and beam control becomes difficult. (Courtesy of Osram-Sylvania.)

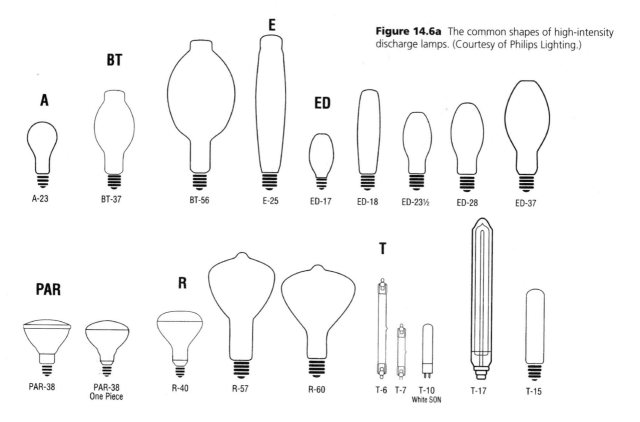

Figure 14.6a The common shapes of high-intensity discharge lamps. (Courtesy of Philips Lighting.)

appropriate for stores, offices, schools, industrial plants, and outdoors where color rendition is important. These lamps are some of the best sources of light today because they combine in one lamp many desirable characteristics: high efficacy (80–125 lumens/watt), long life (10,000–20,000 thousand hours), very good color rendition, and small size for optical control.

High-Pressure Sodium Lamps

When high efficacy (70–140 lumens/watt) and long life are of prime importance, the high-pressure sodium (HPS) lamp group is usually the design choice. Although the color rendition of HPS lamps is not very good, some people find the warm golden-white light acceptable when color is not important. Most of the emitted energy is in the yellow and orange parts of the spectrum (color plate 4).

HPS lighting is most appropriate for outdoor applications, such as lighting for streets, parking areas, sports areas, and building floodlighting. Research has shown, however, that for low-light-level peripheral vision (e.g., to see a deer at the side of the road or a mugger at the edge of a parking lot), cool sources, such as metal-halide lamps, far outperform warm sources, such as high-pressure sodium lamps. Indoor spaces where color rendition is not important can also make use of the lamps' high efficacy. HPS lighting is quite appropriate for many industrial and warehouse spaces.

A low-pressure sodium lamp group also exists. Although it has the highest efficacy of any lamp group (130–180 lumens/watt), its monochromatic yellow light is unacceptable in most applications.

14.7 COMPARISON OF THE MAJOR LIGHTING SOURCES

To help designers choose the best light source for their needs, Table 14.7 compares the major lamp groups by pro-

TABLE 14.7 COMPARISON OF THE MAJOR LAMP GROUPS

Lamp Group	Advantages	Disadvantages	Applications	Efficacy (Lumens/Watt)	Life (Hours)
Incandescent & Halogen	Excellent optical control (e.g., very narrow beams of light are possible) Very good color rendition (especially warm colors and skin tones) Very low initial cost (especially useful when many low-wattage lamps are used) Flexible (easily dimmed or replaced with another lamp of a different wattage) Very small fixtures are possible	Very low efficacy (high energy costs and) burden on the environment Very low lamp life (high maintenance costs) Adds high heat load to buildings, thereby increasing cooling load	For spotlighting, accent, high-lighting and sparkle (residential, restaurants, lounges, museums)	10–25	750–2,500
Fluorescent	Very good for diffused, wide area, low brightness lighting Good color renditions (varies greatly with lamp type) Very good efficacy Long lamp life (Compact fluorescents are efficient replacements for many incandescent-lamp applications.)	Limited optical control (no narrow beams possible) Linear source except CFLs Sensitive to temperature and, therefore, not used outdoors in cold climates	For diffused even lighting of a large area (offices, schools, residential, industrial)	40–90	6,000–20,000
Metal-Halide	Good optical control Very good color rendition (especially of blue, green, and yellow) High efficacy Long lamp life Small fixtures possible	5 to 10 minute delay in start or restart Fairly expensive	For diffused lighting or wide beams (offices, stores, schools, industrial, outdoor)	80–120	9,000–20,000
High-Pressure Sodium	Good optical control Very high efficacy Very long lamp life	Color rendition is poor (mostly orange and yellow) About 5-minute delay in start or restart	For diffused lighting or wide beams where color is not important (outdoor, industrial, warehouses, interior and exterior floodlighting)	80–140	20,000–24,000

viding the advantages, disadvantages, and major applications for each group.

Some of the most important considerations in choosing a lighting system are lighting effect desired, color rendition, energy consumption, illumination level, maintenance costs, and initial costs. When we consider energy consumption and illumination level, the lamp efficacy (lumens/watt) is the prime factor. Typical ranges of efficacy, as well as lamp life, are found in Table 14.7.

14.8 NEW LIGHT SOURCES

Several new light sources might eventually become important providers of light in buildings. Meanwhile, they are appropriate for some limited applications.

Induction Lamp

Also known as the **electrodeless fluorescent lamp** or by the trade names of QL and ICE, the **induction lamp** can last 100,000 hours because it has no electrodes to wear out. It also has a good color rendition index (CRI of 80–84) and very good efficacy (70 to 80 lumens per watt). An induction lamp is ideal for locations where lamp replacement is very difficult.

Sulfur Lamp

Microwaves are focused on a small quartz sphere filled with sulfur and other gases to create a very large amount of high-quality light. For example, one particular sulfur lamp emits 130,000 lumens, which is equivalent to seventy-four 100-watt incandescent lamps. These incandescent lamps require 7,400 watts of power, while the sulfur lamp requires only 1,300 watts (i.e., 100 lumens per watt). Because the light-emitting quartz sphere is about the size of a golf ball, very good beam control is possible, and the lamp is, therefore, frequently used with light-guides, which are described below.

Light-Emitting Diodes (LED)

Solid-state **light-emitting diodes** have been used extensively in electronic equipment because of their very long life. As their efficacy has improved, they have become popular outdoors for traffic lights and indoors for exit signs and strip lighting to define emergency-egress paths. They are appropriate for low-wattage devices since their efficacy is still only about 30 lumens per watt.

A closely related technology is the light-emitting polymer (LEP), which is made of organic semiconducting material while the LED is made of inorganic semiconducting material. Although still experimental, they have the promise of being inexpensive ultra-thin films that could be used as wallpaper.

TABLE 14.9 LIGHTING FIXTURES (LUMINAIRES)

Illustration[a]	Type	
	0%–10% / 90%–100%	*Direct:* Direct lighting fixtures send most of the light down to the workplane. Since little light is absorbed by the ceiling or walls, this is an efficient way to achieve high illumination on the workplane. Direct glare and veiling reflections are often a problem, however. Also, shadows on the task are a problem, when the fixture-to-fixture spacing is too large.
	10%–40% / 60%–90%	*Semi-direct:* Semi-direct fixtures are very similar to direct luminaires except that a small amount of light is sent up to reflect off the ceiling. Since this creates some diffused light as well as a brighter ceiling, both shadows and the apparent brightness of the fixtures are reduced. Veiling reflections can still be a problem, however.
	40%–60% / 40%–60%	*General diffuse:* This type of fixture distributes the light more or less equally in all directions. The horizontal component can cause severe direct glare unless the diffusing element is large and a low-wattage lamp is used.
	40%–60% / 40%–60%	*Direct–indirect:* This luminaire distributes the light about equally up and down. Since there is little light in the horizontal direction, direct glare is not a severe problem. The large indirect component also minimizes shadows and veiling reflections.
	60%–90% / 10%–40%	*Semi-indirect:* This fixture type reflects much of the light off the ceiling and, thus, yields high-quality lighting. The efficiency is reduced, however, especially when the ceiling and walls are not of a high-reflectance white.
	90%–100% / 0%–10%	*Indirect:* Almost all of the light is directed up to the ceiling in this fixture type. Therefore, ceiling and wall reflectance factors must be as high as possible. The very diffused lighting eliminates almost all direct glare, veiling reflections, and shadows. The resultant condition is often used for ambient lighting.

[a]Drawings are from *Architectural Graphic Standards,* 8th ed., Ramsey and Sleeper, © John Wiley & Sons, Inc., 1988.

14.9 LIGHTING FIXTURES (LUMINAIRES)

Lighting fixtures, also called **luminaires,** have three major functions: supporting the lamp with some kind of socket, supplying power to the lamp, and modifying the light the lamp emits to achieve a desired light pattern and to reduce glare. General lighting fixtures are divided into six generic categories by the way they distribute light up or down (Table 14.9).

Luminaires with a sizable direct component are most appropriate when high illumination levels are required over a large area, or when the ceiling and walls have a low reflectance factor. Although their energy efficiency is high, the quality of light usually is not. Direct glare, veiling reflections, and unwanted shadows are all reduced or eliminated by the fixtures with a large indirect component. Task/ambient lighting provides the benefits of both approaches. This lighting is discussed further below.

The quality of the lighting from direct fixtures can be improved by the design of the fixtures. The following section describes the various techniques used to improve these types of luminaines.

14.10 LENSES, DIFFUSERS, AND BAFFLES

The distribution of light from a luminaire (in a vertical plane) is often defined by a curve on a polar-coordinate graph, where the distance from the center represents the candlepower (intensity) in that direction. The candlepower distribution curve of a semidirect lighting fixture is shown in Fig. 14.10a. The up-directed light will reflect off the ceiling to reduce both direct glare and veiling reflections. For the same goal, some direct lighting fixtures are designed to distribute light in a **batwing light pattern** (Fig. 14.10b). The high-angle light that

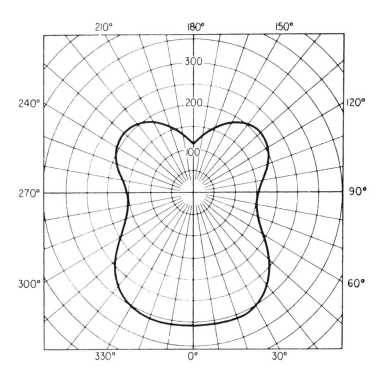

Figure 14.10a Manufacturers generally supply candlepower distribution curves for their lighting fixtures. In this vertical section, the distance from the center determines the intensity of the light in that direction. This curve is for a semidirect lighting fixture. (Courtesy of Osram-Sylvania.)

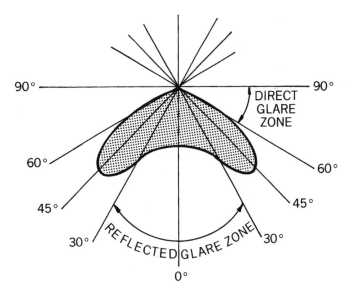

Figure 14.10b Light that leaves the luminaire from 0 degrees to 30 degrees tends to cause veiling reflections, while light in the 60- to 90-degree zone tends to cause direct glare. Fixtures with batwing light-distribution patterns yield a better quality light because they minimize the light output in these problematic zones. However, they are not ideal when computers are used.

causes the direct glare and the low-angle light that causes the veiling reflections are, therefore, avoided as much as possible.

Lenses, prisms, diffusers, baffles, and reflectors are all used in fixtures to control the manner in which light is distributed from the lamps.

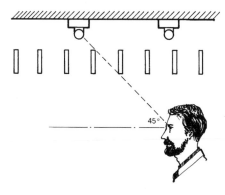

Figure 14.10c Baffles, louvers, and eggcrates are used to shield against direct glare. The direct view of the light sources should be shielded up to at least 45 degrees.

Baffles, Louvers, and Eggcrate Devices

These devices limit direct glare by restricting the angle at which light leaves the fixture (Fig. 14.10c). If these devices are painted white, they, in turn, can become a source of glare. If, on the other hand, they are painted black, much of the light is absorbed and the efficiency of the fixture is very low. These devices can be small and part of the luminaire, or they can be large and part of the architecture (e.g., waffle slab or joists). One-way baffles, such as louvers, joists, and beams, are useful only if viewed perpendicular to their direction (Fig. 14.10d).

Parabolic Louvers

This type of louver is made of parabolic wedges (Fig. 14.10e) with a specular finish. These devices are extremely effective in preventing

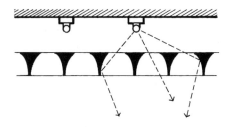

Figure 14.10e Parabolic louvers are very effective in reducing direct glare.

direct glare because the light distribution is almost straight down. Thus, these fixtures have a high visual comfort probability (VCP). They are also very good in preventing veiling reflections in computer monitors and video display terminals (VDTs) (Fig. 14.10f). The penalty for having mostly vertical light is that vertical surfaces are not well illuminated. These louvers also do not solve the problem of veiling reflections on horizontal surfaces.

Figure 14.10d One-way baffles are effective only when people are limited to viewing the ceiling from one direction. For example, in a corridor, the baffles should be oriented perpendicular to the length of the corridor. Use eggcrates when shielding is required in two directions.

Figure 14.10f The left photo shows the room after the luminaire lenses were replaced with parabolic louvers. The reduction in direct glare and veiling reflections in the computer monitor at left and center is very significant. However, notice that while horizontal surfaces are brighter, vertical surfaces are darker (note books in left foreground). (Courtesy of American Louver Company.)

Figure 14.10g Lenses and prisms refract the light down to reduce direct glare.

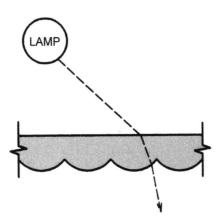
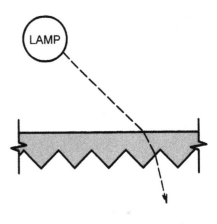

Diffusing Glass or Plastic

Translucent or surface "frosted" sheets diffuse the emitted light more on less equally in all directions. The horizontal component of this distributed light is a cause of significant direct glare. Consequently, these devices have only limited usefulness.

Lenses and Prisms on Clear Sheets

When the surface of clear sheets of glass or plastic is formed into small lenses or prisms, good optical control is possible. The light is refracted so that more of the distribution is down and direct glare is reduced (Fig. 14.10g).

14.11 LIGHTING SYSTEMS

Lighting systems can be divided into six generic types. In many applications, a combination of these basic systems is used.

General Lighting

General lighting consists of more or less uniformly spaced, ceiling-mounted direct lighting fixtures (Fig. 14.11a). It is a very popular system because of the flexibility in arranging and rearranging work areas. Since the illumination is roughly equal everywhere, furniture placement is relatively easy. The energy efficiency is usually low because noncritical work areas receive as much light as the task areas. Light quality, especially veiling reflections, is also a problem, since it is hard to find a work area that does not have a lighting fixture in the offending zone (see Fig. 12.11h).

Localized Lighting

Localized lighting is a nonuniform arrangement, in which the lighting fixtures are concentrated over the work areas (Fig. 14.11b). Fairly high efficiency is possible since nonwork areas are not illuminated to the same degree as the work areas. Veiling reflections and direct glare can be minimized because this system affords quite some freedom in fixture placement. Flexibility in rearranging the

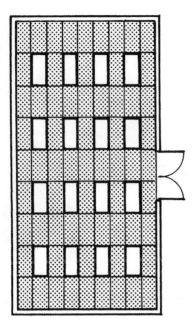

Figure 14.11a This reflected-ceiling plan shows the regular layout of direct luminaires, which is typical of "general lighting systems." This approach is very flexible but not very efficient.

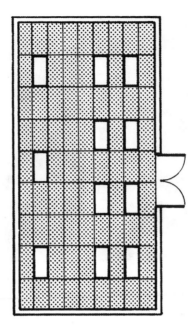

Figure 14.11b This reflected-ceiling plan illustrates what is known as localized lighting. In this system, direct fixtures are placed only where they are needed. It is very efficient but not very flexible.

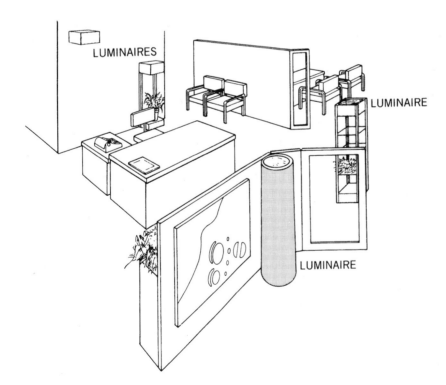

Figure 14.11c Ambient lighting is a soft, diffused light from indirect fixtures. This diagram shows the luminaires mounted either on pedestals (torchieres) or on the wall (sconces). (Courtesy of Cooper Lighting.)

Figure 14.11d Ambient lighting from furniture-integrated lighting fixtures. (Courtesy of Cooper Lighting.)

furniture is more difficult, however, unless track lighting or other adjustable systems are used.

Ambient Lighting

Ambient lighting is indirect lighting reflected off the ceiling and walls. It is a diffused low-illumination level lighting that is sufficient for easy visual tasks and circulation. It is usually used in conjunction with task lighting and is then known as **task/ambient lighting**. Direct glare and veiling reflections can be almost completely avoided with this approach. The luminaires creating the ambient lighting can be suspended from the ceiling, mounted on walls, supported by pedestals, or integrated into the furniture (Figs. 14.11c, 14.11d, and 14.11e). To prevent hot spots, the indirect fixtures should be at least 12 inches below the ceiling, and to prevent direct glare, they should be above eye level (Fig. 14.11d). The ambient illumination level should be about one-third of the task light level.

Task Lighting

The greatest flexibility, quality, and energy efficiency are possible with **task lighting** attached to or resting on the furniture (Fig. 14.11f). Direct glare and veiling reflections can be completely prevented when the fixtures are placed properly (see Fig. 12.13a). Since only the task and its immediate area are illuminated, the energy efficiency is also very high. The individ-

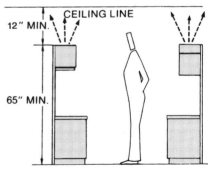

Figure 14.11e Ambient lighting from pendent indirect luminaires. (Courtesy of Peerless Lighting Corporation.)

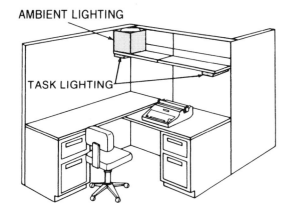

AMBIENT LIGHTING

TASK LIGHTING

Figure 14.11f Note how the task lights are mounted on each side and not in front of the work area because of the problem of veiling reflections. Since an indirect luminaire is included in the office furniture, this system is known as task/ambient lighting. (Courtesy of Cooper Lighting.)

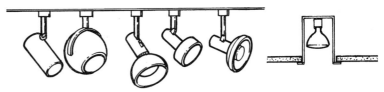

Figure 14.11g Accent lighting is usually achieved with track lighting or canned downlights. To highlight only small areas or objects, low-voltage fixtures with narrow beams of light are especially appropriate. Instead of a centrally located step-down transformer, each luminaire can have its own small transformer.

ual control possible with this personal lighting system can also have significant psychological benefits for workers, who traditionally have little influence oven their environment. To avoid dark surrounding areas and excessive brightness ratios, some background illumination is required. Since indirect luminaires are often used to complement the task lighting, this combination is known as **task/ambient lighting.**

Accent Lighting

Accent lighting is used whenever an object or a part of the building is to be highlighted (Fig. 14.11g). **Accent** illumination should be about ten times higher than the surrounding light level. Since this type of lighting is very variable and is a very powerful generator of the visual experience, designers should give it careful attention.

Decorative Lighting

With a **decorative-lighting** system, unlike all of the others, the lamps and fixtures themselves are the object to be viewed (e.g., chandeliers). Although glare is now called "sparkle," it can still be annoying if it is too bright or if a difficult visual task has to be performed. In most cases the decorative lighting also supplies some of the functional lighting.

14.12 REMOTE-SOURCE LIGHTING SYSTEMS

In remote-source lighting systems, light is efficiently transmitted inside a light guide by the phenomenon of **"total internal reflection."** When the light enters the light guide at a sufficiently narrow beam, the walls of the light guide behave like perfect mirrors. Light guides can be made of hollow plastic pipes, solid plastic rods, or fibers of glass or plastic. Because of the need for narrow beams, sunlight and small, compact light sources are best. The light guides can be designed to be end-emitting or side-emitting to form a linear light source. Remote-source lighting has many benefits: filtering out ultraviolet (UV) and infrared energy, removing the light source from a hazardous or secure area, simplifying maintenance, and reducing energy consumption.

Light pipes are made of a prismatic, plastic film with a narrow-beam light source at one end (Fig 14.12a). For use as a linear diffuser, a mirror is placed at the opposite end, and a special diffusing film is placed where the light is to be extracted. Mirrors are used to make turns in the otherwise straight pipes. Light pipes can deliver large amounts of light to areas difficult or dangerous to reach for relamping. Applications include outlining the top of skyscrapers and illuminating large and high spaces, such as airplane hangars.

Fiber-optics lighting uses special flexible, plastic rods or fibers of plastic or glass to guide the light. A very narrow-beam light source is needed because the light must enter the fibers almost parallel to their length in order for total internal reflections to occur (Fig. 14.12b). The main applications consist of cool, low-wattage light applications, such as jewelry and

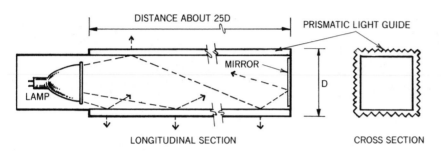

Figure 14.12a Light guides, or light pipes, convey light a distance of about 25 diameters. The light can be conveyed either to the end with small losses, or a linear light source can be created by modifying the light guide to increase the losses along its length.

Figure 14.12b A fiber-optic lighting system consists of a light source (can contain a rotating color wheel), a fiber harness, and a lighting fixture that can deliver the light in any patterns desired from beam to diffused source). (Courtesy Lucifer Lighting Co.)

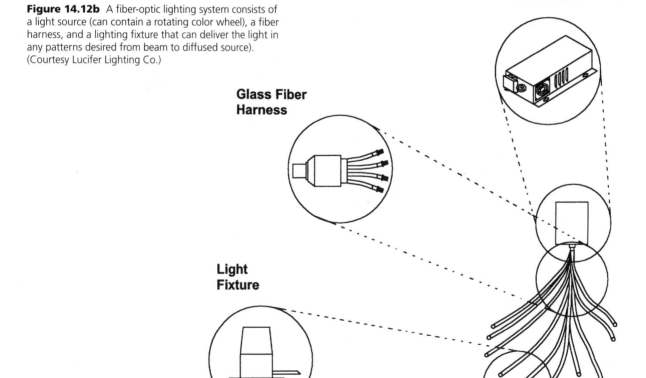

chocolate (read: cool light) display cases. Fiber-optic lighting is also appropriate for decorative or functional lighting where numerous small-light sources and the convenience of one centralized lamp are desired. Troublesome electric wires connected to lamps can be replaced by water-resistant and safe optical fibers. Flexible, plastic rods that look a little like neon lighting are used to show direction, create patterns and signs, and safely illuminate pools.

Light can also be transmitted by the use of mirrors and lenses. This kind of beamed lighting was described in Section 13.16.

14.13 VISUALIZING LIGHT DISTRIBUTION

For both electric lighting and day-lighting, it is very valuable to develop an intuitive understanding of the light distribution from various sources.

Let us first consider how illumination changes with distance from various light sources. For a point light source, the illumination (footcandles) is inversely proportional to the square of the distance (Fig. 14.13a). Notice

that in Fig. 14.13a when the distance doubles (1 to 2 feet), the illumination is reduced to one-fourth (100 to 25 footcandles; see Sidebox 14.13). In most applications, incandescent and high-intensity discharge lamps can be treated as point sources. The main implication of this principle is that point sources should usually be as close as possible to the visual task.

A line source of infinite length is shown in Fig. 14.13b. In this case, the illumination is inversely proportional to the distance. When the distance is doubled (1 to 2 feet), the footcandles are halved (100 to 50). A long string of

fluorescent lamps would create such a situation.

A surface source of infinite area is shown in Fig. 14.13c. In this case, the illumination does *not* vary with distance. A typical example of this kind of light source would be well-distributed indirect lighting in a large room. See Section 13.4 for visualizing the illumination from a finite area source.

The illumination also does not change with distance in a parallel beam of light. It is extremely difficult, however, to create a parallel beam as was explained in Fig. 14.3e. Of common light sources used in buildings

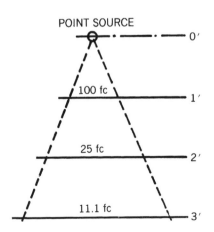

Figure 14.13a The illumination (fc) from a point source is inversely proportional to the square of the distance.

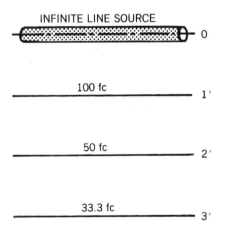

Figure 14.13b The illumination (fc) from a line source of infinite length is inversely proportional to the distance.

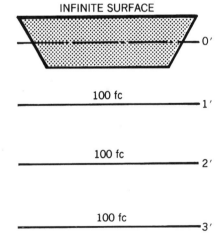

Figure 14.13c The illumination (fc) from a surface of infinite area is constant with distance.

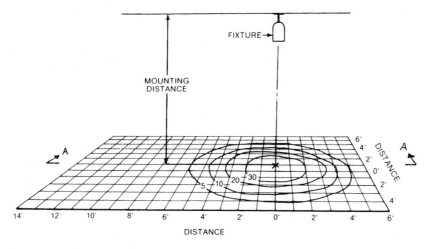

Figure 14.13d This graphic presentation of the illumination pattern is generated from iso-footcandle lines connecting points of equal illumination. (Courtesy of Cooper Lighting.)

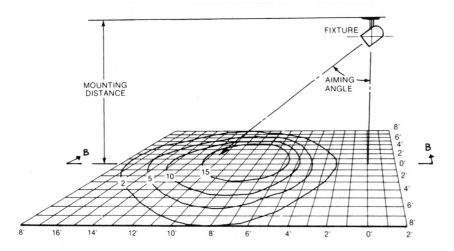

Figure 14.13e When light is not aimed straight at a surface, the isofootcandle lines are elongated. The lines are now of reduced intensity and cover a larger area. (Courtesy of Cooper Lighting.)

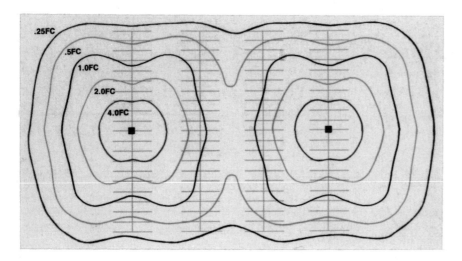

only direct sunlight acts as a beam of parallel light.

The above discussion described how illumination varies with distance from the source. The following discussion will describe how the light, at a fixed distance, is distributed over the workplane. Two major ways exist to graphically display the illumination at the workplane. The first uses points of equal illumination to plot the contour lines of the light pattern in plan. Fig.14.13d illustrates this method for a common light source aimed straight down. Note the concentric pattern of **isofootcandle** rings. In Fig. 14.13e, we see the pattern created when the same source is *not* aimed straight down. Note that the intensities are less, but the area of illumination is greater. This is another example of the consequence of the cosine law, which was explained in Fig. 6.5c. The drawing of Fig. 14.13f illustrates this method as applied to outdoor lighting.

The second graphic method shows a graph of the light distribution superimposed on a section of the room. Fig. 14.13g uses this method to show the same lighting situation as was shown in Fig. 14.13d, but only at section A–A. Similarly Fig. 14.13h shows section B–B of the pattern shown in Fig. 14.13e. When more than one fixture is used, the effect of each is combined for the total because illumination is additive (Fig. 14.13i).

Figure 14.13f Isofootcandle lines used to define the lighting pattern from parking-lot lighting. (Courtesy of Spaulding Lighting, Inc.)

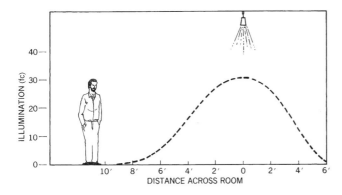

Figure 14.13g In this alternate graphic method of defining the lighting pattern, a curve of the illumination across a room is plotted on top of a section of the space. This diagram, in fact, is section A–A of the room in Fig. 14.13d.

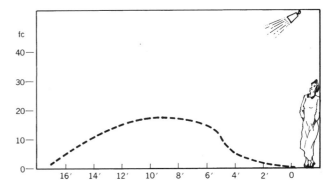

Figure 14.13h This diagram plots the illumination across the room at section B–B of Fig. 14.13e. Again, we can see that when the light source is not aimed normal (perpendicular) to the workplane the maximum illumination is reduced, and the light is spread over a larger area.

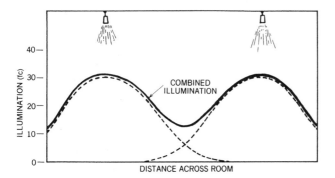

Figure 14.13i When more than one light source is present, the curve defining the combined effect is the sum of the individual curves.

14.14 ARCHITECTURAL LIGHTING

The lighting equipment either can consist of prefabricated luminaires or can be an integral part of the building fabric. In the latter case, it is often known as **architectural lighting.** Ceiling-based systems will be discussed first.

Cove Lighting

Indirect lighting of the ceiling from continuous wall-mounted fixtures is called **cove lighting** (Fig. 14.14a). Besides creating a soft, diffused ambient light, coves create a feeling of spaciousness because bright surfaces (in this case the ceiling) seem to recede. The cove must be designed in such a

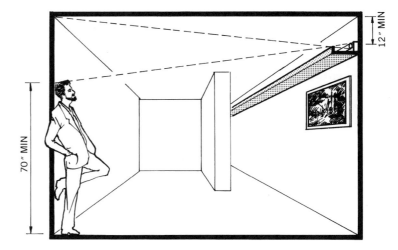

Figure 14.14a Ceilings appear to recede with cove lighting. Lamps must be shielded from view.

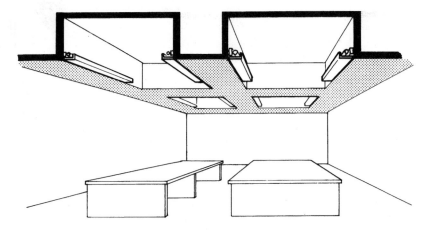

Figure 14.14b Large coffers are often illuminated with cove lighting.

Figure 14.14c Small coffers are best illuminated by direct luminaires in each coffer.

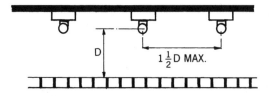

Figure 14.14d The very even brightness of luminous ceilings can create difficulties. Not only is it technically difficult to achieve, but it also tends to simulate a gloomy overcast sky.

manner that a direct view of the light source is not possible, and at the same time it must be far enough from the ceiling to prevent excessive brightness (hot spots) right above the lamps. The inside of the cove, the upper walls, and the ceiling must all be covered with a high-reflectance white paint. Larger rooms require cove lighting on two, three, or four sides.

Coffer Lighting

Coffers (pockets) in the ceiling can be illuminated in a variety of ways. Large coffers often have cove lighting around their bottom edges (Fig. 14.14b), which makes them appear similar to skylights. This technique is sometimes used inappropriately on real skylights as nighttime illumination. In such a design, most of the light is lost through the skylight. Small, square coffers can be illuminated by recessed luminaires (Fig. 14.14c).

Luminous-Ceiling Lighting

The luminous ceiling provides a large area source of uniform illumination by means of diffuser elements suspended below uniformly spaced fluorescent lamps (Fig. 14.14d). The mind often associates this uniform high-brightness ceiling with a gloomy overcast sky. In the worst case, it is similar to being in a fog, where the lighting is so diffused that the three-dimensional world appears rather flat. For these and several other practical reasons, successful luminous ceilings are difficult to achieve.

Wall Illumination

Although most visual tasks take place on the horizontal plane, the vertical surfaces have the greatest visual impact (see Figs. 12.10c and d). When we experience architecture, we are usually viewing vertical surfaces. Functional lighting systems for horizontal workplanes sometimes do not sufficiently illuminate the walls.

Supplementary lighting fixtures mounted on the ceiling or walls can increase the brightness of the walls, emphasize texture, or accent certain features on the walls. Architectural lighting in the form of valances, cornices, and luminous panels are often used to illuminate the walls.

Valance (Bracket) Lighting

Valance (bracket) lighting illuminates the wall both above and below the shielding board (Fig. 14.14e). The placement and proportion of valance boards must result in complete shielding of the light sources as seen from common viewing angles. Valances should be placed at least 12 inches below the ceiling to prevent excessive ceiling brightness. If the valance must be close to the ceiling, a cornice as described below might be more appropriate.

Cornice (Soffit) Lighting

When a valance board is moved up to the ceiling, it is called a **cornice** (Fig. 14.14f). The wall is then illuminated only from above, and the ceiling, which receives no light from the cornice, might appear quite dark. This is called **cornice lighting** or **soffit lighting**. If people are permitted to approach the wall, the light source will be visible unless additional shielding is provided. Cross louvers are quite effective in preventing this direct glare situation (Fig. 14.14g).

Luminous Wall Panels

Luminous wall panels must have a very low surface brightness to prevent direct glare or excessive brightness ratios. Nevertheless, a sense of frustration might exist in the viewer, because the luminous panel implies a window where the view to the outside is denied. The same sense of frustration often exists with the use of translucent glazing in real windows.

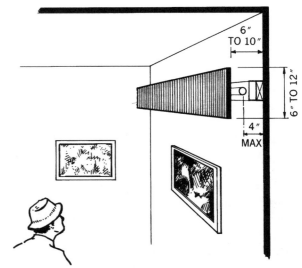

Figure 14.14e Valance lighting can increase the wall brightness, which is so very important in the overall visual appearance of a space. The specific design of a valance depends greatly on the expected viewing angles in a particular room.

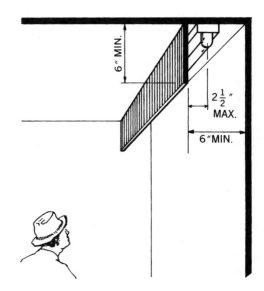

Figure 14.14f Because cornice lighting illuminates only the walls and not the ceiling, excessive brightness ratios can occur.

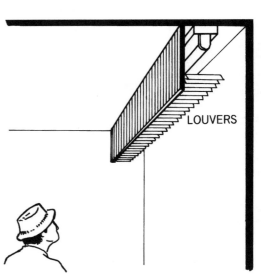

Figure 14.14g In many cases, the viewing angles are such that direct glare will result unless louvers or some other shielding devices are used in the cornice.

14.15 MAINTENANCE

The two main considerations in maintaining a lighting system are the aging of the lamps and the accumulation of dirt on the lamp and fixture.

Fig. 14.15 shows the light output as a function of time for a certain lamp. As the lamp ages, its lumen output depreciates until the lamp fails. The rate of decline and the length of life vary greatly with the specific lamp type, but the general pattern is the same for all. If a large **lamp lumen depreciation** is expected, the initial illumination level is increased to allow for the decline. If lamp life is short, lamp replacement must be made an easy operation. If replacement is difficult, a long-life lamp type should be chosen. Incandescent long-life lamps should generally not be used because their extended life is still very short compared to the normal life of discharge lamps. They also waste a lot of energy.

Light loss due to dirt accumulation is a separate problem. It is a function of the cleanliness of the work area and the design of the luminaire. For example, in dirty areas, such as a woodworking shop, indirect fixtures would not be appropriate because dirt accumulates mostly on the top side of lamps and fixtures. Manufacturers give information on how well their luminaires maintain light output under various levels of dirt accumulation. A **luminaire dirt depreciation factor** can be used to choose the right fixture for a specific environment. Nevertheless, periodic cleaning of lamps and luminaires is required even in clean areas. Thus, easy access for cleaning and "relamping" is an important design consideration.

14.16 SWITCHING AND DIMMING

Properly designed switching can yield functional, aesthetic, psychological, and economic benefits. Switching allows for the flexible use of spaces, as well as the creation of interesting and varying lighting environments. As mentioned before, the definition of personal space and the control over one's environment are important psychological benefits available from individual work area switching. Switching is also one of the best ways to conserve large amounts of energy (money) simply by allowing unneeded lights to be turned off.

Although people can usually be relied on to turn lights on when it is too dark, they almost never turn lights off when they are not necessary. Consequently, to save energy and, therefore, money, it is usually necessary to use automatic devices, such as occupancy sensors, timers, and remote switching equipment. Occupancy sensors can determine if anyone is present in a room and automatically turn the lights off a few minutes after the last person leaves. Photo sensors can respond to the availability of daylighting. Timers can turn lights on and off at a preset cycle or they can be used automatically to extinguish lights a certain amount of time after the lights are turned on manually. Remote-control switching enables people or a computer at a central location to control the lights. This central control of lights is part of what is now often called an **energy management system.** A computer using remote sensors and switches can be programmed to make efficient use of all energy used in a building.

Dimming is another powerful tool for the designer. Incandescent lamps can be dimmed most easily and inexpensively, while most fluorescent lamps can be dimmed at a reasonable cost. Most types of high-intensity discharge lamps can now also be dimmed, but the special equipment required makes it somewhat more expensive. When daylighting is used, switching and dimming are especially important, and were discussed in the previous chapter.

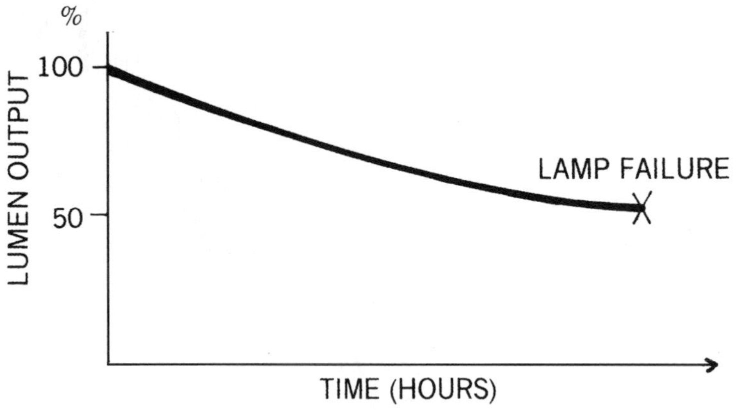

14.17 RULES FOR ENERGY-EFFICIENT ELECTRIC LIGHTING DESIGN

With informed design and efficient equipment, high-quality lighting can be achieved with a power density as low as 1 watt per square foot. This is remarkable when you consider that the norm in the 1980s was about 5 watts per square foot. Use the following strategies to achieve a high-quality, energy-efficient lighting system.

Figure 14.15 Lumen output typically declines over time until the lamp fails.

1. Use light-colored surfaces whenever possible for ceilings, walls, floors, and furniture.
2. Use local or task lighting to prevent the unnecessary high illumination of nonwork areas.
3. Use electric lighting to complement daylighting (see previous chapter for details).
4. Use the lowest recommended light level for the electric lighting. For daylighting, however, the illumination can be set a little higher in the summer and can be much higher during the heating season.
5. Carefully control the direction of the light source to prevent veiling reflections. A small amount of high-quality light can be just as

effective as a large amount of low-quality light.
6. Use high-efficiency lamps (e.g., metal-halide and fluorescent).
7. Use efficient luminaires (e.g., avoid luminaires with black baffles and indirect fixtures in dirty areas).
8. Use the full potential of manual and automatic switching and dimming to save energy and money. Use occupancy sensors, photosensors, timers, and central energy-management systems whenever possible.

Many more specific suggestions for energy-conscious lighting are mentioned throughout the three lighting chapters.

14.16 CONCLUSION

A good lighting design stresses flexibility and quality, not sheer quantity. It must satisfy the biological as well as the activity needs. In doing so, it must prevent direct glare, veiling reflections, and excessive brightness ratios. In addition, in a world of limited resources, good lighting must be accomplished with a minimum of waste. Inefficient lighting systems not only guzzle huge amounts of electrical energy directly, but they also add greatly to the air-conditioning load, which then requires more equipment and still more electrical energy. Good lighting designs support a more sustainable world.

KEY IDEAS OF CHAPTER 14

1. One of the most important characteristics of a lamp is its efficacy (lumens per watt).
2. Another important characteristic of lamps is the color-rendering properties of the emitted light (i.e., whiteness of the light).
3. Incandescent lamps are obsolete for general illumination (even in homes). Use them only for special small-scale applications.
4. Compact fluorescent lamps are a good substitute for most incandescent-lamp applications.
5. Tungsten-halogen lamps should be used only where beam control is vital (e.g., highlighting small objects).
6. Use fluorescent or metal-halide lamps for general illumination where color rendition is important.
7. Use high-pressure sodium lamps where efficacy is more important than color rendition.
8. Task/ambient type of lighting offers both high-efficiency and high-quality lighting.
9. The illumination from a point light source is inversely proportional to the square of the distance.
10. The illumination from a linear light source is inversely proportional to the distance.
11. The illumination from an infinite area light-source does not vary with distance.
12. Use automatic controls, such as occupancy sensors and photocells, to eliminate waste.

Resources

FURTHER READING
(See Bibliography in back of book for full citations. The list includes valuable out-of-print books.)

Brown, G. Z., and M. DeKay. *Sun, Wind, and Light: Architectural Design Strategies.*
Egan, M. D. *Concepts in Architectural Lighting.*

Flynn, J. E. *Architectural Interior Systems: Lighting, Acoustics, Air-Conditioning.*
Gordon, G., and J. L. Nuckolls. *Interior Lighting for Designers.*
Grosslight, J. *Light, Light, Light: Effective Use of Daylight and Electric Lighting in Residential and Commercial Spaces.*
Grosslight, J. *Lighting Kitchens and Baths.*

Futagawa, Y., ed. *Light and Space: Modern Architecture.*
Lam, W. M. C. *Perception and Lighting as Formgivers for Architecture.*
Leslie, R. P., and K. M. Conway. *The Lighting Pattern Book for Homes.*
Leslie, R. P., and P. A. Rodgers. *The Outdoor Lighting Pattern Book.*
Lighting Handbook, IESNA.

Millet, M. S. *Light Revealing Architecture.*

Schiler, M. *Simplified Design of Building Lighting.*

Schiler, M. *Simulating Daylighting with Architectural Models.*

Steffy, G. R. *Architectural Lighting Design.*

Steffy, G. R. *Time-Saver Standards for Architectural Lighting.*

Stein, B., and J. S. Reynolds *Mechanical and Electrical Equipment for Buildings.* A basic resource.

Trost, J. *Electrical and Lighting.*

ORGANIZATIONS
(See Appendix J for full citations.)

IESNA Illuminating Engineering Society of North America

International Association of Lighting Designers (IALD)
www.iald.org

LBNL (Lawrence Berkeley National Laboratory)
http://eande.lbl.gov/BTP/BTP.html

Lighting Design Lab
www.northwestlighting.com

Lighting Research Center, at RPI

National Lighting Bureau

The Thermal Envelope: Keeping Warm and Staying Cool

Waste not, want not, is a maxim I would teach.

Let your watchword be dispatch, and practice what you preach.

Do not let your chances like sunbeams pass you by.

For you never miss the water till the well runs dry.

Rowland Howard, 1876

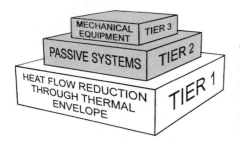

Figure 15.1a The best and most sustainable way to heat a building is to use the three-tier design approach. This chapter covers tier one.

15.1 BACKGROUND

This chapter discusses the creation of an efficient thermal envelope to minimize the heat loss in the winter and the heat gain in the summer. The design of a tight thermal envelope is the first tier of the three-tier design approach to heating a building (Fig. 15.1a).

Suppose we wanted to keep a certain bucket full of water. Our common sense would have us repair the leaks, at least the major ones, rather than just refilling the bucket continually. Yet with regard to energy, we usually keep a leaky building warm by pouring in more heat rather than patching the leaks (Fig. 15.1b). Perhaps if we could see the energy leaking out, we would have a different attitude. Fortunately, such **thermography** does exist now and is very effective in convincing people to upgrade their buildings. In a thermogram of the exterior of a building, hot and cold areas are shown in different shades of grey or colors (Fig. 15.1c). Although this technique is still quite expensive to use on individual buildings, the cost is reasonable in larger projects. Several towns have contracted to have all

Figure 15.1b If we could see heat flowing out of a building as we see water leaking from a container, then our attitudes might be different.

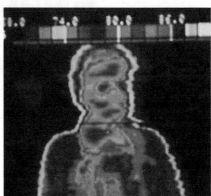

Figure 15.1d Thermogram of the author—a good likeness.

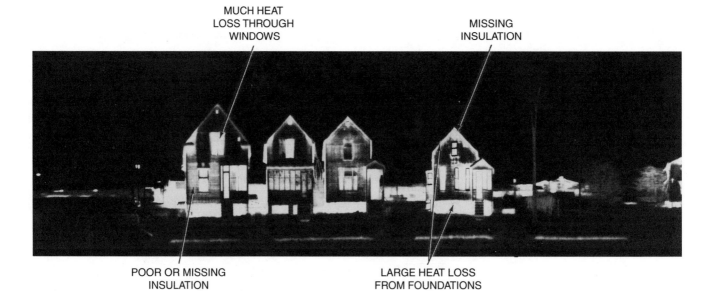

MUCH HEAT LOSS THROUGH WINDOWS

MISSING INSULATION

POOR OR MISSING INSULATION

LARGE HEAT LOSS FROM FOUNDATIONS AND EAVES

Figure 15.1c Thermograms can pinpoint the weakness in the thermal envelope. White indicates the warmest areas, which are a result of the greatest heat loss. [Vanscan (Thermogram) by Daedalus Enterprises, Inc.]

their buildings analyzed by thermography. A thermogram of the author (Fig. 15.1d) proves that he is not hot-headed.

President Jimmy Carter and President Ronald Reagan did not always agree, but they did agree that conservation implied a reduction of comfort. They were both wrong. From experience, we now know that comfort can be *increased* if the proper conservation techniques are used. For example, indoor comfort increases dramatically when insulation is added to the walls, ceiling, and especially the windows. When the author moved into his present home, he was uncomfortably cold even when the thermostat was set at 80°F. The addition of ceiling insulation and insulating drapes oven the windows now allows a thermostat setting of 70°F to provide complete thermal comfort. Thus, insulation not only reduced the energy consumption, but also increased thermal comfort.

Because conservation has such a negative connotation, it is better to talk of **energy efficiency.** It is not only possible, but also more likely to have a higher standard of living through "energy efficiency." After all, higher efficiency will enable us to meet the necessities of heating, cooling, and lighting at less cost, and, consequently, there will be more money left over for other needs.

Making adjustments to new thermal realities has a long history in America. The early settlers in New England found that the wattle-and-daub construction method that they had brought from England was inappropriate in the harsh climate of the Northeast (Figure 15.1e). They quickly exhausted the entire local wood supply in trying to stay warm. Because bringing wood from great distances was expensive, they modified their building method and switched to clapboard siding for a tighter construction and greater comfort. Although this was a great improvement for keeping the cold out, it was not as good as the

Figure 15.1e The traditional wattle-and-daub construction, so popular in old England, was unacceptable in the harsh climate of America. There was little insulation to counter heat flow, and infiltration was a major problem, as can be seen from the cracks.

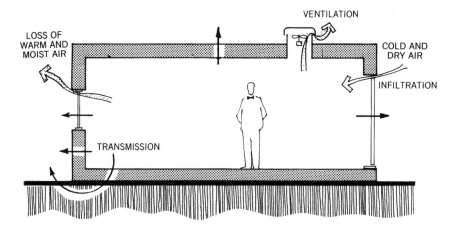

Figure 15.2a The major heat-loss channels are transmission, infiltration, and ventilation.

log-cabin technology that was brought to the United States later by the Swedish immigrants. Compared to the alternatives of those days, the thick logs were good insulation, but the numerous joints were still a significant source of infiltration. It was the invention of that underappreciated material, tar paper, that really cut down on infiltration. By today's standards, wood is also a poor insulator. Today, controlling heat flow is not so much a technical problem as one of will and economics.

15.2 HEAT LOSS

The major channels of heat loss are transmission, infiltration, and ventilation (Fig. 15.2a). Heat is lost by **transmission** through the ceilings, walls, floors, windows, and doors. Heat flow by transmission occurs by a combination of conduction, convection, and radiation. The proportion of each depends mainly on the particular construction system (Fig. 15.2b).

The actual transmission heat loss through a building's skin is a function

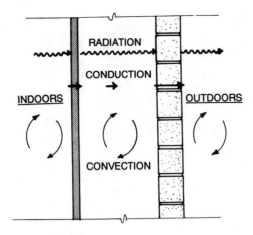

Figure 15.2b Transmission heat flow occurs by a combination of conduction, convection, and radiation. Winter condition is shown.

SIDEBOX 15.2a

Heat Loss Due to Transmission Through Walls, Windows, and Roofs

$$\text{Heat loss per hour by transmission} = \frac{(\text{Area}) \times (\text{Temperature Difference})}{(\text{Thermal Resistance})}$$

$$HL = \frac{A \times (T_i - T_o)}{R_T}$$

where in the inch-pound (I-P) system:

HL is the rate of heat loss in BTU/hour

A is the area in ft^2

T_i is the indoor design temperature in °F

T_o is the outdoor design temperature in °F

R_T is the total resistance in R-value or $\dfrac{(ft^2)(°F)}{BTU/hour}$

or in the SI system:

Watts

square meter

°K (Kelvin)

°K

$\dfrac{m^2(°K)}{W}$

or the more conventional but not as conceptually clear format:

since $U = \dfrac{1}{R_T}$ (see Section 3.17)

$HL = A \times U \times (T_i - T_o)$

where U is the heat-flow coefficient

Example: What is the heat loss through an 8-foot-high and 20-foot-long wall that has a total thermal resistance of 16 (R-value) if the indoor design temperature is 75°F and the outdoor design temperature is 20°F?

$$HL = \frac{(A) \times (T_i - T_o)}{R} = \frac{(20)(8)(75-20)}{16} = 550 \text{ BTU/h}$$

of the area, the temperature difference between the indoors and outdoors, and the thermal resistance (see Sidebox 15.2a). Thus, the heat loss can be minimized by the use of a compact design (minimum area), common walls (no temperature difference across walls), and plenty of insulation (large thermal resistance).

Heat is also lost by the **infiltration** of cold air through joints in the construction, as well as through cracks around windows and doors. The heat loss due to infiltration is a function of how much cold air enters the building and the temperature difference between the indoor and outdoor air. The amount of cold air that infiltrates on the windward side of the building is equal to the amount of hot air that leaves on the leeward side. Thus, counting both the air entering and leaving is an error just like counting both heads and tails to determine the number of coins. Sidebox 15.2b shows how infiltration heat loss is calculated when the air-infiltration rate through cracks around windows and doors is used. A somewhat simpler method of finding heat loss due to infiltration is known as the **air-change method.** If the indoor air is being humidified, latent as well as sensible heat will be lost by infiltration in the winter.

Heat loss due to **ventilation** is very much like infiltration except that it is a controlled and purposeful form of air exchange. Fortunately, heat-recovery devices, as described in Section 16.18, can save as much as 90 percent of the heat, both sensible and latent, that would otherwise be lost due to ventilation.

15.3 HEAT GAIN

Although heat gain in a building is similar to heat loss, some significant differences exist. The similarity is that part of the heat flow through the building envelope is due to the temperature difference between the indoors and outdoors. The differences are primarily due to the load from internal heat sources, the effect of thermal mass, and, of course, the action of the sun (Fig. 15.3).

Depending on the building type, the internal heat sources can be either a major or a minor load. **Internally dominated buildings** are those that have a large amount of heat generated by either people, lights, appliances, or any combination of these. The heat can be in both sensible and latent (water-vapor) form.

Thermal mass can reduce the heat gain when temperatures are fluctuating widely during a day. The insulating effect is most pronounced when the daily temperature range varies from above to below the comfort zone, a situation found in hot and dry climates. (This effect will be explained later in Section 15.10.)

Infiltration is generally less of a problem in the summer than in the winter because of the lower wind velocities. Instead, ventilation is often a major source of heat gain. This is especially true in humid climates because of the large latent-heat component. The same heat-recovery devices, mentioned above for reducing heat loss, can also be used to reduce both sensible and latent heat when ventilation brings in hot, humid summer air. Ventilation and heat-recovery devices are explained more fully in Section 16.18.

For windows, the part of the heat gain due to the temperature difference across the window is calculated the same way as in winter. Of course, summer design temperatures are used. The part of the window heat gain due to solar radiation through the glazing is calculated as shown in Sidebox 15.3a. The second component of solar heat gain is a consequence of the surface heating of opaque surfaces. Dark colors absorb a large amount of solar radiation and get quite hot. This results in a higher temperature differential between indoors and outdoors than can be

---

SIDEBOX 15.2b

Heat Loss Due to Infiltration by the Crack Method

Heat loss per hour by infiltration = (Constant) x (Rate of air loss per minute) x (Temperature difference between indoor and outdoor air)

$$HL = (1.08) \times (CF/M) \times (T_i - T_o)$$

where:

HL	=	Rate of heat loss in BTU/hour
1.08	=	Proportionality constant
CF/M	=	Volume of air gained or lost (not both) in cubic feet per minute
T_i	=	Indoor design temperature in °F
T_o	=	Outdoor design temperature in °F

SIDEBOX 15.3a

Heat Gain Due to Solar Gain Through Glazing

Heat gain per hour by solar gain = (Area) x (Solar Heat Gain Factor) x (Correction Factor due to Shading)

$$HG = A \times SHGF \times SC$$

where in the I-P system:	**or in the SI system:**
HG = Rate of solar heat gain in BTU/hour	Watts
A = Actual area of glazing in ft^2	m^2
SHGF* = Unit solar heat gain through a ft^2 single glazing in $\frac{BTU/h}{ft^2}$	$\frac{W}{m^2}$

SC = Shading coefficient (see Section 9.20 for explanation and table of values)

*SHGF is a function of latitude, orientation of glazing, time of year, and time of day. The table of values is found in the *ASHRAE Handbooks*.

Example: What is the solar heat gain per hour through a clear 4- x 5-foot-double-glazed-window that is 80 percent glass and 20 percent frame? The window is at 40°N latitude and faces south. It is 11 A.M. on March 21.

Solution: Use formula $HG = A \times SHGF \times SC$

where A (of glazing) = $(4 \times 5)(0.8) = 16 \ ft^2$

SHGF from *ASHRAE Handbook* = $\frac{197 \ BTU/hr}{ft^2}$

SC (from Table 9.20) = 0.84

therefore, $HG = (16)(197)(0.84) = 2,648$ BTU/hour

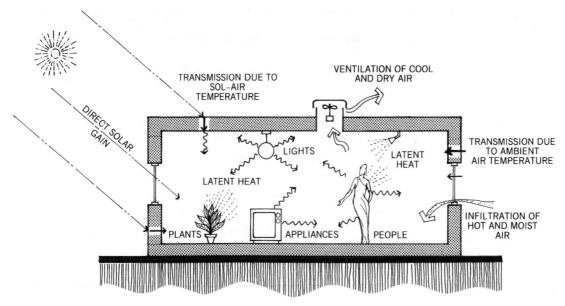

Figure 15.3 Sources of sensible and latent heat gain.

Heat Gain Through the Walls and Roofs

$$HG = \frac{A \times DETD}{R_T}$$

where:

HG = Rate of heat gain in BTU/h

A = Area of wall, roof, opaque door, etc. in ft²

DETD = Design-equivalent temperature difference in °F

R_T = Total thermal resistance of wall or roof in $\frac{(ft^2)(°F)}{BTU/h}$

or $HG = A \times U \times DETD$

where:

U= Heat-flow coefficient = $\frac{1}{R_T}$

accounted for by the actual outdoor-air temperature.

This larger temperature differential is called the **sol-air temperature.** When the insulating effect of thermal mass is also included, the temperature differential used is called the design-equivalent temperature difference. Sidebox 15.3b illustrates how to calculate heat gain through the opaque parts of the building envelope.

Since sol-air temperatures are much lower with light-colored surfaces than with dark-colored surfaces,

one of the most effective and certainly the least expensive ways to reduce heat gain is to specify light-colored building finishes.

15.4 SOLAR REFLECTIVITY (ALBEDO)

The heat gain through a white roof will be about 50 percent that of a black roof. The measure of a surface's reflectivity of solar radiation is called

albedo. It varies from zero to one, where a surface with an albedo of zero would absorb all solar radiation and an albedo of one would reflect all solar radiation. See Table 15.4 for albedo values of typical building surfaces.

A textured, bumpy surface will absorb more radiation than a smooth surface of the same material. Studies done at the Oak Ridge National Laboratory in Tennessee have shown that the temperature of dark-colored roofs routinely exceeds 160°F on sunny summer days, while flat, white surfaces only reach about 135°F, and glossy, white surfaces rarely exceed 120°F (*Cooling Our Communities,* 1992). Other studies have shown that the total air conditioning can be reduced by 20 percent just by changing the albedo of the roof and walls from a typical medium-dark value of 0.3 to a light-colored value of 0.9. During initial design and construction, there is usually no cost penalty for choosing light colors, but the savings go on for the life of the material.

Although polished metal surfaces have a high reflectivity (albedo), they have low emissivity and, therefore, get much warmer than light colors, which have a high emissivity (see

Figure 15.4 White is the traditional and appropriate color in the hot and humid Southeast. This nineteenth century home in Eufaula, AL, uses just about every strategy available for summer comfort.

15.5 COMPACTNESS, EXPOSED AREA, AND THERMAL PLANNING

Until the advent of modern architecture, buildings generally consisted of simple volumes richly decorated. Modern architecture turned that around and created buildings of complex volumes simply decorated. Unfortunately, complex volumes usually result in large surface-area-to-volume ratios. For example, the compact cube and the spread-out alternative of Fig. 15.5a have the same volume and yet the surface area of the dispersed volume is 60 percent greater than that of the cube. In most cases, a building with more surface area requires more resources of every kind for both construction and operation. Throughout history, compact buildings were built not just for the poor and frugal middle class, but also for the rich and powerful (Fig. 15.5b).

A note of caution is in order here because we can easily learn the wrong lessons from the past. For example, our image of a traditional home is one with many wings. This can be a misleading prototype because authentic old homes almost always started out as compact designs.

TABLE 15.4 ALBEDO OF TYPICAL BUILDING SURFACES

Building Surface	Albedo*
White paint	0.5–0.9
Highly reflective roof	0.6–0.7
Colored paint	0.1–0.4
Brick and stone	0.1–0.4
Concrete	0.1–0.4
Red/brown tile roof	0.1–0.4
Grass	0.2–0.3
Trees	0.1–0.2
Corrugated roof	0.1–0.2
Tar and gravel roof	0.05–0.2
Asphalt paving	0.05–0.2

*Values from *Cooling Our Communities,* 1992.

It should be relatively easy to convince people to use light colors in hot climates. Light colors not only save money, energy, and the environment, but also are traditional in hot climates (Fig 15.4). Yet many people are convinced that black roofs are beautiful and that white roofs are ugly or that they will get stained and, therefore, cannot be used.

Section 3.11). Clearly then, light colors should be used in hot climates. In dense urban areas, the paving and other surfaces between buildings should also have a high albedo. Researchers have found that in a typical American city, the value of the albedo can be realistically raised by about 0.15, and this increase in reflectivity will result in a drop of a city's temperature of about 5°F. In such a city, a light-colored building will benefit both from less heat gain from direct solar radiation and because the outdoor-air temperature will be lower.

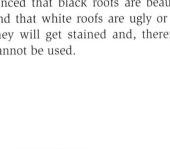

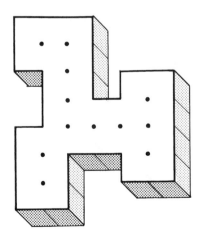

Figure 15.5a Although the volumes are equal the less compact form (right) has 60% more surface area.

Figure 15.5b Like most palaces throughout time, the Andrew Carnegie mansion on Fifth Avenue in New York City is a spacious, ornate, but compact building.

desire for more surface area for daylighting and less surface area for heating and cooling needs.

Ultimately what counts is not total surface but exposed surface. Through the sharing of walls (party walls), great savings in heating and cooling are possible. For example, row housing of four attached units has about 30 percent less surface area than four detached units (Fig. 15.5d).

Exposure can also be reduced by the arrangement of the floor plan for thermal planning. Spaces that require or tolerate cooler temperatures should be placed on the north side of the building. Buffer spaces, such as garages, should be on the north to protect against the cold or on the west to protect against summer heat.

Solar/thermal conditions around a building are shown for both summer and winter in Fig.15.5e. A bubble diagram for a residence based on the solar/thermal conditions is shown in Fig. 15.5f. The kitchen would be on the northeast because it needs little heat, while the breakfast area would be on the southeast to receive morning sun all year. The utility and garage

It was only after many generations of additions that the quaint nostalgic image we have today emerged (Fig. 15.5c).

Some exceptions to the desirability of compact plans exist. When natural ventilation is the dominant cooling strategy and the climate has mild winters, then an open, spread-out plan might be best. If daylighting in a large multistory building has a high priority, a more spread-out plan might also be in order. The glass-covered atrium is usually the result of a simultaneous

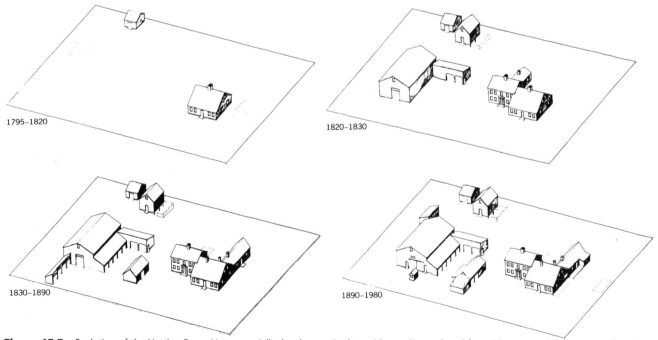

1795–1820

1820–1830

1830–1890

1890–1980

Figure 15.5c Evolution of the Nutting Farm. Note especially the changes in the residence. (Reproduced from *Big House, Little House Back House Barn* by Thomas C. Hubka, by permission of the University Press of New England. Copyright 1984, 1985, 1987 by Trustees of Dartmouth College.)

areas would be on the northwest to buffer against the winter cold and summer heat. Bedrooms are on the north because they are used little during the day and a cool room is better for sleeping.

A plan based on this diagram would be in harmony with our circadian rhythms, thereby promoting both physical and mental well-being (Fig. 15.5g). Consider waking up to sunlight entering through the east windows and then preparing and eating breakfast with bright, cheerful sunlight. You would spend the day in the family room with warm, south lighting, and at night you would sleep in the pleasantly cool bedrooms on the north and northeast (second floor).

Another example of solar/thermal planning is the bubble diagram for an elementary school (Fig. 15.5h). The classrooms face southeast for year-round warm, south lighting from early morning until early afternoon, while the gymnasium is on the north because it needs to be the coolest space.

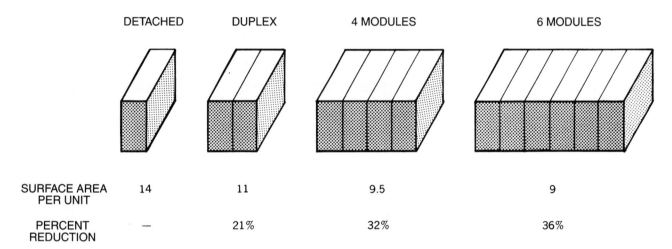

	DETACHED	DUPLEX	4 MODULES	6 MODULES
SURFACE AREA PER UNIT	14	11	9.5	9
PERCENT REDUCTION	—	21%	32%	36%

Figure 15.5d Through the sharing of walls, attached units can significantly reduce the amount of exposed surface area.

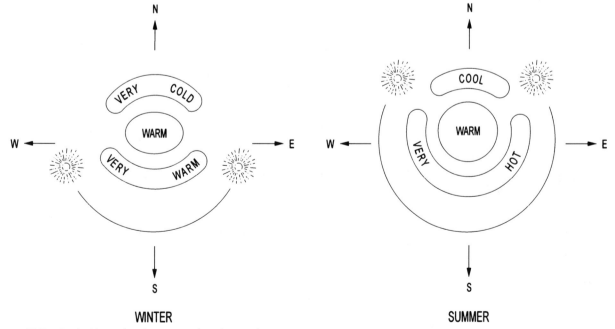

Figure 15.5e A solar/thermal zoning pattern for winter and summer.

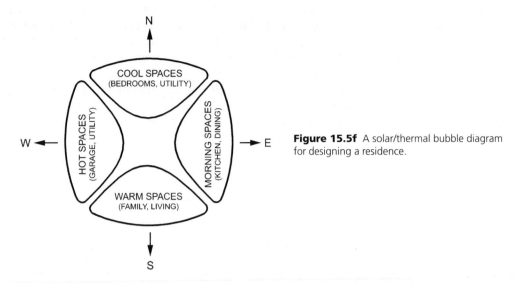

Figure 15.5f A solar/thermal bubble diagram for designing a residence.

Figure 15.5g A conceptual plan for a residence based on solar/thermal planning considerations.

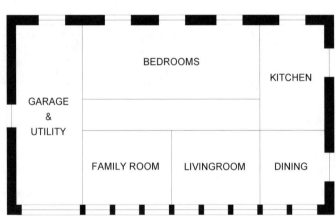

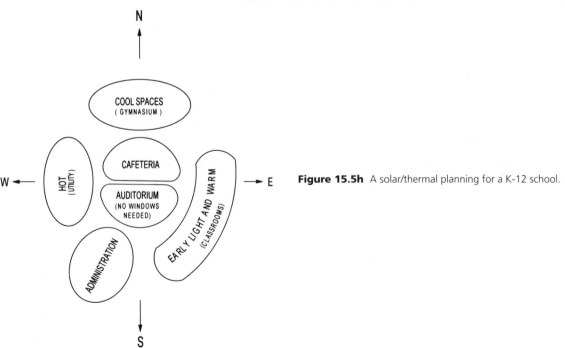

Figure 15.5h A solar/thermal planning for a K-12 school.

15.6 INSULATION MATERIALS

Twenty five years ago many buildings were still built without insulation in the walls. As a consequence of the energy crisis of 1973, the question is no longer should insulation be used but rather which material and how much.

In general, "the more insulation the better" is a good principle to start with for a number of reasons: insulation not only saves money, but also increases thermal comfort; it helps create more sustainable buildings; it is relatively inexpensive; it is very durable; it functions both summer and winter; and it is much easier to install during initial construction than to retrofit later. There is, of course, a limit to how much should be used. The "law of diminishing returns" says that every time you double the amount of insulation, you cut the heat loss in half. This is great the first few times as the heat loss goes from 1 to $1/2$ to $1/4$, etc. Unfortunately, the cost keeps up with the thickness of insulation, while the heat loss decreases by ever smaller amounts (e.g., from $1/32$ to $1/64$ to $1/128$, etc.). Therefore, the optimum thickness is mainly a function of climate and the value (not just cost) of the energy saved. Since future energy supplies and cost are uncertain, it is wise to be conservative and to **use as much insulation as possible.**

The map of Fig. 15.6a gives recommended insulation levels. These values and those required by codes should be considered *minimum* values. Consider that **superinsulated buildings,** which are gaining in popularity, use about twice these levels. The main obstacle to using high levels of insulation is not so much the cost of the insulation, but the need to change construction details to allow the use of thicker insulating materials.

Among the many characteristics of insulating materials, one of the most important is thermal resistance since that will determine the thickness required. The bar chart of Fig. 15.6b compares various insulating materials to each other and to common building materials by showing the thermal resistance of 1-inch-thick samples.

Other important characteristics of insulating materials are: moisture resistance, fire resistance, potential for generating toxic smoke, physical strength, and stability over time. Table 15.6A summarizes the important characteristics of common insulation materials .

Most insulation materials used in buildings fit into one of the following five categories: blankets, loose fill, foamed-in-place, boards, and radiant barriers. Most insulating materials work by creating miniature air spaces. The main exception is **reflective insulation,** which uses larger air spaces faced with foil on one or both sides. This material acts mainly as a **radiation barrier.** The metal foil, which is usually aluminum, is both a poor emitter and absorber of thermal radiation (Fig. 15.6c). The first surface of foil stops about 95 percent of the radiant heat flow and the second surface stops another 4.8 percent for a total of 99.8 percent blockage of radiation. Additional layers of foil help little except to create additional air spaces, which reduce the convection

TABLE 15.6A   INSULATION MATERIALS[a]

Material	Thermal Resistance	Physical Format	Comments on Applications
Fiberglass	3.2	Rolls, batts, and blankets	Good fire resistance Moisture degrades R-value
Rock wool	2.2 4.4	Loose fill Rigid board	Fairly inexpensive
Perlite	2.7	Loose fill	Very good fire resistance
Cellulose	3.2 3.5	Loose fill Sprayed in place	Can be blown into small cavities Requires treatment for resistance to fire and rot Absorbs moisture
Polystyrene (expanded)	4	Rigid board (bead board)	Fairly low cost per R-value Combustible Must be protected against fire and sunlight
Polystyrene (extruded)	5	Rigid board	Very high moisture resistance Can be used below grade Combustible Must be protected against fire and sunlight Good compressive strength Higher cost and R-value than expanded polystyrene
Urethane/ isocyanurate	7.2	Rigid board	Very high R-value per inch Combustible and creates toxic fumes Must be protected against fire and moisture
	6.2	Foamed in place	For irregular or rough surfaces
Reflective foil	Varies widely[b]	Thin sheets separated by air spaces	Effective in reducing summer heat gain through roof Foil must face air spaces Foil should be face down to prevent dust from covering the foil

[a] The thermal resistances are given in R-values per inch thickness. The actual resistance varies with density, type, temperature, and moisture content.
[b] The thermal resistance depends on the orientation of the foil-faced air space and the direction of the heat flow (Table 15.6B).

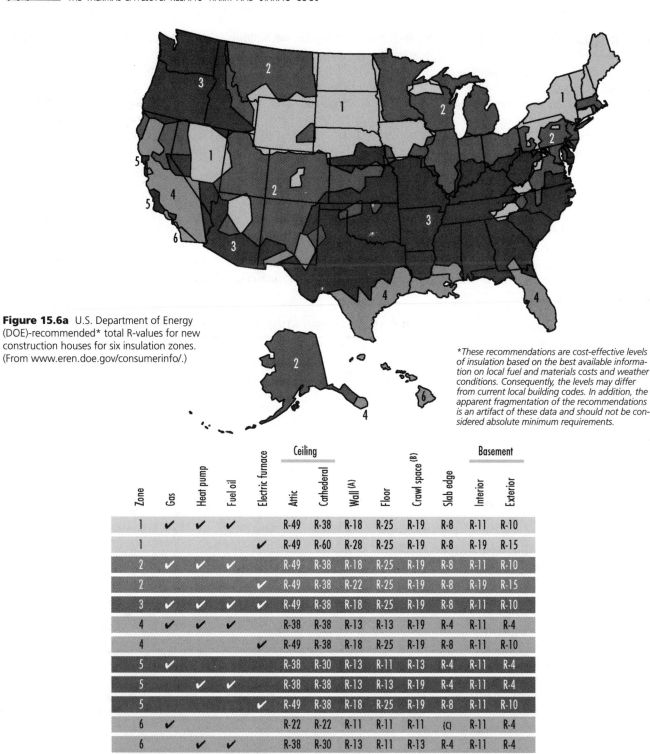

Figure 15.6a U.S. Department of Energy (DOE)-recommended* total R-values for new construction houses for six insulation zones. (From www.eren.doe.gov/consumerinfo/.)

*These recommendations are cost-effective levels of insulation based on the best available information on local fuel and materials costs and weather conditions. Consequently, the levels may differ from current local building codes. In addition, the apparent fragmentation of the recommendations is an artifact of these data and should not be considered absolute minimum requirements.

Zone	Gas	Heat pump	Fuel oil	Electric furnace	Ceiling Attic	Ceiling Cathederal	Wall (A)	Floor	Crawl space (B)	Slab edge	Basement Interior	Basement Exterior
1	✔	✔	✔		R-49	R-38	R-18	R-25	R-19	R-8	R-11	R-10
1				✔	R-49	R-60	R-28	R-25	R-19	R-8	R-19	R-15
2	✔	✔	✔		R-49	R-38	R-18	R-25	R-19	R-8	R-11	R-10
2				✔	R-49	R-38	R-22	R-25	R-19	R-8	R-19	R-15
3	✔	✔	✔	✔	R-49	R-38	R-18	R-25	R-19	R-8	R-11	R-10
4	✔	✔	✔		R-38	R-38	R-13	R-13	R-19	R-4	R-11	R-4
4				✔	R-49	R-38	R-18	R-25	R-19	R-8	R-11	R-10
5	✔				R-38	R-30	R-13	R-11	R-13	R-4	R-11	R-4
5		✔	✔		R-38	R-38	R-13	R-13	R-19	R-4	R-11	R-4
5				✔	R-49	R-38	R-18	R-25	R-19	R-8	R-11	R-10
6	✔				R-22	R-22	R-11	R-11	R-11	(C)	R-11	R-4
6		✔	✔		R-38	R-30	R-13	R-11	R-13	R-4	R-11	R-4
6				✔	R-49	R-38	R-18	R-25	R-19	R-8	R-11	R-10

(A) R-18, R-22, and R-28 exterior wall systems can be achieved by either cavity insulation or cavity insulation with insulating sheathing.
For 2 in x 4 in walls, use either 3-1/2-in thick R-15 or 3-1/2-in thick R-13 fiber glass insulation with insulating sheathing.
For 2 in x 6 in walls, use either 5-1/2-in thick R-21 or 6-1/4-in thick R-19 fiber glass insulation.

(B) Insulate crawl space walls only if the crawl space is dry all year, the floor above is not insulated, and all ventilation to the crawl space is blocked.
A vapor retarder (e.g., 4- or 6-mil polyethylene film) should be installed on the ground to reduce moisture migration into the crawl space.

(C) No slab edge insulation is recommended.

NOTE: For more information, see: Department of Energy Insulation Fact Sheet (D.O.E./CE-0180). Energy Efficiency and Renewable Energy Clearinghouse, P.O. Box 3048, Merrifield, VA 22116; phone: (800) 363-3732; www.ornl.gov/roofs+walls/insulation/ins_11.html or contact Owens Corning, (800) GET-PINK (800-438-7465), www.owenscorning.com

heat flow. Although radiation is independent of orientation, the heat flow by convection is very much dependent on both the orientation of the air space and the direction of heat flow. As a result, the resistance of air spaces and reflective insulation varies greatly with the location in the structure and the time of year. Table 15.6b gives the resistance of air spaces and reflective insulation for different orientations and heat-flow directions. Note the tremendous range of R-values for air spaces that are all about the same size.

Blankets

Although most blankets or batts are made of fiberglass or rock wool, cotton is also being used. The blankets are made to fit between studs, joists, or rafters. Both fiberglass and rock wool are very resistant to fire, moisture, and organic attack. The main health hazard is present during installation and comes from inhaling short fibers.

Loose Fill

Loose-fill materials consist primarily of fiberglass, cellulose (ground-up newspapers), and expanded minerals, such as perlite and vermiculite, that are used mainly to fill masonry-wall cavities. The fiberglass and cellulose are blown into stud spaces and attics. The cellulose is a good, safe product if properly treated with chemicals to make it fire-retardant and resistant to organic attack. Care must be taken to prevent the inhalation of the fine particles both during installation and afterward.

Foamed-in-Place

Most foamed-in-place insulation materials are made of plastics. The main exceptions are foamed glass and Air Krete™ (R = 3.9 per inch), which are foamed minerals that have the advantage of acting as a fire stop and

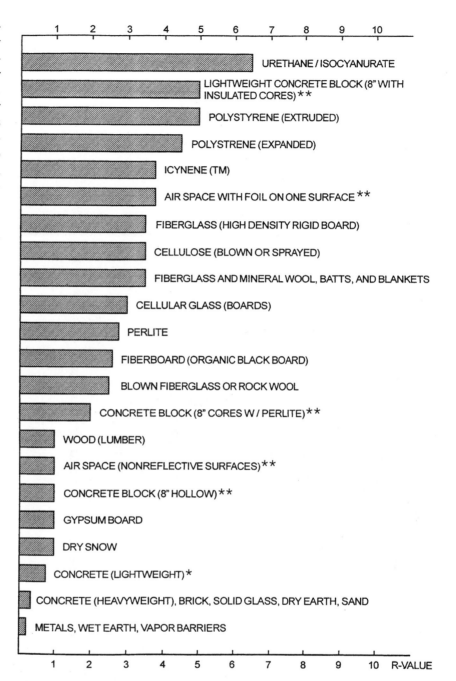

Figure 15.6b A comparison of the thermal resistance of various materials. All values are for 1-inch-thick samples, unless otherwise indicated. The actual resistance of a sample varies with density, temperature, material composition, and, in some cases, moisture content.

*The resistance of lightweight concrete varies greatly with density and aggregate used (R-values vary from 0.2 to 2.0).

**Not per inch but for actual thickness of block or air space.

TABLE 15.6B R-VALUES OF AIR SPACES AND REFLECTIVE INSULATION[a]

Position of Air Space	Air Space (No Reflective Surfaces)	Air Space (with One Reflective Surface)	Reflective Insulation (Two Layers of Foil and Three Air Spaces)
Wall	1	2	11
Ceiling			
winter (heat flow up)	1	2	9
summer (heat flow down)	0.8	3–8[b]	15

[a] R-values are approximate and vary somewhat with thickness of air space and temperature.

[b] Depends greatly on the size of the air space (given range is for ¾-inch to 4-inch air spaces).

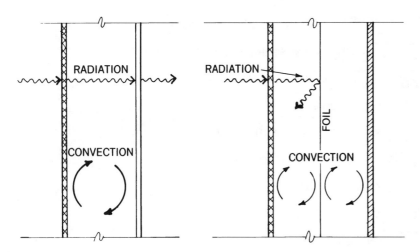

Figure 15.6c Much heat flows across an air space by radiation (left). A radiation barrier blocks heat flow by being both a poor absorber (good reflector) and a poor emitter of radiant energy. When placed in the center of an air space it also slightly reduces heat flow by convection (right).

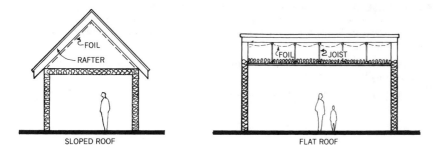

Figure 15.6d Summer heat gain through the roof can be reduced as much as 40 percent by use of a radiation barrier (foil).

of not releasing toxic gases The plastic foams vary tremendously because of both the base materials and the foaming agents. Foams can be sprayed into cavities or on surfaces (e.g., basement walls) both during construction and afterward. In all cases, the cured material needs to be covered to protect both the foam and the occupants. Toxic smoke from burning plastic is a severe hazard, and off-gassing from aging plastic foams is a potential hazard, especially for chemically sensitive people.

Boards

Although most insulation boards are made of foamed plastics, some are made from recycled or waste organic material (e.g., Homosote™). Boards can also be made of fiberglass. Because boards made of polystyrene are very resistant to moisture, they are frequently used to insulate below grade. Extruded polystyrene is both more resistant to moisture migration and has a higher R-value than the less expensive, expanded polystyrene (bead board) material. Boards made of either urethane or isocyanurate have very high R-values but are damaged by moisture and offgassing. For that reason, they are usually covered with a protective foil coating. All plastic insulation boards should be covered to prevent damage to themselves from such forces as ultraviolet (UV) radiation in sunlight, water, and physical attack. They also need to be covered to minimize off-gassing and to prevent toxic smoke from being generated during a fire.

Radiant Barriers

The best application for a radiation barrier is in hot climates just under the roof. Experiments in Florida have shown that the summer heat gain through the roof can be reduced as much as 40 percent. In buildings with rafters, the foil should be attached to the underside of the rafter to create two air spaces each facing a radiant

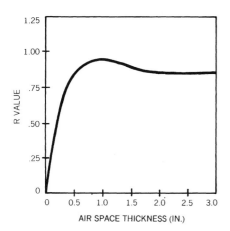

Figure 15.6e The optimum wall or window air space thickness is about ¾ inch. Note that this is for air spaces not faced with reflective material. (After *Climatic Design* by D. Watson and K. Labs, 1983.)

barrier as seen in Fig. 15.6d (left). Since radiant barriers are effective only when the foil faces an air space, a finished attic or vaulted ceiling will not achieve the double radiant-barrier benefit. The foil can also be draped across rafters, joists, or trusses (Fig. 15.6d right). Since working with the thin foil is difficult, most builders use foil pre-applied to sheathing panels.

The resistance of an air space is also a function of the thickness of the air space. As Fig. 15.6e shows, the optimum thickness is about ¾ inch. For thinner spaces, the resistance is less because of greater conduction, and for thicker spaces, the resistance is less because convection currents transfer more heat.

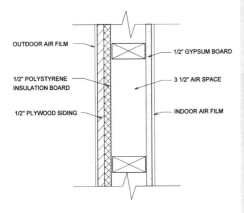

Figure illustrating Sidebox 15.7

15.7 INSULATING WALLS, ROOFS, AND FLOORS

The total thermal resistance of a wall, roof, or floor construction is simply the sum of the resistances of all the component parts. Determining the total resistance of a wall or roof section is useful for comparing alternatives, for complying with codes, and for calculating heat loss. Many codes, company literature, and equations describe the thermal characteristic of a wall or roof by a quantity called the **U-coefficient** rather than the total R-value. The U-coefficient is the reciprocal of the total R-value. The author feels that the U-coefficient is a somewhat counterintuitive concept, and that it is, therefore, usually better to think in terms of total R-value (see Sidebox 15.7 and accompanying diagram).

SIDEBOX 15.7

Total Resistance and U-Coefficient

The total resistance of a wall, roof, window or door equals the sum of all the resistances of the components

$$R_T = \sum R = R_1 + R_2 + R_3 + \dots$$

where:

R_T = the total thermal resistance of the construction detail in R-value

$$\frac{ft^2 \times °F}{BTU/hour}$$

R_1, R_2, etc. = the thermal resistance of each component in R-value

Also, the heat-flow coefficient = the reciprocal of the total resistance of a wall, roof, door, or window

or

$$U = \frac{1}{R_T}$$

where

U = the heat-flow coefficient of the construction detail in U-value or

$$\frac{BTU/hour}{ft^2 \times °F}$$

Example: Find the total resistance and U-value for the wood framed wall detail at left.

Component:	R*
Indoor air film	0.7
½-inch gypsum board	0.45
3½-inch air space	1.0
½-inch polystyrene extruded insulation board	2.5
½-inch plywood siding	0.6
Outdoor air film	0.2
Total Resistance (R_T)	5.45

*Tables of R-values for building components are found in the *ASHRAE Handbooks*

and $U = \dfrac{1}{R_T} = \dfrac{1}{5.45} = 0.183$ U-value

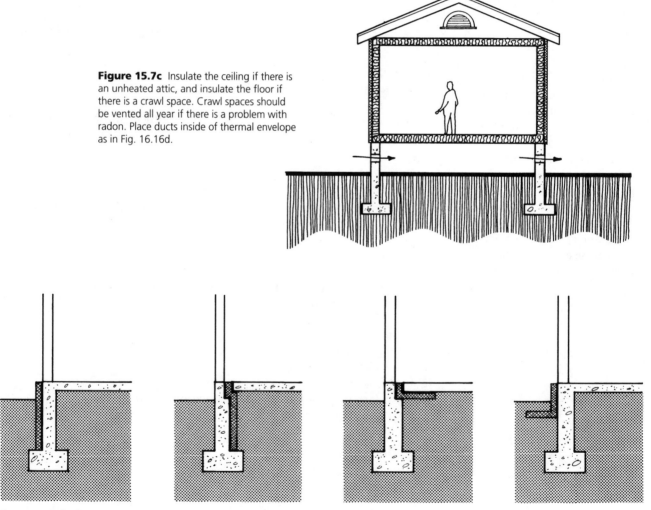

Figure 15.7a Heat bridges caused by studs are greatly reduced by using 2 x 6 studs every 24 inches instead of 2 x 4 studs every 16 inches on center.

RIGID INSULATION

INSULATION BLANKETS

Figure 15.7b Flat-roofed buildings use either rigid insulation on top of roof decks or blankets on top of the suspended ceiling. Insulation on the roof deck is preferred because pipes and ducts should be on the inside of the thermal envelope.

Figure 15.7c Insulate the ceiling if there is an unheated attic, and insulate the floor if there is a crawl space. Crawl spaces should be vented all year if there is a problem with radon. Place ducts inside of thermal envelope as in Fig. 16.16d.

Figure 15.7d Alternative methods for insulating the perimeter of a slab. In all cases, the insulation forces heat to take a long (high-resistance) path through the earth.

As much as possible, a continuous unbroken layer of insulation should surround a building. This thermal envelope should have as few gaps or heat bridges as possible. For example, in framed construction, each stud is a heat bridge, and metal studs are much worse than wooden studs. The heat bridges of 2 x 4 studs can be reduced by using 2 x 6 studs with a wider spacing. This construction detail not only reduces the heat bridges in number and severity, but also allows more insulation in the walls (Fig. 15.7a). Heat bridges also occur where structural elements project through the thermal envelope. Concrete or masonry wing walls, projecting through the building skin, should have thermal breaks. Since concrete and steel beams and slabs cannot have thermal breaks for structural reasons, either minimize the penetrations or encase the projecting elements in insulation. Also, windows with metal frames must have thermal breaks built into them. In summary, to have a good thermal envelope it is necessary to minimize heat bridges and to use thermal breaks wherever conductive materials penetrate the skin.

Roofs

In commercial and institutional buildings, roofs are generally flat, and the insulation can be either on top of the roof deck or resting on the suspended ceiling (Fig. 15.7b). In gabled roof construction where the attic is not used, the insulation is generally in the ceiling (Fig. 15.7c).

Slab-on-Grade

Insulation is not required under slab-on-grade except around the outside edge. Rigid insulation should extend down to the frost line or an equal distance sideways. Thus, the heat flowing through the Earth is forced to take a very long and, therefore, high-resistance path (Fig. 15.7d).

Walls

Although wall details vary tremendously, a few typical details are shown in Fig. 15.7e. Note that most insulation materials should not be exposed on either the indoors or outdoors. Also note that in walls with steel studs, the insulation between the studs is outflanked by the stud heat bridges. Therefore, the R-value of the insulating sheathing is especially important and should be increased.

Structural Insulated Panels (SIP)

Heat bridges are almost nonexistent with **structural insulated panels (SIP),** which are sandwich panels fabricated in a factory, shipped efficiently (no boxes full of air), and erected in as little as one day (Fig 15.7f). The panels connect with splice plates so that there are no studs to act as heat bridges (Fig 15.7g). Thus a 4-inch nominal SIP wall has an R-value of 14, while a similar framed wall has an actual R-value of about 10. Furthermore, there is much less infiltration with the SIP than standard construction.

Structurally, SIP panels are also superior to conventional framing. The facing boards, which can be made of a variety of materials, carry most of the load in the efficient, stressed-skin mechanism. SIP systems also offer great design flexibility. The panels vary in thickness from 2 to 12 inches and can be as large as 8 x 24 feet. Some advocates claim that if you can build it with sticks, you can also build it with panels.

Straw Bales

Although straw bales are not as insulating as many advocates claim, their total R-value of about 28 is excellent for walls. Research at the Oak Ridge National Laboratory in Tennessee has shown straw bales to have an R-value

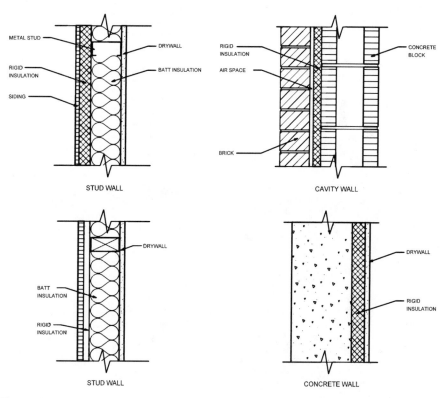

Figure 15.7e Various typical wall details showing type and location of insulation. Note the extra-thick insulating sheathing as the primary thermal envelope component when using steel studs since they are good heat bridges.

Figure 15.7f Structural insulated panels (SIP) create very high-quality thermal envelopes because of high R-value and low infiltration. (Courtesy R-Control Building Systems.)

Figure 15.7g Structural insulated panels (SIP) have eliminated many heat bridges by eliminating all studs. (Courtesy R-Control Building Systems.)

of about 1.5 per inch (*EBN*, October, 1998). Their main virtue is that they are recycled natural materials. When covered with cement stucco, they become a safe and environmentally healthy building system. The bales should be used as an in-fill and not as structural material (Fig. 15.7h). Also see Fig. 17.2e of the Real Goods Solar Living Center.

Crawl Spaces and Basements

If there is a crawl space, the floor should be insulated (Fig. 15.7c). Much more heat is lost through the floor than was previously believed. Also, this approach permits the crawl space to be ventilated of potential moisture and radon problems. Basement walls should be insulated at least to the frost line, but the thickness of the insulation can taper off toward the bottom (Fig. 15.7i). Instead of adding insulation to a concrete wall, many **Insulating Concrete Form (ICF)** systems are available that initially act as formwork for the concrete and steel reinforcing rods and then remain in place as insulation (Fig. 15.7j). The insulating concrete forms (ICF) are either preformed blocks or panels with plastic ties. Most ICFs are made of polystyrene because that plastic is unaffected by water and, therefore, is safe below ground. The ICF systems are also used to build strong, energy-efficient walls above grade. Although such above-grade uses provide much more mass than conventional buildings, they are not well coupled to the indoor air because of the foam insulation. Thus, they do not help with either passive heating or passive cooling.

Care must be taken to protect the foam insulation where it is above ground. Polystyrene insulation, the usual choice, must be protected from UV radiation and physical attack with a protective finish, such as stucco, special paints, sheet metal, or treated wood. Although termites don't eat plastic insulation, they have no trouble making tunnels in it to reach wood.

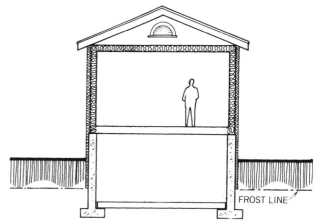

Figure 15.7h The straw bales are generally used as in-fill panels, much as in half- timbered construction.

Figure 15.7i Insulate basement walls at least down to the frost line.

Figure 15.7j Insulating concrete forms (ICF) serve as both formwork and permanent insulation. Although mostly used for foundations, the forms are also used to build full-height walls.

15.8 WINDOWS

To keep drinks hot or cold, a vacuum bottle is unbeatable. The vacuum stops all conduction and convection losses while a silvered coating on the bottle stops most radiant transfer (Fig. 15.8a). The bottle resists being crushed by the one-ton-per-square-foot atmospheric pressure only because of the sharp curvature of the glass. Not having that sharp curvature, double-pane windows could not maintain a vacuum but instead had to be filled with a gas, such as dry air. However, some very recent research is suggesting that windows with a vacuum between the two layers of glass might be possible after all. Since at present they cannot be as good as vacuum bottles, what are the realistic possibilities for the thermal performance of windows?

The bar chart of Fig. 15.8b shows the comparative thermal resistance of different window systems. Note that although double glazing is about twice as good as single glazing in stopping heat flow, it is still only about one-seventh as effective as an ordinary insulated stud wall. It takes about thirteen sheets of glass and twelve air spaces for the window to have the same R-value as a normal framed wall (Fig. 15.8c). It is no wonder, then, that the traditional rule of thumb for standard single glazing was to minimize window area in order to prevent heat loss (Fig. 15.8d). However, new technology has dramatically changed this situation. With high-performance windows, it is no longer necessary to minimize the window area in order to minimize the heat loss.

Consider the fact that switching from R-2 windows to R-4 windows will have the same benefit as switching the walls from R-19 to R-100. This unbelievable situation is correct because both single- and double-glazed windows have essentially been "holes" in the thermal envelope.

Now consider the fact that win-

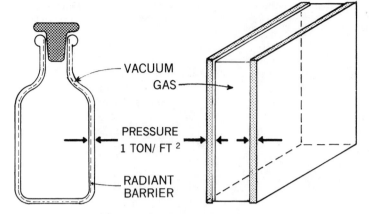

Figure 15.8a A vacuum bottle can stop most heat flow. It resists the 1-ton-per-square-foot atmospheric pressure by its sharp curvature, something a flat window cannot do.

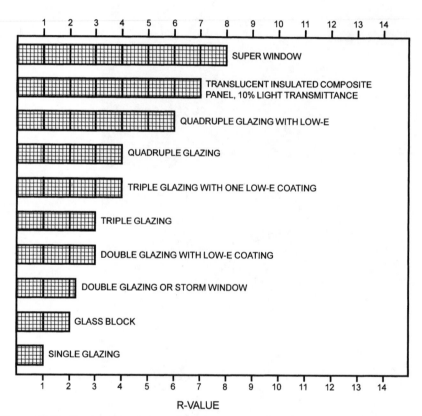

Figure 15.8b The thermal resistance of window systems. The values shown are for the total resistance, which includes the resistance of the air films, air spaces, low-e coatings, and the effect of the frame. Actual R-values vary with temperature, type of glazing, type of coating, thickness of air space, and type of gas in air space. Most plastics are similar to glass when used as glazing.

dows with an R-8 or higher are better than walls with infinitely thick insulated walls. When windows have an R-8 or higher, they collect more heat during the day than they lose during the whole twenty-four-hour period. Actually, south windows need only double or triple glazing to be net gainers of heat, but with super windows even north windows collect more heat than they lose.

The glazing itself, whether glass or plastic, has almost no thermal resistance. It is mainly the air spaces,

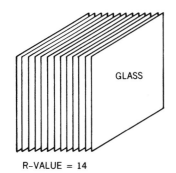
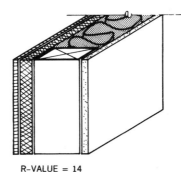

R-VALUE = 14

GLASS

R-VALUE = 14

Figure 15.8c With old technology, it would have taken thirteen sheets of glass and twelve air spaces to develop the same R-value as a standard framed wall. Today, the same R-value can be obtained with two sheets of glass, three plastic films, low-e coatings, krypton gas fill, and efficient, low heat-bridge frames.

One way to insulate.

The way to insulate with a view.

Figure 15.8d Which building will require less heating energy? Until the 1980s, the building on the left would have had the least heat loss, but with new, very high-performance windows, the building on the right needs the least heating. However, a window manufacturer used these images to sell insulating (i.e., double-glazing) windows. Although double glazing was much better than single glazing, the truth was that the windowless building still needed less heating. (Courtesy Anderson Corp., Bayport, MN.)

the surface air films, and low-e coatings that resist the flow of heat (Fig. 15.8e). Single glazing has no air spaces, but it does have the slightly insulating stagnant air films that exist whenever air comes in contact with a building surface. The thermal resistance of a window consists of two parts: the glazing and the frame. In large windows, most of the heat is lost through the glazing, but in smaller windows, the frame becomes very critical. Most frames are made of either wood, plastic, aluminum, or a combination of these materials. Wood has good thermal properties, and when it is protected with vinyl or aluminum it becomes a durable, low-maintenance product. In double or triple glazing, heat is also lost through the edge spacers. High-performance windows use edge spacers with thermal breaks. The edge spacers not only maintain the size of the air space, but

also keep moisture and dirt out in order to prevent condensation (fogging) inside the window. The air films and air spaces control the heat flow by conduction and convection. To reduce heat flow further, radiation must also be considered.

Although clean glass is mostly transparent to solar radiation, it is opaque to heat radiation. Since most of this long-wave infrared (heat) radiation is absorbed, the glass gets warmer, and, consequently, more heat is given off from both indoor and outdoor surfaces (Fig. 15.8f). Thus, in effect, a significant amount of the heat radiation is lost through the glazing. Special coatings on the glass can dramatically reduce this radiant heat flow through the glazing.

Various types of reflective coatings are possible. A silver coating (any polished metal) will not only reflect much of the long-wave heat radiation,

but also much of the solar radiation (visible and short-wave infrared) (see Fig. 15.8g). Because it reflects heat radiation back in during the winter and back out in the summer, it has a higher R-value than clear glazing. This kind of coating is appropriate for

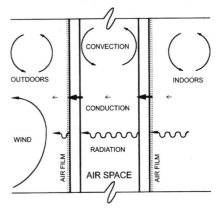

Figure 15.8e Since glass Is a good conductor of heat, most of the resistance to heat flow comes from air films and air spaces (if any) and coatings (if any).

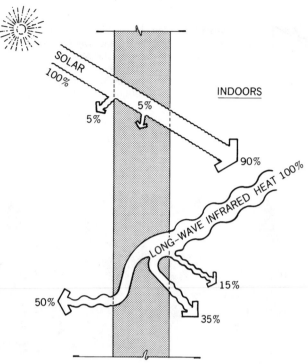

Figure 15.8f Although clear glass transmits most solar radiation, it absorbs most of the long-wave infrared (heat) radiation. Much of this absorbed heat Is then lost outdoors. In the summer, the flow of heat radiation is from the outside in.

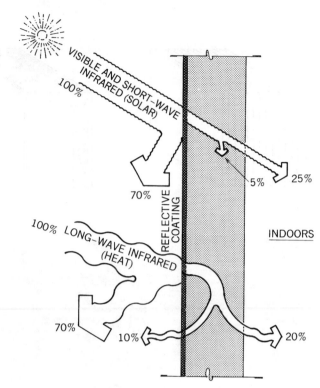

Figure 15.8g A metallic coating reflects most of the solar and heat radiation. Although the summer condition is shown, unfortunately, solar heat gain is also reduced in the winter.

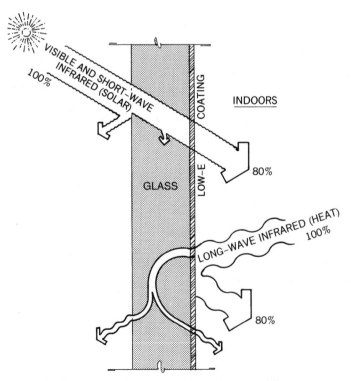

Figure 15.8h Low-e coatings are good for colder climates because they allow high transmission of solar radiation while they reflect heat radiation back inside.

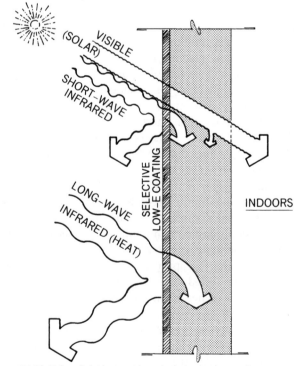

Figure 15.8i When light but not heat is desired, a spectrally selective low-e coating should be used.

buildings that need year-round protection from the sun. However, if winter solar heating is desirable, a different kind of coating will be required.

Special coatings are now available that transmit solar radiation but reflect long-wave infrared radiation. These **low-e** (low-emissivity) coatings are ideal for those buildings that need to reduce winter heat loss and at the same time allow the sun to shine in (i.e., envelope-dominated buildings) (Fig. 15.8h). Because the low-e windows reduce heat flow, they are given a higher R-value. The graph in Fig. 15.8b shows how each low-e coating is about equivalent to an additional pane of glass (and air space) in R-value but without the equivalent increase in the weight or cost of the glass. The cost of double-glazed, low-e glazing is low enough that it should be the minimum standard, except in the mildest climates. The benefits are many: significant energy savings, increased thermal comfort, reduced noise transmission, reduced condensation (fogging), and reduced fading from UV radiation.

In Fig. 13.7 a slightly different low-e coating was described. In those cases where the light but not the heat of the sun is desired, a **selective low-e** coating is used. This type of coating is transparent to visible radiation but reflective to both short- and long-wave infrared radiation. This kind of coating is appropriate for internally dominated buildings, such as large office buildings, in all but the coldest climates (Fig. 15.8i).

In the near future, electrically switchable coatings will become available. A building's operating computer will then automatically select which radiation and how much will be transmitted or reflected from the glazing at any particular time. Meanwhile, movable insulation is a realistic option.

15.9 MOVABLE INSULATION

Movable insulation is a present-day practical technique for achieving some of the future benefits of switchable glazing. These include controlling the light and view through the aperture, reducing heat loss and gain, and creating greater comfort in the winter because of a higher mean radiant temperature (MRT). Since insulating shutters can be made to any thickness, the R-value of shuttered windows can be almost as high as that of the walls. A shutter with 1 inch of extruded polystyrene (R = 5) or 1 inch of urethane/isocyanurate (R = 7) is quite adequate when added to double-glazed, low-e windows (R = 3) to give a total R from 8 to 10. Unfortunately, the shutters are not only an extra expense, but also they take up a lot of indoor wall area dur-

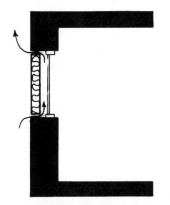

Figure 15.9a Without extra good seals, the wind will short-circuit exterior insulating shutters.

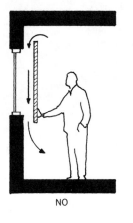
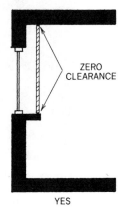

NO   YES

ing the day. Consequently, high-performance (R-4) or super windows (R-7) are attractive alternatives. Outdoor shutters are usually not as effective because wind blowing behind them short-circuits the insulation (Fig. 15.9a).

Drapes with thermal liners are very appropriate since curtains of some kind are often specified anyway for aesthetic and lighting reasons. With an insulating foam or reflective films, drapes or shades can increase the R-value of a window as much as 3 R-units. Care must be taken, however, to prevent the short-circuiting of the insulation by sealing the edges. (Fig. 15.9b). Top and bottom seals are best accomplished by having the drapes extend from the ceiling to the window sill or floor, while magnetic strips or Velcro™ can be used to achieve good edge seals. The drapery should also

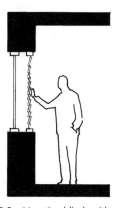

Figure 15.9c Venetian blinds with a reflective coating or insulated louvers can significantly improve the R-value of the windows when they are rotated into the closed position during nights and summer daytimes.

ZERO CLEARANCE

Figure 15.9b Prevent air currents from short-circuiting movable insulation, such as thermal drapery.

contain a vapor barrier to reduce condensation on the windows.

Venetian blinds with a reflective coating or with insulated louvers can effectively control daylight, heat gain, and heat loss (Fig. 15.9c). The Hooker Chemical Co. Headquarters in Niagara Falls, New York, is an excellent case study for this approach (see Fig. 13.11i).

15.10 INSULATING EFFECT FROM THERMAL MASS

The time-lag property of materials can be used to reduce both the peak load and the total heat gain during the summer. Sections 3.18 and 3.19 explained the basic principles behind this phenomenon. The graph in Fig. 15.10a shows the time it takes for a heat wave to flow through a wall or roof. The length of time from when the outdoor temperature reaches its peak until the indoor temperature reaches its peak is called the **time lag**. The graph also shows how the indoor-temperature range is much smaller than the outdoor-temperature range in part because of the moderating effect of the mass. Traditional buildings in hot climates, except for the very humid ones, were usually built of stone, soil, or adobe (Fig. 15.10b).

A structure can use thermal mass to reduce heat gain by delaying the entry of heat into the building until the sun has set. Since each orientation experiences its major heat gain at a different time, the amount of time lag required for each wall and roof is different (Fig. 15.10c). North, with its small heat gain, has little need for time lag. The morning load on the east wall must not be delayed to the afternoon because this would make matters only worse. Consequently, either a very long time lag (more than fourteen hours) or a very short time lag is required on the east. Since mass is expensive, mass with a fourteen-hour time lag is *not* recommended on

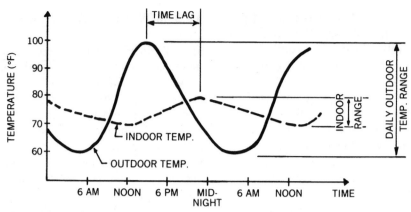

Figure 15.10a The difference between the times when the outdoor and indoor temperatures reach their peaks is called the time lag.

the east. Instead, use either no mass or at least no mass on the outside of the insulation on the east.

Although high sun angles reduce the summer heat gain on the south, the load is still significant. To delay the heat from midday until dark, about eight hours of time lag is recommended for the south wall. Although the west wall receives the maximum heat gain, the number of hours until sunset is fairly short. Thus, the west wall also suffices with about eight hours of time lag. Finally, since the roof receives sunlight most of the day, it would require a very long time lag. Thus, because it is very expensive to place mass on the roof, additional insulation rather than thermal mass is usually recommended there (Fig. 15.10c).

To help in choosing appropriate materials, Table 15.10 gives the time lag for 1-foot-thick walls for a variety of materials.

The time-lag property of materials should not be seen as a substitute for insulation but rather as an additional benefit of massive materials that are used for other purposes, such as structural support. Since it is expensive to add mass to a building, it

Figure 15.10b Adobe or sun-dried mud bricks are being made in this Mayan village in Guatemala. The best adobe bricks are made of a clay- and straw- mixture. The straw gives the dried brick some strength in tension.

TABLE 15.10 TIME LAG FOR 1-FOOT-THICK WALLS OF COMMON BUILDING MATERIALS

Material	Time Lag (hours)
Adobe	10
Brick (common)	10
Brick (face)	6
Concrete (heavyweight)	8
Wood	20[a]

[a]Wood has such a long time lag because of its moisture content.

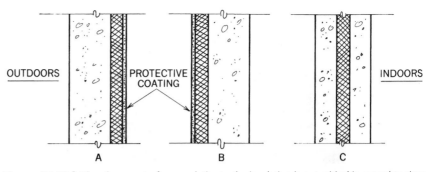

Figure 15.10d The placement of mass relative to the insulation is not critical in regard to time lag but It does have other implications. (A) Mass on outside: good for fire and weather resistance and good for appearance. (B) Mass on inside: good for night-flush cooling and for passive solar heating. (C) Mass sandwich: for some of the benefits of both A and B.

should provide as many benefits as possible.

If mass is used, should it be on the inside on outside of the insulation? For time-lag purposes, the location of the mass is not critical. On the outside of the insulation, the mass can also create an attractive as well as durable, weather-resistant skin. On the inside of the insulation, the mass can support night-flush cooling in the summer and passive solar heating in the winter. In order to achieve all of these benefits, one must often divide the mass by a layer of insulation (Fig. 15.10d).

The importance of light colors in reducing heat gain should not be forgotten. After all, time lag largely postpones heat gain, while light colors significantly reduce the heat gain.

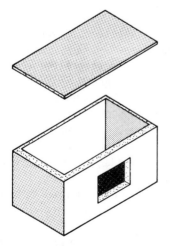

Figure 15.10c In most cases, the south and west walls should have enough mass to yield an 8-hour time lag, while north walls, east walls, and the roof should have little mass for time-lag purposes.

15.11 EARTH SHELTERING

A survey of indigenous underground dwellings around the world shows that most are found in hot and dry climates. In Matmata, Tunisia, chambers and a central courtyard are carved out of the local sandstone, and access to the 30-foot-deep dwellings is by an inclined tunnel. Because of the dry climate, neither flooding nor condensation is a problem. More than 20 feet of rock provides sufficient insulation, time lag, and heat-sink capability to create thermal comfort in the middle of a desert (Fig. 15.11a).

To understand the benefits of earth sheltering, one must understand the thermal properties of soil and rock. First, one must recognize that the insulating value of earth is very poor. It would take about 1 foot of soil to have the the same R-value as 1 inch of wood, and it would take more than 10 feet of earth to equal the R-value of an ordinary insulated stud wall. Thus, earth usually is not a substitute for insulation. What, then, is the main benefit of earth in controlling the indoor environment?

Because of its massiveness, earth can offer the benefits of time lag. In small amounts, the soil can delay and reduce the heat of the day just as massive construction, mentioned previously, does. In large quantities, the time lag of soil is about six months long. Thus, deep in the earth (about 20 feet or more), the effect of summer

heat and winter cold is averaged out to a constant steady-state temperature that is about equal to the mean annual temperature of that climate (see Fig. 10.12a). For example, at the Canadian border, the deep-earth temperature is about 45°F, while in southern Florida it is about 80°F all year. See the map in Fig. 10.12b for deep ground temperature throughout the United States.

The ground is, therefore, cooler than the air in summer and warmer than the air in winter. This is a much milder environment than a building experiences above ground. But the closer one comes to the surface, the more the ground temperature is like the outdoor air temperature. Consequently, the deeper the building is buried in the earth, the greater the thermal benefits. In much of the country, the earth can act as a heat sink to give free cooling because the deep-earth temperature is sufficiently lower than the comfort zone. Also, the heating load is greatly reduced because the deep-earth temperature is much higher than the winter outdoor air temperature.

There are, however, a number of serious implications in underground construction. The biggest problems come from water. Never build below the water table. Have a foolproof way of draining storm water. Wet regions with soils that drain poorly require elaborate waterproofing efforts. In humid climates, condensation can

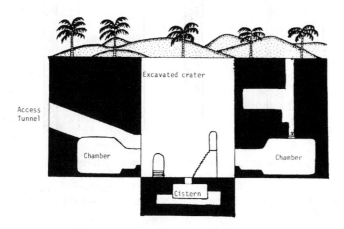

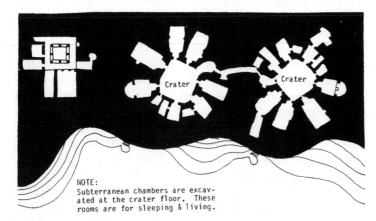

Figure 15.11a In Matmata, Tunisia, chambers and courtyards are cut from sandstone, which functions as both a heat sink and insulation. (From *Proceedings of the International Passive and Hybrid cooling Conference*, Miami Beach, FL, Nov. 6–16, 1981. © American Solar Energy Society, 1981.)

form on the cool walls, and the cooling of humid air encourages the growth of mildew.

Structural problems also increase with the amount of earth cover. The main structural loads to be considered are of three types: weight of earth on roof, soil pressure on walls, and hydrostatic pressure on walls and floor.

Such problems as providing for exit requirements (code) and the psychological needs of people also exist. For example, most people want and need a view of the outdoors.

Where these problems can be resolved, earth-sheltered buildings can offer substantial benefits, the greatest of which is security. By its very nature, an earth-sheltered building will be low to the ground and have a substantial structural system. Thus, it offers good protection against such forces as violent storms (tornadoes, hurricanes, lightning), earthquakes, vandalism, bombs (fallout shelter), temperature extremes, and noise (highway or airport) (Fig. 15.11b). In densely populated areas, the greatest benefit might be the retention of the natural landscape (Fig. 15.11c). Finally, from a heating and cooling point of view, these buildings are very comfortable and require substantially less energy than conventional buildings. For example, an underground factory in Kansas City required only one-third the heating and only one-twelfth the cooling equipment of a comparable above-ground building.

Four major schemes for the design of earth-sheltered buildings exist. The **below-grade scheme** offers the greatest benefits but also has the greatest liabilities (Fig. 15.11d). This type is usually built around sunken atriums or courtyards. The problem of flooding from storms can be partially solved by covering the atriums with domes. In the summer, the earth can act as a substantial heat sink and in the winter as an excellent buffer against the cold.

When an earth-sheltered structure is built on sloping land, the **at-grade scheme** is often the most advantageous since water drains naturally, and there is easy access for people, light, and views (Fig. 15.11e). If built on a south slope, close to 100 percent passive solar heating is possible because of both the small heat loss and large thermal-storage mass of the earth.

On flat land, a mound of earth can be raised to protect a building that is built above grade (Fig. 15.11f). This scheme works well in hot and dry climates where time lag from day to night is very helpful.

Finally, one should consider the **berm-and-sod-roof scheme.** Here, when many openings are required for light and ventilation, earth berms work best (Fig. 15.11g). However, the thermal benefits of earth berms are quite minimal except on west orientations in hot climates and north orientation to deflect the cold wind in cold climates (Fig. 15.11h). Likewise, sod roofs help only a little in cold climates but can significantly reduce the summer heat gain through the roof. Just 1 to 2 feet of earth will furnish sufficient daily time lag to reduce the overheating in hot and dry climates. Plants growing on the sod roof or berm will cool the earth both by shading and evaporation.

If berms are to have any benefit, they must be as continuous as possible. Each penetration of the berm is a major weakness because of the way heat flows through soil. Heat tends to

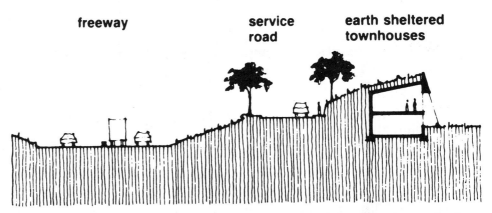

Figure 15.11b In densely populated areas, earth sheltering can help maintain the natural environment as well as protect from noise. (From *Earth Sheltered Housing Code: Zoning and Financial Issues,* by Underground Space Center, University of Minnesota. HUD, 1980.)

Figure 15.11c Earth-sheltered design helps preserve the natural landscape. ("Design for an Earth Sheltered House." Architect: Carmody and Ellison, St. Paul, MN. From *Earth Sheltered Housing Code: Zoning and Financial Issues,* by Underground Space Center, University of Minnesota. HUD, 1980.)

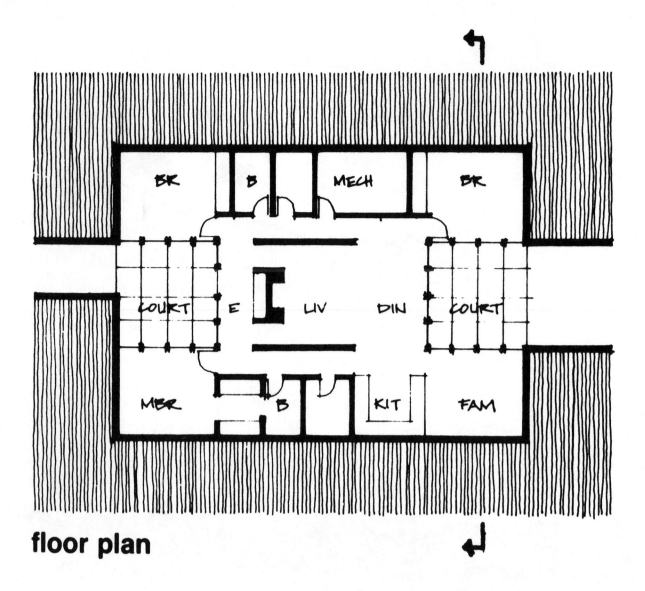

floor plan

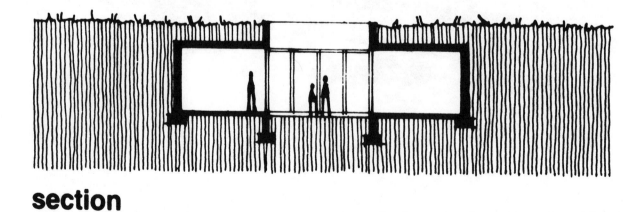

section

Figure 15.11d Below-grade scheme: Rooms are arranged around one or more atriums. Drainage and fire exits are major considerations. (From *Earth Sheltered Housing Code: Zoning and Financial Issues,* by Underground Space Center, University of Minnesota. HUD, 1980.)

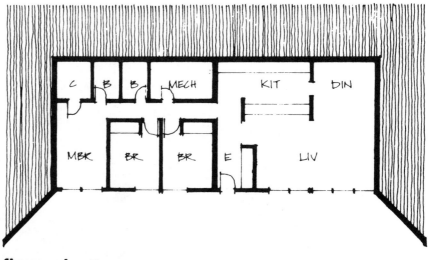

floor plan

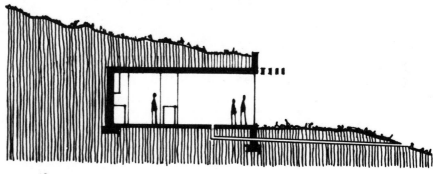

section

Figure 15.11e At-grade scheme:
Drainage, egress, and views are all very good for an earth-sheltered structure built at-grade on a slope. (From *Earth Sheltered Housing Code: Zoning and Financial Issues,* by Underground Space Center, University of Minnesota. HUD, 1980.)

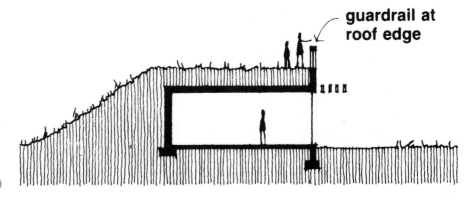

guardrail at roof edge

Figure 15.11f Above-grade scheme: On flat land with poor drainage, an artificial mound might be the best strategy. (From *Earth Sheltered Housing Code: Zoning and Financial Issues,* by Underground Space Center, University of Minnesota. HUD, 1980.)

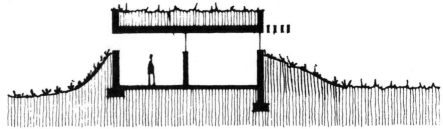

Figure 15.11g Berm-and-sod-roof scheme: When natural ventilation, daylight, and views are important, berms are appropriate. Sod roofs are best for protection from summer heat. (From *Earth Sheltered Housing Code: Zoning and Financial Issues,* by Underground Space Center, University of Minnesota. HUD, 1980.)

Figure 15.11h This highway rest area in Idaho uses earth berms both to deflect the northerly winter winds, and also to deflect the hot summer sun from the east and west facades. South glazing collects winter sun, while a south facing overhang shades the south glazing from the summer sun.

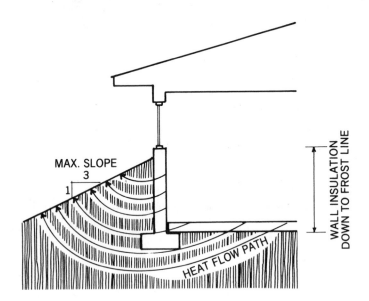

Figure 15.11i Since heat flows through earth in the radial pattern shown, the heat-flow path is quite long at the base of the wall and through the slab edge. Insulation (not shown for clarity) should, therefore, be thickest at the top of the earth-bermed wall.

Figure 15.11j Minimize berm penetration because each opening is a major source of heat loss.

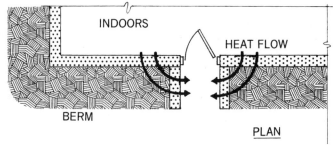

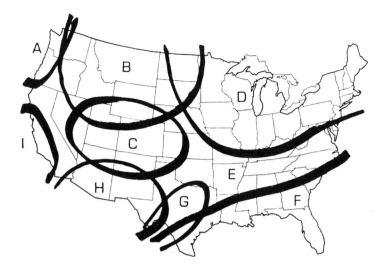

SYNOPSIS OF REGIONAL EARTH TEMPERING ISSUES

A   Cold, cloudy winters maximize value of earth tempering as a heat conservation measure. Cool soil and dry summers favor subgrade placement and earth cover, with little likelihood of condensation.

B   Severely cold winters demand major heat conservation measures, even though more sunshine is available here than on the coast. Dry summers and cool soil favor earth covered roofs and ground coupling.

C   Good winter insulation offsets need for extraordinary winter heat conservation, but summer benefit is more important here than in zone B. Earth cover is advantageous, the ground offers some cooling, condensation is unlikely, and ventilation is not a major necessity.

D   Cold and often cloudy winters place a premium on heat conservation. Low summer ground temperatures offer a cooling source, but with possibility of condensation. High summer humidity makes ventilation the leading conventional summer climate control strategy. An above ground super-insulated house designed to maximize ventilation is an important competing design approach.

E   Generally good winter sun and minor heating demand reduce the need for extreme heat conservation measures. The ground offers protection from overheated air, but not major cooling potential as a heat sink. The primacy of ventilation and the possibility of condensation compromise summer benefits. Quality of design will determine actual benefit realized here.

F   High ground temperatures. Persistent high humidity levels largely negate value of roof mass and establish ventilation as the only important summer cooling strategy. Any design that compromises ventilation effectiveness without contributing to cooling may be considered counterproductive.

G   This is a transition area between zones F and H, comments concerning which apply here in degree. The value of earth tempering increases moving westward through this zone, and diminishes moving southward.

H   Summer ground temperatures are high, but relatively much cooler than air. Aridity favors roof mass, reduces need for ventilation, eliminates concern about condensation. Potential for integrating earth tempering with other passive design alternatives is high.

I   Extraordinary means of climate control are not required due to relative moderateness of this zone. Earth tempering is compatible with other strategies, with no strong argument for or against it.

Figure 15.11k A summary of regional issues in regard to suitability of earth sheltering. (From *Proceedings of the International Passive and Hybrid Cooling Conference,* Miami Beach, FL, Nov. 6–16, 1981. ©American Solar Energy Society, 1981.)

flow in a radial pattern as shown in Fig. 15.11i. A cut in the berm creates a thermal weakness not only at the exposed wall, but also in adjacent parts of the wall (Fig. 15.11j). Because of this heat short-circuiting, there should be as little penetration as possible in any earth cover.

Although many factors determine the appropriateness of earth sheltering, one of the most important is climate. Earth sheltering is most advantageous in hot and dry climates and in regions that have both very hot summers and very cold winters. It is least advantageous in hot and humid regions where water and mildew problems are common and where natural ventilation is a high priority. The map and key in Fig. 15.11k give a more detailed breakdown of regional suitability of earth sheltering.

15.12 MOISTURE CONTROL

Excess moisture not only adds to the cooling load, but also can cause serious problems in buildings. The structure can rust or rot, the insulation can become useless, and the paint can peel. Indoors, windows can fog up, and mildew can grow causing health problems. The moisture can come from the outside or be generated indoors. Although moisture in vapor form can cause mildew to grow, it is mainly water in the liquid form inside the building fabric that causes most problems. Moisture can enter the building envelope either as water or water vapor that then condenses. Moisture can enter the building envelope four ways: bulk moisture, capillary action, air leakage, and vapor diffusion.

1. **Bulk moisture** is liquid water that enters through holes, cracks, or gaps, and it is usually rainwater driven by gravity and wind. Proper design and quality construction will minimize the

amount of bulk moisture penetrating the building skin.

2. **Capillary action** moves liquid water through porous materials and tiny holes by the surface tension on the water. The effect is strong enough to move water vertically against gravity. Capillary action is controlled primarily by fully sealing the porous materials and tiny holes with some material, such as the asphaltic waterproofing on a concrete foundation wall.

3. **Air leakage** carries water vapor through holes and cracks in the building envelope by the action of the wind, fans, or the stack effect.

4. Water vapor also enters the building fabric because of **vapor diffusion,** which is driven by a difference in vapor pressure. Water vapor moves from a higher concentration of moisture (high vapor pressure) to a lower concentration (low vapor pressure).

Water entering the building fabric through bulk-moisture flow or capillary action is immediately ready to cause problems, but water vapor entering through air leakage or vapor diffusion will primarily cause problems if condensation occurs in the wall or roof. Condensation will occur if part of the construction is colder than the dew-point temperature of the moist air. Various ways to prevent this condensation exist: by installing a vapor barrier, by reducing the indoor humidity, and by raising the temperature of the surface where the condensation would have occurred.

Condensation inside the building envelope can be prevented by using a vapor barrier to keep water vapor from migrating to a point in the construction that is cold enough to cause condensation. This phenomenon can be better understood with the concept of the **thermal gradient,** in which the temperatures across a wall (roof, etc.) are graphed on top of a drawing of the wall (roof, etc.), as

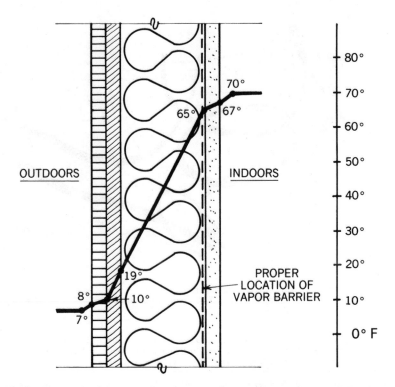

Figure 15.12a The graph of the thermal gradient, which is superimposed on a wall detail, clearly shows the temperature at each layer inside the construction. The dew-point temperature of the indoor air determines where in the wall condensation will occur.

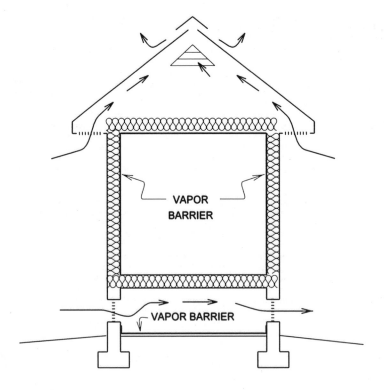

Figure 15.12b Vapor barriers can help prevent certain types of moisture problems. In cold climates, the barriers should be used in the walls on the warm side of the insulation. In crawl spaces, the vapor barriers should cover the ground. They are needed in ceilings only for unheated and unvented attics.

shown in Fig. 15.12a. Thus it is easy to determine the temperature in any part of the construction. For example, the temperature on the indoor side of the Fiberglas batts in Fig. 15.12a is 65°F, while on the outdoor side of the batts it is about 19°F. As the indoor air slowly moves through the wall, its temperature drops as the thermal gradient indicates. As the indoor air cools, its relative humidity (RH) will increase (see Section 4.6). Eventually it will reach 100 percent RH, which is also known as the saturation or dew point. At this point, water condenses out of the air and wets the insulation. Since it is very likely that the indoor air's dew-point temperature is between 19°F and 65°F, condensation will occur if the moist indoor air can get into the insulation. A vapor barrier on the indoor side of the insulation can be used to keep the moist indoor air from reaching the part of the insulation where its dew point is reached, thereby preventing condensation in the wall (Fig. 15.12b).

Note that in very hot and humid climates with mild winters, the problem is reversed. Here, the vapor barrier must be on the outside to keep very moist, summer outdoor air from condensing in the wall cooled from the outside by the air-conditioning system.

Vapor barriers can consist of any low-permeability material, such as closed-cell, plastic foam boards, foil-covered insulation, asphalt-coated Kraft paper on batt insulation, or polyethylene film. Since recent research has shown that air leakage is usually a much larger problem than vapor diffusion, a vapor barrier should have as few joints and holes as possible. Although a continuous sheet of polyethylene is a good choice, holes for electrical outlets are a major problem. In renovations, it is often best to use a vapor-retardant paint.

Water vapor that gets through the vapor barriers should have some way to escape. Thus, only a single vapor barrier should be used, and it should be located on the indoor side of a wall or ceiling in most climates. Many new insulating sheathing boards have a low permeability and, therefore, create, in effect, a vapor barrier on the outside of the wall. When these materials are used, it is important to control the indoor humidity rather than use a vapor barrier on the indoor side of the wall to keep moisture out of the structure.

Attics, roofs, and sometimes walls should be vented to prevent water damage (Fig. 15.12c). These vents

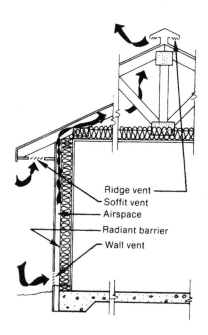

Figure 15.12c Make sure that the air flow from soffit to ridge or gable vents is not blocked by the ceiling insulation. Wall vents are mainly used as a cooling technique. (From *Cooling with Ventilation*. Golden, CO: Solar Energy Research Institute, 1986 (SERI/SP-273-2966; DE8601 701).)

have the additional benefit of allowing hot air to escape in the summer. See Table 15.12 for recommended attic vent areas. For best results, half the vent area should be in the soffit and half at the ridge. Continuous ridge vents, as shown in Fig. 15.12d,

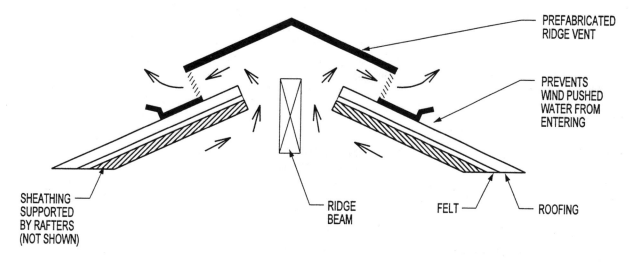

Figure 15.12d Continuous ridge vents efficiently evacuate hot and/or moist air by the combined action of the stack, Bernoulli, and venturi effects (see Fig. 10.5m). Ridge vents must be carefully designed to prevent wind blown water from entering the building.

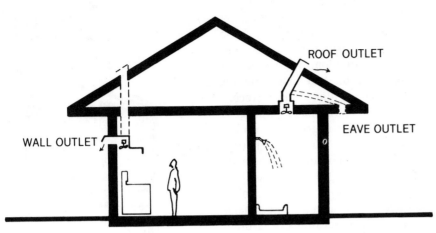

Figure 15.12e Moisture should be exhausted at the source. Vent it outside and never into the attic.

are very effective and highly recommended, as are aerodynamically designed roof vents (see Fig. 10.6t). However, electrically powered roof vents are generally not economical.

It is important to avoid having two different vapor barriers in a wall since moisture that gets in will be trapped. Another way to avoid condensation is to keep the dew-point temperature (humidity) of the indoor air low enough. This is an especially practical method to prevent condensation (fogging) on windows. Thus, if the indoor air is kept dry, its dew-point temperature will be below that of the indoor surface of the glazing, and no condensation will occur. In winter, the outdoor air is quite dry, and the indoor air will be just as dry unless moisture

is added. Exhaust fans in moisture-producing areas, such as bathrooms, kitchens, and laundry areas can prevent adding excessive moisture to the indoor air (Fig. 15.12e). In basements and earth-sheltered buildings, moisture is avoided by good drainage around the walls and a carefully installed waterproofing. In buildings with crawl spaces, the soil should be covered with a polyethylene vapor barrier and the crawl space should be vented (see Table 15.12).

The third method to prevent condensation is to heat the surface where the condensation would occur. This is especially effective for glazing. In automobiles, hot air or electric heating wires heat the glazing to a temperature above the dew point of the indoor air. In buildings, double or triple glazing can produce the same result: a higher temperature of the glass's indoor surface.

TABLE 15.12 RECOMMENDED VENT AREAS

Space Vented	Vent Area/ Floor Area
Crawl Space	
No ground cover	1/150*
With vapor barrier on ground	1/1500*
Attic	
No ceiling vapor barrier	1/150**
With ceiling vapor barrier	1/300**

* At least one vent near each corner of crawl space.

** Half at or near ridge, and half at soffit.

15.13 INFILTRATION AND VENTILATION

In a poorly constructed house with no weatherstripping on doors and windows, 50 percent of the heat loss can be due to infiltration. Good tight construction techniques with quality weatherstripped windows and doors can reduce the loss from infiltration to

about 20 percent of the total building heat loss.

Infiltration is the unplanned introduction of outdoor air through windows, doors, or cracks in the construction due to wind, the stack effect, or the action of exhaust fans. In winter as dry cold air infiltrates, an equal amount of warm moist air leaves the building (Fig. 15.13a). As a result, latent as well as sensible heat is lost. In summer, hot, moist air infiltrates and cool, dry air is lost. Consequently, the cooling load is also both latent and sensible.

Infiltration is controlled first by avoiding windy locations or by creating windbreaks. Minimizing doors and operable windows helps but more important is the seal. A poorly fitted, not weatherstripped window has an infiltration rate five times as great as an average-fit weatherstripped window. In buildings where doors are opened frequently, a vestibule can cut infiltration by 60 percent and revolving doors can cut infiltration by an amazing 98 percent (Fig. 15.13b).

Although a vapor barrier properly installed also acts as a good infiltration barrier, a separate barrier is sometimes used toward the outside of the wall (Fig. 15.13c). Be sure to keep in mind that the infiltration barrier in this position must be **permeable** to enable water vapor to escape from inside the wall. For that reason, most infiltration barriers are made of a tightly woven fabric that can breathe.

Unless one takes precautions, it is possible to make a building too airtight. Fireplaces and gas heating appliances could be starved for air, odors could build up, and eventually a shortage of oxygen for breathing could result. Often, there is also a problem of indoor air pollution. Thus, in very airtight construction, provisions must be made to bring in sufficient fresh air in a controlled manner. This is called **ventilation** and is described in more detail in the next chapter.

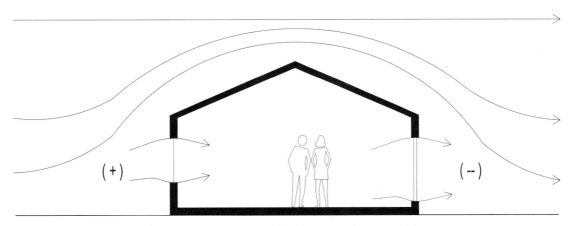

Figure 15.13a Infiltration due to the wind is caused by a push-pull effect from a simultaneous positive pressure on the windward side and a negative pressure on the leeward side.

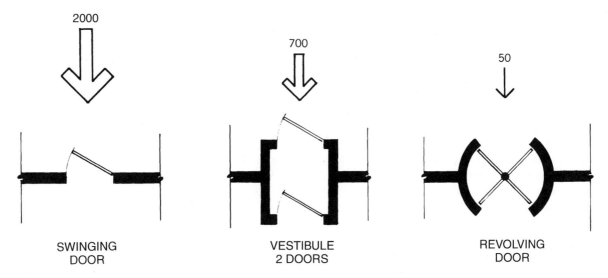

SWINGING
DOOR

VESTIBULE
2 DOORS

REVOLVING
DOOR

Figure 15.13b The numbers indicate the cubic feet of air infiltrating due to one-door operation.

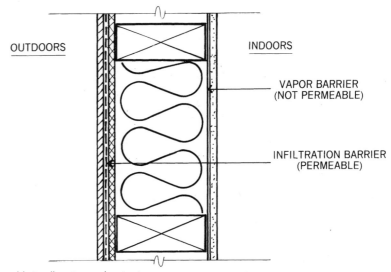

OUTDOORS

INDOORS

VAPOR BARRIER
(NOT PERMEABLE)

INFILTRATION BARRIER
(PERMEABLE)

Figure 15.13c An infiltration barrier must be permeable to allow trapped water to escape.

15.14 APPLIANCES

Appliances vary greatly in the amount of energy they consume and, therefore, the amount of heat they give off. The cost of inefficient appliances is double since first one must pay for using unnecessary energy and then one must pay again for extra cooling to have this unnecesary heat removed from the building.

By law, some appliances like refrigerators must state their efficiency with an Energy Efficiency Ratio (EER) label. In this way, one can make an objective choice. By having an electric ignition, gas appliances can eliminate the wasteful pilot lights. Exhaust fans can eliminate not only sensible and latent heat, but also the polluting products of combustion.

Table 15.14 is included to help focus attention on those appliances that use most energy in the home. As mentioned in Chapter 11, the lighting system is often a major energy user and, therefore, also a major source of heat gain. Thus, in all buildings, but especially nonresidential buildings, the lighting system should have lamps of high efficacy and luminaires of high efficiency.

15.15 CONCLUSION

Every building should be an "energy-efficient building." Such buildings generally cost less initially because their heating and cooling equipment is smaller, and they certainly cost less to operate since their energy bills will be much lower. Not only will owners save money and society save valuable energy, but the architect's desire for a richer visual environment can also be promoted. Money not spent for mechanical equipment or energy is now free to be applied to aesthetic elements.

This discussion on the techniques for keeping warm and staying cool finishes that part of the heating, cooling, and lighting of buildings that is primarily in the domain of the architect. Although the mechanical heating and cooling systems discussed in the next chapter are mainly the responsibility of engineers, architects must still help to integrate these systems properly into the building. It is, therefore, vital for architects to have a general understanding of mechanical systems.

TABLE 15.14 ANNUAL ENERGY REQUIREMENTS OF ELECTRIC HOUSEHOLD APPLIANCES

Appliance	Approximate Kilowatt-Hours/Year
Shaver	1
Clock	20
Fan (circulating)	50
Fan (attic)	300
Oven (microwave)	200
Oven (conventional)	700
Color television	300
Dishwasher	400
Clothes dryer	1,000
Freezer	1,500
Refrigerator	2,000
Hot-water heater	4,500

KEY IDEAS OF CHAPTER 15

1. Designing the thermal envelope is part of the first tier of the three-tier design approach. The better the thermal envelope, the less heating and cooling will be required.
2. Thermography can be use to find holes or weak areas in the thermal envelope.
3. Heat is lost by transmission through walls, roof, floors, windows, and doors; by infiltration through cracks; and purposeful ventilation.
4. Heat is gained from transmission, ventilation, solar radiation, appliances, lighting, and people.
5. In most cases, a compact design will use fewer resources than a spread-out extended design.
6. The extra heating effect of sunlight on a wall or roof is accounted for by the **sol air temperature.**
7. The **design equivalent temperature difference** is used when the insulating effect of mass is also considered.
8. In hot climates use surfaces with high solar reflectivity (albedo).
9. **Thermal planning** arranges building spaces in accordance to their heat and temperature needs as related to solar orientation.
10. Insulation materials consist primarily of very small air spaces separated by a material of low thermal conductivity.
11. Shiny metal surfaces are good radiant barriers because they have both high reflectivity and low emissivity. Radiant barriers are most appropriate under roofs in hot climates.
12. Large air spaces should not be used because their R-value is about the same as small air spaces. Large air spaces should be subdivided into as many small air spaces as possible.
13. Log cabins should not be built because they waste wood and have only mediocre thermal resistance.
14. Avoid creating heat bridges through the thermal envelope.
15. Structural insulated panels (SIP) have both high thermal resistance and great structural strength.
16. Low-e, double-glazed (R-3) windows should be the minimum standard in all but the mildest climates.
17. Use high-performance (R-4 and higher) windows whenever possible.
18. Use **selective low-e** glazing when cool daylight is required.
19. Movable insulation over windows can reduce heat loss during winter nights and heat gain during summer days.
20. In hot climates with medium to large diurnal temperature ranges, massive walls will reduce the heat gain by the phenomenon of the insulating effect of thermal mass.
21. Earth sheltering is most appropriate in climates with very hot summers and very cold winters. It is least appropriate in wet and humid regions.
22. Only use earth sheltering when positive gravity drainage is assured.
23. Grass roofs are mainly beneficial in reducing heat gain.
24. Moisture can enter a building four ways: bulk moisture, capillary action, air leakage, and vapor diffusion.
25. Use vapor barriers to prevent condensation inside walls.
26. Avoid creating excess moisture. Use exhaust fans where large quantities of water vapor are produced.
27. Use eave, ridge, and gable vents to prevent moisture problems in the roof.
28. Use a ground vapor barrier to prevent moisture problems in crawl spaces.
29. Infiltration barriers should stop the wind but not water vapor.

References

Akbari, Hashem, and others, eds., *Cooling Our Communities: A Guidebook on Tree Planting and Light-Colored Surfacing.* Washington, DC: EPA and DOE/Lawrence Berkeley Laboratory, 1992. [Lawrence Berkeley Laboratory LBL-31587]

Environmental Building News, October, 1998.

Resources

FURTHER READING
(See Bibliography in back of book for full citations. The list includes valuable out-of-print books.)

Brown, G. Z., and M. DeKay. *Sun, Wind and Light: Architecure Design Strategies.*
Boyer, L., and W. Grondzik. *Earth Shelter Technology.*
Carmody, R. S. *Earth Sheltered Housing Design.*

Carmody, R.S., S. Selkowitz, and L. Heschong. *Residential Windows: A Guide to New Technologies and Energy Performance.*
Easton, D. *The Rammed Earth House.*
Franta, G., K. Anstead, and G. D. Ander. "Glazing Design Handbook for Energy Efficiency."
Golany, G. *Design and Thermal Performance: Below Ground Dwellings in China.* Newark, DE: University of Delaware Press, 1990.

Hestnes, A. G., R. Hastings, and B. Saxhof, eds. *Solar Energy Houses: Strategies, Technologies, Examples.*

Houben, H., and H. Guillaud. *Earth Construction: A Comprehensive Guide.*

Norton, J. *Building with Earth: A Handbook.*

Lstiburek, J. W. *Exemplary Home Builder's Field Guide.*

Lstiburek, J. W. *Builder's Guide to Cold Climates: Details for Design and Construction.*

Lstiburek, J. W. *Builder's Guide to Mixed Climates: Details for Design and Construction.*

Lstiburek, J. W., and J. Carmody. *Moisture Control Handbook: Principles and Practices for Residential and Small Commercial Buildings.*

King, B. *Buildings of Earth and Straw: Structural Design for Rammed Earth and Straw-Bale Architecture.* Ecological Design Press, 1996.

Krigger, J. T. *Residential Energy.*

Matus, V. "Design for Northern Climates: Cold-Climate Planning and Environmental Design."

Stein, B. and J. S. Reynolds. *Mechanical and Electrical Equipment for Buildings.* A basic resource.

Steven Winter Associates. *The Passive Solar Design and Construction Handbook.*

Underground Space Center. "Earth Sheltered Housing Design: Guidelines, Examples, and References."

Watson, D., and W. M. C. Labs. *Climatic Design.*

ORGANIZATIONS
(See Appendix J for full citations.)

Energy Efficiency & Renewable Energy Clearinghouse (EREC)
Energy Efficient Building Association
Southface Energy Institute

MECHANICAL EQUIPMENT FOR HEATING AND COOLING

"It is not a question of air conditioning versus sea breezes, or fluorescent tubes versus the sun. It is rather the necessity for integrating the two at the highest possible level."

James Marston Fitch
American Building: The Environmental
Forces That Shape It *(p. 237),* © *1972*

16.1 INTRODUCTION

In most buildings, mechanical equipment (tier 3) is required to carry the thermal loads still remaining after the techniques of heat retention or rejection (tier 1) and passive heating or cooling (tier 2) have been applied (Fig. 16.1). With the proper design of the building, as described in the previous parts of this book, the size and energy demands of the heating and cooling

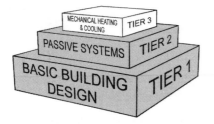

Figure 16.1 The heating and cooling needs of a building are best and most sustainably achieved by the three-tier design approach, and this chapter discusses tier three (mechanical equipment).

equipment can be quite small. Since the heating and cooling equipment is bulky and must reach into every space, it is an especially important concern for the architectural designer.

Although cooling systems are a must for all internally dominated buildings with their high cooling loads from lights, people, and equipment, cooling systems are not necessary for all envelope-dominated buildings. In cold regions with mild or short summers, only heating systems are required. Heating is, therefore, discussed separately from cooling.

16.2 HEATING

Conceptually, heating is very simple: a fuel is burnt and heat is given off. The simplest heating system of all is to have a fire in the space to be warmed.

Until the twelfth century, it was the almost universal practice—even in royal halls—to have a fire in the center of the room with the smoke exiting through the roof or a high window (Fig. 16.2a). This heating method was very efficient, but smoke made the concept of cleanliness inconceivable. Around the Mediterranean Sea and in some other warm climates around the world, small portable heaters, such as **charcoal braziers,** were popular (Fig. 16.2b). The Japanese hibatchi is a similar device. A real exception to these primitive heating systems was the Roman **hypocaust,** where warm air from a furnace passed under a floor and up through the walls (Fig. 16.2c). Traditional Korean buildings use a similar underfloor heating system.

The fireplace came about with the invention of the chimney in the twelfth century AD. Although buildings were now relatively smoke-free, heating them became harder because the efficiency of fireplaces is very low, about 10 percent. The fireplace

Figure 16.2a Some royal halls were still heated by an open fire as late as 1300 AD. The Hall of Penhurst Place. (From *The Mansions of England in Olden Time* by Joseph Nash, Henry Sotheran & Co., 1971.)

Figure 16.2b A portable charcoal brazier.

Figure 16.2c Roman hypocaust heating. (Courtesy of Wirsbo Company.)

remained popular in England because of the relatively mild climate (Fig. 16.2d), but in colder parts of Europe, the ceramic stove with its much higher efficiency (between 30 percent and 40 percent) became popular (Fig. 16.2e).

The English settlers brought the fireplace to the New World where the endless forests could feed the huge appetite of fireplaces in cold climates. Around big cities, like Colonial Philadelphia, the forests were soon cut

Figure 16.2d In England, fireplaces remained popular because of the relatively mild climate. In colder climates, the ceramic stove was preferred. (From *The Mansions of England in Olden Time* by Joseph Nash, Henry Sotheran & Co., 1971.)

down, and an energy crisis developed. Benjamin Franklin responded by inventing a fuel-efficient cast-iron stove.

Franklin realized that the traditional fireplace has several serious deficiencies: it heats only by direct radiation, the hot gases carry most of the heat out through the chimney, and cold air is sucked into the building to replace the warmed room air pulled into the fire to support combustion. Franklin's design addressed all of these issues, and a good modern fireplace must do the same. Today, metal fireplace inserts enable room air to circulate around the firebox (Fig. 16.2f). Sometimes a fan is used to

Figure 16.2e In northern, central, and eastern Europe, masonry and ceramic stoves were used instead of the inefficient fireplace. Cast-iron and steel stoves are even more efficient because they conduct heat faster through the walls; however, the heavy masonry stoves stored heat for all night warmth. In very cold climates like Russia, people would live and sleep on top of very large masonry stoves.

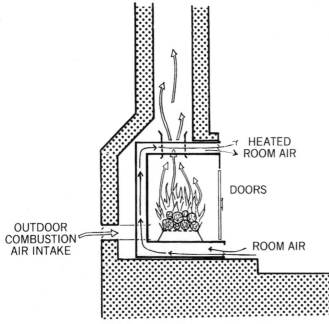

Figure 16.2f A modern efficient fireplace must have doors, outdoor-combustion air intake, and a firebox around which room air can circulate.

increase the heat transfer from the firebox to the circulating room air. A special duct brings outdoor-combustion air to the fireplace. Thus, heated room air is not required to feed the fire. Doors are necessary to prevent any room air from being pulled into the fireplace. Otherwise, even when the fire has died out, the stack effect will continue to pull heated room air out through the chimney. But even with

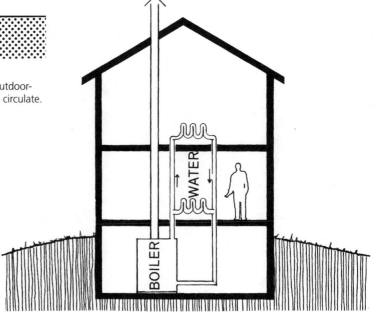

Figure 16.2g Gravity hot air or water systems used natural convection currents to move heat.

these features, modern fireplaces are still only about 30 percent efficient, while a modern metal stove can have an efficiency as high as 70 percent.

Central heating became quite popular in larger buildings in the nineteenth century. Gravity air and water systems worked especially well in multistory buildings with basements. The furnace or boiler, located in the basement next to the wood or coal bin, heated the air or water to create strong natural-convection currents (Fig. 16.2g). By adding pumps and fans, modern heating systems have become more flexible and respond faster.

When choosing or designing a heating on cooling system for a building, one must first know how many different thermal zones are required.

16.3 THERMAL ZONES

Because not all parts of a building have the same heating or cooling demands, mechanical systems are subdivided into individually controlled areas called **zones.** Each zone has a separate thermostat to control the temperature and sometimes a humidistat to control the moisture content of the air. The most common reason for separate zones is the difference in exposure. A north-facing space might require heating while a south-facing space in the same building requires cooling. So, too, a west-facing room might require heating in the morning and cooling in the afternoon, while an east-facing room would experience the reverse situation. Since interior spaces have only heat gains, they require heat removal all year. Thus, a large office building would be divided into at least five zones on the basis of exposure (Fig. 16.3).

Frequently, additional zones are required because of differences in usage. For example, a large conference room requires separate thermal control; otherwise it will be too cold when only a few people are present and too hot when the room is full. A computer room must be on a separate

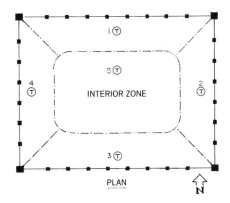

Figure 16.3 A large office building would require at least five zones based on differences in exposure. Each zone will have its own thermostat.

zone for two reasons. It has an unusually high source of internal heat gain, and its hours of operation are quite different from the rest of the building. Buildings are often zoned on the basis of rental areas, too. The number of zones required is an important factor in choosing a particular mechanical system.

16.4 HEATING SYSTEMS

The two major considerations in choosing a heating system are the source of energy (fuel) used and the method of distribution within the building. The choice of a fuel usually depends on both economic factors and what is available. The main choices are gas, oil, coal, electricity, solar energy, and waste-heat recovery. Except in rural areas, wood is too polluting to be a practical fuel.

Oil, coal, wood, bottled gas, and solar energy require building storage space (Fig. 16.4a). Electricity is popular because of its great convenience. Solar energy, the only renewable source in the list, was discussed in Chapters 7 ("Passive Solar") and 8 ("Photovoltaics and Active Solar"). Heat recovery from ventilation is nearly always appropriate, but it can supply only part of the heat required. It will be discussed later in this chapter, along with waste heat from a

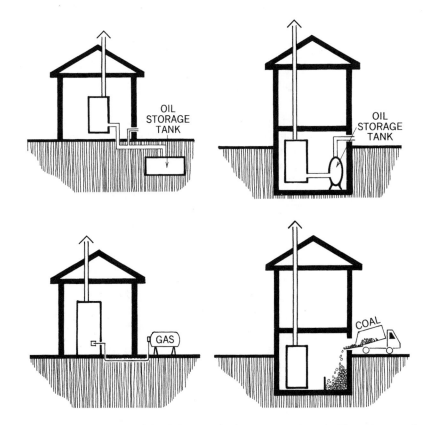

Figure 16.4a Oil, coal, wood, bottled gas, and solar energy require a significant amount of storage space.

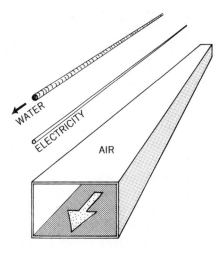

Figure 16.4b An air system requires a substantial amount of a building's volume for ducts and air handling equipment (1–5%).

combined heat and power (CHP) system (see Section 3.21).

Since the distribution system has a great effect on the architecture, one must select it with care. Heat can be distributed in a building by air, water, or electricity. Because of their large size, air ducts require the most forethought, while electric wires require the least (Fig. 16.4b). The space required for ducts and air handling equipment varies between 1 percent and 5 percent of the total volume of a building. The advantages and disadvantages of air, water, and electric distribution systems are summarized in Table 16.4.

16.5 ELECTRIC HEATING

Although many different types of electric-heating devices exist, most use resistance-heating elements to convert electricity directly into heat. The exceptions are the heat pump and heat from the lighting system.

Figure 16.5a illustrates the general types of resistance-heating devices that are available. A great advantage of all the devices shown is that they allow many heating zones to be easily established — each room or part thereof can be a separate zone. Electric boilers or furnaces to heat central hot water or air systems do not have this advantage. Since electric resistance heating is expensive to operate and wasteful of source energy, one should use it only in mild climates or for spot-heating small areas.

The baseboard units heat by natural convection, while the unit heaters have fans for forced convection. Radiant heating is possible at three different intensities. Because of their large areas, radiant floors and ceilings can operate at rather low temperatures (80°F and 110°F, respectively). Radiant panels on walls or ceilings must be hotter (about 190°F) to compensate for their smaller areas. They are used to increase the mean radiant temperature (MRT) near large areas of glazing or other cold spots. High-intensity infrared lamps operate at over 1000°F and, therefore, can be quite small. They look similar to fluorescent fixtures except the linear quartz lamps glow red hot. These high-intensity infrared heaters do not heat air, only solid objects, such as walls, furniture, and people. Therefore, these heaters can be used outdoors for purposes such as keeping people warm in front of hotel or theater entrances. They are also appropriate in buildings, such as warehouses or aircraft hangars, where it is impractical to heat the air. High-intensity infrared heaters can also be powered by gas instead of electricity.

TABLE 16.4 HEATING DISTRIBUTION SYSTEMS

System	Advantage	Disadvantage
Air	Can also perform other functions such as ventilation, cooling, humidity control, and filtering Prevents stratification and uneven temperatures by mixing air Very quick response to changes of temperature No equipment required in rooms being heated	Very bulky ducts require careful planning and space allocation Can be noisy if not designed properly Very difficult to use in renovations Zones are not easy to create Cold floors result if air outlets are high in the room
Water	Compact pipes are easily hidden within walls and floor Can be combined with domestic hot water system Good for radiant floor heating	For the most part can only heat and not cool (exceptions: fan-coil units and valance units) No ventilation No humidity control No filtering of air Leaks can be a problem Slightly bulky equipment in spaces being heated (baseboard and cabinet convectors) Radiant floors are slow to respond to temperature changes
Electricity	Most compact Quick response to temperature changes Very easily zoned Low initial cost	Very expensive to operate (except heat pump) Wasteful, unenvironmental Cannot cool (except for heat pump)

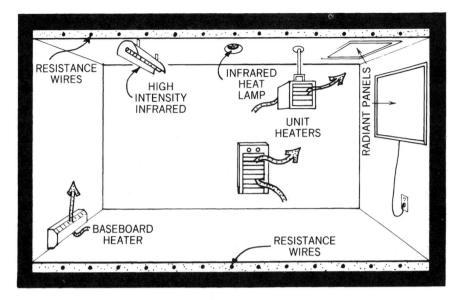

Figure 16.5a Various types of electric resistance heaters.

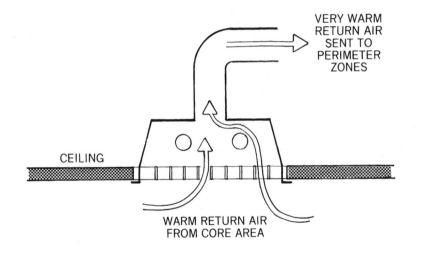

Figure 16.5b Air is heated by returning it through the lighting fixtures.

With the above-mentioned resistance heating devices, 1 BTU equivalent of electrical energy is converted into 1 BTU of heat. However, with a **heat pump,** 1 BTU equivalent of electricity can yield as much as 3 BTUs of heat. The secret of this apparent "free lunch" is that the electricity is not converted into heat, but is used instead to pump heat from outdoors to indoors. Heat is extracted from the cold outdoor air and added to the warm indoor air. Thus, in effect, the heat is pumped "uphill," which is what all refrigeration machines do. A heat pump is a special air conditioner running in reverse during the winter.

Heat pumps are appropriate where both summer cooling and winter heating is required. Since the efficiency of heat pumps drops with the outside temperature, they are not appropriate in very cold climates. The efficiency of heat pumps is described by the **coefficient of performance (COP)** which is defined as

$$COP = \frac{energy\ out}{energy\ in}$$

In mild climates, a COP as high as four can be achieved, while in cold climates it will be under two. Much better efficiencies are possible by coupling the heat pump with the ground water rather than the outdoor air because the ground is much warmer in the winter and is cooler in the summer. Ground-coupled heat-pump systems will be explained after heat pumps are discussed in Section 16.11.

Although heat pumps are the best method of heating with electricity, people sometimes complain of being cold in apparently warm houses. Since heat pumps supply not hot but warm air at approximately 85°F, anyone in the path of the moving air will be chilled by evaporation more than heated by the warm air. Hot-air heating with a gas furnace, on the other hand, supplies much warmer air at approximately 110°F and, therefore, does not cause this cooling effect. The solution for the heat-pump systems is to use larger ducts and more diffusers in order to supply air with a lower velocity, as well as away from people.

Although lights are always a source of heat, they are no more efficient than resistance heating elements (COP = 1). There is a system, however, in which the lighting can be efficiently used for heating. In a large office building a sizable interior zone is lit only by electric lights and requires cooling even in the winter (Fig. 16.3). If the warm return air from the core is further heated by being returned through the lighting fixtures, it will be warm enough to heat the perimeter area of the building (Fig. 16.5b). A side benefit of this system is that the lamps and fixtures last longer because they are cooled by the return air.

16.6 HOT WATER (HYDRONIC) HEATING

Any of the fuels mentioned before can be used to heat water in a **boiler** (Fig. 16.6a). The hot water can distribute the heat throughout the building in several ways.

One of the most comfortable hydronic heating systems is based on the Roman hypocaust. Instead of fire, hot water is pumped through coils of plastic tubing imbedded in the floor (Figs. 16.6b and 16.6c). When concrete slabs are used, the coils can be cast right into the slab. When wooden floors are used, the coils are attached to the subflooring, covered first by lightweight concrete or gypsum, and then covered by wooden flooring or tiles.

A radiant floor will heat the space by both radiation and natural convection. Radiant floors are very comfortable because we enjoy having warm feet and cool heads. For the same reason, radiant ceilings are not very comfortable: they give us a warm head and cool feet. Because the coils are imbedded in thermal mass, there is a long time lag in receiving heat after the system is turned on. This is an advantage in climates and buildings in which the heating demand is fairly constant, but this time lag is a disadvantage when the heating load is intermittent, and the system must respond quickly.

Radiant floor heating is a good match for active solar because solar collectors can efficiently produce the relatively low water temperatures needed (about 90°F) (see Chapter 8).

Most hot-water systems use convectors to transfer the heat from the water to the air of each room (Fig. 16.6d). In the past, hot-water systems used cast-iron radiators, which, in fact, also heated largely by convection (Fig. 16.6e). Today, most convectors consist of fin-tubes or fin-coils to maximize the heat transfer by natural

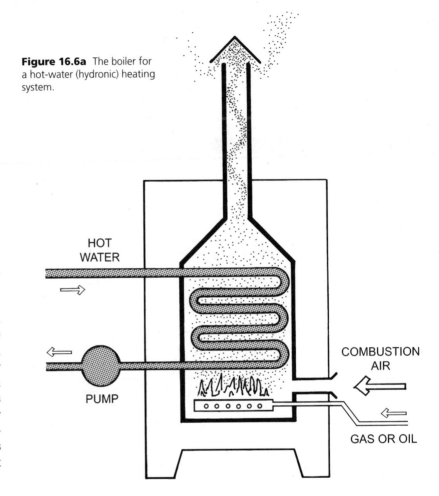

Figure 16.6a The boiler for a hot-water (hydronic) heating system.

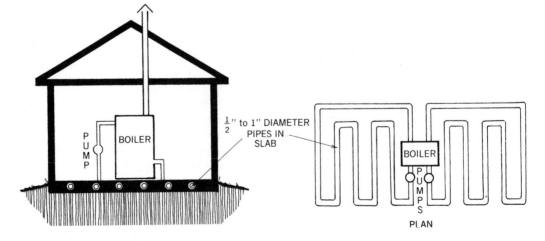

Figure 16.6b Radiant-floor hot-water heating systems often use loops of continuous plastic tubing to minimize joints.

Figure 16.6c For slab-on-grade, radiant floor heating, the concrete is poured over the plastic tubing. The various heating zones are made with continuous tubing with all joints made above the slab to minimize leaks in the concrete.

Figure 16.6d A hot-water system with baseboard convectors.

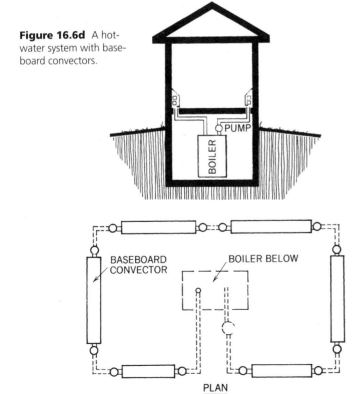

Figure 16.6e Cast-iron radiators heated by both radiation and convection. (From *Architectural Graphic Standards,* Ramsey/Sleeper 8th ed., John R. Hoke, editor, © John Wiley & Sons, Inc., 1988.)

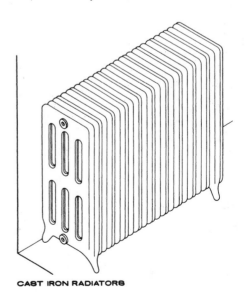

CAST IRON RADIATORS

convection (Fig. 16.6f). Baseboard convectors are linear units placed parallel to exterior walls, while cabinet convectors concentrate the heating where it is most needed — under windows. When there is a large area of glazing from floor to ceiling, a below-floor convector can be used.

Most convectors are designed to be as unobtrusive as possible because it is assumed that mechanical equipment must be ugly. Some architects and manufacturers take a different approach, as can be seen in the elegant radiator/convectors of Fig. 16.6g.

Because convectors rely on natural convection, they must be placed low in a room. However, if a fan is used for forced convection, any mounting position is possible. Because such fan-coil units can also be used for cooling, they are discussed under cooling systems.

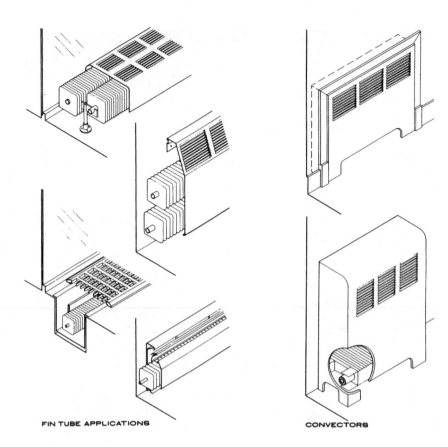

FIN TUBE APPLICATIONS   CONVECTORS

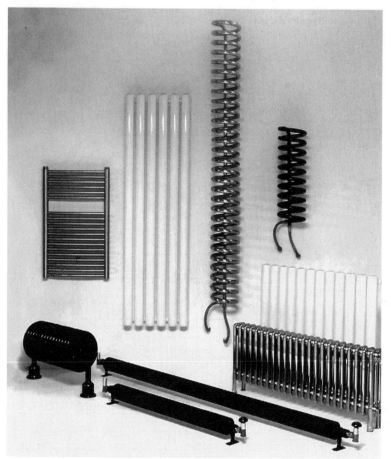

Figure 16.6f Baseboard convectors are unobtrusive but can be blocked by furniture. Cabinet convectors are usually placed under windows. Under-floor convectors are appropriate for areas with floor-to-ceiling glass. (From *Architectural Graphic Standards,* Ramsey/Sleeper 8th ed., John R. Hoke, editor, © John Wiley & Sons, Inc., 1988.)

Figure 16.6g These sculptural, elegant convector/radiators are designed to be exposed to view. (Courtesy 3-D Laboratory, Inc.)

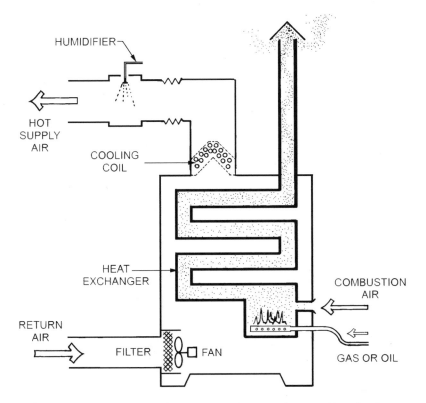

Figure 16.7a A schematic section of a hot-air furnace with optional cooling coils and humidifier.

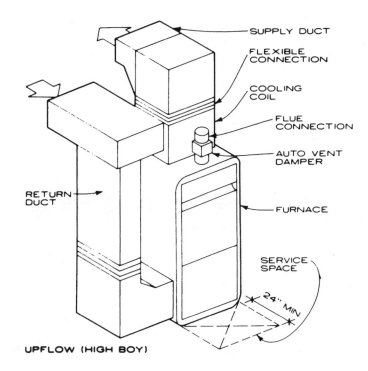

Figure 16.7b The isometric of a hot-air furnace with cooling coil. (From *Architectural Graphic Standards,* Ramsey/Sleeper 8th ed., John R. Hoke, editor, © John Wiley & Sons, Inc., 1988.)

16.7 HOT-AIR SYSTEMS

Air systems are popular because they can perform the whole range of air-conditioning functions: heating, cooling, humidification, dehumidification, filtering, ventilation, and air movement to eliminate stagnant and stratified air layers. Hot-air heating systems are especially popular where summer cooling is also required. Those hot-air systems that supply air at or near floor level around the perimeter of the building are most suitable for cold climates where the heating season is the main consideration. These systems are discussed here, while those more suitable for hot climates will be discussed along with other cooling systems.

A hot-air furnace uses a heat exchanger to prevent combustion air from mixing with room air. A blower and filter are standard, while a humidifier and cooling coil are optional (Figs. 16.7a and 16.7b). Gas furnaces can be very efficient. Some pulse-type gas furnaces convert more than 95 percent of the gas energy into useful heat.

For slab-on-grade construction in cold climates, the **loop-perimeter system** offers the greatest thermal comfort (Fig. 16.7c). The supply air heats the slab where it is the coldest—at the edge. Thus, this system offers both the benefits of hot-air and radiant-slab heating. The main disadvantage is the high initial cost of the system.

The **radial-perimeter system** is a less expensive but also less comfortable way to heat slab-on-grade construction with hot air. This system is more suitable for crawl space construction (Fig. 16.7d). If the crawl space is high enough (about 4 feet), a special horizontal furnace can be used (Fig. 16.7e). The same horizontal furnaces are sometimes also used in attic spaces. The author has crawled though spaces so tight that it was a puzzle how the equipment and ducts were ever installed. Of course, the craftsmanship was very poor and the duct work was very leaky, which

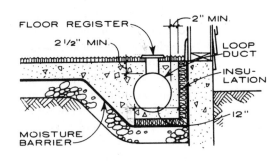
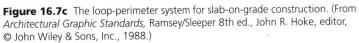
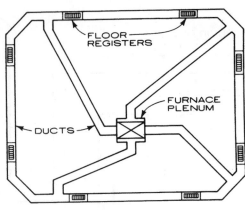

Figure 16.7c The loop-perimeter system for slab-on-grade construction. (From *Architectural Graphic Standards,* Ramsey/Sleeper 8th ed., John R. Hoke, editor, © John Wiley & Sons, Inc., 1988.)

PLAN

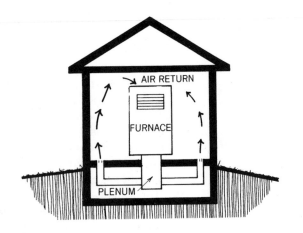

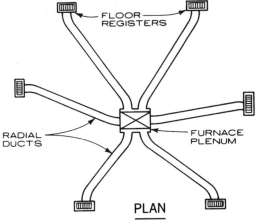

PLAN

Figure 16.7d The radial-perimeter system for crawl space construction. (From *Architectural Graphic Standards,* Ramsey/Sleeper 8th ed., John R. Hoke, editor, © John Wiley & Sons, Inc., 1988.)

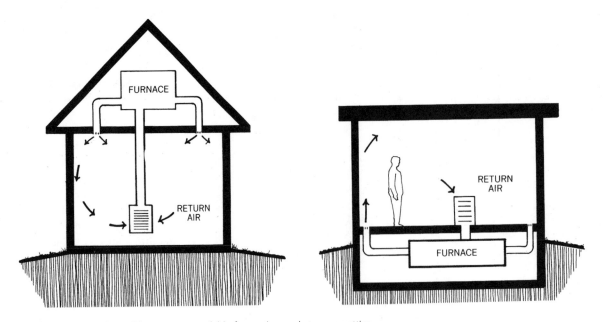

Figure 16.7e Horizontal furnaces are available for use in crawl spaces or attics.

Figure 16.7f The extended-plenum system for basement and crawl-space construction. Supply ducts run parallel and between joists to save on headroom. (Drawing on right from *Architectural Graphic Standards,* Ramsey/Sleeper 8th ed., John R. Hoke, editor, © John Wiley & Sons, Inc., 1988.)

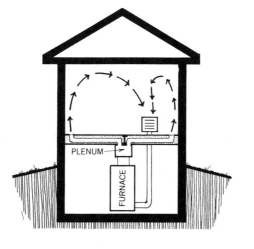
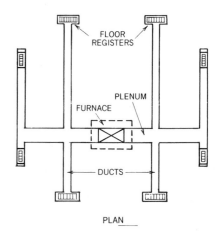

Figure 16.7g Wall furnaces are appropriate for heating single spaces.

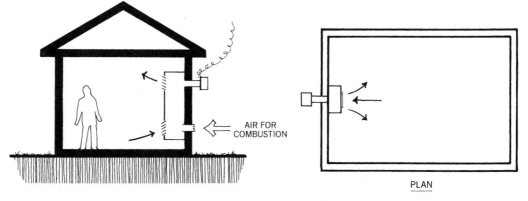

is not just inefficient but unhealthy, as will be explained later. It is imperative that adequate, easily accessed space be provided so that quality installations and servicing can be achieved.

The **extended-plenum system** is most appropriate for buildings with basements because it enables the supply ducts to run parallel and between the joists and, consequently, much space and headroom are saved (Fig. 16.7f).

For heating single spaces, a **wall furnace** can be a practical solution because no duct work is required. When powered by gas, these wall furnaces can draw combustion air and vent directly through the wall on which they are attached (Fig. 16.7g).

In spaces with high ceilings, **unit heaters** powered by gas, electricity, or hot water are often appropriate because they take up no floor area (Fig. 16.7h).

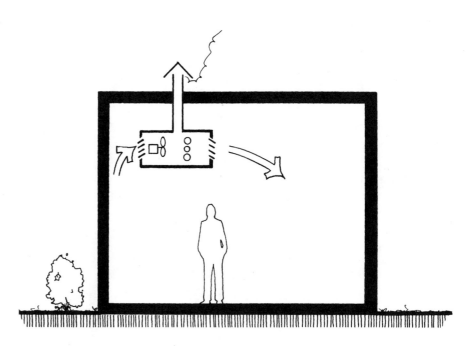

Figure 16.7h Unit heaters can be appropriate for spaces with high ceilings. The heat source can be electricity, hot water, or gas, which, however, must be vented.

16.8 COOLING

Cooling is not as intuitively clear and simple as heating. Cooling, the removal of heat, can be better understood by means of a water analogy. A building in the summer is surrounded by heat trying to get in just as water tries to get into a submerged building (Fig. 16.8a). The water in the analogy is gained both through the envelope and from internal sources. The natural tendency is for the water to flow into the building. Only by pumping it uphill can it be removed again.

In the same way, it is the natural tendency for heat to flow inward when the outdoor temperature is higher than the indoor temperature.

The only way to remove the heat is with a machine that pumps heat, a refrigeration machine (Fig. 16.8b). Before the invention of refrigeration machines about 150 years ago, there was no way to actively cool a building. Although blocks of ice harvested in winter could actually cool a building, the huge amount of ice required made that impractical on all but the smallest scales. One of the big trade items in the nineteenth century was ice, which was harvested from New England lakes in the winter and shipped to the South in the summer. Because of ice's high cost, it was primarily used for cooling drinks.

Until the twentieth century, the only way to achieve some summer comfort was to use the heat-rejection and passive-cooling techniques mentioned previously. Shading, natural ventilation, evaporative cooling, and thermal mass were the main techniques used.

In the 1840s, Dr. John Gorrie, a physician in Apalachicola, Florida, built the first refrigeration machine in an attempt to help his patients suffering from malaria. Although his machine worked, it was not used to cool buildings until the 1920s when a new type of building had a special need for air conditioning. Because movie houses had to close windows to keep the light out, they also kept the cooling breezes out. So most people had their first experience with air conditioning at movie houses, where the marquees announced "air conditioned" bigger than the title of the movies. Although air conditioning was still considered a luxury in the United States in the 1950s, it is now considered a necessity.

16.9 REFRIGERATION CYCLES

The refrigeration machine, a machine that pumps heat, is the critical element of any cooling system. There are basically three refrigeration methods: vapor compression, absorption, and thermoelectric. The compression cycle is the most common, but the absorption cycle is often appropriate when a source of low-cost heat is available. The thermoelectric cycle, which turns electricity directly into a heating and cooling effect, is not used for buildings.

The Compressive Refrigeration Cycle

The **compressive refrigeration cycle** depends on two physical properties of matter:

1. A large amount of **heat of vaporization** is required to change a liquid into a gas. Of course, this heat is released again when the gas condenses back into a liquid.

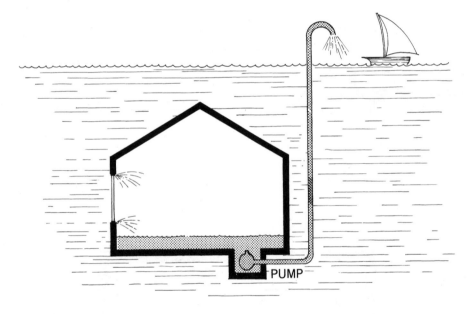

Figure 16.8a A water analogy for cooling: the water that found its way into the submerged building must be pumped "uphill" to get it out.

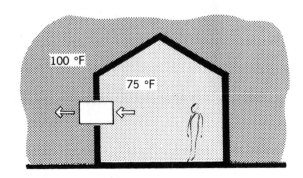

Figure 16.8b A refrigeration machine pumps heat from a lower to a higher temperature.

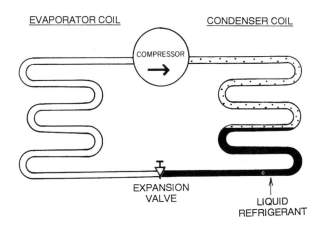

Figure 16.9a The basic components of a compressive refrigeration machine.

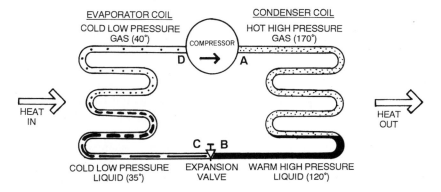

Figure 16.9b Where the refrigerant evaporates it absorbs heat (cools), and where it condenses it gives off heat.

vaporization. Thus, a warm, high-pressure liquid will collect at point "B." The cycle now repeats as a small amount of liquid refrigerant enters the evaporation coil at point "C" and evaporates to collect as a low-pressure gas at point "D."

Almost all refrigerants were made of **chlorofluorocarbons (CFCs)** until 1987 when the world agreed through the Montreal Protocol to phase out these ozone-damaging compounds.

New refrigerants made of **hydrochlorofluorocarbons (HCFCs)** are 90 percent less damaging to the ozone, but are still significant greenhouse gases when they escape from refrigeration machines. Other possible refrigerants include ammonia, propane, and isobutane, but these are all flammable, and ammonia is toxic also.

The Absorption Refrigeration Cycle

The **absorption refrigeration cycle** depends on the same two properties of matter described above for the compressive cycle, as well as a third property:

3. Some liquids have a strong tendency to absorb certain vapors (e.g., water vapor is absorbed by liquid lithium bromide or ammonia).

The absorption refrigeration machine requires no pumps or other moving parts, but it does require a source of heat such as a gas flame or the waste heat from an industrial process. The machine consists of four interconnected chambers of which the first two are shown in Fig. 16.9c.

In chamber "A," water evaporates and in the process draws heat from the chilled water coil (output). The water vapor migrates to chamber "B" where it is absorbed by the lithium bromide. Consequently, the vapor pressure is reduced, and more water can evaporate to continue the cooling process. Eventually, the lithium bromide will become too dilute to further

2. The boiling/condensation temperature of any material is a function of pressure. For example, 212°F is the boiling point of water only at the pressure of sea level (14.7 lb/square inch). When the pressure is reduced, the boiling point is also reduced.

The basic elements of a compression refrigeration machine are shown in Fig. 16.9a. Imagine that the valve is almost closed and the compressor has pumped most of the refrigerant into the condenser coil. Since the valve is only slightly open, only a small stream of liquid refrigerant can enter the par-

tial vacuum of the evaporator coil at point "C" (Fig. 16.9b). The refrigerant boils (evaporates) due to the low pressure. To change state, the liquid will require the large amount of heat called "heat of vaporization." Thus, the evaporator coil will cool as it gives up its heat to boil the refrigerant.

To keep the process going, the compressor continues to pump the refrigerant gas back into the condenser coil. A high pressure gas collects at point "A." Since any gas under pressure heats up, the condenser coil gets hot. As the coil loses heat, the high-pressure refrigerant gas will be able to condense and give up its heat of

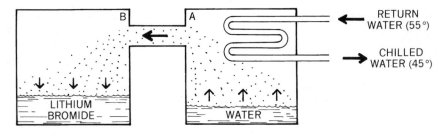

Figure 16.9c The first two chambers of an absorption refrigeration machine.

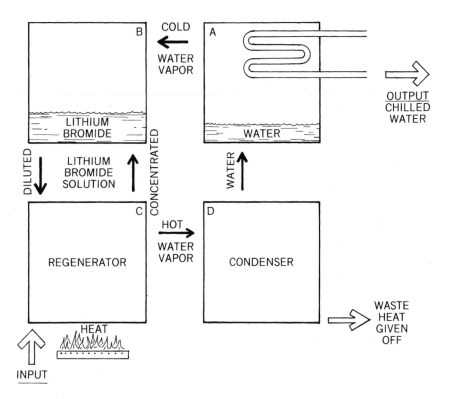

Figure 16.9d The absorption refrigeration cycle has chilled water as an output and a heat source as an input. Waste heat is given off in the process.

absorb water. In chamber "C" (Fig. 16.9d), an external heat source boils the water back off the lithium bromide. The concentrated lithium bromide is returned to chamber "B," while the water vapor is condensed back into water in chamber "D." The last step is to return the liquid water back to chamber "A" so the cycle can continue.

Because the absorption refrigeration-machine cycle is inherently inefficient, the cycle is economical only when an inexpensive source of heat is available. Although the compressive cycle is more efficient, it requires a source of mechanical power, which is most often supplied by an electric motor running on expensive electricity.

In the compressive cycle, the power is required to drive refrigerant pumps that are either of the reciprocating or centrifugal types. The reciprocating type of compressor is most appropriate for small-to-medium-size

buildings while the centrifugal compressor is best for medium-to-large-size buildings. When any of these refrigeration machines are used to chill water, they are known as **chillers.**

Sometimes **evaporative coolers** are included in a discussion of refrigeration machines. Although evaporative coolers often replace air conditioners in dry climates, most types do not remove total heat from a building. Instead, they convert sensible heat into latent heat, which in dry climates creates thermal comfort very economically. Because of their mechanical simplicity, they were discussed with other passive cooling systems in Chapter 10 (see Fig. 10.11a).

16.10 HEAT PUMPS

Every compressive refrigeration machine pumps heat from the evaporator coil to the condenser coil. Fig. 16.10a illustrates a simple through-the-wall or window air-conditioner unit that is essentially a refrigeration machine. One fan cools indoor air by blowing it across the cold evaporator coil, while another fan heats outdoor air by blowing it across the condenser coil.

What would happen if the air-conditioning unit were turned around so that the evaporator coil were outdoors and condenser coil were indoors? The outdoor air would then be cooled, and the indoor air would be heated—just what is needed in the winter.

Instead of turning the whole unit around, it is much easier to reverse just the flow of refrigerant. That also makes it unnecessary to go outside in the winter to reach the controls. A refrigeration machine in which the flow of refrigerant can be reversed is called a **heat pump.** The term is unfortunate because every refrigeration machine pumps heat, even if it is only in one direction. Heat pumps use reversing valves to change the direction of refrigerant flow (Fig. 16.10b).

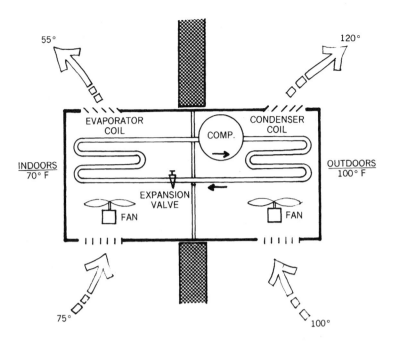

Figure 16.10a A simple through-the-wall air-conditioner unit essentially consists of a compressive refrigeration machine.

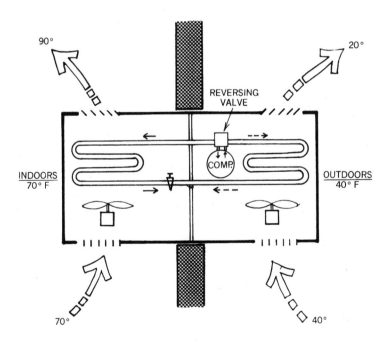

Figure 16.10b In a heat pump, the reversing valve allows the refrigerant to flow in either direction. In the winter condition shown, the outdoor coil becomes the evaporator and the indoor coil, the condenser.

Heat pumps are air conditioners that can switch to heat in the winter. Since, however, they extract heat from outdoor air, their efficiency drops as the outdoor air gets colder. Thus, heat pumps are most appropriate in those climates where summer cooling is required and where the winters are not too cold. Most of the southern half of the United States fits into this category. When heat pumps are coupled with the ground instead of outdoor air, they are called **geo-exchange heat pumps.**

16.11 GEO-EXCHANGE

According to the Environmental Protection Agency (EPA), **geo-exchange** heat pumps are the most energy-efficient, environmentally clean, and cost-effective, active, space conditioning systems available (EPA, 1993). Geo-exchange heat pumps are also known as **geothermal** or **ground-coupled** heat pumps. "Ground-coupled" is a good descriptive term but "geothermal" is not because it is already in use to describe high-temperature heat obtained from deep within the Earth (see the end of Section 2.17). "Geo-exchange" is an excellent term because in the summer, heat pumps move heat from indoors to the ground, which acts as a heat sink, and in the winter, the heat is moved back from the ground and pumped indoors again—a seasonal exchange with the Earth (*geo*).

Water is used to transfer the heat from the heat pump to and from the upper layer of the earth (less than 100 feet). Four different methods are available, and the best choice depends on the local conditions. If a pond is available, it can be the most convenient heat source/heat sink (Fig.16.11a). Where ground water is plentiful and well drilling is easy, an open-loop system can be used; here, the water is pumped out of the ground at one well and returned at another well (Fig. 16.11b). A closed vertical

Pond/Lake

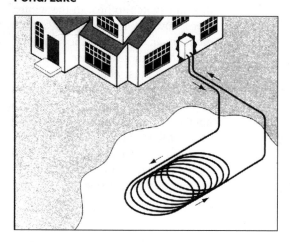

Figure 16.11a A geo-exchange, heat-pump system can use a pond as the heat source/ heat sink. (From U.S. Dept. of Energy, Office of Geothermal Technologies.)

Open-Loop Systems

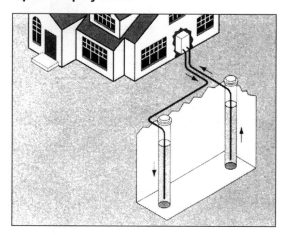

Figure 16.11b A geo-exchange, heat-pump system can use ground water as the heat source/ heat sink. The water should be returned to the ground via a second well. (From U.S. Dept. of Energy, Office of Geothermal Technologies.)

Vertical

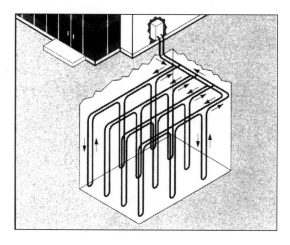

Figure 16.11c In most cases, vertical loops are the best option for geo-exchange heat pumps. (From U.S. Dept. of Energy, Office of Geothermal Technologies.)

Horizontal

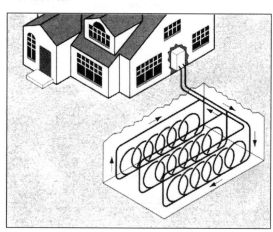

Figure 16.11d When soil conditions make vertical loops impractical, horizontal loops are acceptable. Trenches should be as deep as possible to obtain the best soil temperatures. (From U.S. Dept. of Energy, Office of Geothermal Technologies.)

loop is usually preferred because much less pumping energy is required (Fig. 16.11c).

Vertical closed loops are preferred over horizontal closed loops because the deep ground is warmer in the winter and cooler in the summer than the shallow ground (see Section 10.12). Drilling deep holes can be expensive, however, especially if the ground is rocky. In such a case, if the site is large enough, horizontal loops in fairly deep trenches might be the most cost- effective alternative (Fig. 16.11d).

Since ground-coupled heat pumps are indoors, protected from the elements, they last much longer and require much less maintenance than conventional air conditioners. These pumps also have the aesthetic advantage of being hidden indoors and making almost no noise. During the summer, the heat removed from indoors is first used to heat hot water, in effect, yielding free domestic hot water. A device called a **desuperheater** is added to a standard heat pump to heat the hot water. The rest

of the year, the domestic hot water is heated efficiently from the ground like the rest of the building. The greatest benefit of geo-exchange heat pumps is their high efficiency. They use 44 percent less energy than air-to-air heat pumps and 72 percent less energy than standard air conditioning with electric resistance heating (EPA, 1993). Because ground temperatures are less extreme than air temperatures, geo-exchange heat pumps are appropriate nationwide.

16.12 COOLING SYSTEMS

To cool a building, a refrigeration machine must pump heat from the various rooms into a heat sink. The heat sink is usually the outdoor air but can also be a body of water or the ground (Fig. 16.12a). Cooling systems vary mostly by the way heat is transferred from the rooms to the refrigeration machine and from there to the heat sink (Fig. 16.12b). The choice of the heat-transfer methods depends on building type and size. Cooling systems are often classified by the fluids that are used to transfer the heat from the habitable spaces to the refrigeration machine. The four major categories are direct refrigerant, all-air, all-water, and combination air–water.

Direct Refrigerant Systems

The **direct refrigerant** or DX (direct expansion) **system** is the simplest because it consists of little more than the basic refrigeration machine plus two fans. The indoor air is blown directly over the evaporator coil, and the outdoor air passes directly over the condenser coil (Fig. 16.10a). Direct refrigerant units are appropriate for cooling small-to-medium-size spaces that require their own separate mechanical units.

All-Air Systems

In an all-air system, air is blown across the cold evaporator coil and then delivered by ducts to the rooms that require cooling (Fig. 16.12c). Air

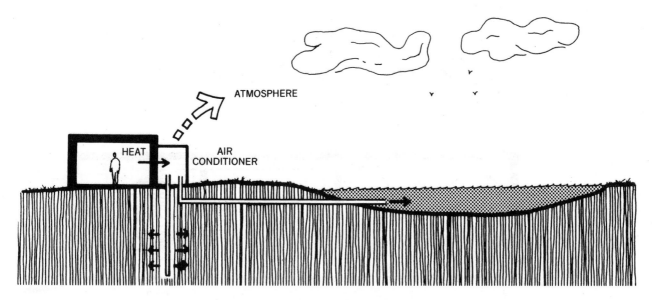

Figure 16.12a Air, water, or the ground can act as the heat sink for a building's cooling system.

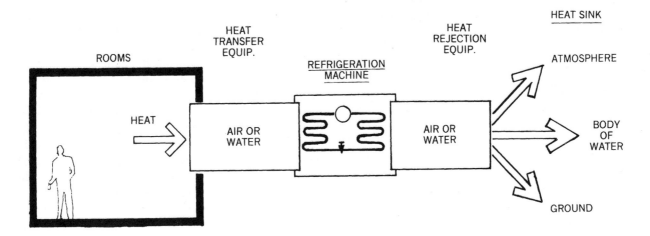

Figure 16.12b Cooling systems vary mainly in how heat is transferred to and from the refrigeration machine.

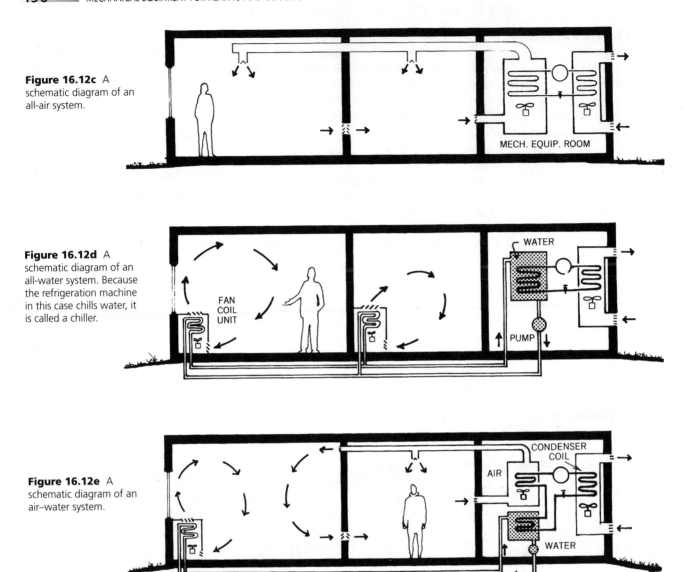

Figure 16.12c A schematic diagram of an all-air system.

Figure 16.12d A schematic diagram of an all-water system. Because the refrigeration machine in this case chills water, it is called a chiller.

Figure 16.12e A schematic diagram of an air–water system.

systems can effectively ventilate, filter, and dehumidify air. The main disadvantage lies in the bulky ductwork that is required.

All-Water Systems

In an all-water system, the water is chilled by the evaporator coil and then delivered to fan-coil units in each space (Fig. 16.12d). Although the piping in the building takes up very little space, the fan-coil units in each room do require some space. Ventilation, dehumidification, and filtering of air are possible but not as effectively as with an air system.

Combination Air–Water Systems

An air–water system is a combination of the above mentioned air and water systems (Fig. 16.12e). The bulk of the cooling is handled by the water and fan-coil units, while a small air system completes the cooling and also ventilates, dehumidifies, and filters the air. Since most of the cooling is accomplished by the water system, the air ducts can be quite small.

The above systems describe how heat is transferred from building spaces to the refrigeration machine.

The following discussion describes how the heat from the refrigeration machine is dumped into the atmospheric heat sink.

In small buildings, the heat given off by a refrigeration machine is usually dumped into the atmosphere by blowing outdoor air over the condenser coil (Fig. 16.12e). To make this heat transfer more efficient, water can be sprayed over the condenser coil. Medium-sized buildings often use a specialized piece of equipment called an **evaporative condenser** (Fig. 16.12f) to dump heat into the atmosphere by evaporating water. A small amount of

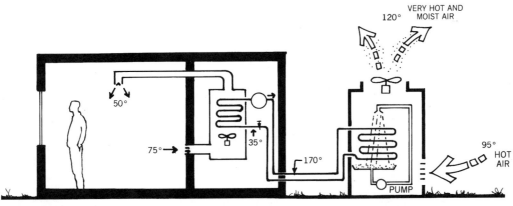

Figure 16.12f In an evaporative condenser, water is sprayed over the hot condenser coils.

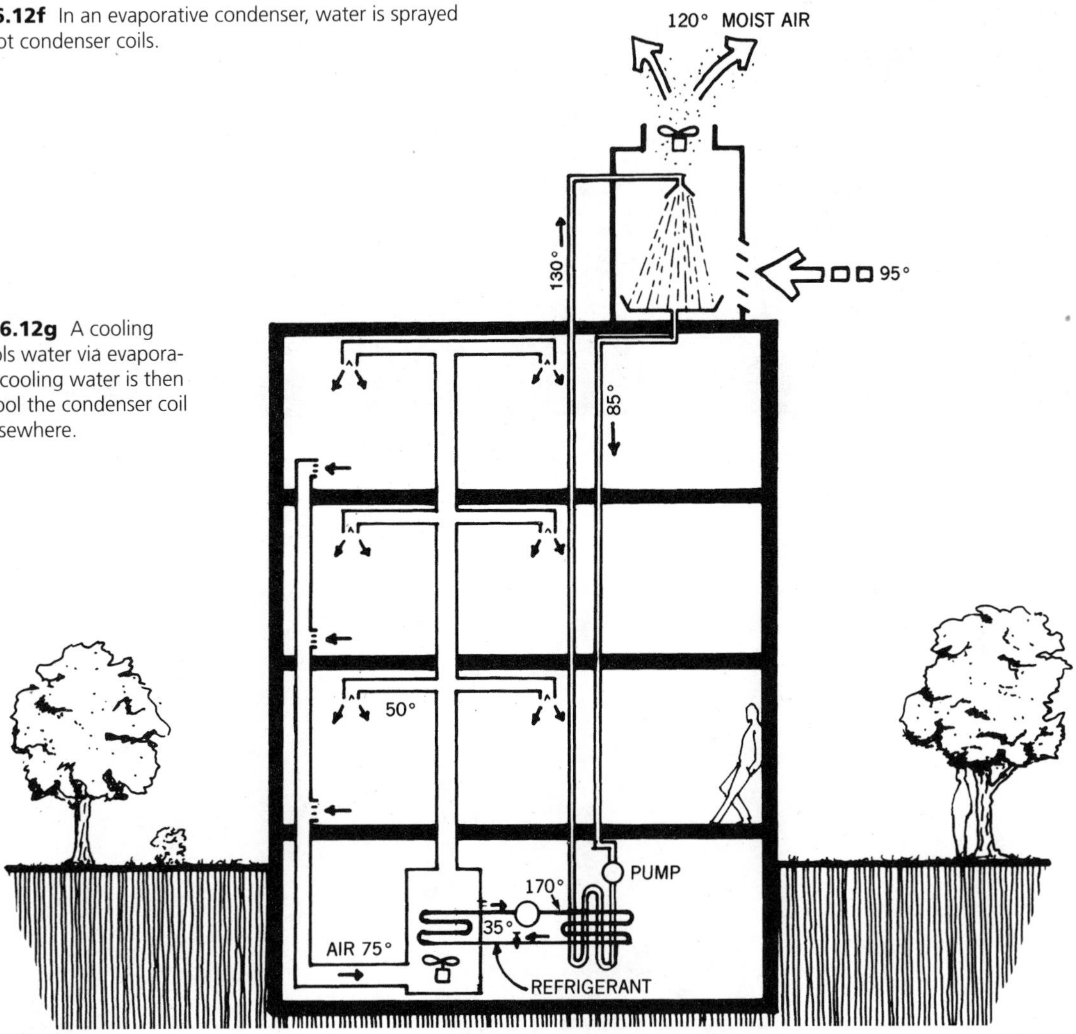

Figure 16.12g A cooling tower cools water via evaporation. This cooling water is then used to cool the condenser coil located elsewhere.

water must be continuously supplied to replace the water lost by evaporation. Since refrigerant lines are limited in length, an evaporative condenser cannot be more than about 60 feet from the compressor and evaporator coil. Thus, for large buildings, cooling towers are frequently a better choice.

A **cooling tower** also dumps heat into the atmosphere via the evaporation of water. However, the evaporating water is used to cool more water rather than refrigerant as in the evaporative condenser. This cooling water is then pumped to the refrigeration machine where it cools the condenser coil (Fig. 16.12g). Although most of the water is recirculated, a small amount of makeup water is again required to replace the water lost by

Figure 16.12h In urban areas, cooling towers are typically located on rooftops.

Figure 16.12i This conical cooling tower is on the roof of one of the World Financial Center buildings in lower Manhattan. The twin towers of the World Trade Center, seen in the background, use the water of the New York City harbor as a heat sink. Large, underground pipes supply and return the warmed, salty water.

Figure 16.12j The small cube at left is a cooling tower for this office building. Blue Cross and Blue Shield Building, Towson, MD.

evaporation. Most cooling towers are placed on roofs (Figs. 16.12h and 16.12i), but when land is available they can be equally well placed at grade (Fig. 16.12j). As a matter of fact, a cooling tower does not have to be a structure at all. A decorative fountain can be a very effective substitute for a cooling tower (Fig. 16.12k).

16.13 AIR CONDITIONING FOR SMALL BUILDINGS

Air conditioning is the year-round process that heats, cools, cleans, and circulates air. It also ventilates and controls the moisture content of the air. The various components of air-conditioning systems have been described above. Some of the most common air-conditioning systems will now be described first for small or one-story buildings and then for large, multistory buildings.

Through-the-Wall Unit

For air conditioning single spaces like motel rooms, a through-the-wall unit is often used. Each of these units essentially consists of a compressive refrigeration machine (Fig. 16.13a). The condenser coil, compressor, and one fan are on the exterior side of an internal partition. The compressor is on the outside because it is the noisiest part of the equipment. On the interior side of the partition, there is the evaporator coil and a fan to blow air over it. As indoor air passes over the evaporator coil, its temperature is often lowered below its dew-point temperature. Consequently, condensation occurs, which must be collected and disposed of. Often, the condensation is used to help cool the condenser coil. An adjustable opening in the interior partition allows a controlled amount of fresh air to enter for ventilation purposes. Return air from the room first passes oven a filter. An electric-strip heater is sometimes supplied

Figure 16.12k These decorative fountains are used in place of a cooling tower. West Point Pepperell factory, Lanett, AL.

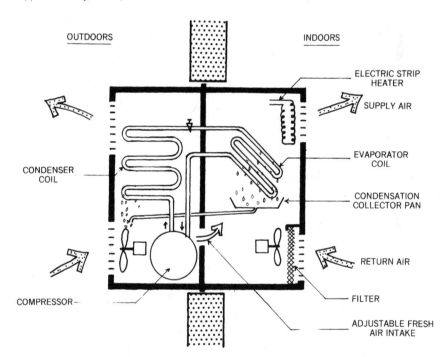

Figure 16.13a A schematic diagram for a through-the-wall air-conditioning unit that can heat as well as cool.

for cold weather. Often, however, a heat pump is used instead of just a refrigeration machine for more efficient winter heating. The electric-strip heaters are then still included but are used for backup heating only when the outdoor temperatures are too low for the heat pump.

Packaged Systems

Like through-the-wall units, **packaged systems** are preengineered, self-contained units where most of the mechanical equipment is assembled at the factory. Consequently, they offer low installation, operating, and

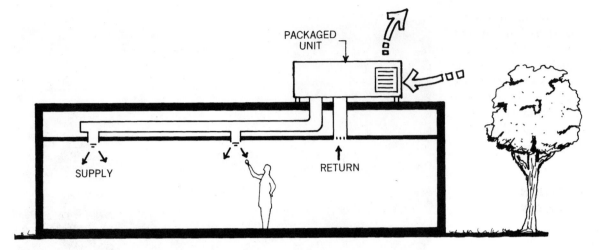

Figure 16.13b A packaged unit can contain both heating and cooling equipment.

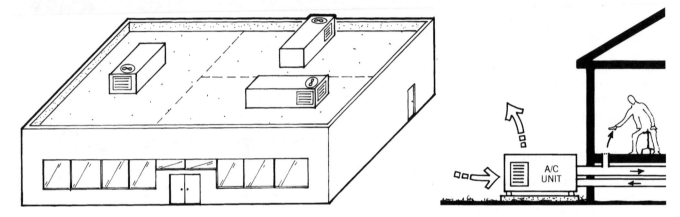

Figure 16.13c Rooftop packaged units are placed over the separate zones that they serve.

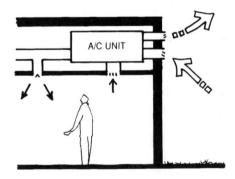

Figure 16.13d A packaged unit designed for crawl-space construction.

maintenance costs. Usually, small buildings are served by one package, while larger buildings get several. Rooftop versions are the most common, with each unit serving a separate zone (Figs. 16.13b and 16.13c). Packaged units are sometimes also used on the ground for buildings with crawl spaces (Fig. 16.13d) or above a suspended ceiling when there is enough space below the roof (Fig. 16.13e).

Packaged units can heat a building via electric strips, a heat pump, or a gas furnace. Electric-strip heaters are appropriate only in very mild climates. As mentioned before, heat pumps are an economical way to heat in the southern half of the United States. In cold climates, gas is the logical source of heat for the packaged units.

Split Systems

Most homes and many other small-to-medium-sized buildings find the **split system** to be most appropriate. In this system, the compressor and condenser coils are outdoors, while the air-handling unit with the evaporator coil is indoors (Fig. 16.13f). As in all cooling systems, condensation from the evaporator coil must be drained away. The air-handling unit also contains the central heating system. As with packaged units, the heating is usually by electric strips, a heat pump, a gas furnace, or an oil furnace.

Figure 16.13e A packaged unit designed for placement above suspended ceiling.

Figure 16.13f A schematic diagram of a split system.

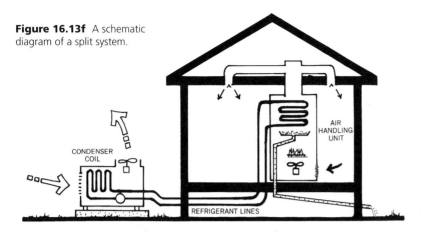

Figures 16.13g and 16.13h illustrate the use of split systems for a small office building. The compressor/condenser units are shown on grade although they could equally well be on the roof if free land were not available. The air-handling units (AHU) with their evaporator coils and heating systems are in a mechanical-equipment room (MER). The supply ducts are above a suspended ceiling but on the indoor side of the roof insulation. Thus, any heat loss from

Figure 16.13g Two split systems for a small office building. TR (top register), CD (ceiling diffuser), M.E.R. (mechanical equipment room), A. H. U. (air handling unit). Double diagonal lines define supply ducts in section, while a single diagonal across a rectangle defines a return duct.

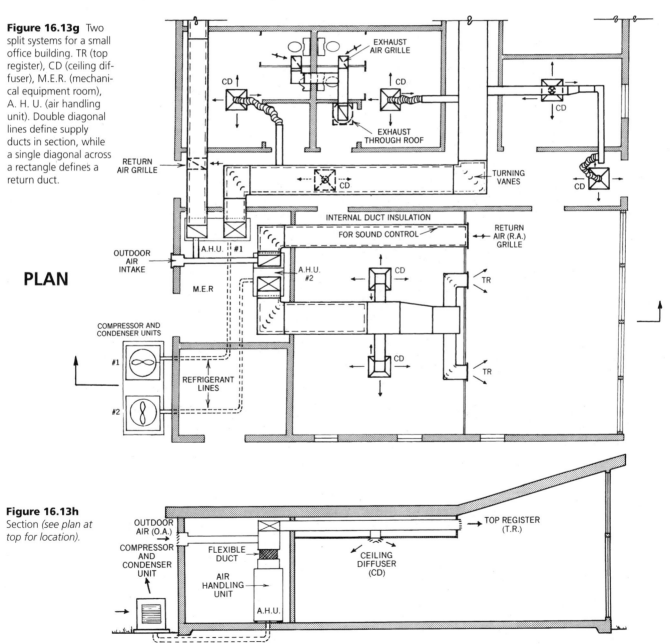

PLAN

Figure 16.13h Section (see plan at top for location).

the ducts is into the air-conditioned space. The air is supplied to each room through a top register (high on wall) or a ceiling diffuser. Return-air grilles and ducts bring the air back to the air-handling units. Instead of grilles, smaller rooms often have undercut doors to allow air to enter the corridor which can act as a return-air plenum (duct). Sections of both supply and return ducts are lined with sound-absorbing insulation to trap noise emitted by the air handling units. A short piece of flexible duct (Fig. 16.13h) prevents vibrations from the air-handling unit from being transmitted throughout the building by the duct system. Ventilation is maintained by means of exhaust fans in the toilets and an outdoor-air intake into the MER.

The split system is very flexible because only two small copper tubes carrying refrigerant must connect the outdoor condenser/compressor with the indoor air-handling unit. However, these units cannot be much more than 60 feet apart. Thus, the split system is appropriate for small-to-medium-sized buildings. Usually separate zones are served by having their own split units (Fig. 16.13i).

Ductless Split Systems

It is often difficult if not impossible to hide ducts when adding air conditioning to an existing building that has either inadequate or no air conditioning. Ducts are especially difficult to conceal in historic-preservation work. Ductless split systems require only two small, copper refrigerant lines between the outdoor compressor/condenser unit and the mini-indoor air-handling units (Fig. 16.13j). The indoor units are compact, unobtrusive, and very quiet. They look and install much like a fluorescent lighting fixture. One condenser/compressor unit can serve up to three mini-air-handling units as much as 160 feet away.

16.14 AIR CONDITIONING FOR LARGE, MULTISTORY BUILDINGS

Most large, multistory buildings use highly centralized air-conditioning equipment. The roof and basement are the usual choice for these **central station systems** (Fig. 16.14a). The basement has the advantage of easy utility connections, noise isolation, not being valuable rental area, and the fact that structural loads are not a problem. The roof, on the other hand, is the ideal location for fresh-air intakes and heat rejection to the atmosphere. Since cooling towers are noisy, produce very hot and humid exhaust air, and can produce fog in cold weather, they are usually placed on the roof. In many buildings, the equipment is split, with some in the basement and some on the roof. To minimize the space lost to vertical air ducts, intermediate mechanical floors are used in very high buildings (Fig. 16.14a). If there is sufficient open land, the cooling tower and some of the other equipment might be placed on grade adjacent to the building (Fig. 16.12j).

In plan, most large, multistory buildings have windows on all four sides and a core area for building services (Fig. 16.14b). Floor-to-floor heights of 13 or 14 feet are often necessary to accommodate the horizontal ducts bringing conditioned air from the core to the perimeter. Lower floor-to-floor heights are possible with special compact mechanical systems or if the perimeter area is supplied with riser ducts at the perimeter (Fig. 16.14c).

When choosing a mechanical system for a building, one must consider the following questions. How much does the equipment cost? How much space does it require? How much energy does it use? How much thermal comfort does it provide? The ther-

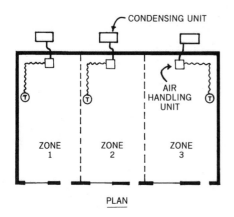

Figure 16.13i Each zone has its own split system controlled by a separate thermostat.

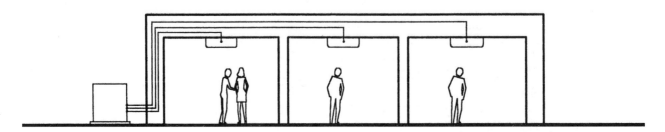

Figure 16.13j Ductless split systems are most advantageous in renovation work where ducts are hard to install or hide. One compressor/condenser unit can feed up to three mini-air handling units.

Figure 16.14a Common locations for centralized mechanical-equipment spaces in large, multistory buildings.

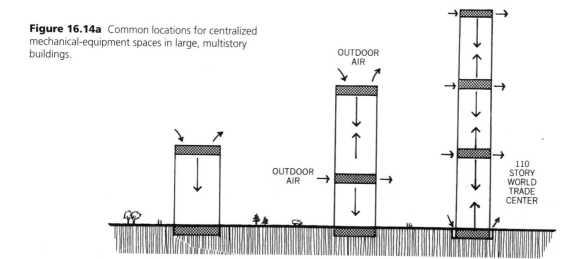

OUTDOOR AIR

OUTDOOR AIR

110 STORY WORLD TRADE CENTER

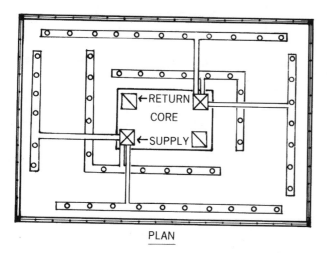

RETURN

CORE

SUPPLY

PLAN

Figure 16.14b A typical air-distribution plan with all risers in the core results in duct crossings and consequently a greater floor-to-floor height.

RETURN RISER

CORE

SUPPLY RISER

PLAN

Figure 16.14c One way to minimize floor-to-floor height is to supply the perimeter zones with separate perimeter risers. These ducts can be expressed on the facade.

mal comfort is mostly a function of the number of zones that can be provided. The more zones there are, the more people will be comfortable. Since zones are expensive, however, this number is usually kept to a minimum. As was explained earlier in this chapter, the average large building has at least five zones because of orientation.

For the purpose of clarity, an "example" building with only three zones will be used to illustrate the major mechanical systems for large, multistory buildings (Fig. 16.14d). A section of this "example" building is shown in Fig. 16.14e. As in most

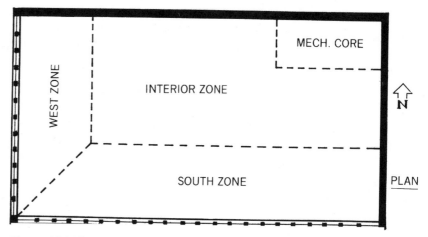

MECH. CORE

WEST ZONE

INTERIOR ZONE

N

SOUTH ZONE

PLAN

Figure 16.14d A special three-zone floor plan is used to illustrate the major mechanical systems available for large buildings. This plan is actually the southwest quadrant of a typical large office building.

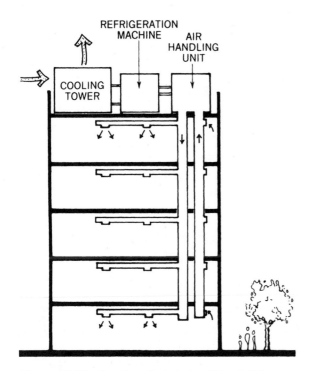

Figure 16.14e A section of a typical multistory building with a rooftop central-station mechanical system. The air-handling unit on the roof serves all floors.

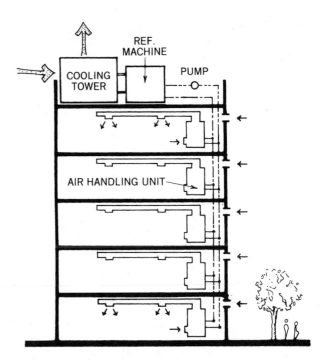

Figure 16.14f A section showing an alternate approach. Although much of the mechanical equipment is still on the roof, each floor has a separate air handling unit. Because the refrigeration machine is now producing chilled water, it is called a chiller.

buildings, the mechanical equipment is shown to be on the roof. This section shows an all-air system served by a single central air-handling unit on the roof. To avoid the large vertical ducts, separate air-handling units can be placed on each floor with only water circulating vertically (Fig. 16.14f). This alternative saves a great deal of energy because moving air great distances requires much more power than moving water.

The major mechanical systems available for large buildings are illustrated by showing a typical floor plan and section of each. The equipment shown in each floor plan is above the ceiling unless otherwise noted. These systems are grouped by the heat-transfer medium used: all-air, air–water, and all-water.

Air systems are of two types: constant-air-volume (CAV) and variable-air-volume (VAV). In CAV systems, the temperature control of a space is achieved by changing the temperature of a constant supply of air. In VAV systems, the temperature control of a space is achieved by varying the amount of supply air delivered to each space at a constant temperature. VAV systems are the more widely used because they are more versatile and efficient than the CAV systems.

All-Air Systems

The great advantage of all-air systems is that complete control over air quality is possible. The main disadvantages are that all-air systems are very bulky and a significant part of the building volume must be devoted to them. They are also less efficient because moving large quantities of air requires a great deal of power. It must be noted that for clarity only the supply ducts are shown on each plan in the following examples. Usually there is also a sizable return-duct system on each floor.

I. *Single-Duct System with Constant Air Volume (CAV):* The single-duct system is basically a one-zone system.

Since a separate supply duct and air-handling unit is required for each zone, this system is most appropriate for small buildings or medium-size buildings with few zones (Fig. 16.14g).

II. *Variable-Air-Volume System (VAV):* This is a single-duct system that can easily have many zones. A variable-volume control box is located wherever a duct enters a separate zone (Fig. 16.14h). A thermostat in each zone controls the air flow by operating a damper in the VAV control box. Thus, if more cooling is required, more cool air is allowed to enter the zone. Since VAV systems cannot heat one zone while cooling another, they are basically cooling-only systems. Because heating is usually required only on the perimeter, a separate heating system can be supplied in conjunction with the VAV system. The low first cost and low energy usage make the VAV system very popular, and it is applicable to almost every building type.

III. *Terminal Reheat System (CAV):* At first, the terminal reheat system looks just like the VAV system previously described, but, in fact, it is very dif-

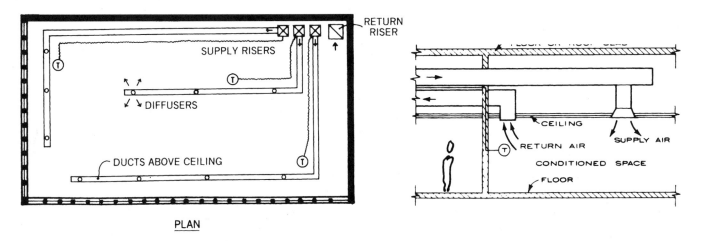

Figure 16.14g A single-duct system. (Section from *Architectural Graphic Standards,* Ramsey/Sleeper, 8th ed., John R. Hoke, editor, © John Wiley & Sons, Inc., 1988.)

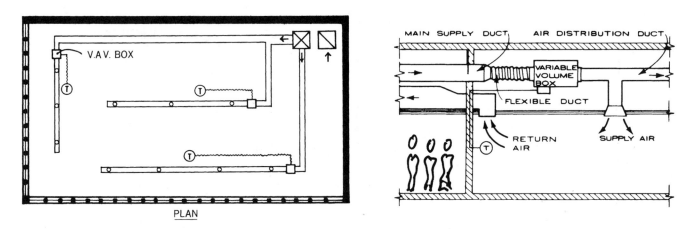

Figure 16.14h A variable-air volume (VAV) system. (Section from *Architectural Graphic Standards,* Ramsey/Sleeper, 8th ed., John R. Hoke, editor, © John Wiley & Sons, Inc., 1988.)

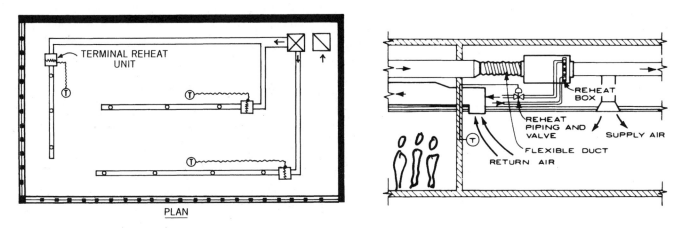

Figure 16.14i A terminal reheat system. (Section from *Architectural Graphic Standards,* Ramsey/Sleeper, 8th ed., John R. Hoke, editor, © John Wiley & Sons, Inc., 1988.)

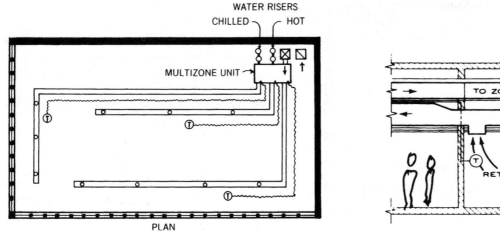
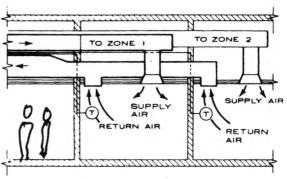

Figure 16.14j A multizone system. (Section from *Architectural Graphic Standards,* Ramsey/Sleeper, 8th ed., John R. Hoke, editor, © John Wiley & Sons, Inc., 1988.)

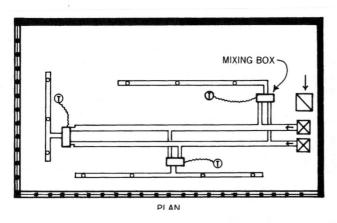
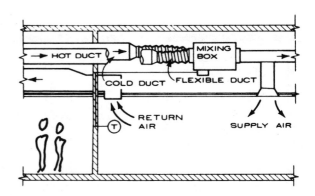

Figure 16.14k A double-duct system. (Section from *Architectural Graphic Standards,* Ramsey/Sleeper, 8th ed., John R. Hoke, editor, © John Wiley & Sons, Inc., 1988.)

ferent. Instead of VAV boxes, this system has terminal reheat boxes in which electric-strip heaters or hot-water coils reheat air previously cooled (Fig. 16.14i). For example, on a spring day the zone with the greatest cooling load will determine how much the air for the whole building is cooled. All other zones will then reheat the cold air to the desired temperature. Thus, most of the building is being heated and cooled simultaneously — a waste of energy. This system was popular in the past because it gave excellent control. It should not be used now except in special cases because of its high energy consumption. Terminal reheat systems can also be of the VAV type. In this case, the terminal reheat box also controls the volume of air. Thus, the control of space temperature is handled efficiently by varying the volume of the supply air, and the reheat function is called for only when some space needs heating while the others all need cooling.

IV. *Multizone System (CAV):* In this mechanical system, every zone receives air at the required temperature through a separate duct (Fig. 16.14j). These ducts are supplied by a special multizone air-handling unit that custom-mixes hot and cold air for each zone. This is accomplished by means of motorized dampers located in the air-handling unit but controlled by thermostats in each zone. Depending on the temperature, the ratio of hot and cold air varies but the total amount of air is constant. The multizone unit is supplied with hot water, chilled water, and a small amount of fresh air.

Each multizone unit can handle up to about eight zones or about 30,000 square feet. Because moderate air temperatures are created by mixing hot and cold air, this system is also somewhat wasteful of energy. First costs are relatively high while the thermal control is relatively poor.

V. *Double-Duct System (CAV):* Like the multizone system, the double-duct system mixes hot and cold air to achieve the required air temperature. Instead of mixing the air at a central air-handling unit, mixing boxes are dispersed throughout the building (Fig. 16.14k). Thus, there is no limit to the number of zones possible.

However, two sets of large supply ducts are necessary.

Although the double-duct system creates a high level of thermal comfort and allows for great zoning flexibility, it is very expensive, requires much building space, and is wasteful of energy.

Air–Water System

The following systems supply both air and water to each zone of a building. Although this increases the complexity of the mechanical systems, it greatly decreases the size of the equipment because of the immense heat-carrying capacity of water as opposed to air. Air is supplied mainly because of the need for ventilation.

VII. *Induction System:* In an induction system, a small quantity of high-velocity air is supplied to each zone to provide the required fresh air and to induce room air to circulate (Fig. 16.14m). Most induction-terminal units are found under windows where they can effectively neutralize the heat gain or loss through the envelope (Fig. 16.14n). As the high-velocity air shoots into the room, it induces a large amount of room air to circulate. This combination of room air (90 percent) and fresh air (10 percent) then passes over heating or cooling coils. Thus, most heating or cooling is accomplished with water while ventilation and air motion are accomplished with a small amount of high-velocity air. Local thermostats regulate the temperature by controlling the

flow of either hot or cold water through the coils.

Unfortunately, high-velocity (up to 6,000 feet per minute) and, therefore, high-pressure ducts are much more expensive than regular-velocity (up to 2,000 feet per minute) ducts. High-velocity air systems also consume much more fan power than normal-velocity systems. Because of these problems, the use of induction systems is limited.

VIII. *Fan-Coil With Supplementary Air:* This system also consists of two separate parts. For ventilation and cooling of the interior areas, there is an all-air system, and for neutralizing the heat gain or loss through the envelope, there are fan-coil units around the perimeter (Fig. 16.14o). The fan-

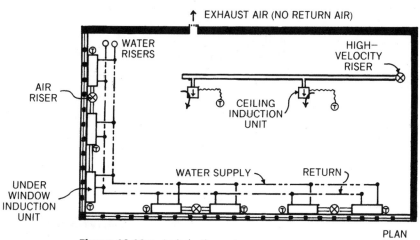

Figure 16.14m An induction system.

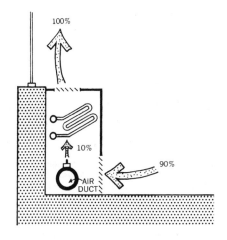

Figure 16.14n An induction unit with cooling or heating coils.

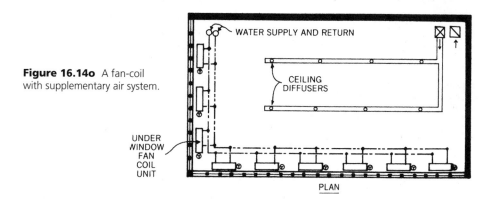

Figure 16.14o A fan-coil with supplementary air system.

coil units are described in more detail below under all-water systems.

VIII. *Radiant Panels with Supplementary Air:* Similar to radiant heating but in reverse, a cool surface can achieve thermal comfort by lowering the mean radiant temperature (MRT). Moderately cool panels will suffice if their area is large. The ceiling is the best surface for radiant cooling because it is large, it is unobstructed by furniture, etc., and it can also cool by convection. Fig. 16.14p shows three radiant cooling systems. Indoor rain from condensation on the ceiling can be prevented by controlling the humidity and panel temperature. The supplementary air is dehumidified in order to control indoor humidity, and the large panels are maintained above the dew-point temperature of the air.

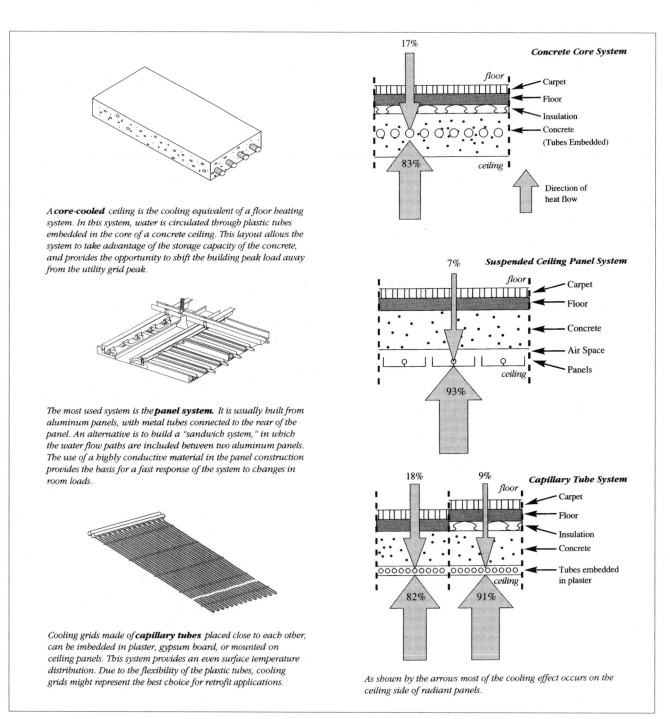

A **core-cooled** ceiling is the cooling equivalent of a floor heating system. In this system, water is circulated through plastic tubes embedded in the core of a concrete ceiling. This layout allows the system to take advantage of the storage capacity of the concrete, and provides the opportunity to shift the building peak load away from the utility grid peak.

The most used system is the **panel system.** It is usually built from aluminum panels, with metal tubes connected to the rear of the panel. An alternative is to build a "sandwich system," in which the water flow paths are included between two aluminum panels. The use of a highly conductive material in the panel construction provides the basis for a fast response of the system to changes in room loads.

Cooling grids made of **capillary tubes** placed close to each other, can be imbedded in plaster, gypsum board, or mounted on ceiling panels. This system provides an even surface temperature distribution. Due to the flexibility of the plastic tubes, cooling grids might represent the best choice for retrofit applications.

As shown by the arrows most of the cooling effect occurs on the ceiling side of radiant panels.

Figure 16.14p   Radiant cooling systems. (From *Center for Building Science News,* Lawrence Berkeley Laboratory, Vol.1, no. 4, Fall, 1994.)

All-Water Systems

Since these systems supply no air, they are appropriate when a large amount of ventilation either is not necessary or can be achieved locally by such means as opening windows.

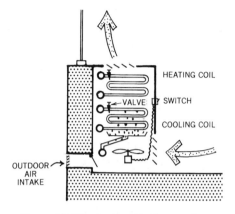

Figure 16.14q A schematic diagram of an under-window fan-coil unit (four-pipe system) with outdoor air intake and condensation drain.

IX. *Fan-Coil System:* The fan-coil unit, as the name implies, basically consists of a fan and a coil within which water circulates. The units are often in the form of cabinets for placement under windows (Fig. 16.14q). The fan blows room air across coils containing either hot or cold water. Thermostatically controlled valves regulate the flow of water through the coils. A four-pipe system, which is shown, has two pipes for hot-water supply and return and another two pipes for cold-water supply and return (Fig. 16.14q). Thus, either heating or cooling is possible at any time of year. In the less expensive but also less comfortable two-pipe system, hot water circulates during the winter and cold water in the summer. In such systems, it is not possible to heat some zones while cooling others. A three-pipe (hot, cold, and a common return) system also exists but wastes energy because hot and cold water return through the same pipe.

Condensation on the cooling coils must be collected in a pan and drained away. When the fan-coil unit is on an outside wall, it is possible to have an outdoor air intake connected to the unit. A three-speed fan switch enables occupants of the zone to have some control over the temperature.

Fan-coil units are most appropriate for air conditioning buildings with small zones (e.g., apartments, condominiums, motels, hotels, hospitals, and schools). Besides the under-window location, fan-coil units are sometimes also located above windows (valance units), in small closets, or in the dropped ceiling above a bathroom or hallway (Figs. 16.14r and 16.14s).

X. *Water-Loop Heat-Pump System:* With this system, each zone is heated or cooled by a separate water-to-air heat pump. A thermostat in each zone determines whether the local heat pump extracts heat from a water loop (heating mode) or injects heat into the water loop (cooling mode) (Fig. 16.14t). The water in the loop is circulated at between 60°F and 90°F. In the summer, when most heat pumps are injecting heat into the loop, the excess heat is disposed of by a cooling tower. In winter, when most heat pumps are extracting heat, a central boiler keeps the water in the loop from dropping below 60°F.

The water-loop heat-pump system really shines in spring and fall or whenever about half the heat pumps are in the cooling mode and the other half are in the heating mode. Here, the heat extracted from the water loop will roughly equal the heat injected, and neither cooling tower nor boiler needs to operate. This system is most appropriate in those buildings and climates where the simultaneous heating of some zones and cooling of others is common. The numerous heat pumps are a major maintenance problem.

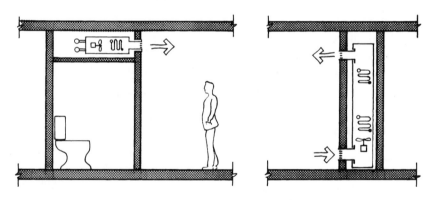

Figure 16.14r Fan-coil units can also be placed above a dropped ceiling or in a small closet.

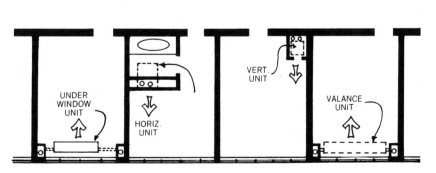

Figure 16.14s Alternate fan-coil systems (plan view). Four different types are shown: under-window unit, above-bathroom ceiling unit, closet unit, and valance (above-window) unit.

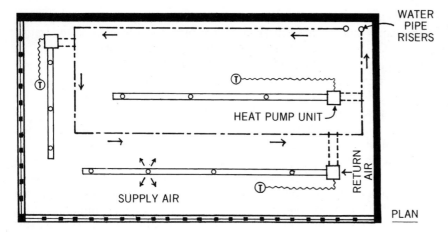

WATER PIPE RISERS

HEAT PUMP UNIT

RETURN AIR

SUPPLY AIR

PLAN

Figure 16.14t Water-loop heat-pump system (all equipment shown in plan is above the ceiling)

16.15 DESIGN GUIDELINES FOR MECHANICAL SYSTEMS

Because the mechanical and electrical equipment requires 6 to 9 percent of the total floor area of most buildings, sufficient and properly located spaces should be allocated for it. By incorporating the following rules and design guidelines at the schematic design stage, one can prevent many serious design problems .

Sizing Guidelines

For the floor-area and ceiling-height requirements of the various parts of a mechanical system see Table 16.15A. Note that the spatial requirements for the air-handling units depend mostly on whether an all-air, air–water, or all-water system is used. Ducts for horizontal air distribution are usually above the ceiling and, therefore, do not use up any of the floor area.

However, since floor-to-floor heights are very much dependent on the size of horizontal ducts, use Table 16.15B for a rough early estimate of duct sizes.

Location Guidelines

I. For small or one-story buildings:
    1. Place the equipment on roof; or
    2. Use a mechanical equipment room (MER) centrally located to minimize duct sizes, and placed along an outside wall for easy servicing (Fig. 16.15a).
II. For large, multistory buildings:
    1. Place the centralized mechanical equipment in basement, on roof, on on intermediate floors (see Fig. 16.14a) but
    2. Cooling tower should be placed on roof or on out-of-the-way adjacent land (Fig. 16.15b).
    3. Any additional MER on each floor should be centrally located to minimize duct sizes. If these MERs require large amounts of outdoor air, they should be located along an outside wall (Fig. 16.15a).

Noise Guidelines

1. Equipment placed outside can be a major source of noise.
2. Surround MER with massive material to stop sound transmission.
3. MERs and ducts should be lined with sound-absorbing insulation.
4. Do not locate an MER near quiet areas like libraries and conference rooms.

Residential Guidelines

1. Use one ton of cooling capacity for each 500 square feet of a standard house.
2. Use one ton of cooling capacity for each 1,000 square feet of a modern, well-designed and well-built house.

TABLE 16.15A SPATIAL REQUIREMENTS FOR MECHANICAL EQUIPMENT

Equipment Type	Floor Area Required[a] (Percentage)	Required Ceiling Height[b] (Feet)
Room for refrigeration machine, heating unit, and pumps	1.5–4	9–18
Room for air-handling units		
All-air	2 to 4	9–18
Air–water	0.5 to 1.5	9–18
All-water	0 to 1[d]	N/A
Cooling tower	0.25	7–16[c]
Packaged (rooftop)	0 to 1	5–10[c]
Split units	1 to 3	8–9

[a] The required floor area is a percentage of the gross building area served (parking is excluded). Use the upper end of the range for small buildings and large buildings with a great deal of mechanical equipment (e.g., laboratories and hospitals).

[b] Use the lower end of the range in ceiling height for smaller buildings.

[c] Since cooling towers and packaged units are usually not roofed over, the heights given are for the actual equipment, not ceiling height.

[d] The required area refers to fan-coil units.

TABLE 16.15B CROSS-SECTIONAL AREA OF DUCTS (HORIZONTAL OR VERTICAL)[a]

System	Cross-Sectional Duct Area per 1,000 Square Feet ofConditioned Space (Square Feet)
All-air	1–2
Air–water	0.3–0.8

[a]When used, return ducts are at least as large as these supply ducts. For high-velocity systems use about one-third these sizes. Use the large end of the range for spaces with large cooling loads or when duct work has many turns. For the vertical-shaft space, use about twice the area of the duct risers.

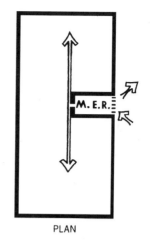

PLAN

Figure 16.15a Usually, mechanical-equipment rooms (MER) should be both centrally located and have access to the outdoors.

General Guidelines

1. The duct layout should be orderly and systematic, like the structural system.
2. Avoid crisscrossing air ducts.
3. Provide adequate access to large equipment that might have to be replaced.

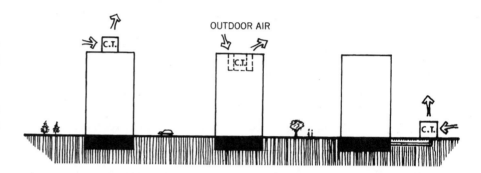

16.16 AIR SUPPLY (DUCTS AND DIFFUSERS)

Figure 16.15b When the mechanical equipment is mainly in the basement, the cooling tower should still be on the roof or on an out-of-the-way location with good outdoor air circulation.

Air is usually supplied to rooms by means of round, rectangular, or oval ducts. However, building elements, such as corridors and hollow beams can also be used. Although round ducts are preferable for a number of different reasons, they require clearances that are not always available. Consequently, rectangular ducts are very popular. However, the ratio of short to long sides, aspect ratio, should not exceed 1:5 because the resulting high airflow friction requires the ducts to have excessively large areas and perimeters (Fig. 16.16a).

At least another 2 inches must be added to the height and width when duct insulation is used. Insulation might be required for three different reasons: to reduce heat gain or loss, to prevent condensation, and to control noise. To prevent condensation, a vapor barrier must be on the outside of

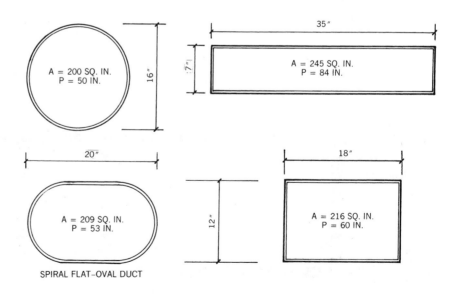

Figure 16.16a All these ducts have the same friction and, therefore, air-flow capacity. The circular duct requires the least volume and material but has the greatest depth. A, area; P, perimeter.

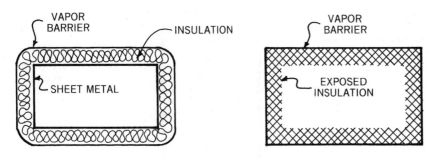

Figure 16.16b To prevent condensation on ducts, a vapor barrier must always be placed on the outside of the insulation. For noise control, a porous Fiberglas insulation must be exposed to the airstream.

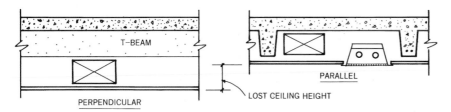

Figure 16.16c Coordinate ducts with beams and lighting fixtures to minimize the space required for the mechanical and electrical systems.

the insulation, while for noise control, the insulation must be exposed to the ainstream (Fig. 16.16b).

Although the size of ducts can be decreased by increasing the air velocity, there are two important reasons for using large ducts and the resultant low-velocity airflow. High-velocity airflow requires significantly more fan power to generate, and it is much noisier than low-velocity airflow.

A duct system is often described as a "tree system" in which the main trunk is the largest duct and the branches get progressively smaller

(see Fig. 16.14e). As much as possible, ducts should run parallel to deep structural elements and lighting fixtures to prevent the space wasted when various building elements cross each other (Fig. 16.16c).

Sometimes the return-air system has a duct "tree" as extensive as the supply-air system. Often, however, the corridor or plenum (space) above a hung ceiling is used instead of return air ducts. To maintain acoustical privacy, short return-air ducts lined with sound-absorbing insulation can connect rooms with the corridor.

Special return-air grilles with built-in sound traps are also available for doors or walls.

Supply ducts are also lined where necessary with sound-absorbing insulation to prevent the short-circuiting of sound from one room to the next. The insulation also reduces the noise transmitted through the air from the air-handling unit (see Fig. 16.13g). To prevent the duct work itself from transmitting noise or vibration, a flexible connection is used in which the main duct connects to the air-handling unit (see Fig. 16.13h).

The supply air enters a room through either a grille, register, or diffuser. A **supply grille** has adjustable vanes for controlling the direction of the air entering a room. A **register** is a grille that also has a damper behind it so that the amount as well as the direction of the air entering a room can be controlled.

In building types or climates where cooling predominates, the air should be supplied high in each room. Although it is convenient to run ducts through an unheated attic or even on the roof, it is not wise from an energy point of view even if the ducts are well insulated. Thus, a dropped ceiling in the corridor is appropriate not only for flat-roofed but also pitched-roof buildings (Fig. 16.16d).

When registers are mounted on upper walls, they are designed to throw the air about three-quarters the

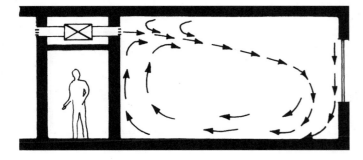

Figure 16.16d Air from an upper-wall register should be thrown about three-quarters the distance across a room before it drops to the level of people's heads.

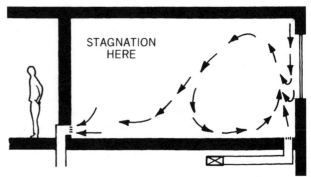

Figure 16.16e Floor registers cannot serve large spaces. They are good at countering the heating or cooling effect of windows.

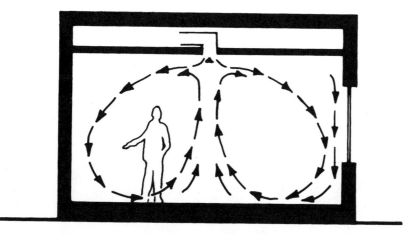

Figure 16.16f The proper air-flow pattern from a ceiling diffuser.

The location of supply-air outlets is very important for the comfort of the occupants. The goal is to gently circulate all of the air in a room so that there are neither stagnant non drafty areas. Make sure, too, that beams on other objects do not block the air supply from reaching all parts of a room (Fig. 16.16h). Where heating is the major problem, the outlets should be placed low.

Diffusers and registers are unnecessary when cloth ducts are used because they are fabricated with materials with different grades of

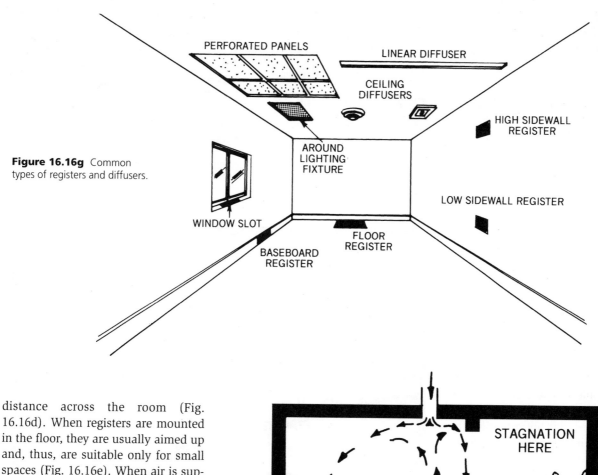

Figure 16.16g Common types of registers and diffusers.

distance across the room (Fig. 16.16d). When registers are mounted in the floor, they are usually aimed up and, thus, are suitable only for small spaces (Fig. 16.16e). When air is supplied from the ceiling, it has to be mixed rapidly with the room air to prevent discomfort for the occupants and, consequently, a **diffuser** is used (Fig. 16.16f). Diffusers can be round, rectangular, on linear. Supply air can also be diffused with large perforated ceiling panels (Fig. 16.16g).

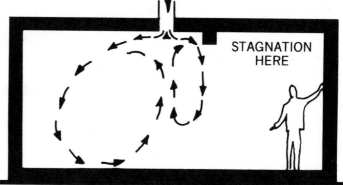

Figure 16.16h Locate air outlets carefully to avoid blocking the air flow with beams or other building elements.

Figure 16.16i No drafts result when cloth ducts are used because the air is supplied along the entire length through the porous fabric. (Courtesy Air Sox, Inc.)

porosity to allow for a range of draft-free air distribution. The sections of lightweight, brightly colored cloth ducts are zippered together and hung from the ceiling (Fig. 16.16i).

In rooms with high ceilings, it is not necessary to cool the upper layers, which, because of stratification, are very warm. Air can be introduced through wall registers just a few feet above people's heads. Stratification, however, is a liability in cold climates. In the winter, the hot air collecting near the ceiling should be brought back down to the floor level by means of antistratification ducts or devices as shown in Fig. 17.7d. In large theaters, good air flow is often achieved by supplying air all across the ceiling and returning it all across the floor under the seats (Fig. 16.16j).

Return air generally leaves a room through a grille, but it can also leave through lighting fixtures, perforated

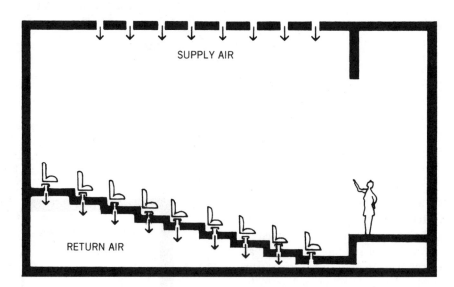

Figure 16.16j In large theaters, the air is often returned under the seating.

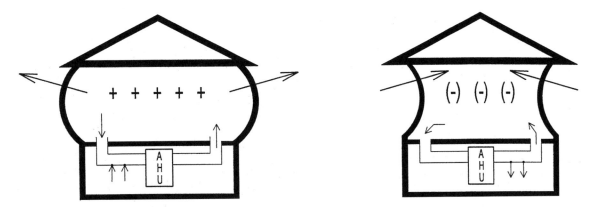

Figure 16.16k If the ducts are outside of the air-conditioned space, it is especially important to have no air leaks. In the worst case, radon and toxic mildew can be sucked into the return duct. (After North Carolina Alternative Energy Corp. (AEC).)

ceiling panels, or undercut doors. The location of the return-air grille has almost no effect on room air motion if supply outlets are properly located. However, to prevent the short-circuiting of the air, do not place return openings right next to supply outlets. Also avoid floor return grilles because dirt is sucked in and small objects fall into them.

Ducts tend to run through the basement, crawl space, or slab, particularly when air is supplied low in the room. Tight-fitting ducts are important; otherwise, the building will come under negative or positive pressure (Fig. 16.16k). If the return ducts leak, toxic material, such as radon or mildew, will be sucked into the return ducts. If the supply ducts leak, the building comes under negative pressure and outdoor air will enter through all the cracks.

The air pressure in these areas should be kept slightly below that of the rest of the building to prevent the contaminated air from spreading. At the same time, the pressure in the rest of the building should be slightly above the atmospheric pressure to prevent the infiltration of untreated air through cracks and joints in the building envelope (Fig. 16.17). These pressures can be maintained by the proper balance between the amount of air that is removed by exhaust fans and the amount brought in through the air-conditioning unit.

Normally, about 10 percent of the air circulated for heating or cooling will be outdoor air. Under certain circumstances, much larger amounts of outdoor air are introduced. For example, the American Society for Heating Refrigeration and Air Conditioning Engineers (ASHRAE)

recommends five times as much outside air for smokers than for nonsmokers. Operating rooms in hospitals use 100 percent outdoor air to prevent the explosion hazard from the anesthetic.

Outdoor air can also be used for cooling when its temperature is at least 15°F below the indoor temperature. Mechanical systems designed to use cold outdoor air for cooling are said to have an **economizer cycle.**

Although mechanical ventilation is a standard part of air conditioning in larger buildings, it is usually left up to natural infiltration and operable windows to ventilate smaller buildings. As long as small buildings were not very airtight, this policy worked quite well. Now, however, many small well-constructed airtight buildings are suffering from indoor air pollution. Even some well-ventilated large build-

16.17 VENTILATION

The excess carbon dioxide, water vapor, odors, and air pollutants that accumulate in a building must be exhausted. At the same time, an equal amount of fresh air must be introduced to replace the exhausted air. The air should be exhausted where the concentration of pollutants is greatest (e.g., toilets, kitchens, laboratories, and other such work areas).

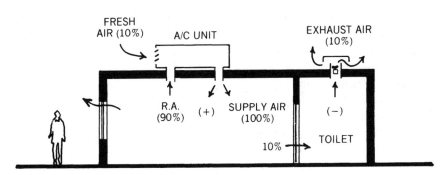

Figure 16.17 A slight positive pressure in the building prevents infiltration A slight negative pressure in toilets prevent odors from spreading to adjacent spaces.

ings have problems problems with indoor-air quality, especially when they are new.

Indoor-air quality (IAQ) is affected by pollution from many sources. Besides the obvious sources from people and their activities, significant sources include unvented combustion, off-gassing of building materials and furnishings, cleaning materials, and the ground below the building. Eliminating the source of the pollutants is the best way to achieve IAQ. What can't be eliminated at the source should be extracted with exhaust fans. Dilution with fresh air should be the last method, although it is the most commonly used method for controlling IAQ.

Air pollution from combustion, which includes the odorless and deadly gas carbon monoxide, comes mainly from unvented kerosene and gas space heaters and attached garages. Unvented carbon monoxide can also be generated by poorly maintained gas appliances, such as kitchen stoves. Tobacco smoke, fireplaces, and leaky wood stoves can also be a problem. Any indoor combustion should be directly vented outdoors.

Off-gasing of **volatile organic compounds** (VOCs), such as formaldehyde from plywood, particleboard, furniture, carpets, and drapes can cause severe reactions in some people. Use materials low in formaldehyde and VOCs and vent buildings well especially when new. If loose asbestos is present, it should be immediately removed.

Radon gas, which is radioactive, odorless, and colorless, occurs naturally in the earth from where it slowly makes its way to the surface. In those locations where there are high concentrations, radon gas must be blocked from entering the building. Radon enters a building mainly through exposed earth as in crawl spaces, cracks, and openings around pipes and electrical wiring.

Moisture in a crawl space or basement encourages mildew growth,

which is highly allergenic. Certain types of microorganisms growing in moist buildings and air-conditioning equipment produce deadly toxins (e.g., like those that resulted in the fatal Legionnaire's Disease). Keeping a building dry and properly maintaining air-conditioning equipment are the best ways to control these microorganisms.

Although indoor green plants can be a source of microorganisms if overwatered, they also scrub out the indoor air. Scientists from NASA have proven that properly maintained plants can clean the air of toxins. Scientist Bill Wolverton predicts that plants will be the technology of choice for improving IAQ in the twenty-first century.

16.18 ENERGY-EFFICIENT VENTILATION SYSTEMS

Required ventilation should not be reduced in order to save energy. After source elimination and extraction have eliminated as many pollutants as possible, ventilation is required to dilute the remaining polluted indoor air. Fortunately, specific techniques

greatly reduce the energy penalties that normally would come with the introduction of outdoor air for ventilation purposes. One system described in Section 8.24 uses a low-cost active solar system to preheat the ventilation air, while other systems recover the heat that is usually lost during ventilation.

Heat Recovery Ventilators

Heat recovery ventilators, also known as **heat exchangers** or **air-to-air heat exchangers,** can capture much of the heat that is ordinarily lost during ventilation. When air is exhausted from a building to make room for fresh outdoor air, a large amount of both sensible and latent heat is lost. In winter, cold, dry outdoor air must be heated and humidified, and in summer, the outdoor air must be cooled and dehumidified. Some heat-recovery ventilators use fixed plates or heat pipes to capture only the sensible heat (Fig. 16.18a). Other ventilators are more sophisticated and recover both latent and sensible heat. The fixed plates in some heat exchangers are made of a special material that enables water vapor to move though the plates (Fig.

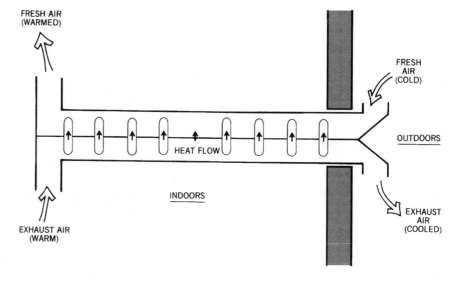

Figure 16.18a Heat-recovery ventilators allow part of the heat in exhaust air to be recovered without cross contamination of the air-streams. Typically, the heat moves through metal partitions, but heat pipes can be used to improve the energy transfer.

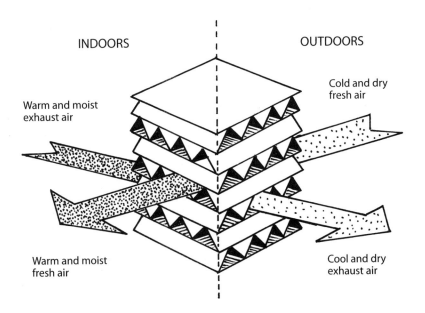

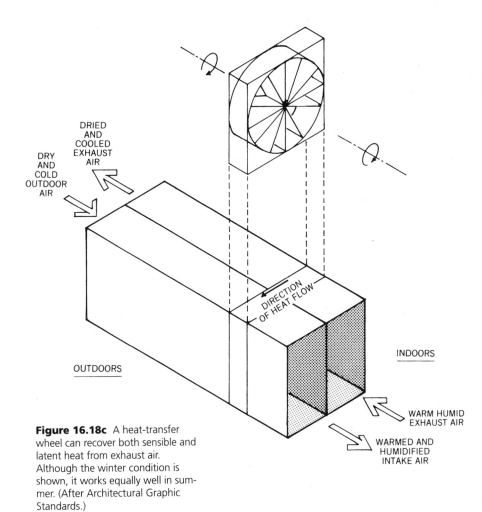

INDOORS   OUTDOORS

Warm and moist exhaust air

Cold and dry fresh air

Warm and moist fresh air

Cool and dry exhaust air

Figure 16.18b Some flat-plate heat exchangers transfer only sensible heat across metal partitions. The one shown uses composite-resin partitions to transfer latent as well as sensible heat (winter condition shown). (After Mitsubishi Electronics, HVAC Advanced Products Division.)

DRIED AND COOLED EXHAUST AIR

DRY AND COLD OUTDOOR AIR

DIRECTION OF HEAT FLOW

OUTDOORS

INDOORS

WARM HUMID EXHAUST AIR

WARMED AND HUMIDIFIED INTAKE AIR

Figure 16.18c A heat-transfer wheel can recover both sensible and latent heat from exhaust air. Although the winter condition is shown, it works equally well in summer. (After Architectural Graphic Standards.)

16.18b). Another type recovers between 70 and 90 percent of the heat by means of a heat transfer wheel covered with lithium chloride, a chemical that absorbs water. As the wheel turns, both sensible and latent heat is transferred from one air stream to another (Fig. 16.18c).

Displacement Ventilation

The typical method of supplying fresh air to spaces is shown in the water analogy of Fig. 16.18d. The polluted, warm water in the pitcher on the top is being diluted by the addition of cool, clean water. A much more effective approach is the method used in the pitcher on the bottom where the cool clean water is displacing the warmer polluted water. Similarly, a **displacement ventilation** system

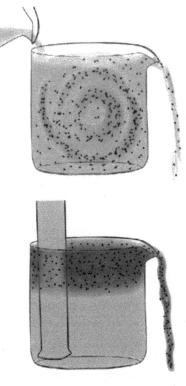

Figure 16.18d Conventional ventilation dilutes polluted air, as in the water analogy on the top, while displacement ventilation eliminates the polluted air more effectively as on the bottom. (Courtesy Air Displacement, Inc.)

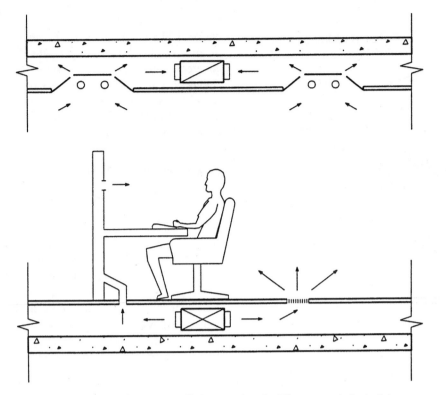

Figure 16.18e Displacement ventilation uses the raised floor to supply fresh air low in the room and sometimes directly to the occupants. The polluted warm air rises and is then exhausted through the lighting fixtures. This arrangement results in exceptionally clean air, prevents the waste heat of the lighting from entering the occupied space, and makes the lighting more efficient by cooling the fixture.

supplies fresh cool air at floor level and exhausts polluted warm air though the ceiling (Fig. 16.18e).

Economizer Cycle

When the outdoor air is sufficiently cool, it can be used instead of the refrigeration equipment to cool a building. Mechanical equipment designed to use ventilation for cooling when the weather is cool enough is said to be using an **economizer cycle.**

16.19 AIR FILTRATION AND ODOR REMOVAL

Besides heating and cooling, the mechanical system must also clean the air. The amount of dust, pollen, bacteria, and odors in air has an effect on the health and comfort of the occupants, as well as the cleaning and re-

decorating costs of the building. Return air should pass through filters first so that coils and fans can be kept free of the dirt that will reduce their efficiency. Among the many types of filters, the most popular are dry filters, electronic filters, water sprays, and carbon filters.

Dry filters are usually made of fiberglass and are thrown away when dirty. Although most dry filters remove only large dust and dirt particles, **high-efficiency particulate (HEPA)** air filters that can remove microscopic particles, such as bacteria and pollen, are available. A very efficient device for removing these very small particles is the electronic air cleaner. Odors can be reduced by passing the air through a water spray or over activated carbon filters. Sometimes, however, the best solution for controlling odors is increased ventilation.

16.20 SPECIAL SYSTEMS

District Heating and Cooling

High efficiency is possible when a complex of buildings can be heated and cooled from a district plant. For example, on many college campuses hot and chilled water is distributed through insulated pipes to all the buildings (Fig. 16.20a). The district plant contains large and efficient boilers, chillers, and cooling towers. Since the equipment is centralized and highly automated, a small staff can easily maintain it.

Combined Heat and Power

As first mentioned in Section 3.21, a **combined heat and power (CHP)** system can be an efficient method of supplying a building or complex of buildings with electricity, hot water, and even chilled water. Through the generation of the electricity on-site, the heat normally wasted can be used for space heating or domestic hot water. The waste heat can also be used to drive an absorption refrigeration machine to generate chilled water for summer cooling. Cogeneration can be used in installations as small as a fast-food restaurant and as large as a university campus.

With the restructuring of the electric power industry, it will be easier and more financially attractive to generate one's own electricity, with the "waste" heat being a bonus.

Thermal Storage

The strategy of saving energy when an excess exists for a time when an energy shortage exists is rapidly gaining in popularity. We have seen this technique used for both passive heating and cooling where the mass of the building is usually used to store heat from day to night. Besides the heat storage of active solar systems, there exists a number of other ways that active mechanical systems can store energy.

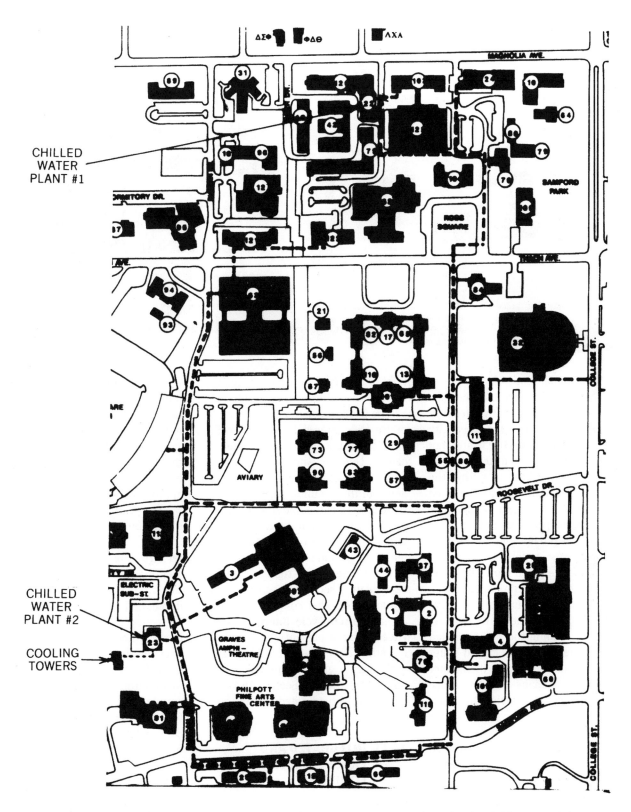

CHILLED
WATER
PLANT #1

CHILLED
WATER
PLANT #2

COOLING
TOWERS

Figure 16.20a Most buildings on the Auburn University Campus are heated and cooled from central plants. Only the district cooling system is shown.

Figure 16.20b This above-ground, thermal/water storage tank is used as an element of the aesthetic design. (Courtesy Chicago Bridge & Iron Company.)

For example, since electric-power companies usually have excess capacity at night, they frequently offer substantially lower rates for electricity during night hours. In the winter, hot water can be generated at night and stored for space heating during the next day. During the summer, chilled water can be produced at night for use during the peak afternoon cooling period the next day. Not only does the electricity cost much less, but the chillers operate much more efficiently during the cool nighttime hours.

The main additional cost for active thermal storage is the water tank required for holding the hot or chilled water. If the size of the chilled water tank is limited, ice rather than water can be stored. Many high-rise office buildings now have swimming-pool-size water tanks in the basement for storing chilled water. Where open land is available, large, insulated tanks can be either partially buried or placed above ground to become part of the architecture (Fig. 16.20b).

Some people have realized that summer heat could be stored for use in the winter. A solar collector operating at peak efficiency in the summer or a condenser coil on an air-condi-

tioning unit could heat water for the winter. In addition, the winter cold can be used to generate chilled water or ice to be stored for the following summer, much like the nineteenth-century New Englanders who harvested ice in the winter and sold it to the South in the summer. When seasonal ice is used, it should be stored in insulated tanks. When water is used, tanks tend to be too large and the water in the ground should be used. Waterproof liners and rigid insulation can isolate a section of earth with its groundwater. This concept is known as an **Annual-Cycle Energy System (ACES),** but at present this technique is still experimental. The geo-exchange, ground-coupled heat pumps described in Section 16.11 use this concept to a limited extent.

16.21 INTEGRATED AND EXPOSED MECHANICAL EQUIPMENT

Usually, mechanical equipment is a completely separate system hidden away behind walls and above suspended ceilings (Fig. 16.21a). Perhaps it is time for the equipment

to come out of the closet. This can happen either by an integration of the mechanical equipment with other building systems, such as the structure, or by exposing the equipment to full view. Many successful buildings exist to illustrate either approach.

The CBS Building, a high-rise office building designed by Eero Saarinen, illustrates how the perimeter structure and mechanical equipment can be integrated (Fig. 16.21b). The appearance of the building is dominated by the triangular columns projecting from the wall every 5 feet. Every other column contains an air riser for an induction unit located on each side of the column. The remaining columns contain either water-supply or return risers. These pipes feed the induction units with hot or cold water depending on the season of the year (Fig. 16.17c).

Although the outside dimensions of the columns remain constant, the inside opening varies with the load. The smallest opening (most concrete) occurs near the ground and the largest opening near the roof. This dovetails nicely with the size of the ducts since they get smaller as they descend from

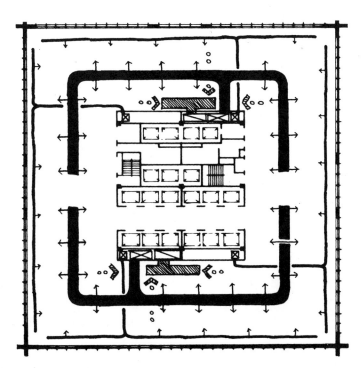

Figure 16.21a Typically, the mechanical equipment is a separate building system hidden from view. (After Guise.)

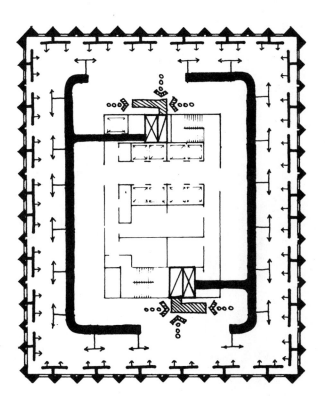

Figure 16.21b The perimeter structure and mechanical equipment are integrated in the CBS Building, New york City. (After Guise.)

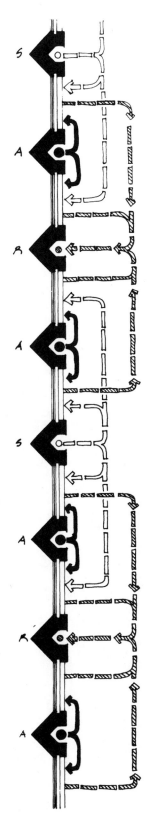

Figure 16.21c Detail of a perimeter structure/mechanical system. A, air-riser ducts; S, water-supply pipe; R, water-return pipe. (From *Design and Technology in Architecture* by David Guise, © John Wiley & Sons, Inc., 1985.)

TOP OF BUILDING

Figure 16.21d The size of both columns and riser ducts varies with height. Moving up the building, the column area gets smaller while the duct area gets larger. (After *Design and Technology in Architecture* by D.Guise.)

BOTTOM OF BUILDING

the mechanical room on the roof (Fig. 16.21d). These columns/mechanical chases make a strong, aesthetic statement on the facade (Fig. 16.21e).

One way to integrate the floor structure and mechanical equipment is illustrated by the Hoffmann-La Roche Building, Nutley, New Jersey. The floor structure consists of an exposed waffle-slab. Lighting fixtures, air-supply outlets, and return grilles are designed to fit into the 2 x 2-foot x 18-inch-deep coffers. All ducts, pipes, and wires run under a raised-floor system that rests on the waffle slab (Fig. 16.21f). The plan in Fig. 16.21g shows how the supply and return ducts are connected to the core of the building.

Instead of separate systems, the integrated approach tries to make each construction element do as many jobs as possible. As Buckminster Fuller urged, "do more with less." Although the integrated approach promises great efficiency and cost reduction, it is a more sophisticated and difficult way to design and build. It requires much more cooperation among the various building professionals than the existing approach of separate systems. The previous two examples were presented to illustrate the concept rather than any particular integrated design.

The most dramatic way to recognize the mechanical equipment as a legitimate part of architecture is to expose it to view. Especially at a time when a more ornate and colorful style is replacing the simplicity and clear lines of modern architecture, the exposed mechanical equipment can add complexity and richness to a building.

An example of this approach is the Occupational Health Center in Columbus, Indiana, where all the pipes and ducts are exposed to view (Fig. 16.21h). Bright colors define and clarify the various systems of supply air, return air, hot-water heating, etc.

Figure 16.21e Eero Saarinen used exposed concrete columns to contain mechanical risers in the CBS office tower in New York City.

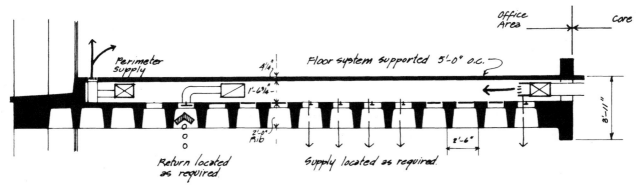

Figure 16.21f A section through the waffle slab of the Hoffmann-La Roche Building. (After *Design and Technology in Architecture* by D. Guise.)

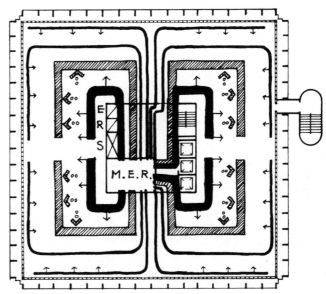

Figure 16.21g Plan of the Hoffmann-La Roche Building. S, supply-riser duct; R, return-riser duct; E, exhaust-riser duct. (After Guise.)

Figure 16.21h Exposing the mechanical equipment can add richness and complexity to architecture. Occupational Health Center, Columbus, IN, by Hardy, Holzman, Pfeiffer. (Courtesy of Cummins Corporation.)

Since exposed ducts must be made of better material and higher-quality work, they cost more than conventional equipment. This higher cost can be offset by savings from the elimination of a suspended-ceiling systems.

At the Centre Pompidou, architects Richard Rogers and Renzo Piano exposed the mechanical equipment not only on the interior but also on the exterior (Fig. 16.21i). Any heat gain or loss to the ducts and pipes exposed on the interior is either helpful or not very harmful. For example, in the winter, any heat lost from ducts on pipes will help heat the interior. This is definitely not the case with exterior ducts or pipes, which are instead exposed to the harsh climate. It is worthwhile to note that in nature no creatures have their guts on the outside of their skin. Such creatures might be born, but they do not survive. The Centre Pompidou might be a great monument, but it should not be a prototype for more ordinary buildings where most of the mechanical equipment should be on the inside of the building skin.

16.22 CONCLUSION

The architect determines the size, and to some extent the design, of the mechanical system during tiers one and two of the three-tier design approach. The architect has the opportunity, if not power, and the responsibility to make the operation of the mechanical system as small a burden as possible on both the building owner and the environment.

Figure 16.21i The Centre Pompidou, Paris, France, by Richard Rogers and Renzo Piano. Much of the mechanical equipment is exposed on the exterior of the building. (Photograph by Clark Lundell.)

KEY IDEAS OF CHAPTER 16

1. Mechanical heating should supply only the small amount of heat still needed after a tight thermal envelope (tier one) and passive solar (tier two) have been fully utilized.

2. Fireplaces are highly inefficient unless they have a heat exchanger, outdoor combustion air, and doors.

3. All but the smallest buildings have more than one thermal zone.

4. The number of required zones has a great impact on the choice of a mechanical system.

5. Heating systems

   a.Hot-water (hydronic) systems are efficient and comfortable but generally are not useful for cooling.

   b. Hot-air systems are appropriate when cooling is also required.

   c. Electric resistance heating should be used as little as possible because it is most wasteful of our energy.

   d. Heat pumps are an efficient way to heat with electricity. Earth-coupled heat pumps are more efficient than air-to-air heat pumps.

   e. **Combined heat and power (CHP) systems** can supply heat at low cost.

6. Cooling systems use refrigeration machines to pump heat from lower temperatures indoors to higher temperatures outdoors.

7. Although the compressive refrigeration cycle is more efficient than the absorption refrigeration cycle, an inexpensive source of heat can make the absorption cycle attractive.

8. In the compressive cycle, the refrigerant moves heat from the evaporator coil to the condenser coil.

9. Ground-coupled heat pumps are now called **geo-exchange** heat pumps. Four energy sources/sinks are possible: ponds, well water, vertical ground loops, and horizontal ground loops.

10. Chillers produce chilled water that is used in air-handling units and fan-coil units.

11. Large quantities of heat are more efficiently dumped into the atmosphere by evaporative coolers and cooling towers.

12. Package units allow a maximum of mechanical fabrication to be performed in the factory.

13. Split systems are, in effect, two package systems.

14. Internal duct insulation and sections of flexible ducts control mechanical noise.

15. **Variable-air-volume (VAV) systems** are efficient but are limited to either heating or cooling a building at any one time.

16. Fan-coil systems provide many zones at modest cost, but they will provide fresh air only if they are located on an outside wall.

17. **Mechanical-equipment rooms (MER)** should be centrally located to minimize the quantity and complexity of the duct work.

18. Round ducts are most efficient, but oblong rectangular ducts require less headroom. Flat-oval ducts are a compromise.

19. Diffusers and registers mix the supply air and room air without causing drafts. Both have dampers to control the volume of air. Grilles might or might not direct the flow of air, and they do *not* contain dampers.

20. **Indoor-air quality (IAQ)** is of the highest priority for health and productivity. Clean, healthy air is achieved by:

    a. Eliminating sources of pollution (e.g. volatile organic compounds, VOCs)

    b. Directly exhausting the source of pollutants

    c. Using ventilation to dilute pollutants

    d. Using living plants to clean air.

21. Heat-recovery units are excellent devices for minimizing heat loss from ventilation. Both sensible and latent heat can be recovered from these air-to-air heat exchangers.

22. Displacement ventilation is more efficient than conventional ventilation in eliminating pollutants.

23. The economizer cycle uses cold outdoor air when available to cool a building.

24. Combined heat and power (CHP) systems, formerly known as **co-generation systems,** can be more efficient than buying electricity and heating energy separately.

25. District heating and cooling systems are more efficient than smaller, individual systems.

26. Thermal-storage systems can take advantage of low night electric rates and the greater efficiency of running refrigeration machines in cool night air. The energy storage (heat sink) can be in the form of chilled water or ice.

27. Exposing mechanical equipment and ducts on the interior is efficient and has aesthetic potential. Exposing mechanical and equipment on the outside of the thermal envelope is very inefficient and should be avoided. *Ducts should be located inside the thermal envelope whenever possible.*

References

Environmental Protection Agency. *Space Conditioning: The Next Frontier.* Office of Air and Radiation, EPA 430-R-93-004, April, 1993.

Resources

FURTHER READING

(See Bibliography in back of book for full citations. This list includes valuable out-of-print books.)

Allen, E. *How Buildings Work: The Natural Order of Architecture.*

Allen, E., and J. Iano. *The Architect's Studio Companion: Technical Guidelines for Preliminary Design.*

ASHRAE. *Air-Conditioning Systems Design Manual.*

ASHRAE. *Terminology of Heating, Ventilation, Air Conditioning, and Refrigeration.*

Banham, R. *The Architecture of the Well-Tempered Environment.*

Bearg, T. W. *Indoor Air Quality and HVAC Systems.*

Bobenhausen, W. *Simplified Design of HVAC Systems.*

Bradshaw, V. *Building Control Systems.*

Dagostino, F. R. *Mechanical and Electrical Systems in Construction and Architecture.*

Egan, M. D. *Concepts in Thermal Comfort.*

Flynn, J. E. *Architectural Interior Systems.*

Guise, D. *Design and Technology in Architecture.*

Knight, P. A. Mechanical Systems Retrofit Manual: A Guide for Residential Design.

Lang, V. P. *Principles of Air Conditioning.*

Lyle, D. *The Book of Masonry Stoves: Rediscovering an Old Way of Warming.*

Mann, P. A. *Illustrated Residential and Commercial Construction.*

Olivieri, J. B. *How to Design Heating-Cooling Comfort Systems.*

Ramsey/Sleeper. Architectural Graphic Standards.

Reid, E. *Understanding Buildings.*

Rush, R. D., ed. *The Building Systems Integration Handbook.*

Stein, B., and J. S. Reynolds, *Mechanical and Electrical Equipment for Buildings.* A basic resource.

Tao, W. K. Y., *Mechanical and Electrical Systems in Buildings.* 2nd ed. Upper Saddle River, NJ: Prentice Hall, 2000.

Traister, J. E. *Residential Heating, Ventilating, and Air Conditioning: Design and Application.*

Trost, J. *Heating, Ventilating, and Air Conditioning.*

Wirtz, R. *HVAC/R Terminology: A Quick Reference Guide*

Wright, L. *Homefires Burning: The History of Domestic Heating and Cooking.*

ORGANIZATIONS

(See Appendix J for full citations.)

American Society of Heating, Refrigerating, and Air-Conditioning Engineers (ASHRAE)

GeoExchange
www.geoexchange.org

The Vital Signs Project
www.ced.berkeley.edu/cedr/vs/

17

CASE STUDIES

"The fact is that our schools have now educated several generations
of architects who depend exclusively on nonarchitectural means of
environmental adaptation. The problems of this approach are now
emerging, partly as an energy-consumption dilemma, but perhaps,
more to my point, problems of architectural expression have
appeared. A paucity of vocabulary has emerged as a professional and
artistic dilemma: If all environmental problems can be handled by
chemical and mechanical means, who needs the architect?"

Ralph L. Knowles,
in **Sun Rhythm Form,** *1981 (p. 151)*

17.1 INTRODUCTION

Ralph Knowles answers the question he raises in the quote on the previous page. He argues that by responding to the energy and environmental problems, the designer enriches architecture. The following case studies

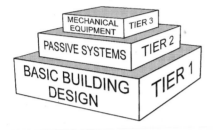

Figure 17.1 The heating, cooling, and lighting needs of buildings are best and most sustainably achieved by the three-tier design approach. The case studies presented in this chapter fully address the appropriate issues at each of the three tiers.

were chosen in part to show how addressing environmental problems can create good and often excellent architecture.

Up to this point, buildings were used to illustrate the particular concept being discussed at the moment. However, since buildings must address many issues simultaneously, they must also incorporate at the same time many different design strategies. The successful integration of many environmental concepts is demonstrated in the following examples.

The buildings presented here as case studies are successful in part because their designers addressed environmental control issues at all three tiers: basic building design, passive systems, and mechanical equipment (Fig. 17.1). The case studies include a variety of building types, sizes, and styles.

17.2 THE REAL GOODS SOLAR LIVING CENTER

Retail Store outside of Hopland, California

Architects: Van der Ryn Architects/ Ecological Design Institute and Arkin Tilt Architects

Landscape designers: Stephanie Kotin and Chris Tebbutt

Built: 1996

The Real Goods Trading Company sells merchandise needed for a sustainable lifestyle and is the leading source of renewable-energy equipment. When new facilities were needed, its president and founder, John Schaeffer, decided it was time to "walk our talk."

A barren, blighted site was chosen in part because of its need for regeneration. The company wanted to

Figure 17.2a The Real Goods Solar Living Center and its elaborate landscaping, as seen from the southeast. Most windows face south, a few face east, and the operable clerestories also face east. (Courtesy Van der Ryn Architects,©Richard Barnes, photographer.)

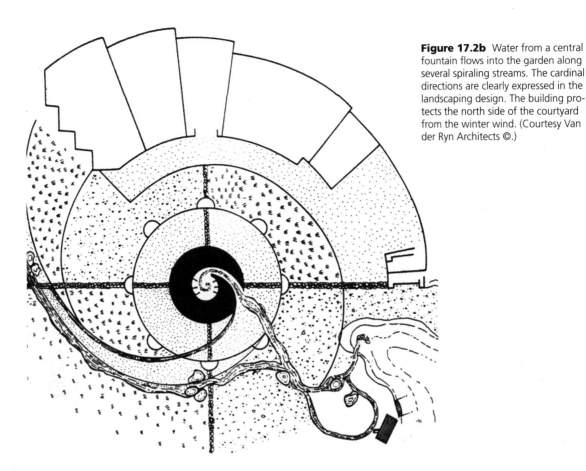

Figure 17.2b Water from a central fountain flows into the garden along several spiraling streams. The cardinal directions are clearly expressed in the landscaping design. The building protects the north side of the courtyard from the winter wind. (Courtesy Van der Ryn Architects ©.)

demonstrate how to restore a rich ecology to land that had been ruined. Unlike the usual situation, the landscaping was started before the building construction, so that the gardens and trees could start their long process of restoring life to this site. Plants were carefully chosen to return nitrates to the ground, thereby making chemical fertilizer unnecessary. Since the site was in a flood plain, ponds were excavated and the fill used to raise the building above the 100-year flood level.

The landscaping was designed to delight the visitors and customers, and to engender the love of nature necessary for the restoration of the whole planet (Fig. 17.2a). The design of the landscape was based on the three themes of "inspiration, production (food gardens), and restoration."

Part of the inspiration came from design elements that reflect the daily and seasonal cycles of our lives. Besides sundials of various sizes, shapes, and materials, fountains that rise and fall with the daily and seasonal cycles of the sun were also added (Fig. 17.2b). The fountains and irrigation systems are powered by photovoltaic (PV) arrays whose output is proportional to the amount of sunshine. Thus, the evaporative cooling from the fountains and irrigation systems rises to a maximum when the need is at a maximum, in the summer. All of the landscaping was carefully designed to moderate the intense summer heat through shading and evaporative cooling.

The Solar Living Center was designed to be energy-independent. Because of the energy-conscious

architecture, it is possible for a small wind machine and seven PV-tracker arrays to generate most of the electricity needed on an annual basis. The building is connected to the power grid mainly to avoid the need for batteries. Power is bought at night, but during a typical sunny, summer day surplus electricity is generated and sold to the power company.

A sophisticated daylighting design enables the electric lighting to be on only about 200 hours instead of the more than 3,000 hours per year of a standard retail or commercial building. Model studies showed that large, south-facing windows with light shelves reflecting off a curved ceiling would create high-quality daylighting (see Fig. 17.2c). The light shelves are translucent so that the area underneath is not in darkness. On winter

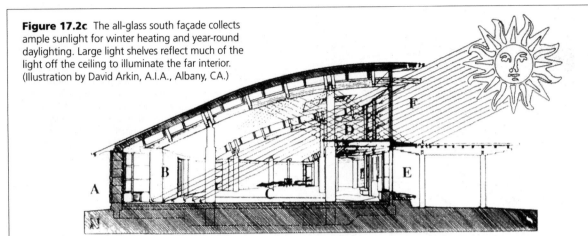

Figure 17.2c The all-glass south façade collects ample sunlight for winter heating and year-round daylighting. Large light shelves reflect much of the light off the ceiling to illuminate the far interior. (Illustration by David Arkin, A.I.A., Albany, CA.)

Cold Day Heating Mode

A The mass of the highly insulated straw-bale and PISE™ walls produce a thermal "flywheel" effect, storing and releasing heat.

B Stored heat in the thermal mass in the concrete floor and columns is also released to maintain comfort.

C Efficient (wood-burning stoves working display models) are the only source of supplemental heat.

D Lightshelves control glare and direct gain, reflecting light onto the curved ceiling.

E Doors and windows can be opened to provide immediate cooling if the temperature indoors is too warm.

F The low winter sun penetrates the building through substantial south-facing glazing.

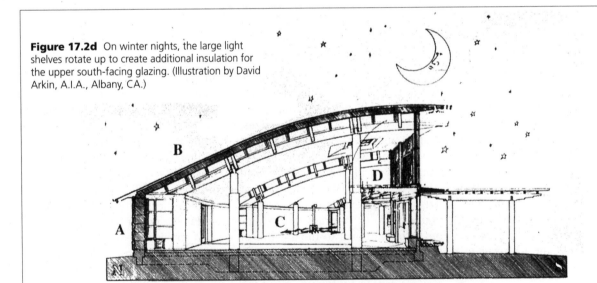

Figure 17.2d On winter nights, the large light shelves rotate up to create additional insulation for the upper south-facing glazing. (Illustration by David Arkin, A.I.A., Albany, CA.)

Night Heating Mode

A The insulation in the straw-bale and roof helps to retain the day's heat inside the building mass.

B The 12" cellulose (recycled newspaper) insulation in the roof provides an R-value greater than 60.

C The wood-burning stoves can be left on a slow burn on the coldest nights, balancing heat loss.

D Lightshelves are hinged and can be folded against the high glazing to reduce heat loss to the night sky.

nights, the light shelves can be swung up to add additional insulation to the glazing (see Fig. 17.2d). Because the roofs were stepped up toward the west, a series of east-facing clerestories pick up additional daylighting. Automatic controls are used to turn off the electric lighting when sufficient daylight is present, and all lighting comes from fluorescent lamps—no incandescent lamps are used at all.

Heat transfer through the thermal envelope was kept to a minimum with generous insulation levels. The roof has 12 inches of loose cellulose with an R-value of about 42. There is also a 1½-inch air space with a radiant barrier that has an R-value of about 8 in summer and 2 in winter. The roofing membrane is made of white Hypalon™, which reflects about 75 percent of the solar radiation. The walls are made of 23-inch-thick rice straw bales with an R-value of about 34 (see Fig. 17.2e). The bales are sprayed using gunite equipment, first with about 1 inch of cement stucco

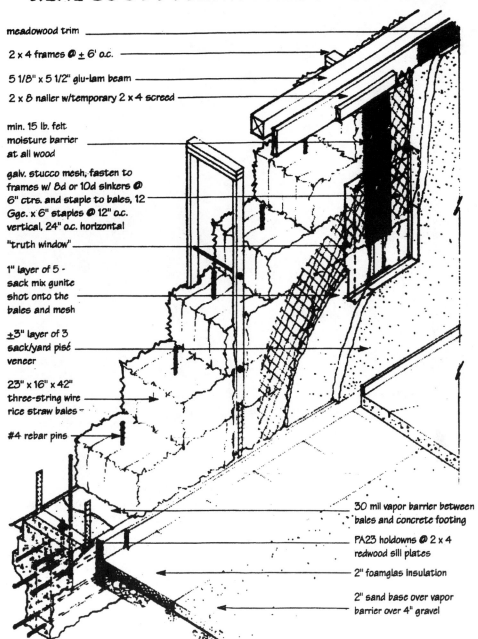

REAL GOODS STRAW BALE WALL SYSTEM

meadowood trim

2 x 4 frames @ ± 6' o.c.

5 1/8" x 5 1/2" glu-lam beam

2 x 8 nailer w/temporary 2 x 4 screed

min. 15 lb. felt moisture barrier at all wood

galv. stucco mesh; fasten to frames w/ 8d or 10d sinkers @ 6" ctrs. and staple to bales, 12 Gge. x 6" staples @ 12" o.c. vertical, 24" o.c. horizontal

"truth window"

1" layer of 5 - sack mix gunite shot onto the bales and mesh

±3" layer of 3 sack/yard pisé veneer

23" x 16" x 42" three-string wire rice straw bales

#4 rebar pins

30 mil vapor barrier between bales and concrete footing

PA23 holdowns @ 2 x 4 redwood sill plates

2" foamglas insulation

2" sand base over vapor barrier over 4" gravel

Figure 17.2e The straw bale walls were reinforced with steel rebar pins and wood frames. The walls are covered with a subcoat of cement stucco and finished with about 3 inches of a soil-cement mixture known as Pise™. (Courtesy Van der Ryn Architects©.)

Figure 17.2f This view from the northwest shows the soil-cement finish on the straw bale walls. (Courtesy Van der Ryn Architects©.)

and then with 3 more inches of Pise™, a soil-cement mixture that gives the building a rich, reddish-brown color (see Fig. 17.2f). The windows, most of which face south, are double-glazed with low-e for an R-value of about 2.8. Thus, little heat is lost in the winter, and very little heat is gained in the summer through the thermal envelope.

The main thermal load is in the summer when temperatures are frequently higher than 100°F. The three-tier approach to cooling design was used. **Heat avoidance** was the first-tier strategy, utilizing high levels of insulation, a very reflective roof membrane, no west windows (see Fig. 17.2g), few east windows, daylighting, and shading for all windows. The

second-tier involved night-flush cooling because the dry climate has a large diurnal temperature range. At night, outdoor air drawn in through openings near the floor cools the massive floor, the walls, and the concrete columns, while the warm air leaves through the operable east clerestories (see Fig. 17.2h). Fans, wind, and the stack effect all help to ventilate the

Figure 17.2g This view from the southwest shows that no windows face west. A sawtooth arrangement on the lower west wall has all glazing on the west façade facing south. (Courtesy Van der Ryn Architects©.)

Figure 17.2h Night-flush ventilation through the operable east clerestories enables the internal mass to cool down in preparation for the next hot day. (Illustration by David Arkin, A.I.A., Albany, CA.)

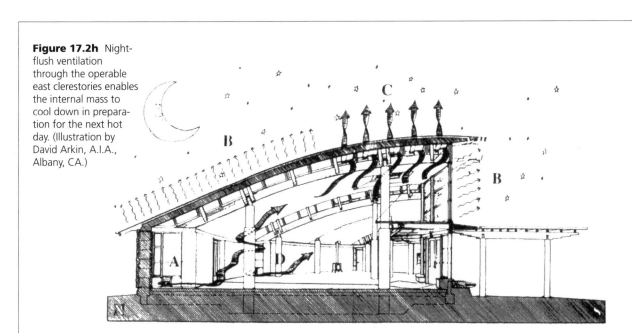

Night Cooling Mode

A Fans in the evaporative coolers run after hot (100° F+) days for additional air changes.

B Roof and high windows radiate heat to the night sky.

C A "stack effect" draws out warm air through clerestory windows and draws in cooler air through openings near the floor.

D Night sky radiation and cool night temperatures are used to "charge" the thermal mass of the building with "coolth" for the next day.

Figure 17.2i The impact of summer heat is prevented by high insulation levels, a very reflective roof membrane, and extensive shading. (Illustration by David Arkin, A.I.A., Albany, CA)

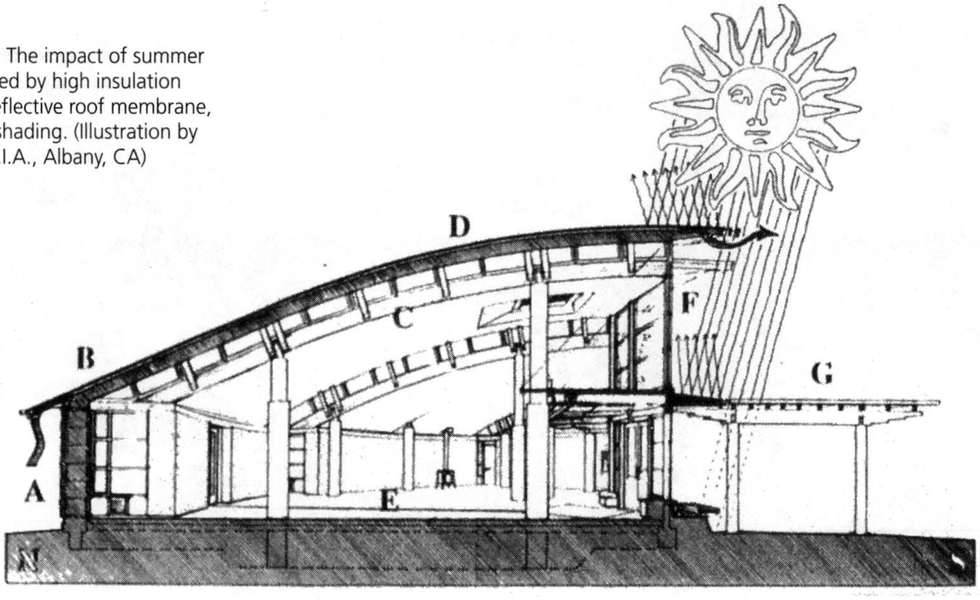

Hot Day Cooling Mode

A The thermal mass and high insulation value of the straw-bale and PISE™ (sprayed soil-cement) wall protects against the thermal transfer of l00°F+ outdoor summer temperatures.

B Air space in roof ventilates heat from radiant barrier over 12" of cellulose insulation (R-60).

C Clerestory windows are closed when outside temperature exceeds interior temperature.

D White Hypalon™ (synthetic rubber) roof membrane reflects gain from solar radiation.

E Thermal mass of concrete floor and columns absorbs heat from people and equipment; stores "coolth" from previous night flush.

F Overhang and awnings shade windows, controlling solar gain. Light shelves and curved white ceiling distribute daylight evenly, reducing need for heat-producing lighting fixtures.

G Trellis shades walls, windows, and walkway in summer; allows gain in winter, and controls glare.

building. During summer days, the windows are closed when the outdoor temperature is higher than the indoor temperature. The top of the south glazing is shaded by overhangs, the middle section by awnings, and the lowest part by a trellis, which also shades the courtyard in front of the building to prevent reflected heat and glare (see Figs. 17.2i and 17.2j). The west wall has a sawtooth arrangement so that windows on the west facade all face south (see again Fig. 17.2f).

For the third tier, evaporative coolers were installed; however, they have never been turned on. Tiers one and two have been so effective that they are all that are needed. A customer once reprimanded the staff members for setting the air conditioning too low. She was astonished to learn that there was no air-conditioning equipment, only heat avoidance and passive cooling. Even on the hottest days, the indoor temperature stays in the 70s (°F).

In winter, the building is heated mainly by direct gain through the all-glass south wall. The large amount of mass is again useful for storing heat. On cold winter mornings, several wood stoves are used as auxiliary heaters, as demonstration items (one

of the products for sale), and for ambiance.

To create such sustainable buildings, the architect Sim Van der Ryn uses a seven-part design process he calls "Eco-Logic" (Schaeffer, 1997):

1. Create program scenarios.
2. Analyze the site and ecological resources.
3. Investigate renewable and sustainable materials and systems.
4. Design the vision.
5. Develop and integrate systems.
6. Complete the detailed design.
7. Facilitate the construction.

Experience has shown that this design process results in buildings that have lower life-cycle costs, a high level of sustainability, a high satisfaction level for the owner and users, and higher productivity for the people working in the buildings.

Van der Ryn has also developed five ecological principles that guide him in his design process (Schaeffer, 1997):

1. The best solution starts from paying attention to the unique qualities of the place.
2. Trace the direct and indirect environmental costs of design decisions: use ecological accounting, or *"environomics."*
3. Mimic nature's processes in design, so that your design fits nature. The four key laws of ecological systems are:
   a. Nature lives off solar income as captured in food webs
   b. All wastes are recycled as food for other processes
   c. Biodiversity promotes stability
   d. Networks of relationships are maintained through feedback.
4. Honor every voice in the design process (local residents and builders, building occupants, maintenance people, and skilled construction people).
5. Making nature visible through design transforms both makers and users.

Most of the information about the Solar Living Center comes from a book that the author highly recommends:

A Place in the Sun: The Evolution of the Real Goods Solar Living Center, Chelsea Green Publishing Co., 1997, by John Schaeffer. (Also recommended is: *Ecological Design,* Island Press, Washington, D.C., 1995, by Sim Van Der Ryn and Stuart Cowan.)

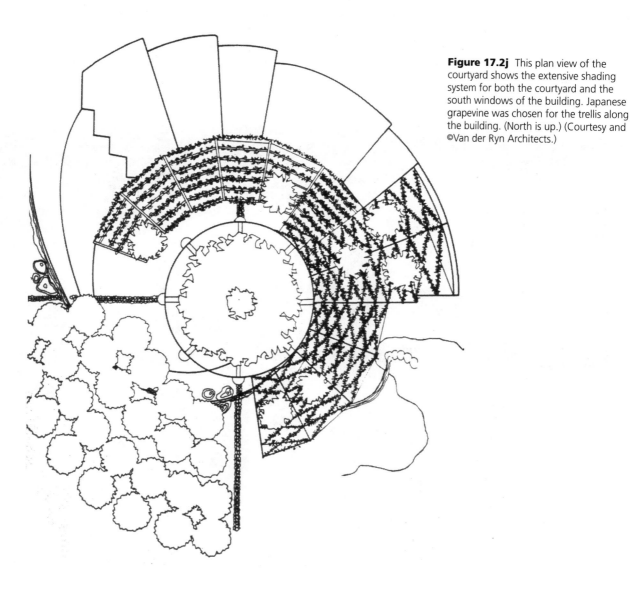

Figure 17.2j This plan view of the courtyard shows the extensive shading system for both the courtyard and the south windows of the building. Japanese grapevine was chosen for the trellis along the building. (North is up.) (Courtesy and ©Van der Ryn Architects.)

17.3 THE URBAN VILLA

Apartment building in
Amstelveen, the Netherlands,
at 52°N latitude

Architect: Atelier Z, Zavrel
Architecten BV, Rotterdam,
Netherlands

Area: about 50,000 square feet
(4,500 square meters)

Built: 1995

The Urban Villa project was built to prove that extreme energy savings are not an obstacle to good architecture in the predominantly cool and cloudy climate of the Netherlands, where the average maximum summer temperature is about 65°F, and the average minimum winter temperature is about 35°F. The design takes advantage of a compact shape and many shared walls to greatly reduce the exterior exposure. The central atrium/sun space acts as an entrance vestibule, as well as a preheater for winter ventilation (Figs. 17.3a and 17.3b).

The design team decided to utilize passive solar and a superinsulated envelope to meet most of the winter-heating requirement. The superinsulated envelope has R-values of:

R-38 for the roof
R-33 for the walls
R-8 for the windows (very high for windows), and
R-1 for the sun-space glazing.

The concrete structure adds thermal mass, and the continuous exterior insulation reduces the number of heat bridges, which are minimized when they do occur. For example, the balconies have a separate structural system so that only a few and small heat bridges are needed to tie the balconies to the main building.

Because of the low winter sun at 52°N latitude (Seattle is only about 48°N latitude), even deep balconies allow ample solar access on the south side for passive solar heating (Fig. 17.3c). Because 40 percent of the south facade is glass, the designers also decided that effective shading and ventilation would be needed to prevent summer overheating. In the summer, however, the balconies are not deep enough to fully shade the windows on hot days. Thus, an exterior roller shade was added at the outside edge of the balconies to automatically descend when indoor temperatures get too high (see again Fig. 17.3c). The sloped glazing of the atrium roof is also shaded on the inside during the summer (Fig.17.3e bottom).

During the summer, both natural and forced ventilation are used for cooling. Cross- ventilation, high and

Figure 17.3a The Urban Villa achieves winter comfort via a compact form, many shared walls, a large south façade that is 40 percent glass, and superinsulation. Summer comfort is achieved through ventilation and shading. Note the vertical, outdoor roll-down shades near the center of the curved façade. (Photograph from CADDET Technical Brochure No. 64.)

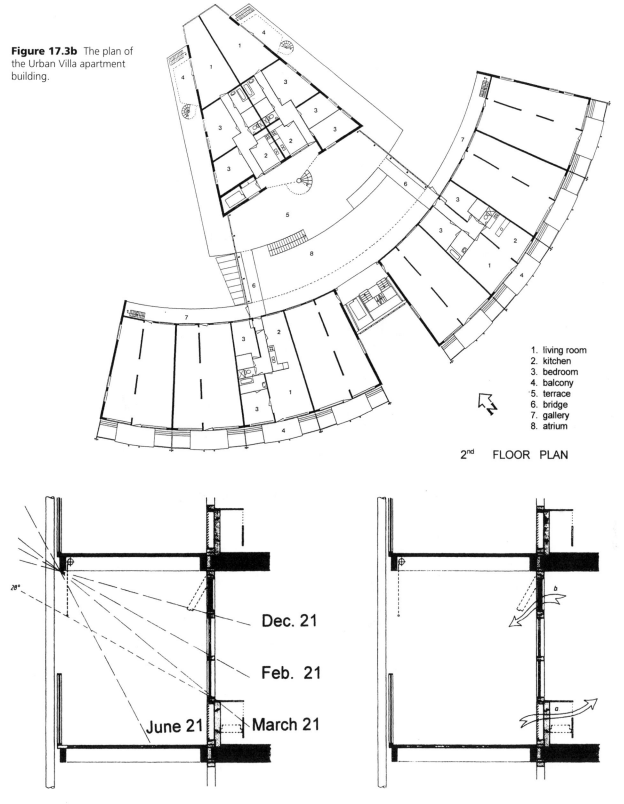

Figure 17.3b The plan of the Urban Villa apartment building.

1. living room
2. kitchen
3. bedroom
4. balcony
5. terrace
6. bridge
7. gallery
8. atrium

2ⁿᵈ   FLOOR   PLAN

28°

Dec. 21

Feb. 21

June 21   March 21

Figure 17.3c A section through the balcony showing the summer and winter sun angles. Note the roll-down shade. (Courtesy Atelier Z, Zavrel Architecten, BV.)

Figure 17.3d Summer ventilation by means of high and low openings. The low openings are through heating convectors. (Courtesy Atelier Z, Zavrel Architecten BV.)

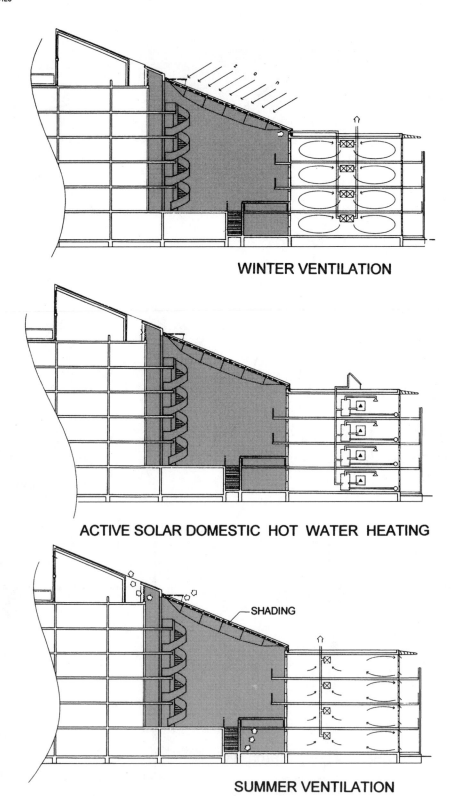

WINTER VENTILATION

ACTIVE SOLAR DOMESTIC HOT WATER HEATING

SHADING

SUMMER VENTILATION

Figure 17.3e Top: Winter passive solar heating for the south facade and the atrium/sun space which preheats the ventilation air for the apartments. Additionally, heat exchangers recover much of the heat from the exhaust air. Middle: Active solar supplies most of the domestic hot water. Each apartment has its own storage tank and auxiliary gas boiler for both domestic-hot-water and space heating. Bottom: Summer shading and ventilation. (Courtesy Atelier Z, Zavrel Architecten, BV.)

low windows (Fig. 17.3d), and exhaust fans remove the hot indoor air. Besides the shading, the atrium is kept from overheating by opening high and low vents (Fig. 17.3e bottom). To maintain indoor air quality in the apartments in the winter, preheated air from the sunspace/atrium passes through heat-recovery units to ventilate each apartment with minimal energy losses (Fig. 17.3e top).

Domestic hot water is supplied primarily from active solar panels on the roof (Fig. 17.3e middle). The sun supplies about 60 percent of the hot water, and high-efficiency (90 percent) gas auxiliary boilers supply the rest, as well as provide space heating when necessary. Building monitoring has shown that energy consumption is 60 percent to 70 percent lower than that of a conventional design, and the users are very satisfied with the buildings.

17.4 THE EMERALD PEOPLE'S UTILITY DISTRICT HEADQUARTERS

Office building near Eugene, Oregon

Architects: Equinox Design and WEGROUP, PC, Architects

Area: 24,000 square feet

Built: 1987

This sophisticated design is a consequence of the collaboration between Equinox Design Inc., WEGROUP PC Architects, and an enlightened electric utility in need of a new headquarters complex, which includes the office building described here (Fig. 17.4a). The climate consists of cool winters, mild summers, and overcast days much of the year. The goal was to create a building that not only created a pleasant and attractive work environment but also was energy-conscious.

With the building designed as a south-facing elongated rectangle, east–west windows could be avoided and no point within the building is far from either north or south windows. All south windows, including the south-facing clerestory, have vine-covered trellises for protection from the high summer sun. In the winter, the deciduous vines lose their leaves and allow solar energy to heat the building (Fig. 17.4b). Concrete-block exterior walls, concrete-block interior fin walls, and concrete ceiling planks provide ample thermal mass.

This same mass is used in the summer for night flush cooling. The interior of the building is flushed with cool night air so that the mass can act as a heat sink the next day. To increase the efficiency of the thermal mass, air is blown through the cavities in the precast, hollow-core con-

Figure 17.4a The Emerald People's Utility District Headquarters (Courtesy of WEGROUP PC Architects and Planners, Solar Strategies by John Reynolds Equinox Design Inc.)

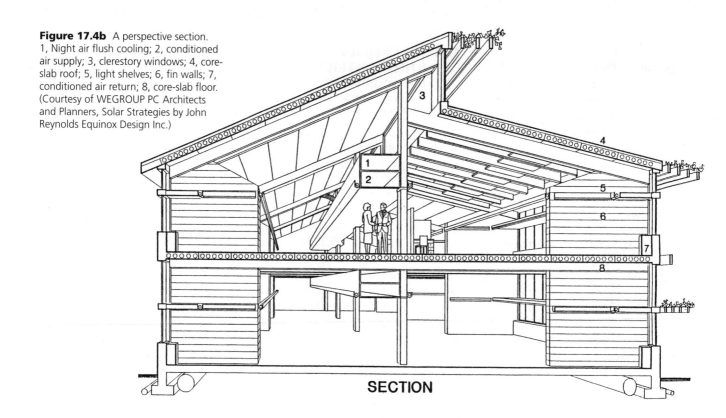

Figure 17.4b A perspective section. 1, Night air flush cooling; 2, conditioned air supply; 3, clerestory windows; 4, core-slab roof; 5, light shelves; 6, fin walls; 7, conditioned air return; 8, core-slab floor. (Courtesy of WEGROUP PC Architects and Planners, Solar Strategies by John Reynolds Equinox Design Inc.)

SECTION

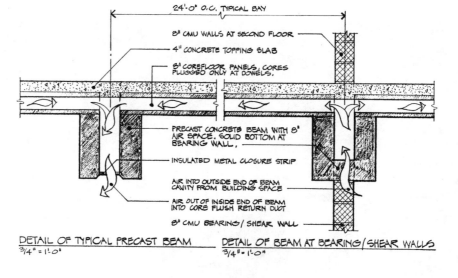

Figure 17.4c Details of air flow through precast, hollow-core concrete slabs. (Courtesy of WEGROUP PC Architects and Planners, Solar Strategies by John Reynolds Equinox Design Inc.)

24'-0" O.C. TYPICAL BAY

8" CMU WALLS AT SECOND FLOOR

4" CONCRETE TOPPING SLAB

8" COREFLOOR PANELS, CORES PLUGGED ONLY AT DOWELS.

PRECAST CONCRETE BEAM WITH 8" AIR SPACE. SOLID BOTTOM AT BEARING WALL.

INSULATED METAL CLOSURE STRIP

AIR INTO OUTSIDE END OF BEAM CAVITY FROM BUILDING SPACE

AIR OUT OF INSIDE END OF BEAM INTO CORE FLUSH RETURN DUCT

8" CMU BEARING / SHEAR WALL

DETAIL OF TYPICAL PRECAST BEAM
3/4" = 1'-0"

DETAIL OF BEAM AT BEARING / SHEAR WALLS
3/4" = 1'-0"

Figure 17.4d "T"-shaped windows place most of the glazing up high for deep daylight penetration. The light shelf reflects the sunlight to the ceiling. (From Solar 88: 13th National Passive Conference. Proceedings. © American Solar Energy Society, 1988.)

534

crete slabs (Fig. 17.4c). Thus, at night the slabs are thoroughly cooled with outdoor air, and during the day the slabs cool the circulating indoor air. Air is also blown through the hollow core slabs on winter mornings to extract the heat stored in the concrete.

Numerous techniques were used to bring quality daylight to all parts of the building. Windows are "T"-shaped so that most of the glazing is placed high (Fig. 17.4d). Both north and south windows are supplied with interior light shelves. The view glazing, which is placed lower on the walls, is shaded by the vine-covered trellises. Venetian blinds provide additional protection from low sun angles. Sunlight entering through the clerestory windows is reflected off the ceiling. Glare from the clerestory windows is avoided by a large air-conditioning duct and by vine-covered trellises (Fig. 17.4b).

The electric lighting is a task/ambient system. Indirect fluorescent fixtures supply a very diffused ambient light ideal for offices, especially when computers are used. The fixtures are arranged in rows parallel to the windows so that light sensors can pro-

gressively turn off rows as more daylight becomes available (Fig. 17.4b).

A variable-air-volume (VAV) system is supplemented by large exhaust fans for summer night flush cooling and by electric resistance heaters under the windows for auxiliary winter heating. An economizer cycle allows outside air to be used for cooling when the outdoor temperature is low enough. Windows that open also allow outdoor air to be used for cooling, especially during the spring and fall.

Because ceilings and walls are hard surfaces that reflect sound, floors are covered with carpets, and acoustical baffles hang from the ceiling. These baffles run perpendicular to the windows to avoid blocking the entering daylight.

The vine trellises on the south-facing windows deserve some extra comments (Fig. 17.4b). The vines add to both the "commodity and delight" of the building. They not only shade the windows but do it in phase with the thermal year. Much of the year, the vine also filters and softens the daylight. Because of the vines, the appearance of the building changes with the seasons. Even the view from the

interior changes radically from the dense green leaves of summer to the colorful leaves of autumn to the leafless vines of winter.

Finally, another special feature of this building was the creation of a *User's Manual.* The owner realized that this building would function properly only with the cooperation of the users. The architect was asked to create a manual to explain the building to the employees working there. Fig. 17.4d is one of the illustrations from that manual.

17.5 HOOD COLLEGE RESOURCE MANAGEMENT CENTER

Academic/residential building at Hood College, Frederick, Maryland

Architect: Burt Hill Kosar Rittelmann Assoc.

Area: 8000 square feet

Built: 1983

Hood College wanted a new building that not only related to the style of the neo-Georgian campus, but also was state-of-the-art, energy-conscious architecture (Fig. 17.5a). The building

Figure 17.5a The Hood College Resource Management Center. Note the solar collectors on the roof. Architect: Burt Hill Kosar Rittelmann Associates. (Courtesy of Burt Hill Kosar Rittelmann Associates.)

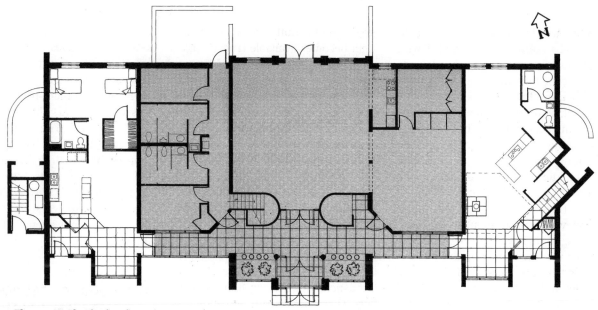

Figure 17.5b The first-floor plan. Academic areas are shaded. (Courtesy of Burt Hill Kosar Rittelmann Associates.)

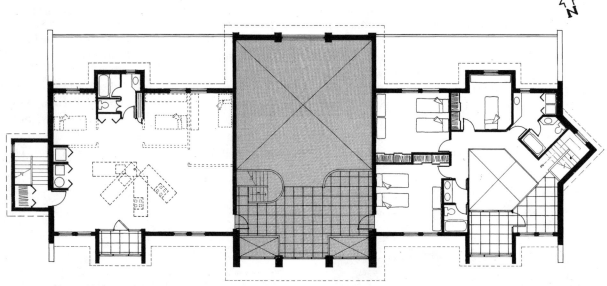

Figure 17.5c The second-floor plan. Academic areas are shaded. (Courtesy of Burt Hill Kosar Rittelmann Associates.)

would support an energy-management curriculum in several ways: by the design of the building itself, by means of academic facilities, and by means of an hands-on experience from living in the experimental residential units of the building.

Since the existing campus plan required the building to be oriented 17 degrees off south, other design features compensated as much as possi-

ble for the deviation from the ideal south orientation. Also, the proposed building location was shifted 20 feet after a site analysis indicated excessive shading from another building. The form is an elongated rectangle oriented approximately east–west. Thus, most windows could face south to capture the winter sun and be well shaded to keep out the summer sun. There are almost no east or west win-

dows and only a few on the north (Figs. 17.5b and 17.5c).

The building is well insulated with 2 x 6 stud walls and brick veneer to allow a wall-insulation level of R-17. Earth berms add extra protection on the east, west, and north sides. The roof is insulated to R-30. Infiltration is minimized with air-lock vestibules and by minimizing non-south-facing windows.

Support functions such as bathrooms, storage, and mechanical spaces are placed mainly along north, east, and west walls to act as buffers.

The extensive south glazing allows a significant amount of passive solar heating. On the first floor, the foyer is a sun space, and on the second floor each apartment also has a sun space. Heat is stored in a number of different areas and materials. Just inside the entrance, vertical, translucent water tubes store heat as well as diffuse the light. Interior brick piers, brick walls, and some concrete-block walls are exposed to the winter sun. Furthermore, all sun-space floor areas are covered with quarry tile for heat-storage purposes (Fig. 17.5b and c).

Cooling is largely accomplished by effective shading and natural ventilation. The basic shading comes from extra-long overhangs and wing walls (Fig. 17.5a and e). Additional shading for the first-floor south windows is provided by trellises covered with deciduous vines. All apartment sun spaces have movable awnings with internal manual controls, while the high windows in the entrance area are shaded by automatically operated internal sun screens. The long and slender form of the building with the mostly north and south windows ensures good cross-ventilation.

Daylighting was an important consideration, especially in the academic areas. High windows, light-colored surfaces, and open trusses were provided for deep light penetration and reduced glare (Fig. 17.5e). Glass doors and a glass balcony railing contribute to the daylighting of the commons area, while the display room and office "borrow" light from the foyer sun space by means of glazed partitions.

Although modern architecture was promoted as being functional, this case study illustrates that other styles can also produce buildings in which the heating, cooling, and lighting are largely accomplished by architectural design.

Figure 17.5d North-south section is at left, and section through sun space adjacent to main entrance is at right. (Courtesy of Burt Hill Kosar Rittleman Associates.)

WATER TUBES

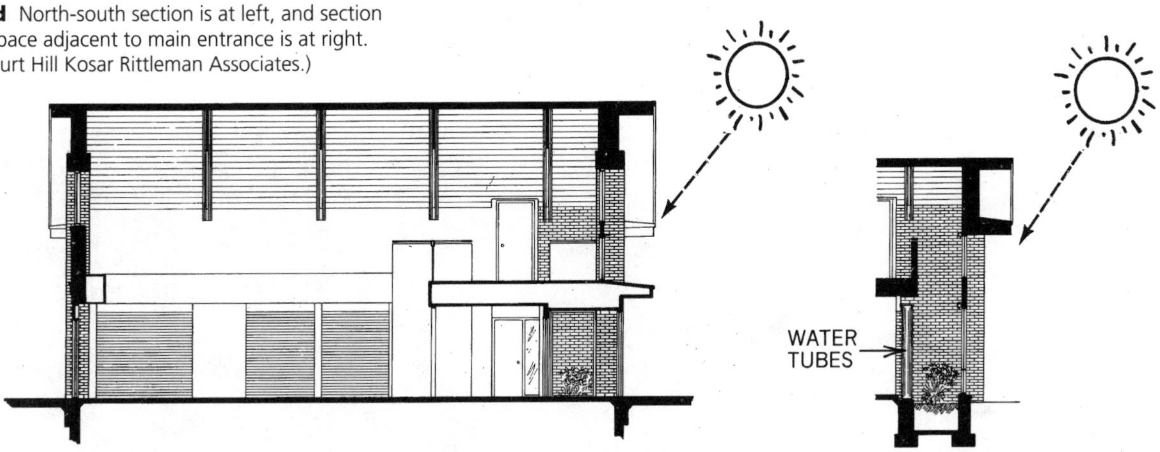

Figure 17.5e A north-south perspective section through the commons area. (Courtesy of Burt Hill Kosar Rittleman Associates.)

HIGH ARCHED WINDOW ON NORTH TO BALANCE LIGHT PENETRATION

HIGH CEILING, LIGHT COLOR FINISHES, OPEN TRUSSES ENHANCE DAYLIGHT

HIGH ARCHED WINDOW FOR DEEP LIGHT PENETRATION

GLASS RAILING FOR LIGHT PENETRATION

LOW NORTH GLAZING TO BALANCE LIGHT AT FLOOR LEVEL

CURVED WHITE SURFACES REDUCE CONTRAST GLARE AT ENTRANCE

TRANSLUCENT WATER TUBES FOR THERMAL STORAGE AND LIGHT DIFFUSION

17.6 COLORADO MOUNTAIN COLLEGE

Blake Avenue College Center, Glenwood Springs, Colorado

Architect: Peter Dobrovolny, A.I.A., of Sunup Ltd.

Area: 32,000 square feet

Built: 1981

This college and community building is located in Glenwood Springs, which lies in a sunny, arid region on the western slope of the Colorado Rockies (Fig. 17.6a). Winters are very cold with a great deal of snow. Warm summer days and cool nights generate high diurnal temperature swings.

Although the climate suggests that the main concern involves winter heating, the fairly large building is mainly an internally dominated type.

Thus, the cooling load from people and lights was more of a concern than originally expected. Through the improvement of the daylighting, the cooling load was reduced sufficiently so that conventional air conditioning was not required.

This three-story building has community-education facilities on the first two floors and offices on the third (Fig. 17.6b). The building has an atrium that acts not only as a central circulation- and meeting-space, but also as a source of solar heating and daylighting (Fig. 17.6c).

A combination of energy efficiency and solar energy furnishes most of the heating needs. The backup, electric-resistance baseboard heaters are needed mainly in the morning until the sun can make itself felt. By build-

ing into the side of a south-facing hill, much of the building is earth-sheltered. Floors, walls, and roofs are very well insulated (R-17, R-26, and R-34, respectively). In addition to the double or triple glazing, all windows are covered with movable insulation at night. All entries into the building are the air-lock vestibule type.

Much of the solar heating occurs via the central sunspace atrium (Fig. 17.6c and d). South-facing, stepped clerestory windows collect both the direct sun and the indirect sun off the highly reflective roof. The heat is then stored in the concrete floor, masonry walls, and the exposed-steel structure. At night, automatic insulating curtains cover the clerestory windows (Fig. 17.6e). The otherwise unheated atrium can be shut off from the conditioned

Figure 17.6a The new center for the Colorado Mountain College. Architect.- Peter Dobrovolny. (© Robert Benson, photographer.)

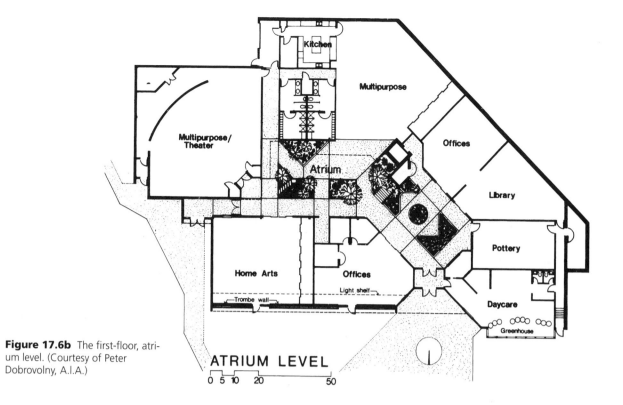

Figure 17.6b The first-floor, atrium level. (Courtesy of Peter Dobrovolny, A.I.A.)

Kitchen

Multipurpose

Multipurpose/ Theater

Offices

Atrium

Library

Home Arts

Offices

Pottery

Light shelf

Trombe wall

Daycare

Greenhouse

ATRIUM LEVEL

0 5 10   20   50

Figure 17.6c A view of the atrium and clerestory windows. (Robert Benson, photographer.)

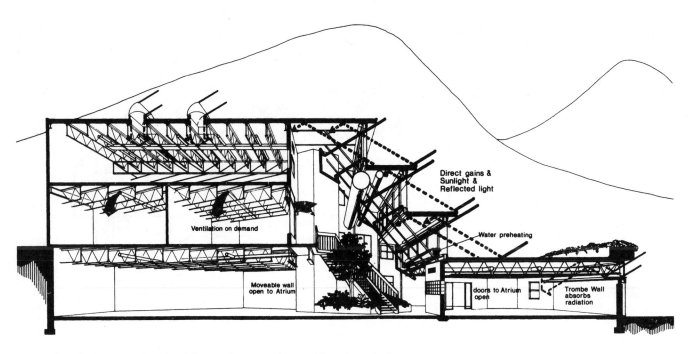

Figure 17.6d The winter-day solar collection. (Courtesy of Peter Dobrovolny, A.I.A.)

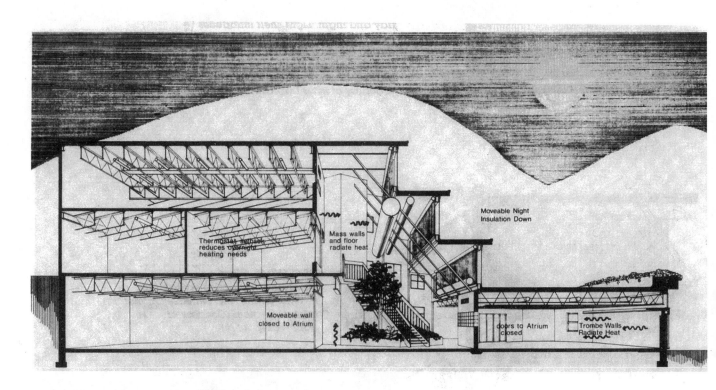

Figure 17.6e Winter night-heating and insulating closures. (Courtesy of Peter Dobrovolny, A.I.A.)

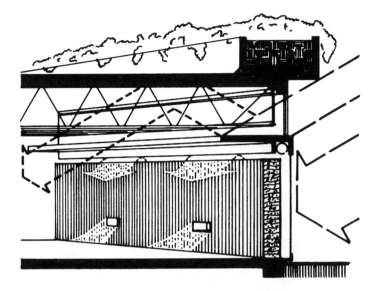

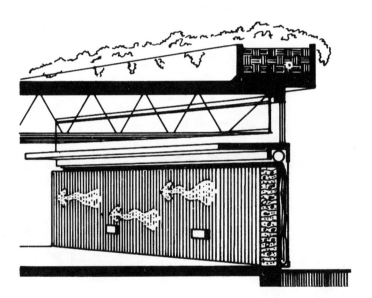

spaces by glass doors and movable walls. Antistratification ducts bring the warm air collecting near the ceiling of the atrium back down to the ground-floor level.

Solar Trombe walls make up most of the south facade that is not covered by windows. Automatic insulating curtains drop between the thermal wall and glazing to greatly reduce nighttime losses (Fig. 17.6f). The day-care center on the ground floor has a sun space filled with water tubes to provide the required thermal mass. A series of skylights introduces heat and light to those parts of the building not facing south walls. Movable reflectors on the skylight increase winter collection while also reducing summer overheating (Fig. 17.6g).

This college building (except computer room) is cooled in summer only by passive means (Fig. 17.6h). The cooling load is minimized by a number of different strategies: daylighting, low electric-light levels, a highly reflective, white roof surface, seasonal awnings over atrium windows, blinds

Figure 17.6f Perspective sections illustrating the action of the Trombe walls during a winter day and night. The light shelf introduces both light and direct-gain heating during the day. Note the insulating curtain between the thermal wall and the glazing during winter nights. (Courtesy of Peter Dobrovolny, A.I.A.)

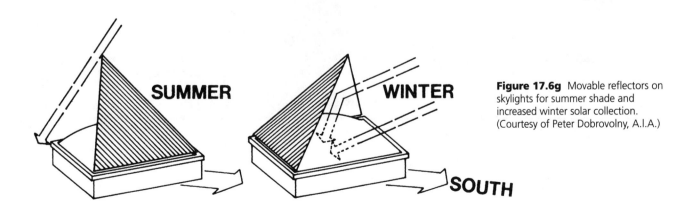

Figure 17.6g Movable reflectors on skylights for summer shade and increased winter solar collection. (Courtesy of Peter Dobrovolny, A.I.A.)

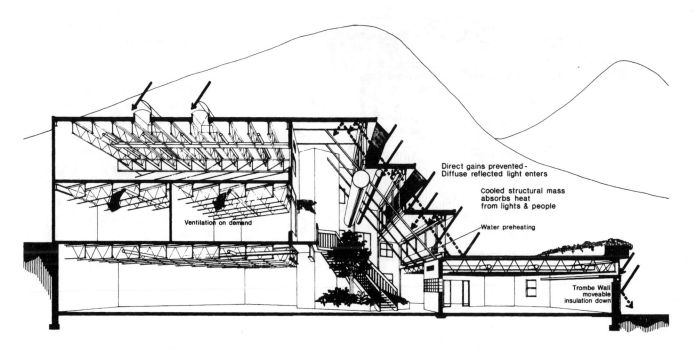

Figure 17.6h Summer-day heat rejection and passive cooling by the heat-sink action of the thermal mass. (Courtesy of Peter Dobrovolny, A.I.A.)

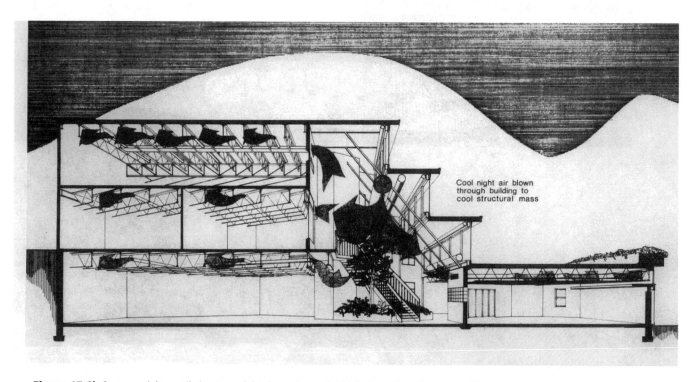

Figure 17.6i Summer-night ventilation to cool the thermal mass (night flush cooling). (Courtesy of Peter Dobrovolny, A.I.A.)

on west windows, reflectors over sky-lights, and movable reflective insulation over the Trombe walls.

The modest cooling load that remains is handled mostly by night flush cooling. Cool night air is brought into the building to cool the thermal mass to about 65°F (Fig. 17.6i). The mass acts as a heat sink to hold the indoor temperature below 78°F on most summer days. Fans and evaporative coolers come on as needed to handle any unusually high cooling loads.

To help minimize the cooling load, both daylighting and task/ambient lighting strategies were used. With the help of physical models, the building was designed so that most interior spaces receive their ambient light from natural sources. Additional task light is supplied as needed. The electric ambient lighting in all spaces can be set at two different levels: low for normal and high for special situations.

On south walls, light shelves introduce quality daylight. Automatic switches turn off the electric lighting when it is not necessary. The highly reflective surfaces on the stepped roofs over the atrium direct much of the light to the atrium ceiling. Adjacent spaces then "borrow" this light from the atrium. As mentioned before, the skylights, which have reversible reflectors, introduce controlled amounts of light.

Domestic hot water is preheated in a series of large tanks just inside the lower-atrium clerestory windows.

The energy consumption is only about one-fifth of that of a conventional building, and about 60 percent of that small amount comes from passive techniques.

17.7 GREGORY BATESON BUILDING

State office building in Sacramento, California
Architect: Office of the State Architect for California
Area: 267,000 square feet
Built: 1981

It is no accident that this office building was named after the noted anthropologist Gregory Bateson. Although the program called for a building that would set an example for energy-conscious design, it also called for a building that would demonstrate the more "humane" values in architecture. Thus, this building is inviting to the public and friendly to its users. The aesthetic is not monumental but informal. Many indentations, setbacks, and terraces break up the facade of this rather large office building (Fig. 17.7a).

Figure 17.7a Each facade of the Bateson Building is somewhat different because the solar impact is different on each orientation of a building. Note the horizontal louvers in a horizontal plane on the south façade and the apparent lack of shading on the west façade in this morning photograph. In the afternoon, roll-down exterior shades protect the west windows just as the east windows are protected in the morning (Fig. 17.7b). Architect: Office of the California State Architect. (Cathy Kelly, photographer.)

Figure 17.7b Compare the north and east facades of this photograph with the west and south facades in Fig. 17.7a. Note how the east windows are protected from the morning sun.

Figure 17.7c The automated fabric roller shades on the exterior of east and west windows are guided by vertical support cables.

In the Bateson Building, architectural design and energy features are very well integrated. The rich, warm, and articulated facades are largely consequences of the various shading elements, and the appearance of the facades varies because the shading needs of different orientations vary (Fig. 17.7b).

Because of its large size and because of the mild climate of Sacramento, the Bateson Building is an internally dominated building. Consequently, cooling rather than heating is the main concern, and daylighting is important in reducing the cooling load. As in any good design, the first priority is to reduce the cooling load from the sun. A trellis system protects the south windows in the summer while allowing the winter sun to enter. Because the building covers a whole city block, east and west windows could not be avoided. Instead,

544

movable, exterior roller shades block the sun on the east in the morning and in the west in the afternoon. These shades automatically glide up and down vertical exterior cables to give east and west windows a clear sunless view for half of each day (Fig. 17.7c).

The atrium, which creates a conceptually clear circulation core and plaza for the workers and the public, also brings daylight to interior offices (Fig. 17.7d).

The atrium-roof glazing is carefully designed to prevent unwanted sunlight from entering. Some of the glazing slopes to the north to capture year-round diffused skylight (Fig. 17.7e). The rest of the roof glazing faces south to capture the winter sun (Fig. 17.7f). In the summer, this south glazing is protected by movable vertical louvers (Fig. 17.7g).

Although the atrium is not air-conditioned, it does act as a buffer space. Thus, much of the building is not exposed to hot or cold outdoor temperatures. The atrium also brings daylight to some of the interior spaces, thereby reducing the cooling load due to lighting (Fig. 17.7h).

Most of the cooling load that remains after the above-mentioned heat-avoidance techniques are em-ployed is handled by night flush cooling. Because Sacramento has a large diurnal temperature range 25°F to 30°F), cool night air is available to flush out the heat from the thermal mass of the exposed-concrete frame building. There are also two 700-ton rock beds under the atrium floor for additional thermal mass. During a summer day, this combined mass of rock beds and building structure can absorb enough heat to accomplish more than 90 percent of the building's cooling needs. The remaining cooling load is served by a conventional chilled-water variable-air-volume system. This air system also circulates

Figure 17.7d A large atrium with south-facing clerestories and north-facing skylights brings light to interior offices in the Bateson Building. The prominent stairs invite people to walk rather than use the elevators. Antistratification tubes hang from the atrium roof. (Courtesy of the Office of the California State Architect.)

Figure 17.7e The steeply sloped clerestories face south (toward the right in the photograph) and the sloped skylights face north. Both the clerestories and skylights have exterior shading devices for solar control. The circles on the atrium roof gables are exhaust fans for night flush cooling. (Courtesy of the Office of the California State Architect.)

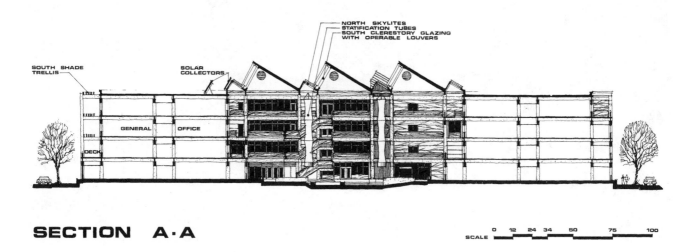

SECTION A·A

Figure 17.7f The north–south section illustrating the sloped south-facing clerestories that capture the winter sun. (Courtesy of the Office of the California State Architect.)

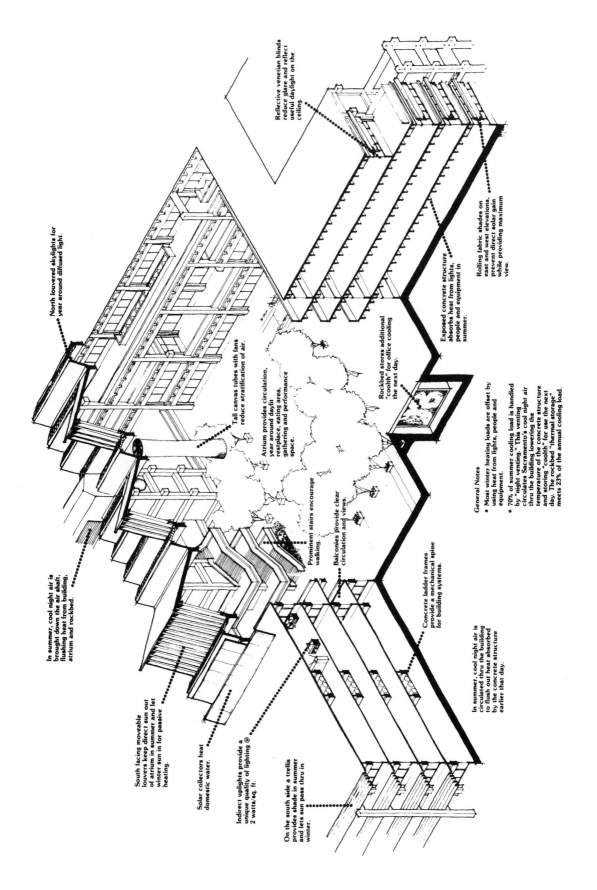

Reflective venetian blinds reduce glare and reflect useful daylight on the ceiling.

Rolling fabric shades on east and west elevations, prevent direct solar gain while providing maximum view.

Exposed concrete structure absorbs heat from lights, people and equipment in summer.

Rockbed stores additional "coolth" for office cooling the next day.

North louvered skylights for year around diffused light.

Tall canvas tubes with fans reduce stratification of air.

Atrium provides circulation, year around daylit restplace, eating area, gathering and performance space.

Prominent stairs encourage walking.

Balconies provide clear circulation and views.

Concrete ladder frames provide a mechanical spine for building systems.

General Notes

• Most winter heating loads are offset by using heat from lights, people and equipment.

• 70% of summer cooling load is handled by "night venting." This venting circulates Sacramento's cool night air thru the building lowering the temperature of the concrete structure and storing "coolth" for use the next day. The rockbed "thermal storage" meets 23% of the annual cooling load.

In summer, cool night air is brought down the air shaft, flushing heat from building, atrium and rockbed.

South facing moveable louvers keep direct sun out of atrium in summer and let winter sun in for passive heating.

Solar collectors heat domestic water.

Indirect uplights provide a unique quality of lighting @ 2 watts/sq. ft.

On the south side a trellis provides shade in summer and lets sun pass thru in winter.

In summer, cool night air is circulated thru the building to flush out heat absorbed by the concrete structure earlier that day.

Figure 17.7g An isometric of the Bateson Building. (Courtesy of the Office of the California State Architect.)

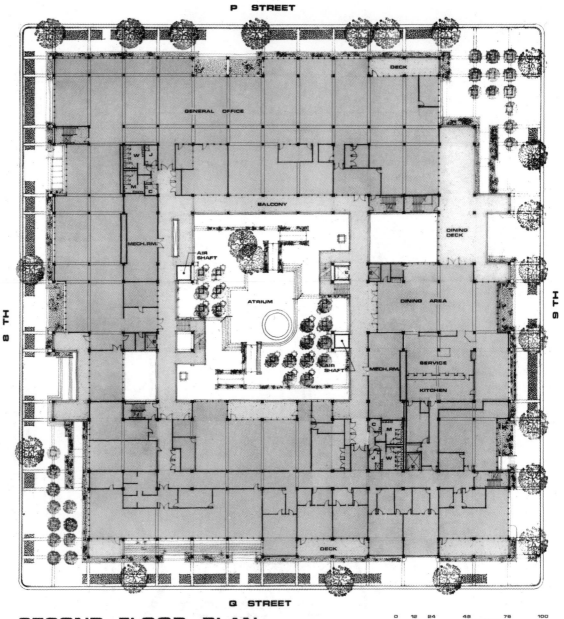

P STREET

8 TH

9 TH

Q STREET

SECOND FLOOR PLAN
ATRIUM AND STREET LEVEL BELOW

SCALE 0 12 24 48 76 100

Figure 17.7h
The second-floor plan. (Courtesy of the Office of the California State Architect.)

the cool outdoor night air throughout the building, bringing it into close contact with the concrete structure to more fully recharge the "thermal batteries." Large exhaust fans are placed at the highest points in the atrium to exhaust the hottest air first (Fig. 17.7e).

The heating load in winter is small because of the climate, the small surface-area-to-volume ratio, and the heat produced by people, lights, and equipment. Most of this small heating

load is handled by the solar energy received through the south windows and the south-facing atrium clerestories. Since the hot air will rise to the top of the atrium (good in summer but bad in winter), antistratification tubes are suspended from the ceiling of the atrium (Fig. 17.7d). These colorful-fabric tubes have fans in their bottom ends that in the winter pull down the warm air collecting near the atrium ceiling. Also, in winter the rock beds can be used to store excess

afternoon heat for use early the following morning. Auxiliary heating comes from perimeter hot-water reheat coils.

By having an atrium, no point in the building is more than 40 feet from a natural-light source. The atrium receives diffused skylight from the north-facing skylights and direct sunlight from the south-facing clerestories. Banner-type screens are lowered in the winter to prevent glare and excessive brightness ratios from the direct sunlight entering the atrium. In the summer, the vertical louvers on the outside of the south-facing clerestories keep out the direct sun but allow a small amount of diffused light to enter. Windows are shielded from sun and glare by means of reflective Venetian blinds.

Task/ambient lighting provides an efficient glare-free visual environment. Indirect fluorescent fixtures provide the soft ambient lighting, and each workstation has locally controlled task lights.

The columns, beams, and floor slabs were left exposed so that they could function as thermal mass. However, the exposed concrete also creates acoustical problems. Carpeting on the floor and vertical white acoustical baffles hanging from the ceiling absorb excess office noise.

Domestic hot water is generated by 2000 square feet of solar collectors located on the roof just south of the clerestories (Fig. 17.7f). The Bateson Building not only uses 70 percent less energy than a conventional building, but much of the electrical power is used during off-peak hours when electricity is plentiful. Much of this efficiency is maintained by a computer-operated energy-management system that senses conditions throughout the building and then decides on the best mode of operation.

The Bateson building is a valuable prototype because it addresses and successfully resolves many of the important issues in architecture. It is a worthy destination for any pilgrim seeking great architecture, especially since it was built at almost the same cost as a conventional building.

17.8 HONGKONG BANK

Hongkong Bank Headquarters, Hong Kong

Architects: Foster Associates

Area: 1,067,000 square feet gross, 47 floors above grade and 4 floors below grade

Built: 1985

For more information, see the following periodicals, each of which devoted an entire issue to this building:

Architectural Review, April, 1986.
Progressive Architecture, March, 1986.

The Hongkong Bank by Norman Foster and Associates is one of the key buildings of the twentieth century (Fig. 17.8a). The building is very striking and attractive, as well as un-usually creative and innovative. Although considered a high-tech building, the innovations are not all technical in nature. The Hongkong Bank is a product of Foster's philosophy, which he articulated when he said that the building "is not arbitrary, nor is it mechanistic. It has logic and it has the social idealism that was—in our terms—manifest in our approach from the outset."

The basic organizational concept is a stacking of separate "villages." The sky lobby of each "village" or bank subdivision is reached by express elevators. Vertical circulation within each "village" is possible only by escalators, which connect the floors both physically and psychologically much better than elevators.

Figure 17.8a The Hongkong Bank Headquarters Building, Hong Kong. Architect: Foster Associates, 1985. (Courtesy Foster Associates. Ian Lambot, photographer. ©)

Both the exterior and interior appearances of the building are largely a result of Foster's desire to "demystify" technology. To enable everyone to understand how the building works, the structure, the circulation system, and much of the mechanical equipment are exposed to view. The mechanisms of the escalators, elevators, and even the automatic doors are enclosed in glass.

The overall form of the building was largely governed by the local building code, which required setbacks to limit the shading of neighboring buildings.

In order to maximize the clear views of the beautiful harbor and surrounding mountains, the building was designed to have almost no columns, open office areas, and floor-to-ceiling windows (Fig. 17.8b). Unlike most other all-glass office towers, the Hongkong Bank has exterior shading devices over all windows (Fig. 17.8c). These horizontal louvered overhangs block both the direct sun and the glare from the bright sky. These sunscreens also act as catwalks for window washing and other maintenance activities. Since much of the solar radiation in this humid climate comes from the hazy sky, there are also miniature Venetian blinds on all windows. To introduce as much daylight as possible into the office areas, the ceiling height is increased around the perimeter of the building (Fig. 17.8g).

To beam daylight into the completely interior atrium (Fig. 17.8d), a sun scoop consisting of a sophisticated system of mirrors is employed (Figs. 17.8e and f). A one-axis tracking mirror on the outside of the building always reflects the direct sun's rays horizontally into the building (Fig. 17.8c). Fixed convex mirrors at the atrium ceiling then reflect the sunlight down into the large interior space.

Daylighting of the public plaza located at street level under the

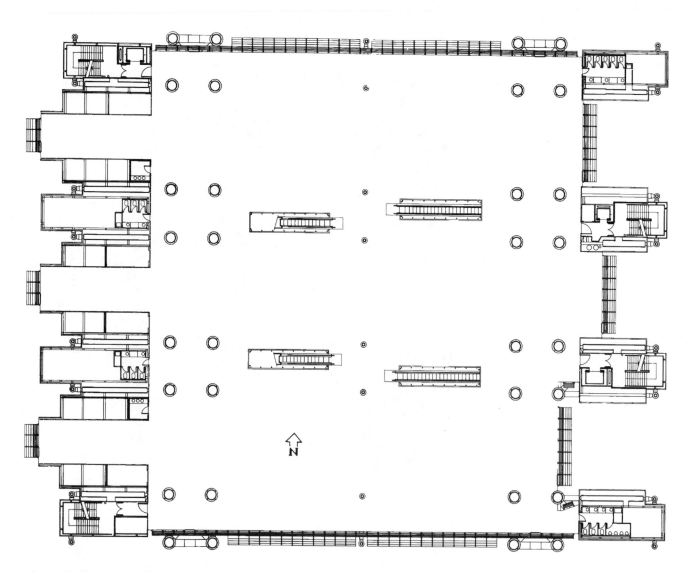

Figure 17.8b A typical office floor. (Courtesy Foster Associates.)

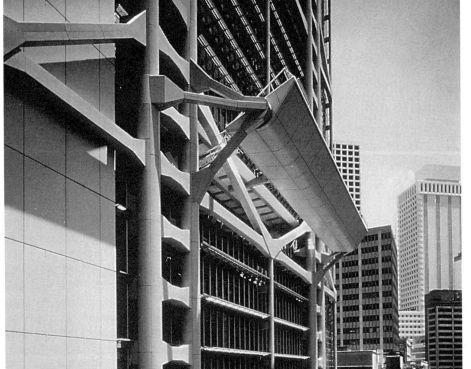

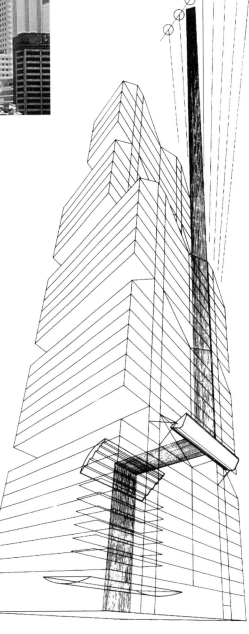

Figure 17.8c Each window is protected by exterior shading devices. A sun scoop beams sunlight into the atrium. (Courtesy Foster Associates. Ian Lambot, photographer.©)

Figure 17.8d The nine-story central atrium. Note the glass underbelly that allows daylight to penetrate to the public plaza below. (Courtesy Foster Associates.)

Figure 17.8e A perspective view illustrating the beamed sunlighting for the atrium and the public plaza below the atrium. (Courtesy Foster Associates.)

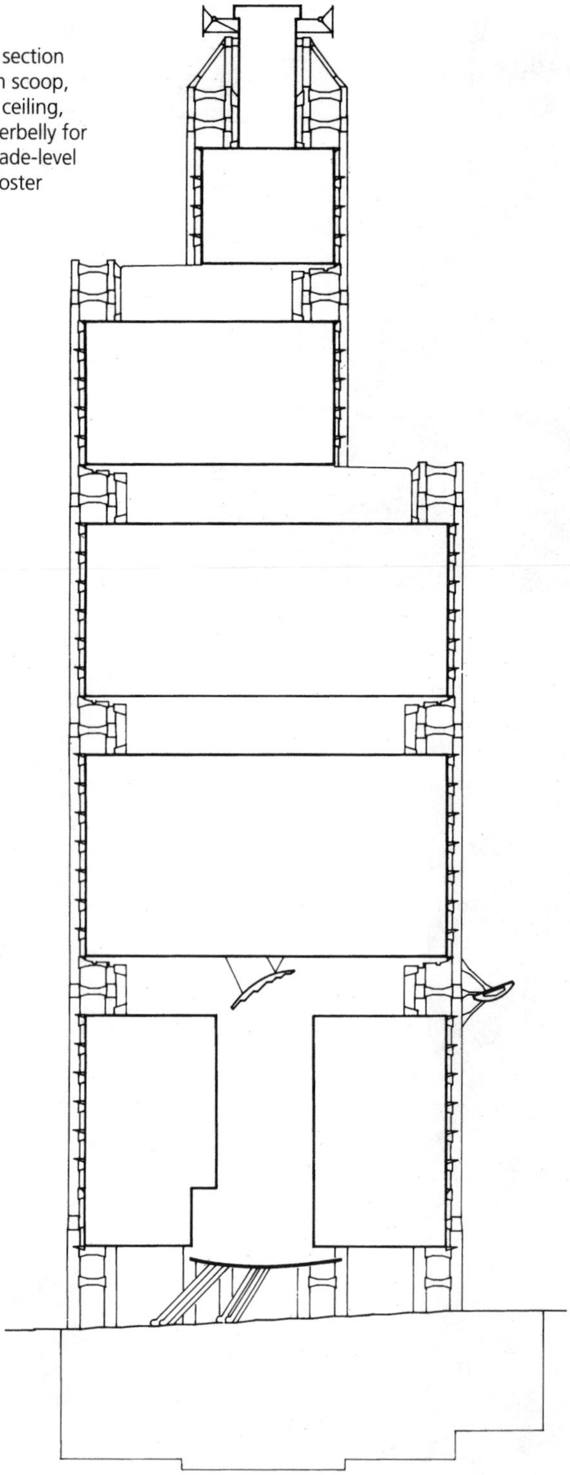

Figure 17.8f A section illustrating the sun scoop, mirrors on atrium ceiling, and the glass underbelly for daylighting the grade-level plaza. (Courtesy Foster Associates.)

strategies are employed to minimize the cooling load. Most of the glazing is located on the north and south facades. The smaller amount of east and west glazing is somewhat protected by the projecting mechanical and stair towers, which act like giant vertical fins (note north arrow on plan of Fig. 17.8b). As mentioned above, all glazing is further protected by external and internal sunscreens (see lower left of Fig. 17.8g).

The cooling load from the electric lighting is minimized by several techniques: some daylighting is used, task/ambient lighting is used, and much of the heat from the fluorescent lamps is vented outside by exhausting air through the lighting fixtures.

Sea water is used as a heat sink for the central chillers located in the basement. Chilled water is then brought to air handlers located in mechanical modules on each floor. These modules were prefabricated in Japan and shipped to Hong Kong. To maximize off-site labor, as much of the mechanical equipment as possible and all toilet facilities were included in the modules. These modules were then stacked and connected to riser shafts.

From the modules, air is circulated under the raised floor and supplied through circular floor diffusers (Fig. 17.8g). Both air and the electrical outlets can be moved for maximum flexibility. Most room air is returned through linear floor grilles to the void below the raised floor, which acts as a return plenum. A small amount of air is exhausted through the ceiling lighting fixtures. Heat gain through the glazing is counteracted by a continuous perimeter supply grille.

The use of the modules and the individual floor diffusers reflects Foster's belief in the concept of decentralization. He hopes that one result of this approach will be individuals' greater understanding and control of their environment.

A building of such complexity and with so much innovation was possible

building was also desired. Consequently, much of the base of the atrium is designed as a glass "underbelly" (Fig. 17.8f) rather than an opaque floor. To maximize its transparency, this hanging "canopy" con- sists of glass sheets supported by a catenary tension structure (see lowermost portion of Fig. 17.8d).

As an internally dominated building in a hot and humid climate, heating is not a concern, but many

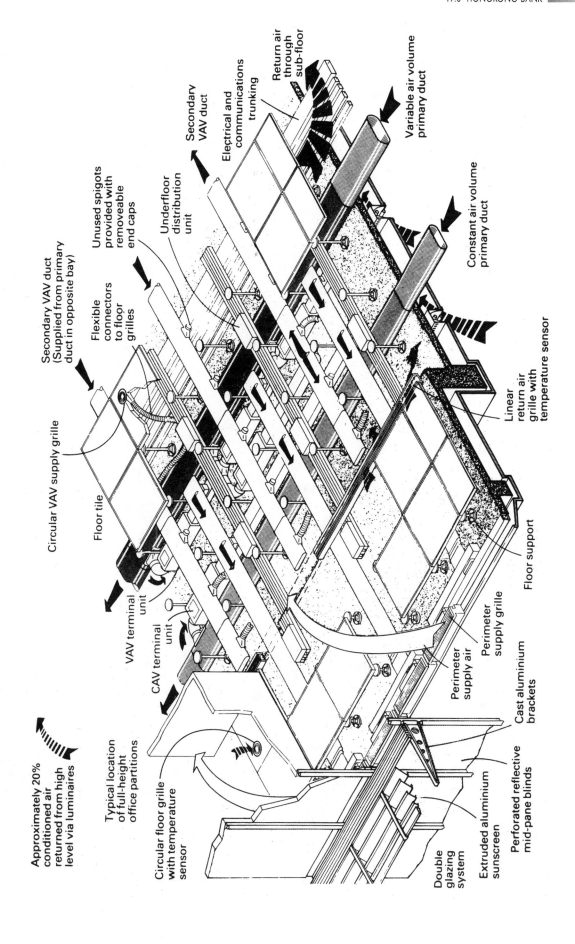

Figure 17.8g An isometric of the modular subfloor-services layout. (Courtesy Foster Associates.)

only by an unusual team approach that included not only the architects, engineers, and contractor, but also manufacturers. Many new products and systems were developed by this successful partnership. The truly international cooperation that created this building may be the greatest contribution of all for the Hongkong Bank Headquarters.

17.9 COMMERZBANK

The Commerzbank Headquarters Building in Frankfurt, Germany

Architects: Foster and Partners, London, United Kingdom

Area: 62 stories, typical floor area is about 8,530 square feet (776 square meters)

Built: 1997

Frankfurt is at about 50°N latitude in a generally cold but still temperate climate, similar to that of Boston. Even though the primary concern is winter heating, buildings also need protection from the summer sun.

One of the major goals for the Commerzbank was to reduce the need for fossil fuels by utilizing natural energies as much as possible. Also, it was important that the building users have access to daylight, views, plants, and fresh air.

The design by Foster and Partners resolved these issues by using sky gardens that open around a central atrium forty-eight stories high (Fig. 17.9a). Because of the stack effect, a horizontal glass diaphragm separates the atrium into twelve-story sections, each of which is surrounded by three sky gardens in a spiral arrangement (Fig. 17.9b). Thus, each twelve-story section has one south-, one west-, and one east-facing sky garden, all open to the central atrium. The sky gardens facing west contain American plants, those facing east contain Asian plants, and those facing south contain Mediterranean plants. Each sky garden is an enclosed sun space because of the four-story-high, dou-

Figure 17.9a The Commerzbank Headquarters Building in Frankfurt, Germany, designed by Foster and Partners. (Courtesy Foster and Partners; Ian Lambot, photgrapher.)

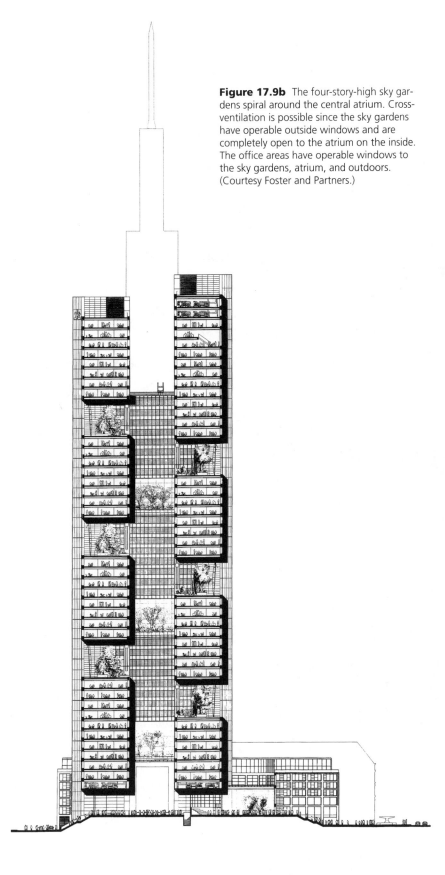

Figure 17.9b The four-story-high sky gardens spiral around the central atrium. Cross-ventilation is possible since the sky gardens have operable outside windows and are completely open to the atrium on the inside. The office areas have operable windows to the sky gardens, atrium, and outdoors. (Courtesy Foster and Partners.)

ble-glazed curtain wall that separates it from the outdoor terrace. To reduce the number of confusing reflections, the curtain wall is canted outward (Fig. 17.19b). The sky gardens feel like outdoor public spaces and are used the same way: for meeting, relaxing, and lunch and coffee breaks.

The corner towers are for wind resistance, vertical support, vertical transportation, and services (Fig. 17.9c). Large riser ducts are not needed because of the extensive natural ventilation available via operable windows to the outdoors, atrium, and sky gardens. The top and a few bottom panels of the sky-garden curtain walls are also operable (Fig. 17.9d). Effective cross-ventilation is possible because all three sky gardens in each twelve-story section are open to each other via the atrium (Fig. 17.9e).

The outdoor windows of the office areas are part of the **climate facade** designed to prevent rain and high winds from entering open windows even on the highest floors (Fig. 17.9f). The climate facade consists of: an outer, fixed glass pane; a 7-inch air space; and an inner, double-glazed, low-e movable sash that tilts in on top. Ventilation slots above and below the outer fixed glass allow controlled ventilation (Fig. 17.9g). Because of the aerodynamic profile of these openings, no air noise is generated. The operable windows allow for a high level of indoor-air quality and, for the occupants, a degree of personal control over their environment.

Heating is achieved with conventional hot-water convectors under the windows, while cooling is achieved by chilled water passing through metal coils above the perforated-metal ceiling panels (Fig. 17.9h). Since the water systems control most of the heat gain and heat loss, the mechanical system supplies only fresh air, and the ducts therefore, are much smaller than those in a conventional office building. This mechanical ventilation system is needed only when the windows cannot be used: on very hot or

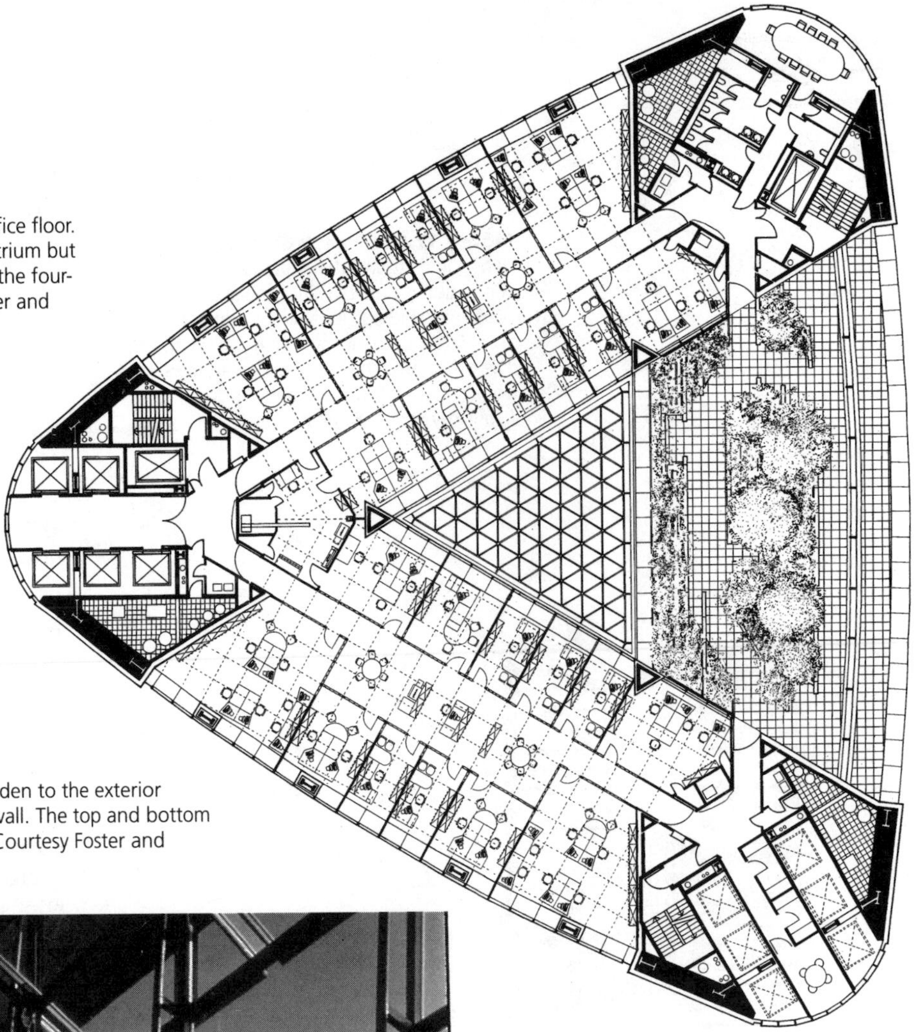

Figure 17.9c The plan of a typical office floor. The sky garden is open to the central atrium but separated from the outdoor terrace by the four-story, glass-curtain wall. (Courtesy Foster and Partners.)

Figure 17.9d The view from sky garden to the exterior through the four-story, glass- curtain wall. The top and bottom panels open for summer ventilation. (Courtesy Foster and Partners; Ian Lambot, photographer.)

very cold days, during storms, or when the outdoor air pollution is very high.

Solar control is achieved via the Venetian blinds located in the 7-inch cavity of the climate facade (Fig. 17.9f). Although the blinds can be controlled locally by the occupants, there is a central, electronic **building management system (BMS)** that can override manual control when necessary. During the summer, the blinds are angled for solar rejection, while in the winter, maximum sunlight is collected and then reflected to the ceiling. Besides the blinds, the BMS also monitors and controls the electric lighting and air-conditioning systems. The BMS also turns on special indicator lights to let occupants know when windows can be opened.

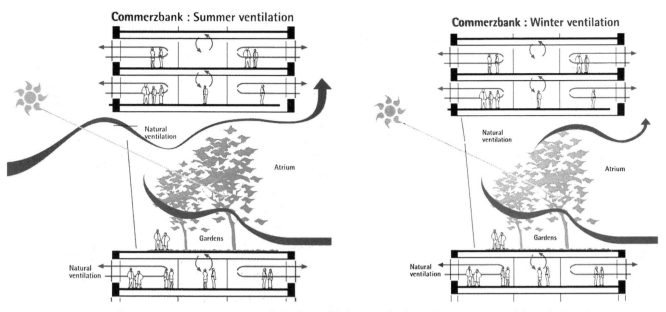

Figure 17.9e Summer cross-ventilation is possible because the sky gardens are connected through the atrium. (Courtesy Foster and Partners.)

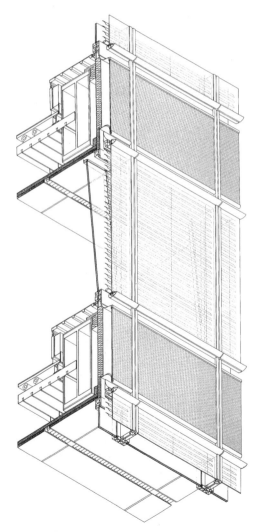

Figure 17.9f The climate façade controls both ventilation and sunlight. (Courtesy Foster and Partners.)

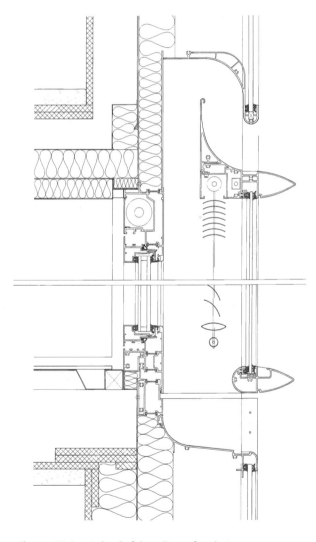

Figure 17.9g A detail of the "climate façade." (Courtesy Foster and Partners.)

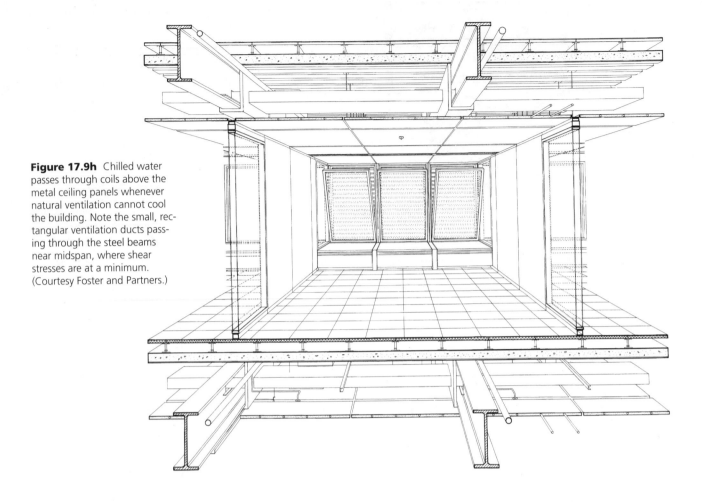

Figure 17.9h Chilled water passes through coils above the metal ceiling panels whenever natural ventilation cannot cool the building. Note the small, rectangular ventilation ducts passing through the steel beams near midspan, where shear stresses are at a minimum. (Courtesy Foster and Partners.)

The Commerzbank was the most "green" skyscraper when it was built in 1997 for more reasons than bringing gardens to the occupants. The building's management estimates that the energy use will be 25 percent to 30 percent less than the already strict German norms, mainly because of the reduced air-conditioning loads.

17.10 PHOENIX CENTRAL LIBRARY

A new central library for Phoenix, Arizona

Architects: William P. Bruder and DWL Architects & Planners

Area: 280,000 square feet

Date: 1995

Arizona's Monument Valley, with its magnificent brown and red mesas, appears to be William P. Bruder's inspiration for the form and color of the Phoenix Central Library. His rational, flexible plan should serve the library well in the rapidly changing future.

Because the climate of Phoenix consists of exceedingly hot summers with temperatures higher than 110°F and very mild winters, heat rejection, shading, and daylighting are the primary passive strategies for this large, internally dominated library. Appropriately, the Phoenix Central Library has no east or west windows, and the all-glass north and south facades are carefully shaded.

The south facade is covered with movable, computer-controlled louvers to regulate both the quantity and quality of the daylighting (Fig. 17.10a). The outdoor louvers are about 3 feet from the glazing to prevent hot air being trapped next to the facade.

Because of the hot climate, even the north glazing is fully shaded with sail-like fins made of a Teflon-coated, perforated acrylic fabric (Fig. 17.10b). The deep but widely spaced fins allow great views of Phoenix and the surrounding mountains (Fig. 17.10c).

The east and west corrugated, copper-covered facades, known as saddlebags, act as buffers against the intense summer sun (see again Fig. 17.10a). They contain the mechanical services, the wind-bracing structure, the fire stairs, and the restrooms (Fig. 17.10d). These service areas are separated from the library spaces by a 12-inch concrete wall whose eight-hour time lag will significantly reduce heat gain. Although the copper facade effectively blocks the sun, tiny perforations let a small amount of daylight through, and at night, the perforated copper facade looks like a veil.

Figure 17.10a The Phoenix Central Library as viewed from the southeast. Computer-operated exterior louvers cover the south façade. Copper-clad saddlebags shield the east and west walls from the intense sun. (Carl Walden Reiman, photographer.)

Figure 17.10c The sail-fins as seen from inside the north facade. (Courtesy william p. bruder, architects, ltd.)

Figure 17.10b The north facade is shaded by the fabric sail-fins. (Courtesy william p. bruder, architects, ltd.)

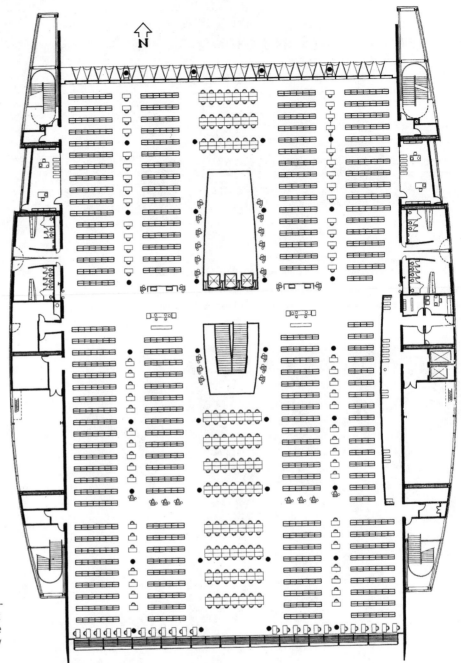

Figure 17.10d The Fifth-Floor Plan. The large, "empty" spaces in each saddlebag are the mechanical-equipment rooms (MERs). All windows are on the north (up) or south facades. (Courtesy william p. bruder, architects, ltd.)

The fifth floor, which is the top level, houses the nonfiction collection and a grand reading room. Directly above the tall, tapered concrete columns are 6 foot diameter skylights right where you would expect the roof structure to rest on the columns (Fig. 17.10e). This structural feat was made possible only because of the tensegrity structural system designed by the very creative engineering firm of Ove Arup & Partners. Cables from the top of the

columns support struts, which in turn support the slightly bowed roof.

The central five-story-high atrium is daylit by nine circular skylights 6 feet in diameter (Fig. 17.10f). Computer-operated specular louvers on top of the skylights project light deep into this "crystal canyon" that contains the grand staircase and glass-enclosed elevators. The highlighted atrium becomes an orientation focus for visitors to the library.

The mechanical-equipment rooms (MER) on each side of the building supply conditioned air to the main supply ducts running north–south just inside the 12-inch concrete wall that separates the library from the "saddlebag" service areas (Fig. 17.10e). These large ducts are in dropped ceilings in order to clear the T-beams that span east–west (Fig. 17.10g). The supply ducts feed the branch ducts above them, which are

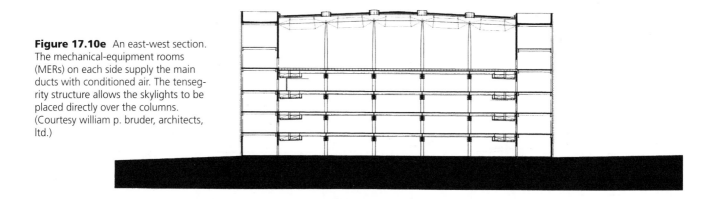

Figure 17.10e An east-west section. The mechanical-equipment rooms (MERs) on each side supply the main ducts with conditioned air. The tensegrity structure allows the skylights to be placed directly over the columns. (Courtesy william p. bruder, architects, ltd.)

Figure 17.10f A north-south section. Nine skylights are located directly over the central circulation spine. Concrete T-beams function as the branch ducts. (Courtesy william p. bruder, architects, ltd.)

Figure 17.10g An east-west-perspective section. The main north-south supply and return ducts are dropped below the east-west-spanning T-beams. Removable panels allow easy access to power and communication cables. (Courtesy william p. bruder, architects, ltd.)

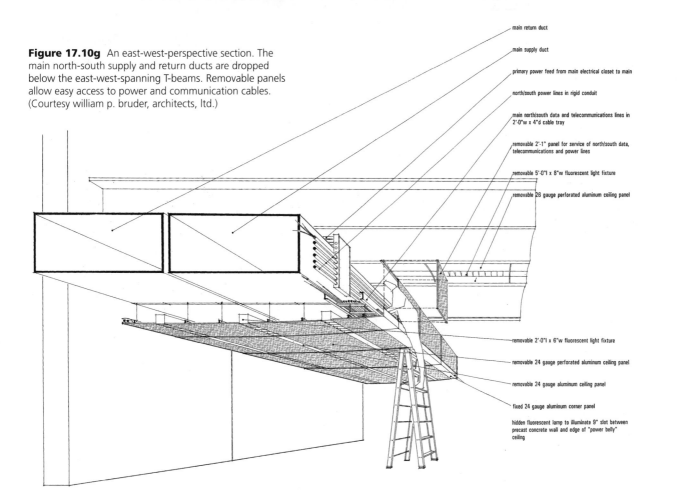

main return duct

main supply duct

primary power feed from main electrical closet to main

north/south power lines in rigid conduit

main north/south data and telecommunications lines in 2'-0"w x 4"d cable tray

removable 2'-1" panel for service of north/south data, telecommunications and power lines

removable 5'-0"l x 8"w fluorescent light fixture

removable 26 gauge perforated aluminum ceiling panel

removable 2'-0"l x 6"w fluorescent light fixture

removable 24 gauge perforated aluminum ceiling panel

removable 24 gauge aluminum ceiling panel

fixed 24 gauge aluminum corner panel

hidden fluorescent lamp to illuminate 9" slot between precast concrete wall and edge of "power belly" ceiling

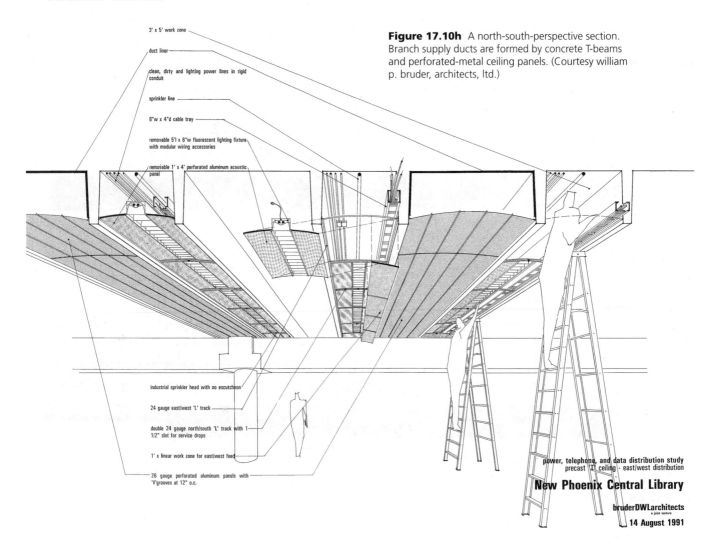

3' x 5' work zone

duct liner

clean, dirty and lighting power lines in rigid conduit

sprinkler line

6"w x 4"d cable tray

removable 5'l x 8"w fluorescent lighting fixture with modular wiring accessories

removable 1' x 4' perforated aluminum acoustic panel

industrial sprinkler head with no escutcheon

24 gauge east/west 'L' track

double 24 gauge north/south 'L' track with 1 1/2" slot for service drops

1' x linear work zone for east/west feed

26 gauge perforated aluminum panels with 'V'grooves at 12" o.c.

power, telephone, and data distribution study
precast 'T' ceiling · east/west distribution

New Phoenix Central Library

bruderDWLarchitects
a joint venture

14 August 1991

Figure 17.10h A north-south-perspective section. Branch supply ducts are formed by concrete T-beams and perforated-metal ceiling panels. (Courtesy william p. bruder, architects, ltd.)

formed by the east–west spanning T-beams (Fig. 17.10h). Perforated-metal panels cover the concrete channels as well as diffuse the air. The power and communication cables follow the same distribution logic. Only the top level has a raised floor for the distribution of air, power, and communication because the ceiling is too high.

William P. Bruder designed the Phoenix Library as a simple rectangular solid in order to save on initial costs and operating costs. The compact design will forever minimize the surface area exposed to the hot Arizona sun and air. Instead of using complex masses, Bruder's design achieves aesthetic interest through the careful choice of color, texture, material, and shading systems.

Resources

FURTHER READING
(See Bibliography in back of book for full citations.)

Davies, C., and I. Lambot.
   Commerzbank Frankfurt: Prototype for an Ecological High-Rise.
"Desert Illumination." *Architecture,* October, 1995; pp. 56–65.
"High Heat, High Tech." *Architecture,* October, 1995; pp. 107–113.
Pepchinski, M. "Commerzbank." *Architectural Record,* 01.98

HORIZONTAL SUN-PATH DIAGRAMS

See Section 6.11 for a discussion of these horizontal sun-path diagrams. Additional sun angle information can be found in Appendix B, *Architectural Graphic Standards* by Ramsey and Sleeper, John Wiley & Sons, Inc., and *The Handbook of Fundamentals* by ASHRAE. All editions of these books cover sun angles.

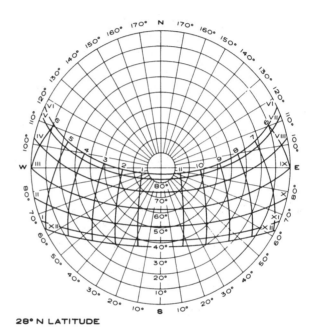

28° N LATITUDE

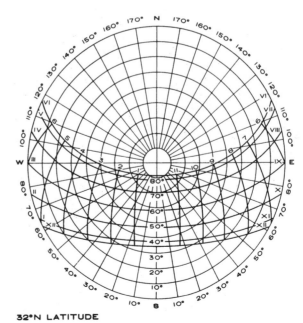

32°N LATITUDE

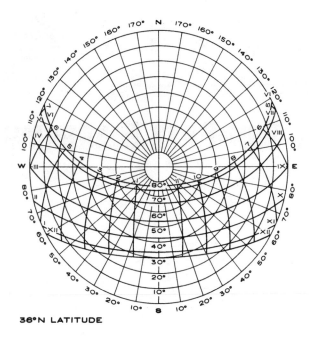

36°N LATITUDE

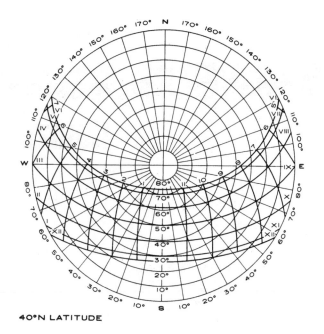

40°N LATITUDE

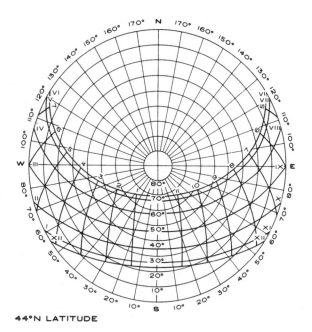

44°N LATITUDE

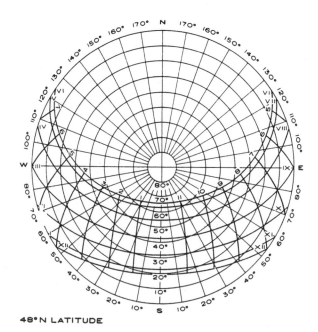

48°N LATITUDE

VERTICAL SUN-PATH DIAGRAMS

See Section 6.12 for a discussion of these vertical sun-path diagrams. These diagrams have been reprinted by permission from *The Passive Solar Energy Book* by Edward Mazria, published by Rodale Press, Emmaus, PA, 1979. A complete set of similar vertical sun paths can be found in *Sun Angles for Design* by Robert Bennet. Additional sun angle diagrams can be found in Appendix A.

Sun Charts

The following sun charts are for all latitudes from 28° to 56°NL at 4° intervals.

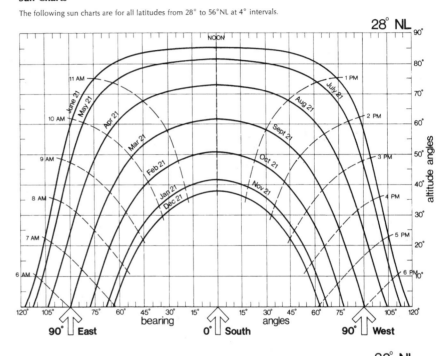

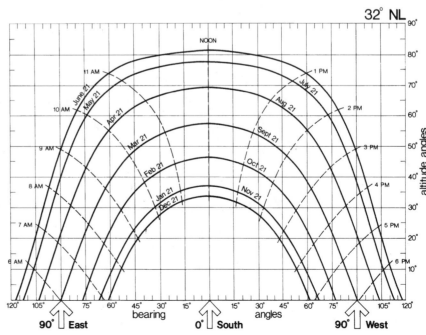

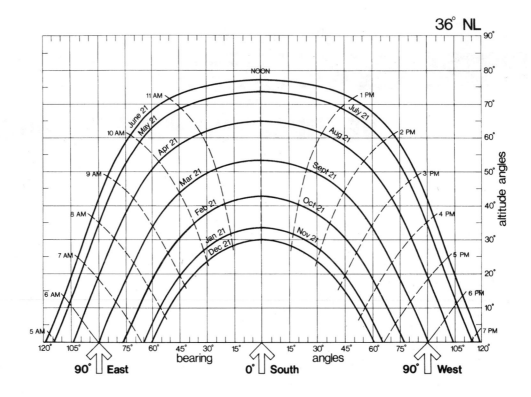

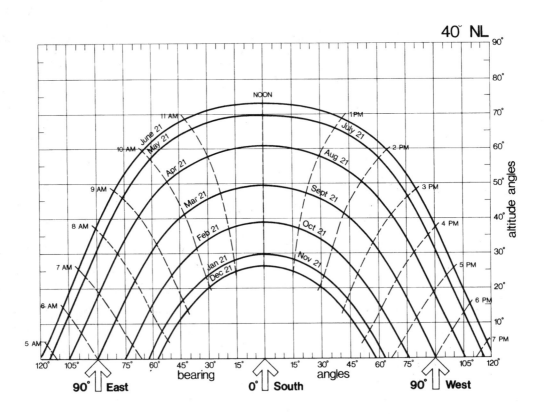

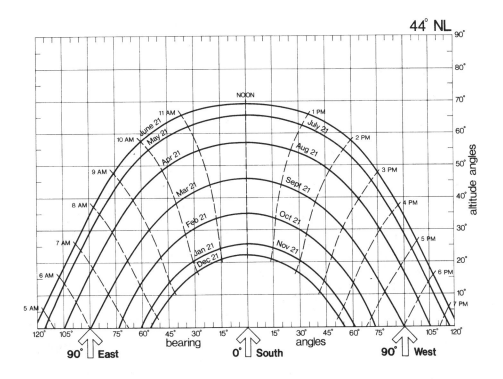

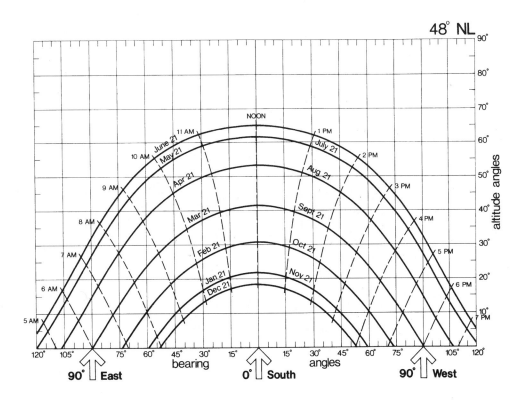

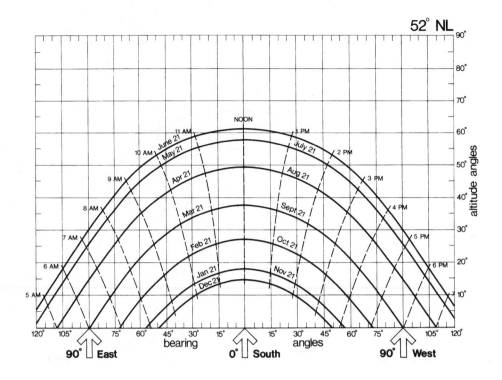

52° NL

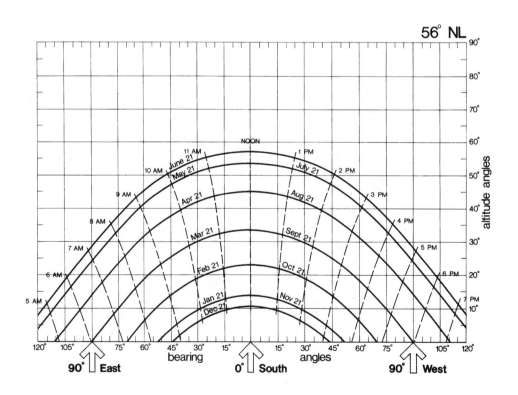

56° NL

SUN MACHINE

C.1 CONSTRUCTION OF SUN MACHINE (HELIODON)

The sun machine shown in Fig. 6.15b consists of three parts:

1. a labeled ribbon, which is taped to the edge of a door
2. the clamp-on lighting fixture, which is supported by the edge of a door
3. the model stand, which rests on an ordinary table

In Fig. C.1a we see the precise spacial relationship of these three parts.

Ribbon

The cloth ribbon should be of a light color, about 2 in. wide and 76 in. long. The locations for the various months should be marked as indicated in Fig. C.1a (e.g., the top end should be labeled as June 21).

Light

Use a 75 or 150 W indoor reflector lamp in a clamp-on lighting fixture. Avoid outdoor type PAR lamps because they are too heavy to be held in a horizontal position. The goal is to get a good quantity of fairly parallel light to shine on the model stand.

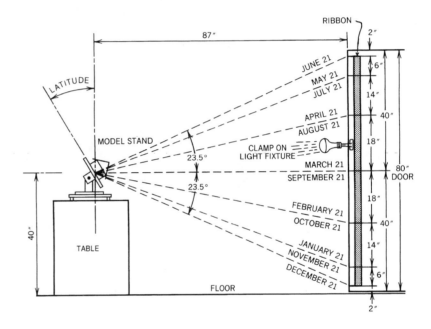

Figure C.1a Sun machine geometry.

Model Stand: (see Fig. C.1b) Parts List

2 pieces of ¾ in. plywood 12 x 12 in.

1 piece of ¾ in. plywood 12 x 10½ in.

2 pieces of wood ¾ x 1½ x 7 in. (Part A)

2 pieces of wood ¾ x 3½ x 7 in. (Part B)

3 carriage bolts ¼ in. diameter 2 in. long with washers and winged nuts

6 wood screws 2 in. long, size #8

4 soft rubber no-slip feet (not gliders)

2 sheets of acetate about 8 x 8 in.

1 wood dowel ¼ in. dia., 1½ in. long

Construction Procedure

Drill a ¼ in. diameter hole in the center of the fixed base, in the corresponding location in the rotating base, and in Parts A and B (Fig. C.1c). Drill a ¾ in. hole in one part "A" as shown in Fig. C.1e. Also drill ³⁄₁₆ in. holes in the rotating base and the tilt table as indicated in Fig. C.1d. Drill all holes as accurately as possible.

Prepare both parts "A" by rounding one end of each. On the part "A" with the ¾ in. hole cut a "V" groove as shown in Fig. C.1e. Glue a red thread into this groove to form a reference line across the hole. Make sure no glue or thread protrudes above the surface. Screw parts "A" to the rotating base as shown in Fig. C.1c. Be sure to drill ³⁄₃₂ in. pilot holes to prevent splitting of the wood. Part A with the ¾ in. hole should be on the west side and have the surface with the thread face inward.

Attach parts "B" to parts "A" with 2 carriage bolts. Then screw the tilt table to parts "B" as shown in Figures C.1c. Again drill ³⁄₃₂ in. pilot holes first in parts "B" to prevent splitting of the wood. Photocopy Fig. C.1f, cut along the dashed lines, and use a paper hole puncher to very carefully punch out the hole for the carriage bolt. Glue this latitude scale to the outside surface of part "B" on the west side. Make sure that the holes are aligned and that the zero line is parallel to the tilt table. Cover the scale with clear plastic or apply several coats of varnish for protection.

On the fixed base label the hours of the day, and on the tilt table, label the cardinal directions of the compass (Fig. C.1d). Attach the four soft rubber feet to the bottom of the fixed base. Use the wood dowel to make a pointer on the rotating base (Fig. C.1d). From the acetate sheets make two washers about 8 in. in diameter with a 1/4 in. hole in the center Assemble the tilt table with the acetate washers between the rotating base and the fixed base. However, there should be no washers between Parts A and B.

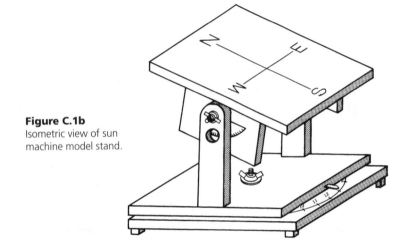

Figure C.1b
Isometric view of sun machine model stand.

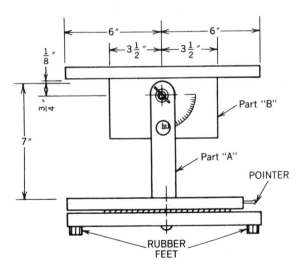

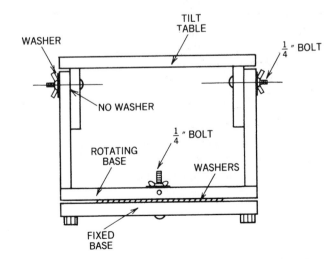

Figure C.1c West side (left); south side (right).

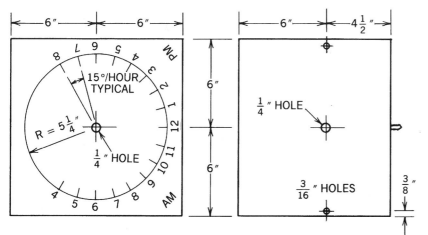

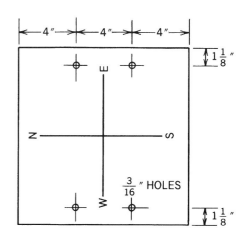

Figure C.1d Fixed base (left); rotating base (center); tilt table (right).

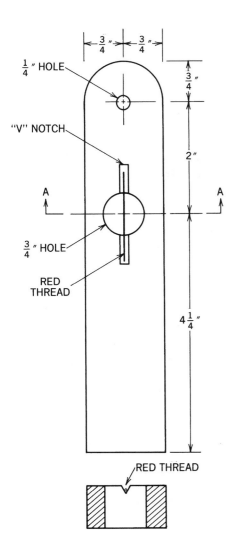

SECTION A–A

Figure C.1e Detail of part A for west side.

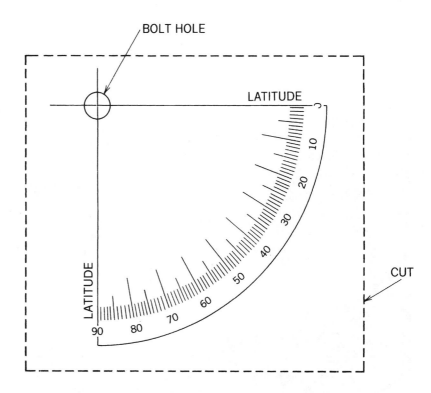

Figure C.1f Sun machine latitude scale.

Make sure that the rotating base can move very freely on the fixed base. It should, however, be possible to completely lock in place the tilt table when the winged nuts on the east and west sides are tightened. Check to make sure that the pointer, 12 noon, and south are all aligned. Also check to make sure that the latitude reads 90° for a horizontal tilt table and 0° when the tilt table is vertical.

C.2 DIRECTIONS FOR INITIAL SET-UP

1. Tape ribbon to the edge of a door as shown in Fig. C.1a.

2. Make sure clamp-on light fixture has a sufficiently long extension cord so that it can be placed anywhere along the vertical edge of the door.

3. Place model stand on a table so that the center of the tilt table is about 87 in. from the door edge and about 40 in. above the floor. Also make sure that 12 noon on the model stand faces the light on the door.

C.3 DIRECTIONS FOR USE

1. Set and fix latitude by adjusting the angle of the tilt table.
2. Attach model to tilt table with push pins or clamps. Align south of the model with south of the tilt table.
3. Set clamp-on light to desired month and aim lamp at model.
4. Turn rotating base to the desired hour of the day.
5. The model will now exhibit the desired sun penetration and shading.

Notes

1. Since the greatest accuracy occurs at the center of the tilt table, small models are more accurate than large models. However, many large models (e.g., site models) can be shifted around so that the part examined is always near the center of the tilt table.
2. The dynamics of sun motion can be easily simulated. Rotate the tilt table about its vertical axis to simulate the daily cycle. Move the light vertically along the edge of the door to simulate the annual cycle of the sun.
3. A correctly constructed sun machine will illuminate the east side of a model during morning hours. Check that the tilt table indicates 0 latitude when it is in a vertical

position. This would be the correct tilt for a model of a building located at the equator. Also make sure that the tilt table is horizontal when the latitude scale reads 90° (north pole).

C.4 ALTERNATE MODE OF USE OF THE SUN MACHINE

For greater accuracy a source with more parallel light is required. Indoors a slide projector at the far end of a corridor would give fairly parallel light rays. The best source of all, of course, is the sun. Since neither of these two sources of light can be moved up or down along the edge of a door, an alternate method of use for the model stand is required. Figure C.4 shows how a sundial is used in this alternate mode. Appendix E describes how sundials can easily be made for various latitudes.

Procedure for Alternate Mode of Use of the Sun Machine

1. Attach the sundial for the appropriate latitude to the model in such a way that the base of the sundial is parallel to the floor plane of the model. Also align the south orientation of the model with that of the sundial.
2. Attach the model to the model stand. In this mode of use the adjustments for latitude and time of day are ignored on the model stand.
3. Tilt and rotate the model stand until the gnomon of the sundial casts a shadow on the intersecting lines of the month and hour desired.
4. The model now exhibits with great accuracy the desired sun penetration and shadows (Fig. C.4).

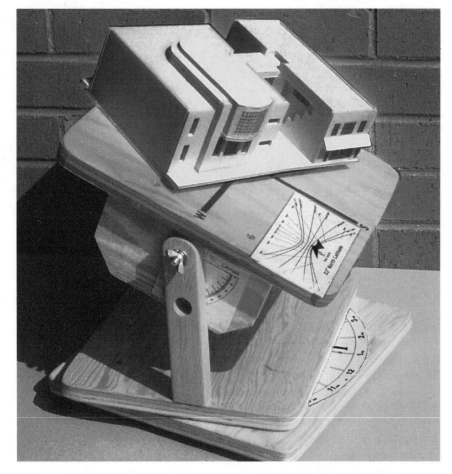

Figure C.4 Alternate mode of use for model stand. Note use of sundial.

METHODS FOR ESTIMATING THE HEIGHT OF TREES, BUILDINGS, ETC.

To determine solar access and shading, one needs to know the approximate height of objects around the site being investigated. This is the case whether model studies or graphical analyses are used. Four methods are described below for finding the height of a tree, but, of course, these methods work equally well for buildings or other objects.

D.1 PROPORTIONAL-SHADOW METHOD

This method can be used only on sunny days during the hours when shadows are fairly long.

Set up a vertical stick so that you can measure its shadow at the same time the shadow of the tree is visible (Fig. D.1). Measure both shadows and the height of the stick. You can then determine the height of the tree by means of the following equation:

$$\frac{H}{S} = \frac{h}{s}$$

or $\quad H = S \times \dfrac{h}{s}$

D.2 SIMILAR-TRIANGLE METHOD

Although this method can be used whether the sun is shining or not, it is important to avoid looking directly into the sun on clear days. Hold a square with one leg horizontal and the other vertical. Use a small level on the horizontal leg or hang a weighted string from the top of the vertical arm to align the square. Sight along the square, and place a finger where the sight line intersects the vertical arm to find the height (h), as seen in Fig. D.2.

Then, by similar triangles:

$$\frac{H}{D} = \frac{h}{d}$$

or $\quad H = D \times \dfrac{h}{d}$

and the Height of the object = H + P

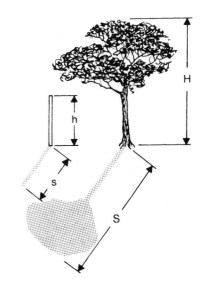

Figure D.1 The proportional-shadow method of finding the height of objects.

Figure D.2 The similar-triangle method of finding the height of objects

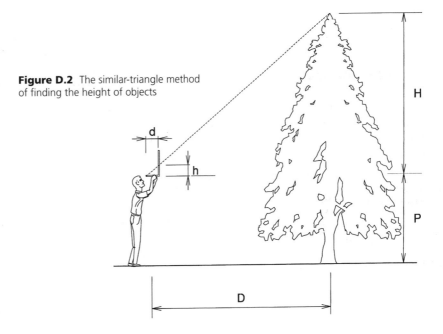

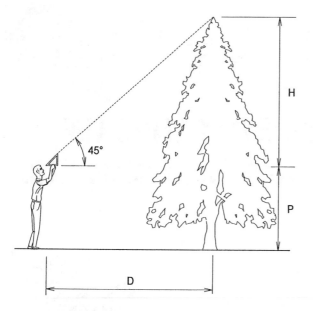

Figure D.3 The 45-degree right-triangle method of finding the height of objects.

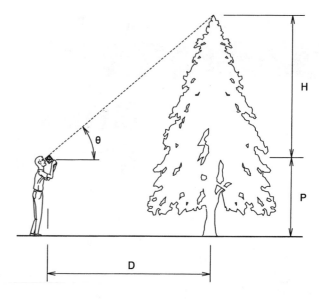

Figure D.4 The trigonometric method of finding the height of objects.

D.3 45-DEGREE RIGHT-TRIANGLE METHOD

This method is a special case of the similar-triangle method. If the site allows one to sight along the hypotenuse of a 45-degree right triangle, the unknown height of the object will simply be the horizontal distance to the sighting point (see Fig. D.3).

$$H = D$$

and the Height of the object = H + P

D.4 TRIGONOMETRIC METHOD

Use a transit or other device to measure the vertical angle "θ" (see Fig. D.4). Commercially available devices and a simple, inexpensive do-it-yourself device to measure this angle are described in Section D.5 below.

Use the following equation to find (H):

$$H = D \tan \theta$$

and the Height of the object = H + P

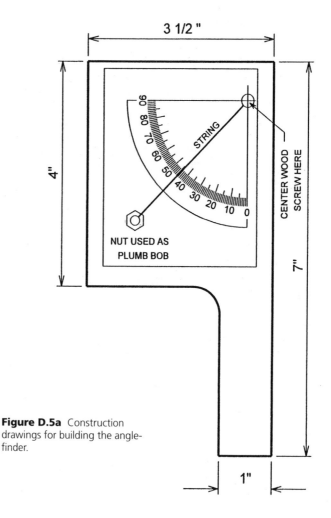

Figure D.5a Construction drawings for building the angle-finder.

D.5 TOOLS FOR MEASURING VERTICAL ANGLES

Of course, a professional transit can be used, but that tool is usually too expensive for this purpose. However, a much less expensive tool can still give precise results. A **clinometer** is a small pocket tool used for finding vertical angles, and it costs about $150 (see Fig. D.5b). It is widely used by foresters for finding the height of trees.*

Also available for about $10 is a special construction protractor, which is meant for finding the slope of pipes or structural members (Fig. D.5c). This tool is used much like the angle-finder, described below.

The angle-finder is a simple and inexpensive do-it-yourself tool for measuring vertical angles, which can be cut from a 1 x 4 board about 7 inches long (see Fig. D.5a). The 90-degree scale can be photocopied from Fig. C.1f in Appendix C. Hang a plumb line from a screw inserted at the point shown in Fig. D.5a. A small machine nut can make a handy plumb bob.

To measure the altitude angle to the top of some object, hold the angle-finder in a vertical plane and sight along its top edge. After the plumb line has stopped swaying, place a finger on the line and measure the angle (Fig. D.5d).

*Clinometers can be obtained from Forestry Supplies, Inc., P. O. Box 8397, Jackson, MS 39284-8397, tel. 1-800-647-5368, www.forestry-supplies.com.

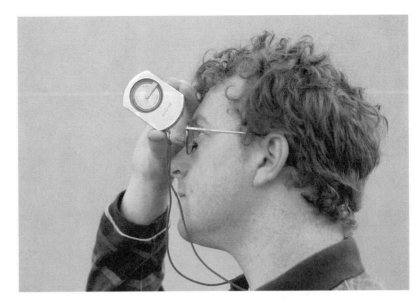

Figure D.5b The commercially available clinometer is used for finding vertical angles.

Figure D.5c This construction protractor was purchased for about $10.

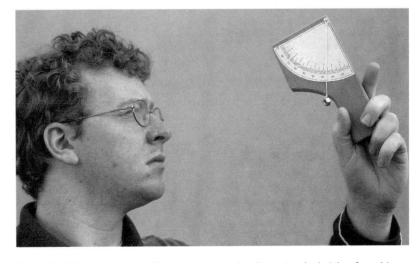

Figure D.5d Here, the angle-finder is being used to determine the height of an object.

SUNDIALS

The following diagrams are used for making sundials that are tools for simulating sun angles in physical models. Their use in conjunction with a sun machine is explained in Appendix C and their use with daylighting models is explained at the end of Chapter 13.

Each sundial requires a peg (gnomon) of a particular height to cast the proper shadow. The length of the gnomon is indicated on each sundial so that enlargements or reductions of the sundials are convenient to make. Copy the sundial that comes closest to the latitude required and glue it on a very thick piece of cardboard or plywood. In the circle (tail of the gnomon height arrow) mount a vertical pin or peg of the proper length to create the gnomon.

These sundial diagrams are derived from the *Solar Control Workbook* by Donald Watson and Raymond Glover.

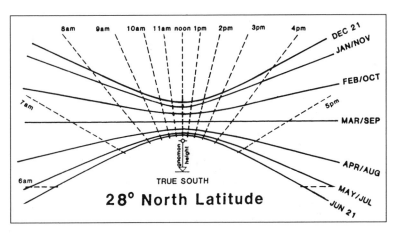

28° North Latitude

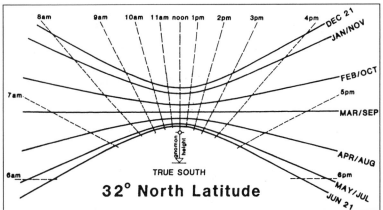

32° North Latitude

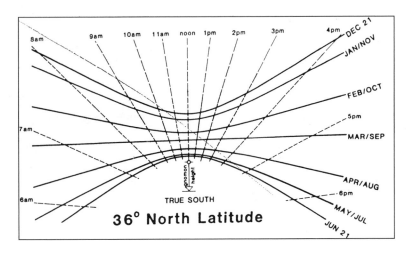

36° North Latitude

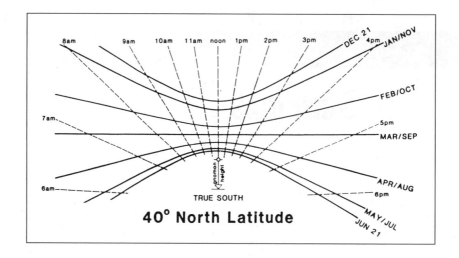

40° North Latitude

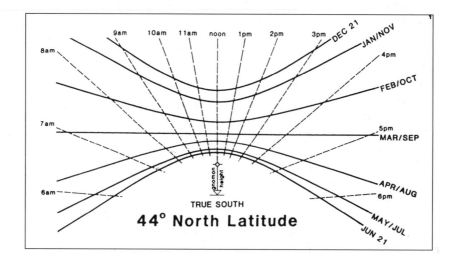

44° North Latitude

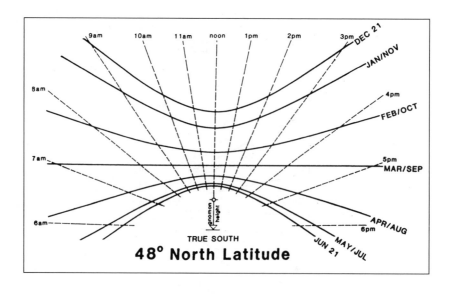

48° North Latitude

SUN-PATH MODELS

Sun path models can be very useful in visualizing the complex motion of the sun. The following diagrams make it quite easy to construct a sun path model for a variety of latitudes. These diagrams are orthographic projections of a skyvault. Normal sun path diagrams as found in Appendix A are not appropriate for this purpose. See Section 6.13 and Fig. 6.13 for an additional discussion of these sun path models.

Section F.1 explains how to build models from the given projections of 0°N, 24°N, 28°N, 32°N, 36°N, 40°N, 44°N, 48°N, 64°N, and 90°N latitude, which are found on the following pages. Section F.2 describes how to make projections for other latitudes.

F.1 DIRECTIONS FOR CONSTRUCTING A SUN-PATH MODEL

Materials List

1. A piece of 4½ x 4½ in. foam-core board at least ³/₁₆ in. thick.
2. A piece of 2 x 4 in. stiff clear plastic film (e.g., acetate)
3. Pipe cleaners, 3 pieces

Procedure

1. To make the base, photocopy the orthographic projection closest to the latitude of interest and glue it on a piece of foam board of the same size. The board should be at least ³/₁₆ in. thick.

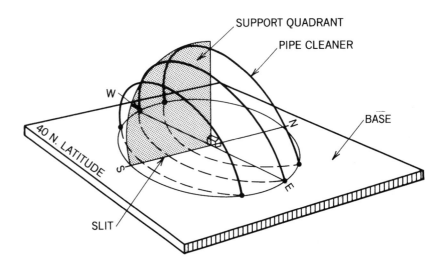

Figure F.1 A sun-path model

2. Cut a ⅛ in. deep and 1½ in. long slit where indicated on the projection (longer slits are required only for 0° and 90° north latitude). See Fig. F.1.
3. Trace the support quadrant on a piece of fairly stiff clear plastic film or photocopy onto clear film specified for copiers (absolutely *not* acetate).
4. Cut out the support quadrant and be sure to cut the three little notches where indicated.
5. Place the support quadrant in the slit so that the zero mark on the quadrant lines up with the zero mark on the base (0 = south).
6. Use a push pin or sharp pencil to make holes at the sunrise and sunset points for each of the three sun paths. The holes should pass

all the way through the base and be angled in the direction of the sun paths.
7. Insert one end of a pipe cleaner in the sunrise hole for June 21. Bend it across the support quadrant and insert the other end in the sunset hole. Pull the pipe cleaner down until it rests in the top notch of the support quadrant. Repeat this procedure for the other two sun paths. Note that the three pipe cleaners should form segments of parallel circles.
8. Glue the pipe cleaners in place and trim off the excess from the bottom of the base.
9. Place a small balsa wood block (less than ¼ in. on a side) in the center to represent a building.

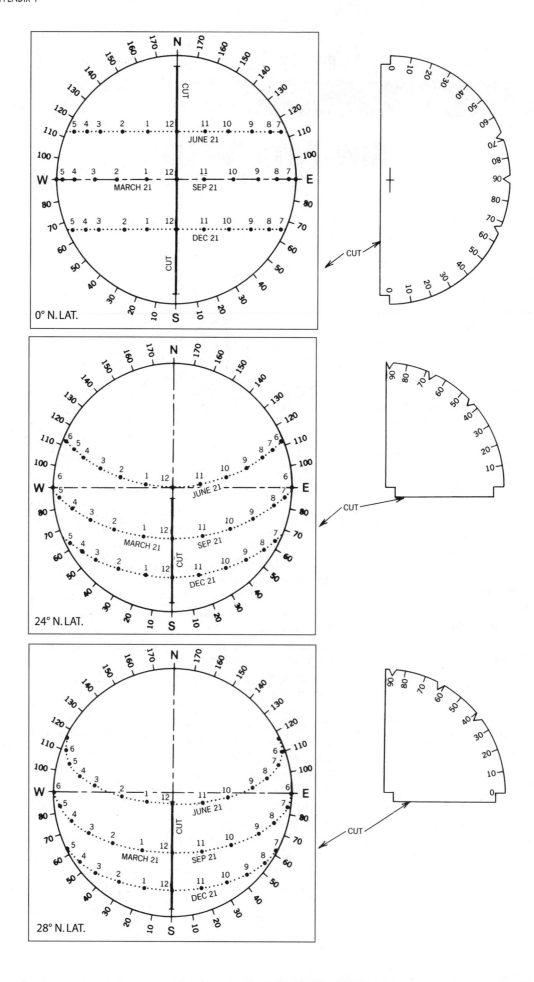

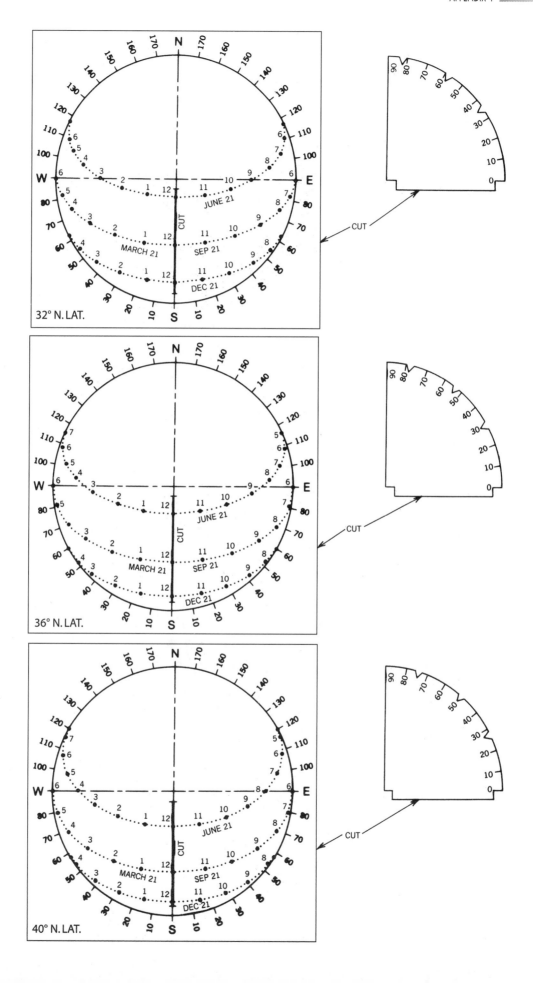

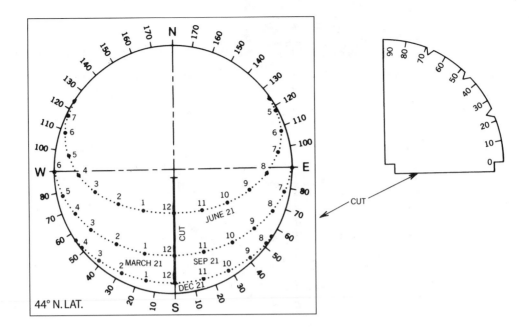

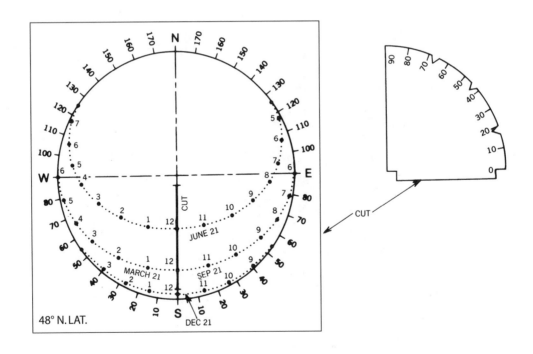

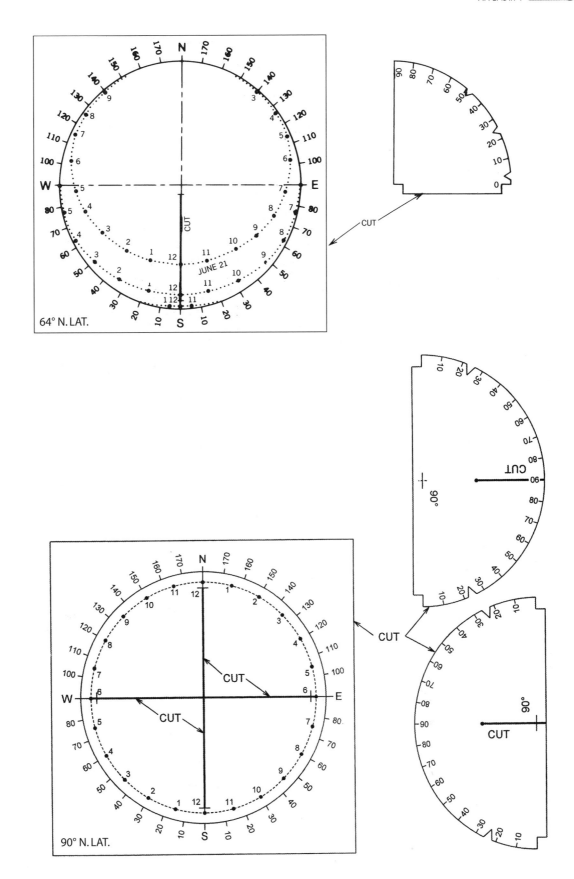

F.2 DIRECTIONS FOR CREATING OTHER ORTHOGRAPHIC PROJECTIONS

Plot the hourly points for the three sun paths of June 21, March/September 21, and December 21 on the projection of the skydome (Fig. F.2a). Connect the points to form the sunpaths. The azimuth and altitude angles for latitudes not given in this Appendix can be found in *Sun Angles for Design* by Robert Bennett and the *ASHRAE Handbook of Fundamentals*.

Use the skydome section of Fig. F.2b to create the pattern for the support quadrant. Mark the altitude angles at 12 noon of December 21, March/September 21, and June 21. Draw small notches at these altitudes to help support the pipe cleaners. Add a tab on the bottom for insertion into the base (see examples on previous pages). Transfer the pattern onto stiff acetate by tracing or by photocopying onto special transparent copying film (*N.B.* Acetate will ruin a copying machine).

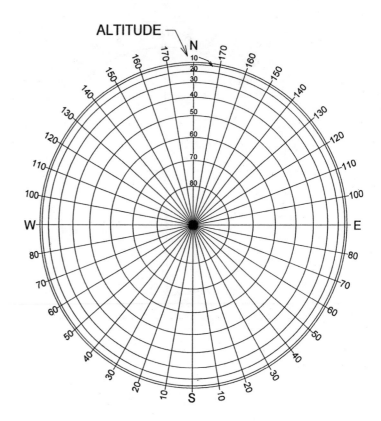

Figure F.2a The orthographic horizontal projection of the sky dome.

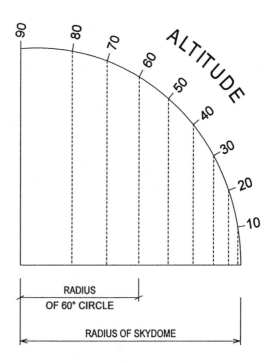

Figure F.2b A vertical section of the sky dome becomes the support quadrant. Radius of 60° circle is included to illustrate how horizontal projection was developed.

COMPUTER SOFTWARE USEFUL FOR THE SCHEMATIC DESIGN STAGE

G.1 UCLA ENERGY DESIGN TOOLS

All the software on the Energy Design Tools web page (www.aud.ucla.edu/energy-design-tools) can be downloaded and used free. These tools were developed specifically for use at the very beginning of the architectural design process. This is the point where most of the decisions are made that will determine the building's final energy performance.

Climate Consultant: At the beginning of a new project, an architect needs to understand the design implications of the local climate. Hourly climate data is available for more than 230 stations from NREL (hot link is on Energy Design Tools web page). ClimateConsultant converts these TMY2 files and displays dozens of different charts analyzing various aspects of the climate.

SOLAR-5 Passive Solar Design Tool: This whole-building energy-analysis tool considers hundreds of different design variables. But at the very beginning of the design process, the architect knows only a few basic facts: the city, the building type, the square footage, and the number of stories. The expert system inside SOLAR-5 uses these four facts to design a simple base-case, energy-efficient building. Then the architect's task is to mold and adapt this basic building to better meet the unique site, the program, and the aesthetic constraints. For each new scheme, SOLAR-5 shows graphically whether the building's energy performance is getting better or worse (Figs. G.1a, G.1b, and G.1c).

SOLAR-2 Window Sunshade Design Tool: This tool lets the architect fine-tune the shading of a window, whether it involves an overhang, fins, adjacent wings of the building, nearby walls, rows of trees, or even large buildings across the street.

OPAQUE Building Envelope Design Tool: Walls and roofs are an assembly of various materials, each with its own thermal properties. This design tool calculates many aspects of the envelope performance and produces dozens of charts and tables, including "Radiation on the Surface," "Surface Temperatures," "Heat Flow Through the Section," and "Solar Angles," plus the information for Form 3R required by the California Energy Code.

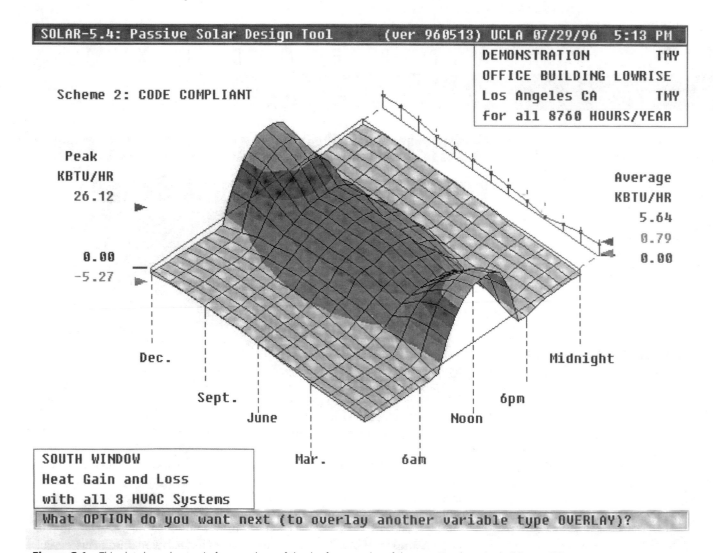

Figure G.1a This plot shows heat gain for every hour of the day for every day of the year. Good passive buildings will be saddle-shaped like this one; in other words, they gain more heat during winter than summer. This plot is for the South Windows (which are always saddle-shaped) and shows that they could collect enough heat in this climate to keep the building warm all winter assuming all other design factors are correctly balanced. (Courtesy Murray Milne.)

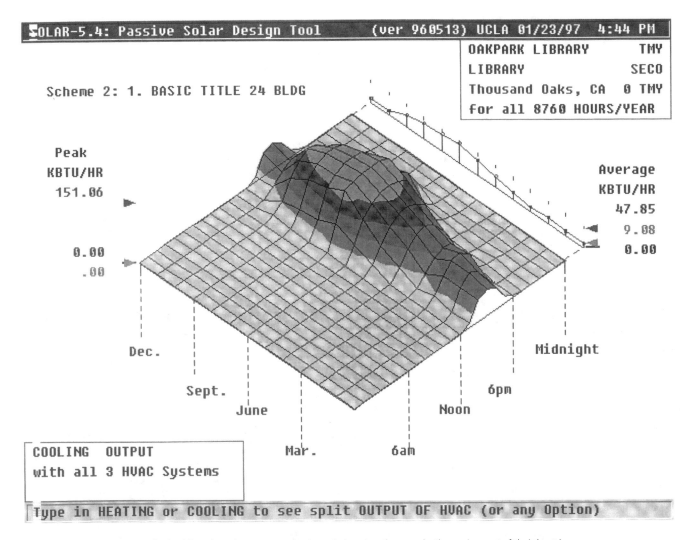

Figure G.1b Bad passive solar buildings have heat-mountain shaped plots; in other words, they gain most of their heat in midafternoon in midsummer, and almost nothing in the winter when it is needed most. This plot shows the cooling output needed from the air conditioner to keep this particular building within the comfort range. (Courtesy Murray Milne.)

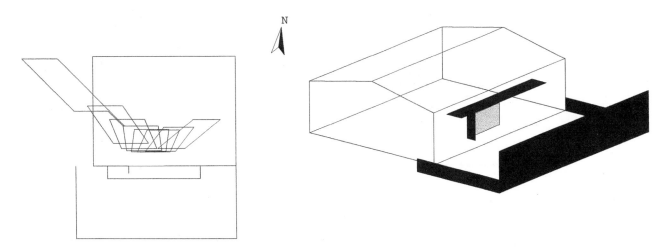

Figure G.1c This three-dimensional sunlight pattern shows the window as "seen" by the sun. In this case, most of the window is exposed to direct radiation in December at 3:00 P.M. (hour 15). The sun patterns on the floor plan show that all day long for this south-facing window, almost all of the sunlight passed over the wall of the adjacent building and under the overhang. Incidentally, in summer this same window is completely shaded and no sun patterns fall on the floor at any time. A dozen other graphs and tables help the architect optimize the design for almost any combination of window, sunshades, and exterior objects. (Courtesy Murray Milne.)

G.2 ENERGY SCHEMING

The Design Process Using Energy Scheming: The process of using Energy Scheming starts with scanning in a drawing (cocktail napkins are fine), copying from another application, or drawing a building in the program. The building surfaces are then taken off with the tape-measure cursor and associated with a specification. The takeoffs, specifications, and other inputs are entirely graphic. No numbers! Next, the building is evaluated and the results shown in a bar graph

and, of course, by Energy Scheming's thermographic images. The bar graph has an optional soundtrack that uses sounds and music to cue the designer to events that are important to a building's energy performance, such as the arrival of occupants in the morning, turning the lights on at sunset, cooling wind velocity, etc. The professor provides advice on how to improve the building's performance. The designer elects to redesign, and the evaluation process begins again (Figure G.2). Energy Scheming 2.5 is available to professionals for $195 and to students

with proof of current enrollment for $49. Site licenses are also available.

Contact: Terry Blomquist
Energy Studies in Buildings Laboratory
Department of Architecture
University of Oregon
Eugene, OR 97403

phone: (541) 346-5647
fax: (541) 346-3626
email: terryb@oregon.uoregon.edu

Checks should be made payable to University of Oregon.

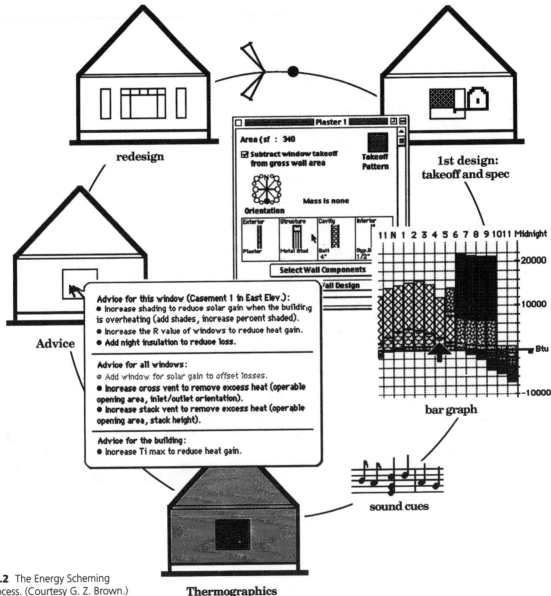

Figure G.2 The Energy Scheming Design Process. (Courtesy G. Z. Brown.)

SITE EVALUATION TOOLS

The construction of two site evaluation tools is described here. For a very quick but approximate site evaluation, use the Sun Locator (Fig. H.1). For a much more precise site evaluation, use the Solar Site Evaluator (Fig. H.2).

H.1 THE SUN LOCATOR

Make an enlarged copy of the sun locator (Fig. H.1), and glue it to a cardboard backing. Trim along the line of the latitude nearest you. Place the locator in a level position at the area where the collectors are to be mounted. Align a compass along the correct magnetic declination line for true north and south. View from the corner over the top of the latitude line from 9 A.M. to 3 P.M. solar time. This is the path the sun will take in mid-winter. If more than 5 percent of the path is blocked, the site might need closer evaluation. Even tree branches without leaves can block a considerable amount of winter sunlight. Consider trimming them if necessary.

H.2 DO-IT-YOURSELF SOLAR SITE EVALUATOR

Adapted from the design by Daniel K. Reif (www.homeplanner.com) from his book *Solar Retrofit: Adding Solar to Your Home*, Brick House Publishing

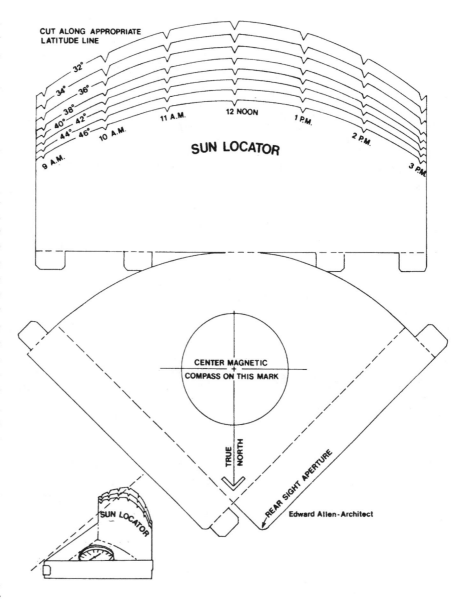

Figure H.1 The Sun Locator. (Courtesy Edward Allen and North Carolina Solar Center.)

Co., Amherst, NH. Also published in the periodical *Fine Home Building*. See Section 6.14 for a general discussion of solar site-evaluation tools.

The Solar Site Evaluator consists of three parts: a semicircular wooden base, a clear plastic mask, and a removable wooden handle (see Fig. H.2). The mask is held to the base via a Velcro™ strip for two reasons. The Velcro™ allows using different masks for different latitudes (4-degree intervals), and it makes the device easier to store and carry in a disassembled state. This Solar Site Evaluator is fairly easy and quick to build and costs only about $15 in materials.

H.3 PARTS LIST

1 pine board, 1 x 8 x 15 (nominal dimensions)
1 wooden dowel, ³/₄ inch diameter minimum and 6 inches long
1 sheet clear acetate, 20 inches x 24 inches, at least 0.005 inch thick
1 compass (inexpensive kind)
1 bull's-eye level
1 "T" nut, ¹/₄ inch diameter and ³/₄ inches long (see fig. H.2)
1 combination wood/machine screw, ¹/₄ inch diameter and 1¹/₂ inches long
1 strip of self-adhesive Velcro™ about 28 inches long and ³/₄ inch wide
1 cotton swab (e.g., Q-Tip™)

H.4 CONSTRUCTION PROCESS

Base

Since a 1 x 8 board has actual width of only 7¹/₄ inches, cut a semi-circle with a 7¹/₄-inch radius from the board. Drill a hole 2 inches from the middle of the straight edge of the base for mounting the "T" nut, as shown in Fig. H.2. About a ¹/₄-inch from the middle of the straight edge, also drill a ³/₃₂-inch hole. Insert a cotton swab in this hole in such a manner that the top of the cotton tip is just above the base. The cotton swab

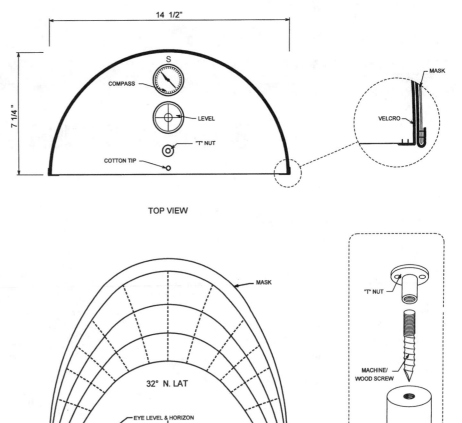

Figure H.2 The Solar Site Evaluator.

is used to prevent potential damage to the eye. Next, glue a Velcro™ strip to the edge of the base. For better holding power, extend the Velcro around the base, as shown in the front view of Fig. H.2, and staple the Velcro at both ends. Finally, glue the compass and level to the top of the base, as shown in Fig. H.2. Make sure that south on the compass is aligned with south on the mask.

Handle

Drill a ³/₁₆-inch hole in one end of the wood dowel, and insert the combination wood/machine screw in such a manner that about ³/₄ inch of the machine screw sticks out of the dowel.

Masks for Latitudes 26° to 46°

Drawings of masks for five different latitudes (28°, 32°, 36°, 40°, and 44°) have been prepared and are shown in Figs. H.4a to H.4e. Use the drawing that is less than 2 degrees from the latitude desired. Enlarge it until the line marked 6 inches is full size, and then transfer the lines onto a

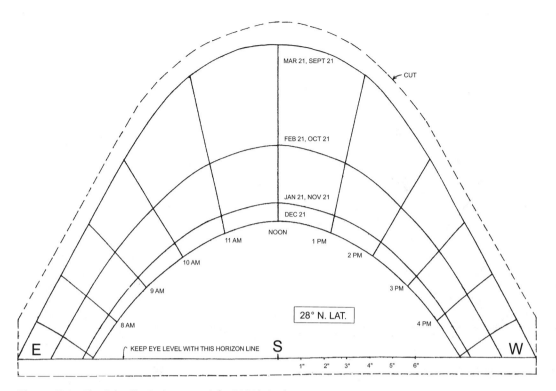

Figure H.4a The Solar Site Evaluator mask for 28°N latitude.

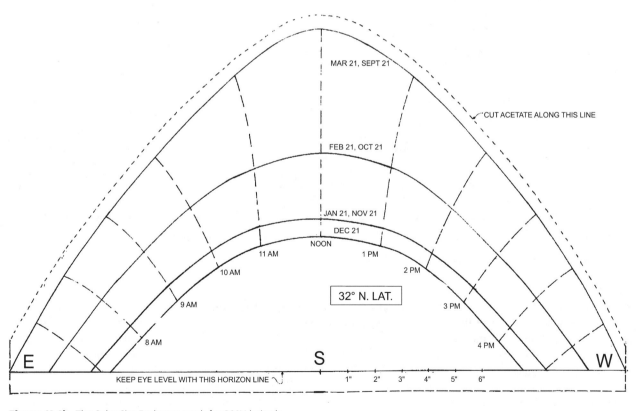

Figure H.4b The Solar Site Evaluator mask for 32°N latitude.

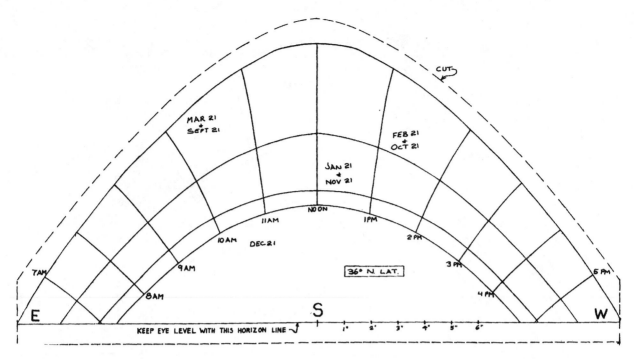

Figure H.4c The Solar Site Evaluator mask for 36°N latitude.

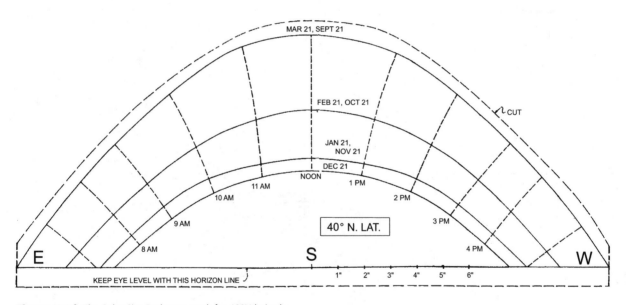

Figure H.4d The Solar Site Evaluator mask for 40°N latitude.

relatively stiff clear film, such as 0.005-inch-thick acetate. Cut the film as shown, and apply along the bottom inside surface of the mask the ¾-inch self-adhesive Velcro™ strip that matches the one on the base. For extra strength, return the Velcro™ strip, as shown in the detail of Fig. H.2.

Masks for Latitudes less than 26° and more than 46°)

Draw the winter sun paths for the required latitude on a **full-scale** version of the altitude/azimuth graph shown in Fig. H.4f. The altitude and azimuth angles for many latitudes are found in Appendices A and B of this book; *Sun* *Angles for Design,* by Robert Bennett; *Architectural Graphic Standards,* by Ramsey/Sleeper; and in the ASHRAE handbooks, such as the 1985 *Handbook of Fundamentals.* Plot the points for each hour of the following sun paths: December 21, November/ January, October/ February 21, and March/September 21. Connect the

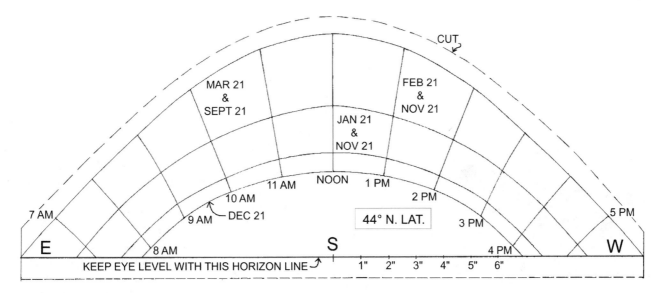

Figure H.4e The Solar Site Evaluator mask for 44°N. latitude.

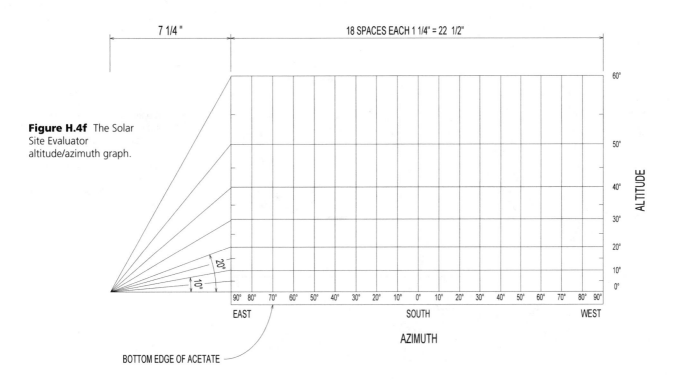

Figure H.4f The Solar Site Evaluator altitude/azimuth graph.

points with solid lines to form the sun paths, and with dashed lines for the hours of the day to form a diagram like that in Fig. H.4d. Be sure to label the mask for the latitude of its sun paths. Then trace onto a sheet of transparent film as described above for "Masks for Latitudes 26° to 46°."

H.5 USING THE SOLAR SITE EVALUATOR

This tool enables the user to determine when and how much a certain location will be shaded from the sun. This is possible because when one looks through the evaluator at the actual site, the sun-path diagrams are superimposed on the view of the site

(see Figs. H.5a and 6.14). It is immediately obvious which trees or buildings are blocking the solar window and by how much. A record of the site condition can be traced on an overlay on the mask (see Fig. 6.12b). This tool is best used on overcast days or on sunny winter days before 9 A.M. and after 3 P.M. so that it is not necessary to look directly into the sun.

Figure H.5a Mounting the Solar Site Evaluator on a camera tripod is especially helpful when tracing a horizon profile.

Steps for Use

1. Attach the mask to the base by means of the Velcro™ strip.
2. Use a camera tripod to support the Solar Site Evaluator if at all possible. The "T" nut will fit a camera tripod. Otherwise, use the handle.
3. At the specific place to be evaluated for solar access, set up the Solar Site Evaluator and level it by means of the bull's-eye level.
4. Orient the tool so that the 0° azimuth reference line faces true south. Use the Magnetic Declination Map in Fig. H.5b to determine how far true north is from magnetic north is your area. At the top of the map, two compasses illustrate the relative position of true north and magnetic north on each side of the zero declination line in the United States. Use the following web site for the exact declination in your area: http://geomag.usg.gov
5. Bring your eye close to the top of the sight (cotton-swab tip), and look through the mask toward the south of the building site.
6. Make an evaluation of the site by determining how much of the solar window is blocked.
7. If a record of the site is desired, draw on the acetate with a washable marker the outline of the objects viewed through the mask. By using clear acetate overlays, you can record any number of sites.

Figure H.5b The magnetic Declination Map of the United States. True south and magnetic south are only aligned along the 0° declination line. Everywhere else, the compass must be rotated east or west according to the declination. For example, in the middle of Massachusetts, the compass must be rotated clockwise 15 degrees. (Data from 1995 U.S. Geological Survey. Map courtesy of North Carolina Solar Center.)

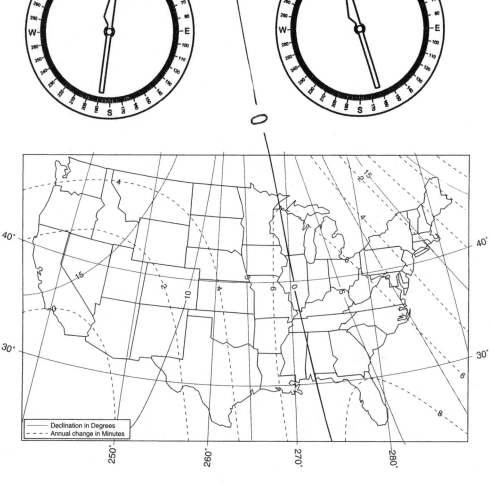

EDUCATIONAL OPPORTUNITIES IN ENERGY-CONSCIOUS DESIGN

The schools listed below have an emphasis in energy-conscious design education. The list has been created as an aid for students wishing to learn more about climatic design, energy-conscious design, environmental design, or sustainable "green" design as it relates to heating, cooling, and lighting.

The following schools have submitted information for this survey:

- Arizona State University
- Cal Poly Pomona
- Cal Poly San Luis Obispo
- Florida A&M University
- Georgia Tech
- Laval University
- New Jersey Institute of Technology
- North Carolina State University
- Rensselaer Polytechnic Institute
- Slippery Rock University
- University of Calgary
- University of California–Berkeley
- University of Florida
- University of Idaho
- University of Louisiana at Lafayette
- University of Michigan
- University of Oregon
- University of Southern California
- University of Texas at Arlington
- University of Texas at Austin
- University of Washington

University: ARIZONA STATE UNIVERSITY
School/College: College of Architecture and Environmental Design
Program: Environmental Design and Planning
Degree(s) offered: Ph.D.
What is offered: The college-wide, interdisciplinary Ph.D. program in Environmental Design and Planning offers concentrations in the following areas based on the research and teaching expertise of participating faculty: design; planning; and history, theory and criticism. Broad in scope, the program involves multidisciplinary research interests at both micro- and macro-scale levels of design and planning. The program provides research experience for students wishing to pursue careers in academia and in industry as members of interdisciplinary-design-and-planning teams on environmental and energy issues, as well as for those wishing to teach in the architecture, design, or planning fields.

Address:	College of Architecture and Environmental Design Arizona State University Tempe, AZ 85287-1905
Web site:	http://www.asu.edu/caed/PHD
Telephone:	(480) 965-4620
E-mail:	CAED.Ph.D@asu.edu
Contact person:	Michael D. Kroelinger

University: CAL POLY POMONA
School/College: College of Environmental Design
Program: Dept. of Architecture and Center for Regenerative Studies
Degree(s) offered: B. Arch./M. Arch., Minor in Regenerative Studies
What is offered: The university-wide Regenerative Studies minor can be subscribed in both the undergraduate and the graduate program. The Architecture, Landscape

Architecture, and Planning departments are major players in the center. Many other departments in the College of Science and the College of Engineering are also involved in the center. The center provides live-in experiences and learning opportunities for sustainable design subjects. With complementary courses in the Architecture Department, architecture students can obtain skills and knowledge in expanded professional design.

Address: College of Environmental Design
California State Polytechnic University
3801 W. Temple Avenue
Pomona, CA 91768
Web site: http://www.csupomona.edu/ ~ env/
http://www.csupomona.edu/ ~ crs/
Telephone: (909) 869-2661 (College)
(909) 869-2683 (Architecture)
(909) 869-5155 (CRS)
E-mail: check in Web pages
Contact person: Hofu Wu, or Brooks Cavin, III
Dr. Hofu Wu, AIA Director, Environmental Design Technology Unit

University: CAL POLY SAN LUIS OBISPO
School/College: Architecture Dept., College of Architecture and Environmental Design
Program: Architecture
Degree(s) offered: B.A. Arch., M.S. Arch.
What is offered: General architecture degree with fifth-year senior-thesis options in adaptive reuse, sustainable design, and housing. The graduate program supports building research in climate-adapted design and sustainability.

Address: 1 Grand Avenue
San Luis Obispo, CA 93407
Web site: www.calpoly.edu/
Telephone: (805) 756-1316
E-mail: mmcdonal@calpoly.edu
Contact person: Margot McDonald

University: Florida A&M University
School/College: Architecture/Landscape Architecture
Program: Graduate
Degree(s) offered: Masters of Architecture, Master of Science in Architecture, Master of Landscape Architecture
What is offered: The student is responsible for choosing the thesis focus in this program. As a result, students with an interest in this area of study have developed theses that center on energy and sustainability. These theses have explored both architectural and urban form as they relate to passive design, daylighting, and urban regeneration. Because approximately 60 percent of the program is thesis-related, the students are expected to be very motivated and self-starters. Interdisciplinary work in Landscape Architecture exploring sustainability, carrying capacity, and ecologically sound decision-making is available. Also available in a coop program with Florida State University are electives in Urban Planning that afford the opportunity to develop expertise in administration, policy-making, and management skills.

Address: School of Architecture
Florida A&M University
Tallahassee, FL 32307
Web site: soa@famu.edu
Telephone: (850) 412-7051
E-mail: malfanojr@aol.com
Contact person: Michael Alfano, Jr.,
AIA Graduate Coordinator

University: GEORGIA TECH
School/College: College of Architecture
Degree(s) offered: Ph.D. and M.S. in Architecture, Field of Study: Building Technology with focus on Environmental Technology
Description
- Integrated systems
- Workplace performance
- Environmentally conscious technologies: The responsive facade
- Thermal and lighting building performance and simulation
- Physical testing and simulation of integrated building components
- Lighting design and simulation
- Climate-responsive building design

Address: Georgia Tech
College of Architecture
Atlanta, GA 30332-0155
Web site: http://murmur.arch.gatech.edu/ ~ tech/
http://murmur.arch.gatech.edu/phd/
faculty/augenbroe/buildingtech.html
Telephone: (404) 894-3838
(404) 894-1686
E-mail: tahar.messadi@arch.gatech.edu
godfried.Augenbroe@arch.gatech.edu
Contact person: Dr. Tahar Messadi, Assistant Professor
College of Architecture
Georgia Tech
Atlanta, GA 30332-0155
(404) 894-3838 (Voice)
tahar.messadi@arch.gatech.edu (email)

Godfried Augenbroe, Associate Professor
College of Architecture
Georgia Tech
Atlanta, GA 30332-0155
(404) 894-1686 (Voice)
godfried.Augenbroe@arch.gatech.edu
(email)

University: LAVAL UNIVERSITY
School/College: School of Architecture
Program: Architecture
Degrees: Bachelor's Degree in Architecture, Master's Degree in Architecture
What is offered:
Bachelor's Degree
  two courses: Daylighting and Physical Ambiences, Climatic Factors and Architectural Design
  two advanced studios: Daylighting in Architecture, A Bioclimatic Approach to Architecture
Master's Degree
  2-year program with thesis

Address:	School of Architecture
	Laval University
	1, cité de le Fabrique
	Quebec City, (Quebec)
	CANADA, G1K 7P4
Web site:	www.arc.ulaval.ca
Telephone:	(418) 656-2543
E-mail:	andre.potvin@arc.ulaval.ca
Contact person:	Andre Potvin

University: NEW JERSEY INSTITUTE OF TECHNOLOGY
School/College: School of Architecture
Degree(s) offered: B. Architecture (five-year), M. Architecture
What is offered:

1. Building Performance, B.A. and M.A. required course
Design decisions that influence the thermal performance of a building are addressed, including comfort, thermal envelope, shading, heating and cooling design loads and energy loads, and complying with energy codes. Emphasis is on shading, with urging to apply shading design in the studio.

2. Environmental Control Systems, B.A. and M.A. required course

The energy part, approximately half of the semester, of the course is on the design and selection of HVAC systems based on the design and energy loads.

3. Passive Solar Design in Housing, elective course, Building Performance required

Each student designs a passive solar heated house based on a client and program defined by an identified client. Sizing and design of: 1) the window area necessary to meat the energy load at design using on-site potential solar energy; 2) the necessary internal thermal mass to store and distribute solar energy; and 3) the shading to control summer solar gain, are used in designing the exterior and interior of the term project house.

Address:	School of Architecture
	NJIT
	Newark, NJ 07102
Web site:	www.njit.edu
Telephone:	(973) 596-30130
E-mail:	bales@megahertz.njit.edu
Contact person:	Erv Bales

University: NORTH CAROLINA STATE UNIVERSITY
School/College: School of Design
Program: Architecture Department
Degree(s) offered: Bachelor of Environmental Design in Architecture (4-year preprofessional undergraduate), Bachelor of Architecture (1-year professional program), and Master of Architecture (1½-year, 2-year, and 3½-year professional graduate programs)
What is offered: A number of courses are offered in which energy-conscious, climatic, and "green"-energy design are the focus. These are ARC 201: Architectural Design Studio: Environment; ARC 211: Natural Systems and Architecture; ARC 414: Environmental Control Systems; and ARC 577 and DDN 777: Sustainable Communities. It is possible to have a concentration at the Master of Architecture level that is focused on issues of sustainability. At the Doctoral (Ph.D. in Design) level, Community and Environmental Design is one of the two areas of substantive knowledge.

Address:	Department of Architecture
	School of Design
	North Carolina State University
	Campus Box 7701
	Raleigh, NC 27695
Web site:	http://design.ncsu.edu/
Telephone:	(919) 515-8350
E-mail:	architecture_design@ncsu.edu
Contact person:	Dr. Fatih A. Rifki

University: Rensselaer Polytechnic Institute
Department/School: School of Architecture
Program: Lighting Research Center
Degree(s) offered: Master of Science in Lighting
What is offered:
Forty-eight credits of project and course work over four semesters covering the physics of light, lighting technology, human factors, and architectural lighting design.

Address:	110 8th Street
	Troy, NY 12180
Web site:	www.lrc.rpi.edu
Telephone:	(518) 276-8716
E-mail:	lrc@rpi.edu
Contact person:	Dan Frering, Manager of Education

University: SLIPPERY ROCK UNIVERSITY
School/College: Department of Parks and Recreation/ Environmental Education
Program: Sustainable Systems Program
Degree(s) offered: Master of Science in Sustainable Systems
What is offered: The multidisciplinary program prepares students to address current environmental challenges by adopting long-term sustainability as the underlying framework for action. Students study sustainability with course work in the built environment, agroecology, and natural resource management. The built-environment emphasis addresses a range of issues associated with environmentally conscious design, including: climate responsive design; healthy building systems and materials; site and building feasibility studies; sustainable landscape design; and design/build. An on-campus demonstration and research facility affiliated with the program provides experiential learning opportunities.

Address: Sustainable Systems Program
PREE/101 Eisenberg
Slippery Rock University
Slippery Rock, PA 16057
Web site: http://www.sru.edu/depts/chhs/pree/ ms3/ms3home.htm
Telephone: (724) 738-2622
E-mail: karen.kainer@sru.edu
Contact person: Karen Kainer, Program Coordinator

University: THE UNIVERSITY OF CALGARY
School/College: Faculty of Environmental Design
Program: Architecture
Degree(s) offered: Master of Environmental Design (M.E.Des.), Master of Architecture (M. Arch.), Doctoral (Ph.D.)
What is offered: The M.Arch. curriculum is accredited and provides a structured introduction for those without a previous architectural education. Students normally complete three years of course and studio work prior to undertaking a Master's Degree Project (MDP). The program includes term-abroad study opportunities, including the Faculty's program in Barcelona. The Architecture Program stresses an understanding of the conceptual issues and operational skills required to participate in the design of the built environment. The study of architecture is pursued through a progressive series of studios, lectures, seminars, ateliers, and Master's Degree Project topics as both a foundation of program specific knowledge and advanced disciplinary and interdisciplinary, research-oriented activities.

Address: 2500 University Drive NW
Calgary, Alberta
T2N 1N4
CANADA
Web site: www.ucalgary.ca/UofC/faculties/EV/

Telephone: (403) 220-6601
E-mail: lee@ucalgary.ca
Contact person: Professor Tang Lee

University: UNIVERSITY OF CALIFORNIA–BERKELEY
School/College: College of Environmental Design
Program: Architecture
Degree(s) offered: B.A. in Architecture, Master of Architecture, Master of Science, Master of Urban Studies, and Ph.D.
What is offered: Beyond the undergraduate survey course on energy and environmental management, there are graduate classes on daylighting, computer-aided design for energy conservation, natural cooling, mechanical systems, evaluation of buildings in use (Vital Signs), research methods, acoustics, and (periodically) design studios with energy or lighting focus.

Address: Wurster Hall
University of California–Berkeley
Berkeley, CA 94720
Web site: http://arch.ced.berkeley.edu/
Telephone: (510) 642-4942
Contact person: Lois Koch, Graduate Office, Department of Architecture. (510) 642-5577

University: UNIVERSITY OF FLORIDA (UF)
School/College: Department of Architecture
Program: Environmental Technology
Degrees offered: Undergraduate, Master's, and Ph.D.
What is offered: The six areas that constitute Environmental Technology are:
1. Energy in architecture
2. Thermal systems in architecture
3. Lighting in architecture
4. Acoustics in architecture
5. Life safety systems
6. Environmental quality

Address: 231 Architecture Building
Post Office Box 115702
Gainesville, FL 32611-5702
Fax: (352) 392-4606
Web site: http://grove.ufl.edu/ ~ mgold/ET_WEB/
Telephone: (352) 392-0205
E-mail: siebein.acoustic@prodigy.net
mgold@ufl.edu
Contact: Gary W. Siebein and Martin A. Gold

University: UNIVERSITY OF IDAHO
School/College: College of Art and Architecture
Program: Professional and Post-Professional Programs in Architecture
Degree(s) offered: M. Arch., M.A. Architecture

What is offered: The professional program offers students a five-year immersion leading to a Master of Architecture degree in architecture in an interdisciplinary-design college. Students are exposed to the basics in energy-conscious design and sustainable architecture in required courses and may specialize in these areas through seminars and studios taught by several professors who specialize in energy, lighting, and sustainable design. The post-professional program (which is also suitable for non-architecture-degree holders) offers intense, self-directed studies with professors who are doing research in the areas of energy, lighting, and sustainability.

Address:	Department of Architecture
	University of Idaho
	Moscow, ID 83844-2451
Web site:	http://www.aa.uidaho.edu/arch/
Telephone:	(208) 885-6781
E-mail:	arch@uidaho.edu
Contact person:	Bruce Haglund

University: UNIVERSITY OF LOUISIANA AT LAFAYETTE
School/College: School of Architecture
Program: Architecture, Industrial Design (emphasis on sustainability)
Degree(s) offered: B. Architecture, B. Industrial Design
What is offered: B. Arch. degree is a typical, five-year NAAB-accredited program with a thesis the last year that allows students to pick their own project—which can be a sustainable, energy-efficient project. Faculty have expertise in mainly hot-humid climates. B. Industrial Design is a four-year accredited degree with the main emphasis on sustainability.

Address:	School of Architecture
	University of Louisiana at Lafayette
	Lafayette, LA 70504-3850
Web site:	www.usl.edu/Academic/Arts/ARCH.html
Telephone:	(337) 482-6225
E-mail:	cazayoux@louisiana.edu
Contact person:	Eddie Cazayoux, AIA

University: UNIVERSITY OF MICHIGAN
School/College: Taubman College of Architecture and Urban Planning
Program: Architecture Program and Doctoral Program in Architecture
Degree(s) offered: B.S., M. Arch., M. Sc., Ph.D.
What is offered: Courses related to sustainable design, intelligent buildings, lighting, building energy conservation, and computer analysis of building environmental performance.

Address:	2200 Bonisteel Boulevard
	Taubman College of Architecture and Urban Planning
	University of Michigan
	Ann Arbor, MI 48109

Web site:	www.caup.umich
Telephone:	(734) 763-3518
E-mail:	daylight@edu
Contact person:	Jong-Jin Kim

University: UNIVERSITY OF OREGON
School/College: Department of Architecture
Degree(s) offered: B. Arch, M. Arch.
What is offered:
ARCH 491/591 Environmental Control Systems I
ARCH 492/592 Environment Control System II
ARCH 493/593 Solar Heating
ARCH 494/594 Passive Cooling
ARCH 495/595 Daylighting
ARCH 407/507 Vital Signs Case Study Investigations
ARCH 407/507 Electric Lighting
IARC 510: The Window

Address:	Department of Architecture
	1206 University of Oregon
	Eugene, OR 97403
Web site:	
	Dept. Home Page:
	http://arch-uo.uoregon.edu/windex.html
	Solar Information Center:
	http://darkwing.uoregon.edu/ ~ sic/
	HOPES
	http://gladstone.uoregon.edu/ ~ hopes
Telephone:	(541) 346-3656
E-mail:	akwok@aaa.uoregon.edu
Contact person:	Alison Kwok

University: UNIVERSITY OF SOUTHERN CALIFORNIA
School/College: School of Architecture
Program: Environmental Controls
Degree(s) offered: Master of Building Science
What is offered: Under the guidance of the faculty, students do examinations into the environmental-control areas of natural lighting, lighting design, energy and water conservation, lifecycle impact, and issues critical in the field. Extensive computer simulations have been developed, as well as methods for examining and testing physical models. Students in the program have published many papers in national and international forums.

Address:	School of Architecture
	Watt Hall 204
	University Park
	Los Angeles, CA 90089-2723
Web site:	www.usc.edu/dept/architecture/mbs/tools/ecstools.html
Telephone:	(213) 740-4591
E-mail:	marcs@rcf.usc.edu
Contact person:	Marc Schiler

University: UNIVERSITY OF TEXAS AT ARLINGTON
School/College: School of Architecture
Program: Architecture
Degree(s) offered: B.S. Arch., M. Arch.
What is offered: One elective Energy Course; two semesters of Environmental Control systems covering Acoustics, Lighting, and Mechanical Systems. NO specialization in E.C. or sustainability.

> **Address:** Box 19108
> University of Texas at Arlington
> Arlington, TX 76019
> **Telephone:** (817) 272-2801
> **Contact person:** C. Lee Wright

University: UNIVERSITY OF TEXAS AT AUSTIN
School/College: School of Architecture
Program: "Design With Climate"
Degree(s) offered: First Professional M. Arch., Emphasis; First Professional M. Arch., Specialization; Post-professional M. Arch.; Master of Science in Architectural Studies (MSAS)
What is offered: The University of Texas at Austin offers four "Design with Climate" graduate architecture-program options. These programs provide a multidisciplinary, focused course of research and study for students and professionals on the salient issues related to the confluence between the built and natural environments. The program focuses on three areas of inquiry: *natural systems, building systems,* and *cultural systems.* The study of natural systems relies upon the disciplines of physics and ecology as they relate to architecture. The study of building systems includes the investigation of those component technologies that are required to construct environmentally responsive architecture. The study of cultural systems requires that natural and building systems be investigated within the complex social and political context of architectural practice. The faculty consists of Francisco Arumi-Noe, Buford Duke, Michael Garrison, and Steven Moore.

> **Address:** School of Architecture
> University of Texas at Austin
> Goldsmith Hall
> Austin, TX 78712-1160
> **Web site:** http://www.ar.utexas.edu/
> **Telephone:** (512) 471-1922
> **Contact person:** Steven A. Moore

University: UNIVERSITY OF WASHINGTON
School/College: College of Architecture and Planning
Program: Architecture
Degree(s) offered: M. Arch. and B.A. in Architectural Studies
What is offered: Lighting certificate (thirty credits) and design studios and courses in climate-responsive design (see Web site).

> **Address:** Box 355720
> University of Washington
> Seattle, WA 98195-5720
> **Web site:** http://www.caup.washington.edu/html/arch/
> **Telephone:** (206) 685-8407
> **E-mail:** see Web site
> **Contact person:** Undergraduate assistant: (206) 685-7236, Graduate assistant: (206) 685-8405

RESOURCES

BOOKS

The following books are highly recommended. Because they cover the whole range of environmental-control topics, they are listed here. (See Bibliography for full citations. The list includes valuable and out-of-print books.)

Anderson, B., ed. *Solar Building Architecture.*

Barnett, D. L., with W. D. Browning. *A Primer on Sustainable Building.*

Brown, G. Z., and M. DeKay. *Sun, Wind, and Light: Architectural Design Strategies,* 2nd edition.

Cofaigh, E. O., J. E. Olley, and O. Lewis. *The Climatic Dwelling: An Introduction to Climatic-Responsible Residential Architecture.*

Crowther, R. L., ed. *Ecological Architecture.*

Daniels, K. *The Technology of Ecological Building: Basic Principles and Measures, Examples and Ideas.*

Flynn, J. E. *Architectural Interior Systems.*

Givoni, B. *Climate Considerations in Building and Urban Design.*

Goulding, J. R., J. O. Lewis, and T. C. Steemers, eds. *Energy Conscious Design: A Primer for Architects.*

Lstiburek, J. W. *Builder's Guide—Hot-Dry & Mixed-Dry Climates; Builder's Guide—Cold Climates; Builder's Guide—Mixed Climates.*

Moore, F. *Environmental Control Systems.*

Olgay, V. *Design with Climate: A Bioclimatic Approach to Architectural Regionalism.*

Pearson, D. *The Natural House Catalog: Everything You Need to Create an Environmentally Friendly Home.*

Ramsey/Sleeper: Architectural Graphic Standards.

Stein, B., and J. S. Reynolds. *Mechanical and Electrical Equipment for Buildings.*

Thomas, R., ed. *Environmental Design: An Introduction for Architects and Engineers.*

Tuluca, A. *Energy-Efficient Design and Construction for Commercial Buildings.*

Watson, D., and K. Labs, *Climatic Design: Energy-Efficient Building Principles and Practices.*

JOURNALS

Energy Design Update. Cutter Information Corp., 37 Broadway, Suite 1, Arlington, MA 02174-5539; 800-964-5125. www.cutter.com.

Environmental Building News. 122 Birge Street, Suite 30, Brattleboro, VT 05301; 802-257-7300. www.buildinggreen.com.

Environmental Design and Construction. 299 Market Street, Suite 320, Saddle Brook, NJ 07663-5312; 415-863-2614. www.edcmag.com.

Home Energy: The Magazine of Residential Energy Conservation. 2124 Kittredge Street, No. 95, Berkeley, CA 94704-9942; 510-524-5405. This source provides information on reducing energy consumption.

Home Power: The Hands-on Journal of Home-made Power. Ashland, OR: Home Power, Inc. P.O. Box 5520, Ashland, OR 97520. 916-475-3179.

Renewable Energy World. James & James Ltd. 35-37 William Road, London, NW1 3ER U.K. +44 171 387 8558, +44 387 8998 (fax). rew@jxj.com, www.jxj.com.

Solar Industry Journal. Solar Energy Industries Association (SEIA), 122 C Street NW, Fourth Floor, Washington, DC 20001-2109. 202-383-2600.

Solar Today. The American Solar Energy Society. 2400 Central Avenue, Unit G-1, Boulder CO 80301. 303-443-3130. Internet: (E-mail) ases@ases.org; (World Wide Web) http:// www.ases.org/solar.

Soplan Review: Journal of Energy Conservation, Building Science, and Construction Practice. The Drawing-Room Graphic Services, Ltd., Box 86627, North Vancouver, BC V7L 4L2 Canada. 604-689-1841.

VIDEOS

Affluenza. Host, Scott Simon. Produced by KCTS/Seattle and Oregon Public Broadcasting (OPB). KCTS Television, 1997. Approximately 57 minutes. 1-800-937-5387. Discusses the environmental impact of a high standard of living.

Arithmetic, Population, and Energy. Dr. Albert A. Bartlett. 1994. 65 minutes.
University of Colorado
ITS-Media Services
Campus Box 379
Boulder, CO 80309-0379
303-492-1857
Fax: 303-492-7017
E-mail:
Kathleen.Albers@colorado.edu

Building Green: Audubon House. Produced by National Audubon Society. Two versions (16 minutes and 28 minutes).
National Audubon Society
700 Broadway
New York, NY 10003
212-979-3000

Case Studies in Sustainable Design. 1 hour, 17 minutes.
The American Institute of Architects
1735 New York Avenue NW
Washington, DC 20006
800-365-ARCH

Ecological Design - Inventing the Future.
From San Luis Video Publishing
P.O. Box 6715
Los Osos, CA 93412
805-528-8322

Environmental Architecture. 30 minutes.
Landmark Films, Inc.
3450 Slade Run Drive
Falls Church, VA 22042
800-342-4336
Fax 703-536-9540

Keeping the Earth: Religious and Scientific Perspectives on the Environment. 1996. 27 minutes.
Publications Department
Union of Concerned Scientists
Two Brattle Square
Cambridge, MA 02238-9105
617-547-5552

Living on the Sun: Case Studies in Sustainable Design. 30 minutes.
Arizona Energy Office
3800 North Central Avenue, Suite 1200
Phoenix, AZ 85012
602-280-1402
Fax 602-280-1445

A Visit with Amory and Hunter Lovins.
Produced by CBS News. Bullfrog Films, 1985, 13 minutes.
Bullfrog Films
P.O. Box 149
Oley, PA 19547
800-543-3764
Fax 610-370-1978
www.bullfrogfilms.com
The super-efficient Lovins house/office is described.

World Population. Produced by Zero Population Growth (ZPG).
ZPG
1400 16th Street NW, Suite 320
Washington, DC 20036
1-800-POP-1956
www.zpg.org.

CD-ROMS

E Build Library: Environmental Building News on CD-ROM. Available from *Environmental Building News.*

Energy and Environmental Technologies: Worldwide Suppliers and Successful Demonstration Projects.
Caddet Center for Renewable Energy
1617 Cole Boulevard
Golden, CO 80401-3393
303-275-4373

Green Building Advisor.
CREST Software Orders
1200 18th Street NW, Suite 900
Washington, DC 20036
1-888-44CREST
Fax 202-887-0497
www.crest.org/software-central

Lior CD-ROM Collection on Renewable Energy - Solar: Bioclimatic Architecture.
LIOR
78 Charlierlaan - B - 1560
Hoeilaart, Belgium
www.lior-int.com

ORGANIZATIONS

Alternative Energy Council
Applied Building Science Center
909 Capability Drive, Suite 2100
Raleigh, NC 27606
919-857-9000
E-mail: moreinfo@aec.ncsu.edu

The Alliance to Save Energy
1200 18th Street, Suite 900
Washington, DC 20036
202-857-0666
www.ase.org

American Council for an Energy Efficient Economy
1001 Connecticut Avenue NW, Suite 9-1
Washington, DC 20036
292-429-8873
www.aceee.org

American Hydrogen Association
1739 West 7th Avenue
Mesa, AZ 85202-190
480-827-7915
Fax 480-967-6601
E-mail: aha@getnet.com
http://www.clean-air.org

American Society of Landscape Architects (ASLA)
636 Eye Street NW
Washington, DC 20001-3736
202-898-2444
Fax 202-898–1185
www.asla.org

American Solar Energy Society
2400 Central Avenue, G-1
Boulder, CO 80301
303-443-3130
http://www.ases.org
National group for renewable-energy education.

American Wind Energy Association
122 C Street NW
Washington, DC 20001

202-383-2505
http://www.econet.org/awea
Great source for wind energy information.

American Society of Heating, Refrigerating, and Air-Conditioning Engineers (ASHRAE)
1791 Tullie Circle NE
Atlanta, GA 30329
www.ashrae.org

Building Codes Assistance Project (BCAP)
1200 18th Street NW, Suite 900
Washington, DC 20036
202-530-2200
www.crest.org/efficiency/bcap

Building Science Corporation
70 Main Street
Westford, MA 01886
978-589-5100
www.buildingscience.com

Caddet Center for Renewable Energy
1617 Cole Boulevard
Golden, CO 80401-3393
303-275-4373
http://www.caddet.co.uk
Summaries of renewable energy projects around the world.

Center for Building Science
www.eande.lbl.gov/CBS/CBS.html

Center for Excellence for Sustainable Development
www.sustainable.doe.gov

Center for Resourceful Building Technology
P.O. Box 100
Missoula, MT 59806
406-549-7678

CREST Center for Renewable Energy and Sustainable Technology
777 North Capitol Street NE, #805
Washington, DC 20002
202-289-5370
E-mail: info@crest.org
http://www.crest.org
Solstice - Web site is a comprehensive resource for searching for renewable energy and sustainable technology.

DOE (see EREC directly below)

E Build, Inc.
122 Birge Street, Suite 30
Brattleboro, VT 05301
802-257-7300
Fax 802-257-7304
E-mail: ebn@ebuild.com
www.ebuild.com

E Source, Inc.
4755 Walnut Street
Boulder, CO 80301-2537
303-440-8500
Fax 303-440-8502
www.esource.com

Ecological Design Institute
245 Gate Five Road
Sausalito, CA 94965
415-332-5806
Fax 415-332-5808
E-mail: ecodes@aol.com

Energy Efficiency and Renewable
    Energy Clearinghouse (EREC)
Energy and Renewable Energy Network
    (EREN)
P.O. Box 3048
Merrifield, VA 22116
1-800-363-3732
Fax 703-893-0400
E-mail: energyinfo@delphi.com
www.eren.doe.gov/erec/factsheets
EREC provides free general and techni-
    cal information to the public on
    many topics and technologies per-
    taining to energy efficiency and
    renewable energy, including photo-
    voltaic systems, solar energy, and
    solar-radiation data.

Energy Efficient BuildingAssociation
490 Concordia Avenue
St. Paul, MN 55103-2441
612-851-9940
www.eeba.org

ENERGY STAR®
Environmental Protection Agency (EPA)
401 M Street SW, (6202J)
Washington, DC 20460
202-564-9190
Toll Free: 888-STAR-YES (782-7937)
Fax 202-564-9569
Fax-back system: 202-233-9659
www.energystar.gov

EPA-Indoor Air Quality Hotline
800-438-4318

EPA-Radon Hotline
404-872-3549; 800-745-0037

Florida Solar Energy Center
1679 Clearlake Road
Cocoa, FL 32922-5703
407-638-1000
Fax 407-638-1010
www.fsec.ucf.edu
A source of information and continuing
    education.

GeoExchange/Geothermal Heat Pump
    Consortium
701 Pennsylvania Avenue NW

Washington, DC 20004-2696
202-508-5500
888-ALL-4-GEO
Fax 202-508-5222
E-mail: info@ghpc.org
www.geoexchange.org

IESNA Illuminating Engineering Society
120 Wall Street, Floor 17
New York, NY 10005
212-248-5000
Fax 212-248-5017/8
iesna@iesna.org
www.iesna.org

International Association of Lighting
    Designers (IALD)
200 World Trade Center
Suite 487, The Merchant Mart
Chicago, IL 60654
www.iald.org

Interstate Renewable Energy Council
    (IREC)
P.O. Box 1156
Latham, New York 12110-1156
617-325-7377
Fax 617-325-6738
E-mail: irec1@aol.com
www.eren.doe.gov/irec

International Solar Energy Association
    (ISES)
Villa Tannheim
Wiesentalstr. 50, D-79115
Freiburg, Germany
+49 761 45906-0
Fax +49 761 45906 99
E-mail: hq@ises.org
www.ises.org

Iris Communications, Inc.
P.O. Box 5920
Eugene, OR 97405-0911
800-346-0104; 541-767-0355
Fax 541-767-0357
www.oikos.com/catalog/

Lawrence Berkeley National Laboratory
    (LBNL)
Building Technologies Program
Energy and Environment Division
Lawrence Berkeley Laboratory,
    Berkeley, CA 94721
510-486-6844
http://eande.lbl.gov/BTP/BTP.html

Lighting Design Lab
400 East Pine Street #100
Seattle, WA 98122
206-354-3864
www.northwestlighting.com

Lighting Resource Center
Rensselaer Polytechnic Institute
Troy, NY 12180-3590
518-276-8716

Fax 518-276-2999
E-mail: lrc@rpi.edu

Minnesota Building Research Center
    (MNBRC)
University of Minnesota
1425 University Avenue SE
Minneapolis, MN 55455
612-626-7819
Fax 612-626-7242
www.umn.edu/mnbrc

National Association of Homebuilders
    Research Centerl (NAHBRC)
400 Prince George's Boulevard
Upper Marlboro, MD 20774-8731
800-638-8556, 301-249-4000
Fax 301-429-3265
www.nahbrc.org

National Audubon Society
700 Broadway
New York, NY 10003
212-979-3000

National Center for Appropriate
    Technology (NCAT)
3040 Continental Drive
Butte, MT 59701
406-494-4572
NCAT's publication *Solar Greenhouses
    and Sunspaces—Lessons Learned*
    describes the experiences of sun-
    space owners and builders during
    the Department of Energy's
    Appropriate Technology Small
    Grants Program.

National Fenestration Rating Council
    (NFRC)
1300 Spring Street, Suite 500
Silver Spring, MD 20910
301-589-NFRC
www.nfrc.org

National Lighting Bureau
Communications Department
8811 Colesville Road, Suite G106
Silver Spring, MD 20910
301-587-9572

National Pollution Prevention Center
    for Higher Education
430 E. University Avenue
Ann Arbor, MI 48109-1115
734-764-1412
Fax 734-647-5841
E-mail: nppc@umich.edu
National Renewable Energy Laboratory
1617 Cole Boulevard
Golden, CO 80401-3393
303-384-7520
Fax 303-384-7540
http://www.nrel.gov
Good source for detailed information
    on renewables.

National Technical Information Service (NTIS)
Technology Administration
U.S. Department of Commerce
Springfield, VA 22161
800-533-6847; 703-605-6000
E-mail: info@ntis.fedworld.gov
www.ntis.gov

Natural Resources Defense Council
40 West 20th Street
New York, NY 10011
212-727-2700
Fax 212-727-1773
www.nrdc.org/nrdcpro/dcofc/dcofcinx.html
"NRDC's Washington DC Eco-Office: Tomorrow's Workplace Today" describes this environmentally concious office in detail.

Negative Population Growth, Inc. (NPG)
1608 20th Street NW, Suite 200
Washington, DC 20009
202-667-8950
Fax 202-667-8953
E-mail: npg@npg.org
www.npg.org

The North American Sundial Society
63 Florence Road
Harrington Park, NJ 07640
E-mail: 75762.3207@compuserve
The North American Sundial Society is dedicated to the study, development, history, and preservation of sundials, including a journal *Compendium.*

North Carolina Solar Center
919-515-3480
E-mail: ncsun@ncsu.edu

Northeast Sustainable Energy Association (NESEA)
50 Miles Street
Greenfield, MA 01301
413-774-6051
Fax 413-774-6053
E-mail: nesea@nesea.org
www.nesea.org

Oak Ridge National Laboratory (ORNL)
Building Envelope Systems and Materials
P.O. Box 2008
Oak Ridge, TN 37831-6070
425-574-4345
www.cad.ornl.gov/kch/demo.html

Pacific Northwest National Labs
800-270-CODE
www.energycodes.org

Rocky Mountain Institute (RMI)
1739 Snowmass Creek Road
Snowmass, CO 81654-9199
970-927-3851
Fax 970-927-3420
www.rmi.org

Sandia National Laboratories
Photovoltaic Systems and Active Solar Departments
Albuquerque, NM 87185-5800
505-844-8159
www.sandia.gov

Solar Energy Industries Association (SEIA)
122 C Street NW, Fourth Floor
Washington, DC 20001-2109
202-383-2600
E-mail: info@seia.org
http://www.seia.org
Trade association for U.S. solar companies.

Southface Energy Institute
241 Pine Street
Atlanta, GA 30308
404-872-3549 x110
Fax 404-872-5009
www.southface.org

Sustainable Buildings Industry Council (SBIC)
1331 H Street NW, Suite 1000
Washington, DC 20005-4706
202-628-7400
Fax 202-393-5043
E-mail: sbicouncil@aol.com
www.sbicouncil.org
SBIC (formerly known as Passive Solar Industries Council) offers workshops around the country on the Builder-Guide computer program and guidelines for passive solar building and remodeling projects. Climate-specific guidelines are available for more than 2,000 cities and towns around the United States. SBIC also provides the building industry with practical, useful information on the use of passive solar technologies in buildings. SBIC developed the "Passive Solar Design Strategies: Guidelines for Home Builders" workshops and the BuilderGuide software.

Union of Concerned Scientists
National Headquarters
Two Brattle Square
Cambridge, MA 02238-9105
617-547-5552
www.ucsusa.org

U.S. Department of Energy's Energy Efficiency and Renewable Energy Clearinghouse (EREC)
P.O. Box 3408
Merrifield, VA 22116
800-363-3732
Fax 703-893-0400
E-mail: doe.erec@nciinc.com
www.eren.doe.gov/erec/factsheets

U.S. Department of Energy
www.doe.gov

U.S. Environmental Protection Agency (EPA)
401 M Street SW (2171)
Washington, DC 20460
202-260-3354
Fax 202-260-0920
www.epa.gov

U.S. Government's Federal Information Network
www.fedworld.gov

U.S. Green Building Council
110 Sutter Street, #906
San Francisco, CA 94104
415-445-9500
Fax 415-445-9911
www.usgbc.org

Utility PhotoVoltaic Group (UPVG)
1800 M Street NW, Suite 300
Washington, DC 20036-5802
202-857-0898
Fax 202-223-5537
E-mail: upvg@ttcorp.com
www.ttcorp.com/upvg

The Vital Signs Project
www.ced.berkeley.edu/cedr/vs/
Features case studies of buildings for students from an energy use and occupants' well-being perspective.

Worldwatch Institute
1776 Massachusetts Avenue NW
Washington, DC 20036-1904
202-452-1999
Fax 202-296-7365
E-mail: worldwatch@worldwatch.org
www.worldwatch.org

Zero Population Growth (ZPG)
1400 Sixteenth Street NW, Suite 320
Washington, DC 20036
202-332-2200; 800-POP-1956
Fax 202-332-2302
E-mail: infor@zpg.org
www.zpg.org

Abrams, Donald W. *Low-Energy Cooling.* New York: Van Nostrand Reinhold, 1986.

AIA Research Corporation. *Solar Dwelling Design Concepts.* For the U.S. Department of Housing and Urban Development Office of Policy Development and Research. Washington, DC: For sale by the Supt. of Docs., U.S. Govt. Printing Office, 1976.

AIA Research Corporation. *Regional Guidelines for Building Passive Energy Conserving Homes.* Written for HUD Office of Policy Development and Research, and the U.S. Dept. of Energy. Washington, DC: Government Printing Office, 1980.

Akbari, Hashem, et al. *Cooling Our Communities: A Guidebook on Tree Planting and Light- Colored Surfacing.* Washington, DC: EPA and DOE/Lawrence Berkeley Laboratory [LBL-31587], 1992.

Allard, Francis, ed. *Natural Ventilation in Buildings: A Design Handbook.* London: James & James, 1998.

Allen, Edward, drawings by David Swoboda and Edward Allen. *How Buildings Work: The Natural Order of Architecture,* 2nd ed. New York: Oxford University Press, 1995.

Allen, Edward, and Joseph Iano. *The Architect's Studio Companion,* 2nd ed. New York: John Wiley & Sons, 1995.

Ander, Gregg D., AIA. *Daylighting Performance and Design.* New York: Van Nostrand Reinhold, 1995.

Anderson, Bruce. *Solar Energy: Fundamentals in Building Design.* New York: McGraw-Hill, 1977.

Anderson, Bruce, ed. *Solar Building Architecture.* Cambridge, MA: MIT Press, 1990.

Anderson, Bruce, and Malcolm Wells. *Passive Solar Energy.* Amherst, NH: Brick House Publishing, 1994, c1981.

Andrews, F. T. *Building Mechanical Systems.* New York: McGraw-Hill, 1977.

ASHRAE. *Passive Solar Heating Analysis: A Design Manual.* Atlanta: ASHRAE, 1984.

ASHRAE. *Air-Conditioning Systems Design Manual.* Atlanta: ASHRAE, 1993.

ASHRAE. *ASHRAE Terminology of Heating, Ventilation, Air Conditioning, and Refrigeration,* 2nd ed. Atlanta: ASHRAE, 1991.

ASHRAE. *ASHRAE Handbook Fundamentals.* Atlanta: ASHRAE, 1997.

Balcomb, J. Douglas. *Passive Solar Buildings.* Cambridge, MA: MIT Press, 1992.

Balcomb, J. Douglas, and Robert W. Jones. *Workbook for Workshop on Advanced Passive Solar Design,* July 12, 1987, Portland Oregon. Sante Fe, NM: Balcomb Solar Association, 1987.

Banham, Reyner. *The Architecture of the Well-Tempered Environment,* 2nd ed. London: The Architectural Press, 1984.

Barnett, Dianna Lopez, with William D. Browning. *A Primer on Sustainable Building.* Snowmass, CO: Rocky Mountain Institute, 1995.

Bartlett, Albert A. "Forgotten fundamentals of the energy crisis." *American Journal of Physics.* 46 (9), September, 1978.

Bearg, David W. *Indoor Air Quality and HVAC Systems.* Boca Raton, FL: Lewis Publishers/CRC Press, 1993.

Bennett, Robert. *Sun Angles for Design.* Bala Cynwyd, PA: Robert Bennett, 1978.

Bobenhausen, William. *Simplified Design of HVAC Systems.* New York: John Wiley & Sons, 1994.

Boonstra, Chiel, ed. *Solar Energy in Building Renovation.* London: James & James, 1997.

Boutet, Terry S. *Controlling Air Movement: A Manual for Architects and Builders.* New York: McGraw-Hill, 1987.

Bowen, Arthur, and S. Gingras. *Wind Environments in Buildings and Urban Areas.* Paper presented at the Sunbelt Conference, December, 1978.

Bowen, Arthur, Eugene Clark, and Kenneth Labs, eds. *Proceedings of the International Passive and Hybrid Cooling Conference,* Miami Beach, FL, Nov. 6–16, 1981. Proceedings. Newark, DE: American Section of the International Solar Energy Society, 1981.

Boyer, Lester, and Walter Grondzik. *Earth Shelter Technology.* College Station, TX: Texas A&MU, 1987.

Bradshaw, Vaughn. *Building Control Systems,* 2nd ed. New York: John Wiley & Sons, 1993.

Brown, G. Z., and Mark DeKay. *Sun, Wind, and Light: Architectural Design Strategies,* 2nd ed. New York: John Wiley & Sons, 1985.

Brown, Robert D., and Terry J. Gillespie. *Microclimatic Landscape Design: Creating Thermal Comfort and Energy Efficiency.* New York: John Wiley & Sons, 1995.

Buckley, Shawn. *Sun Up to Sun Down.* New York: McGraw-Hill, 1979.

Butti, Ken, and John Perein. *A Golden Thread: 2500 Years of Solar Architecture.* New York: Van Nostrand Reinhold, 1980.

Campbell, C. J. *The Coming Oil Crisis.* Multi-Science Publishing Company & Petroconsultants S. A., 1988.

Campbell, Colin J., and Jean H. Leherrére. "The End of Cheap Oil." *Scientific American,* 278 (3), March, 1998.

Carmody, John, and Raymond Sterling. *Earth Sheltered Housing Design,* 2nd ed. New York: Van Nostrand Reinhold, 1985.

Carmody, John, Stephen Selkowitz, and Lisa Heschong. *Residential Windows: A Guide to New Technologies and Energy Performance.* New York: W. W. Norton, 1996.

Carver, Norman F., Jr. *Italian Hilltowns.* Kalamazoo, MI: Documan Press, c1980.

Clegg, Peter, and Derry Watkins. *Sunspaces: New Vistas for Living and Growing.* Charlotte, VT: Garden Way Publishing, Storey Communications, 1987.

Climate Atlas of the United States. Asheville, NC: U.S. Dept. of Commerce Environmental Science Services Administration, Environmental Data Service. For sale by the National Climate Data Center, 1983.

Climate Change: State of Knowledge. Washington, DC: Executive Office of the President of the United States. Office of Science and Technology Policy, 1997.

Cofaigh, Eoin O., John A. Olley, and J. Owen Lewis. *The Climatic Dwelling: An Introduction to Climatic-Responsive Residential Architecture.* London: James & James, 1996.

Commission of the European Communities Directorate-General XII for Science Research and Development. *Daylighting in Architecture: A European Reference Book.* N. Baker, A. Fanchiotti, and K. Steemers, eds. London: James & James, 1993.

Commoner, Barry. *The Politics of Energy.* New York: Knopf, 1979.

Comparative Climatic Data for the United States/through . . ./Prepared at the National Climatic Center. Asheville, NC: U.S. Dept. of Commerce, National Oceanic and Atmospheric Administration. For sale by the National Climatic Data Center. Annual.

Cook, Jeffrey. "Cooling as the Absence of Heat: Strategies for the Prevention of Thermal Gain," a chapter in *Proceedings of the International Passive and Hybrid Cooling Conference,* Miami Beach, FL, Nov, 6–16, 1981. Proceedings. Edited by Arthur Bowen, et al. Newark, DE: American Section of the International Solar Energy Society, 1981.

Cook, Jeffrey, ed. *Award Winning Passive Solar Designs.* New York: McGraw-Hill, 1984.

Cook, Jeffrey, ed. *Passive Cooling.* Cambridge, MA: MIT Press, 1989.

Cooling with Ventilation. Golden, CO: Solar Energy Research Institute, 1986 (SERI/SP-273- 2966; DE96010701).

Crosbie, Michael J. *Green Architecture: A Guide to Sustainable Design.* Rockport, MA: Rockport Publications, 1994. Distributed by North Light Books, Cincinnati, OH.

Crowther, Richard L., ed. *Ecologic Architecture.* Butterworth Architecture, 1992.

Dagostino, Frank R. *Mechanical and Electrical Systems in Construction and Architecture.* Englewood Cliffs, NJ: Prentice Hall, 1995.

Daniels, Klaus. *The Technology of Ecological Building: Basic Principles and Measures, Examples and Ideas.* Trans. by Elizabeth Schwaiger. Basel, Switzerland: Birkhäuser, 1997.

Davidson, Joel. *The New Solar Electric Home: The Photovoltaics How-To Handbook.* Ann Arbor, MI: Aatec Publications, 1987. Available from Aatec Publications, P.O. Box 7119, Ann Arbor, MI 48107, 800-995-1470.

Davies, Colin, and Ian Lambot. *Commerzbank Frankfurt: Prototype for an Ecological High-Rise.* Surrey: Watermark, 1997.

Druse, Ken, with Margaret Roach. *The Natural Habitat Garden.* New York: Clarkson Potter, 1994.

Duly, Colin. *Houses of Mankind.* London: Thames and Hudson, 1979.

Easton, David. *The Rammed Earth House.* White River Junction, VT: Chelsea Green Publishing, 1996.

Egan, M. David. *Concepts in Thermal Comfort.* Englewood Cliffs, NJ: Prentice-Hall, 1975.

Egan, M. David. *Concepts in Architectural Lighting.* New York: McGraw-Hill, 1983.

Ehrlich, Paul R., and Anne H. Ehrlich. *The Population Explosion.* New York: Simon & Schuster Touchtone Books, 1991.

Energy Design Resources. *Skylighting Guidelines.* http://www.energyde-signresources.com/.

Environmental Protection Agency. *Cooling Our Communities.* Washington, DC: U.S. Environmental Protection Agency. Office of Policy Analysis, Climate Change Division. For sale by U.S. GPO Supt. of Docs., 1992.

Evans, Benjamin E. *Daylight in Architecture.* New York: McGraw-Hill, 1981.

Fathy, Hassan. *Natural Energy and Vernacular Architecture: Principles and Examples with Reference to Hot Arid Climates.* Chicago: University of Chicago Press, 1986.

Fitch, James Marston. *American Building: 1. The Historical Forces That Shaped It,* 2nd ed. New York: Schocken, 1973.

Fitch, James Marston. *The Architecture of the American People.* New York: Oxford University Press, 2000.

Fitch, James M., with William Bobenhausen. *American Building: The Environmental Forces That Shape It.* New York: Schocken, 1999.

Fitch, James Marston, consulting ed. *Shelter: Models of Native Ingenuity: A Collection of Essays Published in Conjunction with an Exhibition of Katonah Gallery March 13–May 23, 1982.* The Katonah Galley, 1982.

Flynn, John E., et al. *Architectural Interior Systems: Lighting, Acoustics, Air Conditioning,* 3rd ed. New York: Van Nostrand Reinhold, 1992.

Ford, Robert M. *Mississippi Houses: Yesterday Toward Tomorrow.* Mississippi State University, 1982.

Foster, Ruth S., illustrated by James Lombardi. *Homeowner's Guide to Landscaping That Saves Energy Dollars.* New York: David McKay, 1978.

Franta, Gregory, Kristine Anstead, and Gregg D. Ander. *Glazing Design*

Handbook for Energy Efficiency. American Institute of Architects, 1997.

Futagawa, Yukio, ed. and photography. *Light & Space: Modern Architecture.* Tokyo: A.D.A. EDITA Tokyo Co., 1994.

Gelbspan, Ross. *The Heat Is On: The High Stakes Battle Over Earth's Threatened Climate.* Reading, MA: Addison-Wesley, 1997.

Gipe, Paul. *Wind Power for Home and Business: Renewable Energy for the 1990's and Beyond.* Post Mills, VT: Chelsea Green, 1993.

Gipe, Paul. *Wind Energy Comes of Age.* New York: John Wiley & Sons, 1995.

Givoni, Baruch. *Man, Climate and Architecture,* 2nd ed. New York: Van Nostrand Reinhold, 1976.

Givoni, Baruch. "Integrated-Passive Systems for Heating of Buildings by Solar Energy," *Architectural Science Review,* 24 (2): 29–41, June, 1981.

Givoni, Baruch. *Passive and Low Energy Cooling of Buildings.* New York: Van Nostrand Reinhold, 1994.

Givoni, Baruch. *Climate Considerations in Building and Urban Design.* New York: Van Nostrand Reinhold, 1998.

Golany, Gideon. *Design and Thermal Performance: Below-Ground Dwellings in China.* Newark, DE: University of Delaware Press; London: Associated University Press, 1990.

Golany, Gideon, ed. *Housing in Arid Lands — Design and Planning,* New York: John Wiley & Sons, 1980.

Gordon, Gary, and James L. Nuckolls. *Interior Lighting for Designers,* 3rd ed. New York: John Wiley & Sons, 1995.

Gore, Al. *Earth in the Balance: Ecology and the Human Spirit.* Houghton/Mifflin, 1992.

Goulding, John R., J. Owen Lewis, and Theo C. Steemers, eds. *Energy Conscious Design: A Primer for Architects.* London: Batsford, 1992.

Groesbeck, Wesley, and Jan Stiefel. *The Resource Guide to Sustainable Landscapes and Gardens,* 2nd ed. Environmental Resources, 1995.

Grosslight, Jane. *Lighting Kitchens and Baths.* Tallahassee, FL: Durwood, 1993.

Grosslight, Jane. *Light, Light, Light: Effective Use of Daylight and Electric Lighting in Residential and Commercial Spaces.* Tallahassee, FL: Durwood, 1998.

Guiding Principles of Sustainable Design. Denver, CO: U.S. Dept. of the Interior, National Parks Service, Denver Center, 1993.

Guise, David. *Design and Technology in Architecture.* Rev. ed. New York: Van Nostrand Reinhold, 1991.

Guzowski, Mary. *Daylighting for Sustainable Design.* New York: McGraw-Hill, 1999.

Hawken, Paul. *The Ecology of Commerce.* New York: HarperCollins, 1993.

Hawken, Paul, Amory Lovins, and L. Hunter Lovins. *Natural Capitalism: Creating the Next Industrial Revolution.* Boston: Little, Brown, 1999.

Heinz, Thomas A. "Frank Lloyd Wright's Jacobs II House," *Fine Home Building,* 20–27, June/July 1981.

Helms, Ronald N., and M. Clay Belcher. *Lighting for Energy-Efficient Luminous Environments.* New York: Prentice-Hall, 1991.

Heschong, Lisa. *Thermal Delight in Architecture.* Cambridge, MA: MIT Press, 1979.

Hestnes, Anne Grete, Robert Hastings, and Bjarne Saxhof. *Solar Energy Houses: Strategies, Technologies, Examples.* London: James & James, 1997.

Hightshoe, Gary L. *Native Trees, Shrubs, and Vines for Urban and Rural America: A Planting Design Manual for Environmental Designers.* New York: Van Nostrand Reinhold, 1988.

Hopkinson, Ralph Galbraith, P. Petherbridge, and James Longmore. *Daylighting,* London: Heinemann, 1966.

Hottes, Alfred C. *Climbers and Ground Covers, Including a Vast Array of Hardy and Subtropical Vines Which Climb or Creep.* New York: De La Mare, 1947.

Houben, Hugo, and Hubert Guillaud. *Earth Construction: A Comprehensive Guide.* London: Intermediate Technology Publications, 1994.

Humm, O. *[Photovoltaik und Architekture] Photovoltaics in Architecture: The Integration of Photovoltaic Cells in Building Envelopes.* Basel; Boston: Birkhaüser.

International Energy Agency, *Passive Solar Commercial & Institutional Buildings: A Sourcebook of Examples and Design Insights.* S. R. Hastings, ed. New York: John Wiley & Sons, 1994.

Jaffe, Martin S., and Duncan Erley, illustrated by Dava Lurie. *Protecting Solar Access for Residential Development: A Guidebook for Planning Officials.* Washington, DC: U.S. Department of Housing and Urban Development. Office of Policy Development and Research. For sale by the U.S. Dept. of Docs., U.S. Govt. Print Office, 1979.

Jones, David Lloyd. *Architecture and the Environment: Bioclimatic Building Design.* Woodstock, NY: Overlook Press/Peter Mayer, 1998.

Jones, R. W., and R. D. McFarland. *The Sunspace Primer: A Guide to Passive Solar Heating.* New York: Van Nostrand Reinhold, 1984.

Kachadorian, James. *The Passive Solar House: Using Solar Design to Heat and Cool Your Home.* White River Junction, VT: Chelsea Green Publishing, 1997.

K. Cathcart Anders Architects. *Building Integrated Photovoltaics.* National Renewable Energy Laboratory, 1993. Available from NTIS, NTIS Order No. DE95004056.

Kellert, Stephen R., and Edward O. Wilson. *The Biophilia Hypothesis.* Washington, DC: Island Press, 1993.

King, Bruce. *Buildings of Earth and Straw: Structural Design for Rammed Earth and Straw-Bale Architecture.* Sausalito, CA: Ecological Design Press, 1996.

Knight, Paul A. *Mechanical Systems Retrofit Manual: A Guide for Residential Design.* New York: Van Nostrand Reinhold, 1987.

Knowles, Ralph L. *Energy and Form: An Ecological Approach to Urban Growth.* Cambridge, MA: MIT Press, 1974.

Knowles, Ralph L. *Sun Rhythm Form.* Cambridge, MA: MIT Press, 1981.

Knowles, Ralph L. "On Being the Right Size"; "The Ritual of Space"; "Rhythm and Ritual"; "The Solar Envelope." http://www-rcf.usc.edu/~rknowles.

Kohlmaier, Georg, and Barna von Sartory. *Das Glashaus: Ein Bautypus des 19. Jahrhunderts.* Munich: Prestel-Verlag, 1981.

Komp, Richard J. *Practical Photovoltaics — Electricity from Solar Cells,* 3rd ed. Ann Arbor: Aatec, 1995.

Konya, Allan. *Design Primer for Hot Climates.* London: Architectural Press; New York: Whitney Library of Design, 1980.

Krigger, John. *Residential Energy: Cost Savings and Comfort for Existing Buildings.* Helena, MT: Saturn Resources Management, 1994.

Lam, William M. C. *Perception and Lighting as Formgivers for Architecture.* Christopher Hugh Ripman, ed. New York: Van Nostrand Reinhold, 1992.

Lam, William M. C. *Sunlighting as Formgiver for Architecture.* New York: Van Nostrand Reinhold, 1986.

Lang, Paul V. *Principles of Air Conditioning,* 5th ed. Albany, NY: Delmar, 1995.

Leibowitz, Sandra. *Eco-Building Schools: A Directory of Alternative Education Resources in Environmentally Sensitive Design and Building in the United States.* Eco-Building Schools, 1996.

Leslie, Russell P., and Kathryn M. Conway. *The Lighting Pattern Book for Homes.* New York: McGraw-Hill, 1996.

Leslie, Russell P., and Paula A. Rodgers. *The Outdoor Lighting Pattern Book.* New York: McGraw-Hill, 1996.

Lewis, Jack. *Support Systems for Buildings.* Englewood Cliffs, NJ: Prentice-Hall, 1986.

Lighting Handbook. Mark S. Rea. New York: Illuminating Engineering Society of North America, 1999.

Los Alamos National Lab. *Passive Solar Heating Analysis — A Design Manual.* Atlanta: ASHRAE, 1984.

Lovins, Amory B. *Soft Energy Paths.* New York: Harper Colophon Books, 1977.

Lovins, Amory B., and L. Hunter Lovins. *Brittle Power: Energy Strategy for National Security.* Andover, MA: Brick House Publishing, 1982.

Lstiburek, Joseph W. *Exemplary Home Builder's Field Guide.* Triangle Park, NC: North Carolina Alternative Energy Corporation, 1994.

Lstiburek, Joseph W. *Builder's Guide — Hot-Dry & Mixed-Dry Climates; Builder's Guide — Cold Climates; Builder's Guide — Mixed Climates: A Systems Approach to Designing and Building Homes That Are Healthy, Comfortable, Durable, Energy Efficient and Environmentally Responsible.* Building Science Corporation, 1998.

Lstiburek, Joseph W. *Builder's Guide to Cold Climates: Details for Design and Construction.* Newtown, CT: Taunton Press, 2000.

Lstiburek, Joseph W. *The Builder's Guide to Mixed Climates: Details for Design and Construction.* Newtown, CT: Taunton Press, 2000.

Lstiburek, Joseph W., and John Carmody. *Moisture Control Handbook.* New York: Van Nostrand Reinhold, 1993.

Lyle, David. *The Book of Masonry Stoves: Rediscovering an Old Way of Warming.* White River Junction, VT: Chelsea Green Publishing, 1984.

Lyle, John Tillman. *Regenerative Design for Sustainable Development.* New York: John Wiley & Sons, 1994.

Lyle, John Tillman. *Design for Human Ecosystems: Landscape, Landuse, and Natural Resources.* Washington, DC: Island Press, 1999.

Mann, Peter A. *Illustrated Residential and Commercial Construction.* Englewood Cliffs, NJ: Prentice-Hall, 1989.

Matus, Vladimir. *Design for Northern Climates: Cold-Climate Planning and Environmental Design.* New York: Van Nostrand Reinhold, 1988.

Mazria, Edward. *The Passive Solar Energy Book: Expanded Professional Ed.* Emmaus, PA: Rodale Press, 1979.

McHarg, Ian L. *Design with Nature.* New York: John Wiley & Sons, 1995.

McIntyre, Maureen, ed. *Solar Energy: Today's Technologies for a Sustainable Future.* Boulder, CO: ASES, 1997.

McPherson, E. Gregory, ed. *Energy Conserving Site Design.* Waldorf, MD: American Society of Landscape Architects, 1984.

Miller, Burke. *Buildings for a Sustainable America Case Studies.* Boulder, CO: ASES, 1997.

Millet, Marietta S. *Light Revealing Architecture.* New York: Van Nostrand Reinhold, 1996.

Moffat, Anne Simon, and Marc Schiler, drawings by Dianne Zampino. *Landscape Design That Saves Energy.* New York: Morrow, 1981.

Moffat, Anne Simon, and Marc Schiler. *Energy-Efficient and Environmental Landscaping.* South Newfane, VT: Appropriate Solutions Press, 1993.

Moore, Fuller. *Concepts and Practice of Architectural Daylighting.* New York: Van Nostrand, 1985.

Moore, Fuller. *Environmental Control Systems.* New York: McGraw-Hill, 1993.

Nabokov, Peter, and Robert Easton. *Native American Architecture.* New York: Oxford University Press, 1989.

National Audubon Society and Croxton Collaborative, Architects. *Audubon House: Building the Environmentally Responsible, Energy-Efficient Office.* New York: John Wiley & Sons, 1994.

Norton, John. *Building with Earth: A Handbook.* London: Intermediate Technology Publications, 1996.

Norwood, Ken, and Kathleen Smith. *Rebuilding Community in America.* Berkeley, CA: Shared Living Resource Center, 1995.

NY-Star Builder's Field Guide. Albany, NY: NY-Star, Inc., 1994.

Olgyay, Aladar, and Victor Olgyay. *Solar Control and Shading Devices.* Princeton, NJ: Princeton University Press, 1957.

Olgyay, Victor. *Design with Climate: A Bioclimatic Approach to Architectural Regionalism.* Princeton, NJ: Princeton University Press, 1953.

Olivieri, Joseph B. *How to Design Heating-Cooling Comfort Systems,* 4th ed. Troy, MI: Business News Pub., 1987.

Ottesen, Carole. *The Native Plant Primer: Trees, Shrubs, and Wildflowers for Natural Gardens.* New York: Crown, 1995.

Panchyk, Katherine. *Solar Interior: Energy Efficient Spaces Designed for Comfort.* New York: Van Nostrand Reinhold, 1991.

Pearson, David. *The Natural House Catalog: Everything You Need to Create an Environmentally Friendly Home.* New York: Fireside, 1996.

Pearson, David. *The New Natural House Book-Creating a Healthy, Harmonious, and Ecologically Sound Home.* Rev. New York: Simon & Schuster, 1989; 1998.

Pepchinski, Mary. "Commerzbank." *Architectural Record,* 01.98.

Perlin, John. *From Space to Earth: The Story of Solar Electricity.* Ann Arbor, MI: Aatec, 1999.

Petit, Jack, Debra L. Bassert, and Cheryl Kollin. *Building Greener*

Neighborhoods: Trees as Part of the Plan, 2nd ed. Washington, DC: American Forests: Home Builders Press, 1998.

Photovoltaics in the Building Environment. American Institute of Architects (AIA), 1996. Available from AIA Orders, 2 Winter Sport Lane, P.O. Box 60, Williston, VT 05495-0060. 800-365-2724. ISBN No. 1879304694.

Potts, Michael. *The New Independent Home: People and Houses that Harvest the Sun.* Rev. ed. White River Junction, VT: Chelsea Green Publishing, 1999.

Ramsey/Sleeper Architectural Graphic Standards, 9th ed., John R. Hoke, ed. New York: John Wiley & Sons, 1994 (American Institute of Architects).

Rapoport, Amos. *House Form and Culture* (Foundations of Cultural Geography Series). Englewood Cliffs, NJ: Prentice-Hall, 1969.

Recommendations to the United Nations Commission on Sustainable Development: Final Report on the NGO Renewable Energy Initiative. Freiburg, Ger.: ISES, 1995.

Reid, Esmond. *Understanding Buildings: A Multidisciplinary Approach.* Cambridge, MA: MIT Press, 1984.

Robbins, Claude. *Daylighting: Design and Analysis.* New York: Van Nostrand Reinhold, 1986.

Robinette, Gary O., and Charles McClenon. *Landscape Planning for Energy Conservation.* New York: Van Nostrand Reinhold, 1983.

Robinette, Gary O., ed. *Energy Efficient Site Design.* New York: Van Nostrand Reinhold, 1983.

Rocky Moutain Institute Staff. *Green Development: Integrating Ecology and Real Estate.* New York: John Wiley & Sons & Sons, 1998.

Rudofsky, Bernard. *The Prodigious Builders: Notes Toward a Natural History of Architecture.* London: Secker and Warburg, 1977.

Rudofsky, Bernard. *Architecture Without Architects: A Short Introduction to Non-Pedigreed Architecture.* New York: Museum of Modern Art; Garden City, NJ: Doubleday, 1965.

Ruffner, James A., and Frank E. Bair. *The Weather Almanac,* 2nd ed. Detroit, MI: Gale Research, 1977.

Rush, Richard D., ed. *The Building Systems Integration Handbook.* New York: John Wiley & Sons, 1986.

Santamouris, Mat, and D. Asimakopoulos, eds. *Passive Cooling of Buildings.* London: James & James, 1996.

Scheer, Hermann. *A Solar Manifesto: The need for a total energy supply—and how to provide it.* London: James & James, 1994.

Schiler, Marc. *Simplified Design of Building Lighting.* New York: John Wiley & Sons, 1997.

Schiler, Marc, ed., *Simulating Daylight with Architectural Models.* Los Angeles: University of Southern California, 1985.

Sick, Friedrich, and Thomas Erge, eds. *Photovoltaics in Buildings: A Design Handbook for Architects.* London: James & James, 1996.

Singh, Madanjeet. *The Timeless Energy of THE SUN—for Life and Peace with Nature.* San Francisco: Sierra Club Books, 1998.

Solar Age [This periodical has been discontinued but old issues are still very valuable]. Published by Solar Age Inc., 1976–1986; temporarily continued as *Progressive Builder,* published by International Solar Energy Society, 1986–1987.

Solar Today. American Solar Energy Society. Monthly.

Steffy, Gary R. *Architectural Lighting Design.* New York: John Wiley & Sons, 1990.

Steffy, Gary R. *Time-Saver Standards for Architectural Lighting.* New York: McGraw-Hill, 2000.

Stein, Benjamin, and John S. Reynolds. *Mechanical and Electrical Equipment for Buildings, Ninth Edition.* New York: John Wiley & Sons, 1999.

Stein, Richard G. *Architecture and Energy: Conserving Energy Through Rational Design.* New York: Anchor Books, 1977.

Sternberg, Guy, and Jim Wilson. *Landscaping With Native Trees.* Shelburne, VT: Chapters Publishers, 1995.

Steven Winter Associates. *The Passive Solar Design and Construction Handbook.* Michael J. Crosbie, ed. New York: John Wiley & Sons, 1998.

Strong, Steven J., *The Solar Electric House: Energy for the Environmentally Responsive, Energy-Independent Home.* White River Junction, VT: Chelsea Green, 1993.

Strong, Steven J., with William G. Scheller. *The Solar Electric House: A Design Manual for Home-Scale Photovoltaic Power Systems.* Emmaus, PA: Rodale Press, 1987.

Szokolay, S. V. *Environmental Science Handbook for Architects and Builders.* New York: John Wiley & Sons, 1980.

Tabb, Phillip. *Solar Energy Planning: A Guide to Residential Settlement.* New York: McGraw-Hill, 1984.

Tao, William K. Y. *Mechanical and Electrical Systems in Buildings,* 2nd ed. Upper Saddle River, NJ: Prentice-Hall, 2000.

Taylor, John S. *Commonsense Architecture: A Cross-Cultural Survey of Practical Design Principles.* New York: Norton, 1983.

Thomas, Randall, ed. *Environmental Design: An Introduction for Architects and Engineers.* New York: E & FN Spon, 1999.

Traister, John E. *Residential Heating, Ventilating, and Air Conditioning: Design and Application.* Englewood Cliffs, NJ: Prentice-Hall, 1990.

Trost, J. *Electrical and Lighting.* Upper Saddle River, NJ: Prentice-Hall, 1999.

Trost, J. *Heating, Ventilating, and Air Conditioning.* Upper Saddle River, NJ: Prentice-Hall, 1999.

Tuluca, Adrian. *Energy-Efficient Design and Construction for Commercial Buildings.* New York: McGraw-Hill, 1997.

Underground Space Center, University of Minnesota. *Earth Sheltered Housing Design: Guidelines, Examples, and References.* New York: Van Nostrand Reinhold, 1979.

Van der Ryn, Sim. *The Toilet Papers: Recycling Waste and Conserving Water.* Ecological Design Press, 1995.

Van der Ryn, Sim, and Peter Calthorpe. *Sustainable Communities: A New Design Synthesis for Cities, Suburbs, and Towns.* San Francisco: Sierra Club Books, 1986.

Van der Ryn, Sim, and Stuart Cowan. *Ecological Design.* Washington, DC: Island Press, 1995.

Van der Ryn, Sim, annotator. *Designing Sustainable Systems: A Reader,* Vols. 1 & 2, 1993. Contains many hard-to-find key source articles. [Contact Ecological Design Institute — see Appendix J.]

Vickery, Robert L. *Sharing Architecture.* Charlottesville: University Press of Virginia, 1983.

Watson, Donald. *Designing and Building a Solar House: Your Place in the Sun.* Rev. ed. Charlotte, VT: Garden Way Publishing, 1977.

Watson, Donald, and Kenneth Labs. *Climatic Design: Energy-Efficient Building Principles and Practices.* New York: McGraw-Hill, 1983, [1992?].

Wilson, Edward O. "Is Humanity Suicidal: We're Flirting with the Extinction of Our Species." *New York Times Magazine,* May 30, 1993.

Wirtz, Richard. *HVAC/R Terminology: a Quick Reference Guide.* Upper Saddle River, NJ: Prentice Hall, 1998.

World Commission on Environment and Development. *Our Common Future.* Oxford University, 1987.

Wright, Lawrence. *Home Fires Burning: The History of Domestic Heating and Cooking.* London: Routledge & Kegan Paul Ltd., 1964.

Zeiher, Laura C. *The Ecology of Architecture: A Complete Guide to Creating the Environmentally Conscious Building.* New York: Whitney Library of Design, 1996.

Zweibel, K. *Harnessing Solar Power: The Photovoltaics Challenge.* New York: Plenum Press, 1990.

Aalto, Alvar, 393
Absolute Humidity, *see* Humidity ratio
Absorbtance, 42–44. *See also* Radiation
Absorption refrigeration cycle, 485–486
Acid rain, 25
Active solar, 65, 145, 159, 186–199, 478, 532
   batch type, 195
   collectors, 188–199, 549
   cooling, 187
   cost effectiveness, 187
   definition, 187
   design of 193–195
   domestic hot water, 187–188, 189–190, 194–195, 197–199, 294, 533, 535, 542–543, 549
   efficiency, 187–188
   for ventilation air, 188, 196–197
   history, 188
   hot air collectors, 191–192
   integral collector storage (ICS), 195–196
   orientation, 193–194, 199
   panels, 172
   space heating, 187, 194, 197, 294
   storage tank, 190, 192, 194–196
   swimming pool heating, 187–189, 193, 197
   systems, 189–192, 195–197
   tilt, 194
   ventilation air, 532
Adaptation of eye, 330, 335, 343
Adiabatic change, 61
Adobe, 159, 165, 246, 251, 456–457
Affluence, 15
Afterimage, 338
AIA/Research Corporation, 75
Air change method, 436
Air conditioning: *see also* Cooling; Heating
   systems for large buildings, 496–504
   systems for small buildings, 493–496
Air filtration, 512
Air flow, 56, 255–270
   around buildings, 256–258
   diagrams of, 265–267
   through buildings, 258–270
   types, 255–257
   velocities, 266, 268, 302
Air handling unit (AHU), 494–496, 498
Air pollution, 18, 22, 30, 510–512, 556
Air spaces, 443, 445–447, 453
Air supply, 505–509
Air temperature, 56

Air-to-air heat exchangers, *see* Heat recovery
Air velocities, *see* Air flow
Albedo, 322, 438–439
Alberti, Leone Battista, 141
Alcohol, *see* Ethanol alcohol
Alhambra, Spain, 249
Allee, 315, 320
   pleached, 315, 320
Alternating current, 175
Altitude angle, 128, 131–133
Alumina, activated, 277
Aluminum, 49
Ambient luminescence, 354
American Institute of Architects (AIA), 15
American Society of Heating, Refrigeration, and Air-Conditioning Engineers (ASHRAE), 51, 61–62, 509
American System, 330. *See also* System International
Ammonia, 485
Amoeba analogy, 16–18
Amorphous cells, *see* Photovoltaics cells
Ampere (amp), 174
Amphitheaters, 231
Angle of exposure, 44
Angle finder, 575
Angle of incidence, 238
Annual-cycle energy systems (ACES), 514
Antarctic circle, 129
Antebellum house, 2
Antifreeze, 189, 190
Anti-reflectivity glass, *see* Glass
Antistratification ducts, 541, 545, 548
Apartment building, 530–533
Appliances, 175, 184, 468
APS Building, 181
APS factory, *see* APS Building
Apulia, Italy, 249
Aquatic Center at Georgia Institute of Technology, 181–182, 186
Arbors, 231, 233, 315
Arcades, 202–203
Architectural lighting, 427–429
   cornice (soffit), 429
   cove, 427–428
   luminous-ceiling, 428
   valance (bracket), 429
Arctic circle, 127–129
Arkin Tilt Architects, 522
Array, *see* Photovoltaics

Artificial lighting, 408–409. *See also* Electric lighting
Artificial sky, *see* Daylighting
ASHRAE, *see* American Society of Heating, Refrigeration, and Air-Conditioning Engineers
ASHRAE Standard 90–75, 342
Asian architecture, 204, 472. *See also* Japanese architecture; Chinese architecture; Indian architecture; Middle Eastern architecture
Asphalt, 313
Atlanta, GA, 333
Atmosphere, 21, 30, 68
Atrium, 5–6, 249, 360–362, 440, 458, 460, 530, 533, 538–539, 544–545, 548, 550–552, 554–555, 560
Audubon Society headquarters, 14
Auxiliary furnaces, 191
Awnings, 154, 213–215, 231–232, 316, 319, 528
Azimuth angles, 131–133, 223

Backup power, 175
Baer, Steve, 152, 167
Baffles, 419–420
Balance point temperature (BPT), 216
Balance of system equipment, *see* Photovoltaics
Balcomb house, 146, 159–160
Balcomb, J. Douglas, 159, 162–163
Ballast, 412, 315
   electronic, 354
   magnetic, 354
Bartlett, Albert A., 16, 20
Baseboard convectors, 480
Basilicas, 4–5
Batch type heater, 195. *See also* Active solar, batch type
Bateson State Office building, 214, 270, 273, 374, 378
Batteries, 29, 175–176
Batwing lighting pattern, 419
Bauhaus, 142
"Bay" windows, 2, 248–249, 360
Beamed daylighting, 550–552
Becquerel, 172
Bell Laboratories, 172
Belvederes, 253–254
Berkebile, Bob, 34
Berlin, Germany:
   Integrated Passive and Hybrid Solar Multiple Housing, 146

Berms, 458, 461–463. *See also* Earth sheltering
Bernoulli effect, 257–258
Berry, Wendell, 34
Binoculars, 340
Bioclimatic Chart, *see* Psychrometric-Bioclimatic Chart
Biodiversity, 12, 19
Biological activity, 277
Biological machine, 52
Biological Sciences Building at the University of California at Davis, 241–243
Biophilia, 308
"Black hole" effect, 165
Black light, 328
Blue Cross and Blue Shield Building in Towson, MD, 492
Boilers, 478–479, 494, 532
Books (general resource), 601
Borrani, Odoardo, 337
Bradbury Building, 360–361
Brain, 335, 338, 340
Braungart, Michael, 15
Braziers, 472
Breuer, Marcel, 142, 201
Brightness, 330, 335, 340–341. *See also* Luminance
    constancy, 337
    ratios, 336, 342–343, 355, 372
    relativity of, 337
"Brise soleil," 7, 206–207, 227, 243, 265
British Thermal Unit (BTU), 39, 46
Bronowski, Jacob, 37
Brown, G. Z., 588
Bruder, William P., 558
Brundtland Report, 1987, 11
Building Integrated PV, *see* Photovoltaics
Building management system (BMS), 556
Burt Hill Kosar Rittelmann Association, 525
Butti, Ken, 22

Cabinet convectors, 480
Calcium chloride, 277
Calcium chloride hexhydrate, 165
Callaway Gardens, 264
Calories, 39
Campbell, Colin J., 24
Cancer, 354
Candelas, *see* Candlepower
Candlepower, 328–330, 425
    distribution graphs, 329, 419
Canopy bed, 53
Capillary action, 463–464. *See also* Moisture control
Cappadocia, Turkey, 250–251. *See also* Middle Eastern architecture
Carbon dioxide, 19, 25, 30, 48, 68
Carbon neutral, 30
Carrying capacity, 15
Case studies, 552–562
Cast-iron radiators, 478–479
CBS Building, 514–516
CDD, *see* Degree days, cooling graphs
CD-ROMS, (resource), 602
Ceiling diffusers, *see* Diffusers
Cells, *see* Photovoltaics

Cellulose, *see* Insulation
Center for Environmental Sciences and Technology Management, 182
Center for Regenerative Studies, 14–15
Center le Corbusier, 7
Central station system, 496
Centre Pompidou, 518
Chaco Canyon, NM, 142–143
Chandigarh, India 206. *See also* Indian architecture
Change of state, 38–39
Chapel at Ronchamp, *see* Ronchamp
Charcoal, 472
Charge controller, 176
Charleston, SC:
    vernacular architecture, 3
Chattahoochee River, 312
Chernobyl, 25
Chickee, 252. *See also* Native American architecture
Chilled water, 45
Chillers, 486
Chimney, 472, 474
Chinese architecture, 204
Chlorofluorocarbon (CFC), 21, 485
Churches, 5
Circadian cycles, 352, 354
Cité de Refuge, 204, 206
City Hall of Tempe, Arizona, 237–238
Civil/Mineral Engineering Building at the University of Minnesota, 395
Classical architecture 141, 202. *See also* Greek architecture; Roman architecture
Classical portico, 4, 7, 10, 202–203, 253
Classical revival style, *see* Revivalism, classical
Classrooms, 64
Clean energy, 172
Clerestory, sawtooth, 179
Clerestory windows, 4, 146, 149, 211–212, 294, 522, 525, 538, 546. *See also* Daylighting
Climate, 1, 67–81
    anomalies, 73
    change, *see* Global warming
    cold, 4, 47, 284, 306, 317, 321
    data tables, 74
        explanations, 75–81
        tables, 82–115
    design strategies, 76, 116–123
    hot & dry, 2, 47, 71, 246, 284, 306, 315, 318
    hot & humid, 2–3, 47, 284, 306, 315, 319, 321
    macroclimate, 68–71
    microclimate, 68, 71–75
    mild & overcast, 2, 4
    priorities, 82–114
    regions (U.S.) 74–81
    temperate, 316
    wet, 71
Climate facade, 555–558
Clinometer, 575
Clo, 65
Coal, *see* Energy, coal
Coal liquification, 24
Coefficient of performance (COP), 477

Cogeneration, *see* Combined heat and power
Cold draft, 44
Colleges teaching energy-conscious design, 595–600
Colonial architecture, 144
Colonial Williamsburg, VA, 27, 233, 253–254
Colonnades, 202–203, 282
Color, 330–335, 355. *See also* Daylighting, color
    complementary, 338
    constancy, 338
    rendering index (CRI), 334
    rendition, 334
    temperature (CT), 334
Colorado Mountain College, 169, 538–543
Colosseum, 231
Combined heating and power, 48, 476, 512
Combustion air, 474, 481, 483
Combustion gases, 510
Comfort, *see* Thermal comfort
Comfort range, 56, 78
Comfort zone, *see* Thermal comfort, comfort zone
Commerzbank, 554–558
Committee on the Environment, AIA, 15
Community, 280
Community design, 19
Community planning, 280, 322, 322–323
Compactness, 4, 8–9, 439–440, 562
Composite sandwich panel, 398
Compressive refrigeration cycle, 484–485
Compressors, 485–487, 493–496
Computers, 346
Computer monitors, 351, 420
Computer software, 585–588
    Climate Consultant, 585
    Energy Scheming, 588
    OPAQUE, 585
    SOLAR 2, 585, 587
    SOLAR 5, 585–587
Concentrating solar collectors, 190
Condensation, 277, 464–466, 503
Conduction, 39, 52
Cone cells, 335
Conoco, Inc. Headquarters, 208
Conservation, *see* Energy, conservation; Energy efficiency
Conservatories, 142–143, 158
Conservatory of Princess Mathilda Bonaparte in Paris, France, 143
Constant-air-volume (CAV) system, 498
Contrast, 340–341, 346
Controllers, 184
Convection, 166
    forced, 52
    natural, 52
Convection currents, 68
Convective-loop system, *see* Thermosiphon
Convectors, 478–480
Cool daylight, 238
Cooling, 322, 484–486. *See also* Active solar, cooling; Air conditioning, systems
    history, 484

radiant, 504–505, 555, 558
refrigeration cycles, 484–486
systems, 489–493
tower, 491–493, 496, 498, 504–505, 513
Cooling coils, 481
Copernicus, 128
Coriolis force, 69
Cosine law, 128–129, 166, 283
Courtyards, 249–250
Crawl spaces, 448
Cross-ventilation, 144, 555
Crystal Cathedral, 386–388
Crystal Palace, 54–55, 360
Crystalline silicon cell, *see* Photovoltaics, cells
Cupolas, 254, 263
Current (electric), 174

Daily temperature range, 47
Dams, 31–32
Damper, 191, 481, 506
Davis, CA, *see* Village Homes Subdivision, Davis, CA
Davis house, 167
Daylight, 198, 333
Daylight factor, 368–369, 403
Daylighting, 5–6, 8, 179, 239–240, 294, 312, 314, 356, 523–525, 534–535, 537, 544, 553–558, 558–561
artificial sky, 402–403
availability, 79–80, 82–115
baffles, 385–387, 391–394
beamed, 394–395
biological needs, 365
clear sky, 365–367, 402–404
clerestories, 389–394
orientation, 391
spacing, 391
color, 375
conceptual model, 367–368
cool daylight, 369–371, 383
efficacy of, 369–370
goals of, 371–372
history of, 360–364
illumination levels, 366–368, 371
laws, 360
light scoops, 363, 389, 391–392
light shelves, 379–381, 383, 392–393
light wells, 394–395
louvers, 375, 378–379, 381–382
monitors, 389–390
nature of, 365–367
orientation, 372–373
overcast sky, 365–366, 369, 402–404
physical models, 368–369, 378, 382, 394, 400–404
sawtooth, 389–394
skylights, 361–363, 373, 384–388
spacing, 384–385
strategies, 372–382, 384–397
sunlighting, 370
top lighting, 372–373
tubular, 394–395
why, 364–365
window:
area, 376
location, 375–377

strategies, 375–382
Daylight savings time, *see* Time, daylight savings
Deciduous plants, 213–214
Degree days, 80–81
cooling graphs, 80–81, 83–115
heating graphs, 80–81, 83–115
Degrees Kelvin, 334
Dehumidification, 57–59, 65, 187, 278, 490
Density, 41
Denver Airport, 398
Denver Airport Terminal Building, 397
Depression, 354
Deregulation, 175, 183, 186
Design-equivalent temperature difference (DETD), 438
Design guidelines, *see* Direct gain, design guidelines
Design priorities, 76–77, 82–114
Design strategies, 76. *See also* Climate, design guidelines
Desiccants, 255, 277–278
Desuperheaters, 488
Dew point temperature (DPT), 59, 465, 493
Diamond, Jared, 12
Differential thermostats, 191
Diffusers, 495, 507–508
Direct current, 175
Direct expansion systems, 489
Direct gain, 147–152, 154, 156, 161–162
design guidelines, 150–152
heat, 198
Directionally selective, 239–240
Discharge lamps, 412. *See also* Fluorescent; Mercury; Metal halide; High Pressure
Displacement ventilation, 511–512
District heating and cooling, 512
Diurnal temperatures, *see* Temperature, diurnal range
Dixon, John Morris, 1
Dog trot, 253
Dome, 249–250, 254, 362
Domestic hot-water, 48, 488. *See also* Active solar, domestic hot water
Double-duct system, 500–501
DPT, *see* Dew point temperature
Draft, 56
Drain-back system, 190. *See also* Active solar, systems
Drapes, 164
Dubai, United Arab Emirates:
vernacular architecture 3, 246–247. *See also* Middle Eastern architecture
Duct, 505–509
exposed, 507–508, 517–518
fabrics, 507–508
flexible, 481, 495, 506
installations, 481
insulations, 495, 506
leakage, 509
sizing, 505
systems, 481–483, 490–491, 494–495, 497–501, 504, 560–562
types, 505, 508

Ductless split system, 496
Duct systems, 481–482
Durant Middle School, 389, 391
Dynamic buildings, 9–10

Earth, 487
axis, 126–129
orbit, 126–129, 131
Earth berms, 144, 317–318
Earth sheltering, 251, 457–463, 538
and noise, 458–459
regions, 463
Earth summit in Rio de Janeiro, 15
Earth tubes, 277. *See also* Passive cooling
Easter Island, 12
East Wing, National Gallery of Art, 388
Eco-effectiveness, 15
Ecology, 31
Economizer cycle, 509, 512
Ecosystems, 19
Edison, Thomas, 408
Eddy currents, 255–257
Educational opportunities, 595–600
Efficacy, 409, 417
Eggcrates (shading devices), 210, 214, 227, 239, 345
Egyptian architecture, 246–247, 250, 326, 341. *See also* Middle Eastern architecture
Egyptian hypostyle halls, 389
Ehrlich, Paul, 15
Electrical demand, 364–365
Electrical efficiency, 47–48
Electrical energy, 174, 409
Electrical load, 184
Electric cars, 182
Electric discharge lamps, *see* Lamps
Electricity, 475–477
Electric lighting, *see also* Artificial lighting
controls, 399
and daylighting, 399–400
dimming, 430
occupancy sensors, 430
photo sensors, 430
Electric lighting systems, 421–425, 427–429. *See also* Architectural lighting; Lamps; Luminaires
accent, 423
design rules, 430–431
dimming, 430
general, 421–422
maintenance, 430
remote source, 424–425
switching, 430
task/ambient, 419, 422–423, 549, 552
Electric-resistance heating, 32
Electric strip heater, 493–494, 498
Electrolysis, 34, 176
Electromagnetic radiation, *see* Light; Radiation
Electromagnetic Spectrum, 41–42, 328, 332, Color plate 1, 370
Electronic filters, 512
Elevation, 72
Embodied energy, *see* Energy, embodied
Emerald People's Utility District (PUD) Building, 270, 533–535
Emittance, 42–44. *See also* Radiation

Energy:
    biomass, 29–30
    coal, 25
    consumption, 22
    conversion, 48
    conscious design, 9
    conservation, 9
    crisis of 1973, 22, 34, 243, 364
    efficient design, 9
    embodied, 49
    geothermal, 26, 32–33
    high-grade, 172, 187
    hydroelectric, 29, 30–32
    low-grade, 187
    natural gas, see Gas, natural
    nonrenewable, 22
    nuclear, see Nuclear, energy
    oil, 22, 24, 408
    renewable, 22, 26
    solar, see Solar energy
    storage
    thermal
    wind, 26–29, 48. See also Wind
Energy efficiency, 435
Energy Efficiency Ratio (EER), 468
Energy management system, 430, 549
Engawa, 202, 204, 252
English architecture, 285, 360–361. See
    also specific buildings
Engine-generators, see Generators, engine
Enthalpy, 60–61
Envelope-dominated buildings, 75–76,
    216, 218, 220–222, 225, 284
Environmental Building News (EBN), 12,
    18–19
Environmental Impact, 15
Equator, 129, 134
Equinox, 128–130
Equinox Design, 533
Equivalent spherical illumination (ESI),
    347–348
Ethanol alcohol, 29–30
Eufaula, AL:
    vernacular architecture, 203, 439
Eureka, CA:
    vernacular architecture, 4
Evaporation, 52, 248–249, 276
Evaporative condensers, 490–491
Evaporative coolers, 273, 308, 486
Evaporative cooling, see Passive cooling
Evaporator coils, 485–487, 489–491,
    493–496
Excessive-brightness ratios, 371, 375, 383
Exhaust air, 510–511
Exhaust vents, see Vents
Exponential Growth, see Growth, expo-
    nential
Expo '67, Montreal, see U.S. Pavilion,
    Expo '67, Montreal
Exposure angle, 224, 340
Extended plenum, 483
Eye, 335–337, 342, 353–356, 409. See also
    Adaptation of eye

Fan coil units, 490, 501, 503
Fans, 263, 269–270, 503
    ceiling, 264
    exhaust, 548

whole house, 264
    whole-house attic, 266
Fiberglass, see Insulation
Fiber optics, 395–396, 424–425
Fig ivy, 208
Figure/background effect, 338
Filaments, 332. See also Tungsten fila-
    ments
Filtering, 490
Filters, 493, 512
Fine Arts Center at Arizona State
    University, Tempe, 234
Fins, 210, 214, 219–220, 223–226
    movable, 224
Fin-tubes (coils), 478, 480
Fin wall, 260, 262
Fireplaces, 30, 44, 254, 472–475. See also
    Heating systems
Fitch, James Marston, 471
Flag lots, 292
Flat plate collectors, see Active solar, col-
    lectors
Florida Solar Energy Center, 381, 389
Florida vernacular architecture:
    Native American, 252. See also Native
    American architecture
Flow, 31
Flues, 48
Fluorescence, 328
Fluorescent lamp, 328, 332, 369
Focal glow, 354
Focal point, 411
Footcandle, 329–330, 340–342, 347–348,
    354, 425. See also Lux;
    Illuminance; Illumination
Formaldehyde, 510
Formgivers, 9
Fossil fuels, 22–23, 30, 47–48. See also
    Energy
Foster and Partners, 549, 554
Foster, Richard, 228–229
Fountains, 246, 249, 306, 308, 321, 493,
    523
Four Times Square, 48, 181
Fovea, 335
Foveal surround, 336
Franklin, Benjamin, 474
Freon, 39
Fuel cells, 48–49, 176
Fuels, see Gas; Heating; Oil
Fuller, Buckminster, 228–229, 516
Full shade line, 220–222
Full spectrum, 333–334
Full sun line, 220–222
Furnaces, 481–483, 494. See also Heating,
    systems
Fusion, 126

Galileo, 128
Galleria Vittorio Emanuele, 54
Gamble House, 205
Garbage, 30
Gas:
    furnaces, 481
    natural 20, 24, 30, 33, 48
    methane, 20, 22, 29, 30, 33
    lamps, 408
    landfills, 30

liquified natural gas, 24
Generators, 32
    engines, 176, 183
Geodesic domes, 54, 228
Geo-exchange heat pumps, 32, 487–488
Georgia Tech Natatorium, see Aquatic
    Center at Georgia Institute of
    Technology
Georgia Tech University, 181–182, 186
Geothermal, see Electricity, geothermal;
    Geo-exchange
German architecture, 142, 146
Gestalt theory, 338
Getty Museum, 307
Geysers, 33
Givoni, Baruch, 168
Glare, 344–348, 377–379, 381–382, 385,
    391. See also Veiling reflections
    control, 239
    direct, 344, 346–347, 355, 371–372,
    419–421
    disability, 344
    discomfort, 344
    indirect, 345
Glass, 142, 237. See also Glazing
    anti-reflectivity, 350–351
    clear, 237
Glass block, 239, 360, 383–384, 396–397
Glauber's salt, 166
Glazing, 237–239, 383–384. See also
    Glass
    clear, 237–239
    directionally selective, 239
    dispersed-particle, 239
    electrochromic, 239
    heat absorbing, 237–239
    laminated, 397
    liquid-crystal, 239
    low-e, 239, 383, 452, 454–455
    photochromic, 239
    reflective, 237–239, 454
    responsive, 239, 371, 383
    selective low-e, 409, 454–455
    spectrally selective, 238–239, 370–371,
    383
    tinted, 237–239, 371, 383
    translucent, 383–384
Global Climate Change Treaty: The Kyoto
    Protocol, 15
Global warming, 15, 18–20, 22, 25–26,
    30, 48, 68. See also Climates;
    Greenhouse effect
Gnomon, 137, 154, 156, 577–578
Goddard Institute for Space Studies, 19
Golden Pavilion, 204
Gorrie, Dr. John, 484
Gothic architecture, 5, 6, 360
    cathedrals, 326
Governor's Palace, Colonial
    Williamsburg, Virginia, 233,
    253–254
Graves, Michael, 203, 233
Greek architecture, ancient, 22–23, 142,
    280, 336. See also specific build-
    ings and sites
Greek portico, see Classical portico
Greek revival, see Revivalism, classical
"Green," 9, 14, 18–19

Green and Green (firm), 205
Greenhouse, 142, 158. *See also* Sun
　　spaces
　　effect, 20, 42, 68, 142, 148. *See also*
　　　Global warming
　　gases, 20–21
　　global, 21
　　horticultural, 168–169, 273
Gregory Bateson Building, 543–549
Grid-connected PVs, *see* Photovoltaics
Grid (electric), *see* Power grid
Grille, 495, 506, 509
Groin vaults, 360
Gropius, Walter, 142, 245
Growth, economic, 18
Growth, exponential, 16–18
Growth, population, 15–16, 30
Guggenheim Museum, 362
Gulf Coast:
　　vernacular architecture, 252–253
Gymnasiums, 64

Halogen, 332
Hardwick Hall, 360–361. *See also* English
　　architecture
Hardy Holzman Pfeiffer, 517
Harold Hay house, 167
Hatchobaru power station, 32–33
Hay, Harold, 167, 271
HDD, *see* Degree days, heating
Head (pressure), 31–32
Heat, 38
　　of fusion, 39
　　high-grade, 47
　　latent, 39, 60–61, 273, 278, 510–511
　　low-grade, 47
　　radiant, 38, 41–44
　　sensible, 38, 60
　　total heat, *see* Enthalpy
　　of vaporization, 39, 484–485
Heat avoidance, 246, 270, 526, 545
Heat bridges, 448
Heat capacity, 45
　　by volume, 45–46
　　by weight or mass, 45
Heat-exchanger, 189–190, 474, 481,
　　　510–511, 532. *See also* Heat
　　　recovery
Heat flow, 434–438
　　by conduction, 45–46, 435–437, 453
　　by convection, 40, 45, 435–437, 446,
　　　453
　　by infiltration, 435–438
　　by radiation of heat, 435–439,
　　　446–453. *See also* Heat, radiant;
　　　Radiation
　　by transmission, 435–439
　　transport, 40–41
　　by ventilation, 438–438
Heat-flow coefficient, 46, 438, 447
Heat gain, 437–438. *See also* Heat flow
　　solar, 437–439
Heating, 472–483. *See also* Active solar,
　　　space heating
　　combined heating and power (CHP),
　　　476
　　comparison of systems, 476
　　electric, 475–477

from lights, 477
fuels, 475
heat pump, 476–477
　　resistance, 476–477, 493
history of, 472–475
hot air systems, 477, 481–483, 498
hot water (hydronic), 478–480,
　　498–504
radiant, 476–480
radiant floor, 476–479
systems, 475–476. *See also* Furnaces;
　　Stoves
Heat island, 73, 322
Heat loss, 435–437. *See also* Heat flow
　　human, 52–56
Heat pipes, 510
Heat production in humans, *see*
　　Metabolic rate
Heat pumps, 476–477, 486–488, 494
　　air source, 32
　　geo-exchange, 32, 477, 487–488
　　geothermal, 477, 487–488
　　ground coupled, 477, 487–488
　　ground-source, *see* Heat pump, geo-
　　　exchange
　　refrigeration machines, 45
　　water-loop systems, 503
Heat-recovery, 437, 475, 510–511
Heat rejection, 558
Heat sink, 32, 45, 246, 251, 254–255,
　　269–273, 275, 457–458, 489
Heat-storage materials, 165–166
Hedgerow, 315, 320
Height (finding of), 573–575
　　tools for, 575
Heliocaminus, 142, 280
Heliodon, *see* Sun machine
Heliostat, 394, 396
Hermitage (Andrew Jackson's home), 202
Hibachi, 472
High-efficiency particulate (HEPA) filters,
　　512
Highlighting, 352, 355
High pressure sodium lamp, 328
High velocity supply air, 501, 506
Historic preservation work, 49, 496
Hoffmann-La Roche Building, 516–517
Hogans, 251
Holden, John, 15
Hongkong Bank, 549–554
Hood College Resource Management
　　Center, 154, 169, 535–537
Hooker Chemical, *see* Occidental
Hot air collectors, *see* Active solar, hot air
　　collectors
House of Loreio Tiburtino, Pompeii, Italy,
　　307. *See also* Roman architecture
House size, 13
Housing, 322
　　cluster, 322–323
Houston, TX, 208
Howard, Rowland, 433
Hubbert, M. King, 23
Human Services Field Office in Taos, New
　　Mexico, 145–146
Humidification, 60, 273
Humidifiers, 481
Humidity, 465

Humidity, *see* Relative humidity. *See also*
　　Humidity ratio
Humidity ratio, 56
Hundertwasser, F., 314
Hybrid solar buildings, 159
Hybrid systems, 29, 175–176, 254
Hyderabad, Pakistan:
　　windscoops, 246–247. *See also* Indian
　　　architecture; Middle Eastern
　　　architecture
Hydrides, 34
Hydrochlorofluorocarbons (HCFC), 485
Hydroelectric power, *see* Energy, hydro-
　　electric
Hydrogen, 25–26, 29, 48–49, 176
Hydronic, *see* Heating
Hydro-power, *see* Energy: hydroelectric
Hypalon, 525
Hyperbilirubinemia, 354
Hyperthermia, 52
Hypocaust, 472–473, 478
Hypothermia, 52

IESNA, *see* Illuminating Engineering
　　Society of North America
Illuminance, 329–330. *See also*
　　Footcandle; Lux
Illuminating Engineering Society of North
　　America (IESNA), 342, 348
Illumination, 329–330, 333–334, 340–344,
　　348, 352, 355, 425–427
　　levels, 337, 340–342, 348, 356–357.
　　　See also Daylighting, illumina-
　　　tion levels
Incandescent lamp, 328, 332–333, 338,
　　369
Indian architecture, 206–207, 227, 248.
　　See also Specific buildings
Indirect evaporative coolers, 274
Indoor air quality (IAQ), 196–197, 510
Induction systems, 501
Infiltration, 197, 262, 301, 466–467
Insulating:
　　basements, 450
　　concrete forms (ICF), 450–451
　　crawl spaces, 450
　　effect of thermal mass, 456–457
　　roofs, 448
　　slab-on-grade, 448
　　walls, 448
Insulating effect of mass, 46–47
Insulation, 239, 268, 271–272, 276,
　　443–451
　　Air Krete, 445
　　batts, 465
　　blankets, 465
　　boards, 446
　　cellulose, 443, 445, 524–525
　　comparison of R-values, 445
　　fiberboard, 445–446
　　fiberglass, 443, 445, 465
　　foamed glass, 445
　　foamed-in-place, 445
　　icynene, 445
　　isocyanurate, 443, 446
　　levels, 443–444
　　light weight concrete, 445
　　loose fill, 445

Insulation *(continued)*
 movable, 455–456, 540–541
 perlite, 443, 445
 polystyrene (expanded), 443, 446, 450
 polystyrene (extruded), 443, 446
 radiant barrier, 443, 446–447, 525
 reflective foil, 443, 446–447
 rock wool, 443, 445
 R-values, 444–445
 straw bales, 449–451, 524, 526
 urethane, 443, 446
 vermiculite, 445
Integral collector storage (ICS), 195–196.
  See also Active solar, integral col-
  lector storage
Integrating Sun Machine, 138
Internally dominated buildings, 76,
  216–217, 220, 222, 225, 284, 437
International Association of Lighting
  Designers (IALD), 356
International Panel on Climate Control
  (IPCC), 19
"International style," 7
Inverter, 175–176, 183
Iris, 335
Isocyanurate, *see* Insulation
Isofootcandle, 426

Jacobs II House, 144–145
Japanese architecture, 202, 204–205, 252.
  See also Specific buildings
Japanese Garden, Portland, OR 204, 252
Jasmine, 208
Jefferson, Thomas, 171
Jet lag, 354
Johnson and Burgee, 387
Johnson Wax Building, 362–363
Joule, 39
Journals (resource), 601

Kahn, Louis I., 359, 377, 385–386
Keck, George Fred, 144
Kelly, Richard, 354
Kelvin, *see* Degrees Kelvin
Kerosene lamps, 408
Kilowatt-hours, 468
Kimbell Art Museum, 377, 386
Knowles, Ralph L., 279, 296, 521
Kotin, Stephanie, 522
Kyoto Protocol, *see* Global Climate Change
  Treaty: The Kyoto Protocol

Laherrère, Jean H., 24
Lakeland, Florida house, 13
Lam, William M. C., 325, 351, 407
Lamp-lumen depreciation, 430
Lamps, 408–418. *See also* Electric lighting; Lighting
 cold cathode, 415
 compact fluorescent, 412–415, 417
 comparison of, 417–418
 efficacy, 417
 ellipsoidal (ER), 410–411
 fluorescent, 408–409, 412–415, 417
 halogen, 410–411
 high intensity discharge, 416–417
 high pressure sodium (HPS), 409,
  416–417

 incandescent, 408–413
 induction, 418
 life of, 417
 light-emitting diodes (LED), 418
 light-emitting polymer (LEP), 418
 mercury, 416–417
 metal halide, 409, 416–417
 neon, 415
 parabolic aluminized reflector (PAR),
  410–411
 sulfur, 418
Landfills, 30
Landscaping, 315–321, 522, 523
Lanterns, 254
Law of diminishing returns, 340, 369
Le Corbusier, 7, 125, 204, 207, 209, 227,
  242–243, 265–266, 301, 362–363
Leeward, 256–257
Legionnaires' disease, 510
Lenses, 345
Lewy, Dr. Alfred J., 354
Life-cycle cost, 412–413, 529
Light, 328–358
 distribution, 425–427
 guides, 424–425
 pipes, 424–425
 units, 328–331
Lighting, *see* Daylighting; Electric lighting; Lamps
 accent, 350, 353
 activity needs, 348–351
 bilateral, 376
 biological needs, 351–354
 careers, 356
 certified (LC), 356
 for computer, 351
 design, 355–356
 designers, 356
 and health, 354
 indirect, 351, 353, 355, 357
 of paintings, 350
 of sculpture, 349
 of a task, *see* Task lighting
 of texture, 349
 unilateral, 376
Lighting fixtures, *see* Luminaires
Light meters, *see* Photometer
Light pipes, 395–396
Light shelves, 239, 379, 523–524, 533. *See
  also* Daylighting
Light-to-solar-gain ratio, 371
Lithium bromide, 485–486
Lithium chloride, 511
Log cabins, 253, 435
Loggia, 202–203
Long-wave infrared radiation, *see*
  Radiation
Loop-perimeter systems, 481–482
Lord House in Maine, 173
Louver(s), 210–211, 214, 219, 239–240,
  558. *See also* Daylighting, lou-
  vers
Lovins, Amory, home/office, 8, 14
Low-e coatings, *see* Glazing
Low-voltage, 175, 411–412, 423
Lugano, Switzerland, 73–74
Lumen, 328, 409–410
Luminaire dirt depreciation, 430

Luminaires, 418–425
 generic categories, 418–419
 optical controllers (lenses, louvers,
  eggcrate, etc.), 419–421
 parabolic wedge, 420–421
Luminance, 330, 337. *See also* Brightness
Luminous ceiling, 352. *See also*
  Architectural lighting, luminous
  ceiling
Lux, 330, 340–342. *See also* Footcandle;
  Illuminance; Illumination
Lyle, John T., 14

Madison, WI, 144–145
Magma, 32–33
Maharajah's Palace at Mysore, 206–207.
  See also Indian architecture
Maison Carée, 5. *See also* Roman archi-
  tecture
"Maison d'Homme," 7
Malaria, 484
Mantle, 408
Maps, 275
Marin County Court House, 5–6
Marshal Space Flight Center, 333
Mashrabiya, 248–249
Mass, *see* Thermal mass
Matmata, Tunisia, 457–458
Mayan Indian vernacular architecture,
  268
McDonough, William, 15, 30
Mean radiant temperature (MRT), 44, 56,
  61–63, 268, 269, 455, 476
Mechanical equipment, 8–9, 472–518
 design guidelines, 504–505
 exposed to view, 516–518
 integrated, 514–518
 room, 490, 495, 504, 560–561
 space requirements, 504–505
Melatonin, 351, 354
Menil Collection, 333, 386–387
Mercator projection, 132
Mercury, 412, 414
Mercury lamps, 332–333
Mesa Verde, CO, 251. *See also* Native
  American architecture
Metabolic process, *see* Metabolism
Metabolic ratio, 55
Metabolism, 52
Metal-halide lamps, 332
Metal studs, 449
Methane, *see* Gas, methane
Michelangelo, 5
Microclimate, 54, 75
Middle Eastern architecture, 2, 246–251,
  280–281. *See also* Specific build-
  ings and sites
Milne, Murray, 266–267, 585–587
Misting, 273
MIT Chapel, 356, 362
Models (physical), 136–139, 234–236
 daylighting, 138
Modular, 172
Moisture, 510
Moisture control, 463–467. *See also* Water
  vapor
Mold, 277

Monitor, roof, 258. *See also* Daylighting
Monochromatic, 332
Montreal Expo, *see* U.S. Pavilion, Expo
　　　'67, Montreal
Montreal Protocol on Substances that
　　　Deplete the Ozone Layer (1987),
　　　21, 485
Morocco, 281. *See also* Middle Eastern
　　　architecture
Mt. Airy Public Library, 391–393
MRT, *see* Mean radiant temperature
Multizone system, 500

Nanometers, 332
NASA, 333, 510
Natatorium, *see* Aquatic center
National Gallery of Canada, 394–395
National Oceanic and Atmospheric
　　　Administration, 74, 78–80
Native American architecture, 142–143.
　　　See also Specific buildings and
　　　sites
Natural-convection, 195
Natural gas, *see* Gas, natural
Natural ventilation, 2–3, 78, 202–204,
　　　305–306, 361. *See also* Air flow;
　　　Passive cooling; Windows
Navajo vernacular architecture, 251. *See
　　　also* Native American architec-
　　　ture
Net-metering laws, 175
Neutra, Richard, 209
New England vernacular architecture,
　　　144, 169
New Mexico solar architecture, 146, 152
"Next Industrial Revolution," 15
Nichols, Wayne, 146
Niemeyer, Oscar, 209
Night flush cooling, 527, 533–535,
　　　542–543, 545, 547–548
NOAA, *see* National Oceanic and
　　　Atmospheric Administration
Norman Foster and Associates, *see* Foster
　　　and Partners
North Pole, 134
Notre Dame du Haut at Ronchamp, *see*
　　　Ronchamp
Nuclear:
　　bomb, 24
　　energy, 25
　　fission, 25
　　fusion, 25–26
　　waste, 25
Nutting Farm, 440

Oak Ridge National Laboratory in
　　　Tennessee, 438, 449
Occidental office in Niagara Falls, NY,
　　　381–382, 456
Occupancy sensors, *see* Electric lighting
Occupational Health Center in Columbus,
　　　IN, 516–517
Oculus, 249
Odors, 512
Offending zone, 346–347, 350–351, 421
Off-gassing, 446, 510
Oil, *see* Energy, oil

Oil lamp, 408
Oil shale, 24
Olynthus, Greece, 23, 280
Open plans, 264
Optical control, 411–412
Optical illusions, 336
Orbit, *see* Earth, orbit
Organic waste, 30
Organization of Petroleum Exporting
　　　Countries, 22
Organizations (resource), 602–604
Orthographic projection, 579, 584
Ove Arup & Partners, 560
Overhangs, 210, 212, 214, 219, 220–223,
　　　225, 227, 239, 378
　　movable, 222
Overshot wheel, 31
Ozone:
　　hole, 15, 21
　　layer, 68, 485

Packaged systems, 494–496
Palladio, 142
Pantheon, 249. *See also* Roman architec-
　　　ture
Parabolic mirrors, 190
Parabolic reflectors, 411
Parabolic wedge louvers, 345. *See also*
　　　Luminaires
Parasol roofs, 7, 207
Parochial Church of Riola, 193–194
Parthenon, 336. *See also* Greek architec-
　　　ture
Passive cooling, 8, 39–40, 245–278. *See
　　　also* Air flow
　　by comfort ventilation, 255, 263–264,
　　　266–269
　　by dehumidification, 255, 277
　　earth cooling, 251, 275–277
　　　　direct coupling, 276–277
　　　　indirect coupling, 277
　　by evaporation, 246, 248–249, 251,
　　　255, 273–274, 276
　　　　direct, 273–274
　　　　indirect, 274
　　example of ventilation design,
　　　265–266
　　history of, 246–254
　　in hot and dry climates, 246–251, 270
　　in hot and humid climates, 251–254,
　　　269
　　by misting, 273
　　by radiation, 249, 251, 255, 270–273,
　　　312
　　　　direct, 271–272
　　　　indirect, 271–273
　　types of, 255
　　by ventilation, 251–255
Passive solar, 8, 65, 141–170, 524,
　　　530–541, 546–549. *See also*
　　　Direct gain; Trombe wall; Sun
　　　space
　　comparison of systems, 162
　　definition, 146
　　history of, 141–145
　　orientation, 162–163
　　other systems, 166–169

slope of glazing, 164
PA Technology, 399
Paul Davis house, 166
Paxton, 360
Peak demand, 172–173, 175
Pei, I. M., 388
"Pent" roof, 169
Perception, 330, 335–339, 345, 351
Perdin, John, 227
Perforated metal ceiling panels
　　　(diffusers), 561–562
Pergola, 231, 233, 307, 315
Perlite, *see* Insulation
Permeability, 465–466
Petroleum, *see* Energy, oil
Phase, 212–213
Phase change, *see* change of state
Phase change materials (PCM), 165–166
Phoenix, AZ, 558–561
Phosphor coating, 414–416
Photobiological effects, 354
Photocells, 399
Photoetching, 239
Photography, 401–402, 404
Photometers, 329, 337, 402–404
Photosynthesis, 29
Photovoltaics (PV), 26, 172–186, 198–199
　　array, 173, 175–178, 185, 523
　　balance of system equipment, 29, 176
　　building integrated photovoltaics
　　　(BIPV), 173, 176–182, 185–186
　　cells, 29, 174, 178, 181, 184–185, 228,
　　　230
　　design guidelines, 185–186
　　facades of, 180
　　as glazing, 180–181
　　grid-connected, 175, 183
　　history, 172–173
　　modules, 174, 178, 185
　　orientation of, 179, 185
　　roofing of, 177, 179–180, 184, 198
　　as shading devices, 182
　　shingles, 180, 184
　　sizing PV systems, 183–185
　　stand-alone, 175–176, 183
　　systems, 175–176
　　thin film, 174, 178, 181
　　tilt of 179–180, 182, 185–186
　　tracking, 177
　　　single-axis system, 173
　　　two-axis system, 179
Physical models, 234. *See also*
　　　Daylighting, physical models
Piano, Renzo, 386–387, 518
Pilotis, 265, 301
Pineal gland, 354
Planned Unit Development (PUD), 322
Plantation house, 252
Plants, 72, 215, 308–321. *See also* Trees;
　　　Vines
　　bonsai, 311
　　deciduous, 287–288, 308, 310, 315–317
　　foliation, 308, 310, 388
　　hardiness zones, 311
　　shading, 312
Plato, 22
Plutonium, 25

Polar projection, 132
Polluted air, *see* Air pollution
Polycrystaline cell, *see* Photovoltaic cells
Polyethylene, 32, 465–466
Polystyrene, *see* Insulation
Pompeii, Italy, 307. *See also* Roman architecture
Pools, 306–308, 315, 321
Pool and fountain, 318
Population growth, *see* Growth, population
Portico, *see* Classical portico
Postmodern architecture, 7, 203
Power density, 430
Power grid, 173, 175–176
Pre-Columbian architecture, *see* Native American architecture
Predock, Antione, 234
Price Tower, 243
Prisms, 396–397
Proportional shadow method, 373
Psoriasis, 354
Psychrometric-bioclimatic chart, 65, 76, 78, 82–114
Psychrometric chart, 56–65
Public Utilities Regulatory Policy Act of 1978, 175
Pueblo Bonito, 142–143. *See also* Native American architecture
Pupil, 335, 337
PV, *see* Photovoltaics

Quartz iodine lamp, *see* Halogen

R.H., *see* Relative humidity
Radial-perimeter system, 481–482
Radiant:
    ceilings, 44
    cooling, *see* Cooling; Passive cooling
    environment, 44
    floor heating, 187
    heat, *see* Heat, radiant
Radiation, 41–44, 52, 237
    diffuse-sky, 208–209, 239
    direct, 209
    heat, 239
    infrared, 369–371, 453
    long wave infrared, 21, 41–43, 45, 126, 328
    reflected, 209
    short wave infrared, 41–44, 180
    solar, *see* Solar, radiation
    solar infrared, *see* Radiation, short wave infrared
    ultraviolet, 42–44, 126, 238, 328, 340, 354, 371, 412, 414, 424, 455
    visible, 44, 126
    wavelength, 41–43
Radiators, 270–273
Radon, 277, 448, 509–510
Raised floors, 512, 552
Rancho Seco, CA, 173
Real Goods Solar Living Center, 450, 522–529
Reflectance, 42, 44, 237, 330. *See also* Radiation
    factor (RF), 330, 337, 343, 347, 366
Reflectivity, *see also* Albedo
    solar, 438–439

Reflectors, 149, 411
    specular, *see* Specular reflectors
Refrigerants, 485–487, 491
Refrigeration machines, 39, 45, 484–486, 489–491, 493. *See also* Cooling, refrigeration cycles
Regenerate, 278
Regional design, 2, 7, 10
"Regionalism," *see* Regional design
Registers, 495, 506–507
Relative humidity, 56–65, 73, 78, 274
    graphs, 83–115
Renaissance, 337
Renaissance architecture, 141–142, 360
Resistance, *see* Thermal resistance
Resources, (listing), 601–604
    books, 601
    CD-roms, 602
    journals, 601
    organizations, 602–604
    videos, 601
Retina, 335
Return air, 495–501, 504, 506, 508
Reversing valves, 486
Revivalism, 7
    classical, 7, 202, 253
Revolving houses, 228–230
Rickets, *see* Vitamin D deficiency
Rio de Janeiro, *see* Earth Summit in Rio de Janeiro
Riola, *see* Parochial Church of Riola
Robie House, 204–205, 268–269
Roche, Kevin, 208
Rock bed, 160, 166, 270, 545
Rock storage bins, 191
Rock wool, *see* Insulation
Rod cells, 335–336
Rogers, Richard, 518
Roller shades, 214, 228, 543–549
Roman architecture, 142, 202, 231, 280, 306–307, 360, 472. *See also* Specific buildings and sites
Roman basilicas, *see* basilicas
Ronchamp, 362–364
Roof gardens, 315, 321
Roof ponds, 166, 272, 274
Roof radiation traps, 167
Roof ventilators, 264
Roof vents, 263–264
Rudolph, Paul, 209
"R-value," 46, 398, 524–525, 528, 530, 536, 538

Saarinen, Eero, 356, 362, 514, 516
Sacramento Municipal Utility District, 173, 183, 270
Safdie, Moshe, 394
Saint Maria Degli Angeli, 5
Saltbox, 144, 169
Salt hydrates, 165
San Jose, California;
    fixed outdoor shading system, 232
San Juan Capistrano Public Library, 203, 233
Santa Fe, NM:
    architecture, 159, 282
Satellite images, 13
Saudi Arabia:

vernacular architecture, 2. *See also* Middle Eastern architecture
Sawtooth, 180. *See also* Daylighting
Sawtooth clerestory, *see* Clerestory, sawtooth
Schools, *see* Colleges
Scripps Institution of Oceanography, 19
Sea level rise, 20
Seasonal Affective Disorder (SAD), 354
Selective low-e, *see* Glazing
"Selective" surface, 43–44, 157, 189
Semiconductor materials, 174
Setbacks, 295
Sewage, 30
Shade line, 65, 225–226
Shades, 378, 531, 543–545
Shading, 207, 533, 537, 558
Shading coefficient, 240–241
Shading devices, 219–241
Shading devices (outdoors), 202–244, 549–553. *See also* Awnings; Eggcrates; Fins; Louvers; Overhangs; Plants; Roller shades; Toldos
    fixed, 210, 213
    movable, 212–215
shading by glazing, 27–239, 241
Shading of outdoor spaces, 231–234
Shading periods, 216
Shadow:
    length, 283, 289–290, 296, 299–300
    pattern, 289–295, 299–300
Shelly Ridge Girl Scout Center near Philadelphia, PA, 154–155
Shimane Prefecture, Japan, 282
Short wave infrared radiation, *see* Radiation
Shutters, 246, 253
Sierra Nevadas, 74
Silica gels, 277
Silicon cells, *see* Photovoltaics cells
Similar triangle method, 573
Single-duct systems, 498–499
Site, 280
    design, 280–321
    planning, 292–295
    selection, 283–284
Site evaluation tools, 134–136, 589–594
    construction of, 589–593
    operation of, 593–594
    site evaluator, 589–594
    sun locator, 589
Sky dome, 129–133
Skygardens, 554–557
Skylights, 149, 198, 208, 211–212, 254, 541, 545–546, 560–561. *See also* Daylighting, skylights
Sky vault, 285
Sling psychrometers, 59
Smog, 22
SMUD, *see* Sacramento Municipal Utility District
Socrates, 22
Soil temperature, 275
Sol-air temperature, 438
Solar, *see* Active solar; Daylighting; Passive Solar; Photovoltaics; Reflectivity, solar; Shading

access 280, 285–300, 304, 322
access laws, 296
architecture (historical) 141–145
boundary, 285–288
cells, *see* Photovoltaics cells
constant, 126
data, 80, 82–115
easements, 296
energy, 26, 48
geometry, 126–139
peak values, 83–115
radiation, 21, 42–43, 126, 128–130,
  179, 185, 208, 237, 296
spectrum, 126
window, 130, 134, 136, 285
zoning, 296, 322
  bulk-plane method, 296
  envelope method, 296–298
  fence method, 296, 298
Solar chimneys, 259
Solar collectors, 229. *See also* Active
  solar, collectors
tracking, 228, 230
Solar heat-gain coefficient, 240–241, 371
Solar hemicycles, 144, 149
Solarwalls, 188
Southeast U.S.:
  architecture, 253
  climate, 255
Space heating, *see* Heating
Specific heat, 41, 45. *See also* Heat
  capacity
Specularity, 330–331, 350
Specular reflectors, 164
Splayed:
  roofs, 385
  walls, 377
Split systems, 494–496
Stack effect, 253–254, 257–259, 346, 527,
  554
Stadiums, 231
Stand-alone systems, *see* Photovoltaics
Static pressure, 257–258
Steel, 49
Stoa of Attalos II, 202
Stone, E. D., 209
Storage tanks, 196. *See also* Active solar,
  storage tank
Stoves, 30, 473–474, 524. *See also*
  Heating
Stratification, 2, 40, 249, 258, 508
Strip mining, 25
Structure insulated panels (SIP), 449–450
Sumatra, Indonesia:
  vernacular architecture, 3
Summer solstice, 127, 129–130
Sun, 126
  path diagrams, 131–136
    horizontal, 131–133, 563–564
    vertical, 132–135, 565–568
  path model, 134–135, Appendix F
  paths, 129–136
Suncatcher baffles, 390–391
Sun City, AZ, 322
Sundials, 137–138, 154, 156, Appendix E,
  Appendix C.4
Sundials for modeling, 577–578
Sun-dried mud bricks, 250. *See also*

Adobe
Sun Emulator, 138
Sun-hour, 184–185
Sunlight, 47, 409
Sun machine, 136–139, 234
Sun machine (do-it-yourself):
  construction of, 569–571
  operation of, 571–572
Sun-path models, 579–584
  construction of, 579, 584
Sun room, 142, 280
Sunshine:
  availability, 79–80, 82–115
  graphs, 83–115
Sun spaces, 142, 158–162, 530, 533,
  554–557
  design guidelines, 160–162
Superinsulated, 173
Superinsulation, 443
Surface-area-to-volume ratio, *see*
  Compactness
Sustainability, 9, 12, 15, 172, 529
Sustainable design, 18–19
  energy, 18–19
  issues, 19
Swamp coolers, *see* Evaporative coolers
Swimming pools, 189, 197. *See also*
  Active solar, swimming pool
  heating
System International (SI), 39, 330

Tabb, Phillip, 296
Taliesin West, 306, 308
Tar sands, 24
Task, visual:
  performance of, 339–341, 343–344
  lighting, 342, 356
Tebbutt, Chris, 522
Tempe, Arizona City Hall, 237–238
Temperature, 38–39. *See also* Air temper-
  ature; Color, temperature; Dew
  Point Temperature; Wet Bulb
  Temperature
  annual mean, 73
  conversion, 38
  dew point, 466
  diurnal range, 70–71, 246, 269
  equilibrium, 43–44
  global, 19–21
  graphs, 83–115
  human, 52–53
  range, 456
  swing, 148
Tennessee Valley Authority, 18
Tensegrity structural system, 560–561
"Terminal reheat system," 9
Thermae of Diocletian, 5. *See also* Roman
  architecture
Thermal barriers, 53–55
Thermal breaks, 449
Thermal capacity, 46
Thermal comfort, 51–66, 76, 254. *See also*
  Comfort range
  comfort zone, 61–65, 78, 148
Thermal conditions of the environment,
  56
Thermal energy, 32
Thermal envelope, 4, 443–467

gradient, 464–46
Thermal mass, 65, 144, 148, 150–151,
  153, 157–158, 165, 270,
  456–457, 528, 533–534, 538. *See
  also* Heat-storage materials
Thermal planning, 439
Thermal reheat systems, 498–500
Thermal resistance (R), 45–46, 65, 398,
  436, 438, 443, 447, 449, 455,
  457. *See also* R-value
Thermal storage, 512, 514
Thermal storage wall system, *see* Trombe
  wall
Thermal year, 212–213
Thermodynamics:
  First Law, 47
  Second Law, 47
Thermoelectric effects, 484
Thermography, 434–435
Thermometers, 38–39, 59
Thermosiphons, 166, 195–196. *See also*
  Active solar, systems
Thermostat, 435, 475, 496, 498–501, 503
Thin film PVs, *see* Photovoltaics
Three Mile Island, 25
Three-tier design approach, 8–10, 147,
  183, 187, 246, 280, 327
  one 8–10
  two 8–10
  three 8–10
Through-the-wall unit, 493
Time:
  clock, 131
  daylight savings, 131
  solar, 131
Time-of-day pricing, 175
Time lag, 46–47, 275, 456–457
Time zones, 131
Tocamacho, Honduras:
  vernacular architecture, 283
Toldos, 231–232
Total internal reflection, 424–425
Toxic materials, 196
Tracking solar collectors, *see* Solar
  collector, tracking
Transit, 575
Translucent, 330
  insulation, 452–463
  roofs, 397–398
  walls, 397–399
Transmission (of heat), *see* Heat flow;
  Radiation
  affected by wind speed, 301
  through glass, 42–43
  through opaque systems, 45–47
Transmittance, 237–239
Transoms, 264, 361
Transpiration, 249, 276, 312, 322
Transpired collectors, 196. *See also* Active
  for ventilation air
Trees, 301–302, 305–306, 308–313,
  315–322
Trellises, 208, 225, 231, 233, 311, 314–315,
  319–321, 529, 533, 535, 537, 545
Trigonometric method, 574
Trombe, Felix, 152
Trombe walls, 152–158, 161, 162, 541
  design guidelines, 157–158

Tropic of Cancer, 127, 129, 134
Tropic of Capricorn, 127, 129, 134
Trulli, 249
Tuffa cones, 250–251
Tungsten filaments, 410
Tungsten halogen lamps, *see* Halogen
    lamp
Turbines, 32
Turning vanes, 495
Twain, Mark, 68

U-coefficient, *see* Heat-flow coefficient
Ultra violet radiation, *see* Radiation
Underground buildings, *see* Earth
    sheltering
Undershot wheel, 31
Unite d'Habitation at Marseilles, 265–266
United Nations, 11
United Nations Headquarters in New
    York City, 242–243
United States:
    energy consumption, 34
    nightime, 13
    production of CO_2, 20
United States climate regions, *see*
    Climate, regions (U.S.)
U.S. Department of Energy (DOE), 444
U.S. Department of Housing and Urban
    Development, 296
U.S. Pavilion, Expo' 67, Montreal, 54,
    228–229, 251
Unit heaters, 483
University of Southern California, 296
Uranium, 25
Urban Villa, 530–533
Urethane, *see* Insulation
User manuals, 535

Vacuum, 39, 41
Vacuum bottles, 452
Vacuum-tube solar collectors, 191
Van der Ryn, Sym, 522, 528–529
Vapor barriers, 464–467
Vaporization, 274
Variable-air-volume (VAV) systems,
    498–499, 548, 552–553
VAV control boxes, 498–499
VDT, *see* Computer monitors
Vegetation, *see* Plants
Veiling reflections, 345–348, 350–351,
    355, 371–372, 384, 419–421
Venetian architecture, 203
Venetian blinds, 223, 239–240, 338, 381,
    455–456, 535, 549–550, 556
Vent, 159–160, 228, 250, 252, 465–466

areas 466
    exhaust, 466, 468
    ridge, 465
    soffits, 465
Ventilation, 246, 466–467, 490, 509–512,
    527, 530–532, 537
Ventilation air, 188, 196
Ventilation systems, 197
Ventilators, 258
Ventura Coastal Corp. Administration
    Building, 380–381
Venturi effect, 258
Venturi tubes, 257
Veranda, 203, 208, 252
Vernacular architecture, 2, 10, 202, 280
Vestibule doors, 467
Victorian architecture, 203
Video Display Terminals (VDT), 346, 420
Videos (resource), 601
Village Homes in Davis, CA, 292, 322
Vines, 214–215, 307, 311–312, 314–315,
    320, 533, 535
Visible radiation, *see* Radiation
Visible transmittance, 371
Vision, 335–336
    peripheral, 336, 343
Visual comfort probability (VCP), 345,
    351, 420
Visual performance, *see* Task
Vitamin D deficiency (rickets), 354
Vitruvius, 67–68, 142
Volatile organic compounds (VOCs), 510
Voltage, 174–175, 412, 415. *See also* Low
    voltage
Volts, 174
Volumetric heat capacity, 165. *See also*
    Heat storage materials

Wall furnaces, 483
"Waste equals food," 15
Water, 165
Watermills, 31
Water spray, 39
Water tubes, 154, 156, 537
Wattage, 328
Watt-hours, *see* Electrical energy
Wattle-and-daub, 435
Watts, 174
Wavelength, *see* Radiation, wavelength
Waverly plantation, 253–254
Weather, 68
Weatherstripping, 466
Well water temperatures, 275
West Point Pepperell factory in Lanett,
    AL, 493

Wet-bulb temperature, 59
Wilson, E. O., 308
Wind, 68–69, 78, 266, 280, 301–307,
    315–316, 321. *See also* Energy,
    wind
    resources, 28
    roses, 78–81
    scoops, 2, 3, 246
    speed, 27, 73
        graphs, 83–115
    turbines, 27, 29, 263
    velocity, 301–302, 304
Windbreaks, 280, 282, 301–302, 304, 308,
    316–317, 321
Windchill factor, 56
Windmills, 26,27. *See also* Wind turbines
Windows, 452–456. *See also* Daylighting;
    Glazing
    comparison of R-values, 452
    comparison of types, 454
    and insect screens, 263
    location, 260
    operable, 555–558
    sizing, 262, 268–270
    super, 452
    triple hung, 253, 262
    types, 262
Wind power, 26–29, 175–176, 183
Wind screens, *see* Windbreaks
Wind towers, 247–248
Winter, 126, 128–129
Winter solstice, 127, 130
Wolverton, Bill, 510
Wood, 49
World Congress of Architects, 12
World Financial Center, 492
World Trade Center, 492
Wright, Frank Lloyd, 5–6, 144–145, 149,
    204–205, 243, 268–269, 306,
    308, 325, 362–363. *See also*
    Specific buildings

Xenophon, 22
Xeriscape, 312

Yucatan Peninsula, *see* Mayan Indian
    vernacular architecture

Zeolite, natural, 277
Zone, 501, 503
    thermal, 475–477, 479, 494, 496–504
Zoning, *see* Solar
    solar/thermal, 441–442

NASHVILLE PORTRAITS

Jim McGuire

NASHVILLE PORTRAITS

Legends of Country Music

Sixty Portraits
by Jim McGuire

Introduction
by William R. Ferris

Quotes and interviews
edited by Richard Carlin

Exhibition organized
by Kevin Grogan, Director
The Morris Museum of Art, Augusta, Georgia

THE LYONS PRESS
Guilford, Connecticut
An imprint of The Globe Pequot Press

Photographs Copyright © 2007 by Jim McGuire
Introduction Copyright © 2007 William R. Ferris
Compilation Copyright © 2007 The Morris Museum of Art

For more information about Jim McGuire's photographs, see www.nashvilleportraits.com.

The Morris Museum of Art, located on the Riverwalk in downtown Augusta, Georgia, is home to a broad-based survey collection of Southern art. For more information visit www.themorris.org.

The Lyons Press is an imprint of The Globe Pequot Press.

10 9 8 7 6 5 4 3 2 1

Printed in China

ISBN 978-1-59921-168-8

Library of Congress Cataloging-in-Publication Data is available on file.

To John Foote and Irving Penn

—Jim McGuire

CONTENTS

Preface *by Jim McGuire* viii

Foreword *by Kevin Grogan* ix

Acknowledgments xi

Introduction *by William R. Ferris* xiii

Nashville Portraits: The Exhibition

Ralph Stanley, 1998 2

Guy and Susanna Clark, 1975 4

Vince Gill, 1985 6

Dave Dudley, 1976 8

Tut Taylor, 1974 10

Chet Atkins, 1976 12

Brother Oswald, 1975 14

David Bromberg, 1972 16

Tammy Wynette, 1987 18

Waylon Jennings, 1985 20

J. D. Crowe and The New South, 1975 22

Marty Robbins, 1977 24

Hal Ketchum, 1997 26

Béla Fleck, 1999 28

Emmylou Harris, 1983 30

John Hiatt, 1974 32

John Hiatt, 2004 34

Harlan Howard, 1988 36

Townes Van Zandt, 1990 38

Carole King, 1978 40

Kris Kristofferson, 1990 42

The Dillards, 1977 44

Roy Huskey Jr., 1991 46

Sam Bush, 1981 48

Lyle Lovett and The Boys, 2005 50

Mike Seeger, 2005 52

Rosanne Cash and Rodney Crowell, 1983 54

Bill Monroe, 1989 56

Nanci Griffith, 1980 58

The Highwaymen, 1990 60

Marty Stuart, circa 1978 62

John Hartford, 1972 64

New Grass Revival, 1976 66

Steve Earle, 1975 68

Steve Earle, 1995 70

Jerry Jeff and Susan Walker, 1977 72

John Prine, 1984 74

Mark O'Connor, 2004 .76

Dolly Parton, 1974 78

Junior Brown, 1993 80

Lester Flatt, 1977 82

The Country Gentlemen, 1971 84

Darrell Scott, 1997 86

Cowboy Jack Clement, 1985 88

Roland White, 1975 90

John Jorgenson, 2005 .92

Jack Ingram, 1998 94

Johnny Cash and Dr. Billy Graham, 1978 96

Steve Young, 1976 98

Merle Travis, 1977 100

George Strait, 1982 102

Dale Watson, 2001 104

Vassar Clements, 1972 106

Doc Watson, 1975 108

Travis Rivers, 1971 110

Jerry Byrd and Curley Chalker, 1977 112

Don Williams, 1981 114

Norman and Nancy Short Blake, 1980 116

Keith Whitley, 1984 118

Benny Martin, 1976 120

Recent and Other Work

Mac Wiseman and John Prine, 2007 124

J. D. Souther, 2006 126

Guy Clark with Verlon Thompson
and Shawn Camp, 2007 128

Nanci Griffith, 1997 130

Guy Clark with Django, 1999 132

Exhibition Checklist 134

Index 139

PREFACE

I have always had a thing for black and white. *LIFE* magazine. *On the Waterfront. Blackboard Jungle. Rebel Without a Cause.* These were very powerful images to my young eyes. Even in later years, black-and-white images had a way of staying with me longer and speaking to me louder than anything else.

Color was too literal. You could get all the color you could ever want by just walking out your front door. But when you create a black-and-white image, you are not just recording it on film; you have the obligation to bring your own vision and experience to the process. For me, black and white had a way of making simple, mundane things seem important, and I liked that.

Like many things in my life, the collection of photographs called the "Nashville Portraits" began quite accidentally. From a young age, I was drawn to hillbilly music, to the sounds, the emotion, the honesty, and then, of course, to the people who made it. I can't remember what I had for lunch yesterday, but I can still remember what the room smelled like when I walked in and heard Hank Snow's voice coming from those ratty speakers at a Boy Scout camp in New Jersey, 1953. Discovering country music changed my life in ways I couldn't have imagined.

Unlikely as it sounds, I had the good fortune to spend a year in Vietnam at the beginning of that war. I was twenty-three and in the Air Force. I barely knew how to use a camera, but my commanding officer volunteered me to fly around in a very small plane and shoot aerial photos with a beat-up 35mm camera. I spent the rest of the year doing that and persuaded the local Vietnamese portrait photographer to show me how to develop my first rolls of black-and-white film in an old army tent. And that's where it began for me.

My first job was in New York, working as John Foote's darkroom slave for sixty-five bucks a week. He photographed actors and Broadway singers in black and white. He introduced me to the work of Richard Avedon, Irving Penn, Bert Stern, and Wingate Payne, all working in New York at the time and all at the top of their game. But when I saw Penn's "Small Trades" series, it floored me. He took simple, working people you would probably never notice in your daily travels and presented them in a way that compelled you to look at them, to appreciate them. It was a combination of that beautiful hard lighting, that simple gritty canvas backdrop, and all those glorious tones of gray. They became as important as they were timeless.

Over the past thirty-five years, I have had the good fortune to have met, photographed, and befriended many of my musical heroes. The first musician portraits I shot are in this series. John Hartford, David Bromberg, and Vassar Clements were the first . . . all shot late one night in 1972 after a gig in New York. The hand-painted canvas background I used that night was barely dry and is the same one used in all these portraits, even today.

Most of us have a drawer full of snapshots that remind us of the good times. These are some of mine.

—Jim McGuire
2007

FOREWORD

Nashville, Tennessee, is known for one thing, mainly— country music. It has been one of the country's major centers of music production, manufacture, marketing, and distribution for many years.

For many years, too, Jim McGuire has enjoyed a reputation there as the photographer to get for the career-making album cover photograph. Since 1972, more than five hundred recordings—first LPs and later, CDs—have been graced by McGuire photographs on their covers. The list of prominent artists so recognized continues to grow, but by now it seems as if everyone who matters is included in McGuire's catalog of portraits— an undertaking that evolved from commercial assignment into something highly personal. In the Nashville music community, he has achieved the same level of renown as many of the artists he has photographed.

At first blush, McGuire might seem an unlikely fan of country music. There's nothing about a suburban New Jersey upbringing in the 1950s that would seem to point anyone in that direction. But ever since he first encountered country music through a chance hearing of Hank Snow's ''Spanish Fireball''—he recalls this as taking place while he delivered beer to his counselors at summer camp—this music has been his muse.

In 1961 he enlisted in the U.S. Air Force in search of adventure and, to his mild surprise, became a photographer, honing his skills photographing the American troop buildup in Vietnam before receiving an honorable discharge in 1965. After mustering out of the service, he returned to New Jersey. He commuted to New York City daily to attend a photo technical school for a couple of years before moving to the city in 1967, when he accepted an offer to assist fashion photographer John Foote. This is where McGuire first saw the work of Irving Penn, whose photographs of tradesmen with their tools had a galvanic effect on him. They were also the immediate inspiration for the Nashville Portraits series, which features many of America's most influential singers, songwriters, and musicians who, as often as not, are depicted with the instruments that are the tools of their trade.

After a yearlong stint as a corporate photographer for TWA—an experience that taught him a valuable life lesson: he should never work for anyone else ever again—he decided to go freelance, a decision that gave him the freedom to travel and photograph the bluegrass festivals that were then making their way east and north as part of the folk music revival of the 1960s. This led to his writing a column, ''Riffs,'' for *The Village Voice*. His column and the newspaper credentials that came with it provided him greater access to a world he had come to love.

Eventually, the connections he made and the friendships he had established with bluegrass, folk, and country artists drew him to Nashville's music scene. His first ''Nashville Portrait'' of then-rising star John Hartford— which was actually shot in Greenwich Village—led directly to his move, in 1972, to Nashville, where he began an entirely new phase of his career shooting album covers. By 1974, when he had opened a studio in an old neighborhood grocery store on the west side of town, he had already produced a body of work that was quickly evolving into the Nashville Portraits.

When I first met McGuire more than twenty-five years ago, he was a well-established success—the man whose services everyone hoped to secure. Now, at age sixty-five, he remains the foremost music industry photographer in Nashville and the leading pictorial chronicler of its music community. In Nashville, he is almost as renowned for his love of music and musicians, wry wit, and 1947 Ford station wagon—a magnificent thing, that—as he is for his photographs. But over the years, his photographs have achieved a kind of iconic status and have come to define public perception of artists as unlike one another as Carole King and Doc Watson, Townes Van Zandt and Reba McEntire, Tracy Nelson and Dolly Parton.

McGuire's work, inspired by Irving Penn so many years ago, continues to share certain characteristics with Penn's portraits, particularly his depictions of tradesmen. For one thing, both prefer the controlled, formal conditions provided by a studio setting and a neutral background. (In fact, McGuire today continues to use the same painted canvas backdrop that he used in his first Nashville portrait in 1972.) Clarity, careful composition, and the calculated use of light also characterize the work of both men, and both go well beyond the merely reportorial to produce work of genuine insight, capturing a gesture, a fleeting expression, or mood to reveal something intriguing and unexpected about their subjects. The images that have resulted from their particular combination of technical skill and aesthetic sense constitute a record of a specific aspect of our shared culture.

McGuire and I pored over hundreds and hundreds of photographs to select sixty for inclusion in a traveling exhibition of his work and, subsequently, this book. McGuire himself selected an additional five images for the book, an addendum that includes new photographs taken long after our original selection had been made, as well as a couple of his personal favorites. I wish I could say that there was some scientific method to the selection process. In point of fact, we included those images that both of us could instantly agree upon. The challenge in this case lay in paring down the selection of those things that both of us thought should be in the show and in this book. As is usually the case, the challenge was really in the editing. But in some broad sense, we had an unarticulated goal: breadth of representation and a sampling that represented Nashville and the nature of its music community as it moved from its roots through the era of "The Nashville Sound" and back again to something that more closely resembled the place that it came from.

There are some constants, however, that can be readily discerned throughout this selection. For one thing, the prints are invariably beautifully made. The lessons taught him all those years ago in Vietnam weren't wasted. He is as technically gifted a photographer as I've ever known. Another characteristic that's worth noting, I believe, is the sense of engagement that he achieves. With a very few exceptions—the early portrait in which John Hartford appears to be on the verge of levitation and the much later photograph of the courtly Bill Monroe bonding with his mandolin, to cite just two examples—the sitter is engaged directly with the viewer. McGuire—a consummate craftsman possessed of a Pictorialist sensibility—introduces the viewer to people whose company he is pleased to share. He has spent years painstakingly, unceasingly—well, if you don't count all those days he lost at the racetrack—assembling a body of work that is unlike any other. He has given a face to America's music.

—Kevin Grogan
Director, The Morris Museum of Art

ACKNOWLEDGMENTS

I would like to acknowledge several people for helping bring this work to light. Kevin Grogan, director of The Morris Museum of Art, has been a good friend and a great supporter of this work for many years. He gave me my first one-man show in 1982 when he was director of Cheekwood Fine Arts Center in Nashville. Thank you, Kevin. It's been a long haul but with your help, we did it.

Also a big thanks to Maureen Graney and the folks at The Lyons Press who took the time and made the effort to do this right. Special appreciation to Scott Watrous, president and publisher of Lyons/Globe Pequot, who supported this project from the beginning; and Melissa Evarts, Christine Duffy, and Kevin Lynch for their excellence in production. And kudos to Audrey Domsky and her staff at Smith Kramer, traveling exhibition specialists, for getting this show on the road.

I offer special thanks to my trusty assistants, R. B. "The Guru" Miller and Lynne Cook Oglesby. Without their help and talents through the years, many of these images simply would not have happened. And to my friend John Baeder for teaching me, by example, what art is supposed to be; and John Nikolai, who in his own peculiar way has encouraged me to drag this work from the bowels of my darkroom into the light of day.

A debt of gratitude goes to three of my heroes: Django Reinhardt, for keeping my mind ablaze for those countless hours in the dark; R. Crumb, for teaching me that art is everywhere; and Charles Bukowski, for teaching me proper etiquette at the racetrack.

Thanks also to Wendi Chapman for keeping me between the ditches. And to my main men, Dennis Carney and Jon Hobson, for taking the time and patience to teach an old dog some new tricks. Couldn't have done this without you, boys.

Thank you also to my old friend Guy Clark for showing me that fine craftsmanship and fine art are one and the same, and that you can't have one without the other.

I would also like to acknowledge the people who allowed me to photograph them for this collection. You are the reason I do this.

And especially—thanks to Diesel and Django for their healing powers.

—Jim McGuire

Obviously, this book would not have been possible without the willing participation of Jim McGuire. Fervent thanks are owed him, both for his work and his willingness to share it with others. The social historian Carl Degler once wrote with fascination that the South is a part of the world "where roots, place, family, and traditions are the essence of identity." It is this quality that Jim McGuire has captured in his Nashville Portraits. And if only for that alone, we should all be grateful.

Bank of America has made possible, in part, the traveling exhibition of Jim McGuire's Nashville Portraits that led to this publication, and we are as pleased by their recognition as we are grateful for their support. I am deeply gratified by the participation in this project of Professor William Ferris, author, folklorist, filmmaker, and scholar of Southern culture. His insightful introduction to this book places McGuire and his work in a context that deepens one's appreciation of the photographs themselves and McGuire's achievements as a cultural documentarian.

Richard Carlin, a remarkably resourceful editor, found wonderful comments about McGuire's subjects, the singers, songwriters, and musicians who make up the Nashville Portraits. The special insights offered by the colleagues and friends of those portrayed will significantly enhance the reader's appreciation of McGuire's work. And it will lead one to wonder why some people seem to have all the fun.

When I first approached Scott Watrous, president and publisher of The Globe Pequot Press, with this project, he responded immediately and positively. From the very outset, his support of this project has been critical to its success, as has that of Eugene Brissie, associate publisher of The Lyons Press.

Maureen Graney, the editor-in-chief at The Lyons Press, has shepherded this project through the some- times challenging process of making a good idea a book. I'm grateful for her friendship and her assistance, as well as for the regular day-ending telephone calls that helped to keep it on track.

When my wife, Kate, and I first moved to Nashville in 1980 with our then-young children, we did so in the expectation that it would be something like a Foreign Service duty tour. We anticipated spending three years there before returning to America, but we discovered shortly after our arrival that we'd actually found our way home. During our nearly twenty-year-long residency in that city, we came to appreciate and enjoy, to our pleasant surprise, the musical culture that McGuire has so deftly captured in these photographs. Kate already knows that I'm grateful to her for more than it would be appropriate to note here, but I didn't want to miss this opportunity to express special thanks for the shared experience of Nashville.

—Kevin Grogan
Director, The Morris Museum of Art

INTRODUCTION

Country music opens a fascinating window on the American South. Like patterns in a patchwork quilt, its tunes reflect the region's history. One can trace his or her ancestral roots to Ireland and Great Britain. Country music, the lineal descendant of this music, followed poor, struggling families in the twentieth century as they moved from farms to cities and, often, from the South to the North, giving them much comfort and a reminder of their roots. Predominantly white audiences listened to the music in dance halls, on the radio, on records, and on television.

Black families in the South found their musical home in the blues, a lyrical form with roots in African music. As children, Southern white and black artists learned from each other, and over time their musical traditions mixed, their lyrics and tunes moving across racial lines. Such country music icons as Jimmie Rodgers, Hank Williams, and Johnny Cash each learned to play from older black musicians. And black artists like Charley Pride and DeFord Bailey established successful careers as country music singers.

Country music has come a long way, traveling down that long, lonesome road from the proverbial hills and hollers to the bright lights of Nashville and the Grand Ole Opry. The music's roots lie in the ballad—the song that tells a story. British ballads like "Barbara Allen" traveled across the Atlantic and became a beloved part of American folk music. They inspired homegrown American ballads such as "John Henry" and "Casey Jones." Lost love and tragic train wrecks were familiar themes in these ballads. Their haunting musical tales were first sung a cappella and later with guitar, banjo, and mandolin accompaniment. These instruments added a rhythm to the music that over time achieved breakneck speed in the bluegrass style.

Country music in the twentieth century is inextricably linked with Nashville and its beloved "Opry." Nashville became the Vatican City of country music as its fans made their pilgrimage to hear performances every Saturday night at the Ryman Auditorium, legendary "Mother Church of Country Music" and the home of the Grand Ole Opry from 1943 until 1974. The Opry provided a diverse, fun-filled evening's performance that was and is truly and uniquely American. In "How to Tell a Story," Mark Twain declared that "The art of telling a humorous story—understand, I mean by word of mouth, not print—was created in America and has remained at home." Had he lived well into the twentieth century, Twain would have surely heralded the Opry as an equally authentic American institution.

Country music is homegrown, and performers like Johnny Cash and Dolly Parton on the Grand Ole Opry are as important to America as Luciano Pavarotti and Renata Tebaldi are to Italy. The choice of the name "Grand Ole Opry" is a gauntlet flung in the face of classical opera and European cultural tradition. There, family and friends gather each Saturday night, as worshipful as the most avid audience member at La Scala. Those who are not privileged to hear the Opry in person can listen to the performance on radio station WSM. (The Opry is the longest continuously broadcast program in American radio history, having been broadcast on clear channel WSM since November 28, 1925.) Like a musical beacon, its sound has reached into homes across much of the country. Generations of children, mothers and fathers, and grandparents have gathered around their radios, and their culture has been profoundly shaped and enriched by its broadcasts.

For many years the city of Nashville and the American South—the places that gave birth to country music—have had a complicated relationship with the music. Like its counterpart, the blues, country music reflects the struggles of working-class men and women. Their lives seem far removed, even disconnected, from the world of art museums and studio photography. Bridging these worlds is not an easy task. A significant step in this journey was taken by Walker Evans, whose portraits of white sharecroppers in Hale County, Alabama, shocked readers of *Let Us Now Praise Famous Men* when it was published in 1941. Evans's photographs and James Agee's text celebrate working-class white culture in the South and foreshadow Jim McGuire's *Nashville Portraits: Legends of Country Music.*

Portraits of musicians are an established part of American art. Photographs of these cultural icons become an extension of the artists' musical performances, a part of their personas. The craggy face of Johnny Cash, to cite just one example, is now as familiar a part of our photographic history as it is of our musical history.

This publication, *Nashville Portraits: Legends of Country Music,* marks a major step forward in the photographic exploration of Southern music—really America's music. Images of country, blues, rock and roll, gospel, and other musical artists have long graced the covers of records and CDs, books, and the walls of museums in photographs sometimes taken by fans or by journalists and, often, by studio photographers or

even by the musicians themselves. Country music singer Marty Stuart—one of the musicians featured in *Nashville Portraits*—is himself a fine photographer. Stuart's book *Pilgrims: Sinners, Saints, and Prophets: A Book of Words and Photographs* includes his photographs of Johnny Cash, Merle Travis, Bill Monroe, and many others with whom he has worked. Along with his musical heroes, Stuart acknowledges photographers who inspired his artistic eye, including Alfred Stieglitz, Edward Curtis, and Eudora Welty. He also singles out Jim McGuire "for putting the twang back into black-and-white photography."

This book and the exhibition that spawned it mark a historic moment for both country music and American photography. The American South has been slow to recognize its own traditions in the arts. For many years the predominantly rural region had few art galleries and museums. The few museums that existed in the South did not recognize the region's academically trained resident artists and most definitely did not encourage or support the region's folk artists or their musical counterparts—country and blues musicians. Only in recent years has the South's museum world embraced the region's visual arts as a legitimate part of the history and culture of the region. By extension, other forms of artistic expression have also benefited from this institutional recognition and the legitimization of the visual arts.

The Morris Museum of Art's leadership in recognizing and nurturing Southern art is a central part of this story. The museum's founders, Billy and Sissie Morris, have played a pioneering role through their creation of a fine museum with distinctive collections of depth and quality and wide-ranging public programs that help to create a context for the understanding of the region's rich cultural heritage. Along with its sister institution, the Ogden Museum of Southern Art in New

Orleans, the Morris Museum of Art has raised the bar high. Both museums appropriately embrace photography *and* music as central to their efforts to achieve a more inclusive view of Southern art. While the Morris Museum has focused on "upland styles" of music—bluegrass, country, gospel, and mountain music—the Ogden Museum has explored jazz and blues, music that has been, historically, associated with its home, New Orleans.

Jim McGuire's photographs offer an opportunity for an individual, intimate encounter with artists who perform regularly for audiences that more typically number in the thousands. This experience provides a welcome respite from the settings in which country musicians are usually encountered—whether honky tonks or tent shows, coliseums or the Grand Ole Opry. The magic of their music lies in its ability to make every person in an audience feel a unique bond with the singer through the music. Their songs about leaving home, loneliness, and broken hearts touch the listener in a profoundly personal way. It seems as though the artist is singing their song, just for them, creating a sense of shared experience. Jim McGuire's photographs evoke a similar intimacy. Whether in a quiet museum setting or the confines of this book—the viewer and the performer bond, and country music echoes in the viewer's memory as a result.

Jim McGuire grew up in New Jersey and was attracted to both music and photography at an early age. His father, Felix McGuire, loved jazz. Drawn to country music and the people who created and performed it, McGuire moved to Nashville in the early seventies and never left. He has produced more than five hundred album covers, and his studio is, famously, home to many late-night jam sessions.

Interestingly, Jim McGuire's pantheon of heroes includes both Django Reinhardt and R. Crumb. The Belgian-born gypsy guitarist Reinhardt, a major figure in the explosion of interest in American-style jazz in mid-century Paris, is also an important influence on Mississippi blues musician B. B. King. Cartoonist R. Crumb, subversive satirist and founder of the "underground comix" movement, is a noted record collector and an avid fan of blues and country music—both central themes in his cartoons. In a letter to Greil Marcus and Sean Wilentz, which they include in their recent book *The Rose & the Briar: Death, Love and Liberty in the American Ballad*, Crumb writes, "The huge success of the C.D. of 'Oh Brother' indicates that a lot of people out there are hungry for some music that has some tradition behind it, something deeply rooted."

Like Crumb, Jim McGuire seeks authenticity in music, and he finds it through his photographs of country musicians. McGuire's diverse tastes in photography, literature, and music have prepared him to bring both a trained eye and a sensitive ear to his work.

Jim McGuire's portraits have captured the legends of modern country music since 1972. His images feature such icons as Bill Monroe, Chet Atkins, Dolly Parton, Johnny Cash, Lester Flatt, Merle Travis, Porter Wagoner, Marty Robbins, Ralph Stanley, Tammy Wynette, and Waylon Jennings. Each of these artists pioneered musical sounds and performance styles that have helped to define country music in the public's mind. Bill Monroe, widely considered "the father of bluegrass music" (his band The Blue Grass Boys lent its name to the emerging genre) and what was once a sound unique to his band, which included, among others, Lester Flatt and Earl Scruggs, became a genre when Ralph and Carter Stanley became the first of many bands to adopt that style as their own, launching bluegrass as a movement. Their "sound" has influenced bluegrass musicians around the world.

That first generation of bluegrass musicians dominated the form from its invention in the mid-forties until the mid-sixties, when a new generation of musicians—the so-called "second generations of bluegrass musicians," including The Dillards, Sam Bush, and J. D. Crowe—came to the fore. (At the same time, a group of "progressive bluegrass" bands, including The Country Gentlemen and New Grass Revival, put their own distinctive stamp on the music.)

The success of such pioneers as Mother Maybelle Carter, Kitty Wells, and Loretta Lynn led to the wide, popular appeal of such later stars as Dolly Parton and Tammy Wynette, who were among the first women to lend country music a decidedly feminist tone. Johnny Cash and Waylon Jennings moved from rock and roll in one of its earliest forms—rockabilly (which was itself a combination of hillbilly boogie, Western swing, and blues)—to "country" ballads.

McGuire also pays homage to another generation of performers, including Carole King, David Bromberg, Darrell Scott, Emmylou Harris, Guy and Susanna Clark, Hal Ketchum, John Hartford, John Prine, Sam Bush, Steve Earle, Rosanne Cash, Steve Young, Townes Van Zandt, and Vince Gill. These younger artists have helped to broaden the appeal of country music to their generation. This was especially important after many teenage listeners in the fifties turned from country music and its derivative rockabilly to rock and roll in a less "countrified" form as their principal musical interest. In a voice sometimes compared to that of American folk singer Joan Baez, Emmylou Harris has played traditional ballads with acoustic accompaniment while also performing (along with her "Hot Band") electronic, rock-influenced songs. Townes Van Zandt, a gifted singer-songwriter who helped launch the sound of "alternative country," numbered Lightning Hopkins, Bob Dylan, and Guy Clark among his musical heroes. He left a legacy of great songs, and his Texas-rooted influence stretched well beyond country music and musicians. Rosanne Cash has carried her father's legacy to new audiences through her songwriting and her performances. All of these artists have carried country music to new audiences and help to create new interest in traditional forms of the music for their own generation.

There are also artists such as Norman and Nancy Blake, Jesse Winchester, and Doc Watson, who are associated with traditional "roots music." These musicians have come to Nashville and its country music worlds from many different places and diverse backgrounds—Bashful Brother Oswald from East Tennessee; Béla Fleck from New York City; Don Williams from Floydada, Texas; Jerry Jeff Walker from upstate New York; John Hiatt from Indiana; Marty Stuart from Mississippi; and Mike Seeger from New York City, which certainly says something about the pervasiveness and appeal of the music itself. They even include among their numbers a Rhodes scholar, West Point–educated Kris Kristofferson.

Country music singers of every generation acknowledge their debt to older performers whose music has helped to shape their own. Traditional ballads and hymns from *The Sacred Harp* have influenced virtually every country music singer. These styles of music—one imported, the other, the shape note music found in *The Sacred Harp,* a uniquely American phenomenon—date to colonial-era America and have been recognized as an important influence throughout country music history. At the height of the folk music boom in the early nineteen sixties, Doc Watson, at the suggestion of folk musicologist Ralph Rinzler, replaced his electric guitar with an acoustic instrument and began to play hauntingly beautiful songs and mountain ballads associated with the oral traditions of his home area in and around Deep Gap, North Carolina. Today, Doc

Watson's MerleFest festival draws one hundred thousand people each year to Wilkesboro, North Carolina, to hear musicians perform "roots music" on acoustic instruments. Each of these roots musicians is an important part of this celebration of the various musical styles that have contributed to what we now know as country music.

It is hard to imagine a more diverse group of individuals. What binds them is a deep and abiding respect, a real love for country music. This love is clearly shared by Jim McGuire, whose timeless images capture and hold the many different faces of country music, a "who's who" of its leading writers, producers, and performers. His special gift as a photographer lies in his ability to capture personality, and his special gift to posterity lies in his ability to share that with others. Because of Jim McGuire's *Nashville Portraits*, we don't just know what these people look like. McGuire's artistry has enabled us to know *who* they are. Engaging, generous, good-humored, sometimes self-conscious, always bubbling with ideas, Jim McGuire's subjects are the very picture of creativity.

—William R. Ferris
University of North Carolina at Chapel Hill

NASHVILLE PORTRAITS
The Exhibition

Ralph Stanley

1998

Born Ralph Edmond Stanley on February 25, 1927 in Stratton, Virginia

From 1946 until the death of his guitar-playing brother, Carter, in 1966, Ralph Stanley performed as part of the Stanley Brothers duo, backed by their band, the Clinch Mountain Boys. He has continued to play since then, eventually reviving the Clinch Mountain Boys with whom he continues to tour. His unique style of banjo playing—sometimes called "Stanley style"—and the very traditional bluegrass music he performs, draw on the musical traditions of the area of southwestern Virginia in which he grew up. At the age of seventy-five, he was awarded a Grammy for his performance of "O Death" on the *O Brother, Where Art Thou?* movie soundtrack album.

. . . He's a punk singer or a rock-and-roll singer or a country singer… he's a mountain singer is what he really is. He's way closer to Elvis Presley than the notion of "Dueling Banjos."

—T-Bone Burnett

When I first heard the Stanley Brothers do "Rank Stranger," that really gripped me. I drifted into blue-grass when I was about thirteen or fourteen, and all at once there was the Stanley Brothers, Bill Monroe, Flatt & Scruggs, and then I started discovering folks like the Seldom Scene and Don Stover a few months later. There was just something about it that I can't really put my finger on, but it would electrify me when I'd hear it.

—Jim Lauderdale

The Stanleys didn't have the bands that Monroe and Flatt & Scruggs had, but they had those simple songs about mountain ways and mountain people, and they had those harmonies that really defined the high, lone-some sound . . .

I grew up in those mountains, so I absorbed that same sound. Even when I saw The Beatles on "The Ed Sullivan Show," I said, "They're good, but they're not as good as the Stanley Brothers." During my country years, I'd see Ralph at a bluegrass festival and play him a cassette of our new music, and he'd tell me it was great, like he was giving me fatherly encouragement.

Ralph once told an interviewer, "Ricky's making a name for himself in country music, but I think he'll come back to bluegrass." I think he knew something about me that I didn't even know about myself.

—Ricky Skaggs

His colloquialism and his performances echo the voices that I heard around the kitchen table singing when I was a child. Bill Monroe, of course, is the godfa-ther of bluegrass. But the Stanley Brothers are really the essence of Appalachian mountain music.

—Dwight Yoakam

Guy and Susanna Clark

1975

He: Born November 6, 1941 in Monahans, Texas
She: Born March 11, 1939 in Texarkana, Arkansas

Texans Guy and Susanna Clark, both singer-songwriters, first came to Nashville at the time that McGuire did, in 1972. They became fast friends when McGuire shot the cover photographs for Guy's first studio album, *Old Number One,* which was released by RCA Records in 1975. In the 1970s, when this photograph was taken, the Clarks' Nashville home was a haven for emerging songwriters and musicians. Guy Clark has served as a mentor to many other songwriters, most notably Steve Earle and Rodney Crowell, and numerous artists have recorded Clark-penned songs.

Guy is a songwriter and an actor; he plays a different role every time he sings a song. Guy is one of the people who has really let Nashville and the entire country music community know you can really write intelligent lyrics from the heart and they can turn around and sell records. Not that Nashville didn't know that before, but Guy was always there to remind them.

—Radney Foster

When I think about Guy Clark's music, words come to mind like intelligence. I would apply passion, humor, intensity, vision. Guy's a real artist. He's the consummate artist.

Guy Clark's not just a great Texas songwriter, he's a universal songwriter. I think it is essential for people to know and understand his music, because it will change your consciousness when you get inside it.

—Rodney Crowell

I got a call one night from a friend who was in the music business, who said, there's this guy you need to go hear. I halfheartedly, reluctantly, went down to this little beer joint in downtown Nashville. In a few minutes, this guy came walking out, picked up a guitar, lit up a cigarette, and sat there for about the next hour and proceeded to tear my heart out. I've been a Guy Clark fan ever since.

—Ronnie Dunn

Guy Clark is an original of a certain style of music, of Texas music. I don't know if you would call it country music, or folk music; it's just good music, and it's well written. It's the heart stuff that heroes are made of.

—Nanci Griffith

Vince Gill

1985

Born April 12, 1957 in Norman, Oklahoma

Oklahoma native and guitar virtuoso, Vince Gill has enjoyed a long career with many number-one country records. He began playing in bluegrass bands almost immediately on graduating from high school and debuted on the national scene with the country rock band Pure Prairie League in 1979. A Grammy Award winner, he is a member of the Western Performers Hall of Fame and the Grand Ole Opry. He is well known for his flawless singing and skilled guitar playing. He is married to singer Amy Grant.

I was trying to do the background vocals for "House of Love," and I was mortified at how high they were. I couldn't hit the notes. I [asked Vince Gill], "Will you do a song for me on my record?" and Vince said, "Sure." It amazed me that he didn't even care what the song was about. I said, "It's really high, nosebleed high," and he said, "No problem." I really appreciate his approach to his music, which he loves but doesn't feel like it gives him any more value than the next guy.

—Amy Grant

Vince and I were appearing on a TV special together, and just hanging out with the other musicians. We were joking about how, instead of politicking and brown-nosing [with the show's producers], here we were hanging around the guitar players and the steel players. Vince said, "I guess it's just the musician in us."

I think a lot of times for artists such as me and Vince—and it sounds terrible to say this—being versatile hasn't been that great. Vince can rock out and play his butt off, and I think in a lot of ways that's true of me. Then Vince can turn around and sing you a real coun-

try song and make you cry, and I think a forté of mine is to do something that's pop-oriented and yet emotional and heartfelt.

We both enjoy doing music of all kinds, and I think the fact that the public can't say "This is what Steve is" or "This is what Vince is" has been a real limiting thing.

—Steve Wariner

There's a certain kind of sadness in Vince's song "Young Man's Town," dealing with the natural process of aging and new blood coming in. But in Vince's hands, there's no bitterness. It's reflective, I think, more than anything. There's a real sense of grace about the song.

We tend to think that there's only one place to be, and that's on top, but that's not true. There are all these dimensions out there. Certainly, Vince has been on top, in one sense. And he will always be a star. But there's a certain point where you can't be number one on the chart your entire career. And there are plenty of other great places to be.

—Emmylou Harris

Dave Dudley

1976

Born May 3, 1928 in Spencer, Wisconsin
Died December 22, 2003 in Danbury, Wisconsin

Born David Darwin Pedriska in Spencer, Wisconsin, Dudley is best known for such blue-collar truck anthems as "Trucker's Prayer" and "Truck Drivin' Son-of-a-Gun." In 1963, he had his first hit with "Six Days on the Road," a never-to-be-forgotten truckers' classic, as well as a country hit that has been recorded by many artists over the years. He wasn't just involved in truck-driving songs, but also with other material. (His duet with Tom T. Hall called "Day Drinking" proved this.) However, he is one of the best-known singers of the truck-driving era and came to be regarded as one of the icons in this genre of country music. During the height of his career, the 1960s and '70s, he had more than thirty Top Forty country hits.

If you're in the cab of a truck, Dave Dudley's got to be in there somewhere along the line through the years. . . . Interestingly enough about "Six Days on the Road," it was actually written by a couple of truck drivers in their truck on a six-day run. They brought the song to Jimmy C. Newman. They brought it to Jimmy and Jimmy said, "Ah, you know, this really isn't my cup of tea," and he kind of passed it around to some friends who finally gave it to Dave Dudley, who said, "Oh, this really isn't my cup of tea, either," 'cause Dave was sort of like almost a lounge-ballad-type singer at the time. And Dave put together some money and did a recording session. He was going to do four songs. He had three in the can, and they wanted a fourth song—kind of an up-tempo thing, and he said, "Well, I got this truck-driving thing," and you know, the band heard it and they kind of put it together, and the next thing you know it's a super—well, it's a landmark song. It becomes like the definitive trucking song of all time.

I think he basically gave truckers a persona of somebody that is very hardworking and very patriotic, very steady and loyal. I think he did that. I think he highlighted their good qualities in his music, yeah.

—Dave Nemo

It'd probably be hard to find a truck-stop jukebox in America that doesn't play Dudley's "Six Days on the Road."

—Merle Haggard

When he recorded it in 1963, he launched a whole new breed of music known as "trucker music." And Dudley himself became known as high priest of diesel country.

—Daniel Zwerdling, *All Things Considered* (NPR)

Tut Taylor

1974

Born Robert A. Taylor on November 20, 1923 near Possum Trot, Georgia

Taylor, a self-taught master of the Dobro—which he plays in an unusual style, with a flat pick—is a highly regarded studio musician who is, perhaps, best known as a member of the Aero-Plain Band formed by John Hartford in 1971. It's groundbreaking album of 1971, *Aero Plain,* was an inspired combination of traditional bluegrass musicianship (of a very high order, given the band's personnel) and the hippie spirit of the times.

Tut has always been there, right in the middle of things.

—Norman Blake

. . . Although best known as the "flat-pickin' Dobro man," Tut is a fine mandolinist, and his Tut Taylor Music also built some mighty fine mandolins. He played mandolin with Roy Acuff at the last Opry performance at the Ryman, he picked with Bill Monroe and Roland White; he was business partner with Randy Wood in the Old Time Pickin' Parlor. . . . Tut Taylor should be declared a National Treasure.

—David Drake

Tut Taylor was a professional sign painter and musician in Milledgeville, Georgia. He had a woodworking shop along with the sign business. . . . In 1969 Tut received a call from Gibson offering the prospect of carving work for the newly proposed All-American and Florentine banjos. Tut contacted Randy and me to gauge our interest in participating. . . . Tut and Randy saw it as a possible foundation for a new business venture. . . . It was a major move for all of us. For Tut it involved closing his business in Milledgeville and moving his sizable family to Nashville for a less than totally secure future. . . .

We found a building . . . at 111 4th Avenue North. . . .

It wasn't much of a place. It measured 20 by 60 feet total and was in poor condition, but Tut and Randy quickly set to work to build walls dividing the building into a showroom, a small office, and a workshop. . . .

My personal overhead was quite low . . . but Tut Taylor had a large family to feed. Those early days at the shop remain vivid in my mind and are a source of great nostalgia. It was pure magic to have Norman Blake there in our little office/studio and to be able to go back and forth from our shop and the backstage area of the Johnny Cash TV show and meet world-class musicians, but I also recall that there were some days when we took in as little as five dollars for the entire day's proceeds. We didn't enter the business with delusions of grandeur. It was evident from the start that we were not going to become millionaires overnight. . . .

One day . . . I opened the store and found a note from Tut to me taped on the inside door leading back to the repair shop. Tut felt that it was time for him to move on . . . to pursue his own musical career.

—George Gruhn

Tut always tried to help me along into the music. All I can say is, "Wow, Amazing, and Thank You, Jesus."

—Nancy Blake

Chet Atkins

1976

Born Chester Burton Atkins on June 20, 1924 in Luttrell, Tennessee
Died June 30, 2001 in Nashville, Tennessee

**Known as "Mister Guitar," Atkins was a trailblazer who is widely credited
for the creation of the so-called "Nashville Sound." One of the most influential
and best-loved guitarists in the history of the instrument, he became
the president of RCA Records and produced many classic country albums.**

[Chet] taught me to be a forward thinker. Even though this business can dash you against the rocks, there's always this hope that I'm going to hit another lick. He looked forward to the next lick he was going to hit.

—Suzy Bogguss

Chet picked really great musicians [for my recordings]. He knew they could play, so he let 'em play. He would have a general idea of how he wanted the song to go. If he had a lick or riff in mind, he'd go out and show it to the musicians and then sit back and let them go. He was an expert at staying out of the way.

I'd just go hang out with him. We would be talking and he would be noodling around on the guitar. I would forget he was the greatest guitar player in the world. . . . He was bigger than life, and he so underplayed it. He was there to play and that was it. He would much rather concentrate on playing than entertaining an audience. But he got pretty good at the entertaining part, too.

Chet made our dreams possible . . . I really don't think I could have accomplished what I did with anybody else but Chet.

—Bobby Bare

Chet made fun of stuffed shirts and hypocrites and hadn't much time for blowhards or preachers. He loved Mark Twain, and he was a fine storyteller, though he never would tell one on the show because all his best stories were a little ribald or irreverent. . . . He was an artist and there was not a bit of pretense in him. He never waved the flag, he never held up the cross, he never traded on his own sorrows. . . . His humor was self-deprecating. He was always his own best critic. And he inspired all sorts of players who never played anything like him.

—Garrison Keillor

He was the first person I looked up when I moved to Nashville in '64.

—Dolly Parton

I see him as a father figure in terms of what he did for me and how he changed my life. When he went out to the meeting [with RCA] he played my music and didn't tell them what color I was. He just let them hear my voice and wanted to know what they thought. . . . I've been around presidents and a lot of other [important] people, but there only have been two people on this earth that I was nervous around: Chet Atkins and Mickey Mantle. It's because of the respect I have for them.

—Charley Pride

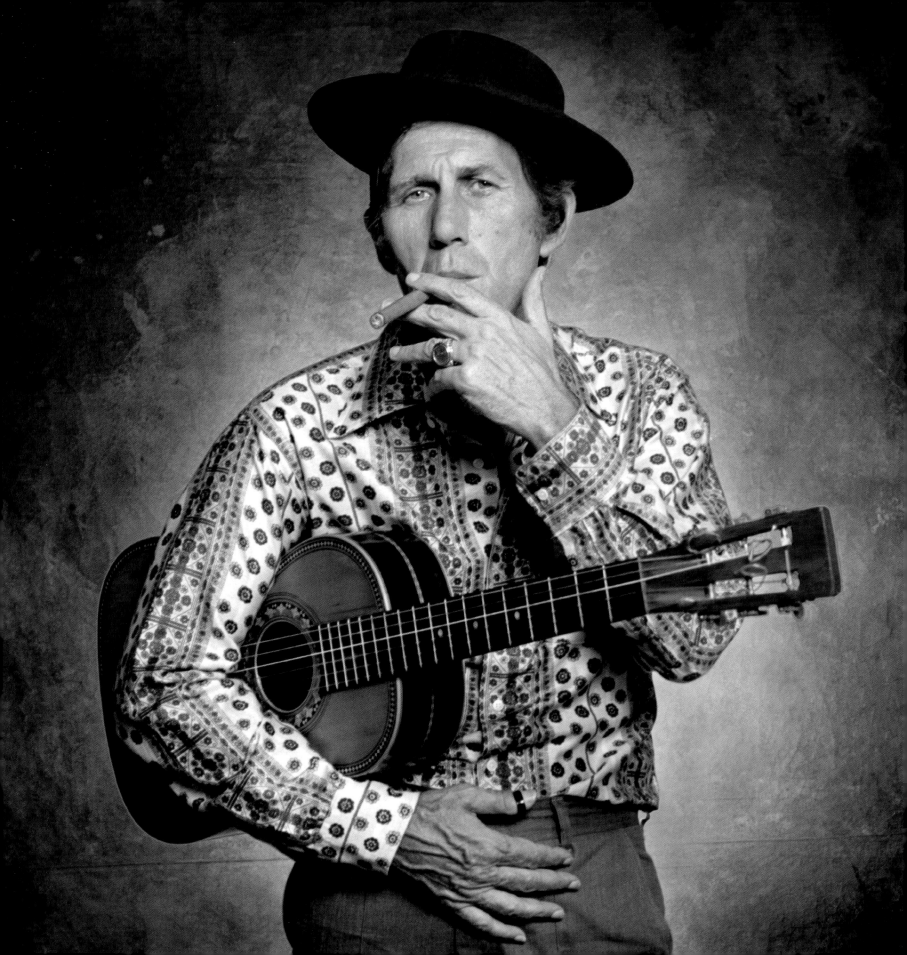

Brother Oswald

1975

Born Beecher Ray [Pete] Kirby on December 26, 1911 in Sevierville, Tennessee
Died October 17, 2002 in Madison, Tennessee

Hailing from East Tennessee, Brother Oswald played Dobro and served as comic relief with the great Roy Acuff and his bands for more than fifty years.

Oswald was the glue that held the Roy Acuff Show together.

—Opry announcer Hairl Hensley

I'd say Oswald was one of a kind. He learned from a Hawaiian boy by the name of Rudy Waikiki when he was in the North playing clubs. A guy came in one night and said there's a crowd going down the street because there's a Dobro player, and Os said, "Well, [there'll] be a Dobro player here tomorrow night." So he went and got him a Dobro, and started to learn it the hard way. He come across Rudy Waikiki, and he learned a lot of things from him. Os said he was really a great player, a very good picker, and he learned from him. That started his Dobro playing.

We just enjoyed each other on the stage. I enjoyed his work because he was so professional, and it was always consistent. No matter how bad he felt, he never let it show through. He always come off to the audience like [he was] having fun, this is it.

When you're together that much, you're really closer than brothers and sisters a lot of times. We traveled with each other on the road, and we understood each other quite a bit. To me, he was a part of country music history.

You're just lucky to be around a person like that; it's been a blessing in my life.

—Charlie Collins

David Bromberg

1972

Born September 19, 1945 in Philadelphia, Pennsylvania

A musician's musician and an eclectic singer, songwriter, and multi-instrumentalist who has performed bluegrass, blues, folk, country, and rock and roll with equal facility, Bromberg is especially well known for his quirky, humorous lyrics. He went from being a top studio player to a touring sideman with, among others, Jerry Jeff Walker and Bob Dylan. He now lives in Wilmington, Delaware, where he deals in vintage violins.

I called him the kamikaze bluesman. He'd play whatever came into his head. He was fearless that way.

—John Sebastian

David is a skillful professional going about his business, the business of making music.

—Roger Dietz

For the nearly forty years I have known David . . . I am still mesmerized by his wryly humorous, technically accomplished presentation. Bromberg is an intense, savvy, seemingly enigmatic artist. The guitar wizard labored long and hard to assimilate and perform the music he performs. . . . I find it amusing and ironic that listeners caught up in a frenzy of notes often project superhuman attributes onto the serious artist that in the final analysis is more restrained than one can imagine. . . .

He has a completely personal style of playing the guitar so that just from his finger work, you know it's David Bromberg. The tone, the attack, the phrasing . . . It's very bluesy and gutsy. But he also has this sweet side of playing. . . . As a bandleader, he was very inspiring. The ethos of the band was, if it excited us, if it turned us on, he encouraged that.

—Jay Ungar

Tammy Wynette

1987

Born Virginia Wynette Pugh on May 5, 1942 near Tremont, Mississippi
Died April 6, 1998 in Nashville, Tennessee

Once known as "The First Lady of Country Music," Tammy Wynette was a talented singer-songwriter who, during the 1960s and '70s, simply dominated the country music charts. In 1968 and '69, she had five number-one hits, among them three of her signature songs, "Stand by Your Man," "D.I.V.O.R.C.E.," and "Take Me to Your World." In all, she had seventeen number-one songs, and her work, along with that of Loretta Lynn, Dolly Parton, and Lynn Anderson, helped to define the role of the female country singer, influencing many of the younger singers who followed them. Many of her songs seemed, at least to her fans, to be influenced by her tumultuous private life. Married five times—productively in the case of her marriage to singing partner, George Jones—she was often ill, surviving numerous hospitalizations and surgeries. She succumbed, unexpectedly, to a cardiac arrhythmia at the age of fifty-six.

Tammy has always had that different-type voice. When I first heard "Apartment #9" [Wynette's 1966 debut single], the roof came off. I thought, "Well, now, there's a girl that can sing!"

—George Jones

The first time I saw her in person was in the early '70s at one of my father's "guitar pulls" in his living room, when a lot of musicians and songwriters previewed their new work. I was about nineteen years old, with purplish hair and insouciance to spare, and the honored guests were George Jones and Tammy Wynette. I sat slack-jawed and transfixed as they sang "(We're Not) the Jet Set" ("No, We're not the Jet Set / We're the old Chevrolet set / Our steak and martini / Is draft beer with weenie"). She sat on the plush blue antique sofa, hair poufed out to here, with nails, makeup, and outfit perfectly coordinated. She looked like a lotus blossom sitting next to George Jones, a perfect foil, but completely herself. It was the most relaxed I was ever to see her.

Tammy was sweet, in the way that only Southern women are sweet, and a bundle of nerves. I don't ever think she got over her ascendancy from the beauty parlor. She was a vehicle for her Voice, and it seemed to have ambition of its own, sometimes overreaching her personal understanding or goals. . . . I think of her—proud but not egotistical (a feat in itself), delicate and strong—and how the world will never be innocent enough again to produce a Tammy Wynette.

—Rosanne Cash

The first time I heard her sing live, I was just a kid. She hit that high F in "Stand by Your Man," and I just fell out of my chair. I thought, "Man, that chick is wailin'!"
. . . Tammy was a landmark—one of those people we navigate by.

—Pam Tillis

Waylon Jennings

1985

Born June 15, 1937 in Littlefield, Texas
Died February 13, 2002 in Chandler, Arizona

A Texas native and self-taught guitarist, Waylon Arnold Jennings was a disc jockey when Buddy Holly recruited him to play bass in his band. He was famous for giving up his seat to Ritchie Valens on Holly's fatal wintertime flight to Mason City, Iowa. He became a Nashville legend and is generally credited with starting the "outlaw" movement in country music—the title track of his 1972 RCA release "Ladies Love Outlaws" lent the movement its name. With his friend and close collaborator, Willie Nelson, he took a career-defining step with the release of "Wanted: The Outlaws!" in 1976. Later, he and Nelson collaborated on their biggest hit, "Mamas, Don't Let Your Babies Grow Up to Be Cowboys," and then, along with Johnny Cash and Kris Kristofferson, formed the supergroup, The Highwaymen. Plagued by addiction and ill health, he left the road in 2001 and succumbed to diabetic complications at the age of sixty-four.

Waylon was a true original. He came along and wasn't afraid to do his own thing and changed the sound of country music. He didn't just open doors—he knocked down doors for all artists who want to break the mold and have their own sound. . . . The spirit of who he was and what he stood for will be with us forever.

—Martina McBride

Doing things his own way, speaking his mind regardless of the consequences, and reaching out and touching the true inner feelings of his audience are all things that Waylon stood for throughout his career. I shared a real kinship with him.

At times he was shunned by many in those establishments for his "outlaw" honesty. In spite of this, his monumental contribution to our industry will be felt for decades to come. I . . . hope to carry on the legacy that he inspired.

—Travis Tritt

My dad would talk about retiring, but he never could. He said the road just keeps calling you back. Sharing his music was what he loved.

—Shooter Jennings

He proved you could go outside the lines, stick to your guns, and be successful. Any time any artist wonders about their path, they just need to think about Waylon Jennings, because he definitely did it his own way.

—Kenny Chesney

He was more of a pussycat than he was an outlaw.

—Bobby Bare

J. D. Crowe and The New South

1975

Born James Dee Crowe on August 27, 1937 in Lexington, Kentucky

J. D. Crowe has toured with various versions of his bluegrass band, The New South, for more than thirty years. An influential banjo player, he got his start playing with Jimmy Martin's Sunny Mountain Boys while still in his teens. In 1975, he put together what is widely thought to be one of the most influential bluegrass bands of all time. Depicted here, it consisted of (left to right): Ricky Skaggs on mandolin and fiddle, Bobby Sloan on acoustic bass, J. D. Crowe on banjo, Jerry Douglas on Dobro, and Tony Rice on guitar and vocals.

Even after more than fifty years in the business, J. D. Crowe still has a lot to say musically.

—Eddie Stubbs, WSM Grand Ole Opry

The first [bluegrass] CD [I heard] was *J. D. Crowe & the New South* with Tony Rice, Ricky Skaggs, Jerry Douglas, and J. D. Crowe. And that is *the* bluegrass album. To many bluegrass fans, that's the record right there. When I heard those harmonies and that music, I thought, "Man! This is it!"

—Bradley Walker

When they came in the studio Monday morning, these guys had no idea what they were going to be playing and singing, so it's pretty much all off-the-cuff. But we all came from the same school of music—we all grew up on Bill Monroe, Flatt & Scruggs, the Stanley Brothers, the Osborne Brothers, J. D. Crowe, and so on—and we all kind of have the same feel and the same groove.

—Jim Mills

There is no way on God's green earth that I would be here playing music for anyone here if it weren't for J. D. Crowe; he is my all-time hero.

—Dan Tyminski, Union Station

Play it like J. D. did.

—Jimmy Martin's advice to each of his banjo players who followed J. D. Crowe in his Sunny Mountain Boys band

Marty Robbins

1977

Born Martin David Robinson on September 26, 1925 in Glendale, Arizona
Died December 8, 1982 in Brentwood, Tennessee

One of the most popular and successful country and western singers of his era, Robbins grew up in a suburb of Phoenix, Arizona, where, while still a kid, he became a huge fan of the legendary singing Cowboys. (Western music remained a constant throughout his forty-year career.) He learned to play guitar while in the navy during World War II. He returned to Phoenix after the service and established himself locally as a performer and the host of his own radio and, later, television programs. Little Jimmy Dickens, a one-time guest on his television program, recommended him to Columbia Records, leading to his move to Nashville where he became one of the most popular members of the Grand Ole Opry.

Marty Robbins was a pretty soulful cat.

—Hoyt Axton

I've known some great ones, and he was one of the greatest. He was such a great performer and he could write what he could perform. He had it all.

—Chet Atkins

My basic roots were Ernest Tubb, Hank Williams Sr., and Marty Robbins. I also liked some of the early rockers like Elvis and Little Richard, and also the folk-blues guys like Leadbelly and Big Bill Broonzy. Usually I just follow my own taste and not gear it to the marketplace. My music is a hodgepodge of country, blues, folk, and rock 'n' roll. I've got sort of a nasal timbre to my voice that people seem to like.

—Jimmie Dale Gilmore

"El Paso" by Marty Robbins, whose music I still love, is one of my favorite songs. There was something about his gunfighter and cowboy ballads that was like poetry to me. They're what made me realize that you could create characters and tell stories in songs.

—Bernie Taupin

Hal Ketchum

1997

Born Hal Michael Ketchum on April 9, 1953 in Greenwich, New York

Ketchum, a singer-songwriter from upstate New York who arrived in Nashville via Austin, Texas, made his living around Nashville as a fine carpenter before signing with Curb Records. (This portrait was shot at a photo session for his first album for Curb.) Though he'd been recording since the mid-1980s, he did not have a major hit until the release of his album *Past the Point of Rescue* in 1991. He continues to record and release new material.

I was in Nashville, presenting at the CMT awards, and my buddy said, "Hey, we should try to make a record." I got some great writers—Hal Ketchum wrote two songs, Bernie Taupin. I financed the whole thing myself. . . .

Hal Ketchum is a buddy of mine. He said, "Don't do it"[sign with a major label]. His label hasn't let him make a record in four years.

—John Corbett

Hal has an incredibly beautiful voice, and I wanted an introspective, quiet work to go between the more upbeat and louder pieces. . . . His music is my music, just as *Swan Lake* was Tchaikovsky's music. There's no reason I shouldn't express myself through the music of my life.

—Paul Vasterling, Nashville Ballet

Hal is one of those guys who I think does exactly what he wants to do. He's much like a Mary Chapin Carpenter who is always gonna remain true to what they do, and occasionally that will fall into the realm of mainstream, and they'll get the success. Hal really impresses me as someone who has his own agenda, and if radio and the industry can work into it, great. If they can't, well, so be it.

—Eddie Haskell

Béla Fleck

1999

Born Béla Anton Leos Fleck on July 10, 1958 in New York, New York

Born in New York City, but a resident of Nashville for more than twenty years, Fleck is an acclaimed instrumentalist who has revolutionized the banjo, taking it from traditional bluegrass to jazz and beyond. The founder of the Flecktones, Fleck is the only musician to be nominated for Grammy Awards in jazz, bluegrass, pop, country, spoken word, Christian, composition, and world music categories.

I drifted into bluegrass when I was about thirteen or fourteen. . . . There was just something about it that I can't really put my finger on, but it would electrify me when I'd hear it. I started playing the banjo and did that for several years. . . . Later, I started playing Dobro when I was about seventeen, and then rhythm guitar. A few years later, my banjo playing started decreasing. . . . I just had the feeling that I was never going to be an innovator on the banjo or somebody that would bring something new, or I'd never be as clean as Earl Scruggs or as innovative as Béla Fleck.

—Jim Lauderdale

These bluegrass guys are phenomenal musicians that have exceptional ears. They can hear the music I'm writing in their own way and add their own personality to the music. [My composition] "Snap Dragon" features [Fleck's] boldest playing. . . . I wrote a lead line to the song, created a bridge, and then in the middle I said,

"Everybody drop out except for me and Béla." It was a blend of the two musics, because you had the standard banjo sound in bluegrass with the picking sound and the tenor saxophone blasting over the top.

—Bill Evans

For most people when they play the banjo, the thumb sounds quite different from the fingers. Béla [has achieved] an evenness between the thumb and fingers and all over the instrument, and a level of control that [is] more typical for jazz or classical players. . . . We keep each other on our toes.

—Edgar Meyer

Béla is taking the instrument and the tradition of the banjo and bringing it up as a seriously virtuosic instrument. When I first listened to the Flecktones, I heard a completely fresh sound.

—Chick Corea

Emmylou Harris

1983

Born April 2, 1947 in Birmingham, Alabama

A Grammy Award–winning country music singer, songwriter, and musician, Emmylou Harris is also a much-sought-after backing vocalist and duet partner in addition to her work as a solo artist and bandleader. Since 1968, she has worked with numerous artists, including Ryan Adams, Willie Nelson, Roy Orbison, Bob Dylan, and, perhaps most notably, Dolly Parton and Linda Ronstadt. This photograph was shot for the cover of her *White Shoes* album on location at the Tennessee Fine Arts Center at Cheekwood (now the Cheekwood Museum of Art) in Nashville.

Emmylou's got something happening with her voice. You don't hear people of her experience singing a lot, so there's emotion in her voice that is so rare. I'm hearing things in her voice that I've never heard before. I get really big chills. She's tapped into stuff that you have to have lived her life to tap into. You can't fake that stuff. She's got this soul in her voice that gets deeper and deeper.

—Patty Griffin

In the '60s, women country singers did not own their intelligence or their integrity so much. Our society being what it is, women were kind of directed by men. And Emmy was kind of the first country artist to stand up with some sort of integrity and not be directed by men; she certainly collaborated with men, but she wasn't directed by them. Maybe the first feminist, in a way. She doesn't lead with that one thing.

Emmylou Harris is a supreme artist. In her way, even as an interpreter, she's a poet. She's very sensitive in communications. She communicates with her songs, poetically. The songs she chooses, the language, the atmosphere. It's her vision.

—Rodney Crowell

Emmy has the best collection of material always. I've never known her to have less than twenty new songs that I've never heard before. I always just figured she stayed up later than I did. Any chance to sing publicly or privately with Emmylou I'll take any day.

—Linda Ronstadt

In my writing, I'd been dealing a fair bit with the male-female shape. With that amazing voice, Emmylou, now more than ever, seems to capture a kind of universal female experience, something that speaks to all ages.

—Mark Knopfler

John Hiatt

1974

Born John Robert Hiatt on August 20, 1952 in Indianapolis, Indiana

**Originally from Indianapolis, Indiana, Hiatt, a rock guitarist, pianist, singer,
and songwriter, moved to Nashville in the early 1970s to write songs
and to find his musical voice. After his song "Sure as I'm Sitting Here"
(recorded by Three Dog Night) became a Top Forty hit, he was signed to a
recording contract by Epic Records. This portrait was shot just before
the release of his first solo album, recorded in 1974.**

He's like the Barry White of singer-songwriters.

—Mary Chapin Carpenter

John likes a lot of the stuff I like, such as R&B and blues,
so we have similar taste. He knows what that music is
all about. . . . And he's an excellent guitar player and a
great singer; a fabulous singer. I think he's got the best
white voice I've heard in years. I'd picked up his
records, and I liked his songs okay, but I liked him the
best—his voice, his playing, and his approach. It
seemed like he was good all by himself, apart from
just being a writer.

I figured the guy would get onto what kind of thing
I was up to, and figure out a way to write something for
me. . . . And ["The Way We Make a Broken Heart" is] a
good song. It doesn't have any particular identification
about it. It could be a country song, a ballad, or anything. It's just the kind of tune I like.

—Ry Cooder

John Hiatt

2004

John has released twenty albums, and his songs have been covered by Bob Dylan, Bonnie Raitt, Eric Clapton, B. B. King, Joan Baez, and Jimmy Buffett, to cite just a few.

I have known John a long time, and we are friends. He used to live out here [in Los Angeles] and I used to see him quite a lot back in the early 1980s and the 1970s when he was collaborating, and I have the greatest respect for his musicianship and his writing and as a guy . . . He's a great guy and, you know, a formidable musician.

—Eric Clapton

His songs are terrific and it's like a homecoming to work with John and to get that old magic going again. John is the consummate songwriter, and I think there's a unique edge, something quite wonderful, that comes through [when we play together].

—Sonny Landreth

Harlan Howard

1988

Born Harlan Perry Howard on September 8, 1927 in Detroit, Michigan
Died March 3, 2002 in Nashville, Tennessee

Howard, a native of Detroit, was one of the most prolific and successful of the "old school" Nashville songwriters. While living and working in Los Angeles, he enjoyed a few minor successes, before the breakout hits "Pick Me Up on Your Way Down" and "Heartaches by the Number" led him to move to Nashville in 1960. There, he became an immediate success, his songs recorded by innumerable country artists. Among his biggest hits was the Patsy Cline classic, "I Fall to Pieces." A member of the Nashville Songwriters Hall of Fame and the Country Music Hall of Fame, he was a force in country music until his death.

In the universe of country music, Harlan is as relevant as the Ryman.

—Kris Kristofferson

When I first started going to Nashville, I was living with Harlan Howard. I had met him in Austin and he invited me. He said, "I know you're coming up. You're trying to get a publishing deal or a record deal. So just stay at my house."

So I'm sitting in the living room at night with Harlan and Waylon Jennings and Allen Reynolds and Jim Rooney, like a pride of lions. They said, "Son, listen twice and talk once; maybe you'll learn something here."

I was sitting around at Harlan's and I played a song called "Someplace Far Away," a folk song. And he looked at me and said, "That's it; that's where you need to go." Prior to that, I'd played him two or three things that were just clever. And he said, "One thing you need to bear in mind as a songwriter is that it's all been said before. If you can just learn to say it from your own perspective in some kind of honest fashion, people will gravitate to it. Because we're all telling the same story, but if you do it from your own heart and your own perspective, people will get it."

—Hal Ketchum

If you were a songwriter and you were working hard at it, that's all it took to get his respect. He had this great love of writers, and he loved to share his own view.

—Pat Alger

Townes Van Zandt

1990

Born John Townes Van Zandt on March 7, 1944 in Fort Worth, Texas
Died January 1, 1997 in Mount Juliet, Tennessee

One of the great tragic figures of country music, Townes Van Zandt was a folksinger, songwriter, performer, and poet. He was particularly influential in the emergence of alternative country in the '70s. Steve Earle described him as the greatest songwriter who ever lived, and his influence was felt by many other artists, including Emmylou Harris, Nanci Griffith, and Lyle Lovett. Bob Dylan refers to this Texas native as his favorite songwriter. He wrote hundreds of haunting songs that have been widely recorded, perhaps most notably "Pancho and Lefty," which was a number-one hit for Willie Nelson and Merle Haggard in 1983.

[I once said] "Townes Van Zandt's the best songwriter in the world, and I'd stand on Bob Dylan's coffee table in my cowboy boots and say that." And I ran into Townes, you know, after it, and he said, "Man, that's really a nice quote, but you know, I've met Bob Dylan's bodyguards, and I don't think it's a really good idea."

—Steve Earle

One of the things Townes was so good at, was in his writing, he allowed you to associate with his song, or be part of it without demanding your participation. You participated in it whether you liked it or not, you know, and it [could] scare you or make you chuckle or whatever, inspire you, but made you feel part of it, somehow.

—Guy Clark

He was our great poet of loss and loneliness and rust. His favorite words were American place names. To hear Townes say "Mojave" or "Comanche" or "Badlands" or "Cleveland" was to connect with something lost long ago, to a forgotten time when the mythic Mike Fink ran keelboats down the Mississippi, and the golden spike of the Transcontinental Railroad still hadn't been hammered into the great salt desert.

—Douglas Brinkley

Townes always greeted me with kindness. [When I first saw him at Gerde's Folk City in New York decades ago], I thought, "My God, it's the ghost of Hank Williams!" But now I know Townes was just forging his own path down the lost highway.

—Emmylou Harris

I think of Townes Van Zandt as being the greatest folk songwriter that my native state of Texas ever gave birth to. His song "Tecumseh Valley" that I recorded and still sing every night that I perform is about this woman named "Caroline." And my middle name is "Caroline." And singing that song every night and her story has saved me from making the same mistakes that Caroline made in her life.

—Nanci Griffith

Carole King

1978

Born Carole Klein on February 9, 1942 in Brooklyn, New York

**Active as a singer, songwriter, and pianist since the 1960s, King
has been inducted into the Songwriters Hall of Fame
and the Rock and Roll Hall of Fame. McGuire photographed her
during a recording session in Austin, Texas.**

Everybody knew everybody. One of the bands playing around was the Middle Class, protégés of Carole King and Gerry Goffin. Charlie [Lackey, the bassist], I think, brought Carole to see Flying Machine, a band James Taylor and I had, at the Night Owl Café. After that, I was amazingly fortunate that Carole started calling me to do some demos, which was a tremendous education in how to play on records from her and Gerry Goffin. I was very fortunate, at an early age, to run into Carole and Gerry, who had already had massive experience with all the hits they had. What great teachers!

—Danny Kortchmar

At times, it seems that Carole King has written and/or recorded every other popular song since the early 1960s: "Will You Still Love Me Tomorrow," "So Far Away," "It's too Late, Baby," "Loco-Motion," "You Make Me Feel Like a Natural Woman."

—Scott Simon, NPR

Back when we lived out in the Hill Country, in Boerne, we had so much land that I ran wild. I never really got into listening to music. When I got older I started listening to Hanson and the Spice Girls, and then my mom made me listen to all of her old music, like Janis Joplin, Carole King, Bob Dylan, the Rolling Stones. I hated them then; I'd cover my ears and yell. But now I love them.

—Hilary Duff

We wrote "If It's Over" in a little over an hour. It seemed very natural. For someone who's basically a legend, she's totally unaffected.

—Mariah Carey

Kris Kristofferson

1990

Born June 22, 1936 in Brownsville, Texas

An influential singer, songwriter, and actor, Kristofferson came to Nashville in 1965, fresh out of the army, to pursue his dream of writing country songs. Although he is best known for such songs as "Me and Bobby McGee" and "Help Me Make It Through the Night," he became a well-respected, much-in-demand film actor after his debut in Dennis Hopper's *The Last Movie* in 1971. He concentrated on film acting for a time, but in the early 1980s he, along with Willie Nelson, Johnny Cash, and Waylon Jennings, formed the superb group, The Highwaymen, with whom he toured and recorded. A member of the Nashville Songwriters Hall of Fame and the Country Music Hall of Fame, he continues to write and record his songs.

I was a songwriter myself back then and when I heard Kris, I said, "Oh, so *this* is the way you're supposed to write a damn song."

—Donnie Fritts

Kristofferson is unusual, maybe even unique, in the pantheon of country songwriters. In terms of sheer numbers of good to great songs written, Kristofferson can't touch Harlan Howard, Hank Williams, Jimmie Rodgers, Dolly Parton, Merle Haggard, Bill Anderson, Mel Tillis, Roger Miller, and quite a few others. But the small handful of great songs that he did write had more cultural impact than many of those other folks combined—and not just on country music culture but on America. I think this impact is mostly due to the way Kristofferson was open, in ways his country colleagues weren't, to incorporating the ideas and issues of the counterculture.

—David Cantwell

Kris said I looked like a black snake. He looked in my face and saw I was dying and wondered why. Then he wrote me a song. The song was "To Beat the Devil," a story about a poet who meets the devil in a bar, drinks his beer, then steals his song. It's the kind of thing that William Blake might have written if he'd had Kristofferson's sense of humor.

—Johnny Cash

It was the first time I'd seen him play solo, just him standing up there with an acoustic guitar and singing those songs. And he's amazing. As it happened, Kris was coming to the studio just to visit that day. He just walked through the door, and we recorded the whole album in about an hour and a half.

—Don Was

He could dig for the simple truth of a character. He's smart, and you would sit up and take notice when he entered a room. Kris Kristofferson knows how to wear the boots.

—John Sayles

The Dillards

1977

Douglas Flint Dillard: Born March 6, 1937 in Salem, Missouri
Rodney Dillard: Born May 18, 1942 in Salem, Missouri

**Douglas and Rodney Dillard formed the original bluegrass band,
The Dillards, in the mid-1960s. Although they were a huge influence on the
group of musicians who were central to the development of the country
rock movement in Southern California in the '60s, they are, perhaps, best
known as regulars on *The Andy Griffith Show* between 1963 and 1966, playing
a fictional local bluegrass band in Mayberry known as The Darlings.**

The Dillards have a drive and enthusiasm that shine through the hard, sparkling brightness of their music. They call it "Back Porch Music" and they enjoy the making of it so obviously that the listener finds himself caught up and carried away. If in listening you find yourself ruminating on bygone times, sensing perhaps the musical feeling and creativeness of your ancestors long since finished with their part in the making of America, if you get the feeling of the campfires and the log cabins and the lonely candlelight burning in the wilderness, you will have amply rewarded The Dillards.

—Jim Dickinson

I started playing a little bit when I was seventeen, but then I saw a group called The Dillards, and that really turned my crank. That gave me a passion for playing. I followed them for two years, attending every show I could. Doug Dillard's banjo playing was a real inspiration.

—John McEuen

When banjo legend Doug Dillard and criminally underrated mandolinist Dean Webb tear into a number, laughs are replaced by awe. Dillard's banjo has retained its splashy, brittle tone, with the speediest passages sounding like a riled rattlesnake.

Webb's solos featured phrases that attacked from the middle of the beat, and his ability to toss chords into frantic improvisations is as thrilling today as it was in the days of Sheriff Andy Taylor and deputy Barney Fife.

Rodney Dillard's quavering yet potent tenor is among the most affecting sounds in bluegrass.

—David Duckman

Roy Huskey Jr.

1991

Born December 17, 1956 in Nashville, Tennessee
Died September 6, 1997 in Nashville, Tennessee

Son of famed Nashville session player Junior Huskey (he had played on the Nitty Gritty Dirt Band's legendary *Will the Circle Be Unbroken* album in the early 1970s), Roy was, at the time of his untimely death, widely considered Nashville's premier acoustic bassist. He was in constant demand for all kinds of acoustic music, ranging from bluegrass to the "unplugged" work of Steve Earle. He was a member of Emmylou Harris's hot acoustic band, The Nash Ramblers. The impact of his early demise on the acoustic music community was deeply felt.

Roy Huskey was a pure, original, state-of-the-art, bona fide, textbook example of a bass-playing genius, truly irreplaceable.

—Marty Stuart

Roy was the best doghouse bass player who ever lived. I was asked to sing at his funeral where I was a pallbearer. I scanned a couple of ideas about what I was going to sing but I wasn't happy with any of them. So I finally decided to write something ["Pilgrim"]. Everyone on that track is someone who worked with Roy or was close to Roy.

—Steve Earle

I remember the night that I first met him. He asked me where I was from and what instrument I played. I told him I was an upright bass player. Well, I guess he was a bit eccentric, and he talked with his eyes shut. He turned to me, and with closed eyes he said: "Well, that's good, because there needs to be more of us in the world."

—Dennis Crouch

I don't remember the first time I ever met Roy Huskey Jr., but I do remember, as clear as day, the first time I saw him in full force with the rest of The Nash Ramblers. It was as if grace itself had come to earth in the form of those hands slapping that upright bass. I had never seen anything so elegantly beautiful breeding something so meaty and rich, a pool of sound that enveloped me and drove the rest of the music that pranced on top of it. . . .

Whenever I think of him now, I see his standard plaid flannel shirt, blue jeans, and green army jacket.

—Meghann Ahern (Emmylou
Harris's daughter)

He was the heartbeat of our band.

—Emmylou Harris, on Roy Huskey's role
in The Nash Ramblers

Sam Bush

1981

Born on April 13, 1952 in Bowling Green, Kentucky

Sam Bush is a renowned mandolin player and accomplished bluegrass vocalist originally from Bowling Green, Kentucky. He formed the improvisational bluegrass band, the New Grass Revival, in 1971. Over the years of its existence, the band went through many changes of personnel, and when it disbanded for good in 1989, Bush was the sole remaining founding member of the group. He joined Emmylou Harris's Nash Ramblers, and, in 1995, worked as a sideman with Lyle Lovett and Béla Fleck's Flecktones. He has been touring and recording with his own band since the mid-'90s.

Has Sam Bush ever played something that didn't feel perfect? I seriously doubt it. Sam Bush's blood pulses with the rhythm of the gods.

—Chris Thile, Nickel Creek

Sam Bush is a bluegrass legend, and it's not unfair to consider him a Bill Monroe for our times—a visionary who reveres traditions but doesn't let that stop him from expanding the form even further.

—Gary Graff

Hearing his incendiary picking and sawing is a bit like watching the special effects in *Terminator 2*: Bush is so adept at doing the impossible on his instruments that one soon takes the technique for granted and just enjoys the musical ride, which is as it should be.

—Jim Washburn

Sam has been my hero since I saw New Grass Revival at the University of Chicago Folk Festival in 1972. When I went to my first lesson with Jethro [Burns] and he asked me what I wanted to learn, I asked him to show me how to play like Sam! In retrospect, this seems tantamount to going to [classical guitarist Andrés] Segovia and asking him to show you how to play like [Jimi] Hendrix.

—Don Sternberg

In both the New Grass Revival and his own band, Bush has been the Keith Richards of the mandolin, a genius at using chords to suggest both melody and rhythmic propulsion.

—Geoffrey Himes

Lyle Lovett & The Boys

2005

These four alternative country singer-songwriters toured together
between 2005 and 2007.

**Traveling and performing concerts together are four premier Americana songwriters
and fast friends (left to right): Lyle Lovett, John Hiatt, Joe Ely, and Guy Clark.**

We have no set list, we have no agenda, and we have no fear.

—Guy Clark

If there was a just god overseeing the world of country music, Guy Clark, Lyle Lovett, John Hiatt, and Joe Ely would be selling out stadiums and Toby Keith would be selling used stereos out of the back of his pickup truck.

—Eric Volmers

John Hiatt is such a marvelous guitar player, he can play all the beautiful parts with you. He and Lyle do some marvelous stuff together.

I remember one night I sang a song about a dog. Joe's sitting next to me and says, "I gotta song about a dog." Then John Hiatt says, "I do, too." That kind of stuff just happens.

—Guy Clark

If we were just getting together in a living room to tell stories and sing songs, this is the closest to that. It's very similar to just sitting around in the living room, swapping songs. We all get on stage and somebody will tell a story and play a song, and that will remind somebody else of another story.

—Joe Ely

We're arranged alphabetically. Coming after "C" and "E" and "H," I always feel like I end up sounding like "L."

—Lyle Lovett

Mike Seeger

2005

Born August 15, 1933 in New York City

A folk musician and folklorist, Mike Seeger was born into the world of music in New York City. (His parents worked with the famous musicologists John and Alan Lomax, and his half brother is famed folksinger Pete Seeger.) At the age of twenty-five, he became a founding member of the original New Lost City Ramblers, a trio that recorded traditional, old-time string band music. He is a singer with a distinctive, personal style and a multi-instrumentalist who plays autoharp, banjo, fiddle, dulcimer, mouth harp, mandolin, guitar, and Dobro with equal authority. His well-documented influence on the American folk music scene has been profound.

We were offended by more commercial acts like the Kingston Trio, and wanted to make this more authentic approach to the music available. It hadn't been heard in the cities, at least not played by live, young musicians. Nobody was playing fiddle in these folk groups at the time, and Mike [Seeger] bringing in the autoharp—this was all new.

—John Cohen

Just as the musical traditions of rural Southerners are rich, deep, and varied, so too have been the life, career, and contributions of our foremost champion of Southern folk music—Mike Seeger. And, just as Southern rural music has had an impact on the larger musical world in ways untold, Mike Seeger's influence has been broad and pervasive in more ways and on more levels than probably even he realizes.

—Paul F. Wells

Rosanne Cash and Rodney Crowell

1983

She: Born May 24, 1955 in Memphis, Tennessee
He: Born August 7, 1950 in Houston, Texas

Rosanne Cash, the daughter of Johnny Cash and his first wife, Vivian, was married to Texas singer-songwriter Rodney Crowell from 1979 until 1992. A successful, Grammy Award–winning singer-songwriter in her own right, she has had more than twenty Top Forty country singles. Her work draws from many genres, including pop, folk, blues, and rock and roll. She is also a published author and respected amateur painter. Considered a part of both the mainstream and alternative country traditions, Rodney Crowel, originally from Houston, is closely associated as a songwriter with his contemporary, Steve Earle. He performed as part of Emmylou Harris's Hot Band in the 1980s before striking out on a successful solo career. Now, he is also a successful record producer.

Rosanne and I have always had this thing: anything is fair as long as it's honest and it rhymes.

—Rodney Crowell

Rosanne's one of those who have forcefully expanded the idea of what "country music" can be, and who it can reach—a rare accomplishment and (in her case, anyway) an admirable one.

—Robbie Fulks

I said, "I'm gonna write a song that's made up of dialogue: Two people who had a relationship ten years ago run into each other on the streets of New York. And the lyrics of the song are what they say back and forth to each other."

I wrote "It's Such a Small World," and I liked it.

I knew it was a man-and-woman song, and I knew I didn't want to give it away to, like, a Dolly Parton and Kenny Rogers kind of situation. I wanted to record it myself, and I knew if I recorded it with a woman, there was no other woman that I really had as much credibility with as I did with Rosanne.

So I just held onto it, and finally I felt confident enough that I could make a record with her and not have it seem like I was trying to use her to achieve my own ends.

—Rodney Crowell

Rosanne's writing is perfectly in sync with her life. What makes her special as a writer is what makes her special as a person—she's thoughtful, sensitive, perceptive. She writes about things that are real to her, and that's why you don't just listen to her songs, you feel them.

—Lyle Lovett

54

Bill Monroe

1989

Born William Smith Monroe on September 13, 1911 in Rosine, Kentucky
Died September 9, 1996 in Nashville, Tennessee

Monroe, considered by all to be the father of bluegrass music, toured with his band, the Blue Grass Boys, for more than fifty years before his death in 1996. During his lifetime, he was inducted into the Country Music Hall of Fame, the Bluegrass Hall of Fame, and the Rock and Roll Hall of Fame—the only artist so honored.

He put his whole life in the music.

—Earl Scruggs

I believe he was the single most prolific musician who ever lived. Here is a guy who started playing music as a teenager, who never did anything much but play. He was able to enjoy the better part of seventy years as a musician. Beyond the pure talent, he had the work ethic. At the end he was still a phenomenal musician, a really powerful man. I think in a hundred years his music will be as immortal as any music. Monroe was an experience. . . . You didn't play with him, you played *against* him. He was like a rock. You can ask any musician. There has never been a rhythm player who was his equal.

— Butch Robins

The only living American to have originated an American music form.

—President Ronald Reagan

I worked for [Monroe] for a year, and I really learned to realize then just what a great musician and singer and writer he was. He kind of formed a music. It's hard for us to see anybody today that invented a style of music, but he did. If it was accidental, it don't matter. He did it! [laughs] I have to say, it was such a training ground for so many musicians. . . . I think he liked taking on new musicians and singing with them and playing with them. I think he really enjoyed that. He was not only a bandleader and great musician and singer; he was also a training ground for other musicians.

—Del McCoury

Bill Monroe was not a traditional artist. In his day, he was cutting edge. He was a radical. I mean, he was a jazz musician. He could have played with Charlie Parker or Louis Armstrong or John Coltrane. Musically, he was such an innovator and a classic example of God-given raw talent who could play anything. But we look back now, forty or fifty years' time, and we see him as a really traditional bluegrass guy. Well, he wasn't, if you go back and look at where he came from and the music he was playing prior to discovering bluegrass.

—Ricky Skaggs

Nanci Griffith

1980

Born Nanci Caroline Griffith on July 6, 1953 in Seguin, Texas

Austin, Texas, singer, guitarist, and songwriter (now living in Nashville), Nanci Griffith has mined numbers of different musical genres, including folk, country, and a hybrid she calls "folkabilly." She won the 1994 Grammy for Best Contemporary Folk Album with *Other Voices—Other Rooms,* but her work has sometimes enjoyed greater success through the performances of other artists. (Kathy Mattea and Suzy Bogguss have both enjoyed big hits with her songs, for example.) She has toured widely and recorded collaboratively with many artists. She is particularly popular in Ireland and has recorded with The Chieftains, the leading proponents of traditional Irish folk music.

I was about as big as Nanci was [in the '60s when she first heard me play]. And somehow, later on, when we met at Kerrville [Folk Festival], God gave it to me to remember exactly who she was. When I decided to resume touring in the late '70s, I asked Nanci to be my opening act. I don't think she had played outside of Texas at that time, [but] she did beautifully, just like I knew she would.

—Carolyn Hester

My goal when I signed her was not to change her, 'cause I really liked what she did. I was thinking country listeners would fall for her, 'cause she's just so refreshing. I don't think she signed with me in Nashville specifically to be a country star. I think she knew that her music is not necessarily pop or country, but she thought I was someone who probably could relate to her as an artist.

—Tony Brown

The Highwaymen

1990

A country music "supergroup," The Highwaymen included, from left to right, Waylon Jennings, Willie Nelson, Kris Kristofferson, and Johnny Cash, all renowned for their influence on the so-called outlaw movement. Formed originally in 1985, before the group had a collective name, Jennings, Nelson, Kristofferson, and Cash adapted the name of the album, *Highwayman*, to serve the group as its own. The group recorded together for ten years and two of their live performances were filmed, the last not long before the death of Waylon Jennings in 2002.

Willie's the old coyote, Waylon's the riverboat gambler, I'm the radical revolutionary, and Johnny's the father of our country.

—Kris Kristofferson

They were dark horses who came in and were standing there on the horizon, never backing down from the system. They were road-worn, these loners telling you stories about their lives. If they carried guns, if they did drugs, they weren't going to apologize for it.

—Shooter Jennings

Johnny Cash is up there singing his greatest hits, and so is Waylon and Kris, and some nights I really get [so] into watching them that I forget that I am supposed to be putting on a show myself.

—Willie Nelson

There hasn't been a new Kristofferson in twenty years, or a new Nelson or Jennings . . . someone who made a difference in the business . . . someone who came along, stepping out where no man has gone before in our business.

I think it is the "Urban Cowboy" syndrome, Nashville [record executives] still trying to sell country records to the people who are buying cowboy boots in New York City. I worry about it a lot, but don't like to think of this Highwaymen tour as a last roundup. I'd like to believe it's an inspiration, a new beginning.

—Johnny Cash

I think I've gone through a lot of my performing knowing I didn't really deserve to be on the same stage with Willie [Nelson], Waylon [Jennings], and Johnny [Cash], but I was there.

—Kris Kristofferson

Marty Stuart

circa 1978

Born September 30, 1958 in Philadelphia, Mississippi

Almost completely preoccupied with country music from a very young age, Stuart taught himself how to play the guitar and mandolin. He emerged as a fourteen-year-old instrumentalist, so accomplished that he was invited to join Lester Flatt's band. After ill health forced Flatt to break up his band, he worked as a sideman with Vassar Clements and Doc Watson before joining Johnny Cash's band. He stayed with Cash until 1985, before embarking on his own successful recording and performance career in Nashville. A member of the Grand Ole Opry since 1993, he remains an avid collector of country music memorabilia and a fan himself. He has served on the board of the Country Music Foundation (for a time as its president), and has written numerous magazine articles related to music.

Marty is our baby brother now. He's a family member. We all cried that night [that I gave Marty my father's old guitar]. It was like the Lord sent us over there to lift up our baby brother.

—Mavis Staples

Marty's a mighty fine musician.

—Ralph Stanley

I've found a new friend and a brother in Marty Stuart. I never wanted to throw my support toward anybody else in this business who was in my category or close to my age—never wanted to do it at all—until Marty and I started working together. We actually like each other. We really do. Marty and I share a lot of the same influences: bluegrass and contemporary and gospel and rockabilly and straight-ahead country.

Marty was on his way to visit some friends and stopped the first night of my show to say hello. I said, "Man, come out and sing 'The Whisky Ain't Workin' with me." When he walked out on that stage there was some kind of magic. The crowd went crazy. At Fan Fair, people came up to his booth and my booth all week long after that and said, "What a great thing!" We knew right then that it had a special spirit that made it different.

—Travis Tritt

Marty, now that Cash has passed on, carries on that same kind of artistic integrity. He's a treasure and a musicologist. He knows more about the Grand Ole Opry and about musicians and about instruments than probably anybody else that's ever been in the country music industry.

—Tony Brown

Marty Stuart is like a brother to me. Not only is he amazingly talented and inspired, but he truly cares about the history and importance of music.

—John Carter Cash

John Hartford

1972

Born John Harford on December 30, 1937 in New York City
Died June 4, 2001 in Nashville, Tennessee

Hartford, a highly original singer-songwriter who was known for his witty lyrics and unique vocal style, made his mark early, writing "Gentle On My Mind" for Glen Campbell. In 1971, he formed a band with Vassar Clements, Tut Taylor, and Norman Blake that released the album *Aero Plain,* which became a classic for fans of newgrass music; the band has since been known as the Aero-Plain Band. He played his own style of banjo and fiddle, often while clog dancing, with a number of different ensembles, but is perhaps best remembered for his solo performances in which, from one song to the next, he would switch from banjo to guitar to fiddle. His knowledge of river lore was well known, and he was musically obsessed by steamboats and life along the Mississippi.

John's music and his life were totally intertwined.

—Emmylou Harris

John Hartford was one of the rarest of musical birds. He had one foot deeply rooted in the past and the other always at least a few steps into the future—and both were dancing.

—Larry Groce, *Mountain Stage*

People in my age group were severely affected by his music and his attitude about music. He's the guy who told me, "Don't ever get famous for something you don't like doing." Nobody played any wilder music than John Hartford.

—Tim O'Brien

He's finally accomplished something he himself dodged all his life, putting one icon after another on a pedestal. Whether it was a riverboat captain or an unknown fiddle player or Bill Monroe and Earl Scruggs, he studied and admired them all. And I think it's his turn to go to the pedestal now. Hartford's body of work is a whole new chapter . . . in American folk music.

—Marty Stuart

You didn't want to follow John. If he was playing from nine to ten, you could forget about playing after that, because the crowd was his.

—Ricky Skaggs

When he's looking at his washing machine, he could write a song about the washing machine. Whatever interested him, he could write a song about it or write a poem about it.

—Tommy Smothers

New Grass Revival

1976

This was an original, innovative bluegrass band from Bowling Green, Kentucky, whose founding members consisted of, from left to right, Courtney Johnson (banjo, guitar), Curtis Burch (guitar, Dobro), Sam Bush (mandolin, fiddle, guitar), and John Cowan (bass). They took traditional bluegrass in a totally new direction, bringing a "jam band" sensibility to a very tradition-bound genre. (They were as likely to play songs by Bob Marley or The Beatles as they were Earl Scruggs or Reno and Smiley.) The band toured with Leon Russell and performed with The Grateful Dead before breaking up for good in 1989.

WHEREAS, newgrass music, also known as progressive bluegrass music, is one of two major subgenres of bluegrass music that drew attention in the 1970s; and . . .

WHEREAS, in 1972, Sam Bush, an accomplished bluegrass musician from Bowling Green, Kentucky, became the founding member of the New Grass Revival along with Courtney Johnson, Ebo Walker, and Curtis Burch from Louisville, Kentucky, and nearby areas of Western Kentucky; and . . .

WHEREAS, Sam Bush has credited Ebo Walker with coining the term, "newgrass"; and . . .

WHEREAS, the New Grass Revival never played traditional bluegrass music but instead brought elements of rock and roll, jazz, and blues to the group's sound; and . . .

WHEREAS, the New Grass Revival, with occasional changes in members, experienced a successful performing and recording career; and . . .

NOW, THEREFORE,
Be it enacted by the General Assembly of the Commonwealth of Kentucky: . . .
(1) Bowling Green, Kentucky, is hereby designated as the birthplace of newgrass music; and
(2) Sam Bush, born on April 13, 1952, in Bowling Green, Kentucky, is hereby designated the Father of Newgrass Music in the Commonwealth of Kentucky.
—General Assembly of the Commonwealth of Kentucky, Act HB 294

We had a large impact in that we made it more acceptable to do a long jam and improvise. We showed young people that there was a different way to play bluegrass. We were a band for an audience that may not have liked straight bluegrass. People would like the way we would play a Bill Monroe tune, then ask us whose song it was. Then later they would discover that they liked the way Bill Monroe played it too.

—Sam Bush

Steve Earle

1975

Born January 17, 1955 in Fort Monroe, Virginia

Though born in Virginia and a resident of Nashville since the mid-1970s, Steve Earle, a singer-songwriter, country rocker, and political activist, is chiefly associated with Texas and such Texas artists as Guy and Susanna Clark, with whom he worked closely on first moving to Nashville. He's been involved in various political causes from the early days of his career, when he allied himself with anti-Vietnam campaigners.

I first became aware of Steve Earle when we'd just signed to Chrysalis Records. John Williams, who was responsible for us getting signed and produced our first record, said: "Have you heard this guy from Texas?" His album *Guitar Town* was just out, and John said, "You should get this." So we got a copy and I was just blown away. I think it's the best record of the 1980s.

It's down to the power of the lyrics and the imagination and the concern for those who don't make it in the American Dream. He maybe writes lyrics better than anyone in the last thirty years. He's never been a big commercial act, but "My Old Friend the Blues" probably stands up as well as any of the songs he's done. I read somewhere that he said it was one of his favorites; one of the most complete things he'd written.

It's a classic song: that image of one's only solace being the pain. Women leave you, friends leave you, things go wrong, but you're guaranteed that's going to be there. It's a kind of antidote to "I Saw The Light" by Hank Williams, where the constant is the figure of God—the power that you're convinced is there, that can guide you and help you. The opposite of that is the pain that never leaves you; and I think he put it so well in that song. We don't play it at every show, but during every tour we've done it at some stage.

I've followed Steve Earle's career very closely since then. I've seen him play many times, solo and with a band. The power of his performance is unbelievable. And I don't think anybody who listens to the lyrics, who understands what it is he's singing about, who gets it, wouldn't be moved by him.

—Charlie Reid, The Proclaimers

Steve Earle

1995

Steve Earle continues to perform original songs often dealing with strong political and social issues. His 2002 album *Jerusalem* was chiefly inspired by the U.S.-led war on terrorism. He is the subject of a documentary film, *Just an American Boy*, which explores his political views as well as his music.

Like all great protest singers, Earle looks for the stories that haven't been told in the mainstream media, and confronts an America looking for scapegoats in the wake of September 11. His *Jerusalem* is the bravest album of a career not lacking in courage.

—Greg Kot

He knows how close he came to never writing another song, never performing in public again. He's one of those rare persons whose sobriety has given him a second chance. All that being the case, the fact that he's chosen to make his politics the central point of his music means that he's risking his life for something that's important to him.

—Grant Alden

Jerry Jeff and Susan Walker

1977

He: Born Ronald Clyde Crosby on March 16, 1942 in Oneonta, New York
She: Born August 14, 1948 in Vernon, Texas

Originally from south-central New York state, singer-songwriter Ronald Crosby adopted the stage name of Jerry Jeff Walker in 1966 and has become an icon in Austin, Texas, since moving there in the early 1970s. "Mr. Bojangles" is, perhaps, his best-known work, and it has been recorded by dozens of artists, ranging from Bob Dylan to Nina Simone and Philip Glass to the Nitty Gritty Dirt Band. After a lengthy series of records for MCA and Elektra, produced after his move to Austin, he gave up on the mainstream record industry and founded his own independent label, Tried and True Music. His wife, Susan, is its president and manager. Jerry Jeff and his work are chiefly associated with the country-rock outlaw scene that also included such artists as Waylon Jennings, Willie Nelson, Guy Clark, and Townes Van Zandt. His birthday bash has become a near-legendary event that attracts top musicians and thousands of fans to Austin.

Jerry Jeff has bounced in and out of my life like the Mad Hatter on the way to the tea party. On some of our misadventures there were people we encountered who surely would have been justified in having both our heads. I am glad we lived through it.

—Jimmy Buffett

Mesmerizing on stage, with just him and his guitar.

—Guy Clark

In the years that I've known Jerry Jeff, every single time I've been to his house he's had a guitar in his hand. . . . What I learned from him is the love of music and the craft.

—Pat Green

I used to follow Jerry Jeff around like a Deadhead. . . . When I first saw Jerry Jeff, I thought, "That's me. Just a total gypsy."

—Todd Snider

In high school, I started seeing Willie and Jerry Jeff. . . . That kind of stuff changed my life, you know. It was so exciting and so good that I figured I wanted to do something like that.

—Jack Ingram

John Prine

1984

Born October 10, 1946 in Maywood, Illinois

Prine, a country/folk singer-songwriter originally from Illinois, has achieved critical and commercial success since his move to Nashville in the early 1970s. His grandfather played guitar with Merle Travis, and he took up the instrument himself at the age of fourteen. He was a postman in Chicago and had served in the military before beginning his musical career. Already a star in Chicago's folk music scene, he was discovered in a local club by Kris Kristofferson. He is known for his wildly imaginative songs and unusual voice and singing style. His 2006 release, *Fair & Square,* was awarded the Grammy for Best Contemporary Folk Album.

I have been an admirer of John Prine since the early '70s, when his first album came out. . . . John Prine has taken ordinary people and made monuments of them, treating them with great respect and love. . . . He is a truly original writer, unequaled, and a genuine poet of the American people.

—Ted Kooser, 2005 Pulitzer Prize–winning poet

[When I first heard John Prine in 1971] he was unlike anybody I'd ever seen—such a young kid, and yet he's writing songs like "Hello in There." I can't imagine myself at that age writing anything remotely that good. John was singing some of the best songs I've ever heard, and they still are the best songs I've ever heard. . . . The best of his songs are timeless. They're like folk music, completely original and unpredictable.

—Kris Kristofferson

"Angel from Montgomery" is still one of the most powerful songs I've ever heard. It was so moving and so heartbreaking, especially for me as a young woman. . . . I loved John's story—that he was a mailman, that he'd been in the army, that he was obviously from the Southern part of the country and moved up. He's salt of the earth—in the old days, Will Rogers was our John Prine.

—Bonnie Raitt

I think the best writers are those who write about their life experience and don't try to come up with something tricky. That's what I love about John's songs— they're just his observations on life. He has an amazing ability to write songs that are very emotional and can make you cry, and yet they're funnier than hell. John's about the best songwriter out there.

—Billy Bob Thornton

Mark O'Connor

2004

Born August 5, 1961 in Seattle, Washington

O'Connor is widely considered to be the most important fiddler of his generation. He started out on the guitar, turning to the fiddle at the age of eleven. Mentored by Texas fiddler Benny Thomasson, by the time he was fourteen he had won three national championships. He has successfully moved among several musical genres, composing, arranging, and performing bluegrass, classical, and jazz. He has appeared in concert and recorded with many artists, including Béla Fleck, Yo-Yo Ma, James Taylor, Stéphane Grappelli, Edgar Meyer, and Tony Rice.

Mark is a supreme master at what he does. Between thought and execution there are no impediments, zero. But what really interests me is that this is a person in constant development, hungry to develop. The intriguing question is, "Where is Mark O'Connor going?"

—Yo-Yo Ma

He has made fiddling a classical art form. He will draw people to this show who normally wouldn't attend a classical performance because he has such wide appeal. But, number one, the quality is there. He wouldn't even be in my mind-set if it weren't for the great quality he offers. Our hall is open to a new way of listening, a new way of communicating. The bottom line is, good stuff is good stuff.

—Sherryl Nelson, executive director, Spivey Hall, Atlanta, Georgia

Mark's style is of course informed by the music he studied and absorbed as he grew up. It's a very fresh and invigorating style. And I think that it brings a fasci-nating perspective to the classical world. . . . People often use the tired term "crossover" to describe Mark's artistry, but I think that what we're seeing is the beginning of a new kind of music. Mark is one of several musicians—Edgar Meyer is another—for whom improvisation is a recaptured art. Improvisation was a normal part of one's musicianship until the early 1800s. So that Mozart and Beethoven, for example, were superb improvisers. It's interesting that it has come back to us through popular music.

—Mark Watt, dean, Blair School of Music, Vanderbilt University

Mark can hold his own with any classical violinist—he's a fiddler not to be believed. I worked at the Santa Fe Chair Music Festival when he came there, and when I saw the reaction on classical fiddlers' faces when they saw him play, which was one of true amazement, I had no qualms at all about putting him in the concert hall.

—Santa Fe Symphony's general manager, Daniel Kosharek

Dolly Parton

1974

Born Dolly Rebecca Parton on January 19, 1946 in Locust Ridge, Tennessee

Now a huge star, Dolly Parton started performing as a child, singing on local radio and television in East Tennessee. She first performed at the Grand Ole Opry at the age of thirteen and came to Nashville for good right out of high school in 1964. Her initial success came as the writer of hit songs for Hank Williams Jr. and Skeeter Davis, among others. Her own stardom was assured when she was teamed with Porter Wagoner in 1967. She has since established herself in concert performance, movies, and television. The winner of seven Grammys and innumerable other awards, she is a member of the Nashville Songwriters Hall of Fame, as well as the Country Music Hall of Fame.

I can't imagine anybody, especially in country, who doesn't try to emulate Dolly in some way.

—Emmylou Harris

I am the biggest Dolly Parton fan. I love her songwriting. She's a great singer, a great performer. She can act. She's the whole package, and the personality is the icing on the cake. I think she's great, and I can only hope to follow in her footsteps. . . . I think it's great that someone like her who has had to overcome family obstacles like she has—life obstacles that she has overcome—still walks around with a smile on her face.

—Kellie Pickler

I hated the song "Islands in the Stream." I had been doing it for three days. And it was Barry Gibb producing it with the Bee Gees, and I sounded like one of the Bee Gees. I just said, "Barry, I don't even like this song anymore." He said, "You know what we need? We need Dolly Parton."

From the moment she walked in the room, it was a totally different song. It just had a lilt to it. It had fun. It had energy. It was playful—all of the things it wasn't when I did it by myself. When she came in, it had a dif-ferent purpose. It was people singing together. It was like communication, as opposed to me just singing about something.

—Kenny Rogers

I just remember everybody always loving Dolly, like anytime she came on television. It was always, "Oh look, there she is!" I listened to mostly bluegrass growing up, so I didn't have country records I was studying like my bluegrass records. When I was nineteen, I sang on her *Eagle When She Flies* album. To hear her over the headphones . . . was much different than hearing a record or seeing someone on television. . . . This was before they mixed it, so it was bare. It was completely bare. And I was like, "Oh, my goodness, there is nothing like that!" . . . That blew me away, and ever since then, I'm like, that's a completely different animal.

She's just so loving and encouraging just by the way she says hello to you. . . . I gotta be all right if she's saying hello to me like that. I've never seen her treat anybody but that way. Anybody and everybody, just like that. So amazing and very sweet.

—Alison Krauss

Junior Brown

1993

Born Jamieson Brown on June 12, 1952 in Kirksville, Indiana

Junior Brown is a renowned instrumentalist who first learned to play piano from his father ("before I could talk"), later turning to guitar. His career began in the 1960s, and he spent that decade developing his astonishing musical skills. Later, he settled in Austin, Texas, where the lines between musical styles and genres are sometimes indistinct. There, he and his wife, "the lovely Miss Tanya Rae," set up shop at the legendary Continental Club where, when not on tour, they perform "music for everybody." His off-the-wall, original, country-rock songs—his admiration for Ernest Tubb inspired "My Baby Don't Dance to Nothing But Ernest Tubb"—are almost as well known as his cherry red "guit-steel," a double-necked guitar he calls "Big Red."

You don't find stylists today like you did in the Ernest Tubb era. That's what sets Junior aside from a lot of other artists. He's got his own style. And his rapport with his fans—he draws a vast audience, from young college kids up to the older, traditional country music fans.

—David McCormick, owner,
Ernest Tubb Record Shop, Nashville

People just looked at us and thought we belonged together. It was like he had this person he could train to play just the type of rhythm he needed for his sound. It's really important in Junior's music, and a lot of people don't play that style of rhythm anymore. It's aggressive, very percussive. For a long time we didn't even have a drummer. I took the place of a drummer and a guitar.

—Tanya Rae Brown,
band member and Junior's wife

Junior Brown is not your typical Fender-toting cowboy. One, he's too old; he's forty-six. Two, he doesn't wear jeans, he wears slacks. And three, between his gravelly, baritone drawl and the way he mixes idioms, from Hawaiian luau to rockabilly, western swing to soulful blues, he doesn't sound like anybody Nashville's produced.

—Anne Goodwin Sides, NPR

Junior told me once that he was very impressed by Jimi Hendrix, by the way he was very wild, and yet he was very controlled. He knew where everything was going in a solo, and he [Junior], I think tries for the same effect, and I think he hits it virtually every time.

—Mitch Mitchell, drummer, Jimi Hendrix Experience

Sure, he's the missing link (but between what?), and yeah, his virtuosity is as good for a gasp as his taste is for a laugh (of glee, derision, disquiet, your choice). Ultimately, though, he's one of those artists, almost impossibly rare, whose ideas just can't be predicted. What he really does is just flat git it when the going gets strange.

—Patrick Carr

Lester Flatt

1977

Born Lester Raymond Flatt on June 19, 1914 in Overton County, Tennessee
Died May 11, 1979 in Nashville, Tennessee

One of the pioneers of bluegrass music, Flatt possessed one of the most recognized voices in all of music. He first came to prominence in the 1940s, playing and singing as a member of Bill Monroe's Blue Grass Boys. In 1948, he formed a partnership with banjo wizard Earl Scruggs, also a Blue Grass Boys alumnus, and formed a new group, the Foggy Mountain Boys, that toured and recorded for the next twenty years. In 1969, Flatt started a new band, Nashville Grass, hiring most of the Foggy Mountain Boys, and he continued to record and perform with that group until his death. He was a posthumous honoree, elected as a member of the first group of inductees into the International Bluegrass Music Hall of Honor in 1991.

Lester heard me play, and he offered me a job. I was in the ninth grade in Mississippi one day, and the next weekend I was a Grand Ole Opry performer and playing poker with Roy Acuff backstage. What could be finer?

Lester used to preach to me about building a slow foundation so you have something to fall back on in a cold season. I thought it best that I become a band member and do this right, and grow up and get the knowledge from the people that invented this music around here. So perhaps when I'm twenty-eight, thirty years old, I can give my solo career a chance. It won't be burned out. And that's kind of the way it worked out.

—Marty Stuart

I remember going to a drive-in movie theater in Pikeville with my father when I was five years old and seeing, between the two films that were playing that evening, a show by bluegrass masters Lester Flatt and Earl Scruggs. There was something about the music that just reached down inside me and spoke to me about things I felt that I couldn't even explain. I knew then that was what I needed to do.

—Patty Loveless

I went to see Lester—I don't know how many days it was before he passed away, but he was really in bad shape. He was in the Baptist Hospital here in Nashville. He could hardly talk loud enough for me to tell what he was sayin'. He wanted to know if we could play some reunion dates together. And my answer immediately was, "Lester, number one, I want you to get well. Number two, yes we'll play dates together when you get well. But my biggest concern now is for you to get more strength and get to feelin' better; then we'll talk about doin' reunions." So that was kind of the way we left it.

—Earl Scruggs

The Country Gentlemen

1971

Seminal bluegrass group originally formed in Washington, D.C., in the late 1950s.

The Country Gentlemen started, auspiciously, on July 4, 1957, as a last-minute replacement for Buzz Busby's Bayou Boys when several members of that band were injured in a car accident. After some early changes in the makeup of the group, they continued to perform together for more than forty-five years, the permanent lineup of the band having become known as the "classic" Country Gentlemen. From left to right: Doyle Lawson, Bill Emerson, Charlie Waller, and Bill Yates.

Although they were respectful of tradition, The Country Gentlemen were also one of the most important early innovative bands in bluegrass, taking the genre into new arenas of repertoire and stylistic performance while steadfastly using acoustic instruments.

—Bill Malone

My dad had a stroke in 2000 and Ronnie Davis asked me to fill some dates. But it scared me. I knew I could do it, but not how it would go over. . . . Christmas 2002 Dad gave me [his 1937 Martin] guitar . . . and I went up to him and said, "Listen to this tape. . . . You were in the hospital, and I was in Savannah playing with The Country Gentlemen." That's when he said, "You could do this!"

—Randy Waller, Charlie's son

After Charlie's passing, when Randy Waller was replacing the members of the Country Gentlemen band after he'd been in there awhile, Jimmy Gaudreau worked a date with him, filling in on mandolin and vocals. By the way, Jimmy was who I hired to replace John Duffey when Duffey left the Gents in '69. Anyway, they were playing somewhere recently, and Jimmy kicked off "Two Little Boys," one of our old classic songs, and the crowd just went wild. They tore the place up and Jimmy said to Randy, "Gosh, what would they do if Adcock was here, too?" So those two cooked it up and came to me and we talked to Tom Gray, too, and we decided we'd play a few dates. I guess it's like bringing things full circle at this point, to have us old guys up there who kind of started the thing, and have a fellow there who helped carry it on, and then to include the son of one of the original guys. It means a lot to me to keep that memory alive, no kidding. And it's a great thing to give the fans something again that meant so much to them.

—Eddie Adcock, original banjo player,
The Country Gentlemen

Darrell Scott

1997

Born James Darrell Scott on August 6, 1959 in London, Kentucky

Born on a tobacco farm in London, Kentucky, Scott is a singer-songwriter who has written several mainstream country hits recorded by such artists as Travis Tritt, Brad Paisley, Patty Loveless, and The Dixie Chicks, among others. His songs are touched by the Appalachia of his youth; his mastery of most stringed instruments has established him as one of Nashville's premier instrumentalists and made him a much-sought-after studio musician.

. . . that soulful voice and the gift of playing any instrument in the room—exceedingly well—and you've got the makings of museum-quality work every time the man takes the notion.

—Rodney Crowell

His playing is unreal. He's just limitless in his approach and knowledge of instruments.

—Sam Bush

Darrell is clearly at the top of his game as a writer, instrumentalist, singer, and producer. Nobody else can do what he does.

—Nick Forster

He is one of the great writers of our time. He brings a complete package through perfectly crafted lyrics, groove, and melody. When you listen to his songs, you can tell he's lived them.

—The Dixie Chicks

This may sound way over the top, folks, but for me, I feel this strong about this piece of music. *Theatre of the Unheard* has impacted me in the same way Joni Mitchell's *Court and Spark*, Miles Davis's *Sketches of Spain*, or The Beatles' *Revolver* records did upon initial listening. I think that *Theatre of the Unheard* is that important of a record, I really do.

—John Cowan

Cowboy Jack Clement

1985

Born Jack Henderson Clement on April 5, 1931 in Whitehaven, Tennessee

Singer, songwriter, and producer "Cowboy" Jack Clement—he picked up his nickname while playing steel guitar during his student days at Memphis State University—is one of the seminal figures in both rock and roll and country music. He was a producer and engineer for Sam Phillips at Sun Records studio in Memphis, where he worked with such future stars as Roy Orbison, Johnny Cash, Carl Perkins, and Jerry Lee Lewis. He moved to Nashville in 1959 and went to work for RCA, making records for such stars as Charley Pride and Ray Stevens. A member of the Nashville Songwriters Hall of Fame, his songs have been recorded by Johnny Cash, Dolly Parton, Ray Charles, and Elvis Presley, among many others.

I was very fortunate to have a great record producer in Jack Clement. What Jack did was to give me the courage to go back and back to the songs until the cuts were the very best I could do. I've always been given artistic control of my projects, but I haven't always known just what to do with it.

—John Hartford

I knew he was the only one who'd see how "Ring of Fire" could work. There wasn't any point in even discussing it with anyone else.

—Johnny Cash (autobiography)

To imagine the world of country music without Jack Clement would be to imagine the world without flowers or birds. There would have been no record of "Whole Lot of Shakin' Going On" by Jerry Lee Lewis, one of many records Jack introduced and engineered while at Sam Phillips's Sun Records in his hometown of Memphis. We would not have heard Johnny Cash sing "Ballad of a Teenage Queen" or "I Guess Things Happen That Way," two of Jack's early hit songs (to say nothing of Cash's renditions of "Dirty Old Egg-Suckin' Dog" or "Flushed from the Bathroom of Your Heart").

—Jim Rooney

Roland White

1975

Born April 23, 1938 in Madawaska, Maine

**Roland White moved from Maine to California with his family while a
teenager. With his brothers, Clarence and Eric, he formed a band, Three Little
Country Boys, that in late 1962 became renowned as The Kentucky Colonels;
the group broke up with the effective end of the folk boom in 1965. He
played guitar in Bill Monroe's band until 1969 before joining Lester Flatt's
band, Nashville Grass, and, in 1973, Country Gazette. From 1988 until 1998,
he played mandolin in the Nashville Bluegrass Band before leaving
to form his own band. He now lives in Nashville, plays concerts with his
own band, and teaches mandolin.**

For those of us who love bluegrass music, it's always been the same. It's understood that Bill Monroe is the father of it all . . . but time has a way of crawling along and taking the responsibility for the future to other souls. As long as I know that Roland has a voice in the inner circle, I know that everything's going to be all right concerning the tradition and future of bluegrass.

—Marty Stuart

In 1964 Bob Dylan was just emerging on the music scene, and folk music was everywhere. We got out of our car in Berkeley; the nearby Cabale Creamery was small and unadorned, with a small stage and maybe a couple dozen small tables for the audience. . . . I wasn't expecting anything special, but, to make a long story short, when the [Kentucky] Colonels came out and played their opening set, I was wiped out! They were great! As far as their stage personalities, Billy Ray [Latham] was the band's "bad boy" and clown; Roger Bush was the straight man/bass player/MC; Roland and Clarence White had (especially Clarence) nonspeaking roles. . . . Roland was the leader of the band. They were each excellent musicians . . . .

—Stan Wolfe

John Jorgenson

2005

Born July 6, 1956 in Madison, Wisconsin

One of the pioneers of the American gypsy jazz movement, John Jorgenson has performed as a solo artist and has collaborated with other musicians all over the world. His broad musical palette has enabled him to play with artists as diverse as Elton John, Luciano Pavarotti, Bonnie Raitt, and Benny Goodman. He is an A-list session player, performing and recording in Los Angeles, Nashville, and London, who has appeared on numerous platinum-selling and Grammy-winning recordings. This photograph is a portrait of John dressed and made up as Django Reinhardt, whom he was playing in the film *Head in the Clouds.*

To say that John is a superlative guitarist is an understatement. He is simply great.

—Earl Scruggs

John has mastered many styles down to the letter, but what blows me away is the way he can honor the style and honor the tune, but still leaves his thumbprint—you can tell it is him, it is identifiable. For me that is the icing on the cake for any player. With Earl [Scruggs], John plays electric on "Foggy Mountain Breakdown," and he will knock you a new one every time he takes a solo. He is a fun cat to play with.

—Brad Davis

I've had a lot of great times with John, but my favorite memory is of the time he and Stan Lynch were recording with me for *The Byrds* boxed set in Nashville. John played bass that night. The groove we got was incredible! We all just looked at each other and smiled. John was truly a Byrd!

—Roger McGuinn

John is not only one of my favorite musicians but also one of the nicest people in the industry. He has contributed his musicianship to several of my albums as well as some of the most memorable and important live performances I have ever done. He is one of very few musicians who is capable of pulling off performances with the likes of Earl Scruggs as well as Elton John. I appreciate him for his talent, positive attitude, and professionalism.

—Travis Tritt

J. J. is not only a musician of vast experience and knowledge, he is also a powerful composer and arranger. I love his music; it has a depth and freshness that stands out in our world today. What a Groovehound!!!!

—Tommy Emmanuel

Jack Ingram

1998

Born November 15, 1970 in Houston, Texas

A Texas-based singer, songwriter, and country rocker, Ingram is chiefly associated with the Red Dirt country music scene. Having made a name playing the college and festival circuits and performing club dates in Dallas and Houston, he released several albums on independent labels before signing with Warner Music. He has toured with Brooks & Dunn, as well as Sheryl Crow, and now records for Big Machine Records.

If there's ever been a Texas artist determined to punch through Music Row's mainstream, it's Dallas's Jack Ingram—a veteran of three Nashville labels, nominal success, and a never-say-die attitude. That resilience, toughness, and learning curve netted the rocker-writer his first number-one hit on country radio in 2006—the sweeping mid-tempo "Wherever You Are"—and a stint opening for Brooks & Dunn.

With the defiance singular to those Texas acts, Ingram might be willing to do what's necessary to get on the radio, but the white-hot flame of his live show isn't something he'll turn down under any circumstances. With thin-legged rockers in calf-gripping boots backing him up, it was a raw intensity to the playing that brought the songs of the past fifteen years to life on stage.

—Holly Gleason

We want to live on the edge of the mainstream; that way we can kind of lead the mainstream. It's kind of the Wayne Gretzky approach: Let's be there before the puck gets there. Jack fit in perfectly with that because he is such a great live artist with a great live base. He's already self-sufficient. . . . We didn't have to compromise Jack to get on the radio.

—Scott Borchetta

On the big shows, it doesn't matter who you are . . . you're just rattled. You can't help it. There's just too much energy out there to tap into, whether you like it or not....

My hands shake. It's visible. You can see it. It's just natural. My friend, Jack Ingram, it doesn't matter if he's playing for 10 or 10,000 people, his hands are shaking the entire night because that's just what he is—a bottle of nervous energy.

—Pat Green

When you were in school, there were guys who thought they were cool and guys that were just cool. Jack is the kind of guy who doesn't have to try to be cool, but is just the coolest.

—Paul Williams

Johnny Cash and Dr. Billy Graham

1978

Mr. Cash: Born John R. Cash on February 26, 1932 in Kingsland, Arkansas
Died September 12, 2003 in Nashville, Tennessee

Dr. Graham: Born William Franklin Graham Jr. on November 7, 1918,
near Charlotte, North Carolina

**Two legends in their own fields of endeavor, they were great
friends for many years. "The Man in Black," Cash was a legendary singer-
songwriter with hits including "I Walk the Line" and
"Folsom Prison Blues." Here he poses with his spiritual inspiration
and close friend, Dr. Billy Graham.**

The integrity he gave to his art is the same integrity that informed him as a parent. He taught me what it meant to be a good parent: trust, respect, and a wide-open mind.

—Rosanne Cash

It was electric, like shaking hands with lightning. . . . John was already so much larger than life. He was John Wayne, Abraham Lincoln—everything, you know? He was like Muhammad Ali: People loved him because he loved them. . . . The more that I got to know him, the more larger-than-life he seemed to me. . . . They say familiarity breeds contempt. But with him, it bred respect. He never got smaller.

—Kris Kristofferson

With Johnny Cash, his voice always intrigued me because it's got so much power in it. It gets on the tape, and you can put symphony orchestras with it or a roomful of banjos or a roomful of horns or a whole bunch of rhythm guitars—whatever you want—and it doesn't drown him out. I always called him "Captain Decibel" for that reason. The loudest recording voice I ever heard. Just thick, full. It's like a great solo instrument.

—Cowboy Jack Clement

So many people think we spent ten years creating that style. It was there the first eight bars we played, and we spent the next four years trying to get rid of it.

—Marshall Grant,
original member, The Tennessee Two

He had his own Folsom Prison inside, and time kept dragging on. But he found a way out. . . . He felt deeply the hunger and suffering of other people. That's what made his songs great, and that's also what made him great.

—Al Gore

He was more than wise. In a garden full of weeds—the oak tree.

—Bono

Steve Young

1976

Born on July 12, 1942 in Newnan, Georgia

A pioneer of the Americana, country rock, and alternative country sounds, this singer-songwriter was also a vital force in the development of the "outlaw movement" that developed in Nashville during the 1980s. Waylon Jennings recorded an entire album of his songs, kicking off the movement, and Hank Williams Jr., as well as many others, have recorded his powerful songs. His "Lonesome, On'ry, and Mean" became something of an anthem for the movement, and "Seven Bridges Road," perhaps his best-known song, was a major hit for the Eagles.

For that voice, that guitar, and those songs to come together in one person is a wonder.

—Townes Van Zandt

In 1963, Steve and I met. We were rich with optimism. Folk music was just about to plug in, to amplify its central concern for civil rights, our generation's cause célèbre. Civil rights drove the coffeehouse lyrics of every folksinger. Music veterans of the McCarthy era (who'd seen the witch hunt of idea-suppression) mixed easily with the new radical-chic that was to shake the very foundations of American politics. It took courage. Steve had it. . . .

Audiences were nailed to the floor with Steve's incredible voice, incendiary guitar virtuosity, birthright to the blues, and vise-like grip of Scotch/Irish traditions. He'd come to Los Angeles informed by a life rich in adversity, balancing conflicted faiths. He still has more questions than answers. In that, he stands apart from the crowd (as he did when we met). Back then, I took his sense of uncertainty as a sign of sanity. It was not temporary.

—Van Dyke Parks

Most of my heroes have been those people who just did what they did, without too much regard for what other people might like. From Mozart to Waylon and all points in between. I guess you'd have to take my dad into account, too. I certainly grew up watching him insist on being an individual. We're a lot alike in that way.

—Jubal Young, Steve Young's son and also a singer-songwriter

As a writer, Steve is in a league with Dylan and Hank Williams—and he sings like an angel.

—Lucinda Williams

Merle Travis

1977

Born Merle Robert Travis on November 29, 1917 in Rosewood, Kentucky
Died October 20, 1983 in Tahlequah, Oklahoma

Arguably one of the most influential guitar stylists of the twentieth century, Merle Travis emerged from western Kentucky to become a prominent singer and songwriter. The coalfields of his youth and the exploitation of miners inspired some of his greatest hits, including "Sixteen Tons" and "Dark as a Dungeon." He developed his own unique guitar picking style that became known as "Travis picking," which had a profound influence on, among others, Chet Atkins and Hank Thompson. Longtime fan Doc Watson named his son, Merle Watson, in Merle Travis's honor. He is a member of both the Nashville Songwriters Hall of Fame and the Country Music Hall of Fame.

Merle Travis could write you a hit song and sing it; he could draw you a cartoon, play you a great guitar solo, or fix your watch.

—Chet Atkins

[In 1971], Rollin' Rock Records was founded by Rockin' Ronny Weiser, with the purpose of keeping alive traditional forms of rock 'n' roll . . . "Merle Travis is back in town!" I related with excitement to Ronny. "Let's see if he'll record with us at Rollin' Rock!" After talking with Merle, he agreed to do it. . . . Finally the big night arrived. I drove my yellow 1966 Cadillac convertible to Merle's house. . . . When Merle saw my car he got excited and called his wife to come have a look at it. "Back in 1954 we had a yellow Cadillac convertible," he exclaimed. "This one is just like it!" A great guitarist inspired by Merle called Thom Bresh, who, it transpired, was Merle's biological son . . . joined us and rode to the studio also.

We arrived at Ron Weiser's house, and I introduced Thom and Merle to Ron, who had the recorder humming, and immediately Merle got out of the case his beautiful Gibson with his name on the neck and soon set himself to work. The first song he played on was "Guitar Rag." We did twelve tunes in total, including a blues instrumental with no vocal.

After finishing the tunes around 11 P.M., Merle, Thom, Ron, and I went to a nearby restaurant for a bite to eat. Ron and I were enthralled that we had just put on tape the great Merle Travis, a hero in our eyes, but in his own, a simple Kentucky boy with some talent and lots of good luck.

—Ray Campi

No guitarist in the music world was more original. Merle made those six strings sound like an entire orchestra.

—Cliffie Stone

George Strait

1982

Born May 18, 1952 in Poteet, Texas

Known for his unique style of western swing music, George Strait began his solo recording career at MCA Records in 1981. A member of the Country Music Hall of Fame, he has been nominated for more Country Music Association awards than any other artist. He holds the record for the most number-one songs on the Billboard Country Music Charts, and over the length of his long career has had more albums certified gold, platinum, or multi-platinum than any other artist except Elvis Presley and The Beatles. It is unlikely that his record of fifty-four number-one records will ever be broken.

I couldn't imagine anything more thrilling than to be on a tour with George Strait. Just last week, I was at the CMA Awards watching him be inducted into the Hall of Fame, and this week I received word that I would be on the tour with him. This is surreal! I keep pinching myself to make sure this isn't just a dream.

—Taylor Swift

I literally used to sit on a stool on a stage in a room a lot smaller than this one, in a bar a lot smaller than the room we're in now, and I sang a lot of George Strait songs over and over again. I know I've won entertainer of the year from the CMA twice, but if someday there's a kid out there who I mean half as much to as George Strait does to me, then I would consider myself a success despite any of this. I'm honored to know him. I'm honored to be his friend. He's my hero.

—Kenny Chesney

[George] was the person that really made me fall in love with music, basically. When I first started listening to his music, that's when it really hit me that I wanted to pursue this more. I can pretty much safely say that without him, without his music, I probably wouldn't be doing this. After I really got into his music, it just branched off from there. That's where it all started for me.

—Blaine Larsen

George Strait!!! He's the king of country music, and I get to take part in his arena tour (not to mention stare at him in those Wranglers). I'm so honored that George would take a chance on a new artist like me. I've admired him my whole life and, as a singer-songwriter from Texas, it is like winning the lottery.

—Miranda Lambert

Dale Watson

2001

Born Kenneth Dale Watson on October 7, 1962 in Birmingham, Alabama

**A Texas honky-tonk fixture who has been playing the bars and working
on the road for more than twenty-five years, Austin-based Watson has
recorded hundreds of original, hard-core country songs that have also been
recorded by other artists. Hank Williams III has dubbed him
"the savior of traditional country music."**

With his tattooed arms and Mohawk pompadour, Dale Watson cuts an impressive figure. But it's his music that makes a real impression. Watson does honky-tonk the old-fashioned way, without regard for radio, only for real-life themes (truckers, cheaters, sausage and eggs) and instrumental solos as authentically roadhouse as sweat-stained shirts and a neon tan.

—Alanna Nash

There's no pop-music style more anachronistic than country-and-western truck-driving tunes. Despite the well-intended efforts of small big-rig labels like Diesel Only, when it comes to songs about eighteen-wheelers and the open road, authenticity isn't overrated. Dale Watson is a totally legit heir to kings of the road like Dave Dudley.

—Wayne Robins

Vassar Clements

1972

Born April 25, 1928 in Kinard, South Carolina
Died August 16, 2005 in Nashville, Tennessee

Vassar Clements, a legendary Nashville fiddler and session player, was entirely self-taught. Over the length of a fifty-year career, he played on more than 2,000 albums and performed with a wide range of artists—from Woody Herman and the Nitty Gritty Dirt Band to The Grateful Dead and Jimmy Buffett. His career began when he joined Bill Monroe's Blue Grass Boys as a teenager in 1942. Though he was a major influence in the developing "new-grass movement," he cited big-band swing music as a significant influence over his own style and musical development.

It was the most lonesome, scary sound coming out of a fiddle I'd ever heard. I played the mandolin, and once I heard this music, I ditched everything I ever knew and went back and tried to play mandolin like Vassar played fiddle. Years later, I played the Opry, and I saw this man playing fiddle. He stood straight, with his eyes closed, and he was playing the prettiest music you could ever imagine. It froze me on the spot.

—Marty Stuart

The worst swear word he ever said was "dad bime," as in "Dad bime, that fiddle won't tune!" It was his own word—I never heard it anywhere else. I don't think I ever heard him say a bad word about anyone, and he was more interested in you than you might have thought, learning about the people he was with.

—John McEuen

I was *flabbergasted* the first time I heard Vassar Clements play the fiddle. I was a sophomore in high school and rushed out to get a copy of the album *Will the Circle Be Unbroken* as soon as I heard the first

notes, I was hooked. Never before had I heard fiddle playing of such incredible, expressive power, dark, brooding, bluesy, and wonderfully inventive.

—Matt Glaser

When Vassar Clements reemerged on the music scene in 1971 [with John Hartford], he really raised the bar for my own musicianship in that his playing inspired me to want to be that good. Through the years, Vassar always continued to inspire me, and I was enormously blessed to have him play on many of my own records as well as on collaborations with other artists. . . . The man was my hero [and] I am so grateful for all of wonderful times and laughter we shared. . . . Vassar's music will live on for all of us to enjoy.

—Tony Rice

The Miles Davis of bluegrass fiddle. An amazing free-associative player, totally spontaneous. Very pure musical force. Somebody I feel lucky to have played with.

—Béla Fleck

Doc Watson

1975

Born Arthel Lane Watson on March 3, 1923 in Stoney Fork Township, North Carolina

An accomplished instrumentalist, singer, and songwriter, this native of Deep Gap, North Carolina, has been performing professionally since childhood. He lost his sight before his first birthday, but that did not prevent him from becoming a proficient guitar player and banjoist. He developed his own "Watson style" of traditional guitar flat-picking and has had a great influence on the work of such other star instrumentalists as Tony Rice and Clarence White. He was awarded the National Medal of Arts by President Clinton in 1997.

One of the Top Ten traditional artists of the twentieth century. Doc's the one who made the flat-picked guitar legitimate and magnificent around the world.

—Dan Crary

Many youngsters and old-timers came to listen and show appreciation. . . . The sitting room at Doc's house was crowded, and songs were remembered that had been laid away years since.

—Ralph Rinzler

I'm intimidated by him, but it's a real thrill to play with him. He teaches me something new every time we sit down together. We did the whole record [*Third Generation Blues*] in first takes—one day, two sessions—and I was pretty nervous. There was a lightning storm during one of the songs, and a CB-radio signal ruined another take, but those were the only songs we repeated.

—Richard Watson (his grandson)

I first saw Doc perform a brilliant solo concert back in 1963, at Club 47 in Cambridge, around the time he appeared at the Newport Folk Festival. A year later, my late father became his representative, and began the forty-year association between our families. For me, it has been a tremendously rewarding relationship.

Doc's goal was to take his Appalachian heritage—often disparaged in mainstream culture—and share it with and thus have it claimed by the rest of America and the rest of world. He found tremendous satisfaction when people who had grown up in very different backgrounds—from Ivy League universities to rural Africa—would respond to and be moved by songs that Doc had learned through his family, church, and community.

—Mitch Greenhill

Travis Rivers

1971

Born Friday morning, 12:04 A.M. December 5, 1941 at Brackenridge Hospital,
Room 202, window bed in Austin, Texas

**A widely traveled figure in the music industry, Texan Travis Rivers managed
and produced Nashville-based blues singer Tracy Nelson in the 1970s and
'80s. (Nelson had been the vocalist for Mother Earth, a San Francisco band
of the late 1960s; Rivers had been the band's manager.) Over the years, he has
helped many other fledgling artists get started, including John Hiatt and
Emmylou Harris, and, famously, he was the man who, some years earlier,
had driven Janis Joplin from Port Arthur, Texas, to San Francisco.**

My next break came at Sam Hood's "Gaslight" club in Greenwich Village. One day I went to hang an exhibit of my work in the lobby. Travis Rivers, manager and producer of Nashville blues singer, Tracy Nelson, came in to meet with the club owner. I was on a ladder hanging prints when he came through the lobby, and as I looked down, he nodded and walked downstairs. An hour later on his way out, he paused and was looking at the prints, and asked me "Who's the photographer?" I told him "Jim McGuire," and his response was flattering: "Great work; where can I find him?" I came down off the ladder and introduced myself and he invited me to come to Tracy's show that night. We became and still remain great friends. Not long after that he invited me to visit him in Nashville and shoot some photographs of Tracy at her farm (my first Nashville album cover)."

—Jim McGuire

I got a call from Roy [Gaines] saying Little Richard was coming out of retirement to play rock and roll again and was holding band auditions. I went down and made the first cut and knew I'd nailed the gig, and ran into Jessie Hill there, the singer who'd done "Ooh Poo Pah Doo." He tried to get me to start a band with him, saying, "It's a big old world out there. Forget Richard." Before I went back for the second round, I got a call from Travis Rivers saying he had a band in San Francisco called Mother Earth, and they needed a guitarist and bassist. Powell St. John was one of the singers, and when Bob and I went up there, Powell was really cooking. Tracy Nelson was the other vocalist, and we got the gig. The next day we started recording.

—John "Toad" Andrews

How I happened to join Big Brother? Well, Chet Helms sent Travis Rivers to get me. What I usually say is that I wanted to leave Texas, but that's not what really happened. I didn't want to leave. But . . . how could I not go [with Travis]!

—Janis Joplin

Jerry Byrd and Curley Chalker

1977

Jerry Byrd: Born Gerald Lester Byrd, March 9, 1920 in Lima, Ohio
Died April 11, 2005 in Honolulu, Hawaii
Curley Chalker: Born Harold Lee Chalker, October 22, 1931 in Enterprise, Alabama
Died April 30, 1998 in Nashville, Tennessee

**Byrd and Chalker were two of the most renowned and revered steel guitar
players ever to play and record in Nashville. Both played on hundreds of
early Nashville recording sessions. Byrd eventually retired to
Hawaii where he spent his later years teaching young Hawaiians how
to play old-style, Hawaiian steel guitar.**

I met Curley in the late '60s when he moved to Nashville. . . . My bass player, Buck Evans, would go down to Printer's Alley and play with Curley [and] I would come by the club and listen. Curley got to asking me to get up and sit in on vocals. . . . Then one night he asked me to learn "The Shadow of Your Smile" so we could do it on stage. Well, the first night we tried it on stage, and he went into the jazz-swing turnaround and just played his everlovin' butt off, I was hooked.

A couple of years later . . . I decided to bring in Curley [to record] "The Shadow." . . . We dimmed the lights real low and Curley touched those strings with his magic hands and we started the song. . . . What followed was a special moment in our country music. . . . Curley played a classic turnaround; every steel guitar player who has ever heard it still just shakes their head in wonder. . . .

I was booked for a concert in St. Louis, Missouri, in 1970, when "The Shadow" reached number one. . . . Well, I was sweating it 'cause I knew folks would be wanting to hear the song . . . and we couldn't play it without Curley. . . . The crowd started hollering for "The Shadow." I was just into the explanation about how it was impossible to do our number-one record . . . when from behind the curtain came the sweetest sound . . . the steel guitar intro to my song. The curtain parted and there Curley sat, grinning from ear to ear. We did "The Shadow" three times before the crowd would let us be. . . .

—Stan Hitchcock

One day while en route to a local record shop with my dad, a new sound blasted through our 1940 Chevy's radio speaker. . . . a song called "Moonland" and played by an "unnamed artist." It was several weeks before the DJ played another tune by this same guitarist, and it was "Steelin' Is His Business." The lyrics to that song, sung by Rex Allen, included the line, "He plays the steel guitar and his name is *Jerry Byrd.*"

—Ray Montee

Jerry Byrd came along and played with great tone and originality. He was the first steel player I heard who was a true artist, and most players today consider him the godfather of the modern steel guitar.

—Lloyd Green

Don Williams

1981

Born May 27, 1939 in Floydada, Texas

Williams spent seven years with the Texas-based folk-pop group The Pozo-Seco Singers before coming to Nashville and embarking on a solo career in 1971. His smooth, romantic voice and straightforward vocals, combined with his imposing build, earned him the nickname "the gentle giant of country music." He retired in 2006 after a career that included seventeen number-one hits. His songs have been recorded by many other artists, including Johnny Cash, Eric Clapton, Lefty Frizzell, Sonny James, Charley Pride, and Pete Townshend.

The only man who can make 'em stand up when he sits down and plays.

—Doug Stone

Don Williams is a guy who probably makes it look easier than anybody in the world. I watched him do a show in London, and I loved it. Those people were just in a trance with him.

—Randy Travis

The best thing about Don Williams is that it's so hard to peg him against this, that, or the other country music era. Hits like "Amanda," "You're My Best Friend," "I Believe in You," and "Good Ole Boys like Me". . . are so understated it's as if they float on top of Nashville history.

—Dan Cooper

Norman and Nancy Short Blake

1980

He: Born March 10, 1938 in Chattanooga, Tennessee
She: Born June 11, 1952 in Independence, Missouri

Norman Blake is a Grammy-nominated singer-songwriter and a multi-instrumentalist who is widely considered one of the best flat-pick acoustic guitar players of his time. He has played in a number of folk and bluegrass groups and has played backup for Bob Dylan, Kris Kristofferson, Joan Baez, and June Carter. He is particularly well known for his work with John Hartford, Tony Rice, and his wife, Nancy Blake. He was featured on the multi-platinum *O Brother, Where Art Thou?* album, and participated in the Down from the Mountain tour that followed its release. Norman and Nancy Blake have lived in an old farmhouse in north Georgia for many years, not far from Sulphur Springs where Norman was raised. Playing mostly traditional guitar and cello duets, they travel widely, performing in concert and at folk and bluegrass festivals.

As the sands of time erode the rocks that are America's musical treasures, one name keeps cropping up: Norman Blake. Anyone familiar with him knows he's never strayed far from his birthplace of Chattanooga, Tennessee—in nineteen hundred and thirty-eight (as Norman might say). Nor has he strayed musically, focusing on preserving a form of American music that might otherwise be lost.

—Chris Frank, efolkmusic.org

For many years I have dreamed of doing one more project with my friend and hero guitar picker, Norman Blake. Although we have played a lot together over the years, we haven't recorded since the days of the *Aero Plain* album [in 1971]. Well, it finally happened. For some time Nancy [Blake] wanted us to do a project, so we went back to a wonderful period in our lives and *Shacktown Road* appeared [in 2006] in sepia tone.

—Tut Taylor

When you work with guys like Norman Blake, you see how when they play, they're not thinking about what lick or run they can throw into their own part, they're processing music on a global level and trying to contribute to the greater musical good. As I've grown as a player, I've tried to take a more communal approach too. I realized it's okay not to play every single note I know in every solo. I no longer feel like I have to show off, but it's still taken me time to build up to that level of comfort where I can play simply. Being accepted by my heroes actually boosted my confidence in that respect.

Norman always reminds me of the beauty of simplicity, the beauty of allowing the tone of the guitar to speak. And I love the way he moves from chord to chord. Norman's got a certain pocket rhythmically that I love to try to get.

—Bryan Sutton

Keith Whitley

1984

Born Jessie Keith Whitley on July 1, 1955 in Sandy Hook, Kentucky
Died May 9, 1989 in Goodlettsville, Tennessee

Originally from Sandy Hook, Kentucky, Whitley grew up playing bluegrass music with Ricky Skaggs and played in various bluegrass bands through the 1970s and '80s. Chiefly associated with the Clinch Mountain Boys, he was signed as a solo country artist to RCA Records in 1984. He was married to fellow country music star Lorrie Morgan. A longtime alcoholic, his career was tragically cut short by his death from alcohol poisoning at the age of thirty-four. Now, nearly twenty years later, his influence can still be heard in the work of such contemporary stars as Tim McGraw.

I love people that sing about broken hearts, whether it's Frank Sinatra or George Jones or Keith Whitley or Merle Haggard or anybody who's had their heart broken and sing from that spot. Those are my favorite singers.

—Dierks Bentley

I always smoke a cigarette for him, right down to the butt, and then I gently place the butt right by his tombstone. For the first year after Keith passed away, every song I was pitched was just morbid, and people get tired of hearing it. And I get tired of feeling it. I live it every day anyway—through my kids and Keith's family and his mom and my own memories. I don't want to live it twenty-four hours a day. My music has got to be something that makes me happy. People are going to think what they want to think, but I did not intend for them to think that. If they want to think it, and it makes them feel better, that's fine.

—Lorrie Morgan, ex-wife

Growing up my dad always liked Keith Whitley, and I listened to every record he had. It was such an influence that I tried to sing like him every chance I could.

—Joe Nichols

Benny Martin

1976

Born Benny Edward Martin Sr. on May 8, 1928 in Sparta, Tennessee
Died March 13, 2001 in Nashville, Tennessee

Martin hitchhiked to Nashville, clutching his fiddle, at the age of thirteen and went on to a long career as a sideman in bands fronted by Bill Monroe, Hank Williams, Flatt & Scruggs, Roy Acuff, and John Hartford.

His fiddle was like an extension of his arms. He played with his whole body. I was thunderstruck. There was so much going on that I've been the rest of my life grasping it all. It was the beginning of a hopeless addiction. . . .

He told me many times that fiddling is *all* in the bow...he'd rather have a great bow and a mediocre fiddle than the other way around . . . his bow licks, his timing, and his syncopation are the key to what he's doing. I think he underestimates what he does with his left hand. His selection of notes that describe the melodies he's playing is totally his own, and it's hard to hear them any other way once you've heard his setting. His accenting and his slides are very important.

There are two kinds of artists in this business. There are those who get famous in their lifetime, and there are those whose fame doesn't come till after they've passed awa y . . . Mozart, Beethoven, Bach, Charlie Parker . . . and Benny's also in that class. We're gonna hear a lot about him in the time to come—this is his time.

For those who were lucky enough . . . no video or tape recorder or record could do justice to hear him play—in person—sitting in your living room or on stage. To hear him play—there's no way that could have ever been captured. It was like sitting on the head of a locomotive . . . it was the most powerful thing you ever heard.

He said that when he passed away, he wanted to be buried with a little tiny air conditioner, a guitar capo, and a set of jumper cables.

—John Hartford

RECENT and OTHER WORK

Mac Wiseman and John Prine

2007 studio portrait

Nashville, Tennessee

In spring 2007, at age eighty-two, the legendary bluegrass singer Mac Wiseman joined
John Prine to record a series of duets for a CD titled *Standard Songs for Average People,*
a very sweet and sentimental recording by two of the most distinctive voices in
bluegrass and contemporary folk music. I've photographed John many times over
the years, but this was the first time I'd met Mac. It was a real pleasure for me to
finally get him on the canvas and hear his many stories.

—Jim McGuire

J. D. Souther

2006 studio portrait
Nashville, Tennessee

Originally from Detroit, J. D.'s stature as a solo performer and songwriter developed
in Los Angeles, where he released several solo albums and cowrote many of the Eagles'
biggest hit singles. He has lived on a farm outside of Nashville since 2002. I had met
him previously at a party, but this was the first chance I had to work with him.
He came into the session with some pretty specific ideas about how he wanted to be
photographed, but this was my favorite image from the session.

—Jim McGuire

Guy Clark with Verlon Thompson and Shawn Camp

2007 studio portrait

Nashville, Tennessee

From Arkansas, Texas, and Oklahoma, respectively, these are three of Nashville's
most respected singer-songwriters. Each has his own solo career, but they often play
shows together. This is what happens when you get three very funny guys
together that are good friends and ask them to "do something" for the camera.
I'm always delighted when musicians surprise me.

—Jim McGuire

Nanci Griffith

1997 studio portrait
Nashville, Tennessee

Nanci is a longtime Nashville resident. This was shot during a photo session
for her 1997 CD *Blue Roses from the Moons.* Nanci usually comes to photo sessions with
very good ideas about what she wants to come away with. This is somewhat
of a departure for her, as it's more glamorous than usual, and I think represents her as
an artist more established in her career than some of the earlier portraits.

—Jim McGuire

Guy Clark with Django

1999 studio portrait

Nashville, Tennessee

Django is an Australian shepherd named after one of my heroes, Django Reinhardt.
She has lived with me since she was a pup. This is just Guy having a very
private conversation with Django, who was hanging on every word. This photograph
eventually made it to the cover of Guy Clark's 1999 CD, *Cold Dog Soup*.

—Jim McGuire

EXHIBITION CHECKLIST

Chet Atkins, 1976
Studio portrait/Nashville
Silver print
Image Size: 14" x 14"
Framed: 21 ¼ x 21 ¼ x 1 ½"

Norman and Nancy Short Blake, 1980
Studio portrait/Nashville
Silver print
Image Size: 14" x 14"
Framed: 21 ¼ x 21 ¼ x 1 ½"

David Bromberg, 1972
Studio portrait/New York
Silver print
Image Size: 14" x 14"
Framed: 21 ¼ x 21 ¼ x 1 ½"

Junior Brown, 1993
Studio portrait/Nashville
Silver print
Image Size: 14" x 14"
Framed: 21 ¼ x 21 ¼ x 1 ½"

Sam Bush, 1981
Studio portrait/Nashville
Silver print
Image Size: 14" x 14"
Framed: 21 ¼ x 21 ¼ x 1 ½"

Jerry Byrd and Curley Chalker, 1977
Studio portrait/St. Louis
Silver print
Image Size: 14" x 14"
Framed: 21 ¼ x 21 ¼ x 1 ½"

Johnny Cash and Dr. Billy Graham, 1978
Studio portrait/Nashville
Silver print
Image Size: 14" x 14"
Framed: 21 ¼ x 21 ¼ x 1 ½"

Rosanne Cash and Rodney Crowell, 1983
Studio portrait/Nashville
Silver print
Image Size: 14" x 14"
Framed: 21 ¼ x 21 ¼ x 1 ½"

Guy and Susanna Clark, 1975
Location portrait/Nashville
Silver print
Image Size: 14" x 14"
Framed: 21 ¼ x 21 ¼ x 1 ½"

The Country Gentlemen, 1971
Studio portrait/New York
Silver print
Image Size: 14" x 14"
Framed: 21 ¼ x 21 ¼ x 1 ½"

Cowboy Jack Clement, 1985
Studio portrait/Nashville
Silver print
Image Size: 14" x 14"
Framed: 21 ¼ x 21 ¼ x 1 ½"

Vassar Clements, 1972
Studio portrait
Silver print
Image Size: 14" x 14"
Framed: 21 ¼ x 21 ¼ x 1 ½"

J. D. Crowe and The New South, 1975
Studio portrait/Nashville
Silver print
Image Size: 14" x 14"
Framed: 21 ¼ x 21 ¼ x 1 ½"

The Dillards, 1977
Studio portrait/Nashville
Silver print
Image Size: 14" x 14"
Framed: 21 ¼ x 21 ¼ x 1 ½"

Dave Dudley, 1976
Studio portrait/Nashville
Silver print
Image Size: 14" x 14"
Framed: 21 ¼ x 21 ¼ x 1 ½"

Steve Earle, 1975
Studio portrait/Nashville
Silver print
Image Size: 14" x 14"
Framed: 21 ¼ x 21 ¼ x 1 ½"

Steve Earle, 1995
Studio portrait/Nashville
Silver print
Image Size: 14" x 14"
Framed: 21 ¼ x 21 ¼ x 1 ½"

Lester Flatt, 1977
Studio portrait/Nashville
Silver print
Image Size: 14" x 14"
Framed: 21 ¼ x 21 ¼ x 1 ½"

Béla Fleck, 1999
Studio portrait/Nashville
Polaroid transfer print
Image Size: 14" x 14"
Framed: 21 ¼ x 21 ¼ x 1 ½"

Vince Gill, 1985
Studio portrait/Nashville
Silver print
Image Size: 14" x 14"
Framed: 21 ¼ x 21 ¼ x 1 ½"

Nanci Griffith, 1980
Studio portrait/Nashville
Silver print
Image Size: 14" x 14"
Framed: 21 ¼ x 21 ¼ x 1 ½"

Emmylou Harris, 1983
Location portrait/Nashville
Silver print
Image Size: 14" x 14"
Framed: 21 ¼ x 21 ¼ x 1 ½"

John Hartford, 1972
Studio portrait/New York
Silver print
Image Size: 14" x 14"
Framed: 21 ¼ x 21 ¼ x 1 ½"

John Hiatt, 1974
Studio portrait/Vanderbilt
Silver print
Image Size: 14" x 14"
Framed: 21 ¼ x 21 ¼ x 1 ½"

John Hiatt, 2004
Studio portrait/Nashville
Silver print
Image Size: 14" x 14"
Framed: 21 ¼ x 21 ¼ x 1 ½"

The Highwaymen, 1990
Studio portrait/Nashville
Silver print
Image Size: 14" x 14"
Framed: 21 ¼ x 21 ¼ x 1 ½"

Harlan Howard, 1988
Studio portrait/Nashville
Silver print
Image Size: 14" x 14"
Framed: 21 ¼ x 21 ¼ x 1 ½"

Roy Huskey Jr., 1991
Studio portrait/Nashville
Silver print
Image Size: 14" x 14"
Framed: 21 ¼ x 21 ¼ x 1 ½"

Jack Ingram, 1998
Studio portrait/Nashville
Silver print
Image Size: 14" x 14"
Framed: 21 ¼ x 21 ¼ x 1 ½"

Jerry Jeff and Susan Walker, 1977
Studio portrait/Nashville
Silver print
Image Size: 14" x 14"
Framed: 21 ¼ x 21 ¼ x 1 ½"

Waylon Jennings, 1985
Studio portrait/Nashville
Silver print
Image Size: 14" x 14"
Framed: 21 ¼ x 21 ¼ x 1 ½"

John Jorgenson, 2005
Studio portrait/Nashville
Silver print
Image Size: 14" x 14"
Framed: 21 ¼ x 21 ¼ x 1 ½"

Hal Ketchum, 1997
Studio portrait/Nashville
Silver print
Image Size: 14" x 14"
Framed: 21 ¼ x 21 ¼ x 1 ½"

Carole King, 1978
Studio portrait/Austin
Silver print
Image Size: 14" x 14"
Framed: 21 ¼ x 21 ¼ x 1 ½"

Kris Kristofferson, 1990
Studio portrait/Nashville
Silver print
Image Size: 14" x 14"
Framed: 21 ¼ x 21 ¼ x 1 ½"

Lyle Lovett and The Boys, 2005
Studio portrait/Nashville
Digital print
Image Size: 14" x 14"
Framed: 21 ¼ x 21 ¼ x 1 ½"

Benny Martin, 1976
Studio portrait/Nashville
Silver print
Image Size: 14" x 14"
Framed: 21 ¼ x 21 ¼ x 1 ½"

Bill Monroe, 1989
Studio portrait/Nashville
Digital print
Image Size: 14" x 14"
Framed: 21 ¼ x 21 ¼ x 1 ½"

New Grass Revival, 1976
Studio portrait/Nashville
Silver print
Image Size: 14" x 14"
Framed: 21 ¼ x 21 ¼ x 1 ½"

Mark O'Connor, 2004
Studio portrait/Nashville
Silver print
Image Size: 14" x 14"
Framed: 21 ¼ x 21 ¼ x 1 ½"

Brother Oswald, 1975
Studio portrait/Nashville
Silver print
Image Size: 14" x 14"
Framed: 21 ¼ x 21 ¼ x 1 ½"

Dolly Parton, 1974
Studio portrait/Vanderbilt
Silver print
Image Size: 14" x 14"
Framed: 21 ¼ x 21 ¼ x 1 ½"

John Prine, 1984
Studio portrait/Nashville
Silver print
Image Size: 14" x 14"
Framed: 21 ¼ x 21 ¼ x 1 ½"

Travis Rivers, 1971
Studio portrait/New York
Silver print
Image Size: 14" x 14"
Framed: 21 ¼ x 21 ¼ x 1 ½"

Marty Robbins, 1977
Studio portrait/Nashville
Silver print
Image Size: 14" x 14"
Framed: 21 ¼ x 21 ¼ x 1 ½"

Darrell Scott, 1997
Studio portrait/Nashville
Silver print
Image Size: 14" x 14"
Framed: 21 ¼ x 21 ¼ x 1 ½"

Mike Seeger, 2005
Studio portrait/Nashville
Silver print
Image Size: 14" x 14"
Framed: 21 ¼ x 21 ¼ x 1 ½"

Ralph Stanley, 1998
Studio portrait/Nashville
Silver print
Image Size: 14" x 14"
Framed: 21 ¼ x 21 ¼ x 1 ½"

George Strait, 1982
Studio portrait/Nashville
Silver print
Image Size: 14" x 14"
Framed: 21 ¼ x 21 ¼ x 1 ½"

Marty Stuart, circa 1978
Studio portrait/Nashville
Silver print
Image Size: 14" x 14"
Framed: 21 ¼ x 21 ¼ x 1 ½"

Tut Taylor, 1974
Studio portrait/Nashville
Silver print
Image Size: 14" x 14"
Framed: 21 ¼ x 21 ¼ x 1 ½"

Merle Travis, 1977
Studio portrait/Nashville
Silver print
Image Size: 14" x 14"
Framed: 21 ¼ x 21 ¼ x 1 ½"

Townes Van Zandt, 1990
Studio portrait/Nashville
Silver print
Image Size: 14" x 14"
Framed: 21 ¼ x 21 ¼ x 1 ½"

Dale Watson, 2001
Studio portrait/Nashville
Silver print
Image Size: 14" x 14"
Framed: 21 ¼ x 21 ¼ x 1 ½"

Doc Watson, 1975

Studio portrait/Nashville

Silver print

Image Size: 14" x 14"

Framed: 21 ¼ x 21 ¼ x 1 ½"

Roland White, 1975

Studio portrait/Nashville

Silver print

Image Size: 14" x 14"

Framed: 21 ¼ x 21 ¼ x 1 ½"

Keith Whitley, 1984

Studio portrait/Nashville

Silver print

Image Size: 14" x 14"

Framed: 21 ¼ x 21 ¼ x 1 ½"

Don Williams, 1981

Studio portrait/Nashville

Silver print

Image Size: 14" x 14"

Framed: 21 ¼ x 21 ¼ x 1 ½"

Tammy Wynette, 1987

Studio portrait/Nashville

Silver print

Image Size: 14" x 14"

Framed: 21 ¼ x 21 ¼ x 1 ½"

Steve Young, 1976

Studio portrait/Nashville

Silver print

Image Size: 14" x 14"

Framed: 21 ¼ x 21 ¼ x 1 ½"

INDEX

Boldfaced page numbers indicate featured biographies and portraits.

A

Acuff, Roy, 10, 14, 82, 120
Adams, Ryan, 30
Adcock, Eddie, 84
Aero Plain (Aero-Plain Band), 10, 64, 116
Aero-Plain Band, 10, 64
Ahern, Meghann, 46
Alden, Grant, 70
Alger, Pat, 36
Ali, Muhammad, 96
Allen, Rex, 112
alternative country music, xvi, 38, 50, 54, 98
"Amanda," 114
Americana, 98
Anderson, Bill, 42
Anderson, Lynn, 18
Andrews, John "Toad," 110
The Andy Griffith Show, 44
"Angel from Montgomery," 74
"Apartment #9," 18
Armstrong, Louis, 56
Atkins, Chet, **12–13,** 24, 100
Austin, Texas, 26, 72, 80, 104
Axton, Hoyt, 24

B

Bach, Johann Sebastian, 120
Back Porch Music, 44
Baez, Joan, 34, 116
"Ballad of a Teenage Queen," 88
banjo music, 2, 22, 28, 52, 64
"Barbara Allen," xv
Bare, Bobby, 12, 20
Bayou Boys, 84

The Beatles, 2, 66, 86, 102
Bee Gees, 78
Beethoven, Ludwig van, 76, 120
Bentley, Dierks, 118
Big Brother, 110
Big Machine Records, 94
Billboard Country Music Charts, 102
Blake, Nancy Short, 10, **116–17**
Blake, Norman, 10, 64, **116–17**
Blake, Williams, 42
Blue Grass Boys, xvii, 56, 82, 106
Bluegrass Hall of Fame, 56
bluegrass music, xv–xvi, 2, 10, 16, 22, 28, 44, 46, 48, 56, 66, 82, 84, 90, 116, 124
Blue Roses from the Moons (Griffith, N.), 130
blues, 16, 54
Bogguss, Suzy, 12, 58
Bono, 96
Borchetta, Scott, 94
Bresh, Thom, 100
Brinkley, Douglas, 38
Bromberg, David, **16–17**
Brooks & Dunn, 94
Broonzy, Big Bill, 24
Brother Oswald, **14–15**
Brown, Junior, **80–81**
Brown, Tanya Rae, 80
Brown, Tony, 58, 62
Buffett, Jimmy, 34, 72, 106
Burch, Curtis, **66–67**
Burnett, T-Bone, 2
Burns, Jethro, 48
Busby, Buzz, 84
Bush, Roger, 90
Bush, Sam, **48–49, 66–67,** 86

Byrd, Jerry, **112–13**
The Byrds (The Byrds), 92

C

Camp, Shawn, **128–29**
Campbell, Glen, 64
Campi, Ray, 100
Cantwell, David, 42
Carey, Mariah, 40
Carpenter, Mary Chapin, 26, 32
Carr, Patrick, 80
Carter, June, 116
Carter, Mother Maybelle, xvi
"Casey Jones," xv
Cash, John Carter, 62
Cash, Johnny, 20, 42, 54, **60–61,** 62, 88, **96–97,** 114
Cash, Rosanne, 18, **54–55,** 96
Cash, Vivian, 54
Chalker, Curley, **112–13**
Charles, Ray, 88
Chesney, Kenny, 20, 102
The Chieftains, 58
Christian music, 28
Clapton, Eric, 34, 114
Clark, Guy, **4–5,** 38, **50–51,** 68, 72, **128–29, 132–33**
Clark, Susanna, **4–5,** 68
Clement, Cowboy Jack, **88–89,** 96
Clements, Vassar, 62, 64, **106–7**
Clinch Mountain Boys, 2, 118
Cline, Patsy, 36
Clinton, Bill, 108
Cohen, John, 52
Cold Dog Soup (Clark, G.), 132
Collins, Charlie, 14
Coltrane, John, 56

composition music, 28
Continental Club, 80
Cooder, Ry, 32
Cooper, Dan, 114
Corbett, John, 26
Corea, Chick, 28
Country Gazette, 90
The Country Gentlemen, **84–85**
Country Music Association Awards, 102
Country Music Foundation, 62
Country Music Hall of Fame, 36, 42, 56, 78, 100, 102
country rock music, 44, 80, 98
Court and Spark (Mitchell, J.), 86
Cowan, John, **66–67,** 86
Cowboys, 24
Crary, Dan, 108
Crosby, Ronald. *See* Walker, Jerry Jeff
Crouch, Dennis, 46
Crow, Sheryl, 94
Crowe, J. D., **22–23**
Crowell, Rodney, 4, 30, **54–55,** 86
Crumb, R., xvii

D

"Dark as a Dungeon," 100
The Darlings, 44
Davis, Brad, 92
Davis, Miles, 86, 106
Davis, Ronnie, 84
Davis, Skeeter, 78
"Day Drinking," 8
Dickens, Little Jimmy, 24
Dickinson, Jim, 44
Dietz, Roger, 16
The Dillards, **44–45**

"Dirty Old Egg-Suckin' Dog," 88
"D.I.V.O.R.C.E.", 18
The Dixie Chicks, 86
Django (dog), **132–33**
Dobro music, 10, 14, 52
Douglas, Jerry, **22–23**
Down from the Mountain (tour), 116
Drake, David, 10
Duckman, David, 44
Dudley, Dave, **8–9,** 104
Duff, Hilary, 40
Duffey, John, 84
Dunn, Ronnie, 4
Dylan, Bob, 16, 30, 34, 38, 40, 72, 90, 98, 116

E
Eagles, 98, 126
Eagle When She Flies (Parton), 78
Earle, Steve, 4, 38, 46, 54, **68–69, 70–71**
"El Paso," 24
Ely, Joe, **50–51**
Emerson, Bill, **84–85**
Emmanuel, Tommy, 92
Evans, Bill, 28
Evans, Buck, 112
Evans, Walker, xiv

F
Fair & Square (Prine), 74
Fan Fair, 62
"The First Lady of Country Music." *See* Wynette, Tammy
Flatt, Lester, 62, **82–83,** 90
Flatt & Scruggs, 2, 22, 120
Fleck, Béla, **28–29,** 76, 106
Flecktones, 28, 48
"Flushed from the Bathroom of Your Heart," 88
Flying Machine, 40
Foggy Mountain Boys, 82

"Foggy Mountain Breakdown," 92
folk music, 4, 16, 52, 54, 74, 124
"Folsom Prison Blues," 96
Forster, Nick, 86
Foster, Radney, 4
Frank, Chris, 116
Fritts, Donnie, 42
Frizzell, Lefty, 114
Fulks, Robbie, 54

G
Gaines, Roy, 110
Gaudreau, Jimmy, 84
"Gentle On My Mind," 64
Gibb, Barry, 78
Gill, Vince, **6–7**
Gilmore, Jimmie Dale, 24
Glaser, Matt, 106
Glass, Philip, 72
Gleason, Holly, 94
Goffin, Gerry, 40
Goodman, Benny, 92
"Good Ole Boys Like Me," 114
Gore, Al, 96
Graff, Gary, 48
Graham, Billy, **96–97**
Grammy Awards, 2, 6, 28, 58, 74, 78, 92
Grand Ole Opry, xiii, 6, 10, 24, 62, 78, 82
Grant, Amy, 6
Grant, Marshall, 96
Grappelli, Stéphane, 76
The Grateful Dead, 66, 106
Gray, Tom, 84
Green, Lloyd, 112
Green, Pat, 72, 94
Greenhill, Mitch, 108
Gretzky, Wayne, 94
Griffin, Patty, 30
Griffith, Nanci, 4, 38, **58–59, 130–31**
Groce, Larry, 64
Gruhn, George, 10
"Guitar Rag," 100
Guitar Town (Earle), 68
gypsy jazz movement, 92

H
Haggard, Merle, 8, 38, 42, 118
Hall, Tom T., 8
Hansen, 40
Harris, Emmylou, 6, **30–31,** 38, 46, 48, 54, 64, 78, 110
Hartford, John, 10, **64–65,** 88, 106, 116, 120
Haskell, Eddie, 26
Head in the Clouds (film), 92
"Heartaches by the Number," 36
"Hello in There," 74
Helms, Chet, 110
"Help Me Make It Through the Night," 42
Hendrix, Jimi, 48, 80
Hensley, Hairl, 14
Herman, Woody, 106
Hester, Carolyn, 58
Hiatt, John, **32–33, 34–35, 50–51,** 110
The Highwaymen, 20, 42, **60–61**
Highwaymen (The Highwaymen), 60
Hill, Jessie, 110
Himes, Geoffrey, 48
Hitchcock, Stan, 112
Holly, Buddy, 20
Hood, Sam, 110
Hot Band, xvi, 54
"House of Love," 6
Houston, Texas, 94
Howard, Harlan, **36–37,** 42
Huskey, Junior, 46
Huskey, Roy, Jr., **46–47**

I
"I Believe in You," 114
"I Fall to Pieces," 36
"If It's Over," 40
"I Guess Things Happen That Way," 88
Ingram, Jack, 72, **94–95**
International Bluegrass Music Hall of Honor, 82

Irish folk music, 58
"I Saw The Light," 68
"Islands in the Stream," 78
"It's Such a Small World," 54
"It's Too Late, Baby," 40
"I Walk the Line," 96

J
J. D. Crowe & the New South (J.D. & the New South), 22
James, Sonny, 114
jazz, 28
Jennings, Shooter, 20, 60
Jennings, Waylon, **20–21,** 36, 42, **60–61,** 72, 98
Jerusalem (Earle), 70
John, Elton, 92
"John Henry," xv
Johnson, Courtney, **66–67**
Jones, George, 18, 118
Joplin, Janis, 40, 110
Jorgenson, John, **92–93**
Just an American Boy (documentary film), 70

K
Keillor, Garrison, 12
Keith, Toby, 50
Kentucky, General Assembly of the Commonwealth of, 66
The Kentucky Colonels, 90
Kerrville Folk Festival, 58
Ketchum, Hal, **26–27,** 36
King, B. B., 34
King, Carole, **40–41**
Kingston Trio, 52
Kirby, Beecher Ray "Pete." *See* Brother Oswald
Knopfler, Mark, 30
Kooser, Ted, 74
Kortchmar, Danny, 40
Kosharek, Daniel, 76
Kot, Greg, 70
Krauss, Alison, 78
Kristofferson, Kris, 20, 36, **42–43, 60–61,** 74, 96, 116

L

Lackey, Charlie, 40
"Ladies Love Outlaws," 20
Lambert, Miranda, 102
Landreth, Sonny, 34
Larsen, Blaine, 102
The Last Movie (film), 42
Latham, Billy Ray, 90
Lauderdale, Jim, 2, 28
Lawson, Doyle, **84–85**
Leadbelly, 24
*Let Us Now Praise Famous
    Men* (Walker), xvi
Lewis, Jerry Lee, 88
Lincoln, Abraham, 96
Little Richard, 24, 110
"Loco-Motion," 40
Lomax, John and Alan, 52
"Lonesome, On'ry, and
    Mean," 98
Loveless, Patty, 82, 86
Lovett, Lyle, 38, 48, **50–51,** 54
Lynch, Stan, 92
Lynn, Loretta, xvi, 18

M

Ma, Yo-Yo, 76
Malone, Bill, 84
"Mamas, Don't Let Your
    Babies Grow Up to Be
    Cowboys," 20
mandolin music, 48, 52, 62, 90
"The Man in Black." *See*
    Cash, Johnny
Mantle, Mickey, 12
Marcus, Greil, xv
Marley, Bob, 66
Martin, Benny, **120–21**
Martin, Jimmy, 22
Mattea, Kathy, 58
McBride, Martina, 20
McCormick, David, 80
McCoury, Del, 56
McEuen, John, 44, 106
McGraw, Tim, 118
McGuinn, Roger, 92
"Me and Bobby McGee," 42

MerleFest, xvii
Meyer, Edgar, 28, 76
Middle Class, 40
Miller, Roger, 42
Mills, Jim, 22
"Mister Guitar." *See*
    Atkins, Chet
Mitchell, Joni, 86
Mitchell, Mitch, 80
Monroe, Bill, xv, 2, 10,
    48, **56–57,** 64, 66, 82, 90,
    106, 120
Montee, Ray, 112
"Moonland," 112
Morgan, Lorrie, 118
Morris Museum of Art, xiv
Mother Earth, 110
Mountain Stage (Groce), 64
Mozart, Wolfgang Amadeus,
    76, 98, 120
"Mr. Bojangles," 72
"My Baby Don't Dance to Noth-
    ing But Ernest Tubb," 80
"My Old Friend the Blues," 68

N

Nash, Alanna, 104
The Nash Ramblers, 46, 48
Nashville, Tennessee, xiii–xiv,
    12, 46, 106, 112
Nashville Bluegrass Band, 90
Nashville Grass, 90
Nashville Songwriters Hall of
    Fame, 36, 42, 78, 88, 100
Nashville Sound, 12
National Medal of Arts, 108
Nelson, Sherryl, 76
Nelson, Tracy, 110
Nelson, Willie, 20, 30, 38, 42,
    60–61, 72
Nemo, Dave, 8
newgrass music, 64, 66, 106
New Grass Revival, 48, **66–67**
New Lost City Ramblers, 52
Newman, Jimmy C., 8
The New South, **22–23**
Nichols, Joe, 118
Night Owl Café, 40

Nitty Gritty Dirt Band, 46, 72,
    106

O

O'Brien, Tim, 64
O Brother, Where Art Thou?
    (movie soundtrack), 2, 116
O'Connor, Mark, **76–77**
"O Death," 2
Ogden Museum, xiv–xv
Old Number One (Clark, G.), 4
"Ooh Poo Pah Doo," 110
Orbison, Roy, 30, 88
Osborne Brothers, 22
Other Voices—Other Rooms
    (Griffith, N.), 58
outlaw country music, 20, 60,
    72, 98

P

Paisley, Brad, 86
"Pancho and Lefty," 38
Parker, Charlie, 56, 120
Parks, Van Dyke, 98
Parton, Dolly, 12, 18, 30, 42,
    54, **78–79,** 88
Past the Point of Rescue
    (Ketchum), 26
Pavarotti, Luciano, 92
Perkins, Carl, 88
Phillips, Sam, 88
Pickler, Kellie, 78
"Pick Me Up on Your Way
    Down," 36
"Pilgrim," 46
*Pilgrims: Sinners, Saints,
    and Prophets: A Book of
    Words and Photographs*
    (Stuart), xiv
pop music, 28, 54
The Pozo-Seco Singers, 114
Presley, Elvis, 2, 24, 88, 102
Pride, Charley, 12, 88, 114
Prine, John, **74–75, 124–25**
Printer's Alley, 112
Pugh, Virginia Wynette. *See*
    Wynette, Tammy
Pure Prairie League, 6

R

Raitt, Bonnie, 34, 74, 92
"Rank Stranger," 2
RCA Records, 12
Reagan, Ronald, 56
Reid, Charlie, 68
Reinhardt, Django, xv, 92, 132
Reno and Smiley, 66
Revolver (The Beatles), 86
Reynolds, Allen, 36
Rice, Tony, **22–23,** 76, 106,
    108, 116
Richards, Keith, 48
"Ring of Fire," 88
Rinzler, Ralph, xvi, 108
Rivers, Travis, **110–11**
Robbins, Marty, **24–25**
Robins, Butch, 56
Robins, Wayne, 104
rockabilly, xviii
Rock and Roll Hall of Fame,
    40, 56
rock and roll music, 16, 54, 88
Rodgers, Jimmie, 42
Rogers, Kenny, 54, 78
Rogers, Will, 74
Rolling Stones, 40
Rollin' Rock Records, 100
Ronstadt, Linda, 30
Rooney, Jim, 36, 88
*The Rose & the Briar: Death,
    Love and Liberty in the
    American Ballad* (Marcus
    and Wilentz), xv
Russell, Leon, 66

S

The Sacred Harp, xvi
Sayles, John, 42
Scott, Darrell, **86–87**
Scruggs, Earl, 28, 56, 64, 66,
    82, 92
Sebastian, John, 16
Seeger, Mike, **52–53**
Seeger, Pete, 52
Segovia, Andrés, 48
Seldom Scene, 2

"Seven Bridges Road," 98
Shacktown Road (Blake, Blake, and Taylor), 116
"The Shadow of Your Smile," 112
Sides, Anne Goodwin, 80
Simon, Scott, 40
Simone, Nina, 72
Sinatra, Frank, 118
"Six Days on the Road," 8
"Sixteen Tons," 100
Skaggs, Ricky, 2, **22–23**, 56, 64, 118
Sketches of Spain (Davis, M.), 86
Sloan, Bobby, **22–23**
Smothers, Tommy, 64
"Snap Dragon," 28
Snider, Todd, 72
"So Far Away," 40
"Someplace Far Away," 36
Songwriters Hall of Fame, 40
Souther, J. D., **126–27**
Spice Girls, 40
spoken word (music category), 28
St. John, Powell, 110
Standard Songs for Average People (Wiseman and Prine), 124
"Stand by Your Man," 18
Stanley, Carter, 2
Stanley, Ralph, **2–3**, 62
Stanley Brothers, 2, 22
Staples, Mavis, 62
"Steelin' Is His Business," 112
Sternberg, Don, 48
Stevens, Ray, 88
Stone, Cliffie, 100
Stone, Doug, 114
Stover, Don, 2
Strait, George, **102–3**

Stuart, Marty, xiv, 46, **62–63,** 64, 82, 90, 106
Stubbs, Eddie, 22
Sunny Mountain Boys, 22
Sun Records, 88
"Sure as I'm Sitting Here," 32
Sutton, Bryan, 116
Swan Lake (Tchaikovsky), 26
Swift, Taylor, 102

"Take Me to Your World," 18
Taupin, Bernie, 24, 26
Taylor, James, 40, 76
Taylor, Tut, **10–11,** 64, 116
"Tecumseh Valley," 38
The Tennessee Two, 96
Theatre of the Unheard (Scott), 86
Thile, Chris, 48
Third Generation Blues (Watson, Doc and Richard), 108
Thomasson, Benny, 76
Thompson, Hank, 100
Thompson, Verlon, **128–29**
Thornton, Billy Bob, 74
Three Dog Night, 32
Three Little Country Boys, 90
Tillis, Mel, 42
Tillis, Pam, 18
"To Beat the Devil," 42
Townshend, Pete, 114
Travis, Merle, 74, **100–101**
Travis, Randy, 114
Tried and True Music, 72
Tritt, Travis, 20, 62, 86, 92
truck-driving music, 8, 104
"Truck Drivin' Son-of-a-Gun," 8
"Trucker's Prayer," 8
Tubb, Ernest, 24, 80
Tut Taylor Music, 10
Twain, Mark, 12

"Two Little Boys," 84
Tyminski, Dan, 22

U

Ungar, Jay, 16

V

Valens, Ritchie, 20
Van Zandt, Townes, **38–39,** 72, 98
Vasterling, Paul, 26
Volmers, Eric, 50

W

Wagoner, Porter, 78
Waikiki, Rudy, 14
Walker, Bradley, 22
Walker, Ebo, 66
Walker, Jerry Jeff, 16, **72–73**
Walker, Susan, **72–73**
Waller, Charlie, **84–85**
Waller, Randy, 84
"Wanted: The Outlaws!", 20
Wariner, Steve, 6
Warner Music, 94
Was, Don, 42
Washburn, Jim, 48
Watson, Dale, **104–5**
Watson, Doc, 62, 100, **108–9**
Watson, Merle, 100
Watson, Richard, 108
Watt, Mark, 76
Wayne, John, 96
"The Way We Make a Broken Heart," 32
Webb, Dean, 44
Weiser, Rockin' Ronny, 100
Wells, Kitty, xvi
Wells, Paul F., 52
"(We're Not) the Jet Set," 18
Western Performers Hall of Fame Awards, 6
western swing music, 102

"Wherever You Are," 94
"The Whisky Ain't Workin' ," 62
White, Barry, 32
White, Clarence, 90, 108
White, Eric, 90
White, Roland, 10, **90–91**
White Shoes (Harris), 30
Whitley, Keith, **118–19**
"Whole Lot of Shakin' Going On," 88
Wilentz, Sean, xv
Williams, Don, **114–15**
Williams, Hank, xiii, 24, 38, 42, 68, 98, 120
Williams, Hank, III, 104
Williams, Hank, Jr., 78, 98
Williams, John, 68
Williams, Lucinda, 98
Williams, Paul, 94
Will the Circle Be Unbroken (Nitty Gritty Dirt Band), 46, 106
"Will You Still Love me Tomorrow," 40
Wiseman, Mac, **124–25**
Wolfe, Stan, 90
Wood, Randy, 10
world music, 28
WSM (radio station), xiii–xiv
Wynette, Tammy, **18–19**

Y

Yates, Bill, **84–85**
Yoakam, Dwight, 2
"You Make Me Feel Like a Natural Woman," 40
Young, Jubal, 98
Young, Steve, **98–99**
"Young Man's Town," 6
"You're My Best Friend," 114

Z

Zwerdling, Danile, 8